Springer Series in Materials Science

Volume 324

The Springer Series in Materials Science covers the complete spectrum of materials research and technology, including fundamental principles, physical properties, materials theory and design. Recognizing the increasing importance of materials science in future device technologies, the book titles in this series reflect the state-of-the-art in understanding and controlling the structure and properties of all important classes of materials.

More information about this series at https://link.springer.com/bookseries/856

Bernd Rauschenbach

Low-Energy Ion Irradiation of Materials

Fundamentals and Application

Springer

Bernd Rauschenbach
Felix Bloch Institute for Solid State Physics
University Leipzig
Leipzig, Germany

Innovative Surface Technologies GmbH (IOT)
Leipzig, Germany

Leibniz Institute of Surface Modification (IOM)
Leipzig, Germany

ISSN 0933-033X ISSN 2196-2812 (electronic)
Springer Series in Materials Science
ISBN 978-3-030-97276-9 ISBN 978-3-030-97277-6 (eBook)
https://doi.org/10.1007/978-3-030-97277-6

This Springer imprint is published by the registered company Springer Nature Switzerland AG
The registered company address is: Gewerbestrasse 11, 6330 Cham, Switzerland

This book is dedicated to my wife, Christa,
to my daughter, Katrin, and to my son,
Stephan

Preface

The book focuses on the interaction of low-energy ions with surfaces and how these processes affect the surface and near-surface regions of materials. These processes are of outstanding importance for numerous modern industrial applications and products, especially in microelectronics, nanotechnology, mechanical engineering and the optical industry. Due to the complexity of the interaction of low-energy ions with solid surfaces, only single aspects have been addressed in the published literature regarding the effect of these processes for the evolution of topography, the figuring and smoothing of surfaces, the formation of nanostructures and the fabrication of thin films. Therefore, the book is an attempt of coherent representation of these research areas.

The book starts with the introduction to the basis of the classical ion–solid interaction at low ion energies. The first four chapters of the first part of the book deal with collision processes (Chap. 2), the energy loss of energetic ions in solids and the depth distribution of implanted particles (Chap. 3), the damage structure from a single defect to a completely amorphized layer after low-energy ion irradiation (Chap. 4) and the sputtering of atoms from the surface and near-surface regions (Chap. 5). The aim of these chapters is to give the reader an overview of the interaction processes on the one hand and to provide the elementary basis for understanding the following chapters on the other hand. The mathematics is kept to a minimum to represent the underlying physics.

In the second part of this book, individual aspects of the interaction of low-energy ions with solid surfaces are discussed in detail. The changes in the topography of surfaces (Chap. 6) are often the first observable modification after low-energy ion irradiation. These topography changes after ion beam bombardment can be described based on different theoretical methods. In addition to the erosion of the surface by ion bombardment, secondary ion-beam-induced effects, such as surface diffusion, re-deposition and viscous flow, also contribute to the modification of the surface. In the next chapter (Chap. 7), the technologies of the ion beam figuring and ion beam smoothing are introduced to polishing of surfaces or to correct the surface shape in a predetermined and controlled fashion. Chapter 8 deals with a new self-organization-based method to fabricate ordered patterns of ripples and holes on the nanometer scale. In a first approximation, the periodic modulation of the surface under low-energy ion bombardment can be described by an interplay between curvature-dependent sputtering and surface diffusion. Chapter 9 describes some aspects of the deposition of ionized atoms or molecules with hyperthermal energy to form thin films on substrates. This has some key advantages such as the high purity of the deposited material, the low temperatures during deposition, the anchoring of the layers in the substrate, the high lateral layer uniformity and the possibility to control the film properties. This technology is also successfully used for cleaning surface. The ion beam-assisted deposition, a method characterized by independently controllable and simultaneous flux of ions and atoms to the substrate surface, is presented in Chap. 10. The physical principles and the application of this process for the production of high-quality coatings are discussed in detail. In the last chapter (Chap. 11), a fascinating method to produce nanostructures based on the low-energy ion beam-assisted glancing angle deposition is introduced. By controlling of the substrate polar and azimuthal rotations, during sputtering-induced deposition, a large variety of the dimension and the shape of nanostructures can be realized.

The book is designed to be suitable for undergraduate and graduate students as well as for scientists in research institutes and practitioners in industry.

The idea for this book arose from courses on ion–solid interaction that I have offered since the 1990s at the universities of Hamburg, Augsburg, Leipzig and Wuhan for undergraduate and graduate students in experimental physics and materials science, as a natural outgrowth of the emerging research activities in this field.

In contrast to irradiation with electrons or photons, accelerated ions can transfer significant energies and momentums when interacting with atoms of the target material. Therefore, ion beam techniques have become an integral part of modern high technology in recent decades. For example, effective doping of semiconductors, deposition and etching of layers in the semiconductor industry, smoothing of optical surfaces or the application of energetic ions for surface and thin film analysis would not have been possible without the application of low-energy ions. Modern ion beam technology is currently facing new challenges. This includes promoting the development of new components for computer and communications technology, new biomaterials or materials for energy conversion. Ion beams are increasingly used for high-precision surface processing, nanostructure fabrication and efficient space propulsion. In particular, low-energy ions, i.e., ions with energies up to a few kiloelectron volts, are the focus of interest. Ions in this energy range are easy to generate, their characteristic parameters such as ion energy and ion current density are straightforward to control, and the ion beam systems are inexpensive to implement and can be easily integrated into other research and production systems.

The roots for the engagement of the subject to this book go back to my activity as a young scientist in the seventies and the possibility to accept invitations for research stays in different countries after the peaceful revolution at the end of the eighties and the beginning of the nineties of the last century.

I am indebted to many friends, colleagues and collaborators. The collaboration with Rainer Grötschel, Karl-Heinz Fährmann, Viton Heera, Karl Hohmuth,

Andreas Kolitsch, Edgar Richter, Matthias Posselt and Kurt Helming from the Helmholtz-Zentrum Dresden-Rossendorf (formerly known under Zentralinstitut für Kernforschung Rossendorf), with Natalie Moncoffré and Gilbert Marest from the Université de Lyon and Gerhard Linker and Otto Meyer from the Forschungszentrum Karlsruhe (formerly known under Kernforschungszentrum Karlsruhe) was a period of intense cooperation in the use of nuclear methods to elucidate solid-state physics issues, the study of ion beam-induced phenomena and the focused applications of ion beams to modify surface properties.

Most of what I know about the topic of this book I have learned from my colleagues over the years. I would like to take this opportunity to express my gratefulness working with them, especially Wilhelm Hinz (Zentralinstitut für Anorganische Chemie Berlin), Gerhard Blasek (Technische Universität Dresden), Bernd Stritzker, Jörg Lindner, Helmuth Karl and Matthias Schreck (Universität Augsburg), Bernd Abel, Frank Frost, Jürgen W. Gerlach, Thomas Hänsel, Andriy Lotnyk, Stephan Mändl, Darina Manova, Stefan G. Mayr, Horst Neumann and Andreas Nickel (Leibniz-Institut für Oberflächenmodifizierung, Leipzig), Marius Grundmann and Rüdiger Szargan (Universität Leipzig), Wolfgang Ensinger (Technische Universität Darmstadt), Walter Assmann (Ludwigs-Maximilians-Universität München), Stephan Facsko, Karl-Heinz Heinig and Wolfhard Möller (Helmholtz-Zentrum Dresden-Rossendorf), Ulrich Gösele, Johannes Heydenreich, Wilfried Erfurth, Gerhard Kästner and Christian Dietsch (Max-Planck-Institut für Mikrostrukturphysik Halle/S.), Carsten Riedel and Reiner Mehnert (IOT GmbH Leipzig), Frank Richter (Technische Universität Chemnitz), Bernd Meyer (Technische Universität Hamburg-Harburg), Hans-Florian Zeilhofer and Hans-Henning Horch (Technische Universität München), Ibrahim Abdulhalim and Alina Karabchevsky (Ben-Gurion University of the Negev, Beer Sheva), Agustin R. González-Elipe (University Sevilla), Yuri Trushin (Saint Petersburg Academic University) and Petar A. Atanasov (Bulgarian Academy of Sciences, Sofia).

I would also like to thank my Ph.D. students, Christian Rathje (Technische Universität Hamburg-Harburg), Liu Chang, Stephan Geier, Sascha Henke, Claus Hammerl, Josef Hartmann, Axel Königer, Uwe Preckwinkel, Sigmar Schosser, Stefan Sienz, Stephan Six, Hermann Wengenmair, Axel Wenzel, Siegmar Schoser, Michael Zeitler, (Universität Augsburg) and Mario Behrens, Vikas Baranwal, Marina Inés Cornejo, Annemarie Finzel, Isom Hilmi, Chinmay Khare, Susan Liedtke-Grüner, Jan Lehnert, Jan Lorbeer, Marisa Mäder, Michael Mensing, Lena Neumann, Christian Patzig, David Poppitz, Ulrich Roß, Philipp Schumacher, Xinxing Sun, Marc Teichmann, Erik Thelander, Bashkim Ziberi, (Leibniz-Institut für Oberflächenmodifizierung Leipzig), for their direct and indirect contributions to the subject matter of this book.

I am grateful to the many colleagues who provided diagrams, figures, micrographs or unpublished papers from their own work, in particular Wolfgang Bolse (Universität Stuttgart), Alexey Guglya (National Scientific Center Kharkov), Jozef Keckes (Montanuniversität Leoben), Stefan Linz (Westfälische Wilhelms-Universität Münster), Andreas Mutzke (Max-Planck Institut für Plasmaphysik, Greifswald), Edmund G. Seebauer (University of Illinois), Peter Sigmund (University of Southern Denmark, Odense), Roger E. Stoller (Oak Ridge National Laboratory), Werner

Wesch (Friedrich-Schiller-Universität Jena) and Dexian Ye (Rensselaer Polytechnic Inst. Troy, New York).

I am also grateful to colleagues who have contributed in various ways to the preparation of this book. These colleagues, Christoph Grüner and Axel Schindler (Leibniz-Institut für Oberflächenmodifizierung Leipzig), Tansel Karabacak (University of Arkansas at Little Rock), Thomas Michely (Universität zu Köln), Stephan Rauschenbach (University Oxford and Max-Planck-Institut für Festkörperforschung Stuttgart), read drafts of various chapters of the book and offered helpful recommendations and suggested a number of clarifications and improvements in the manuscript.

Many results were achieved in the context of numerous approved projects, for whose financial support I would like to express my sincere thanks to the Deutsche Forschungsgemeinschaft (DFG), Leibniz-Gemeinschaft (WGL), Deutschen Akademischen Austauschdienst (DAAD), Bundesministerium für Bildung und Forschung, Sächsisches Staatsministerium für Wissenschaft und Kultur and numerous companies.

I would like to thank Sabine for critically reviewing parts of the manuscript. I am grateful to Zachary Evenson from Springer-Verlag for the pleasant cooperation in completing the book.

The author (bernd.rauschenbach@iot-gmbh.de) would be grateful to the reader for comments and suggestions to improve and correct this book.

Leipzig, Germany Bernd Rauschenbach

The original online version of this Back Matter was revised: the list of material for this title has been published excluding the page number indices. Now, it is incorporated into this element.

Contents

Abbreviations

ADF	Angle annular dark-field
AES	Auger electron spectroscopy
AFM	Atomic force microscopy
amu	Atomic mass unit
ATP	Adenosine triphosphate
BCA	Binary collision approximation
bcc	Body-centered cubic
BRF	Beam removal function
BSA	Bovine serum albumin
CNC	Computerized numerical control
DC	Diode or direct current
DLC	Diamond-like carbon
DNA	Deoxyribonucleic acid
dpa	Displacements per atom
EBL	Electron beam lithography
EBSD	Electron backscatter diffraction
ECR	Electron cyclotron resonance
EDX	Energy dispersive X-ray spectroscopy
EELS	Electron energy loss spectrometry
EQE	External quantum efficiency
ES barrier	Ehrlich–Schwoebel barrier
ESD	Ion-induced effective surface diffusion
ESI	Electrospray ionization
ES-IBD	Electrospray ion beam deposition
EUV	Extreme ultra-violet
fcc	Face-centered cubic
FET	Field-effect transistor
FFT	Fast Fourier transform
FIB	Focused ion beam
FP	Frenkel pair
FT	Fourier transform

FTIR	Fourier-transform infrared spectroscopy
FWHM	Full width at half maximum
GISAXS	Small angle X-ray scattering under grazing incident angles
GLAD	Glancing angle deposition
Hb	Hemoglobin
HbA1c	Glycated hemoglobin
hcp	hexagonal closed-packed
HPOG	Highly oriented pyrolytic graphite
HR-TEM	High-resolution transmission electron microscopy
HSFR	High-spatial frequency roughness
IBAD	Ion beam-enhanced deposition
IBA-PLA	Ion beam-assisted pulsed laser deposition
IBA-SD	Ion beam-assisted sputter deposition
IBA-TD	Ion beam-assisted thermal deposition
IBD	Ion beam (direct) deposition
IBF	Ion beam figuring
IBS	Ion beam smoothing
ISS	Ion scattering spectroscopy
JCPDA	Joint committee on powder diffraction standards
KS equation	Kuramoto–Sivashinsky equation
LAFR	Low-spatial frequency roughness
LEED	Low-energy electron diffraction
LNT	Temperature of liquid nitrogen
LSPR	Localized surface plasmon resonance
LSQR	Iterative method for the solution of large linear systems of equations and least squares problems
LSS theory	Lindhard–Scharff–Schiøtt theory
MALDI	Matrix-assisted laser desorption/ionization
MATLAB	Programming and numeric computing platform
MBE	Molecular beam epitaxy
MC simulation	Monte Carlo simulation
MD simulation	Molecular dynamic simulation
MEMS	Micro-electro-mechanical systems
MFSR	Mid-spatial frequency roughness
ML	Monolayer
MOKE	Magneto-optic Kerr effect
MOSFET	Metal-oxide semiconductor field-effect transistor
MS-IBD	Ion beam deposition with mass separation
NRT model	Norgett–Robinson–Torrens model
NSL	Nano-sphere lithography
OAD	Oblique angle deposition
ODF	Orientation density function
PKA	Primary knock-on atom
PL	Photoluminescence
PLD	Pulsed laser deposition

PMMA	Polymethylmethacrylate
PSD	Power spectral density
pv	Peak-to-valley
PVD	Physical vapor deposition
RBS	Rutherford backscattering
RF	Radio frequency
RFA	Retarding field analyzer
RHEED	Reflection high-energy electron diffraction
R-IBD	Reactive ion beam deposition
rms	Root mean square
RT	Room temperature
SAD	Selected area diffraction
SEF	Surface-enhanced fluorescence
SEIRA	Surface-enhanced infrared absorption
SEM	Scanning electron microscopy
SERS	Surface-enhanced Raman spectroscopy
SF	Stacking faults
SIMS	Secondary ion mass spectrometry
SQUID	Superconducting quantum interference device
SRIM	Stopping and ranges of ions in matter
STEM	Scanning transmission electron microscopy
STM	Scanning tunneling microscopy
TEM	Transmission electron microscopy
TOF-SIMS	Time-of-flight secondary ion mass spectrometry
TRIM	Transport of ions in matter
ULE	Ultra-low expansion glass
Vg	Vitellogenin
XEDS	Energy-dispersive X-ray spectroscopy
XPS	X-ray photoelectron spectroscopy
XRD	X-ray diffraction
YSZ	Yttria-stabilized zirconia
ZBL	Ziegler–Biersack–Littmark

Physical Constants

Name	Numerical value	Unit
Atomic mass unit	$amu = 1.6605 \times 10^{-24}$	kg
Avogadro number	$N_A = 6.0221367 \times 10^{23}$	molecules/mol
Bohr radius	$a_o = 0.052917725$	nm
Bohr velocity	$v_B = 2.1877 \times 10^8$	$cm \cdot s^{-1}$
Boltzmann constant	$k_B = 1.380658 \times 10^{-23}$ $= 8.617385 \times 10^{-5}$	J/K eV/K
Elementary electric charge	$e = 1.60218 \times 10^{-19}$ $e^2 = 1.4399651$	c eV·nm
Mass of electron	$m_e = 9.1093897 \times 10^{-28}$	g
Molar gas constant	$R = 8.314510$	J/mol·K
Planck's constant	$h = 6.62607 \times 10^{-34}$	J·s
Reduced Planck's constant	$\hbar = h/2\pi = 6.5822 \times 10^{-16}$	eV·s
Rydberg energy	$E_R = 13.606$	eV
Velocity of light	$c = 2.9979 \times 10^8$	$m \cdot s^{-1}$
Stefan–Boltzmann constant	$\sigma_s = 5.67051 \times 10^{-8}$	$W/m^2 \cdot K^4$
Universal gas constant	$R = 8.3145$	$J \cdot g^{-1} \cdot mole \cdot K$
Vacuum dielectric permittivity	$\varepsilon_o = 8.854 \times 10^{-12}$	$AsV^{-1}m^{-1}$

Conversions

1 atomic unit (a.u.) = 27.210 eV
1 eV/atom = 23.069 kcal/mol = 96.521 kJ/mol
1 eV = 1.6022×10^{-19} J
1 calorie = 4.184 J
1 sccm = 1.69×10^{-3} Pa·m^3/s = 2.27×10^3 mol/s
1 μA/cm^2 = 6.25×10^{12} ions/cm^{2}s

Chapter 1
Introduction

Abstract The first chapter places the modification of surfaces by low-energy ion bombardment in the context of modern surface modification techniques, formulates the requirements for a low-energy ion bombardment facility, outlines the main physical processes induced by low-energy ions, and points out some recent applications with low-energy ions that are beyond the scope of the book.

Precise control of the structure, topography, and composition of the surface and near-surface regions is a mandatory requirement for modern surface processing technologies. A general strategy for the development of surface modification techniques is based on kinetically limited processes instead of equilibrium processes. One of these techniques is low-energy ion irradiation, which has been an attractive area of research for several decades. From the beginning, research and development work in the field of interaction of low-energy ions with solids has been closely related to practical applications. Compared to other surface modification techniques, the high energy density after penetration of an energetic ion into the solid has a significant impact on the resulting atomic structure and consequently on the properties of the surface and near-surface regions.

The energy range of low-energy ions is generally between a few electron-volts and a few kilo- electron-volts. The lower energy limit can be defined either by an ion energy higher than the typical thermal energy of an atomic particle, or by the kinetic energy of an ion on the order of the binding energies. In this lower energy range, the interaction of ions with the solid is characterized by lattice distortions and phonon-induced excitations. The low-energy range up to about 100 eV is also called hyperthermal. The upper energy limit is less precisely definable as inelastic interactions and deviations from the simple binary collision process become increasingly important.

Numerous designs for direct ion irradiation of surfaces or ion beam-assisted deposition of thin films with low-energy ions are described in the literature. In general, an equipment for low-energy ion irradiation should satisfy the following requirements:

B. Rauschenbach, *Low-Energy Ion Irradiation of Materials*, Springer Series in Materials Science 324, https://doi.org/10.1007/978-3-030-97277-6_1

- capability to generate stable ion beams from a variety of gases and solid materials,
- possibility of controlling the ion energy, ion current density and the angle of ion incidence,
- performance of the ion beam experiments under HV and UHV conditions,
- temperature control of the substrate during the ion bombardment, and
- beam impurities arising either from neutral atoms or molecules exiting the source thermal energy or from charge exchange in the beamline should be negligible.

For fundamental studies on the elementary process of interaction of low-energy ions with surfaces, further requirements have to be considered:

- presence of mass separation facility in the beam line,
- the capability of the ion beam system to provide mass-analyzed ion currents at very low energies and with low energy spreads (beam divergence), and
- the possibility of in-situ analysis of composition and crystallinity of the substrate surface or growing film.

Ion sources are the most important and indispensable components of ion irradiation systems. For the wide variety of applications in the field of the low-energy ion irradiation, there are many types of ion sources with different working mechanisms. The variety of the ion sources arises from the different ways of ion generation and also the variety of generating plasma such as AC discharge, DC discharge, arc discharge, micro-wave discharge and laser driven plasmas [1]. Typical requirements are high beam intensities, small beam divergence, and high stability. In the past, the main disadvantage of using these ion sources was that these ion sources could not generate sufficient ion current densities at low ion energies. A response to the need for higher ion currents at low ion energies was the development of the broad beam ion sources by Kaufman et al. [2] in the seventies of the last century. Later a variety of ways were used to increase the current capacity of broad beam ion sources, including ion sources without gratings and multiple gratings (up to three gratings) as well as decreasing the grating spacing and increasing the number of apertures.

In general, a distinction is made between two principles for transporting ions at low energies from the ion source to the target due to the space-charge effect. On the one hand, the ions are accelerated to high energies after the ion source, mass analyzed, and then decelerated to low energies immediately before the target [3]. On the other hand, the generation of low energy and hyperthermal ions in an ion beam source is used in combination with a compact quadrupole mass analysis system equipped with entry and exit ion optics without deceleration before the target [4].

Why low energies? To answer this question, different aspects have to be considered. Numerous ion irradiation experiments have shown that low-energy ions in particular have a significant effect on the structure of the near-surface layer and the growth of the layers on the surface. Experimental investigations and theoretical studies have shown many times that low-energy ion irradiation results, among other things, in creation of mobile vacancies and interstitial atoms, local relaxation of the crystal lattice, enhanced diffusion, sputtering of surface, formation of defects, atomic mixing between thin films and the underlying substrate, film densification, preferred grain

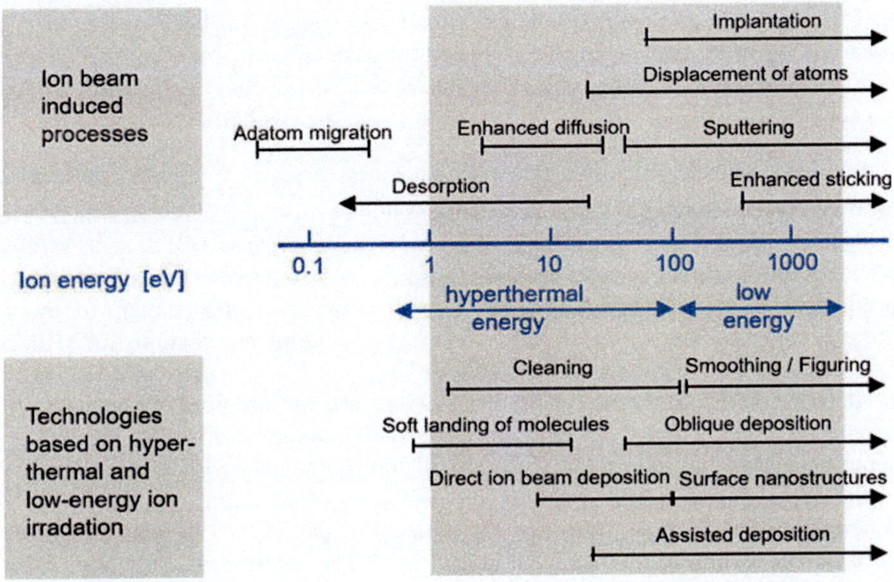

Fig. 1.1 Hyperthermal and low-energy ion range corresponding to some selected ion beam induced processes and engineering applications

orientation, build-up of stress states, stimulating chemical reaction, tailoring film composition, low-temperature epitaxial growth, synthesis of metastable phases and nanostructures etc. [5, 6]. Figure 1.1 shows the energy range of low-energy ions together with some selected ion beam-induced processes and the resulting ion beam stimulated engineering processes.

From the point of view of the application of these surfaces and thin films, these changes in properties induced by ion irradiation can be very beneficial but also undesirable. A key advantage of low-energy ion irradiation is the ability to distinguish between beneficial and undesirable effects by the choice of ion energy and beam intensity, i.e. to control the ion beam induced changes in properties, topography and composition.

In contrast to high-energy ion irradiation (ion energies higher than a few tens of keV), the costs for construction, operation and maintenance of a low-energy ion beam facility are comparatively low. Consequently, the application of low-energy ions to modify surfaces, to support the deposition of thin films or grow of artificially structured material surfaces has found its way into a number of industrial application. For example, ion implantation, plasma processes techniques, plasma immersion ion implantation, sputtering and reactive ion beam etching are routinely employed in the manufacture of microelectronic components, in mechanical engineering or for processing optical surfaces. Another field of application of low-energy ions is surface analysis with the aim, for example, to determine the chemical composition sensitively as a function of the sample depth.

Despite the wide range of applications of low-energy ion beam technologies, it can be expected that further very successful developments based on knowledge of interaction between low-energy ions and solids will become known in the next decade. Indicators for this assumption are the impressive technological evolutions in the past decade.

For example, the modern focused ion beam (FIB) technology is becoming widely available in semiconductor research and processing environments, as well as in failure analysis and chip-design centers. It is expected that FIB systems will be successfully advanced in the field of mask repair, the inspection of integrated circuits and electronic devices, the localized milling and also the preparation of specimens for transmission electron microscopy as well as in micromachining applications for MEMS [7].

Other examples of the use of low-energy ions arise in the field of deposition of nonvolatile organic molecules or nanoparticles. Electrospray ionization assisted-ion beam deposition [8] as a method to deposit molecular ions with kinetic energies in the hyperthermal range onto surfaces providing an essential tool for the non-destructive deposition of large organic molecules, particularly proteins, peptides viruses, saccharides as well as DNA and viruses. The recent development of high-flux ionization sources has opened up new opportunities for the precisely controlled preparation of both two-dimensional and three-dimensional architectures of these molecules by ion soft landing.

Also the deterministic placement of single atoms by low-energy ion implantation up to energies of some keV can be a key capability for the fabrication of nanometer scale and single atom solid-state devices [9]. Examples are devices include donors linked to quantum dots or single color centers in diamond (referred to as nitrogen-vacancy centers or NV centers), which can be used in quantum computers in the future.

It should also be noted that the tool for generating low-energy ions, the ion source, has enormous potential for development [10]. For example, the use of ion sources as electric propulsion for moving satellites and spacecraft in space has experienced a renaissance since it was possible to increase available power on spacecraft, as demonstrated by the very recent appearance of all-electric communication satellites, significantly.

References

1. B.H. Wolf, K.N. Leung, B. Sharkow, R. Becker, T. Jolly, G. Alton, J. Ishikawa, Characteristics of ion sources, in *Handbook of Ion Sources*, ed. by B. Wolf (CRC Press, Boca Raton, 1995), pp. 23–329
2. H.R. Kaufman, J.M.E. Harper, J.J. Cuomo, Developments in broad beam, ion sources technology and applications. J. Vac. Sci. Technol. **21**, 764–767 (1982)
3. A.H. Al-Bayati, D. Marton, S.S. Todorov, K.J. Boyd, J.W. Rabalais, D.G. Armour, J.S. Gordon, G. Duller, Performance of mass analyzed, low-energy, dual ion beam system for materials research. Rev. Sci. Instr. **65**, 2680–2692 (1994)
4. J. W. Gerlach, P. Schumacher, M. Mensing, S. Rauschenbach, I. Cermak, B. Rauschenbach, Ion mass and energy selective hyperthermal ion-beam assisted deposition setup. Rev. Sci. Instrum. **88**, 063306 (2017)
5. T. Takagi, lon-surface interactions during thin film deposition. J. Vac. Sci. Technol. A **2**, 382–399 (1984)
6. J.E. Greene, S.A. Barnett, J.-E. Sundgren, A. Rockett, Low-energy ion/surface interactions during film growth from the vapor phase: effect on nucleation and growth kinetics, defect structure, and elemental incorporation probabilities, in *Plasma-Surface Interactions and Processing of Materials*, ed. by O. Auciello, A. Gras-Marti, J.A. Valles-Abarca, L. Flamm (Kluwer Academic Publishers, Dordrecht, 1990), pp. 281–311
7. J. Orloff, M. Utlaut, L. Swanson, *High Resolution Focused Ion Beams: FUIB and Its Applications* (Springer Science+Business Media, New York, 2003)
8. S. Rauschenbach, G. Rinke, R. Gutzler, S. Abb, A. Albarghash, D. Le, T.S. Rahman, M. Dürr, L. Harnau, K. Kern, Two-dimensional folding of polypeptides into molecular nanostructures at surfaces. ACS Nano **11**, 2420–2427 (2017)
9. K. Groot-Berning, T. Kornher, G. Jacob, F. Stopp, S.T. Dawkins, R. Kolesov, J. Wrachtrup, K. Singer, F. Schmidt-Kaler, Deterministic single-ion implantation of rare-earth ions for nanometer-resolution color-center generation. Phys. Rev. Lett. **123**, 106802 (2017)
10. S. Mazouffre, Electric propulsion for satellites and spacecraft: established technologies and novel approaches. Plasma Source Sci. Technol. **26**, 033002 (2016)

Part I
Fundamentals

Chapter 2
Collision Processes

Abstract Collisions between, on the one hand, the incident ion and the target atoms and, on the other hand, between target atoms themselves are the fundamental processes of ion–solid interaction. In a collision with low and medium ion energies, the interaction between two particles can be described by a screened Coulomb potential. In detail, various screened potentials and their screening function are presented in a summarized form. Interactions between particles in the very low or hyperthermal energy range can, on the one hand, be represented approximately by a Born–Mayer potential and, on the other hand, it is to be taken into account that the attractive part of the interaction potential cannot be ignored any more. In the following, the classical description of the collision processes in both the laboratory system and the center-of-mass system is briefly presented, the collision parameters are introduced, and the scattering angles for the colliding and collided particles are formulated. Subsequently, the total scattering cross section, a quantity indicating the probability of the interaction between ions and target atoms, and the differential scattering cross-section, indicating the number of ions scattered into a differential solid angle at given polar and azimuthal angles, are introduced. The method proposed by Lindhard et al. to represent the differential cross-section as a function of a single parameter for the screened Coulomb interaction is also presented.

The description of the interaction of an energetic ion with target atoms is based on the consideration of the forces of attraction or repulsion between the partners of interaction. In the simplest case of a collision event, two atomic nuclei, which are surrounded by a cloud of electrons, interact with each other. A comparison of the velocity of the outer shell electron of a target atom, called Bohr velocity v_B, in order of 10^8 cm/s, with the projectile (ion) velocity, v, can be helpful in determining the interaction potential during a collision.

At a collision with low energies $v_B \gg v$, the electrons follow the particle movement and again take their initial configuration after the collision. This process can be referred to be adiabatic and the collision can be described using quasi-molecular potential formed during the collision. Consequently, the initial and the final energy are the same, i.e. the collision is elastic. At higher energies $v_B \ll v$, the electrons are not capable to follow the changes of the nuclei and electron cloud positions of

B. Rauschenbach, *Low-Energy Ion Irradiation of Materials*, Springer Series in Materials Science 324, https://doi.org/10.1007/978-3-030-97277-6_2

the other impact partners. This results in an extreme charge distribution state that occurs when the particles are at their closest approach. There, it's highly probable that inelastic processes are triggered.

2.1 Interaction Potentials

In general, the interatomic potential can be determined by solving the Schrödinger equation for all atomic positions. For this purpose, so-called 'ab-initio' methods are used like Hartree–Fock method, density functional theory, configuration interaction or coupled cluster approaches [1]. These 'ab-initio' methods provide self-consistent field equations which are dependent on the parameters of the atomic positions. The solution of the self-consistent field equations is limited, because the computer calculation time increases extremely with the number of electrons in the considered system. Consequently, semi-empirical methods are used, which try to fit a generated interatomic potential to experimental or 'ab-initio' data.

According to Bohr [2], a classical consideration may be executed when the de Broglie wavelength is small in comparison to the impact parameter b in a head-on collision. This condition is fulfilled for the most frequently applied ion-target atom combinations in the middle- and high-energy ranges. The interaction between two atoms or molecules is strongly depends on the inter-nuclear distance r between these particles and their charges (see Fig. 2.1).

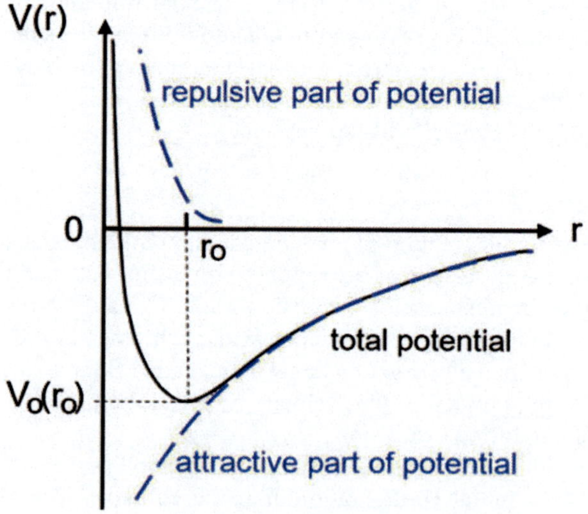

Fig. 2.1 Schematic presentation of the interatomic potential as function of the inter-nuclear distance r, with r_0 corresponding to the equilibrium distance (in crystal lattice the nearest neighbor distance)

For large distances, the forces of attraction prevail. For example, these attractive forces can be Van der Waals forces, but also forces between dipoles. These forces are responsible for generation of a liquid or a solid. If the particles are brought closer together, the repulsive forces become quickly significant and exceed the attractive forces (see Fig. 2.1).

The potential energy or simply the potential V(r) can be expressed by the force F. When the force is conservative (meaning that work done in moving a particle between two different points is independent of the taken path), the potential is given by

$$V(r) = -\int_{\infty}^{r} F(r)dr, \tag{2.1}$$

where r is the interatomic distance. It follows that the force is given by

$$F(r) = -\frac{dV(r)}{dr}. \tag{2.2}$$

If is assumed that only pairwise interactions are involved (for example at the interaction of two atoms) so-called pair potentials are used. For example, the Coulomb force describes the attraction of particles due to their electric charge along the connection-line between the two particles and can expressed by

$$\mathrm{F(r)} = \frac{1}{4\pi\epsilon_o\epsilon}\frac{\mathrm{Z_1Z_2}e^2}{\mathrm{r}^2}, \tag{2.3}$$

where ε_o is vacuum dielectric permittivity (= 8.854×10^{-12} As V^{-1} m^{-1}), ε is the dielectric permittivity, Z_1 and Z_2 are atomic numbers of the interacting particles and e (= 1.60218×10^{-19} C) is the elementary electric charge (e^2 = 1.439 eV nm). Then, the potential energy or the Coulomb interatomic potential which is necessary to approach both particles from infinity to the distance r is given by

$$\mathrm{V(r)} = \frac{1}{4\pi\epsilon_o\epsilon}\frac{\mathrm{Z_1Z_2}e^2}{\mathrm{r}}. \tag{2.4}$$

The exact determination of the potential for the different particle–particle (atom–atom, atom-ion, etc.) interaction is possible only for simple cases. The assumption that the particles are hard and non-deformable spheres represents the simplest case. There, the potential function V(r) is zero for distances r larger than the radius of the particle and is infinite for distances smaller than the particle radius. In general, it is easy to introduce such potential.

It must be noted that no single potential function is capable to describe the interaction between accelerated particles for all distances of the separation r. With increasing energy of the particles from hyperthermal energies (some tens of electron-volts) to very high energies, the magnitude of the overlapping electronic clouds of two particles can be predicted. Therefore, two contributions to the repulsive part of the potential between two particles can be expected. On the one hand, the contribution to the repulsive part of the potential between two particles is increased with interaction energy, on the other hand an increased energy contribution to lift the electrons in a higher energetic level of the atoms is necessary without violating the Pauli exclusion principle. In the latter case, the overlapping of electrons from the colliding particles followed by promotion of electrons to higher, unoccupied atomic levels. With the approach of the particles towards each other, the energy increases, because a larger number of the electrons are affected.

2.1.1 *Repulsive Potentials*

According to Fig. 2.1, the interaction between accelerated particles is determined by the repulsive part of the potential. In the following the different contributions to the repulsive potential are introduced. While the interaction between two particles with very high energy can be described with the Coulomb potential, the interaction with high and middle energies must be described with a screened Coulomb potential. Interactions between particles in the low or hyperthermal energy ranges can be approximately depicted by a Born–Mayer potential.

2.1.1.1 Screening Coulomb Potential

In the collision regime, the energy difference between the kinetic energy/velocity of the projectile and the energy of the target electrons strongly determines the choice of the potential. If the kinetic energy of the projectile exceeds the velocity of the outer shell target electrons then the interaction is repulsive only, and a Coulomb potential between the two particles can be applied [3]. On the other hand, for lower kinetic energies of the projectile, quantum mechanical codes are necessary to evaluate the potential because here attractive interactions are more significant. Between these two extreme energy regions, the kinetic energy of the projectile is comparable to or less than the energy of the target electrons. The screening of the nucleus by electron clouds influences the interaction potential [2]. The simplest way to include this effect is to multiply the Coulomb potential with a so-called screening function $\chi(r/a)$, which leads to the two-body screened Coulomb potential

$$V(r) = \frac{Z_1 Z_2 e^2}{r} \chi(r/a), \tag{2.5}$$

where a is the screening length and r is the distance between the two atoms. The screening function tends to be 1 if r = 0 and to be 0 if r = ∞. The term $1/4\pi\epsilon_0\epsilon$ in this equation is set equal 1 [Gaussian system of units, c.f. (2.4)]. Therefore, the screening function can be interpreted as the ratio of the real potential to the Coulomb potential. One of the earliest theoretical expression of the screened Coulomb potential has been proposed by Bohr [2], where the screening functions consist only on an exponential term

$$V(r) = \frac{Z_1 Z_2 e^2}{r} \exp\left(-\frac{r}{a_B}\right), \tag{2.6}$$

where the negative sign demonstrates that the screening is a function of the distance r. This potential quickly decreases with increasing distance r. The Bohr screening length, a_B, is given by

$$a_B = \left(\frac{9\pi^2}{128}\right)^{1/3} \frac{a_o}{\left(Z_1^{2/3} + Z_2^{2/3}\right)^{1/2}} \approx 0.88534138 \frac{a_o}{\left(Z_1^{2/3} + Z_2^{2/3}\right)^{1/2}}, \tag{2.7}$$

where a_0 is the Bohr radius (physical constant, radius of hydrogen atom in its ground state) and given by $a_0 = \hbar^2/e^2 m_e = 0.052918$ nm, with m_e the mass of the electron). The application of this potential is limited on the case that $r < a_o$, that means for high energy collisions only (in general >100 keV).

Later, Firsov [4] has determined a screening function $\chi(r/a)$ on the basis of overlapping Thomas–Fermi atoms. In the Thomas–Fermi model of an atom it is assumed that the electrons are distributed uniformly in a small volume element, but the electron density can still vary from one small volume element to the next. The derivation of the Thomas–Fermi equation for an isolated atom is shown in Appendix A, (A.14), and given by

$$\frac{d^2\chi}{dx^2} = \frac{\chi^{3/2}}{x^{1/2}}, \tag{2.8}$$

where $x = r/a_{TF}$. The Thomas–Fermi or Firsov screening length, a_{TF}, is given by

$$a_{TF} = \frac{3}{\left(32\pi^2\right)^{2/3}} \left(\frac{\hbar^2}{2e^2 m_e Z^{1/3}}\right) = 0.8853 \frac{a_o}{Z^{1/3}}, \tag{2.9}$$

where the average atomic number Z of both particles is used in several forms. Firsov [5] has used

$$Z^{1/3} = \left(Z_1^{1/2} + Z_2^{1/2}\right)^{2/3}, \tag{2.10}$$

and Ziegler et al. [6] have applied

$$Z^{1/3} = \left(Z_1^{0.23} + Z_2^{0.23}\right). \tag{2.11}$$

The Thomas–Fermi differential equation, (2.8), cannot be analytically solved. Therefore, a numerical calculation of the screening function $\chi(r)$ becomes necessary. A number of approximations for the screening function are presented in the literature (see e.g., [3, 6–12]). Some of the most popular approximations of the screening functions are summarized in Table 2.1.

The screening functions are commonly developed by fitting a proposed functional form to available data. This information can be obtained from either first-principle calculations or experimental measurements (e.g., measurement of the scattering cross-section or by channeling experiments).

For interatomic distances < 0.1 nm, potentials by Moliére [14], Lenz and Jensen [9, 10], Ziegler et al. [6], the Kr–C-potential [15] or the Nakagava–Yamamura potential [12] are suggested. Likely, the ZBL potential is one of most applied potentials. In comparison to the rapid decrease of the Bohr screened two-body potential with increasing distance r, the potentials based on the Thomas–Fermi theory decrease more slowly. Consequently, these potentials can be advantageously used in the intermediate distance range. For comparison, in Fig. 2.2 several screening functions in dependence on the reduced distance r/a are summarized. When empirical and measured interatomic potentials are compared, discrepancies are sometimes found. Therefore, correction factors to the screening length of the form

$$f = \alpha\left(Z_1^{1/2} + Z_2^{1/2}\right) + \beta \tag{2.12}$$

have been introduced, where α and β are listed as fitting parameters [16].

2.1.1.2 Born–Mayer Potential

The exponential form of the potentials shown above is based on quantum mechanical considerations of a moderate interpenetration of the closed electron shells [7, 17]. In addition to the Coulomb potentials with exponentially screened terms, there is a two-parameter potential for low collision energies, so-called hyperthermal energy (energies between some ten and a few hundred electron-volt) which can be expressed by

$$V(r) = A_{BM}\exp\left(-\frac{r}{a_{BM}}\right), \tag{2.13}$$

where energy parameter A_{BM} and the screening length a_{BM} are empirical parameters and have to be determined for each particle combination. This potential was originally introduced by Born and Mayer [17] to calculate the lattice energy of crystalline ionic compounds. The Born–Mayer potential describes the situation where the distance between two atoms is somewhat smaller compared to the equilibrium distance (see schematic in Fig. 2.3), i.e. the repulsion between the nuclei is small because the positive charges of the nuclei are nearly screened by the electrons. Figure 2.3 also

Table 2.1 Summary of most common screening functions $\chi(r/a)$, where a_B, a_L, a_{TF}, a_{ZBL}, and a_{NY} are the scattering radii by Bohr [3], Lindhard [11], Firsov [5], Ziegler et al. [6] and Nakagawa and Yamamura [12], respectively

Screening function $\chi(r/a)$	Screening length
Bohr [3]: $\chi\left(\frac{r}{a}\right)=\exp\left(-\frac{r}{a_B}\right)$	$a_B=\frac{a_o}{\left(Z_1^{2/3}+Z_2^{2/3}\right)^{1/2}}$
Lenz [9] and Jensen [10]: $\chi\left(\frac{r}{a}\right)=\exp\left(-3.11\sqrt{\frac{r}{a}}\right)\left(1+3.11\sqrt{\frac{r}{a}}+3.234\left(\frac{r}{a}\right)+1.4586\left(\frac{r}{a}\right)^{3/2}+0.248\left(\frac{r}{a}\right)^2\right)$	$a_L=\frac{0.8853a_0}{\left(Z_1^{2/3}+Z_2^{2/3}\right)^{1/2}}$
Sommerfeld [13] $\chi(r)=\left[1+\left(r/12^{2/3}\right)^{\lambda}\right]^{-3/\lambda}$ with $\lambda=0.772$	
Moliere [14]: $\chi\left(\frac{r}{a}\right)=0.35\exp\left(-\frac{0.3r}{a_F}\right)+0.55\exp\left(-\frac{1.2r}{a_F}\right)+0.1\exp\left(-\frac{6r}{a_F}\right)$	$a_{TF}=\frac{0.8853a_0}{\left(Z_1^{1/2}+Z_2^{1/2}\right)^{2/3}}$
Wilson, Haggmark, Biersack (Krypton–Carbon potential) [15]: $\chi\left(\frac{r}{a}\right)=0.191\exp\left(-\frac{0.278r}{a_F}\right)+0.474\exp\left(-\frac{0.63717r}{a_F}\right)+0.335\exp\left(-\frac{1.919r}{a_F}\right)$	

(continued)

Table 2.1 (continued)

Screening function $\chi(r/a)$	Screening length
Ziegler, Biersack, Littmark (ZBL) [6]: $\chi\left(\frac{r}{a}\right) = 0.18175\exp\left(-\frac{3.1998r}{a_{ZBL}}\right) + 0.50986\exp\left(-\frac{0.9423r}{a_{ZBL}}\right) + 0.28022\exp\left(-\frac{0.4029r}{a_{ZBL}}\right) + 0.02817\exp\left(-\frac{0.20162r}{a_{ZBL}}\right)$	$a_{ZBL} = \frac{0.8854a_0}{Z_1^{0.23}+Z_2^{0.23}}$
Nakagawa and Yamamura [12]: $\chi\left(\frac{r}{a}\right) = \exp\left[-A_1\left(\frac{r}{a_{NY}}\right)\right] + A_2\left(\frac{r}{a_{NY}}\right)^{1.5} - A_3\left(\frac{r}{a_{NY}}\right)^2$ with $A_1 = 1.51$, $A_2 = 0.763\,(Z_1^{0.169} + Z_2^{0.169})/(Z_1^{0.307} + Z_2^{0.307})$ $A_3 = 0.191\,[(Z_1^{0.0481} + Z_2^{0.0481})/(Z_1^{0.307} + Z_2^{0.307})]^{4/3}$	$a_{NY}\, \frac{0.8854a_0}{(Z_1^{0.307}+Z_2^{0.307})^{2/3}}$

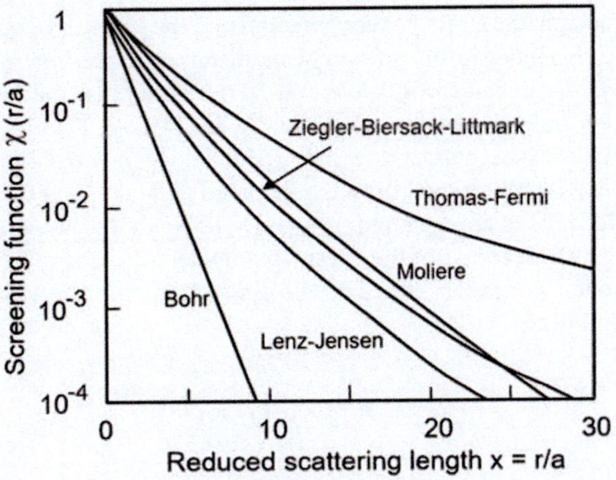

Fig. 2.2 The universal scattering function by Ziegler et al. [6] versus the reduced scattering length $x = r/a_{ZBL}$. For comparison other screening functionare shown

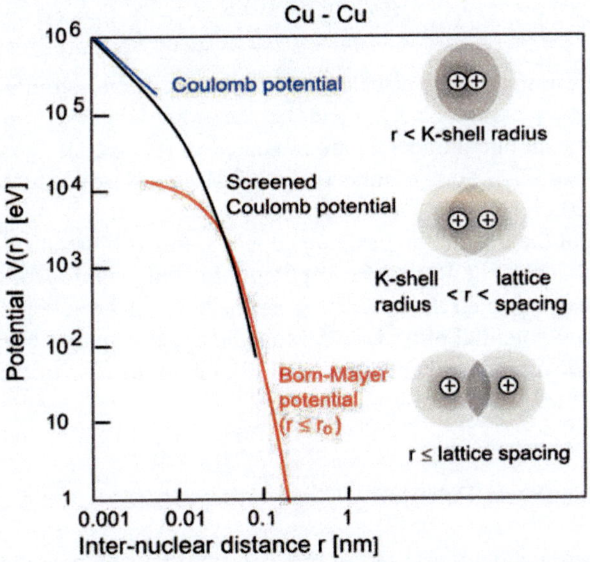

Fig. 2.3 Left: potential function for the interaction between copper atoms. Depending on the inter-nuclear distance r the total potential function is composited by the Coulomb potential, the screened Coulomb potential and the Born–Mayer potential functions. Right: schematic presentation of the applicability of the different repulsive interatomic potentials. The light gray regions depict the radii between the innermost electronic shells and the ionic radius. The dark gray areas represent the overlap of the electron clouds of the atoms during the collision processes. The lattice spacing of Cu crystal is about 0.36 nm (figures adapted from [18] and modified)

illustrates the fact that the Born–Mayer potential and the screened Coulomb potential show the same dependence on inter-nuclear distance in the intermediate region. Extrapolation of the linear part of potential V(r) reveals that the interval of inter-nuclear distance r ranged between $1.5a_0$ and $6\ldots8a_o$ [19].

For homo-nuclear pair interactions (interaction between two particles of the same chemical species), Abrahamson [19] has calculated and tabulated the value for A_{BM} and a_{BM}. While the screening length a_{BM} only depends to a small extent on the nuclear spacing r, the increase of the energy parameter A_{BM} correlates strongly with the atomic number. For hetero-nuclear pair interactions the Born–Mayer potential must be modified [8]

$$V(r) = A_{BM}^{Z_{12}} \exp\left(-\frac{r}{a_{BM}^{Z_{12}}}\right), \tag{2.14}$$

where the following combinations rules are used

$$A_{BM}^{Z_{12}} = \sqrt{A_{BM}^{Z_1} A_{BM}^{Z_2}} \quad \text{and} \quad a_{BM}^{Z_{12}} = 2\frac{a_{BM}^{Z_1} a_{BM}^{Z_2}}{a_{BM}^{Z_1} + a_{BM}^{Z_2}}. \tag{2.15}$$

Usually, the parameters can be calculated from the compressibility and the elastic moduli. The energy parameter $A_{BM}^{Z_i}$ and the screening length $a_{BM}^{Z_i}$ of the chemical species i with the atomic number Z_i are tabulated in [8]. A more generalized set of parameters $A_{BM}= 52(Z_1Z_2)^{3/4}$ eV and $a_{BM} = 0.0219$ nm were proposed by Andersen and Sigmund [20].

An example of the interaction between copper atoms is illustrated in Fig. 2.3. It is clearly shown that that the Born–Mayer potential is only significant for low energy or larger distances between the interacting partners. It has been demonstrated that a Born–Mayer type potential provides a reasonable approximation of the interatomic potential in the near threshold energy range [20], i.e. in the case of the hyperthermal ion bombardment.

2.1.1.3 Inverse Power Potential

As first described and demonstrated in Fig. 2.3, no repulsive potential exists which is valid for all inter-nuclear distances r and can be applied successfully. Logically, the potential function must be composed piecewise.

Lindhard et al. [11] have proposed to approximate the interatomic potential by simple analytic expressions. It is suggested that the screening function $\chi(r/a)$ in (2.5) can be approximated piecewise, assuming that the potential V(r) is proportional to r^{-s}, i.e. $V(r) \propto C/r^s$, where C is a constant and s = 1/2, 1, 2, 3. Thereby, the Thomas–Fermi screening function yields to

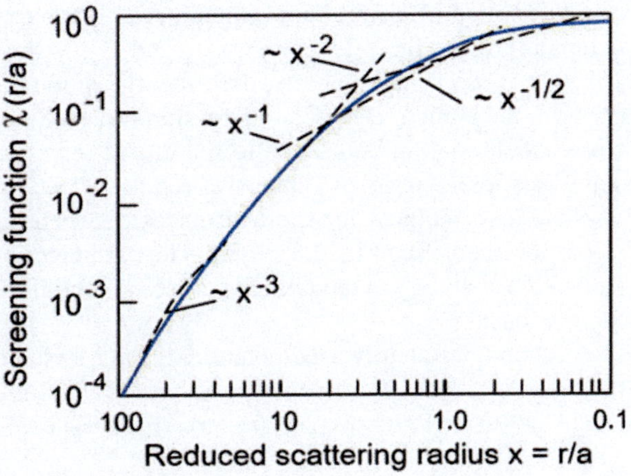

Fig. 2.4 Thomas–Fermi screening function and the piecewise approximation of this function by the power function x^{-s} with $x = r/a_{TF}$

$$\chi(x) = \frac{k_s}{s}\left(\frac{a_{TF}}{r}\right)^{s-1}, \tag{2.16}$$

where k_s is an adjustable parameter. In Fig. 2.4 the Thomas–Fermi screening function and its piecewise approximation in limited intervals of $x = r/a_{TF}$ are shown. For example, the power s must be 1 for low values of r (unscreened Coulomb potential) and 2 or 3 for larger values of r. Then, the screening function is given by

$$\chi(x) = a_1 \exp(-b_1 x) + a_2 \exp(-b_2 x) + \cdots + a_n \exp(-b_n x). \tag{2.17}$$

The advantages of this power law potential description is (i) to find simple approximations of the screening function over defined ranges of r/a_{TF} for s = 1/2, 1, 2 and 3 [19] and (ii) the description of the differential cross section as a function of only one parameter (see Sect. 2.2.5).

2.1.2 *Attractive Potentials*

When the energy of the interacting particles is comparable with the hyperthermal energy, i.e. the particle separation is around between 0.05 and 0.3 nm, the attractive part of the interaction between the atoms in the potential cannot be ignored (see Fig. 2.1). Here the Lennard–Jones potential

$$V(r) = 4E_{LJ}\left[\left(\frac{r_{LJ}}{r}\right)^{12} - \left(\frac{r_{LJ}}{r}\right)^{6}\right] \tag{2.18}$$

can be applied in a first approximation [22], where r_{LJ} is the distance where the potential crosses the zero energy axis and E_{LJ} is the well depth (minimum at $r = 2^{1/6} r_{LJ} \approx 1.12\, r_{LJ}$). The r^{-12} term describes the short range electronic repulsion and the r^{-6} term the attractive London dispersion forces. Small distortions in the electronic closed shell configuration lead to the dipole interactions proportional to r^{-6} (Van der Waals interaction). An intensive repulsion between two atoms proportional to r^{-12} can be calculated for short distances between both atoms (overlapping of the electronic shells, see Fig. 2.3, right). The parameter of the equilibrium spacing and well depth of energy in the Lennard–Jones potential must be determined by ab-initio calculations.

A second, frequently used potential is the Morse potential [23] given by

$$V(r) = E_M \exp[-2a_M(r - r_M)] - 2E_M \exp[-a_M(r - r_M)], \qquad (2.19)$$

where E_M is the well depth, a_M is a parameter which determines the width of the potential barrier and r_M is the equilibrium distance between two particles. Similar to the Lennard–Jones potential, the Morse potential is also a combination of a short-range repulsion term (first term) and a long-range attractive term (second term).

A major drawback of these combined potentials is that they have a very limited range of applicability.

2.2 Collisions Between Ions and Atoms

The interaction of an incident projectile which could be an ion or atom with a stationary target atom can be described as a binary collision event. Both particles, incident ion and target atom interact with each other through a potential. The collision can be characterized by loss of the kinetic energy and a change of the direction of propagation of the incident atom and a gain in the kinetic energy of the target atom, which is recoiled from the initial position. The collision is termed to be elastic under the assumption that electronic excitations during the collision process can be neglected. This condition implies that the total momentum of the collision partners during the collision is constant and also the total energy is conserved during the collision, which leads to the assumption that the total kinetic energy is conserved before and after the collision. Although an elastic collision is an idealized approximation of real interactions, there are many applications where this approximation is appropriated. The theoretical background of the collision processes and the scattering phenomena have been described in numerous textbooks [24–32].

For the following considerations, nuclear reactions and relativistic collisions are not taken into account. Under the assumption that the nuclear radii from target and projectile particles are smaller than the distance of the closest approach (impact parameter, see Sect. 2.2.2) no nuclear reactions are to be expected. Non-relativistic collisions are given, if the collision velocity is considerably smaller compared to the speed of light.

2.2.1 *Kinematic of Binary Elastic Collisions*

The usual description of elastic collisions is based on the application of the conservation laws, in particular conservation of energy, as well as linear and angular momentums. In this case, the momentum and the total kinetic energy are conserved, whereas in an inelastic collision there is a transfer between kinetic energy and internal energy of the collision partners.

Collision problems can be commonly discussed in two different coordination systems, the laboratory (lab) and the center-of-mass (cm) systems. For an observer in the laboratory, the lab-system is the frame to describe collision experiments. Frequently, the detailed mathematical analysis of collision events is simpler in the cm-system. The cm-coordination system moves with respect to the lab-coordination system so that the origin is concordant with the center-of-mass. In Fig. 2.5 the collision between two particles and the definition of the angles (scattering angle θ and recoiling angle φ) are schematically shown in a *laboratory system*. It is assumed, that an incident particle of mass M_1 and kinetic energy E (or velocity v) strikes an individual atom with the mass M_2 which is at rest ($v_2 = 0$). As result of the collision, the moving particle will be deflected from the initial direction (on the right side of Fig. 2.5) by an angle θ and acquires a velocity v_1, while the other atom will scatter by an angle φ relative to the direction of the initial particle motion.

The following equations reflect the conservation of the energy and the momentum parallel and perpendicular to the direction of the incident particle

$$\tfrac{1}{2}M_1 v^2 = \tfrac{1}{2}M_1 v_1^2 + \tfrac{1}{2}M_2 v_2^2, \tag{2.20}$$

$$M_1 v = M_1 v_1 \cos\theta + M_2 v_2 \cos\varphi, \tag{2.21}$$

$$0 = M_1 v_1 \sin\theta - M_2 v_2 \sin\varphi. \tag{2.22}$$

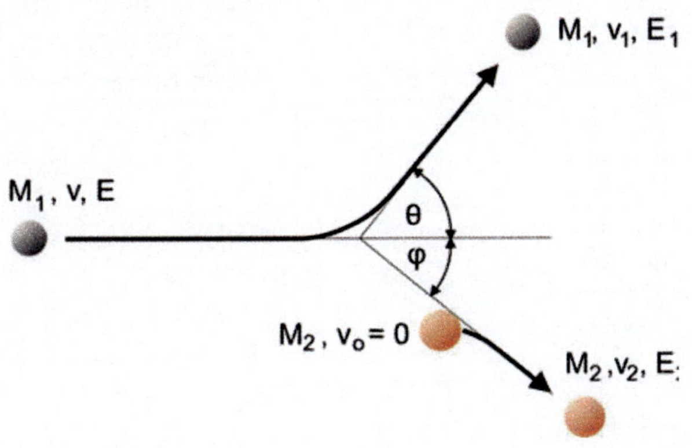

Fig. 2.5 Velocities and energies before and after a binary collision in the lab-system

Using these equations, the final velocities v_1 and v_2 can be determined depending on the initial velocity v and the scattering angles θ and φ, respectively,

$$\left(\frac{v_1}{v}\right)^2 - 2\left(\frac{v_1}{v}\right)\left(\frac{M_1}{M_1+M_2}\right)\cos\theta - \left(\frac{M_2-M_1}{M_1+M_2}\right), \tag{2.23}$$

$$v_1 = v\left(\frac{\sqrt{M_2^2 - M_1^2\sin^2\theta} + M_1\cos\theta}{M_1+M_2}\right)^2, \tag{2.24}$$

$$v_2 = v\frac{2M_1}{M_1+M_2}\cos\varphi. \tag{2.25}$$

Consequently, the energy of the scattered particle and the transferred energy on the recoil particle after the collision are given by

$$E_1 = E\left(\frac{\sqrt{M_2^2 - M_1^2\sin^2\theta} + M_1\cos\theta}{M_1+M_2}\right)^2, \quad (0 \le \theta \le 180°) \tag{2.26}$$

and

$$E_2 \equiv T = \gamma E\cos^2\varphi = T_{\max}\cos^2, \quad (0 \le \varphi \le 90°) \tag{2.27}$$

with the transfer energy efficiency factor

$$\gamma = \frac{4M_1M_2}{(M_1+M_2)^2}. \tag{2.28}$$

In a head-on collision, the projectile is scattered either through an angle $\theta = 180°$ for $M_2 > M_1$ or through $\theta = 0°$ for $M_2 < M_1$.

On the base of relations (2.25)–(2.27), the following cases can be discussed:

Mass ratio	Scattering angle	Transferred energy	Sign	Comment
$M_1 < M_2$ or $M_1/M_2 < 1$	$0° \le \lvert\theta\rvert \le 180°$	$\frac{T}{E_1} = \frac{2}{\pi}(1-\cos\theta)$	+ sign is taken in (2.26)	From no scattering up to total back-scattering
$M_1 = M_2$ or $M_1/M_2 = 1$	$0° < \lvert\theta\rvert < 90°$	$\frac{T}{E_1} = \sin^2\theta$		

(continued)

(continued)

Mass ratio	Scattering angle	Transferred energy	Sign	Comment
$M_1 > M_2$ or $M_1/M_2 > 1$	$0° < \|\theta\| < 90°$	$\frac{T}{E_1} = \left(\frac{M_1}{M_2}\right)\theta^2$	+ and – signs are taken in (2.26)	Forward scattering, only

The scattering angle φ of the recoiled particle varies between 0° and 90° independent of mass.

It should be noted, that (2.26) and (2.27) are the fundamental relations for surface analysis by Rutherford backscattering, ion-scattering spectrometry and elastic recoil spectrometry.

Using (2.26)–(2.28), the relation between both scattering angles can be determined to

$$\tan\theta = \frac{M_2}{M_1 - M_2}\frac{\sin 2\varphi}{\cos 2\varphi}. \quad (2.29)$$

In Fig. 2.6 the quotients v_1/v $(= E_1/E)$, termed as the kinematic factor and the ratio E_2/E, are plotted as function of the scattering angle θ (Fig. 2.6, left) and recoiling angle φ (Fig. 2.6, right) for different values of the mass ratio M_2/M_1, respectively.

The collision process in the *cm-coordination* system is schematically shown in Fig. 2.7. In contrast to the laboratory system, where one particle (projectile) moves initially and the other particle is stationary, in the center-of-mass system a point CM is defined where the mass of the two particles may be considered to be concentrated and where external forces may be applied. The concept of the center-of-mass is that of an average of the masses determined by their distances from a reference point

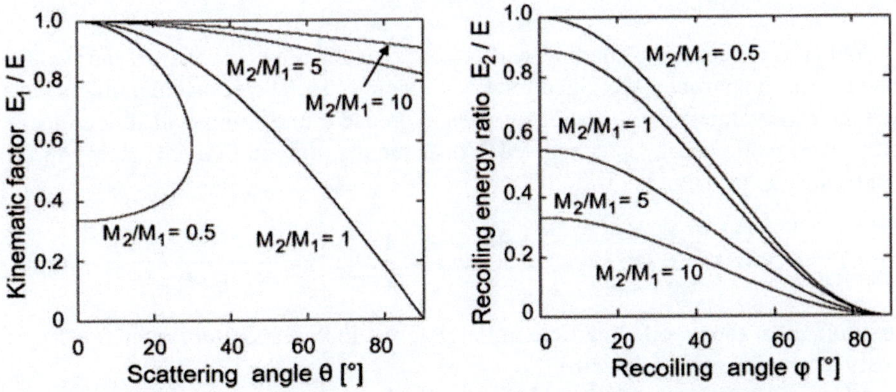

Fig. 2.6 Scattering energy ratio (kinematic factor) for the scattered particles with the mass M_1 as function of scattering angle θ (left) and the recoiling energy ratio for the recoiled particles with the mass M_2 as function of the recoiling angle φ (right). The parameter is the mass ratio M_2/M_1 different values of the mass ratio M_2/M_1, respectively

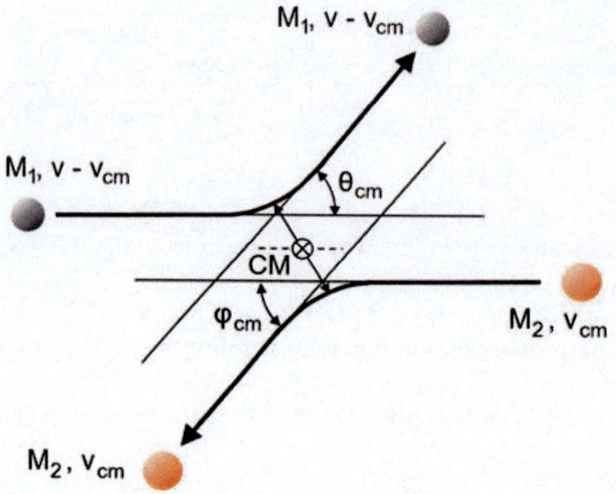

Fig. 2.7 Schematic of a binary elastic collision in the center-of-mass system between a projectile with the mass M_1 and a resting atom with the mass M_2. The center-of-mass is marked by CM

(see Fig. 2.7). From the conservation of the momentum consideration, the following equation can be obtained

$$(M_1 + M_2)v_{cm} = M_1 v \quad \text{or} \quad v_{cm} = \frac{M_1}{M_1 + M_2} v = \frac{M_r}{M_2} v, \tag{2.30}$$

where a reduced mass can be defined as

$$M_r \equiv \frac{M_1 M_2}{M_1 + M_2}. \tag{2.31}$$

With that the center-of-mass velocity v_{cm} of a system of two particles is the average velocity of the particles weighted relative to their mass. The collision partners have the same the velocity in the center-of-mass reference frame before and after collision since the kinetic energy is conserved. Consequently, the total kinetic energy in the center-of-mass system is given by

$$E_c = \frac{M_1}{2}(v - v_{cm}) + \frac{M_2}{2} v_{cm} = \frac{1}{2} M_r v^2 = \frac{M_2}{M_1 + M_2} E, \tag{2.32}$$

assuming that the recoiled particle initially at rest in the laboratory reference frame (barycentric coordinate system).

In a similar fashion, relationships between the scattering angles in both systems can be obtained:

$$\tan\theta = \frac{M_2 \sin\theta_{cm}}{M_1 + M_2 \cos\theta_{cm}}, \tag{2.33}$$

and

$$\varphi = \frac{1}{2}(\pi - \theta_{cm}). \tag{2.34}$$

Combining (2.27) and (2.34), the transferred energy on the particle with mass M_2 is given by

$$T = \gamma E \sin^2 \frac{\theta_{cm}}{2}, \tag{2.35}$$

where for the head-on collision ($\theta_{cm} = 0$). The maximum energy which may be transferred is

$$T_{max} = \frac{4M_1M_2}{(M_1 + M_2)^2} E = \gamma E, \tag{2.36}$$

where γ transfer energy efficiency factor. Corresponding to the maximum energy transfer, the minimum energy transfer is given by

$$T_{min} = \frac{(M_1 - M_2)^2}{(M_1 + M_2)^2} E, \tag{2.37}$$

which can be carried away by the incident particle. A great difference in the masses of the collision partners leads to a reduction of the energy transfer. If $M_2 \gg M_1$ then is $\theta = \theta_{cm}$ and in the case that $M_2 = M_1$ then is $\theta = \theta_{cm}/2$.

All of these relations are independent of the character of the interaction between the collision partners, since the only conservation of momentum and energy are considered.

2.2.2 *Dynamics of the Binary Elastic Collisions*

The determination of the scattering angle θ_{cm} and the transformation in the laboratory angles are aims of numerous scattering studies. In order to calculate the scattering angles, the trajectories of the scattered and recoiled particles after a binary collision process must be known. Depending on the incident kinetic energy, the nature of the interaction potential V(r) and the magnitude of the impact parameter b, calculations of the trajectories were performed in a cm-coordination system (see e.g., [27, 32] and Appendix B).

The interaction of two particles in the scattering process can be simplified significantly by application of the concept of motion of a particle with the reduced mass M_r (see 2.31). In the central force field, which is centered at the origin of center-of-mass coordinates (see Fig. 2.8), it is assumed that two particles interact during the collision process by the central collisions force, i.e. the force acts on the line connecting the

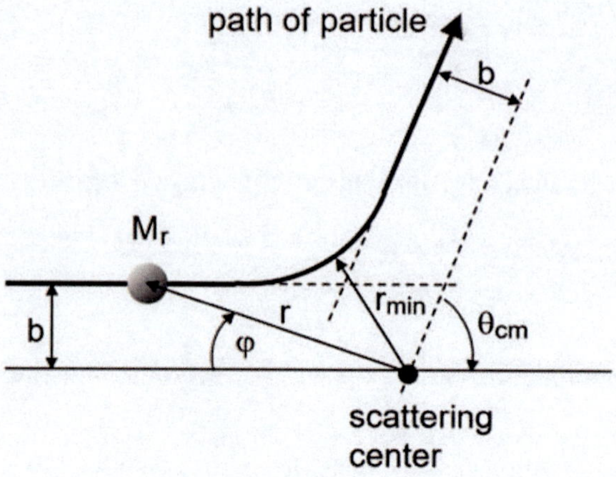

Fig. 2.8 Trajectory of a particle with the reduced mass M_r in a central force field at the point of the closest approach (turning point). φ is the angle between the line linking the scattering center and the path asymptote is called impact parameter and gives the vertical distance between the scattering center and the incident particle. The parameter r_{min} is the closest approach during the scattering process

centers of these particles. In the following, the two-body collision problem is reduced to a one-body collision problem. This is possible because in the cm-system linear momentum of particles within the collision process is always zero.

According to (B.7) in Appendix B, the magnitude of angular momentum at the motion of a particle in the center force field is conserved and can be described by

$$L = M_r r^2 \frac{d\varphi}{dt} \quad \text{or} \quad L = M_r b \frac{dr}{dt} = b\sqrt{2M_r E_c} = M_r v b, \quad \text{i.e.} \ \frac{d\varphi}{dt} = \frac{vb}{r^2}, \tag{2.38}$$

because the angular momentum is independent of time and v is the initial velocity of the particle with the reduced mass.

In Appendix B, the formal expression (B.12) for motion of a particle with reduced mass M_r in the central field under the condition of the conservation of energy and the angular momentum, the so-called radial velocity, is given to

$$\frac{dr}{dt} = \sqrt{\frac{2}{M_r}[E_c - V(r)] - \left(\frac{L}{M_r r}\right)^2} = v\sqrt{1 - \frac{V(r)}{E_c} - \left(\frac{b}{r}\right)^2}, \tag{2.39}$$

where E_c is the total kinetic energy (note that in cm-system $E_c = E_r$, see Appendix B). By integration of (2.39), the radial position in dependence of time, the time integral [see (B.13)], can be obtained. In the experiments, this information cannot be determined. Scattering experiments deliver the deflection angle. This angle can be calculated by integration of (2.39)

$$\varphi = \int \frac{Ldr}{r^2\sqrt{2M_r[E_c - V(r)] - \left(\frac{L}{r}\right)^2}} + const.$$

$$= \int \frac{bdr}{r^2\sqrt{1 - \frac{V(r)}{E_c} - \left(\frac{b}{r}\right)^2}} + const., \tag{2.40}$$

where (B.7) is used in the form $d\varphi = \left(L/M_r r^2\right)dt$.

At the point of the closest approach (turning point), the distance between the scattering center and the particle reaches a minimum ($r = r_{min}$). In this case (see Fig. 2.8), where the energetic term in (2.40) is zero, follows

$$1 - \frac{V(r_{min})}{E_c} - \left(\frac{b}{r_{min}}\right)^2 \equiv 0, \tag{2.41}$$

because radial velocity $dr/dt = 0$. When $dr/dt = 0$ for each impact parameter b, the distance of the closest approach is given by

$$r_{min} = \frac{b}{\sqrt{1 - \frac{V(r_{min})}{E_r}}}. \tag{2.42}$$

With knowledge of the potential (see Sect. 2.1) and with (2.42) the distance of the closest approach r_{min} for $b = 0$ (head-on collision) can be estimated to

$$r_{min} = \frac{Z_1 Z_2 e^2}{E_o}\frac{M_1 + M_2}{M_2}, \tag{2.43}$$

where a Coulomb potential is assumed.

An aim of scattering experiments is the measurement of the asymptotic scattering or deflection angle θ_{cm} of a particle trajectory after collision in the central force field with a potential V(r) as function of the impact parameter and the energy E_c. According Fig. 2.8, the trajectory r(t) is symmetric about the connecting line between the point of closest approach and the scattering center (center-of-mass system). Therefore, an integration of (2.40) from r_{min} to infinity should result in the determination of the scattering angle (see Appendix B)

$$\varphi = \int_{r_{min}}^{\infty} \frac{bdr}{r^2\sqrt{1 - \frac{V(r)}{E_c} - \left(\frac{b}{r}\right)^2}}. \tag{2.44}$$

With the substitution of relation between the lab- and cm-systems, (2.34), follows

$$\frac{1}{2}(\pi - \theta_{cm}) = b\int_{r_{min}}^{\infty} \frac{dr}{r^2\sqrt{1 - \frac{V(r)}{E_c} - \left(\frac{b}{r}\right)^2}}, \tag{2.45}$$

which can be transformed to scattering integral or deflection function

$$\theta_{cm} = \pi - 2b \int\limits_{r_{\min}}^{\infty} \frac{dr}{r^2\sqrt{1 - \frac{V(r)}{E_c} - \left(\frac{b}{r}\right)^2}}. \tag{2.46}$$

This equation delivers the asymptotic scattering angle for the case of the two-body central force scattering in dependence on the energy in the center-of-mass E_c, the impact parameter b, and the interatomic potential V(r). The choice of the potential determines the solvability of the scattering integral equation. For example, the equation of the scattering integral for Coulomb and hard-sphere potentials can be solved analytically [29]. Generally, more realistic potentials are very complicated. This equation can then only be solved approximately by numerical methods. For large and small scattering angles, θ_{cm}, approximation is applied. For large scattering angles (collisions with the tendency to be head-on collisions) hard-sphere approximations and for small scattering angles (i.e. large E_c and b), the integral in (2.47) must be solved by the so-called momentum approximation (see e.g., [30]).

Now, the transferred energy for a head-on collision, given by (2.27), can be rewritten as

$$\frac{T}{E} = \gamma \cos^2\left(\int\limits_{r_{\min}}^{\infty} \frac{b\,dr}{r^2\sqrt{1 - \frac{V(r)}{E_c} - \left(\frac{b}{r}\right)^2}}\right). \tag{2.47}$$

2.2.3 *Scattering Cross-Section*

The result of the interactions between an incident particle beam and the particles in the target is that some of the particles in the beam are deflected and emerge from the target traveling along a direction with the scattering angles θ and φ in regard to the original beam direction. On the other side, some of the beam particles leave the target non-scattered. The number of scattered particles along a well-defined direction of the scattering angles are counted in a detector (schematically illustrated in Fig. 2.9).

It can be assumed that if a target is bombarded by an uniform particle flux with identical kinetic energy (velocity), the scattering cross-section σ gives the probability that a scattering event takes place between two collision partners. The scattering cross-section (measured in barn, where 1 barn ≡ 10^{-28} cm^2) can be defined by the ratio of the number of collisions per target scattering center and time unit to the number of projectiles per time and area (i.e. per ion current density). It is the simplest assumption that the effective cross-section can be interpreted as the (circular-) area with the radius b (impact parameter) which must be met by the primary particle to trigger the corresponding collision process. Then, the scattering cross-section is given by

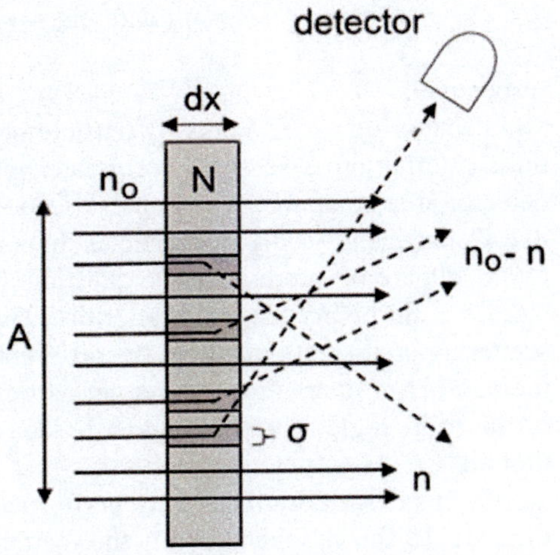

Fig. 2.9 Schematic representation of the scattering process (N is atomic number density)

$$\sigma = \pi b^2 \tag{2.48}$$

and the total scattering cross-section is given by the sum over the individual scattering cross-sections

$$\sigma = \sum_i \sigma_i. \tag{2.49}$$

The probability P that an incident particle sustains a collision process with a target atom is given by the total scattering cross-section per cross-section area A of the incident particle beam (Fig. 2.8) and can be expressed by $P(E) = N\sigma dx$. When it is assumed that the fraction of scattered particles, d(n_o-n), is scattered in the infinitesimal layer thickness dx of a thin layer with a total thickness d, then the fraction of the scattered particles is given by

$$\frac{d(n_o - n)}{n_o} = -\frac{NA\sigma}{A}dx = -N\sigma dx, \tag{2.50}$$

where N is the atomic number density of the target and can be calculated by $N = \rho \cdot N_A/M$ [ρ is the mass density, M is atomic weight and N_A is the Avrogadro number (= 6.022 × 10^{23} atoms/mol)]. The number of the scattered particles can be calculated by integration of (2.50) over the total thickness

$$\int_{n_o}^{n} \frac{d(n_o - n)}{n_o} = -N\sigma \int_0^d dx, \tag{2.51}$$

where n_o is the number of the incident particles. Then, the number of the transmitted and scattered particles can be expressed by

$$n = n_o \exp(-N\sigma d) \quad \text{and} \quad n_o - n = n_o\left[1 - \exp(-N\sigma d)\right], \tag{2.52}$$

respectively.

In scattering experiments by particle bombardment of thin targets the differential scattering cross-section is of greater interest, because a detector with a limited detector area is capable to establish a scattered particle only in a solid angle element dΩ. Consequently, only the particles that are scattered into an infinitesimal angular range, dθ, are of interest. So the question arises, which range of the impact parameter db is necessary for this scattering? This can be described with the differential scattering cross-section, where the differential scattering cross-section is defined as the number of interactions per target particle that lead to scattering into an element of the solid angle at a given angle divided by the number of incident particles per unit area.

For this description is usually performed in the center of mass reference frame. Figure 2.10 shows schematically the scattering of a particle by a particle at rest as a result of a repulsive force.

According to the above definition, the incident particles, n_o, (see Fig. 2.9) passing the infinitesimal area dσ (black area in Fig. 2.10, left) would to be scattered into the infinitesimal solid angle dΩ (black area in Fig. 2.10, right), i.e. the differential scattering cross-section is given by the ratio dσ/dΩ.

Based on a simple geometric consideration, the number of incident particles on the infinitesimal area is given by (see Fig. 2.10)

$$n_o \cdot d\sigma = n_o b d\varphi db = 2\pi n_o b db \tag{2.53}$$

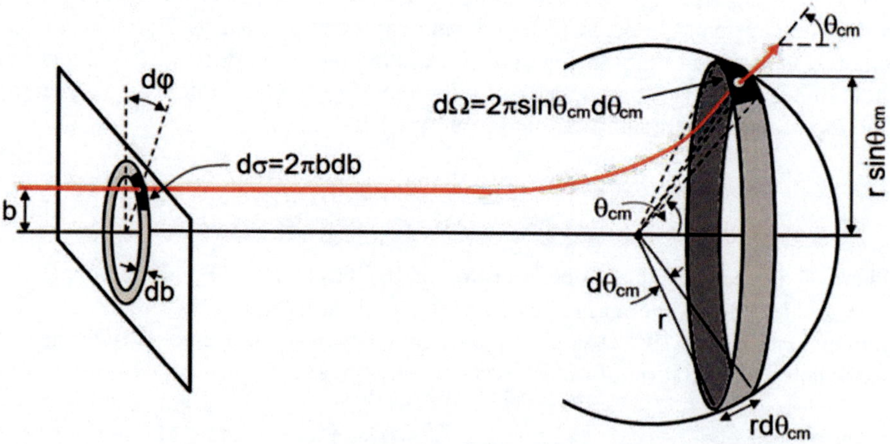

Fig. 2.10 The trajectory of an incident particle at repulsive interaction with a stationary target particle is schematically shown in the center-of-mass system, where the impact parameter b indicates the point of the closest approach. The area of the grey ring (left) is $2\pi bdb$. The radius of the grey ring (right) is $r \cdot \sin\theta_{cm}$, the circumference $2\pi r \cdot \sin\theta_{cm}$ and the width is $r \cdot d\theta_{cm}$ (see Appendix C)

or

$$d\sigma = 2\pi b db, \tag{2.54}$$

where the azimuthal angle, φ, can be eliminated, because the force is considered as a central force and thus has an axial symmetry (i.e. $\varphi = 2\pi$).

The area of grey ring (Fig. 2.10, right) is $2\pi r \cdot \sin \theta_{cm} \cdot r d\theta_{cm} = 2\pi r^2 \cdot \sin \theta_{cm} d\theta_{cm}$ and the solid angle is Ω = area/$r^2 = 2\pi \sin \theta_{cm} d\theta_{cm}$ (see Appendix C). Then, the number of particles scattered into the solid angle is given by

$$n_o \cdot \frac{d\theta_{cm}}{r^2} = n_o \cdot d\Omega = n_o \frac{(r d\theta_{cm})(r \sin \theta_{cm} d\varphi)}{r^2} = n_o \sin \theta_{cm} d\theta_{cm} d\varphi \tag{2.55}$$

or

$$d\Omega = 2\pi \sin \theta_{cm} d\theta_{cm} \tag{2.56}$$

Consequently, the differential scattering cross-section per unit solid angle in the center-of-mass system can be defined by

$$\frac{d\sigma}{d\Omega} = \frac{2\pi b db}{2\pi \sin \theta_{cm} d\theta_{cm}} \tag{2.57}$$

or

$$\frac{d\sigma}{d\Omega} = \frac{b}{\sin \theta_{cm}} \left| \frac{db}{d\theta_{cm}} \right|, \tag{2.58}$$

where the absolute value sign is used to demonstrate that the cross-section is positive. As an example, Fig. 2.11 shows the calculated scattering cross-section for Ar ion implantation in two different metals. The scattering cross-section decreases with increasing scattering angle and energy of the incident ions.

In the laboratory coordination system, the differential scattering cross-section can be written without explicit description by

$$\left(\frac{d\sigma}{d\Omega}\right)_{lab} = \frac{d\sigma}{d\Omega} \cdot \frac{\left(M_r^2 + 2M_r \cos \theta_{cm} + 1\right)^{3/2}}{M_r^2 \cdot |M_r + \cos \theta_{cm}|}. \tag{2.59}$$

Now, the total scattering cross-section can be calculated by integration over all solid angles

$$\sigma_{tot} = \int \frac{d\sigma}{d\Omega} d\Omega = \int_0^{2\pi} d\varphi \int_o^{\pi} \sin \theta d\theta \frac{d\sigma}{d\Omega}. \tag{2.60}$$

Fig. 2.11 Scattering cross-sections for Ar ion implantation into Au and Cu as function of the scattering angle (for calculation a ZBL potential is used)

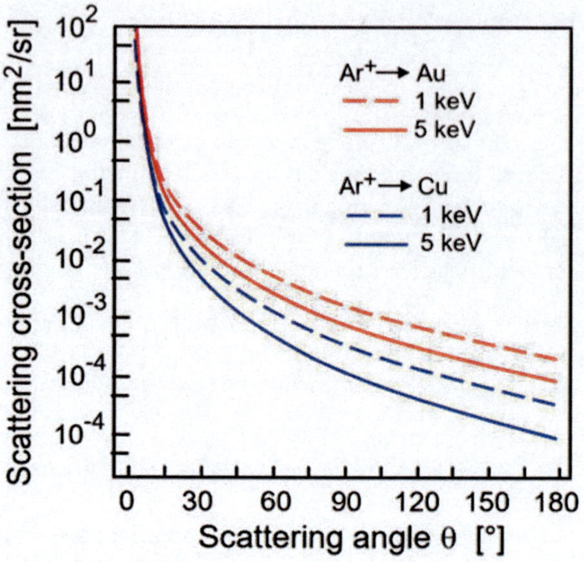

In contrast to the differential cross-section, the total cross-section is in both frames (center-of-mass and lab-systems) the same, because the total numbers that are involved are not depended on the frame.

Then, the impact parameter can be obtained by integration of (2.58)

$$b^2 = 2\int_{\theta_{cm}}^{\pi} \frac{d\sigma}{d\Omega} \sin\theta_{cm} d\theta_{cm}. \tag{2.61}$$

Example 1: Scattering Cross-Section of Hard Spheres

As an example, the differential and total cross-section for elastic scattering of hard spheres should be calculated. The angle between the direction of impact and the surface normal, α, is the same as between recoil direction to this normal. Figure 2.12 illustrates the scattering, where θ is the scattering angle. According Fig. 2.12, the impact parameter b can be expressed by

$$b = R\sin\alpha = a\sin\left(\frac{\pi - \theta_{cm}}{2}\right) = -a\cos\left(\frac{\theta_{cm}}{2}\right) \tag{2.62}$$

where R $= r_1 + r_2$ and $\alpha = (\pi - \theta_{cm})/2$ and therefore $b = R\sin\left(\frac{\pi - \theta_{cm}}{2}\right) = R\cos\left(\frac{\theta_{cm}}{2}\right)$.

Then, term $|db/d\theta_{cm}|$ in (2.58) can be expressed by

Fig. 2.12 Geometry of the collision of hard spheres with the diameters r_1 and r_2 in the center-of-mass frame

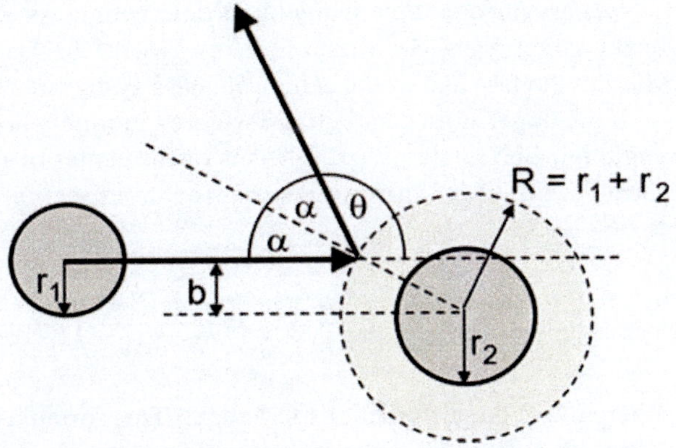

$$\left|\frac{db}{d\theta}\right| = \frac{R}{2}\sin\left(\frac{\theta_{cm}}{2}\right). \tag{2.63}$$

The differential cross-section is given by

$$\frac{d\sigma}{d\Omega} = \frac{1}{\sin\theta_{cm}} \cdot \frac{R}{2}\sin\left(\frac{\theta_{cm}}{2}\right) = \frac{R^2}{2}\frac{\cos\left(\frac{\theta_{cm}}{2}\right)\sin\left(\frac{\theta_{cm}}{2}\right)}{\sin\theta_{cm}} = \frac{R^2}{4}, \tag{2.64}$$

and the total cross-section [see (2.60)], i.e. the integral of the differential cross-section over all angles must be equal to the total or integral cross-section (where $\theta = \theta_{cm}$)

$$\sigma_{tot} = \int d\Omega \frac{d\sigma}{d\Omega} = \int_0^{\pi} \sin\theta d\theta \int_0^{2\pi} d\varphi \frac{d\sigma}{d\Omega} = \frac{R^2}{4}\int_0^{\pi} \sin\theta d\theta \int_0^{2\pi} d\varphi = \frac{R^2}{4}4\pi = \pi R^2. \tag{2.65}$$

The expression represents the area of a sphere of radius *R* projected onto a plane perpendicular to the incoming particle.

Example 2: Rutherford Scattering Cross-Section
The attraction or repulsion of particles because of their electrical charges is called Coulomb or electrostatic interaction. The description of two interacting particles within the formalism of two-body problem involves the motion of two oppositely charged masses separated by a distance r. This interaction can be reduced to a one-dimensional motion of a single particle with the reduced mass M_r and a Coulomb potential. The differential Rutherford scattering cross-section [see Appendix D, (D.14)] is given by

$$\frac{d\sigma}{d\Omega} = \left(\frac{Z_1 Z_2 e^2}{4E}\right)^2 \frac{1}{\sin^4\left(\frac{\theta}{2}\right)}. \tag{2.66}$$

It is assumed that the incident particle with mass M_1 is scattered on an infinitely heavy target mass M_2, which is at rest ($M_1 \ll M_2$). Under this condition, (2.66) is valid in the lab- and in the center-of-mass systems.

If the target atom cannot be considered infinitely heavy, the energy and scattering angle must be reinterpreted in terms of the center-of-mass system. Then, the above formula, Eq. (2.66), gives the scattering cross section in the center-of-mass system to be

$$\left(\frac{d\sigma}{d\Omega}\right)_{cm} = \left(\frac{Z_1 Z_2 e^2}{4E_c}\right)^2 \frac{1}{\sin^4\left(\frac{\theta_{cm}}{2}\right)}, \tag{2.67}$$

where dΩ is the differential solid angle. This formula can be analytically converted to the lab-system for the scattered incident particle

$$\begin{aligned}\left(\frac{d\sigma}{d\Omega}\right)_{lab} &= \left(\frac{d\sigma}{d\Omega}\right)_{cm} \frac{\left[\cos\theta + \sqrt{1-\left(\frac{M_1}{M_2}\right)^2 \sin^2\theta}\right]^2}{\sqrt{1-\left(\frac{M_1}{M_2}\right)^2 \sin^2\theta}} \\ &= \left(\frac{Z_1 Z_2 e^2}{4E}\right) \frac{\left[\cos\theta + \sqrt{1-\left(\frac{M_1}{M_2}\right)^2 \sin^2\theta}\right]^2}{\sin^4\left(\frac{\theta}{2}\right)\sqrt{1-\left(\frac{M_1}{M_2}\right)^2 \sin^2\theta}}\end{aligned} \tag{2.68}$$

and for the recoiled target atom

$$\left(\frac{d\sigma}{d\Omega}\right)_{lab} = \left(\frac{Z_1 Z_2 e^2}{4E_c}\right)^2 \frac{1}{\cos^3(\varphi)}. \tag{2.69}$$

2.2.4 Energy-Transfer Cross-Section

In the previous subsections, the question of the probability with which an ion with energy E transfers an amount of energy between T and T + dT to a target atom, i.e. the question of energy-transfer cross-section, remained unanswered.

It is assumed that the probability of scattering of an ion with energy E passing through a target is given by P(E) = Nσ(E)dx [see (2.50)], where σ(E) is the total scattering cross section. Then the probability P(E,T) that this ion undergoes an energy transfer into the range T and T + dT by a collision in dx is given by

$$\frac{dP(E)}{dT}dT = Ndx\frac{d\sigma(E)}{dT}dT = \frac{1}{\sigma(E)}\frac{d\sigma(E)}{dT}dT \equiv P(E,T)dT. \quad (2.70)$$

Similarly, the probability, $P(\theta_{cm}) = \sigma(\theta_{cm})Ndx$, that an incident ion is deflected by the angle θ after interacting with the scattering target atom can be described as follow

$$\frac{dP(\theta_{cm})}{dp} = N\frac{d\sigma(\theta_{cm})}{dp}dpdx = \frac{1}{\sigma(\theta_{cm})}\frac{d\sigma(\theta_{cm})}{dp}dp \equiv P(\theta_{cm}, b)db, \quad (2.71)$$

where p is the linear momentum (see, e.g., Appendix D) of the particle, θ_{cm} is the scattering angle in the center-of-mass system (see Fig. 2.7) and b is the impact parameter. The total energy-transfer cross-sections can be obtained if the probabilities in (2.70) and (2.71) are set equal to unity [30], i.e.

$$\sigma(E,T) = \int_{T_{min}}^{T_{max}} \frac{d\sigma(E)}{dT}dT \quad \text{and} \quad \sigma(\theta_{cm}) = \int_{b_{max}}^{0} \frac{d\sigma(\theta_{cm})}{db}db, \quad (2.72)$$

where T_{max} is the maximum transferred energy (see 2.36) and T_{min} is the lower limit for the energy transfer process. It should be noted that the two expressions in (2.72) are equivalent.

2.2.5 Simplified Version of the Differential Scattering Cross-Section

It is evident, that the deflection function, (2.46), and with it also the differential cross-section, (2.58), are dependent on six parameters (Z_1, Z_2, M_1, M_2, E_c, and θ_{cm}). Lindhard, et al. [11, 33] have suggested a procedure that allows to depict the differential cross-section as a function of a single-parameter t for screened Coulomb interaction. They introduce the dimensionless parameter

$$t \equiv \varepsilon^2 \frac{T}{T_{max}} = \varepsilon^2 \sin^2 \frac{\theta_{cm}}{2}, \quad (2.73)$$

where the dimensionless energy is called reduced energy and can be expressed by

$$\varepsilon = \frac{a_{TF}E_c}{Z_1Z_2e^2} = \frac{a_{TF}}{Z_1Z_2e^2}\frac{M_2}{M_1+M_2}E = 0.03255\frac{E}{Z_1Z_2}\frac{M_2}{M_1+M_2}, \quad (2.74)$$

where the ion energy is given in units of (eV).

Then, a universal one-parameter differential scattering cross-section equation (c.f. 2.57) in reduced notation was given by Lindhard et al. [33] as

$$d\sigma = -\frac{\pi a_{TF}^2}{2t^{3/2}} f\left(t^{1/2}\right) dt, \tag{2.75}$$

where this relationship is a function of the parameter t, only. The parameter t is proportional to the transferred energy times the initial energy. This so-called 'magic formula', (2.75), is based on a momentum approximation of the scattering process. For a Thomas–Fermi atom, the scaling function $f(t^{1/2})$ as a function of the parameter t is calculated and tabulated in [11].

Lindhard et al. considered $f\left(t^{1/2}\right)$ to be a simple scaling function and the variable t to be a measure of the depth of penetration into the atom during a collision (see Fig. 2.3), with large values of t representing small distances of approach. A generalized expression of the function $f\left(t^{\frac{1}{2}}\right)$ is given by

$$f\left(t^{1/2}\right) = \lambda_m t^{1/2-m}\left[1 + \left(2\lambda_m t^{1-m}\right)^q\right]^{-1/q}, \tag{2.76}$$

where the power law fitting variables m, λ_m and q for the different scattering functions are compiled. For example, the scattering function $f(t^{1/2})$ according to Thomas–Fermi potential can be fitted by $\lambda_m = 1.309$, m = 1/3 and q = 2/3. For six Coulomb potentials with various scattering functions, Winterbon has given the parameter values m, λ and q by last-square fittings (see Appendix of [34]). The value $f(t^{1/2})$ grows with increasing the dimensionless energy ε, the scattering angle θ_{cm}, and the impact parameter b. The importance of this equation is proven by the fact, that this allows a simple analytic solution of the integral equations for the determination of the range and damage distributions (see Chaps. 3 and 4).

A more generalized expression of (2.76) for power law scattering is given by

$$f\left(t^{1/2}\right) = \lambda_m t^{1/2-m} \tag{2.77}$$

and describes the scattering in the presence of a potential of the form $V(r) \propto r^{-1/m}$, approximately, where λ_m is a dimensionless power law fitting variable and given in Table 2.2 for different m values [11].

It should be noted that

(i) for low energies t is small, i.e. m = 1/3 and $V(r) \propto r^{-3}$, yielding a $t^{1/6}$ dependence,
(ii) for high energies is the scattering effect insignificant, i.e. m = 1 and $V(r) \propto r^{-1}$, yielding a $t^{-1/2}$ dependence, and
(iii) for intermediate energies is m = 1/2 and $V(r) \propto r^{-2}$, yielding not dependence, i.e. the cross-section is independent of the energy.

Table 2.2 Values of the constant λ_m for different values m (=1/s)

m	0	1/5	1/4	1/3	1/2	1
λ_m	24	2.92	1.4	1.309	0.327	0.5

Lindhard et al. [11] have applied a power-law approximation to the Thomas–Fermi cross-section with the aim to obtain approximate values of the energy-transfer differential cross-section. Using (2.75) and (2.76), the power-law energy-transfer differential cross-section is given by

$$d\sigma(E, T) = C_m E^{-m} T^{-1-m} dT, \quad 0 \leq T \leq T_{\max}, \tag{2.78}$$

where m is a number between 0 and 1. In detail, Sigmund [35] suggests m = 1 for Rutherford scattering, m = 1/2 for medium-mass ions in the keV range and m = 1/3 in the lower keV and upper eV region. According to Winterbon et al. [36], following values for m are proposed for different ranges of the dimensionless energy ε: m = 1/2 for ε ≤ 0.2, m = 1/3 for 0.08 ≤ ε ≤ 2 and m = 1 for ε ≥ 10 (Rutherford scattering). In the low-energy region (hyperthermal ions) a Born–Mayer potential is used instead of the Thomas–Fermi-potential. In this case m is close to zero [32].

The power law constant in (2.78) is given by

$$C_m = \frac{\pi}{2} \lambda_m a_{TF}^2 \left(\frac{M_1}{M_2}\right)^m \left(\frac{2Z_1 Z_2 e^2}{a_{TF}}\right)^{2m}. \tag{2.79}$$

The following table, Table 2.3, summarizes the energy-transfer differential cross-section for unscreened and screened Coulomb potentials, where $0 < T < T_{max}$ and the screening radius is a Thomas_Fermi screening radius a_{TF}, [see (2.9)].

Assuming a head-on-head collision, the energy transfer is maximum, T_{max}, and is given by (2.36). Then, for collisions which are described with a screened Coulomb potential (e.g., m = 1/2 in the power law approximation) the cross-section is given by [37]

$$d\sigma(E, T) = \frac{Z_1 Z_2 \pi^2 e^2 a_{ZBL}}{4\sqrt{T_{\max}}} \frac{M_1}{M_1 + M_2} T^{-3/2} dT, \tag{2.80}$$

where a_{ZBL} is the Ziegler–Biersack–Littmark screening radius (see Table 2.1).

Table 2.3 Summary of the power law potential energy-transfer cross-section for unscreened and screened potentials

Potential	Power law variable	Energy-transfer differential cross-section $d\sigma(E,T)/dT$
Unscreened Coulomb potential (Rutherford scattering)	m = 1 $\lambda_m = 0.5$	$=(Z_1 Z_2 \pi^2 e^2)^2 \left(\frac{M_1}{M_2}\right) E^{-1} T^{-2}$
Screened Coulomb potential	m = 1/2 $\lambda_m = 0.327$	$\approx Z_1 Z_2 e^2 a_{TF} \left(\frac{M_1}{M_2}\right)^{1/2} E^{-1/2} T^{-3/2}$
	m = 1/3 $\lambda_m = 1.309$	$\approx 2 a_{TF}^{4/3} (2 Z_1 Z_2 e^2)^{2/3} \left(\frac{M_1}{M_2}\right)^{1/3} E^{-1/3} T^{-4/3}$

A Thomas–Fermi screening radius, a_{TF}, is assumed and the power law variables according Winterbon et al. [36] are used (see Table 2.2)

For low-energy collisions the Born–Mayer interatomic potential and m = 0 is assumed. Then the cross-section constant is

$$C_o = \frac{\pi}{2}\lambda_0 a_{BM}^2, \tag{2.81}$$

with $a_{BM} = 0.0219$ nm and $\lambda_{m=0} = 24$ [20].

2.3 List of Symbols

Symbol	Notation
A_{BM}	Born–Mayer energy parameter
C_m	Power law constant
E	Energy of particle
E_c	Total kinetic energy in the center-of-mass
E_{LJ}	Lenard-Jones energy parameter
E_M	Morse energy parameter
E_r	Total energy in cm-system
F	Force
L	Angular momentum
M	Atomic weight
M_i	Mass of particle i
M_r	Reduced mass
N	Atomic number density
P	Probability
T	Transferred energy
T_{max}	Maximal transferred energy
T_{min}	Minimal transferred energy
V(r)	Potential energy, potential
Z_i	Atomic number of particle i
a	Screening length
a_o	Bohr radius
a_B	Bohr screening length
a_{BM}	Born–Mayer screening length
a_F	Firsov screening length
a_L	Lindhard screening length
a_M	Parameter of the Morse potential well

(continued)

(continued)

Symbol	Notation
a_{NY}	Nakagava–Yamamura screening length
a_{TF}	Thomas–Fermi scattering length
a_{ZBL}	Ziegler–Biersack–Littmark (universal) scattering length
b	Impact parameter
d	Thickness
e	Elementary electric charge
$f(t^{1/2})$	Lindhards's scaling function
m	Power law variable (=1/s)
m_e	Mass of electron
n	Number of transmitted particles
n_o	Number of scattered particles
p	Momentum
r	Distance
r_{min}	Distance of the closest approach
s	Power variable in the inverse power potential
t	Dimensionless collision parameter
v	Ion velocity
v_B	Bohr velocity
v_{cm}	Velocity in the center-of-mass system
v_i	Velocity of particle i
Ω	Solid angle
γ	Transfer energy efficiency factor
ε	Reduced energy
θ	Scattering angle
θ_{cm}	Scattering angle in the center-of-mass system
λ_m	Power law fit variable
ρ	Mass density
σ	Scattering cross-section
φ	Scattering angle of the recoiled particle, recoiling angle
φ_{CM}	Scattering angle of the recoiled particle in the center-of-mass system
χ	Screening function

References

1. T. Helgaker, P. Jorgensen, J. Olsen, *Molecular Electronic-Structure Theory* (Wiley, Chichester, 2002)
2. N. Bohr, The penetration of atomic particles through matter. Det. Kgl. Dan. Vid. Selskab. Mat.-Fys. Medd. **XVIII**(8) (1948)
3. M.A. Kumakhov, F.F. Komarov, *Energy Loss and Ion Range in Solids* (Gordon and Breach Science Publishers, New York, 1981)
4. O.B. Firsov, Interaction energy of atoms for small nuclear separations. JETP J. Exp. Theoret. Phys. **5**, 1192–1196 (1957)
5. O.B. Firsov, Calculation of the interaction potential of atoms. JETP J. Exp. Theoret. Phys. **33**, 696–699 (1957)
6. J.F. Ziegler, J.P. Biersack, U. Littmark, *The Stopping and Range of Ions in Solids* (Pergamon Press, New York, 1985)
7. I.M. Torrens, *Interatomic Potentials* (Academic Press, New York, 1972)
8. W. Eckstein, in *Computer Simulation of Ion-Solid Interactions.* Springer Series in Materials, vol. 19 (Berlin, 1991)
9. H. Jensen, Die Ladungsverteilung in Ionen und die Gitterkonstante des Rubidiumbromids nach der statistischen Methode. Z. Phys. **77**, 722–745 (1932)
10. W. Lenz, Über die Anwendbarkeit der statistischen Methode auf Ionengitter. Z. Phys. **77**, 713–721 (1932)
11. J. Lindhard, V. Nielsen, M. Scharff, Approximation method in classical scattering by screened coulomb fields—notes on atomic collision I. Det. Kgl. Danske Vid. Selskab. Mat. Fys. Medd. **36**(10) (1968)
12. S.T. Nakagawa, Y. Yamamura, Interatomic potential in solids and its applications to range calculations. Rad. Eff. **105**, 239–256 (1988)
13. A. Sommerfeld, Integration der Differentialgleichung des Thomas-Fermischen atoms. Z. Phys. **78**, 283–308 (1932)
14. G. Moliére, Theorie der Streuung schneller geladener Teilchen I. Z. Naturforsch. A **2**, 133–145 (1947)
15. W.D. Wilson, L.G. Haggmark, J.P. Biersack, Phys. Rev. B **15**, 2458–2468 (1977)
16. D.J. O'Conner, J.P. Biersack, Comparison of theoretical and empirical interatomic potentials, Nucl. Instr. Meth. Phys. Res. B **15**, 14–19 (1986)
17. M. Born, J.E. Mayer, Zur Gruppentheorie der Ionenkristalle. Z. Phys. **75**, 1–18 (1932)
18. D.R. Olander, *Fundamental Aspects of Nuclear Reactor Fuel Elements* (Technical Information Center, Oak Ridge, 1976)
19. A.A. Abrahamson, Born-Mayer-type Interatomic potential for neutral ground-state atoms with $Z = 2$ to $Z = 105$. Phys. Rev. **178**, 76–79 (1969)
20. P. Andersen, Sigmund. Nucl. Instr. Meth. **38**, 238–240 (1965)
21. P.J. Van den Hoek, A.D. Tenner, A.W. Kleyn, E.J. Baerends, Hyperthermal alkali-ion scattering from a metal surface: a theoretical study of the potential. Phys. Rev. B **34**, 5030–5042 (1986)
22. J.E. Lennard-Jones, On the determination of molecular fields. Proc. Roy. Soc. A **106**, 463–477 (1924)
23. P.M. Morse, Diatomic molecules according to the wave mechanics. II. Vibrational levels. Phys. Rev. **34**, 57–64 (1929)
24. P. Sigmund, Collision theory of displacement damage, ion range, and sputtering. Rev. Roum. Phys. **17**, 823–870, 969–1000, 1079–1106 (1972)
25. L. Landau, E. Lifshitz, *Mechanics* (Butterworth-Heinemann, Oxford, 1982)
26. R.E. Johnson, *Introduction to Atomic and Molecular Collisions* (Plenum Press, New York, 1982)
27. E.S. Mashkova, V.A. Molchanov, *Medium-Energy Ion Reflection from Solids.* Modern Problems in Condensed Matter Sciences, vol. 11 (Elsevier, Amsterdam, 1985)
28. E.W. McDaniel, J.B.A. Mitchell, M.E. Rudd, *Atomic Collisions, Heavy Particles Projectiles* (Wiley, New York, 1993)

29. R.S. Williams, Quantitative intensity analysis of low-energy scattering and recoiling from crystal surfaces, in *Low Energy Ion-Surface Interactions*, ed. by J.W. Rabalais (Wiley, Chichester, 1994), pp. 1–54
30. M. Nastasi, J.W. Mayer, J.K. Hirvonen, *Ion-Solid Interactions: Fundamentals and Applications* (Cambridge University Press, Cambridge, 1996)
31. R. Smith, M. Jakas, D. Ashworth, B. Oven, M. Bowyer, I. Chakarov, R. Webb, *Atomic and Ion Collisions in Solids and at Surfaces* (Cambridge University Press, Cambridge, 1997)
32. H. Goldstein, C.P. Poole, J.L. Safko, *Classical Mechanics* (Addison-Wesley, San Francisco, 2002)
33. J. Lindhard, V. Nielsen, M. Scharff, P. Thomsen, Integral equations governing radiation effects—notes on atomic collisions III. Det. Kgl. Danske Vid. Selskab. Mat. Fys. Medd. **33**, 1–42 (1963)
34. K.B. Winterbon, Heavy-ion range profiles and associated damage distributions. Rad. Eff. **13**, 215–226 (1972)
35. P. Sigmund, Theory of sputtering I. Sputtering yield of amorphous and polycrystalline targets. Phys. Rev. **184**, 383–416 (1969)
36. K.B. Winterbon, P. Sigmund, J.B. Sanders, Spatial distribution of energy deposited by atomic particles in elastic collisions. Det. Kgl. Danske Vid. Selskab. Mat.-Fys. Medd. **37**(14) (1970)
37. L.C. Feldman, J.W. Mayer, *Fundamentals of Surface and Thin Film Analysis* (North-Holland, New York, 1968)

Chapter 3
Energy Loss Processes and Ion Range

Abstract Energetic ions are mainly subject to two loss mechanisms during interaction with target atoms, the nuclear energy loss and the electronic energy loss. The dominating interaction processes of the movement of low-energy particles through matter are elastic collisions with target atoms. In this chapter, the different approaches to determine the nuclear energy loss per unit length, also referred to as nuclear stopping power, are presented in detail. The electronic energy loss, which is less significant for low-energy ions, is briefly described. Based on the knowledge of nuclear and electronic energy loss, the range of incidence particles can be approximately determined as a function of acceleration energy. It can be shown that by use of the projected range and its standard deviation, three-dimensional concentration distribution of the implanted ion species below the surface can be determined. Since often the experimentally determined concentration profiles deviate considerably from a simple expected Gaussian distribution, higher order moments of the Gaussian distribution must be included in the calculation of the concentration distribution. Computer simulations can be used to calculate not only the concentration distribution, but also the trajectories of the particles, the number of reflected ions and sputtered atoms, the distribution of vacancies and interstitials, etc. In this chapter, the two most common computer simulation codes for the analysis of particle-solid interactions, molecular dynamical code and the Monte Carlo code, are briefly presented.

A brief overview of the energy loss and the ion range distribution as well as the computer simulation of ion–solid interaction is presented in this chapter. More details to these topics are available in [1–9].

When an energetic particle (ion, atom) with the kinetic energy E penetrates a material, this particle loses its energy along its trajectory until it rest inside the material. For example, Fig. 3.1 displays energy loss of a He ion penetrates into a Ni target (simulation by Monte Carlo program TRIM, version SDTrimSP). The color indicates the decrease of the energy along its trajectory. The energy loss is the result of the interaction of the particle with the nuclei and electrons of the target material. A distinction is usually made between two loss mechanisms, nuclear energy loss and electronic energy loss [1]. The nuclear energy loss is caused by elastic collisions of the incident particle with target atomic nuclei and can result in the displacement of

B. Rauschenbach, *Low-Energy Ion Irradiation of Materials*, Springer Series in Materials Science 324, https://doi.org/10.1007/978-3-030-97277-6_3

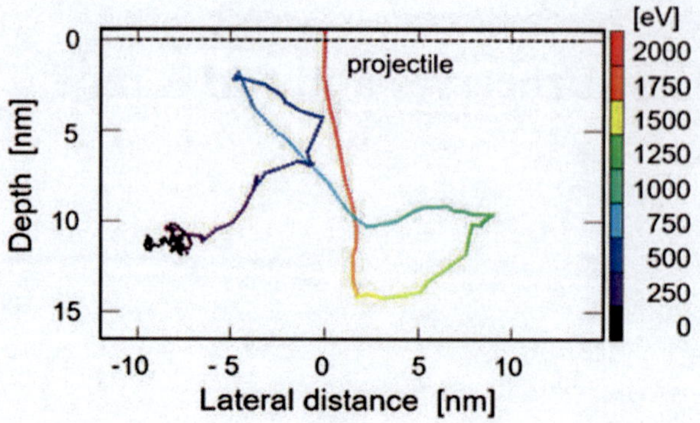

Fig. 3.1 Trajectory of a 2 keV He ion penetrating a Ni target at normal incidence. The He atom is stopped if its energy is smaller than 1 eV (figure is taken from [10])

atoms, while the electronic energy loss is attributed to the inelastic collision with the electrons of the target material due to excitation and ionization.

Accordingly, the average energy loss per unit path length in the material with an atomic number density N is defined by the stopping power S(E). The total energy loss by both collision phenomena can be expressed by

$$S(E) = S_n(E) + S_e(E) = \frac{1}{N}\left(\frac{dE}{dx}\right)_n + \frac{1}{N}\left(\frac{dE}{dx}\right)_e, \tag{3.1}$$

where x is the distance along the particle trajectory (note that the stopping power is a force). It is obvious that the energy loss is determined by incident particle (atomic number, mass) and the target material (atomic number, mass and density).

3.1 Nuclear Energy Loss

The dominating interaction processes of the movement of low-energy particles through matter are elastic collisions with target atoms. These collisions result in an energy transfer to the target atoms. In order to analyze the individual collision processes in detail, different approaches to determine the nuclear stopping power (nuclear energy loss per unit length) are known.

3.1.1 Nuclear Stopping Power Using the Thomas–Fermi Screening Function

Nuclear stopping provides a direct transfer of momentum and energy from the projectile to a target nucleus. The probability P of a particle of energy E to undergo a collision within an infinitesimal distance dx in the bombarded substrate, which leads to a loss of energy between T and T + dT, can be described by (2.70) [5, 11], where the differential energy-transfer cross-section, dσ/dT, was defined in (2.78). By multiplying (2.70) with the transferring energy, the average energy loss for the distance dx follows to

$$\langle dE\rangle = \int T\frac{dP(E)}{dT}dT = N\int_{T_{\min}}^{T_{\max}} T\frac{d\sigma}{dT}dT\,dx \tag{3.2}$$

or

$$\left(\frac{dE}{dx}\right)_n = N\int_{T_{\min}}^{T_{\max}} T\frac{d\sigma}{dT}dT \tag{3.3}$$

for infinitesimal distances. With help of (3.1) follows for the nuclear stopping cross-section

$$S_n(E) = \frac{1}{N}\left(\frac{dE}{dx}\right)_n = \int_{T_{\min}}^{T_{\max}} T\frac{d\sigma}{dT}dT. \tag{3.4}$$

Using (2.78) for the differential energy-transfer cross-section with $T_{\min} = 0$, the nuclear stopping cross-section can be written as

$$\begin{aligned} S_n(E) &= \int_0^{T_{\max}} C_m E^{-m} T^{-1-m} T\,dT \\ &= C_m E^{-m}\int_0^{T_{\max}} T^{-1-m} T\,dT = \frac{C_m E^{-m}}{1-m} T^{1-m}\Big|_0^{T_{\max}} \\ &= \frac{C_m E^{1-2m}}{1-m}\gamma^{1-m} \end{aligned} \tag{3.5}$$

or expressed with (2.36)

$$S_n(E) = \frac{C_m E^{1-2m}}{1-m}\left[\frac{4M_1M_2}{(M_1+M)^2}\right]^{1-m}, \tag{3.6}$$

where for m = 1/3 and m = 1/2 the ranges of validity are given for $\varepsilon \leq 0.2$ and $0.08 \leq \varepsilon \leq 2$, respectively [12]. Assuming m = 1/2 and the constant C_m given in (2.77), the (3.6) can be written by

$$S_n(E) = 4\pi\lambda_{1/2}a_{TF}e^2NZ_1Z_2\frac{M_1}{M_1+M_2}. \tag{3.7}$$

Lindhard et al. [2, 11] have derived the differential cross-section for the screened Coulomb interaction between the incident particle and target atom (see Sect. 2.2.5). For this purpose, the dimensionless reduced energy, ε, was introduced (see 2.74). This allows to define the reduced or normalized stopping cross-section [8], given by

$$s_n(\varepsilon) = \frac{d\varepsilon}{d\rho_L}, \tag{3.8}$$

and also the reduced length, ρ_L, given by

$$\rho_L = 4\pi a_{TF}^2 N\frac{M_1M_2}{M_1+M_2}x. \tag{3.9}$$

Here, a_{TF} is the Thomas–Fermi screening length for the interaction potential (see 2.9) and x is a measure of the distance. On the base of an algebraic transformation shown in Appendix E, the power law approximation to the normalized nuclear stopping cross-section, E.6, is given by

$$s_n(\varepsilon) = \frac{\lambda_m}{2(1-m)}\varepsilon^{1-2m}, \tag{3.10}$$

where λ_m is the power law fitting parameter given in Table 2.2. Lindhard et al. [2] have detailed discussed the behavior of the normalized nuclear stopping cross-section versus the reduced energy ε. For intermediate energies (m = 1/2 and $\lambda_m = 0.327$, see Table 2.2) the normalized stopping cross-section $s_n(\varepsilon) = 0.327$, i.e. the $s_n(\varepsilon)$ is independent of energy. For low energies (m = 1/3) $\lambda_m = 1.309$ and $s_n(\varepsilon) = 0.981\varepsilon^{1/3}$ and for high energies (m = 1, Rutherford energy range) $\lambda_m = 0.5$ and $s_n(\varepsilon)$ drops proportional to 1/ε.

Now, the nuclear stopping power from (3.6), $S_n(E)$, can be expressed for all projectile-target combinations by means of the reduced energy ε and the normalized nuclear stopping power $s_n(\varepsilon)$, (3.10), by

$$S_n(E) = 4\pi a_{TF}Z_1Z_2e^2\frac{M_1}{M_1+M_2}s_n(\epsilon). \tag{3.11}$$

3.1.2 Empirical Relations of the Normalized Nuclear Stopping Power Function

The proposed relation for the normalized elastic stopping cross-section by Lindhard et al. [11] was approximated by different empirical expressions with the aim to improve the accuracy of the calculated values in comparison with the experimental results. For example, the empirical relations of the normalized stopping cross-section based on Thomas–Fermi potential are given by Matsunami et al. [13]

$$s_n^{TF}(\varepsilon) = \frac{3.441\varepsilon^{1/2}\ln(\epsilon + 2.718)}{1 + 6.355\epsilon^{1/2} + \epsilon\left(6.882\epsilon^{1/2} - 1.708\right)}, \tag{3.12}$$

by Strydom and Gries [14]

$$\begin{aligned} s_n^{SG}(\varepsilon) &= -0.0205(\log\varepsilon)^3 - 0.106(\log\varepsilon)^2 - 0.0188(\log\varepsilon) + 0.435 \\ &\quad for\; 0.001 \le \varepsilon < 0.2 \\ s_n^{SG}(\varepsilon) &= 0.0583(\log\varepsilon)^3 - 0.11(\log\varepsilon)^2 - 0.175(\log\varepsilon) + 0.356 \\ &\quad for\; 0.2 \le \varepsilon < 10 \\ s_n^{SG}(\varepsilon) &= 0.5[\ln(1.294\varepsilon)] \\ &\quad for\; 10 \le \varepsilon < 100, \end{aligned} \tag{3.13}$$

by Kalbitzer et al. [15]

$$s_n^{Kal}(\varepsilon) = \frac{1.7\sqrt{\varepsilon}\cdot\ln(\varepsilon + 2.718)}{1 + 6.8\epsilon + 3.4\epsilon^{3/2}}, \tag{3.14}$$

by Wilson et al. [16], especially for low-energy interaction,

$$s_n^{WHB}(\varepsilon) = \frac{0.5\cdot\ln(1+\varepsilon)}{\epsilon + A\epsilon^B}, \tag{3.15}$$

where fitting constants are tabulated in [16]. A Kr–C potential instead a Thomas–Fermi potential given by

$$s_n^{KrC}(\varepsilon) = \frac{0.5\cdot\ln(1 + 1.2288\epsilon)}{\epsilon + 0.1728\epsilon^{1/2} + 0.008\epsilon^{0.1504}} \tag{3.16}$$

was applied by Eckstein et al. [17].

3.1.3 Nuclear Stopping Power Based on the Universal Screening Function

For a more precise calculation of the nuclear stopping power, Ziegler et al. [3] have proposed an approach which based on the universal or ZBL screening function with the scattering length a_{ZBL} (see Table 2.1). Using (3.4), the normalized stopping power (E.6 or E.9) can be given in the generalized form by

$$s_n(\varepsilon) = \frac{\varepsilon}{\pi a_{ZBL}^2 \gamma E} \int_0^{T_{max}} T \frac{d\sigma(E)}{dT} dT, \tag{3.17}$$

where a_{ZBL} is the universal screening function [3] given in Table 2.1. This expression can be rewritten with the help of (2.35) and (2.72) to

$$s_n(\varepsilon) = \frac{\varepsilon}{a_{ZBL}^2} \int_0^{\infty} \sin^2 \frac{\theta_{cm}}{2} d(b^2) \tag{3.18}$$

(details, e.g., see [5]).

The integral in (3.18) was numerically calculated by Ziegler et al. [3]. A practicable relationship for the reduced stopping power cross-section could be obtained by analytical fitting of the results after integration

$$s_n^{ZBL}(\varepsilon_{ZBL}) = \frac{\ln(1 + 1.1383\varepsilon_{ZBL})}{2\left(\varepsilon_{ZBL} + 0.01321\varepsilon_{ZBL}^{0.21226} + 0.19593\varepsilon_{ZBL}^{0.5}\right)} \quad for\ \varepsilon_{ZBL} < 30$$

and

$$s_n^{ZBL}(\varepsilon_{ZBL}) = \frac{\ln(\varepsilon_{ZBL})}{2\varepsilon_{ZBL}} \quad for\ \varepsilon_{ZBL} > 30, \tag{3.19}$$

where ε_{ZBL} is the ZBL reduced energy

$$\varepsilon_{ZBL} = 0.03253 \frac{M_2}{Z_1 Z_2 (M_1 + M_2)\left(Z_1^{0.23} + Z_2^{0.23}\right)} E. \tag{3.20}$$

Then, the nuclear stopping power $S_n(E)$ is expressed as

$$S_n(E) = 8.462 \times 10^{-15} \left(\frac{M_1}{M_1 + M_2}\right)\left(\frac{Z_1 Z_2}{Z_1^{0.23} + Z_2^{0.23}}\right) s_n(\varepsilon_{ZBL}) \tag{3.21}$$

in units of [eV cm^2/atom], where the incident ion energy E is given in units of [eV]. Ziegler et al. [3] have investigated the accuracy of the relationships for nuclear and

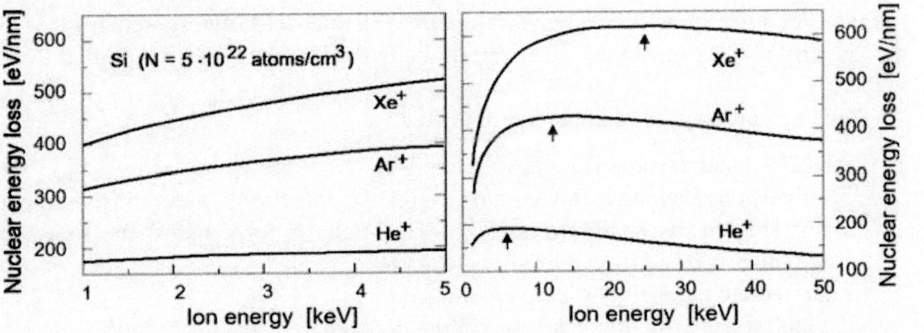

Fig. 3.2 Nuclear energy loss of He, Ar and Xe ions in silicon as function of the incident ion energy. (left for energies up to 5 keV and right for energies up to 50 keV). Arrows indicate the maxima of the energy losses

electronic stopping power (see next subchapter) by comparing results obtained with the above equation on over 13,000 experimental data points from over 1000 published articles. As an example, in Fig. 3.2 the nuclear energy loss of several noble gas ions in dependence of the ion energy after perpendicular implantation in silicon is shown. As anticipated, the energy loss increases with increasing ion mass. The maximum of the nuclear energy loss (arrows in Fig. 3.2, right) is shifted in accordance with ion mass to higher ion energies.

According to Ziegler et al. [3] the second order energy straggling Ω^2 and also the is the third order energy straggling Λ^3 are given by

$$\Omega^2 = 2.601 \times 10^{-13}\left(\frac{2Z_1Z_2}{M_1+M_2}\right)^2 \Omega^2(\varepsilon) \tag{3.22}$$

with $\Omega^2(\varepsilon) = 1/\left(4 + 0.197\varepsilon^{-1.6991} + 6.584\varepsilon^{-1.0494}\right)$ and

$$\Lambda^3 = 1.599 \times 10^{-11}\left(Z_1^{0.23} + Z_2^{0.23}\right)\left(\frac{2Z_1Z_2}{M_1+M_2}\right)^3 \Lambda^3(\varepsilon) \tag{3.23}$$

with $\Lambda^3(\varepsilon) = \left[\left(\frac{1}{0.33495\varepsilon^{2.4145}}\right)^{1/2} + \left(\frac{1}{0.12377\varepsilon}\right)^{1/2}\right]^2$, respectively.

3.2 Electronic Stopping Power for Low Incident Energies

The inelastic or electronic energy loss is the result of the energy loss due to collisions with the target electrons. At high incident particle energies, the inelastic losses are particularly predominant because the target electrons are capable to pick up substantial amounts of energy from the incident particle. The key point in the analysis of the inelastic interaction processes is the ratio of the particle velocity (energy), v, to the velocity of electrons in the first orbit of the Bohr's atom model, v_B, (called Bohr velocity).

Three velocity regions can be distinguished:

1. Small incident particle velocities or energies, i.e. $v \ll Z_1^{2/3} v_B$ and $v > v_B Z_1{}^{1/2}$: positively charged ions tend to be neutralized by electrons captured from target atoms (region of the Firsov and Lindhard-Scharff formalisms, the inelastic energy loss is proportional to ion velocity),
2. Intermediate incident particle velocities, i.e. $v \approx Z_1^{2/3} v_B$: the inelastic energy loss has a maximum at about $v = 3 v_B Z_1{}^{2/3}$, and
3. High incident particle energies, i.e. $v \gg Z_1^{2/3} v_B$: the inelastic energy loss decreases because the interaction time during collisions becomes shorter (region of the Bethe-Bloch formalism)

For ion incident energies lower a few 10 keV, the first velocity range (region 1) is of interest, only. In this energy region two theoretical considerations on the inelastic energy loss by Firsov [18] and also by Lindhard and Scharff [19] have been proposed.

According to Firsov [18], each binary collision is considered as leading to a superposition of the electron orbits of incident particle and target atom and the formation of a quasi-molecule [18]. This quasi-molecule is composed of electrons from both atoms crossing the boundary between these atoms. The electron transfer causes a transfer of momentum and leads to energy loss of the incident particle. This geometrical consideration of momentum exchange between the incident particle and the target atom during the pervasion of both electron clouds provides a relationship for the energy loss to electrons per collision with a target atom which is proportional to the ion velocity.

The electronic energy loss of a projectile in a solid is obtained by means of the following relation by numerical integration from zero to infinity (details see [5,18])

$$S_e(E) = 2\pi \int_0^\infty T_e b db = 2\pi k_F e^2 a_0 (Z_1 + Z_2) \frac{v}{v_B}, \tag{3.24}$$

where the constant $k_F = 1.08$, a_o is the Bohr radius ($\equiv 0.052918$ nm) and the energy loss of electrons per collision is given by

$$T_e = 0.35 \frac{(Z_1 + Z_2)^{5/3} \hbar / a_0}{\left[1 + 0.16 (Z_1 + Z_2)^{1/3} b / a_0\right]^5} v. \tag{3.25}$$

For calculation of the electronic stopping power, (3.24) can be transformed to

$$S_e(E) = 3.25 \times 10^{-17} (Z_1 + Z_2) \sqrt{\frac{E}{M_1}}, \tag{3.26}$$

where the energy E, the mass M_1 and the electronic stopping power $S_e(E)$ are given in units of [eV], [g] and [eV cm^2], respectively.

The second model by Lindhard and Scharff [19] is based on the idea of elastic scattering of free target electrons (model of the free electron gas) in the field of the screened point charge. It is assumed that free electron gas consists of electrons at zero temperature on a fixed uniform positive background with overall charge neutrality. During the passage of the incident particle, the center of the negatively charged cloud shifts a little behind moving ion. This shifting process results in a separation of the charges. Consequently, this charge separation generates an electrical field, which brakes the moving ion and leads to an inelastic energy loss. Also, in this case, it could be demonstrated that the velocity of the incident particle is proportional to the ion energy and can be given to

$$S_e(E) = 3.83 \times 10^{-15} Z_1^{1/6} \frac{Z_1 Z_2}{\left(Z_1^{2/3} + Z_2^{2/3}\right)^{3/2}} \left(\frac{E}{M_1}\right)^{1/2}, \tag{3.27}$$

where again the ion energy E, the mass M_1 and the electronic stopping power $S_e(E)$ are given in units of [eV], [g] and [eV cm^2], respectively. The normalized electronic stopping power can be expressed as

$$s_e(\epsilon) = \left(\frac{d\epsilon}{d\rho}\right) = k_{LS}\varepsilon^{1/2} = 0.0794 \frac{Z_1^{2/3} Z_2^{1/2}}{\left(Z_1^{2/3} Z_2^{2/3}\right)^{3/4}} \frac{(M_1 + M_2)^{3/2}}{M_1^{3/2} M_2^{1/2}} \varepsilon^{1/2}, \tag{3.28}$$

where k_{LS} is Lindhard's normalized or reduced electronic stopping factor.

Lindhard and Scharff [20] have given the electronic stopping power without a derivation. Later, Sugiyama [21] has shown that the difference between the Lindhard and Scharff model and the Firsov model lies in the choice of the interatomic potential. In comparison with experimental results it has to be pointed out that the Lindhard-Scharff model provides a reasonable accordance, whereas the Firsov model can only be applied for Z_1/Z_2 or $Z_2/Z_1 < 4$. In Fig. 3.3 (left), the linear dependence of the electronic energy loss on the quadratic root of the ion energy over a large energy range is demonstrated for the case of keV-implantation of several ion species in GaN [22].

According to (3.2), the total energy loss is given by the sum of the nuclear and electronic energy loss. Figure 3.3 (right) shows the energy losses in silicon as function of the Ar ion energy. This example demonstrates that the nuclear interaction (nuclear stopping power) is the dominant energy loss mechanism for low ion energies (here up to an ion energy of about 10 keV is nuclear energy loss about one order of magnitude larger than the electronic energy loss).

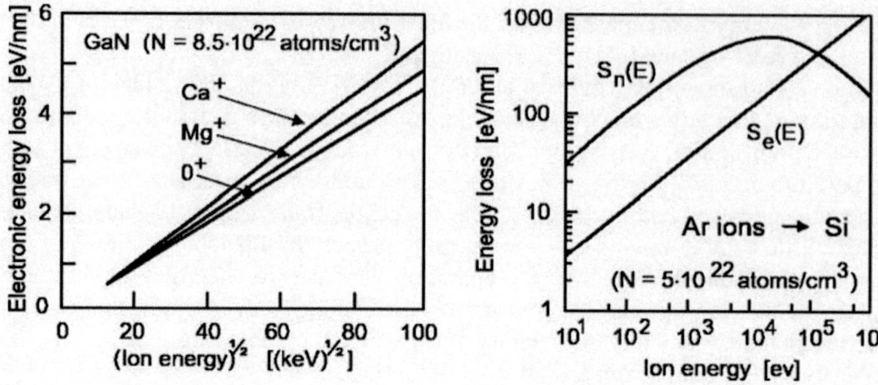

Fig. 3.3 Left: electronic energy loss $(dE/dx)_e = NS_e(E)$ as function of the ion energy calculated for GaN after implantation induced n- and p-type doping with O^+, Ca^+ or Mg^+ ions, respectively. Right: comparison of nuclear and electronic energy loss for the case of Ar ion irradiation of silicon as a function of ion energy

3.3 Stopping Power for Compounds

The calculation of the energy loss in targets with more than one component (element) is based on the application of the Bragg's rule, proposed by Bragg and Kleeman [23]. It is assumed that the interaction processes between the incident ion and a target atom is independent of all other surrounding target atoms. According to the Bragg's rule, the stopping power of a compound is calculated by the linear combination of stopping powers of the individual elements.

$$S^c(E) = \sum_i n_i S^i(E), \tag{3.29}$$

where $S^c(E)$ and $S^i(E)$ are the stopping cross-sections of the compound c, and the element i, respectively and n_i is the molar fraction of the element i, so that $\sum_i n_1 = 1$.

Deviations from the Bragg's rule can be expected mostly at comparatively low energies and for light elements, when the relative contributions of valence electrons to the stopping power becomes relatively large. Experimentally, deviations from Bragg's rule in the order of 10–20% for the stopping maximum for light gases and solid compounds containing heavier elements were found [24]. Ziegler et al. [3] have proposed a core- and bond- model, where each molecule is described as a set of atomic cores and bonds, corresponding to the non-bonding core and bonding valence electrons, respectively. When applying the model to the calculation of the energy loss of compounds it can be expected that the deviations from the Bragg's rule are less than 10%.

3.4 Ion Range

The stopping powers are frequently used to calculate the range of implanted ions into a solid. According (3.2), the energy loss of an incident particle on its way through a solid is the sum of the nuclear and electronic stopping. Information on the stopping powers $S_n(E)$ and $S_e(E)$ allows the slowing-down of a particle in a solid to be determined, quantitatively. Rearranging (3.2) and integration over the energy of the incident particle E yields to the total range R(E) of this particle, expressed as

$$R(E) = \int_o^E \frac{dE}{\left(\frac{dE}{dx}\right)_n + \left(\frac{dE}{dx}\right)_e} = \frac{1}{N}\int_o^E \frac{dE}{S_n(E) + S_e(E)}, \tag{3.30}$$

or with the dimensionless Lindhard units (see 3.9) by

$$R = \frac{\rho_L}{4\pi a_{TF}^2 \gamma N}. \tag{3.31}$$

Because of multiple collision events, the ion range normal to the surface, the so-called mean projected ion range R_p(E,) is smaller than the total ion range R(E). Also a lateral spreading (straggling) of the ions into the targets can be expected.

Statistical fluctuation in the stopping processes leads to a spread of the ion range. Lindhard et al. [2] have developed an approach, the LSS theory, to determine the mean projected ion range $\overline{R_p}$ and the mean projected range straggling (standard deviation of the projected range), $\overline{\Delta R_p}$, quantitatively. The theory uses probability functions instead of tracing the paths of individual particles. The result is an integral equation that matches the implanted concentration distribution. Based on approximations, which have to be made to solve this integral equation, the precision is not so high compared with results calculated by computer simulations (see Sect. 3.6).

For low energies of the incident particles, the LSS theory can be used to make approximate statements about the dependence of the projected range on the mass ratio and about the relationship between the projected range and range straggling:

$$\begin{aligned} &\text{for } M_2/M_1 = 2 \text{ is } R/R_p \approx 2.2,\\ &\text{for } M_2/M_1 = 1 \text{ is } R/R_p \approx 1.6,\\ &\text{for } M_2/M_1 = 1/2 \text{ is } R/R_p \approx 1.2. \end{aligned} \tag{3.32}$$

In the case that $M_1 > M_2$ and the nuclear stopping dominates the range the straggling is given by

$$2.5\Delta R_p \cong \frac{R}{1 + (M_2/M_1)} \tag{3.33}$$

or

$$\Delta R_p \cong R_p/2.5. \tag{3.34}$$

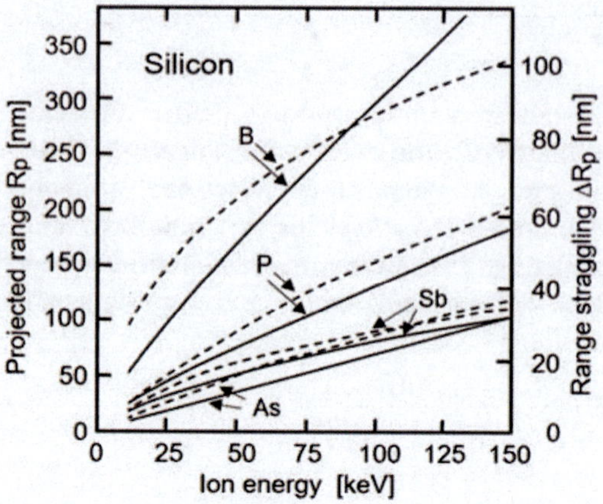

Fig. 3.4 Projected range R_p (solid line) and range straggling ΔR_p (dotted line) of B, P, Sb and As ions perpendicularly implanted in silicon, calculated with SRIM simulation program

Winterbon et al. [12] have proposed a more precise relation between the range, mean projected ion range and the mean projected range straggling as function of the mass ratio M_2/M_1. It can be noted that the LSS theory allows to calculate the range of ions in matter with about 20% accuracy.

A large number of computer programs have been developed in the field of the ion range calculations and their derivations, as partly introduced in Sect. 3.6. As an example, in Fig. 3.4 the calculated projected range R_p and the standard deviation ΔR_p of several ion species implanted in silicon with ion energies up to 150 keV are shown.

3.5 Range Distribution

As a result of the nuclear and electronic stopping processes, implantation of ions in the solid state leads to a three-dimensional concentration distribution of the implanted ion species below the surface. The path of each ion in a solid varies randomly because of the stochastic nature of these stopping processes. Therefore, instead of the total path length, the measured projected quantities range, R_p, and straggling, ΔR_p, are used to describe the implanted concentration distribution. Figure 3.5 shows schematically that the concentration distribution must be considered perpendicular and parallel to the implanted surface. That means, it must be distinguished between the vertical (plane A in Fig. 3.5) and the lateral concentration distributions (plane B in Fig. 3.5). For calculation of the spatial distribution of the implanted concentration, the spatial probability distribution of the point of rest p(X,Y) of the implanted atom must be known. Lindhard et al. [2] have developed two integro-differential equations for $\langle R_p \rangle$ and $\langle R_\perp \rangle$. An analytical solution of these equations is not possible. But, a conversion into three 2nd order differential equations by Taylor expansion is a way to solve these equations, where new quantities for $\langle R_p \rangle$, $\langle R_c \rangle$ and $(\langle R_p^2 \rangle - \langle R_\perp^2/2 \rangle)$ with $\langle R_c^2 \rangle = \langle R_p^2 \rangle + \langle R_\perp^2 \rangle$ are introduced. In general, the range distribution of implanted species

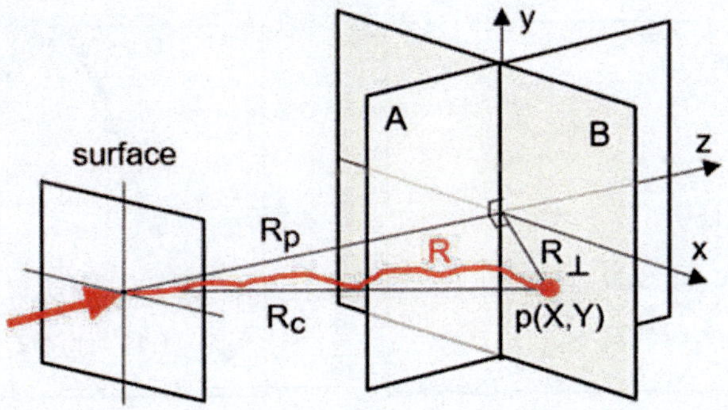

Fig. 3.5 Schematic picture for the definition of the parameters. R_p and $R_\perp$ are the projected ranges perpendicular and parallel to the implantation plane, respectively. A is plane of the vertical concentration distribution and B is plane of the lateral concentration distribution

is assumed to be a symmetrical Gaussian distribution as a first approximation (see Appendix F).

This distribution is characterized by two moments, the mean projected range R_p and the range straggling or standard deviation ΔR_p. The first moment is the average penetration depth und does not influence the shape of the distribution. The second moment is a measure of the width of the Gaussian distribution. Hence, the concentration profile of the implanted species is given by

$$N(x) = \frac{\Phi}{\sqrt{2\pi}\,\Delta R_p} \exp\left[-\frac{1}{2}\frac{(x - R_p)^2}{(\Delta R_p)^2}\right] \tag{3.35}$$

under the condition that all implanted ions are retained (see Appendix F). Φ is the ion fluence (number of particles per unit area) and x is the normal distance from the target surface. In Fig. 3.6 (left) the Gaussian shape of calculated concentration distribution after As ion implantation in Si is exemplarily demonstrated as function of the ion energy. Also shown in Fig. 3.6 (right), the nitrogen concentration profile is measured by Auger electron spectroscopy (AES) after nitrogen ion implantation in iron at low temperature [25]. The measured profile can be approximately described by a Gaussian distribution. It is evident, that a Gaussian distribution function is convenient to describe concentration profiles because its first two moments can be easily used with acceptable accuracy. The fluence related to the given concentration profile can be calculated by

$$\begin{aligned}\Phi &= \int_{-\infty}^{\infty} N(x)dx = \int_{-\infty}^{\infty} \frac{\Phi}{\sqrt{2\pi}\,\Delta R_p} \exp\left[-\frac{(x - R_p)^2}{2\Delta R_p^2}\right]dx \\ &= \int_{-\infty}^{\infty} \sqrt{2}\Delta R_p \frac{\Phi}{\sqrt{2\pi}\,\Delta R_p} \exp(-y^2)dy,\end{aligned} \tag{3.36}$$

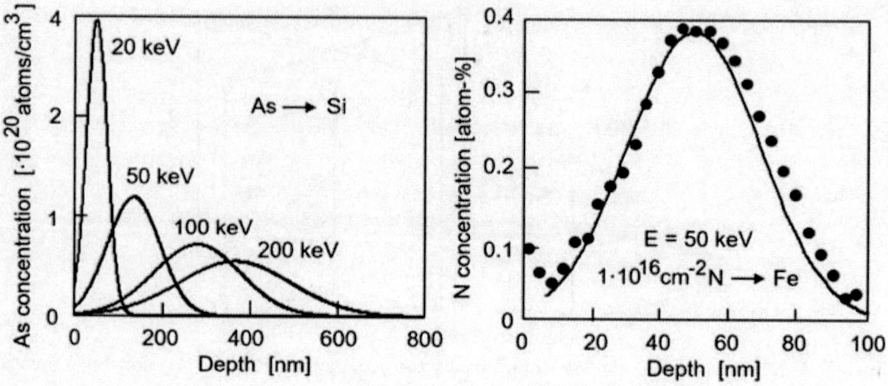

Fig. 3.6 Left: calculated Gaussian concentration profiles after implantation of As ions in Si as function of the ion energy. Right: comparison of the calculated Gaussian concentration profile (solid line) with the measured concentration distribution (dots) after 50 keV implantation of 1×10^{16} N ions/cm^2 in iron at the temperature of liquid nitrogen. The concentration distribution was measured by Auger electron spectroscopy and calibrated by nitrogen concentration standards. From the measured profile the projected range $R_p = 47$ nm and longitudinal straggling $\Delta R_p = 19$ nm were determined in good agreement with a SRIM simulation

where $y = (x - R_p)/2\Delta R_p$ and the maximum concentration at $x = R_p$ is given by N_{max} and can be expressed by

$$N_{\max} = \frac{\Phi}{\sqrt{2\pi}\Delta R_p} \cong 0.4\frac{\Phi}{\Delta R_p}. \tag{3.37}$$

With knowledge of the atomic number density (number of particles per unit volume), N, the implanted concentration

$$C = \frac{N_{\max}}{N_{\max} + N} \tag{3.38}$$

at the peak of the concentration distribution can be determined. The integration of (3.36), the so-called Gaussian integration, can be realized with the help of the following integral [9]

$$\int_{o}^{\infty} \exp(-ax^2)dx = \frac{1}{2}\sqrt{\frac{\pi}{a}}. \tag{3.39}$$

Consequently, the integration leads to

$$\Phi = \sqrt{2\pi}\Delta R_p N_{\max}. \tag{3.40}$$

If the point of rest of the implanted ion is given by the coordinates x and y (see Fig. 3.5), $\langle \Delta X \rangle$ and $\langle \Delta Y \rangle$ are the standard deviations, respectively, where

$$\langle \Delta X \rangle = \langle \Delta Y \rangle = \sqrt{R_{\perp}^2/2}, \tag{3.41}$$

since $\langle X^2 \rangle$ is equal to $\langle Y^2 \rangle$ in an amorphous solid. The spatial distribution of the implanted ions can by described by the distribution function

$$N(x, y, z) = \frac{\Phi}{(2\pi)^{3/2} \Delta R_p \langle \Delta X \rangle \langle \Delta Y \rangle} \exp\left[-\frac{x^2}{2\langle \Delta X \rangle^2} - \frac{y^2}{2\langle \Delta Y \rangle^2} - \frac{(z - R_p)^2}{2(\Delta R_p)^2} \right], \tag{3.42}$$

where the spatial concentration distribution forms an ellipsoid having a center at (0, 0, R_p) on the x, y, z coordinates.

Frequently, experimentally measured concentration profiles differ significantly from a simple expected Gaussian distribution. This means that the construction of the depth distribution with two parameters R_p and ΔR_p is not sufficient for a complete description of the concentration profiles. High-order moments of the Gaussian distribution are required to verify the measured profiles more accurately (see Appendix F). Consequently, in addition to the first two moments, which are given by the mean projected range R_p and the range straggling ΔR_p, the third and the fourth moments of the probability function are used to describe in detail the asymmetry of a measured concentration profile (see Appendix F). The third moment, called skewness, characterizes the tilt of the distribution away from the symmetric distribution and the fourth moment, called kurtosis, is a measure of the peak sharpness of the distribution.

Accurate reproduction of concentration profiles generally requires consideration of both higher order moments in addition to the first two moments of the distribution.

3.5.1 *Modified Range Distribution by Sputtering and Annealing*

The concentration distribution can be influenced by sputtering (see Chap. 5). Specially, a modified concentration distribution can be expected for low implantation energies and high ion fluences. The intensive sputter-induced surface erosion sets a limit to the number of ions that are implanted. Thus, an equilibrium between the implanted species and the eroded atoms is achieved, which is independent of the maximum ion fluence. According to Ryssel and Ruge [6], the change of the concentration profile caused by sputtering can be evaluated under the assumption, that (i) sputtering yield Y is constant, (ii) the distribution of implanted species is given by a Gaussian distribution in absence of sputtering and (iii) density changes due to ion bombardment induced damages are neglected.

Then, the concentration profile influenced by sputtering, $N_s(x)$, can be determined by integration of (3.35) over the implantation time t by

$$N_s(x) = \int_0^t N(x + v_o t, t) dt, \tag{3.43}$$

where $v_o \cdot t = Y \cdot J/N \cdot t = \Phi Y/N$ is the eroded depth (see 5.2), N is the atomic number density, v_o is the erosion velocity and t is the time of ion bombardment. In detail, the concentration distribution is given by

$$N_s(x) = \frac{N}{2Y}\left[erf\left(\frac{x - R_p - \Phi \frac{Y}{N}}{\sqrt{2}\Delta R_p} \right) - erf\left(\frac{x - R_p}{\sqrt{2}\Delta R_p} \right) \right]. \tag{3.44}$$

For $t \to \infty$ follows the saturation concentration profile

$$N_s(x) = \frac{N}{2Y} erf\left(\frac{x - R_p}{\sqrt{2}\Delta R_p} \right) \tag{3.45}$$

and the maximum concentration

$$N_{s,\max}(x) = \frac{N}{2Y} erf\left(\frac{-R_p}{\sqrt{2}\Delta R_p} \right) \approx \frac{N}{Y} \quad \text{for } R_p > 3\Delta R_p. \tag{3.46}$$

Consequently, the maximum is independent of the ion fluence, but dependent on the relation between the atomic number density, N, of the target material and the sputtering yield Y. It means that for high-fluence ion implantation, the maximum concentration is not influenced, but the Gaussian distribution of the implanted species is asymmetrically expanded.

The concentration profile of the implanted ion species can be also modified by the temperature during a post-annealing at higher temperatures after the implantation, significantly. The mathematical treatment based on one-dimensional diffusion (Fick's second diffusion law) and the assumption that (i) the initial concentration distribution after ion implantation can be described by a Gaussian profile and that (ii) the implantation induced damages can be neglected. Then, the concentration profile by annealing is given by

$$N(x,t) = \frac{N_{\max}}{\sqrt{2\pi\left(\Delta R_p^2 + 2Dt\right)}} \exp\left[-\frac{1}{2} \frac{\left(x - R_p\right)^2}{\left(\Delta R_p^2 + Dt\right)} \right], \tag{3.47}$$

where D is the constant diffusion coefficient and t is the annealing time (see Appendix G). The influence of the surface can be considered for two limiting cases. On the one hand no flux of the diffused species through the surface is possible (species are 'reflected' at the internal surface) or otherwise the surface is open for the diffused species (surface act as a sink). Under these boundary conditions the concentration distribution can be expressed by

$$N(x,t) = \frac{\Phi}{\sqrt{2\pi\left(\Delta R_p^2 + 2Dt\right)}}\left\{\exp\left[-\frac{1}{2}\frac{(x - R_p)^2}{(\Delta R_p^2 + Dt)}\right]\right.$$

$$\left.\pm\exp\left[-\frac{1}{2}\frac{(x + R_p)^2}{(\Delta R_p^2 + Dt)}\right]\right\}, \qquad (3.48)$$

where the algebraic sign of the last term in (3.48) is positive for the first case and negative for the second case. It is important to note that the modification of the Gaussian distribution by diffusion as result of the post-annealing procedure becomes significant, when $\sqrt{2Dt} \geq \Delta R_p$.

3.5.2 *Lateral Distribution*

If it assumed that all implanted ions enter the surface (see Fig. 3.5) at a single point (0, 0, 0) and stopped at a certain point (x, y, z), then the Gaussian distribution around y and z coordinates (plane B in Fig. 3.5) is given by

$$N(x,y,z) = \frac{N(x)}{2\pi\Delta R_\perp^2}\cdot\exp\left[-\frac{y^2 - z^2}{\sqrt{2\pi}\,\Delta R_\perp}\right], \qquad (3.49)$$

because it is reasonable to assume that a lateral profile can be represented as a Gaussian distribution, where $\Delta R_\perp$ is the longitudinal or lateral straggling. In the case of the implantation along the line (0, 0, z), this distribution is clearly given by

$$N(x,y) = \frac{N(x)}{\sqrt{2\pi}\Delta R_\perp}\cdot\exp\left[-\frac{y^2}{2\Delta R_\perp}\right] \qquad (3.50)$$

after integrating over z from $-\infty$ to $+\infty$. This two-dimensional concentration distribution is formally constructed by the product of vertical (depth) and the lateral distributions. From the point of view of the application of ion implantation, e.g., for doping semiconductors, this hypothetical distribution is often combined with a masking function (step function) [25]. This can be considered as a window with an opening width $y = \pm a$ over which the substrate to be implanted. Then, the two-dimensional concentration distribution is given by

$$N(x,y) = \frac{N(x)}{2}\left[erfc\left(\frac{y-a}{\sqrt{2}\Delta R_\perp}\right) - erfc\left(\frac{y+a}{\sqrt{2}\Delta R_\perp}\right)\right]. \qquad (3.51)$$

It is obvious, that with the influence of the lateral straggling becomes increasingly relevant with decrease of the mask width and increase of the ion energy.

3.6 Simulation of Ion Range and Range Distribution

Basically, two groups of computer simulation codes to analyze the particle-solid interactions can be distinguished. On the one hand, there are molecular dynamical (or classical) codes. Here, the collision dynamics of a particle with a system of many atoms in the bombarded solid is developed with knowledge of the interacting forces in defined time steps. On the other hand, the Monte Carlo (MC) codes study the results of collisions by a series of successive binary collision events. Extensive summaries of the computer simulation of ion solid state interaction are known [26–30].

3.6.1 Molecular Dynamic Simulations

Conceptually, the approach of the classic molecular dynamics (MD) method is based on the numerical solution of Newton's equations of motion for interacting many-particle system (details see e.g., [31]). Molecular dynamic computer codes describe the motion of each particle in a volume containing n particles. The movement of a particle is caused by the forces F, which are exercised by all other particles, i.e. $n - 1$ particles. Then the movement of every particle i can be determined consecutively by the integration of the Newtonian equation of motion for the time

$$M_i \frac{d^2 r_i}{dt^2} = M_i \frac{d\boldsymbol{v}_i}{dt} = \sum_{j=1, i \neq j}^{n} \boldsymbol{F}_{ij}. \tag{3.52}$$

Consequently, an exponential increase in the number of pairwise calculations can be expected for such n^2 particles. Therefore, the computation time increases significantly, because a large number of particles is necessary to get reliable information about the simulated system.

The interaction forces are controlled by the potential V(r) (see Sect. 2.1), where the force $F_{ij} = -\nabla V(r_{ij})$ and r_{ij} is the distance between two particles. In general, empirical potentials or the potentials determined by ab-initio calculations are used. Calculation of the interatomic forces is especially time consuming with MD simulations. Consequently, the system to be examined is often small and normally does not enclose more than some hundred to some tens of thousands particles.

Commonly, the integration of (3.52) is based on the use of numerical one-step methods (Euler-Cauchy-, Heun- or Runge–Kutta-approaches). With the aim to reduce the number of forces, calculations based on the Verlet algorithms are more and more applied to integrate Newton's equations of motion

$$\frac{d\boldsymbol{v}_i}{dt}\left(t+\frac{\Delta t}{2}\right)=\boldsymbol{v}_i(t)+\frac{\Delta t}{2M_i}\boldsymbol{F}_i(t) \tag{3.53}$$

and

$$\boldsymbol{r}_i(t+\Delta t)=\boldsymbol{r}(t)+\Delta t\boldsymbol{v}_i\left(t+\frac{\Delta t}{2}\right). \tag{3.54}$$

After this calculation the new positions of the particles are known as starting point of the integration of (3.54)

$$\boldsymbol{v}_i(t+\Delta t)=\boldsymbol{v}_i\left(t+\frac{\Delta t}{2}\right)+\frac{\Delta t}{2m_i}\boldsymbol{F}_i(t+\Delta). \tag{3.55}$$

In order to discretize and numerically solve the initial value problem, time steps Δt are chosen which are of the order of femtoseconds (smaller than the period of the lattice vibration in the bombarded solid). Figure 3.7 illustrates the MD algorithm, schematically. Initially, the simulation cell with the side lengths L_x, L_y and L_z is defined, where the origin of the simulation cell frequently is $L_x/2$, $L_y/2$ and $L_z/2$. The size of the cell is defined using rules of thumb. Periodic boundary conditions are used for simulation in an infinite medium. When simulating processes in a crystalline

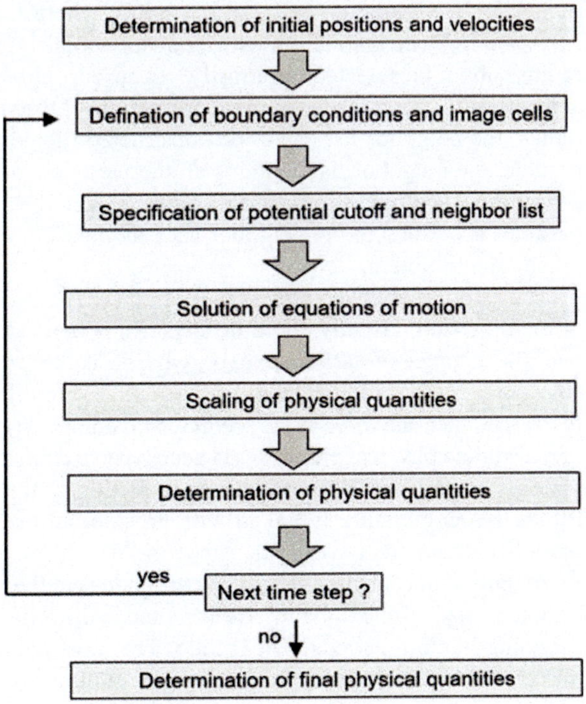

Fig. 3.7 Schematic of the MD algorithm

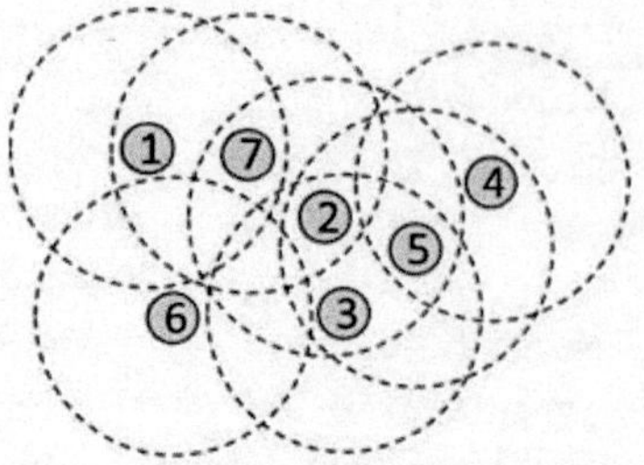

	Particle							Index
	1	2	3	4	5	6	7	
Neighbors	7	3	5	5	2	-	2	1
		5	2		3			2
		7						3

Fig. 3.8 Example of a neighbor list

material, the unit cell is the simplest way to define the simulation cell due to the three orthogonal axes of such unit cells. As a consequence, the simulation cell consists of multiples of the unit cell. The initial velocity of the atoms in the target corresponds to the temperature of the target. Thus, the probability distribution of the velocity components is given by the Maxwell–Boltzmann equation. Under periodic boundary conditions, the initial simulation cell is surrounded by an infinite, so-called image cell, which is formed by multiplying the simulation cell. The application of a cutoff radius and a neighbor list can be a way to manage the time of the pairwise force calculation. The cutoff radius limits the interaction of an atom with surrounding atoms, i.e. the interaction of the atom with particles outside the cutoff radius is ignored. If a cutoff radius is defined, then the sphere with cutoff radius r_{cut} should be inside the simulation cell, i.e. $r_{cut} < L/2$, where L is the smallest value of $L_x/2$, $L_y/2$, and $L_z/2$ (no interaction with its own periodic boundary conditions). The cutoff radius cannot be selected, arbitrarily, because a cut of the energy potential function at an arbitrary point leads to unrealistic results of the simulation. Based on the cutoff radius, the neighbor list has to be constructed (demonstrated in Fig. 3.8). For each particle, the neighbor list contains all the particles (neighbor particles) that are at a certain distance (or less) from this particle. Calculation of the forces between two particles according to the neighbor list can make computational scaling simulations much more efficient.

The solution of the equations of motion (see 3.53–3.55), starts with the definition of the time step. Usually, the time step, Δt, is defined by the travel time of an atom over the distance of a/25 up to a/10, where a is the lattice parameter. The numerical solution of the equations of motion (see above) provides the positions and velocities, firstly, and then the force at the predicted positions. After these calculations, a scaling of the studied physical properties is necessary to control the conformity of the desired system. An important point is the introduction of the temperature into the system. Either the temperature is linked with an external reservoir or the velocity of each particle is rationed at every time step.

At the end of each time step, the actual values of the most quantities of interest may be determined. Typically, runs are executed in the order of $10^3 \ldots 10^7$ time steps corresponding in nanoseconds of real time. The most popular MD simulation programs in the field of the ion–solid interaction are summarized in the Table 3.1.

The determining disadvantage of the MD simulations is the restricted time and spatial scales. For the description of many processes at the interaction of energetic ions with solid state surfaces longer periods and a significantly bigger number of atoms would be necessary.

For example, Fig. 3.9 shows the evolution of the collision cascade induced by a 10 keV recoil in Fe at an ambient temperature of 0 K [30]. It is shown that a very large number of atoms is initially displaced. When the cascade cools down, almost all of them return to perfect crystal positions.

3.6.2 Monte Carlo Simulations

The Monte Carlo (MC) method is a stochastic technique that relies on random numbers and probability statistics to sample the conformational space when it is impossible to compute an exact result with a deterministic algorithm. In the field

Table 3.1 Frequently used computer codes for the MD simulations of the ion–solid interactions

Computer code	References
QDYN	D.E. Harrison, M.M. Jakas, Rad. Eff. 99 (1986) 153
SPUT3	M.H. Sapiro, T.A. Tombrello, Nucl. Instr. Meth. B30 (1988) 152
PARASOL	G. Betz et al., Rad. Eff. and Defects in Solids 130/131 (1994) 251
MODYSEM	V. Konoplev, A. Gras-Marti, Phil. Mag. A 71 (1995) 1265

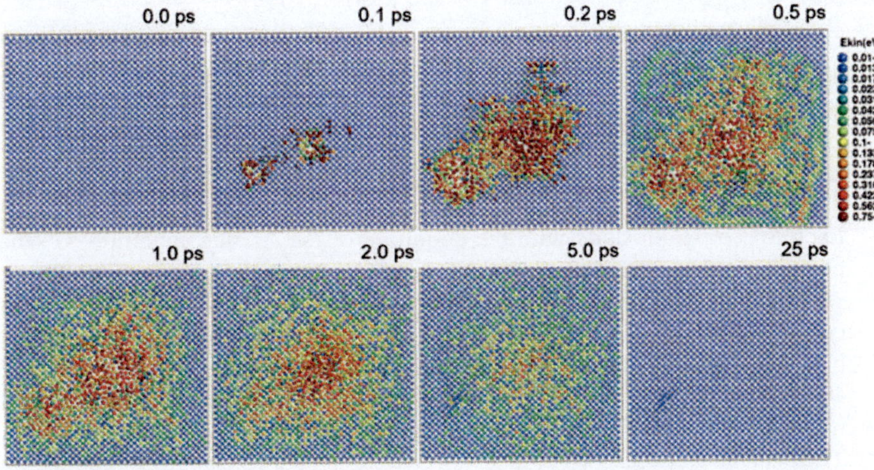

Fig. 3.9 Molecular dynamic simulation of the collision cascade induced by a 10 keV Fe recoil in Fe at an ambient temperature of 0 K. The circles indicate atom positions in a 2-unit cell (4 atom layer) thick cross-section through the center of the simulation cell, and the color scale the kinetic energy of the atom (figure is adapted from [30])

of MC simulations, with the aim of determining the ion range, range distribution, number of sputtered particles, etc., the trajectory of each particle is tracked in time. The entire process is divided into time steps. At each time step, a decision must be made that determines the state of the next time step. For example, after each time step, the question of whether or not a collision will occur must be answered. The random numbers used in MC simulation are generated by a pseudo-random number generator. The random number enables the reproduction of the irregular behavior of system particles and describes this behavior in a compressed form. Thus, with a large number of projectiles and target atoms, the path of all system atoms can be maintained and the next collision partner can be reliably selected. Concluding the MC simulation gives a statistical analysis of the particle trajectories.

Generally, MC simulations are based on the generation of a sequence of uniformly distributed random numbers in the interval [0, 1] by a random generator. It is the aim to find a relation between the irregularly distributed parameter x inside the limits x_1 and x_2 with the distribution function f(x) and an assigned random number R. The condition of the probability

$$p(R) = \begin{cases} 1 & \text{for } 0 \leq R \leq 1 \\ 0 & \text{for } R < 0 \text{ and } R > 1 \end{cases} \tag{3.56}$$

and,

$$f(x) = 0 \text{ for } x < x_1 \text{ and } x > x_2$$

leads by means of integration, $\int_0^R p(R)dR = \int_{x_1}^x f(x)dx$, and thus to random number $R = \int_{x_1}^x f(x)dx$ or (symmetry of the R distribution)

$$R = \int_x^{x_2} f(x)dx. \tag{3.57}$$

MC calculations can be considered as consecutive step sequences:

1. Each MC program starts with the definition of the starting condition, i.e. the position on the surface, the initial energy and the initial directions $\Omega = (\theta, \varphi)$, where θ is polar scattering angle and φ is azimuth scattering angle. After each collision with atoms of the bombarded solid, the next position of the incident particle can be obtained by saving all positions of the involved atoms (see MARLOW code) or by determination of the impact parameter (see Sect. 2.2.2) using a random number (see TRIM or ACAT codes).
2. The distance between two collisions (mean free path λ) can be determined on the base of the assumption that this mean free path is constant and given by

$$\lambda = N^{-1/3}, \tag{3.58}$$

where N is the atomic number density. A second possibility is based on the assumption that the collision probability is dropped proportional to path distance. Then, the probability of a particle to carry out a first collision is given by

$$p(\lambda) = \int_0^{\lambda} f_{mfp}(\lambda)d\lambda = 1 - \exp(-N\sigma\lambda), \tag{3.59}$$

where σ is the scattering cross-section (see Sect. 2.2.3) and $f_{mfp}(\lambda)$ is the distribution function of the mean free paths. Consequently, the mean distance to the first collision, λ_o, is

$$\lambda_0 = \int_0^{\infty} \lambda f_{mfp}(\lambda)d\lambda = 1/N\sigma. \tag{3.60}$$

With (3.56) the relation between the random number and the collision probability is given by

$$R_{mfp} = p(\lambda) = 1 - \exp(-\lambda/\lambda_o) \tag{3.61}$$

and then the mean free path can be expressed by

$$\lambda = -\lambda_o \ln\left(1 - R_{mfp}\right). \tag{3.62}$$

Because R_{mfp} is uniformly distributed between 0 and 1, (3.62) can be redrafted to

$$\lambda = -\lambda_o \ln\left(R_{mfp}\right).$$

3. Consequently, each particle moves in steps of fixed distance. The energy loss of an incident particle along the path L is

$$\Delta E = \int_0^L \left[\left(\frac{dE}{d\lambda}\right)_n + \left(\frac{dE}{d\lambda}\right)_e\right] d\lambda \approx \left[\left(\frac{dE}{d\lambda}\right)_n + \left(\frac{dE}{d\lambda}\right)_e\right]\lambda$$
$$= \left[\left(\frac{dE}{d\lambda}\right)_n + \left(\frac{dE}{d\lambda}\right)_e\right] N^{-1/3} \tag{3.63}$$

4. After the distance λ, the next collision occurs, where the impact parameter b is determined with a random number R_b distributed in the interval between 0 and 1, uniformly. The second random variable is the polar scattering angle θ (see Sect. 2.2.2). Assuming the differential scattering cross-section is $2\pi b db$ (see

2.54), then the probability function $f_b(x) = d\sigma/\sigma$ can be expressed by

$$f_b(b) = \frac{2\pi b}{\pi b_{max}^2} = \frac{2b}{b_{max}^2}, \tag{3.64}$$

where b_{max} is a maximum impact parameter which corresponds to a minimum scattering angle θ. According (3.57) the random number R_b is linked with b by

$$R_b = \int_0^p f_b(b)db \quad \text{or} \quad b = b_{max}\sqrt{R_b} \tag{3.65}$$

The random number R_b can be defined in a different manner. For example, in the case of a cylindrical volume with the length λ_o and the impact radius is

$$b_{max} = \frac{1}{\sqrt{\pi\lambda N}} = \frac{1}{\sqrt{\pi\sqrt[3]{N}}} \tag{3.66}$$

or in the case of a spherical volume with the radius r_{max} is

$$b_{max} = \frac{1}{\sqrt[3]{\frac{4}{3}\pi N}} \approx 0.6204\sqrt[3]{N}. \tag{3.67}$$

5. Further, the azimuth scattering angle φ must be calculated after the scattering event. The distribution function due to planar scattering geometry is given by

$$f_\varphi(\varphi) = \frac{1}{2\pi} \quad \text{or} \quad \varphi = 2\pi R_\varphi \tag{3.68}$$

6. The energy loss, (3.63), is calculated with the help of equation (2.35) and then subtracted from particle energy before the collision.
7. If the energy of the particle is lower than the so-called cutoff energy (in the order of magnitude of binding energy of the bombarded solid) the particle is stopped.

On the one hand, the algorithm described here must be repeated several times for a large number of projectiles to serve as a basis for later statistical analysis. On the other hand, the algorithm is formulated in computer codes in such a way that they can be used by operators without major inconvenience. Examples for frequently used computer codes are summarized in Table 3.2.

The different simulation programs are capable to determine the particle trajectories, the cascade development, the ion range and their spatial distribution, the distribution of defects, the sputtering yield, backscattering and reflection coefficient, etc. Figure 3.10 demonstrates the application of the MC simulation code SRIM for the calculation of the ion range and the depth distribution. When the assumption of

Table 3.2 Frequently used computer codes for the MC simulation of the ion–solid interactions

MC code	References
MARLOWE	M.T. Robinson, I.M. Torres, Phys. Rev. B9 (1974) 5008
TRIM	J.P. Biersack, L.G. Haggmark, Nucl. Instr. Meth. 174 (1980) 257
ACAT	W. Takeuchi, Y. Yamamura,, Rad. Eff. 71 (1983) 53
SASAMAL	Y. Miyagawa, S. Miyagawa, J. Appl. Phys. 54 (1983) 7124
TRIM.SP	J.P. Biersack, W. Eckstein, Appl. Phys. A34 (1984) 73
Crystal-TRIM	M. Posselt, Rad. Eff. & Def. Solids 130/131 (1994) 87
SRIM	J.F. Ziegler, J.P. Biersack, M.D. Ziegler, see [3]
Dynamical MC code	References
EVOLVE	M.L. Roush, T.S. Andreadis, O.F. Goktepe, Rad. Eff. 55 (1981)
TRIDYN	W. Möller, W. Eckstein, Nucl. Instr. Meth. B 59–60 (1991) 814
dynamic-SASAMAL	Y. Miyagawa, et. al. J. Appl. Phys. 70 (1991) 7289

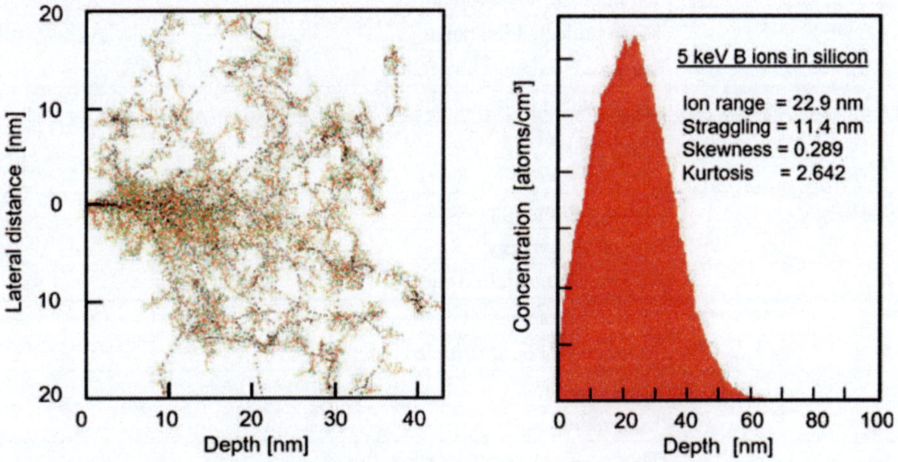

Fig. 3.10 Left: trajectories of atoms displaced in a collision cascade as calculated with the computer simulation program SRIM for normal incidence of twenty-two 5 keV boron ions on silicon. The surface is located at zero depth. Right: concentration distribution of 65,500 boron ions implanted into amorphous silicon with an energy of 5 keV

independent motion of the particles is no longer guaranteed, attempts are made to describe the resulting nonlinear phenomena by means of dynamic MC programs (see Table 3.2). Currently, MC codes are often preferred over MD codes because they are more flexible and require fewer assumptions in the application.

3.7 List of Symbols

Symbol	Notation
C	Concentration
C_m	Power law constant
D	Diffusion coefficient
E	Energy of particle
F	Force
L	Path
M_i	Mass of particle i
N	Atomic number density
N(x)	Concentration distribution
N_{max}	Maximum concentration
R	Ion range or random number
R_p	Mean projected ion range
ΔR_p	Projected ion range straggling
$\Delta R_\perp$	Longitudinal or lateral straggling
S(E)	Stopping power
$S_e(E)$	Electronic stopping power
$S_n(E)$	Nuclear stopping power
T	Transferred energy
T_{max}	Maximum transferred energy
T_{min}	Minimum transferred energy
Z_i	Atomic number of particle i
a	Lattice spacing
a_o	Bohr radius
a_{TF}	Thomas–Fermi scattering length
a_{ZBL}	Ziegler–Biersack–Littmark (universal) scattering length
b	Impact parameter
b_{max}	Maximum impact parameter
k_{LS}	Lindhard's reduced electronic stopping factor
m	Power law variable (= 1/s)
n	Number of particles
v_o	Erosion velocity
Λ^3	Third order energy straggling
Φ	Ion fluence
Ω^2	Second order energy straggling
$s_e(\varepsilon)$	Normalized or reduced electronic stopping power

(continued)

(continued)

Symbol	Notation
$s_n(\varepsilon)$	Normalized or reduced nuclear stopping power
v	Particle velocity
v_B	Bohr velocity
v_o	Erosion velocity
γ	Transfer energy efficiency factor
ε	Reduced energy
ε_{ZBL}	Reduced ZBL energy
θ_{cm}	Angle of incidence in center-of-mass system
λ	Mean free path
λ_m	Power law fitting parameter
ρ	Mass density
ρ_L	Reduced length
σ	Scattering cross-section

References

1. N. Bohr, The penetration of atomic particles through matter. Det. Kgl. Danske Vid. Selskab. Mat.-Fys. Medd. **XVIII**(8) (1948)
2. J. Lindhard, M. Scharff, H.E. Schiøtt, Range concepts and heavy ion ranges—notes on atomic collision II. Det. Kgl. Danske Vid. Selskab. Mat.-Fys. Medd. **33**(10) (1963)
3. J.F. Ziegler, J.P. Biersack, U. Littmark, *The Stopping and Range of Ions in Solids* (Pergamon Press Inc., New York, 1985)
4. E. Rimini, *Ion Implantation: Basics to Device Fabrication* (Kluwer Academic Publisher, Boston, 1995)
5. M. Nastasi, J.W. Mayer, J.K. Hirvonen, *Ion-Solid Interactions* (Cambridge University Press, Cambridge, 1996)
6. H. Ryssel, I. Ruge, *Ionenimplantation* (Akademische Verlagsgesellschaft Geest & Portig K.-G., Leipzig, 1978)
7. P. Sigmund, *Particle Penetration and Radiation Effects* (Springer, Berlin, 2006)
8. G.S. Was, *Fundamentals of Radiation Materials Science* (Springer, Berlin, 2007)
9. K. Suzuki, *Ion Implantation and Activation*, vol. 1 (Bentham Science Publishers, Oak Park, 2013)
10. A. Mutzke, R. Schneider, W. Eckstein, R. Dohmen, K. Schmidt, U. von Toussaint, G. Badelow, SD TrimSP Version 6.00, IPP Report 2019-02 (2019)
11. J. Lindhard, V. Nielsen, M. Scharff, Approximation method in classical scattering by screened coulomb fields—notes on atomic collision I. Det. Kgl. Danske Vid. Selskab. Mat.-Fys. Medd. **36**(10) (1968)
12. K.B. Winterbon, P. Sigmund, J.B. Sanders, Spatial distribution of energy deposited by atomic particles in elastic collisions. Det. Kgl. Danske Vid. Selskab. Mat.-Fys. Medd. **37**(14) (1970)
13. N. Matsunami, Y. Yamamura, Y. Itikawa, N. Itoh, Y. Kazumata, S. Miyagawa, K. Morita, R. Shimizu, H. Tawara, Energy dependence of ion-induced sputtering yields of monoatomic solids. Atomic Data Nucl. Data Tabl. **31**, 1–80 (1984)

14. H.J. Strydom, W.H. Gries, A table of normalized sputtering yields for mono-elemental polycrystalline targets. Report SMAT (1984)
15. S. Kalbitzer, H. Oetzmann, H. Grahmann, A. Feuerstein, A simple universal fit formula to experimental nuclear stopping power data. Z. Physik A **278**, 223–224 (1976)
16. W.D. Wilson, L.G. Haggmark, J.P. Biersack, Calculations of nuclear stopping, range, and straggling in the low-energy region. Phys. Rev. B **15**, 2458–2468 (1977)
17. W. Eckstein, C. Garcia-Rosales, J. Roth, W. Ottenberger, Sputtering data. Report IPP-9/82 (1993)
18. O.B. Firsov, A quantitative interpretation of the mean electron excitation energy in atomic collision. Sov. Phys. JETP **36**, 1076–1080 (1959)
19. J. Lindhard, M. Scharff, Energy dissipation by ions in the keV region. Phys. Rev. **124**, 128–130 (1961)
20. J. Lindhard, M. Scharff, Energy loss in matter by fast particles of low charge. Det. Kgl. Danske Vid. Selskab. Mat.-Fys. Medd. **27**(15) (1953)
21. H. Sugiyama, Electronic stopping power of atomic particle. Rad. Effects **56**, 205–212 (1981)
22. B. Rauschenbach, Ion implantation, isolation and thermal processing of GaN and related Materials, in *III-Nitride Semiconductors: Electrical, Structural and Defects Properties*, ed. by O. Manasreh (Elsevier, Amsterdam, 2000), pp. 193–244
23. W.H. Bragg, R. Kleeman, On the alpha particles of radium, and their loss of range in passing through various atoms and molecules. Philos. Mag. **10**, 318–340 (1905)
24. J.F. Ziegler, J.M. Manoyan, Stopping of ions in compounds. Nucl. Instr. Meth. B **35**, 215–228 (1988)
25. B. Rauschenbach, Microstructural investigations of iron implanted with nitrogen ions at the temperature of liquid nitrogen. Nucl. Instr. Meth. Phys. Res. B **18**, 34–46 (1986)
26. S. Furukawa, H. Matsumura, H. Ishihara, Theoretical consideration of lateral spread of implanted ions. Jap. J. Appl. Phys. **11**, 134–142 (1972)
27. W. Eckstein, *Computer Simulation of Ion-Solid Interactions*. Springer-Series in Material Science, vol. 10 (Springer, Berlin, 1991)
28. H.H. Andersen, Computer simulations of atomic collisions in solids with emphasis on sputtering. Nucl. Instr. Meth. B **18**, 321–343 (1987)
29. R. Smith, *Atomic and Ion Collisions in Solids and at Surfaces* (Cambridge University Press, Cambridge, 1997)
30. K. Nordlund, S.J. Zinkle, A.E. Sand, F. Granberg, R.S. Averback, R.E. Stoller, T. Suzudo, L. Malerba, F. Banhart, W.J. Weber, F. Willaime, S.L. Dudarev, D. Simeone, Primary radiation damage: A review of current understanding and models. J. Nucl. Mater. **512**, 450–479 (2018)
31. D. Frenkel, B. Smit, *Understanding Molecular Simulation from Algorithms to Applications* (Academic Press, San Diego, 2002)

Chapter 4
Ion Beam-Induced Damages

Abstract Bombardment of surfaces with accelerated ions leads to the formation of defects, where lattice atoms can be displaced after collision with the incident ion or a higher order recoil atom if the transferred energy is higher than the displacement threshold energy. The Kinchin-Pease model, modified Kinchin-Pease model, and the Norgett–Robinson–Torrens model are presented and the calculation of the defect generation rate is demonstrated. Of particular importance is the determination of the number of displacements per atom (dpa), since this number is the benchmark for quantifying radiation-induced changes of material properties after ion bombardment. Assuming that elastic collisions are dominant, the spatial distribution of the deposited energy is discussed. If the assumption of subsequent binary collisions is not valid, the formation of high energy density cascades can be expected. Both the displacement spike and the thermal spike are introduced and the mean features of these cascades are discussed. Then, the behaviour of irradiation-induced point defects as a function of irradiation time (fluence) and the temperature during ion bombardment is described and the main analytical relationships are given. As the concentration of ion beam-induced defects increases, the probability of formation of an amorphous layer also increases. The formation of an amorphous phase for a given material depends on the ion species, ion energy, fluence, ion current density, and irradiation temperature. The various models proposed in the literature for the ion beam-induced amorphization process are briefly summarized.

The scope of this chapter focuses exclusively on the low-energy ion beam-induced formation of defects resulting from the bombardment of surfaces with accelerated ions. Many comprehensive books, overviews and magazine articles deal with theoretical and experimental aspects of evolution of the radiation damage (see e.g., [1–8]).

If an ion hits the surface and comes to the rest in the solid, it is subject to a number of collision processes. Particle traveling is assumed to be particles moving along straight trajectories between two collisions. These two-body collision events are the basic requirement of collision processes by binary collision approximation (BCA). A further (second) condition is that the moving atoms primarily collide with

B. Rauschenbach, *Low-Energy Ion Irradiation of Materials*, Springer Series in Materials Science 324, https://doi.org/10.1007/978-3-030-97277-6_4

particles at rest. Based on these conditions, the collision process is described as a linear cascade.

The first target atom which is displaced after the collision with the incident particle is the so-called primary knock-on atom (PKA). Typically, such events take place in about 1 fs (=10^{-15} s). The kinetic energy of a PKA dissipates by two distinct processes. One part of the initial kinetic energy is lost by excitation of electrons and the other part is transferred in elastic collisions with other target atoms. If the transferred PKA energy is higher than the so-called threshold displacement energy, target atoms can be displaced from their lattice sites. Commonly, the damages created by ion irradiation of a target are characterized by the number of displacements per atom (dpa). One dpa represents the fluence at which every target atom in the irradiated volume has been knocked from its lattice place on average. The displaced target atoms are capable to trigger a series of further knock-on lattice atoms, when the transferred energy is high enough. Consequently, a branching tree-like structure, called displacement cascade, is generated within a time scale of about 100 fs. The displaced atoms in the cascade are moving with hyperthermal energies (about 1 eV up to 100 eV) accompanied by high-excited electrons. As example, Fig. 4.1 shows a simulation of 2 keV He ion which penetrates into a Ni target at normal incidence (calculated by means of the Monte Carlo simulation program TRIM, version SDTrimSP [9]). In this figure the same trajectories as in Fig. 3.1 are displayed together with the generated recoils of the first (red) and second generation (blue). After completion of the PKA induced displacement cascade (after about 100 fs), the generated damages are rapidly rearranged and form point defects or small clusters. Under special conditions in the ion bombardment (energy of the PKA's is in the range of 1 keV up to any 100 keV), a splitting of the displacement cascade can be observed, which leads to sub-cascades.

Radiation damages after ion bombardment have been intensively studied for many decades. Diversely theoretical models were developed to describe the different

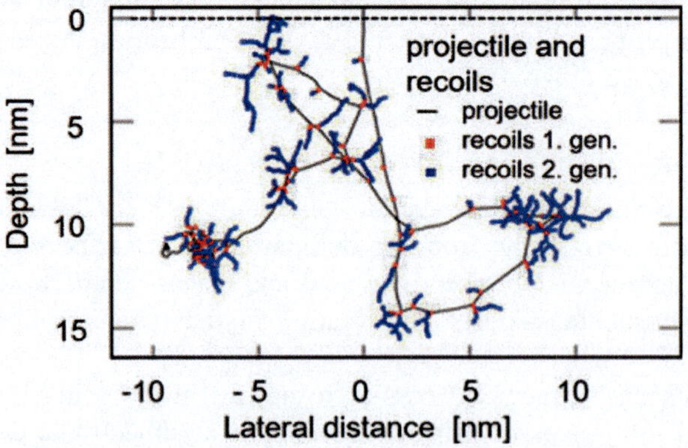

Fig. 4.1 Trajectory and generated recoils of a 2 keV He ion penetrating a Ni target at normal incidence (figure is taken from [9])

damage events as well as the effect of the ion beam-induced damages on the material properties. The primary tool for investigating of damages is the MD simulation, because the time and the length scales correspond to the process times of the collision cascade. The experimental works to study the ion beam-induced damage formation and evolution are based on electrical resistivity measurements and application of the high-resolution transmission electron microscopy. In addition, analysis techniques such as X-ray scattering, positron annihilation spectroscopy, field ion microscopy and small angle neutron spectroscopy are applied.

4.1 Threshold Displacement Energy

A lattice atom can be displaced after collision with the incident ion or a higher order recoil atom if the transferred energy is higher than the displacement threshold energy E_d. This energy is defined as the minimum kinetic energy needed to displace an atom in the bombarded material in order to generate a stable Frenkel pair (FP), which is a vacancy and an interstitial. If the energy less than E_d, the struck atom excites large lattice vibrations, but remains at its stable lattice position. These vibrations are transmitted to surrounding atoms and results in a local heating. If $E > E_d$ the struck atom is capable to overcome the potential barrier as a displaced atom.

Experimentally, the threshold displacement energy can be determined e.g., by measurement of the resistivity of the material at very low temperatures, by transmission electron microscopy or by positron annihilation spectroscopy after particle (electron, ion) bombardment. Typically, the threshold displacement energy in solids ranges between 10 eV and 50 eV [10], i.e. E_d is about five up to ten times higher than the thermally activated formation energy of a Frenkel pair (between 5 eV and 10 eV). For example, Fig. 4.2 illustrates the measured damage function of copper

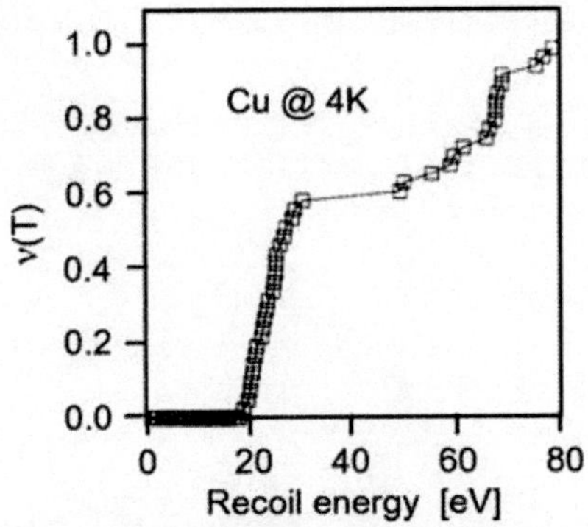

Fig. 4.2 Damage function versus the PKA energy (recoil energy) with a plateau at 0.6 Frenkel pairs. The presentation in this figure based on in-situ electron resistivity damage measurements of single Cu crystals at T = 4K (figure is taken from [11])

in dependence of PKA energy (see Sect. 4.2). The threshold energy in Cu ranges in dependence of the crystal direction between 19 eV (lowest crystal direction) and 30 eV. From the theoretical point of view, the displacement energy can be calculated at detailed knowledge of the interaction potential between the atoms in the material by computer simulation (e.g., molecular dynamic simulations).

Simple approximations have been proposed to determine the threshold energy based on the correlation between the enthalpy or heat of sublimation ΔH_{sub} (sum of the enthalpies of fusion or melting and vaporization) and the threshold displacement energy E_d, because the sublimation enthalpy represents the energy required to transfer an atom or molecule directly from the solid to the gaseous state without the intermediate liquid state at given temperature and pressure. Consequently, the sublimation enthalpy of an atom is defined as the energy required to move from its position in the solid to infinity and decreases linearly with the increase of temperature and pressure. According to Mitchell et al. [12], the threshold displacement energy should be between $4\Delta H_{sub}$ and $5\Delta H_{sub}$. In contrast, many experiments have demonstrated an interval for the displacement energy between $4\Delta H_{sub}$ and $10\Delta H_{sub}$.

It is well-known that for a solid with a crystalline structure the potential barrier is dependent on the crystallographic direction. Consequently, it can be expected that the heights of the potential barriers vary in relation to the crystal direction and, therefore, the displacement energy of a lattice atom should also be strongly dependent on the crystallographic direction. That means, that each crystal direction has an individual threshold displacement energy, $E_d \langle hkl \rangle$, along principal crystallographic axes. For example, Fig. 4.3 shows the calculated distribution of the threshold energy as a function of the crystallographic direction in iron (bcc lattice). In dependence on the crystallographic direction the threshold displacement energy in iron ranges between 15 eV and 65 eV. For this reason, it can be distinguished between a minimum and a maximum of the displacement energy and a weighted average energy $E_{d(av)}$ over all

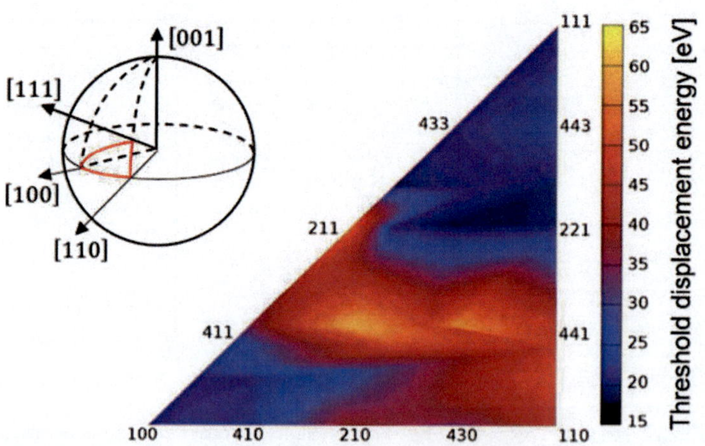

Fig. 4.3 Calculated threshold displacement energy distribution in iron as function of crystallographic directions (figure adapted from [13])

directions. In general, this average threshold displacement energy is larger than the minimum threshold displacement energy $E_{d(\mathrm{min})}$ by a factor of up to two. In the end, the large variation of the threshold displacement energy is caused by superposition of crystallographic dependence of this energy barrier and the statistical nature of the collision processes. It should be noted that the production probability of a defect is also strongly influenced by thermal recombination.

The damage mechanism of Frenkel pair production at low energies is generally assumed to be focused replacement collisions. These focused collision sequence transfers energy along close packed directions of the lattice. Thermal vibrations will tend to defocus such focused replacement collision sequences and hence can only account for an increase in the threshold displacement energy E_d.

Thompson [1] has proposed an analytical expression for the threshold displacement energy at low-energy ion bombardment

$$E_d = \frac{A_{BM}}{2} \exp\left(-\frac{a}{2a_{BM}}\right), \tag{4.1}$$

where A_{BM} and a_{BM} are the coefficients of the Born–Mayer potential (see Sect. 2.1.1.2) and a is the lattice plane spacing. It is obvious that the effect of the increasing temperature on the lattice spacing and the elastic constants will result in lower displacement energies.

In Appendix H the threshold displacement energies for carbon, iron and silicon are summarized as example. This survey in Appendix H demonstrates large variations in the threshold displacement energy. In principle, it must be established that the magnitude of the threshold displacement energy displays significant variance as a function of the crystallographic direction, temperature during ion bombardment and stress state of the bombarded material.

4.2 Primary Knock-On Atom

The primary knock-on atom (PKA) is the first atom that is displaced from is lattice place as result of the collision with the incident particle. The incident particle partially transferred its kinetic energy onto the PKA. If the transferred energy is high enough, a series of further knock-on lattice atoms are produced, resulting in a linear collision cascade. If the transferred energy to the struck atom, T, is higher than the displacement energy E_d, a displacement cascade is created which is characterized by a large number of permanent Frenkel pairs (see Fig. 4.4, left). If the energy of the struck atom is too low ($T < E_d$), no permanent displaced atoms are generated, i.e. these lattice atoms are only deflected from their lattice places and then returned (Fig. 4.4, right). Lattice vibrations are stimulated which induce a warming of the bombarded material. A replacement collision sequence along a close packed crystallographic direction without permanent displaced lattice atoms is called a focusing collision chain or focuson (see also Fig. 4.12). Generally, this appears as a special behavior when the

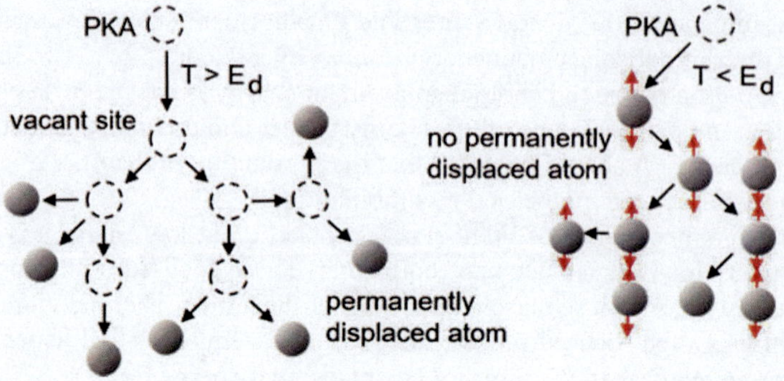

Fig. 4.4 Schematic presentation of a replacement collision sequence or collision cascade. Left: subsequent collision cascade with permanently displaced atoms. Right: a cascade of non-permanently displaced atoms. The red arrows symbolize the excited lattice vibrations

incoming particle is travelling parallel to a close packed direction (e.g., [100] in fcc lattice or [111] in bcc lattice). At the end of a replacement collision sequence a Frenkel pair is generated.

It is assumed that PKA of kinetic energy T hits a stationary lattice atom. During collision, a part of this energy, $\epsilon - E_d$, is transferred on the lattice atom. The energy of this knocked-on lattice atom is assumed to be ϵ, because $E_d \ll T$, i.e. $\epsilon - E_d \approx \epsilon$. The kinetic energy of the PKA atom after collision is $T - \epsilon$ (see Fig. 4.5).

Then, the number of displaced atoms, can be given by the following balance equation

$$N_d(T) = N_d(T - \epsilon) + N_d(\epsilon). \tag{4.2}$$

This equation cannot be used to determine the number of displaced atoms, because the transferred energy to the collision partner is not specified, i.e. T can vary between 0 and T.

Fig. 4.5 The change in the PKA energy after the first collision in a cascade (figure adapted from [14] and modified)

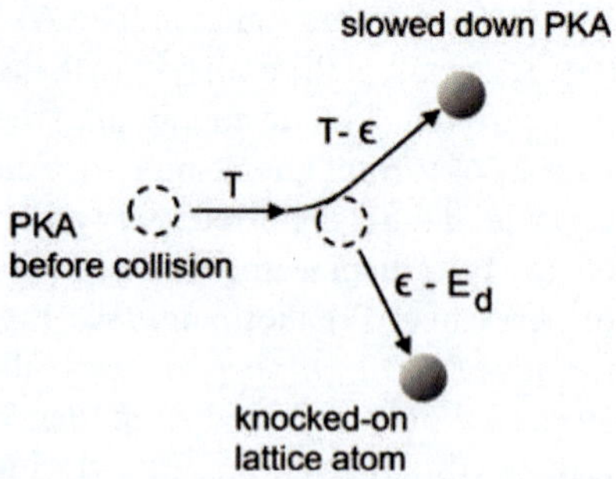

4.2.1 The Kinchin–Pease Model

An approach to quantify the number of displaced atoms was proposed by Kinchin and Pease as a function of the kinetic energy T of the PKA [15] and is based on the following assumptions [14, 15]:

(i) The collision cascade is generated by sequences of two-body elastic collisions between atoms.
(ii) The energy transfer cross-section is given by energy transfer between hard spheres.
(iii) The displacement probability is 0 for $T \leq E_d$ and 1 for $T \geq E_d$.
(iv) Energy is partitioned in a collision between the arriving atom, $T - \epsilon$, and the recoil with the energy ϵ. No energy is dissipated in the lattice (the energy E_d is neglected in the energy balance).
(v) Energy loss by electronic stopping is determined by a cutoff energy level, E_{cut}. If PKA energy > E_{cut}, then no additional displacements occur until energy has been reduced below the cutoff.
(vi) Effects of the crystal structure are neglected.

The differential energy-transfer cross-section and the probability that a PKA of the energy T transfers energy in the range (ϵ, d ϵ) to the struck atom is given by

$$\sigma(T,\epsilon) = \frac{\sigma(T)}{T} \quad \text{and} \quad \frac{\sigma(T,\epsilon)}{\sigma(T)}d\epsilon = \frac{d\epsilon}{T}, \tag{4.3}$$

respectively, where the transfer energy efficiency factor γ is assumed to be one (details see [6, 8]).

In the Kinchin–Pease model, a sharp energy cutoff E_{cut} is assumed, so that below E_{cut} the energy loss is caused by elastic collisions and above E_{cut} by electronic loss. For $T < E_{cut}$ the number of the displaced atoms can be determined by integration of (4.2)

$$N_d(T) = \frac{1}{T}\int_0^T [N_d(T-\epsilon) + N_d(\epsilon)]d\epsilon. \tag{4.4}$$

The integration about integrals on the right hand of (4.4) leads to $\int_0^T N_d(\epsilon)d\epsilon$. Thus, the equation governing $N_d(T)$ is given by

$$N_d(T) = \frac{2}{T}\int_0^T N_d(\epsilon)d\epsilon \tag{4.5}$$

If this energy T is lower than E_d, no displacement atom can be anticipated. If it is lower than $2E_d$, exactly one displaced atom can be expected, i.e. $N_d(T) = 0$ if

Fig. 4.6 Number of displaced atoms as a function of PKA energy according to the Kinchin–Pease model

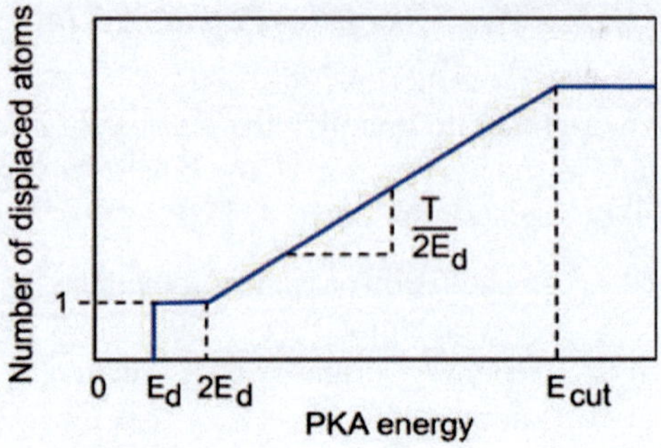

$E_d \leq T < 2E_d$. The PKA induces at least one displacement, i.e. $N_d(T) = 1$. If the initial PKA energy is higher than $2E_d$, the PKA is capable to cause more than one displacement. Consequently, after n collisions 2^n lattice atoms are displaced and the transferred energy is $T/2^n$. Under this condition, the solution of the integral equation (see e.g. [4, 6]) leads to

$$N_d(T) = \frac{T}{2E_d} \tag{4.6}$$

Accordingly, the average number of displacement atoms is given by

$$N_d(T) = \begin{cases} 0 & if\, 0 < T < E_d \\ 1 & if\, E_d \leq T < 2E_d \\ \frac{T}{2E_d} & if\, 2E_d \leq T < E_{cut} \\ \frac{E_{cut}}{2E_d} & if\, T \geq E_{cut} \end{cases}. \tag{4.7}$$

A schematic of the average number of the displaced atoms as a function of the PKA energy is shown in Fig. 4.6. The number of displaced atoms, as determined by the Kinchin–Pease relationships, (4.7), is characterized by a linear relation between the number of displaced atoms and the PKA energy T.

4.2.2 The Norgett-Robinson-Torrens Model

It is well-known, that an energetic recoil atom that slows down in a solid loses a certain energy amount. This energy partially dissipates in electronic excitations (inelastic interactions). Consequently, the remaining energy, called damage energy, is only available for the generation of atomic displacements [16]. Lindhard et al. [17] have derived a comprehensive analysis of the partition of energy between electronic and nuclear processes (called Lindhard's energy partition theory). In the framework of the Thomas–Fermi approximation, Lindhard and coworkers have assumed that the energy T of the PKA can be portioned into the inelastic energy loss in the collision cascade by electron excitation, η, and the elastic PKA energy loss, υ, i.e.

$$T = \upsilon(T) + \eta(T), \tag{4.8}$$

where it is assumed that (i) electrons do not produce recoil atoms, (ii) energy transfer to the electrons is small in relation to the atoms, (iii) the atom binding energy can be neglected, and (iv) $T \leq$ the kinetic energy of the projectile ion.

Its common knowledge that the number of displacements according to the Kinchin–Pease formalism is overestimated. Particularly, (i) the application of more realistic potentials instead of the hard sphere potential and (ii) the implementation of the electronic losses should lead to an improved accuracy of the Kinchin–Pease damage function. Based on a variation of these assumptions in the Kinchin–Pease model, Norgett et al. [18] have proposed a model (NRT model) to predicted numbers of atomic displacements (number of Frenkel pairs) generated by a self-ion PKA as function of the total PKA energy to

$$N_d(T) = \xi \frac{\upsilon(T)}{2E_d} = \xi \frac{T - \eta(T)}{2E_d} \quad \text{for } T > 2E_d/\xi, \tag{4.9}$$

where $\nu(T)$ is the damage energy, i.e. the energy available to induce displacements of atoms.

In the NRT model, the damage energy is given by

$$\nu(T) = \frac{T}{1 + k_{cas} g(\epsilon)}, \tag{4.10}$$

where the parameter k_{cas} and remaining function $g(\varepsilon)$ [16] are given by

$$k_{cas} = 0.1337 Z_1^{1/6} \sqrt{Z_1/M_1} \quad \text{and} \quad g(\varepsilon) = \varepsilon + 0.40244\varepsilon^{3/4} + 3.4008\varepsilon^{1/6}, \tag{4.11}$$

respectively. The reduced energy ε is given by (2.74).

According to Robinson [19], the linear proportionality between the number displaced atoms and T/E_d should be corrected by a factor ξ, called displacement efficiency, with the objective to find a realistic potential between the hard sphere potential and the unscreened Coulomb potential (Rutherford collision).

The solution of the integral equation governing the production of Frenkel pairs in the cascade [19–21] suggests a value of the factor $\xi = 0.8$ independent of the PKA energy, target material and temperature. According to Robinson [19], the damage efficiency decreases tardily with PKA energy, and decreases more steeply for lighter elements than heavier ones. Following this, (4.7) can be simplified by

$$N_d(T) = \xi \frac{T}{2E_d} = 0.4 \frac{T}{E_d}. \tag{4.12}$$

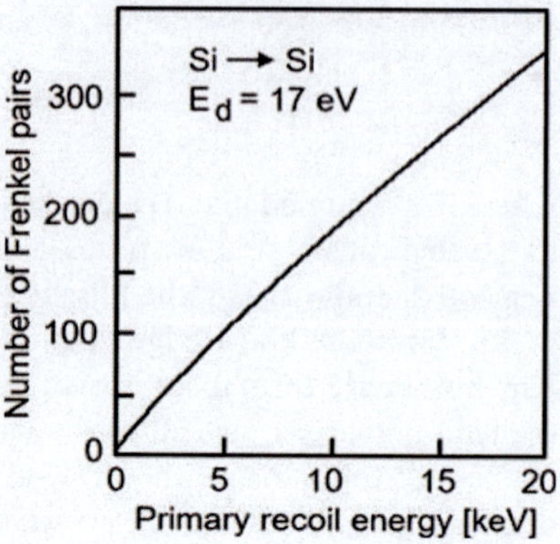

Fig. 4.7 Average number of Frenkel pairs in Si as function of the primary recoil energy

On the base of the assumption by Lindhard et al. [17] that $v(T)$ can be interpreted as part of the energy transferred into nuclear collision processes, a modified Kinchin–Pease relation [19] can be given by

$$N_d(T) = \begin{cases} 0 & if\, 0 < T < E_d \\ 1 & if\, E_d \leq T < 2E_d/\xi. \\ \xi \frac{T}{2E_d} & if\, \frac{2E_d}{\xi} \leq T < \infty \end{cases} \tag{4.13}$$

In general, this formula is applied as the international standard for quantifying irradiation damages. As an example, Fig. 4.7 shows the number of Frenkel pairs (PKA's) in silicon as function of the primary Si ion energy.

The modified Kinchin–Pease relation doesn't take into account collisions between atoms of different masses and the variation of displacement energy with crystal orientation. Both the Kinchin–Pease model and the Norgett–Robinson–Torrens model anticipate that the damage production occurs linear to the deposited energy. It is also known, that these models overestimate the number of Frenkel pairs (see Sect. 4.5). Experimental studies and MD simulation have demonstrated that the damage is often less than expected if the deposited energy is well above the threshold displacement energy E_d [5], because the displaced atoms have sufficient energy so that they are capable to recombine.

For very low ion energies, i.e. $2E_d \leq T \cong E \leq 100\, E_d$, Sigmund [22] has shown that the number of displaced atoms is given by

$$N_d(T) = \frac{6}{\pi^2}\frac{T}{U}\ln\left(1 + \frac{U}{E_d}\right), \tag{4.14}$$

where U is the binding energy lost by a lattice atom when leaving the lattice site (in a first approximation equal the number of breaking bonds times the displacement energy). This is an upper limit for the number of displaced atoms since loss of defects by replacement collisions is neglected. Since the displacement processes in metals and semiconductors are substantially different, these materials must be considered, separately. For semiconductors the displacement energy is determined by the number of atomic bonds which must be interrupted (e.g., $E_d = 4\,E_B$ for Si and Ge, where E_B is the atomic binding energy). With $U = E_d = 4\,E_B$ is the number of displaced atoms $N_d(E) = 0.42\,(E/E_d)$, which is the same results given by the modified Kinchin–Pease relation. For metals, the displacement energy, E_d, can be neglected compared to the binding energy U. Consequently, (4.14) predicts the number of displacement atoms to $N_d(E) = 0.61\,(E/E_d)$.

In praxis, Monte Carlo simulations are applied to calculate the displacement damages. These simulations are based on the binary collision approximation, where collision events are considered to be independent from each other and no collisions between moving species are allowed. All generated Frenkel pairs are considered to be stable, although a particular pair could be annihilated if the distance between vacancy and interstitial sites is less than the given recombination radius. The recombination radius is a free input parameter in these simulations. Monte Carlo simulations provide detailed information on the resulting spatial damage distribution. The most used approaches are the TRIM, SRIM and MARLOWE simulation codes (see Sect. 3.6).

4.3 Displacement Generation Rate

In the previous subchapter the number of displacements produced by a single PKA was determined. But, an implanted ion produces numerous PKA's of different energies on the path through the solid until it comes to rest. The total number of displacements per volume unit and time, called the displacement generation rate, is therefore of particular interest.

The displacement rate, $R_D(x)$, in units of [displaced atoms/cm^{3}s], i.e. the number of PKA's generated per unit volume and unit time by an ion with energy E′ in the energy interval between T and T + dT at depth x is given by [6, 8]

$$NJ\left[\frac{\sigma_D(dE')}{dT}\right]dT, \tag{4.15}$$

where $\sigma_D(dE')$ is the displacement cross-section and can be interpreted as the probability of a particle of energy E′ impacting a sessile lattice atom with a recoil energy T. It should be noted that ions lose energy due to electronic excitations in addition to elastic collisions as they travel through the solid. Consequently, the energy of the incident ions must be corrected with respect to the distance traveled, i.e. E′ is the energy of the ion at depth x. This cross-section is given by [see (2.72)]

$$\sigma_D(E) = \int_{T_{min}}^{T_{max}} N_d(T)\frac{\sigma(E', T)}{dT}dT, \tag{4.16}$$

and $N_d(T)$ is the number of displaced atoms [see, e.g., modified Kinchin–Pease relationship, (4.13)], and $\sigma(E')$ dT is the differential energy transfer cross-section for a moving particle with the energy E′. Then, the displacement generation rate is given by

$$R_D(x) = NJ \int_{T_{min}=E_d}^{T_{max}=\gamma E} N_d(T)\frac{\sigma(E')}{dT}dT = NJ\sigma_D(E'), \tag{4.17}$$

where T_{max} is maximal transferred energy [see (2.36)] and γ is the transfer energy efficiency factor [see (2.28)]. As is well known, the electronic energy loss is given by $(dE/dx)_e = k_{LS}E^{1/2}$, where E is the incident ion energy and k_{LS} is Lindhard‘s electronic stopping factor [see (3.28)]. Then,

$$E' = E'(x) = E - \int_0^x \left(\frac{dE}{dx}\right)_e dx = E - \int_0^x k_{LS}E^{1/2} = (E - k_{LS}x)^2 \tag{4.18}$$

The number of the displaced atoms at a depth x, $R_D(x)/NJ$, in units of [dpa/(ion/cm^2], is given by [6]

$$\frac{R_D(x)}{NJ} = \frac{\pi}{4}\frac{Z_1^2 Z_2^2}{E_d E}\varepsilon^4\left(\frac{M_1}{M_2}\right)\ln\left(\frac{\gamma E}{E_d}\right), \tag{4.19}$$

if it is assumed that $\sigma(E, T)$ can be described by Rutherford scattering and $N_d(T)$ can be taken from the Kinchin–Pease relationship (details see [6]). Note that the displacement generation rate, R_D, is referred to as K_o in Sect. 4.8.

4.4 Displacements Per Atom

In the previous chapters it could be ascertained that the NRT relations are widely used as standard to evaluate the damage induced to structural material properties by radiation. With the objective to quantity irradiation induced changes in materials, the measure displacements per atom (dpa) was introduced, where one dpa means that on average one ion in the irradiated volume displaces exactly one target atom from its lattice site. This measure is related to the number of the Frenkel pairs (FP’s) in the irradiated material, which significantly affect the macroscopic material properties. A helpful conversion from the total number of Frenkel pairs per incident ion and unit depth to displacements per atom can be given by

$$[dpa] \equiv \left[\frac{displacements}{atom}\right] = \left[\frac{Frenkel\ pairs}{atom}\right]$$
$$= \left[\frac{Frenkel\ pairs}{nm \cdot ion}\right] \cdot \frac{fluence\ in\ [ions/cm^2]}{atomic\ density\ in\ [atoms/cm^3]} \cdot \left[\frac{10^7 nm}{cm}\right]. \quad (4.20)$$

Based on this definition, the dpa at depth x can be expressed by means of (4.17)

$$\frac{R_D(x)}{N} t = \Phi \int_{T_{\min}=E_d}^{T_{\max}=\gamma E} N_d(T) \frac{\sigma(E')}{dT} dT \equiv dpa(x), \quad (4.21)$$

where the ion fluence $\Phi = Jt$. Frequently, this relation is approximated by assuming that the number of displacements in the depth x, $N_d(x)$, is given by the modified Kinchin–Pease relationship, (4.13), i.e.

$$\frac{N_d(x)}{\Phi} = \xi \frac{F_D(x)}{2E_d}, \quad (4.22)$$

where damage energy is replaced by the deposited energy $F_D(x)$, (see next Sect. 4.5). Then, the dpa as function of the depth can be expressed by

$$dpa(x) \cong 0.4 \frac{F_D(x)}{E_d N} \Phi. \quad (4.23)$$

It is also of interest to know the total number of dpa over the range of recoils, R. Based on the modified Kinchin–Pease relation. The total number of dpa can be approximately determined by

$$dpa \cong \xi \frac{\nu(T)}{2E_d} \frac{\Phi}{NR} = 0.4 \frac{\nu(T)}{E_d} \frac{\Phi}{NR}, \quad (4.24)$$

where the $F_D(x)$ is replaced by the damage energy $\nu(T)$, see (4.9).

It should be recalled that the dpa parameter is the standard to quantifying radiation-induced changes in material properties after ion bombardment, as well as after neutron and electron irradiation. However, it was recognized from experimental studies and computer simulations that the dpa value can be both overestimated and underestimated. For example, the calculated number of ion beam induced defects in metals was a factor of 3 - 4 greater than the experimentally determined values [7, 23, 24]. Otherwise, the number of atoms which are permanently displaced from their lattice position with the objective to replace an atom in another lattice position, the so-called atomic mixing value, was smaller than experimentally verified [23, 25].

Consequently, a athermal recombination-corrected displacement per atom model (arc-dpa model) was proposed by Nordlund et al. [26, 27] to overcome the failing of the other models and to realistically describe the number of generated Frenkel pairs (FP's). For this purpose the relative damage efficiency, $\xi_{arc}(T)$, is introduced, which gives the ratio between the number of FP's to the number of calculated Frenkel pairs according to the NRT model.

Assuming a strong in-cascade recombination significantly reduces the number of surviving defects. This recombination is a preferentially athermal recombination, i.e. without diffusion of the defects. Consequently, it can be expected that in the result of the recombination the number surviving defects is smaller than calculated by the Kinchin–Pease or the NRT models. Therefore, a relative damage efficiency was introduced [26, 27] by

$$\xi_{arc}(T) = \frac{1 - c_{arc}}{\left(\frac{2E_d}{0.8}\right)^{b_{arc}}} T^{b_{arc}} + c_{arc} \approx \frac{1 - c_{arc}}{(2.5E_d)^{b_{arc}}} T^{b_{arc}} + c_{arc}. \tag{4.25}$$

That describes the influence of the athermal recombination, where b_{arc} and c_{arc} are material constants which can be determined by MD simulations or experiments for a given material (tabulated in [28] from atom mass of lithium up to atom mass of uranium). Herewith, the athermal recombination corrected-displacement damage (arc-dpa) can be expressed by

$$\langle N_d(T) \rangle = \begin{cases} 0 & if\, 0 < T < E_d \\ 1 & if\, E_d \le T < 2E_d/0.8 \\ 0.8 \frac{\nu(T)}{2E_d} \xi_{arc}(T)\; if & \frac{2E_d}{0.8} \le T < \infty . \end{cases} \tag{4.26}$$

On the base of this relationships, the well-known fact that displacement per atom (dpa) overestimates the damage production in metals under energetic displacement cascade conditions can be corrected. Figure 4.8 compares the original NRT damage function for dpa calculation with the function that accounts for athermal recombination (arc-dpa). The damage production in the cascade exhibits a weak sublinear behavior with damage energy.

Similar to the arc-dpa function, a replacements per atom (rpa) function, $\xi_{arc}(T)$, an be derived with the objective to determine the number of atoms that are displaced from their initial lattice site in a cascade and end up in another site (replaced atom) [26]. On this basis, the amount of atomic mixing in energetic displacement cascade (number of defects per displaced atom versus damage energy) can be more realistically determined.

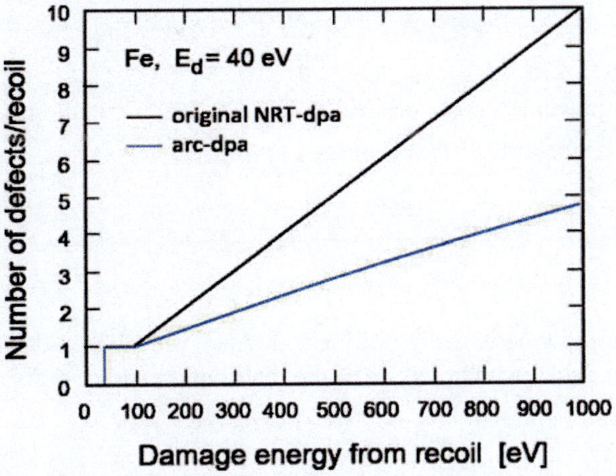

Fig. 4.8 Damage function of iron according to the NRT model and to the model of athermal recombination model (arc-dpa). The constants for Fe are $b_{arc} = 0.568$ and $c_{arc} = 0.286$ (figure is adapted from [26])

4.5 Spatial Distribution of the Deposited Energy

The spatial distribution of the density of energy deposited in both the nuclear collision and electronic processes is of fundamental importance to the physical properties of materials modified by ion beams. Various methods have been developed to determine the distribution of the energy (damage).

Winterbon et al. [29] have developed a simple approach to determine the deposited energy for purely elastic stopping. The energy deposited in primary collisions on the path element dx is given by

$$dE = NS_n(E(x))dx = F_D(x)dx, \tag{4.27}$$

where x is the path length of an ion traveled from the initial kinetic energy E down to energy E(x). Consequently, $F_D(x)$ is the depth distribution function of the energy loss, while neglecting the electronic energy losses. Under the assumption that the power law cross-section is given by (2.76) and inserting (3.5) and (3.27) into (4.27), the depth distribution function of the deposited energy (referred to as deposited energy) can be expressed by

$$F_D(x) = \frac{E}{2mR}\left(1 - \frac{x}{R}\right)^{\frac{1}{2m}-1}, \tag{4.28}$$

where m = 1/s is the power law variable (see Table 2.2) and R(E) is the ion range, which was determined in frame of this theory to be

$$R(E) = \frac{(1-m)}{2m}\frac{\gamma^{m-1}}{NC}E^{2m}, \tag{4.29}$$

where C_m is the power law constant [see (2.79)]. For this distribution, the total amount of the deposited energy is given by

$$\int_{-\infty}^{+\infty} F_D(x)dx = E, \tag{4.30}$$

and identical to the kinetic energy of the ion. Also, the average damage depth $\langle X\rangle$ and the average damage width $\langle \Delta X\rangle$ of the distribution can be derived to

$$\langle X\rangle = \frac{1}{E}\int_{-\infty}^{+\infty} xF_D(x)dx = \frac{2m}{1+2m}R \quad \text{and} \quad \frac{\langle \Delta X^2\rangle}{\langle X\rangle^2} = \frac{\langle X^2\rangle - \langle X\rangle^2}{\langle X\rangle^2} = \frac{1}{1+4m}. \tag{4.31}$$

The major simplifications in the last two equations are that the energy transfer from the incident particle to the recoil atoms is neglected, and that the ion path length is identical with the range of the energy deposition. Different approaches have been developed to calculate the deposited energy by taking into account the electronic energy loss, the energy loss straggling as well as the path length corrections.

Brice [30] has proposed an analytical method for direct calculation of the spatial distribution of energy deposition into damage or ionization for ions implanted into solid targets by considering the multiple scattering of the incident ion beam and the partitioning of the energy of the recoiling target atoms. In this calculation, the energy transport within the target by recoiling target atoms is not considered. Because of this approximation, a lower energy limit exists below which calculations are not valid.

The method of moments (moments of the energy distribution) by Winterbon [31] involves a solution to the coupled set of integral equations which govern the spatial moments of the energy-deposition profiles. Energy transport by recoiling target atoms is implicit in these equations. Subsequently, the distributions of the deposited energy can be calculated from a finite set of moments. For relatively symmetric distributions, such as the depth distribution of implanted ions, only the first few moments are required for an accurate construction of the profile. Because the distributions are generally far from symmetric, however, a large number of moments are required for their accurate description.

Approaches to determine the energy (damage) distribution based on the transport equation assume binary collisions and an integro-differential equation, which determine the evolution of the damage distribution. Winterbon et al. [29] have given an analytic solution of the integro-differential equation for the damage distribution, which uses the power cross-section [32]. This theory, referred to as the WSS theory, gives the average longitudinal, $\sqrt{\delta_x\langle \Delta X^2\rangle}$, and transverse straggling, $\sqrt{\delta_y\langle Y^2\rangle}$, corresponding to a large number of cascades. The quantities δ_x and δ_y are contraction parameters established to correct the statistical cascade to an individual cascade [33], since the moments estimated by the transport theory or computer simulations refer only to the assembly averages. In dependence of the mass ratio M_2/M_1 between 0.075 and 25, the contraction parameters vary between 1 and 0.25 for m = 1/2.

This leads to a collision cascade with spatial dimensions that can exceed those of a single cascade. Figure 4.9 shows the calculated first and second moments as function of the mass ratio M_2/M_1. The average damage depth $\langle X \rangle$ is always smaller than the path length R of the ion. The ratio $\langle X \rangle/R$ decreases with increasing the mass ratio because for $M_1 \ll M_2$ the ions undergo many large-angle deflections. With the increase of the mass ratio the distribution is broadened even more. For $M_2/M_1 \gg 1$, the distribution of the deposited energy shows the tendency to be spherical. Unfortunately, the depth distribution of the damages cannot be expressed by an analytical formula. In fact, Monte Carlo simulations (see Sect. 3.6.2) are frequently used to calculate the spatial distributions of damage energy, vacancies and interstitials as result of ion bombardment. Then, the complete history of a collision cascade is obtained by statistical averages of the quantities of interest over a very large of number of particles to obtain the desired distributions. For a many applications, the first order moment and the second order moments, i.e. the individual cascade longitudinal and transversal straggling of the damage distribution, provide a reasonable approximation of the distribution within a collision cascade in the shape of a spheroid (see schematic in Fig. 4.10).

According to Sigmund [34], the distribution of the deposited energy, $F_D(x, y, z, E)$, can be figured out in terms of moments $\langle X^n \rangle$ (see Appendix F) under the assumption that elastic collisions are dominant by

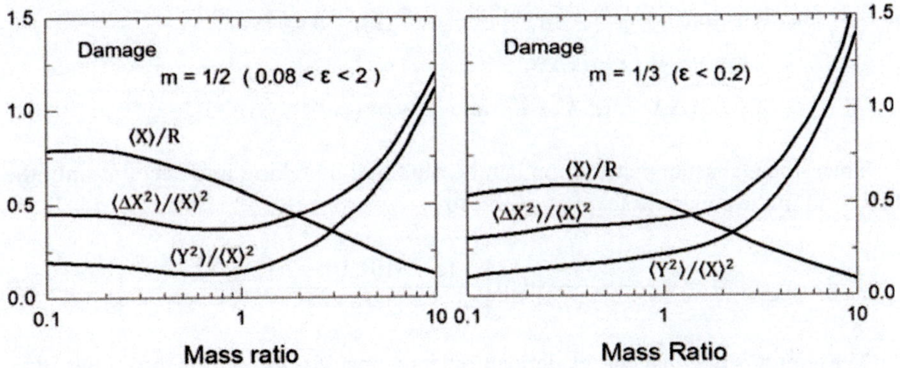

Fig. 4.9 Ratio of the average damage depth $\langle X \rangle$ to the path length of the ion as functions of the mass ratio M_2/M_1, where $\sqrt{\langle \Delta X^2 \rangle}$ and $\sqrt{\langle Y^2 \rangle}$ are average longitudinal and transverse straggling, respectively. Left for m = 1/2, and right for m = 1/3 (adapted from [29])

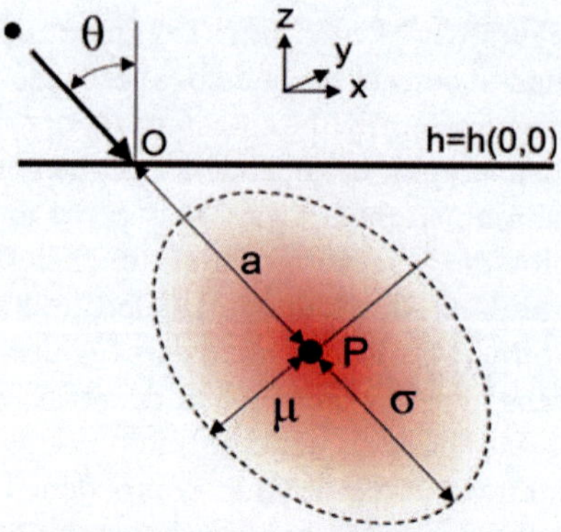

Fig. 4.10 Schematic of the damage distribution in an elastic collision cascade after ion impact under an angle θ, where the longitudinal and transversal straggling of the damage distribution are $\sigma = \sqrt{\delta_x\langle\Delta X^2\rangle}$ and $\mu = \sqrt{\delta_y\langle Y^2\rangle}$, respectively. The kinetic energy dissipates inside the near-surface region forming a collision cascade around the ion track which is generally of ellipsoidal shape. The point Q indicates the point of the maximal energy deposition at the surface and P the point of the maximal energy deposition

$$F_D(x, y, z, E) = \frac{\upsilon(E)}{\langle\Delta X^2\rangle^{\frac{1}{2}}}\left[\varphi_o(\xi) - \frac{\Gamma_1}{6}\varphi_3(\xi) + \left(\frac{\Gamma_2}{24}\varphi_4(\xi) + \frac{\Gamma_1^2}{72}\varphi_6(\xi)\right) + \cdots\right], \tag{4.32}$$

where

$$\langle\Delta X^n\rangle = \langle(\mathrm{x} - \langle X\rangle)^n\rangle \quad \text{with the order of moments n} = 2, 3, \ldots,$$
$$\varphi_n(\xi) = (2\pi)^{-1/2}\left(d^n/d\xi^n\right)\exp(-\xi^2/2), \quad \mathrm{n} = 1, 2, \ldots,$$
$$\xi = (\mathrm{x} - \langle X\rangle)/\langle\Delta X^2\rangle^{1/2},$$
$$\Gamma_1 = \langle\Delta X^3\rangle/\langle\Delta X^2\rangle^{3/2} \quad \text{and} \quad \Gamma_2 = \langle\Delta X^4\rangle/\langle\Delta X^2\rangle^2 - 3.$$

Then, the Gaussian distribution can be obtained by taking into account only the first term in the square bracket of (4.32) to

$$F_D(x, y, z, E) = \frac{E}{(2\pi)^{3/2}\sigma\mu^2}\exp\left\{-\frac{[z - h(0,0) + a]^2}{2\sigma^2} - \frac{x^2 + y^2}{2\mu^2}\right\}. \tag{4.33}$$

The contours of equal energy deposition are ellipsoids. According to Bradley [35], the average energy density at each point (x, y, z) within the solid can be given by means of the momentums, p_{II}, $p_\perp$, parallel and perpendicular to the x–y plane as follows

$$F_D(x, y, z, E) = \frac{E}{(2\pi)^{3/2}\sigma\mu^2} \exp\left\{-\frac{p_{II}^2}{2\sigma^2} - \frac{p_\perp^2}{2\mu^2}\right\}$$
$$= \frac{E}{(2\pi)^{3/2}\sigma\mu^2} \exp\left\{-\frac{(a - x\sin\theta + z\cos\theta)^2}{2\sigma^2} - \frac{(x\cos\theta + z\sin\theta)^2 + y^2}{2\mu^2}\right\}, \tag{4.34}$$

for normal ion impact in the x-direction, where a is the mean distance between the point of impact and the maximum of deposited energy. The damage energy is approximately replaced in (4.33) by the ion incident energy E. As an example, the 2-dimensonal energy distribution due to 1500 eV ion bombardment into Ge and Si in dependence of the mass of the incident ion is shown Fig. 4.11 [36]. The MD simulations are carried out using the Stillinger–Weber potential for the Si–Si and Ge–Ge interaction, and a Moliere potential for the inert gas ion interaction with Ge and Si. As anticipated, the energy deposition distribution is remarkably ellipsoidal for heavier ions. Heavier ions penetrate deeper into the target, making the axial dimension of the energy profile larger than its lateral dimension. For the same ion energy and ion species, the axial dimension is smaller in the denser Ge target than in the less dense Si target.

Hubler et al. [37] have tested the validity of the assumption that spatial distribution can be approximated by a Gaussian and found that the Sigmund predictions do not describe the skewness of the spatially resolved sputter yield for very low ions and high ion energies. It should be noted, that Feix et al. [38] have also observed significant deviations from the Gaussian distribution of the deposited energy. In particular, they have found a minimum of the energy distribution near the position where the ion penetrates the surface, and the decay of energy deposition with distance to ion trajectory is exponential rather than Gaussian.

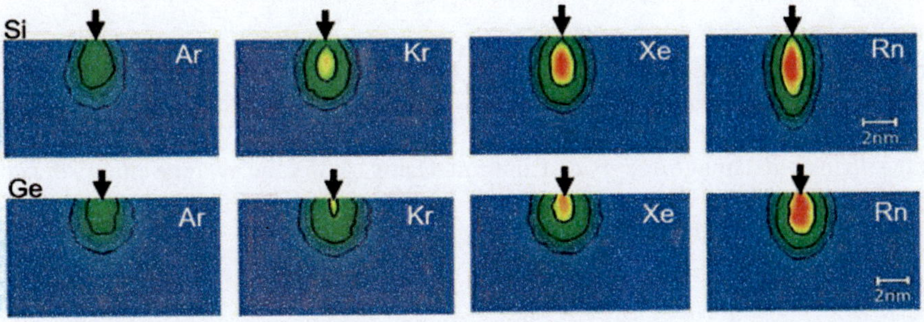

Fig. 4.11 Energy distribution for 1500 eV rare gas ion bombardment of Si and Ge. The contour levels represent the energy levels of 1.0, 0.5 and 0.25 after Rn, Xe and Kr ion bombardment, while for Ar ion bombardment the contour levels represent the energy levels of 0.5 and 0.25 eV. The ion directions are indicated by black arrows (figures adapted from [36])

In general, $F_D(x,E)$ cannot be directly deduced from SRIM simulations [exception: the version SDTrimSP V5.05, is capable to calculate all parameter of (4.32)]. Bolse [39] has proposed an approach to determine deposited energy by

$$F_D(x, E) = n_v(x)\frac{v(E) \cdot E}{N_v}, \tag{4.35}$$

where N_v is the total number of vacancies and n_v (x) is the vacancy density produced by primary and secondary collisions, which can be calculated by means of MC simulations. The mean energy deposited per target atom can be expressed by $F_D(x, y, z, E)/N$ [29, 33], where N is the atomic number density of the target material.

It should be noted that the experimental determination of N_d by channeling or backscattering is difficult, because (i) the annihilation of Frenkel pairs at low temperatures and (ii) defects collapse into vacancy dislocation loops. Consequently, the number of measured defects is frequently less than expected (calculated).

4.6 Non-linear Cascades

In the above consideration all expressions are linear to the PKA energy. If the mass of both collision partners are large, then the average number of displaced atoms increases rapidly. Thus, the assumption of subsequent binary collisions cannot be applied any longer, i.e. a non-linear behavior of the deposited energy density and a dependence of the number of displaced atoms on the ion energy can be expected. Such cascades are generally referred to high-energy density cascades.

According to Averback [41], the volume of the cascade can be determined by using the second moments of the energy distribution and the contraction parameters δ_x and δ_y as established by Sigmund et al. [33] (see Sect. 4.5), to

$$V_{cas} = \frac{4}{3}\pi\left[\langle\delta_x \Delta X\rangle^2 + 2\langle\delta_y Y\rangle^2\right]^{3/2}, \tag{4.36}$$

approximately, where $\langle\Delta X^2\rangle$ and $\langle Y^2\rangle$ are the longitudinal and transversal moments of the deposited damage energy distribution. Using (4.35), the average energy density in the cascade can be estimated as

$$\langle\theta\rangle \cong \frac{\nu(E)}{N V_{cas}}. \tag{4.37}$$

With this relationship, the energy density of a cascade can be approximately estimated with energy densities in excess of about 1 eV/atom being typical for target materials with high atomic numbers.

Usually, the time sequence of the interaction processes of an incident ion with a kinetic energy higher than the thermal energy with the initially stationary substrate atoms is subdivided into individual successive processes. Initially, the atoms of the substrate are displaced and excited by the incoming ion. If the transferred energy during the primary knock-on collision process is higher than the displacement energy, the PKA with the residual energy collides with other atoms thus giving rise to higher generations of collisions. These collisions produce a multitude of low-energy recoils in all directions. A so-called isotropic *collision cascade* is formed. For light incidence ions the distance between two successive collisions (mean free path-length) is larger than the interatomic distance. Therefore, the collision cascade appears as a diluted distribution of defects. In contrast, for heavy low-energy ions the mean free path-length is comparable to the interatomic distance and thus a very dense collision cascade is generated. In this case, a many displaced atoms in a very small region are participating in the dissipation of the incident energy. Hence, the concept to describe the slowing down process by a sequence of successive binary collisions with a fixed threshold energy cannot be applicable when each atom receives about 1 eV/atom. The duration of the interaction between the energetic particle and the atoms of the bombarded substrate atoms can be estimated on the base of (3.1) to

$$t = \frac{1}{N}\int_0^E \frac{dE}{S(E).v} = \frac{1}{N}\int_0^E dE\frac{1}{S(E)}\sqrt{\frac{2E}{M_1}}, \tag{4.38}$$

where the $v = 2E/M_1$ is the particle (ion) velocity and M_1 the particle (ion) mass. For low kinetic energies (< some keV) this time is in the order of 10^{-13} s or 0.1 ps. Note that a single collision process between two atoms takes about 10^{-15} s. Sigmund [34] has estimated the slowing down time of a heavy projectile in an elastic collision cascade as

$$t = \int_0^E \frac{dE}{S(E)\cdot v} \cong \frac{2mR(E)}{(2m-0.5)v}, \tag{4.39}$$

or rather for a power low variable m = 1/2 to

$$t = \frac{2R(E)}{\sqrt{2E/M_1}}, \tag{4.40}$$

where R(E) is the total path length of the incident ion. For example, the slowing down time after Te ion bombardment of Ag (m = 1/2) varies between 10^{-13} s and 10^{-12} s for energies between 40 and 150 keV [34].

In the time regime up to about several hundred femtoseconds, isolated recoils (interstitials) and vacancies or displacement (thermal) spikes are formed as the result of this slowing down process. The total time scale of all interaction processes ranges from femtoseconds up to seconds, i.e. over 10^{15} magnitudes.

Table 4.1 Chronological sequence during the slowing down of an energetic PKA [8]

Duration (10^{-12} s)	Processes	Results
~10^{-3}	• Energy transfer of the incident particle	• Formation of knock-on atom (PKA)
~10^{-3}…0.2	• Slowing down of PKA • Generation of a thermal and displacement spike	• Formation of recoil atoms and vacancies • Formation of subcascades
~0.2…3.0	• Cool down of the thermal spike	• Stable interstitials • Interstitials clusters • Athermal recombination
~3.0…10	• Cascade cooling to ambient target temperature	• Depleted zone in the displacement cascade
>10	• Thermal migration of defects • Interaction of defects with sinks	• Evolution of the microstructure (e.g., segregation, precipitation, second phase dissolution, dislocation, etc.,)

According to Seitz and Koehler [41], two types of spikes can be distinguished, the displacement spike and the thermal spike. The displacement spike is characterized by a sequence of atomic displacements happening in such striking distance that all atoms in a very small volume are displaced nearly at the same time. In the thermal spike the local energy after ion bombardment is lower so that they are not displaced, but vibrational energy is created and with it the temperature is raised locally. In Table 4.1 the chronological sequence during the slowing down process of energetic PKA's is summarized. The incidence particle settles down approximately within a femtosecond of entering the target and generates a knock-on atom. During the slowing down process the knock-on atom is able to displace other substrate atoms from the lattice sites (displacement cascade) or excite lattice vibrations (thermal spike). This is followed by a cooling process of the cascade. The cooling process is sealed off after about several tens of picoseconds. In low-energy ion irradiation experiments typical ion beam current densities between a few ten μA/cm^2 and some mA/cm^2 are used. There, it can be expected that the evolution of a spike and its disassembly is an isolated event in time, because the time interval between existence of a cascade and the next ion impact is many magnitudes larger than life time of a cascade (1mA/cm^2 = 6.24×10^{15} ions/cm^2 s = 62.4 ions/nm^2 s).

4.6.1 *Displacement Spike*

Brinkman [42] and later Seeger [43] have developed a model of the displacement spike, with the aim to describe the displacement cascade and also the mean free path-length between collisions. Figure 4.12 shows the displacement spike, schematically, characterized by a depleted zone in the center of the spike and an enrichment of

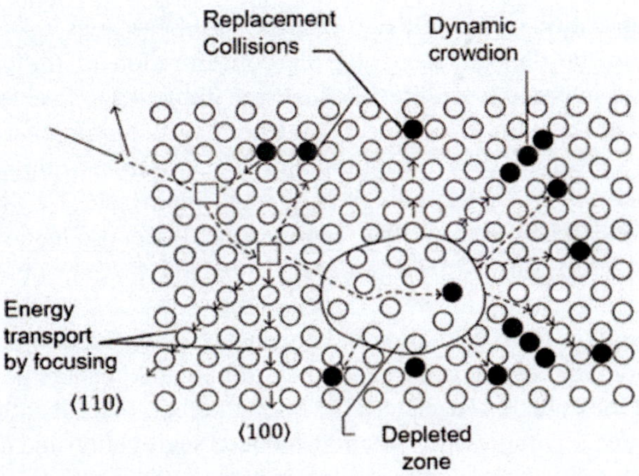

Fig. 4.12 Schematic picture of the displacement spike (adapted from [43]). Solid line is the path of the incident particle, dashed line the path of the PKA, dark points are interstitial atoms and open quadrates are vacancies

interstitials around this zone. Based on the assumption that the mean free path-length λ is given by $\lambda = 1/N\sigma$ [see (I.2)] and the cross-section is given in terms of the differential energy-transfer cross-section, Olander [14] has estimated λ for transfer energies higher than the displacement energy E_d, where a Born–Mayer potential was applied. For example, in the case of Cu atoms moving in Cu the mean free path-length in the displacement cascade rapidly droops down up to about 0.2 nm for energies of the moving particle higher then E_d and smaller than 100 eV. In this range, closely spaced displacement collisions can be expected. According to Stoller [7] the development of displacement cascade is similar for all ion energies and characterized by a highly energetic, disordered core region during the initial phase of the cascade. The number of interstitials and vacancies is increased. In this phase of the cascade development, the number of the interstitials and vacancies are generated in the same manner. Figure 4.13 demonstrates the time evolution of a cascade in iron, which

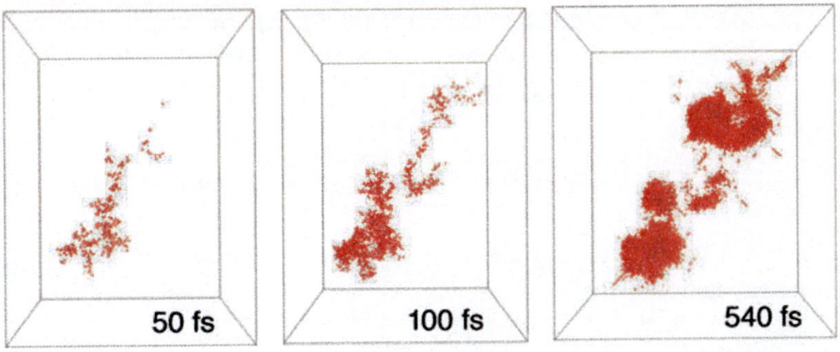

Fig. 4.13 MD simulation of the evolution of a 40 keV cascade in iron at 100 K (adapted from [7])

results in the separation of the collision cascade in subcascades.

Depending on the deposited energy, the concentration of the created Frenkel defects reaches a maximum within 0.1–2 ps (see Table 4.1). This leads to a significant reduction of the number of defects by athermal recombination because of the extremely high spatial vacancy-interstitial pair concentration within a time interval of some picoseconds. The PKA energy is of vital importance for the formation of the displacement cascade. At low PKA energies the defect production is dominated by simple binary collisions, while a cascade-like behavior can be observed at higher energies.

The most frequent form of radiation damages by ion bombardment of materials are isolated point defects (vacancies and interstitials) and clusters generated during the evolution of the collision cascade. These point defects significantly influence the material properties. Examples are radiation-induced segregation and hardening, void swelling and also stable point defect clusters in semiconductor (e.g., so-called {311} like-defects or dislocation loops). Immediately after the formation of a cascade, interstitials and vacancies can recombine, diffuse to sinks (surfaces, interfaces, grain boundaries, dislocations, etc.) and become trapped or escape from the immediate region, where the annihilation by defect sinks, recombination or clustering limits the concentration of defects. The result of the interaction of ion beam-induced defects with the sinks in the bombarded material leads to a steady-state concentration of defects that is higher than the equilibrium concentration of the non-bombarded material.

The energy of the generated recoils is higher than the threshold displacement energy (most recoils are formed at energies near E_d). According to Andersen [10], a collision cascade-induced random walk process of the generated recoils can be assumed. Then, the effective diffusivity within a collision cascade can be expressed by

$$D_{cas}t_{cas} = dpa\frac{\langle\delta^2\rangle}{6}, \tag{4.41}$$

where $\langle\delta^2\rangle$ is the mean squared range of the displaced atoms (see also Sect. 8.2.1.3), t_{cas} is the life time of the cascade and dpa is the number of displacements per atom given by (4.24). When combining both (4.41) and (4.23), the total number of recoils produced by each ion is 0.067 E/E_d and the effective diffusion coefficient due to the ballistic collisions processes is given by [44]

$$D_{cas}t_{cas} = 0.067\frac{F_D\delta^2}{E_dN}\Phi, \tag{4.42}$$

where F_D is the damage energy per unit length given by (4.33), N is the atomic number density and Φ is the ion fluence.

4.6.2 Thermal Spike

In contrast to the displacement spike, in thermal spikes the local atomic energy deposition is lower so that lattice atoms are not displaced. This situation can be the case at the end of the collision cascade, when the energy transfer can be realized neither displacements nor replacements. The concept of the thermal spike by Seitz and Koehler [41] is based on the assumption that the deposited energy is dissipated by intensification of the lattice vibration amplitudes. Within the first 10^{-13} to 10^{-11} s after the collision, the involved atoms are in an extreme energetic state which avoids a description with the laws of the equilibrium thermodynamics. Only after about 10^{-11} s the increase of the local temperature in the spike volume can described by the classic laws of the thermal conduction. Hence, the thermal spike can be defined as a very small region in which the deposited energy causes heating or in some cases melting within an extreme short time scale ($\approx 10^{-11}$ s). Therefore, the temperature, T, as function of the time, t, and location/position, r, can be determined by the heat conduction equation for isotropic medium $\partial T/\partial t = D_{th}$T, where $D_{th} = \kappa/c\cdot\rho$ is the thermal diffusivity, ρ is the mass density, κ is the thermal conductivity, and c is the specific heat capacity.

If the ion energy deposited at a point or along a straight line then spherical and cylindrical thermal spikes can be expected, respectively. For the determination of the temperature distribution and the total number of rearrangement processes in spike, the heat conduction equation was solved for the spherical spikes by Seitz and Köhler [41] and for the cylindrical spike by Vineyard [45] and Hofsäss et al. [46]. In general, it is assumed that a fraction, Q, of the ion incident energy deposited in the spike is converted into thermal energy (i.e. lattice vibrations) and the subsequent energy dissipation results in a temperature distribution. The generated heat diffuses thermally outwards in the radial direction of the spherically or cylindrically symmetrical spike. The local temperature increase due to the heat input from the collision leads to a number of atomic jumps.

According to Vineyard, the temperature distribution, T(r, t), for a spherical thermal spike can be determined if the heat conduction equation is solved assuming a specific power law dependence on temperature for the thermal conductivity and the heat capacity (i.e. $\kappa = \kappa_o T^{n-1}$ and $c = c_o T^{n-1}$ where $n \geq 1$ and k_o and c_o are constants). Then, the equation of heat conduction can be expressed by

$$\nabla \cdot \kappa \nabla T = c\rho \frac{\partial T}{\partial t} \tag{4.43}$$

and the temperature in radial distance r for a spherical spike (point source) is given by

$$T_{sph}(r,t) = \left[\frac{nQc_o^{1/2}}{(4\pi\kappa_o t)^{3/2}}\right]^{1/n} \exp\left(\frac{-c_o r^2}{4n\kappa_o t}\right)$$

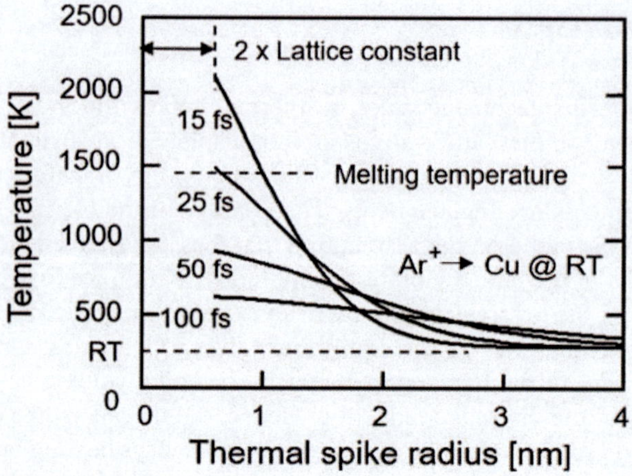

Fig. 4.14 Temperature distribution versus thermal spike radius for four different times after 5 keV Ar-ion bombardment of copper at room temperature (RT). The deposited thermal energy Q ($\approx$ nuclear energy loss) is estimated to be 1050 eV/nm. A thermal diffusivity of 1×10^{-4} m^2/s and the starting time ($\approx a^2/4D_{th}$) are used, where a is the lattice constant of Cu

or for temperatures (n=1) given by

$$T_{sph}(r,t) = \frac{Q}{8\pi^{3/2} c_o \rho (D_{th}t)^{3/2}} \exp\left(\frac{-r^2}{4D_{th}t}\right), \tag{4.44}$$

for times not too close to t = 0 and for distances not to close r = 0. The macroscopic ambient target temperature is taken as absolute zero. It is known that the product $c_o\rho$ in the last equation varies only less sensitively depending on the material than $\kappa_o \kappa_o$ (in a rough approximation is the ratio $\kappa_0^{metal}/\kappa_0^{isulators} \approx 10^2 \ldots 10^3$). Thus, it is obvious that the heating by the thermal spike will be two to three orders of magnitude more intense in insulators than in metals.

As an example, Fig. 4.14 shows the radial temperature distribution in Cu after 5 keV Ar-ion bombardment. In a rough approximation, the thermal energy Q is assumed to be identical to the nuclear deposited energy (stopping power) of the 5 keV Ar ions in Cu. The maximum temperature can be expected close to the ion track. If the temperature is higher than the melting or the evaporating temperature, the thermal spike region can be liquidated or evaporated. The energy is deposited into a spherical volume with a radius larger than the lattice constant of Cu. In larger distances the temperature dissipates in order of some lattice vibration periods.

It should be noted, that (4.44) can be applied when the diffusion length, $\sqrt{D_{th}t}$, is significantly larger than atomic distances. Because of its simplicity, this model is frequently used to estimate the temperature of the thermal spike.

During the time of existence of a thermal spike (up to 1 ps), the atoms are subject to atomic rearrangement processes. Based on the assumption of Vineyard [45] that the atomic motion in generated spike is thermally driven, it can be assumed that the thermally activated jumps obey a directional Arrhenius law. Then, the jump rate (total number of jumps in thermal spike per unit length and unit time) is given by

$$\upsilon(r,t) = \upsilon_o \exp\left(-\frac{E_s}{k_B T(r,t)}\right), \tag{4.45}$$

where E_s is the activation energy (threshold energy) for the jump in the thermal spike and ν_o is a temperature-independent constant in order of magnitude of the phonon frequency ($\approx 10^{13}\ s^{-1}$). Then the total number of these rearrangement processes in the thermal spike (total number of jumps per unit length) is given by

$$\eta = N \int_0^\infty dV \int_0^\infty \nu(r,t)dt, \tag{4.46}$$

where N is the atomic number density. For a spherical spike, η_{sph} is given by

$$\eta_{sph} = N \int_0^\infty 4\pi r^2 dr \int_0^\infty \nu_o \exp\left(-\frac{E_s}{k_B T(r,t)}\right)dt. \tag{4.47}$$

Based on Vineyard's analysis [45] of the thermal spike and the assumption that the thermal conductivity and the heat capacity are temperature-dependent magnitudes (see above), the integration (4.47) by changing the variables yields to the solution [45]

$$\eta_{sph} = \frac{\sqrt{3/5}}{10\pi} N \frac{\nu_o n Q^{5/3}}{\kappa_o c_o^{2/3}} \left(\frac{k_B}{E_s}\right)^{5n/3} \Gamma\left(\frac{5n}{3}\right), \tag{4.48}$$

where $\Gamma(5n/3)$ is the gamma function evaluated at 5n/3.

Seitz and Koehler [41] applied a simplification to integrate (4.47). It was assumed that the temperature in the spike is higher than zero only for spike lifetimes $t > r^2/4D_{th}$. Then, the temperature is given by the pre-exponential term of (4.44), i.e. $T_{sph}(r,t) \approx Q/8\pi^{3/2} c_o \rho (D_{th} t)^{3/2}$. Thus, the number of rearrangement processes in the thermal spike is given by

$$\begin{aligned}\eta_{sph} &\approx N \int_0^\infty 4\pi r^2 dr \int_{r^2/4D_{th}}^\infty \nu_o \exp\left(-\frac{E_s}{k_B T_{sph}}\right)dt \\ &\approx N \int_0^\infty 4\pi r^2 dr \int_{r^2/4D_{th}}^\infty \nu_o \exp\left(-\frac{E_s}{k_B Q} 8\pi^{3/2} c_o \rho (D_{th} t)^{3/2}\right)dt.\end{aligned} \tag{4.49}$$

The integration of (4.49) can be achieved by choosing appropriate substitutions (see e.g., [41, 46]). An approximated expression for the total number of displacements in the spherical thermal spike follows to

$$\eta_{sph} \approx 0.016\,\delta^{*}\left(\frac{Q}{E_s}\right)^{5/3}, \tag{4.50}$$

where δ^* is a coefficient depending on material parameters, such as the thermal diffusivity and atomic volume as well as on the number of symmetrically equivalent neighbors with which an atom can interact, i.e. describes the effectiveness of the local energy dissipation (should lie between 1 and 10 [34], where $\delta^* = 1$ is frequently used). Seitz and Koehler [41] have discussed $Q \approx 50$ eV and $E_s \approx 3$ eV as typical values for a spherical thermal spike. Kelly and Naguib [47] have assumed that the rearrangement processes in the thermal spike can occurs rapidly above a critical associated temperature (is identified with a recrystallization temperature T_C). According to these authors, the activation energy is given by $Q = 35 \pm 5\, k_B T_c$.

For a line source of the length R, the deposited ion energy dissipates in a cylindrical volume $V(t) = \pi r^2(t) R$ and the radius $r(t) = 2(D_{th}t)^{1/2}$ and is given by

$$T_{cyl}(t, r) = \frac{Q/R}{4\pi c\rho}\frac{1}{D_{th}t}\exp\left(\frac{-r^2}{4D_{th}t}\right), \tag{4.51}$$

where inside the cylinder a homogeneous temperature is expected. According to Vineyard [45], the number of displacement processes for the simplest case, where the thermal conductivity and the heat capacity are constant, is given by

$$\eta_{cyl} = N\int_0^\infty 2\pi r dr \int_0^\infty \nu_o \exp\left(-\frac{E_s}{k_B T(r,t)}\right)dt \approx \frac{1}{8\pi} N \frac{\nu_o}{\kappa_o c_o}\left(\frac{E_s}{Q}\right)^2. \tag{4.52}$$

Consequently, the number of displacements (defects) created per unit track length of the incident ion is proportional to the phonon frequency, inversely proportional to the thermal conductivity and the thermal heat capacity. Commonly, the defect density predicted by (4.52) is smaller than experimentally observed. Therefore, it is assumed [48] that strongly reduced values for thermal conductivity as well as the specific heat must be applied (both physical magnitudes fall with increasing temperature for most materials).

The thermal spike considerations discussed above are valid if it can be assumed that the evolution of a spike is dominated mainly by elastic interactions, where inelastic interactions can be neglected (i.e. $S_e(E) \approx 0$). For ion energies exceeding the hyperthermal energy range, electronic interactions must be increasingly included in the considerations. Assuming that the energetic ion is capable to induce nuclear collisions and electronic excitations in the solid along the ion path. This interaction must be described by two coupled differential equations governing the energetic processes in the lattice and the electron system and their coupling. Lifshitz et al. [49] have proposed a coupled differential equation system to describe the time-dependent transient process for a cylindrical geometry, which is as follows

$$c_e(T_e)\frac{\partial T_e}{\partial t} = \frac{\partial}{\partial t}\left[\kappa_e(T_e)\frac{\partial T_e}{\partial t}\right] - g(T_e - T) + A_E(r, t) \quad (4.53)$$

$$c_n(T_n)\frac{\partial T_n}{\partial t} = \frac{\partial}{\partial t}\left[\kappa_n(T_n)\frac{\partial T_n}{\partial t}\right] - g(T_e - T), \quad (4.54)$$

where g is the electron–phonon coupling factor related to the rate of energy exchange between the electrons and the lattice. T_e, T_e, c_n, c_e and κ_e, κ_n are the temperature, specific heat and the thermal conductivity for the electronic and nuclear (lattice) system, respectively. $A_E(r, t)$ is the energy density per unit time supplied by the incident ions to the electronic system at radius r and time t such that $\iint 2\pi r A_E(r, t) dr dt = Se(E)$ [50]. It must be noted that the parameters are again non-linear (c.f. Vineyard approach). The coupled differential equation system can be numerically solved (for this purpose, several methods have been proposed in the literature).

The two-temperature model accounts for the energy transfer from hot electrons to the lattice vibrations due to electron–phonon interaction and the electron heat conduction from the ion path to the bulk of the target. The main effect of the energy deposition and subsequent temperature rise is the energy transfer from the excited electrons to the lattice, as well as cooling of the thermal spike due to the electron heat conduction. While an ion moves along the track in the bombarded material, the heat can induces melting with corresponding changes in structure. Quenching rates in the thermal spike (temperature in the spike divided by the life time of the spike) between 10^{10} and 10^{12} K/s can be estimated, approximately. These quenching times are far shorter than those reached for splat cooling (between 10^4 and 10^8 K/s) or pulsed laser irradiation (between 10^5 and 10^8 K/s). Consequently, when an original crystalline material cools down again, the quenching time is too short for a full reconstruction the crystalline structure. In agreement with experimental observations a significant number of structure defects remains.

4.7 Reactions of Radiation Induced Point Defects

After ion bombardment with energetic ions different types of irradiation induced defects are simultaneously generated. Various reactions take place between these defects, resulting in changes in the distribution of the different defect species. Examples of these reactions are

- formation of dislocation loops by collapsing of vacancy-rich cores of displacement spikes,

- defect annihilation by recombination of vacancies and interstitials with each other,
- formation of clusters by interaction of point defects of the same species (e.g., void formation by vacancy clustering or interstitial clustering),
- formation of new phases (e.g., amorphous phases, see Sect. 4.9),
- diffusion of isolated defects to clusters, voids, dislocation loops and also preexisting lattice defects as dislocations or grain boundaries and
- absorption of interstitials and vacancies at extended sinks as dislocations, voids, precipitates, grain boundaries, etc.

The evolution of the concentration distribution of the particle bombardment induced vacancies and interstitials is primarily determined by two reaction types, (i) the defect-defect reactions (recombination of vacancies and interstitials) and (ii) the defect-sink reactions. In addition to voids, grain boundaries, dislocations and dislocation loops also the free surface act as sink. Such sinks can be three-dimensional sinks (e.g., voids or precipitates), two-dimensional sinks (e.g., grain boundaries or the free surface) or one-dimensional line sinks (e.g., dislocations or dislocation loops). For example, the free surface acts as a sink for interstitials and vacancies. Vacancies disappear on the surface and interstitials form small imperfections on the surface.

The *defect-defect reaction* between the interstitial, *i*, and vacancy, *v*, is characterized by the rate constant K_{iv} and is given in units of [reactions per atom and per unit time]. As result of this reactions, for example, a divacancy is formed by two vacancies or point defects (vacancy and interstitial) are annihilated by recombination. It is assumed that the recombination rate constant K_{iv} is independent of both defect concentration as well as defect generation and can be written as

$$K_{iv} = \xi_{iv}(D_i + D_v), \tag{4.55}$$

where ξ_{iv} is a geometric factor and D_i and D_v are the defect diffusion coefficients of the interstitials and vacancies, respectively.

In the second case, the *defect-sink reaction*, the defects disappear by absorption into sinks. For example, both point defects types (vacancies and interstitials) can be absorbed into irradiation-induced precipitates and reduce the strain state energy or the mismatch of the precipitates with the surrounding matrix. The rate of point defect absorption into a sink is given by

$$K_{is} = \xi_{is} D_i \quad \text{and} \quad K_{vs} = \xi_{vs} D_v, \tag{4.56}$$

where the geometrical factors ξ_{is} and ξ_{vs} represent the interaction radii for the reaction between the sink and the point defects (interstitial or vacancy). The derivations and discussion of the different rates are given in detail (e.g., [6, 8, 51]).

The evolution of the spatially averaged defect concentration of vacancies and interstitials generated by irradiation, C_v and C_i, can be considered as result of a competition between the local rate of defect generation and the recombination/annihilation and also diffusion into or out of the local volume. Therefore, the point defect balance can be given by [8]

$$\begin{pmatrix} Generation \\ rate \end{pmatrix} = (Diffusion) + (Recombination) + \begin{pmatrix} Annihilation \\ at\,sinks \end{pmatrix} + (Clustering). \quad (4.57)$$

In the context of the classical rate theory (see e.g., [6, 51–53]) these reactions are modeled as first-order 'chemical' reactions by a system of non-linear partial differential equations, called point defect balance equations

$$\frac{\partial C_v}{\partial t} = K_o - K_{iv} C_i C_v - K_{vs} D_v C_s \quad (4.58)$$

and

$$\frac{\partial C_i}{\partial t} = K_o - K_{iv} C_i C_v - K_{is} D_i C_s, \quad (4.59)$$

where K_o (= ξ dpa/time of irradiation) is the defect production (generation) rate (compare with defect production rate R_D in units of [number of defects per unit volume and unit time], Sect. 4.4), ξ is the damage cascade efficiency, C_i is the concentration of interstitials, C_v is the concentration of the vacancies, K_{iv}, K_{vs} and K_{is} are rate constants for the reactions indicated by the suffix combination. A clustering of point defects is neglected. The defect generation rate K_o can be approximately calculated by means of (4.19). Another way is the approximation

$$K_o = \sigma_d J, \quad (4.60)$$

where J is the ion current density or ion flux in units of (cm^{-2} s^{-1}). The total displacement cross-section σ_d in units of (cm^2) can be calculated with Monte Carlo simulation programs for example (see Sect. 3.6.2).

The steady-state condition, $\partial C_v/\partial t = \partial C_i/\partial t = 0$ yields an equation system

$$K_o = K_{iv} C_i C_v + K_{vs} D_v C_s \quad (4.61)$$

and

$$K_o = K_{iv} C_i C_v + K_{is} D_i C_s, \quad (4.62)$$

which demonstrates that the defect generation rate K_o is given by rate of point defect absorption into sinks and the annihilation rate by recombination. Consequently, it can be distinguished between two regimes that limit this accumulation process in dependence on the temperature, the defect generation rate and the material [8]: the sink-dominated regime and the recombination-dominated regime. In the sink-dominated regime, the defects are annihilated by long-range diffusion to extended sinks. In this case the steady-state concentration is proportional to K_o. In the second case, in the recombination-dominated regime, the defects are annihilated in clusters and the steady-state concentration is proportional to $K_o^{1/2}$. Under the steady-state condition, $\partial C_v/\partial t = \partial C_i/\partial t = 0$, it is also possible to determine the atomic fractions of defects [54].

At low temperatures (homologous temperature $T_H \leq 0.25$), the thermal formation rate of vacancies can be neglected. The defect loss is determined by the direct defect recombination. The atomic fractions are given by

$$C_v = \sqrt{\frac{K_o D_i}{\alpha D_v}} \quad \text{and} \quad C_i = \sqrt{\frac{K_o D_v}{\alpha D_i}}, \tag{4.63}$$

where recombination coefficient α is proportional to K_{iv}.

At moderate temperatures ($0.25T_M < T < 0.6T_M$), the defect concentrations are determined by the defect loss due to fixed sinks and can be expressed by

$$C_v = \frac{K_o}{k_v^2 D_v} \quad \text{and} \quad C_i = \frac{K_o}{k_i^2 D_i}, \tag{4.64}$$

where k_v and k_i are sink strengths for vacancies and interstitials, respectively.

At higher temperatures ($T_H > 0.6$), it is assumed the irradiation induced defects are annealed out. Consequently, the vacancy concentration is in the thermal equilibrium and determined by thermal formation rate of vacancies.

The temporal evolution of the defect concentration during particle bombardment can be obtained by solving the coupled non-linear partial equation system, (4.58) and (4.59). This equation system must be solved numerically, because no analytical solution is possible without substantial constraints. Sizmann [51] discussed this equation system under appropriate boundary initial conditions. In Fig. 4.15 the evolution of the defect concentration as function of the bombardment time in the recombination-dominated regime for the case of low temperatures (e.g., if the homologous temperature $T_H < 0.5$) and low sink concentration C_s is shown. At the beginning of the particle bombardment, the material is free from defects. Then, the isolated irradiation induced interstitials and vacancies (Frenkel pairs) are generated by particle bombardment (stage I). With increase of irradiation time the point defect starts to accumulate proportional to rate $K_o t$. After a time t_1 the recombination is in balance with the bombardment induced defect generation. Consequently, no further increase in the defect concentration can be observed (stage II). At time t_2, the concentration of the interstitials C_i decreases significantly, because the faster defects (in contrast to the vacancies) arrive at the sinks and are absorbed (stage III). For bombardment times above t_3 (stage IV), the slower vacancies arrive also at the sinks. This means that the concentration of vacancies cannot rise further. It can be concluded that the defect concentration is proportional to the inverse ratio of their diffusivities, i.e. $D_v C_v = D_i C_i$. If the sink concentration and the temperature are high (defects are very mobile), the time t_2 is shorter than time t_1. The sink-dominated regime according to Sizmann [51] is schematically demonstrated in Fig. 4.16. At a time t_2, a significant part of the

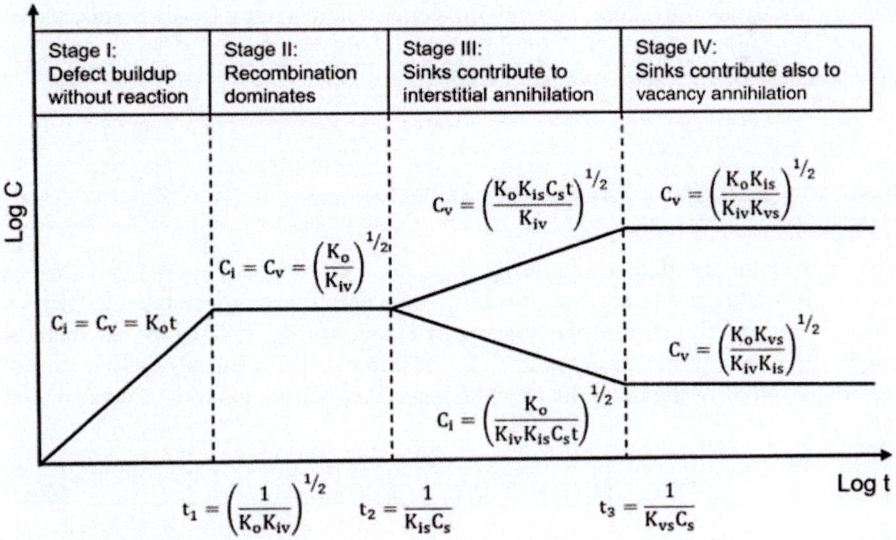

Fig. 4.15 Defect kinetic in homogenous solid during irradiation as function of the bombardment time in the recombination-dominated regime for the case of low temperatures and low sink concentrations (figure adapted from [51])

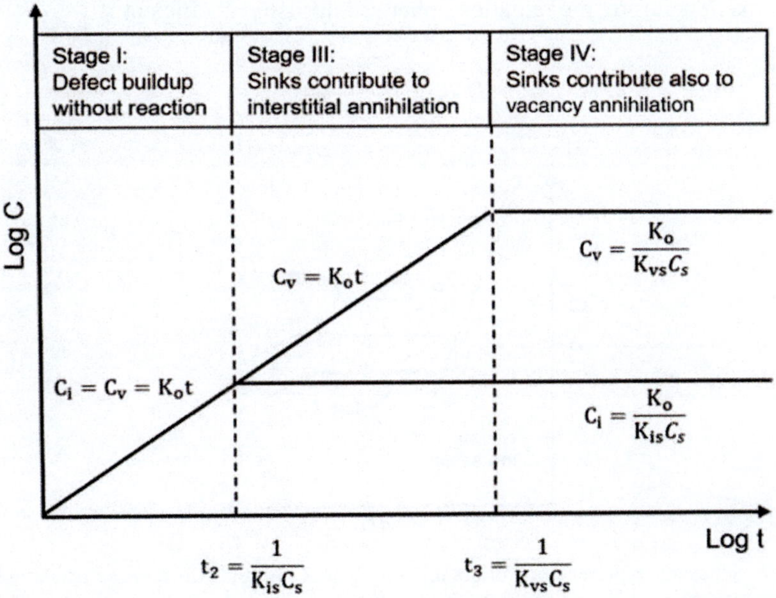

Fig. 4.16 Defect kinetic in homogenous solid during irradiation as function of the bombardment time in the sink-dominated regime for the case of high temperatures and high sink concentrations (figure adapted from [51])

interstitials arrive at the sink, i.e. their concentration reaches a steady-state concentration (generation of defects = annihilation of defects at the sinks). In contrast, the concentration of vacancies increases until t_3.

4.8 Ion Radiation Enhanced Diffusion

The analysis of the defect kinetic by Sizmann [51] has also shown that the ion bombardment induced increase in defect concentrations is accompanied by an enhancement of the atomic movement. In detail, Russel [54] supposed that the contributions to the radiation enhanced diffusion in a solid (metal) are the sum of the contributions of the individual components (vacancies, interstitials, divacancies, etc.), i.e.

$$D = f_v D_v C_v + f_i D_i C_i + f_{2v} D_{2v} C_{2v} + \cdots, \tag{4.65}$$

where f is the correlation factor. According to Russel [54], three different temperature regimes for the diffusion coefficient can be distinguished (Fig. 4.17). At low temperatures the concentration of vacancies produced by ion bombardment is high enough for the vacancy-interstitial recombination to dominate over the diffusion of fixed sinks. Therefore, the radiation enhanced diffusion coefficient is

$$D \approx \sqrt{\frac{D_v D_i K_o}{\alpha}}, \tag{4.66}$$

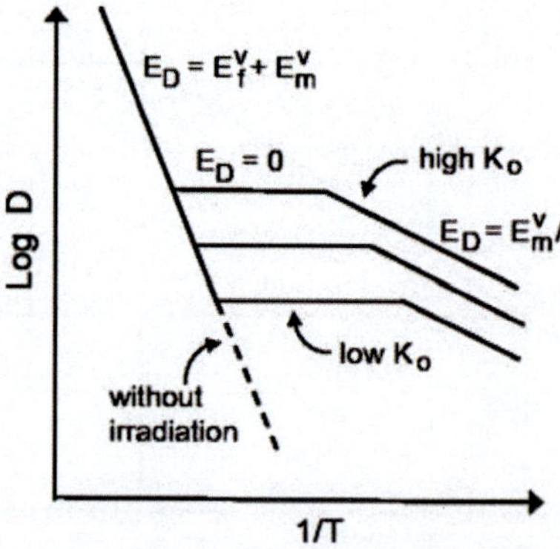

Fig. 4.17 Schematic representation of the different regimes for the diffusion coefficient in irradiated metal at different displacement rates K_o (figure adapted from [54]). In addition, the diffusion coefficient versus the reciprocal temperature is shown for non-irradiated metal (dashed line)

where α is the recombination coefficient. At moderate temperatures the loss of vacancies and interstitials will be determined by fixed sinks. This leads to a diffusion coefficient of

$$D \approx \frac{2K_o}{k_i^2}, \tag{4.67}$$

whereby the diffusion coefficient is nearly independent of the temperature. At higher temperatures ($T > 0.6T_M$), the defect mobility is so high that the generated defects immediately annihilate. Consequently, the concentration of defects is determined by the vacancy concentration in the thermodynamic equilibrium. This means that the diffusion is independent of the radiation-induced defects. The diffusion behavior can be described by vacancy diffusion and has an activation energy composed by the activation energy of vacancy formation E_f^v the activation energy of the vacancy migration E_m^v.

4.9 Ion Beam-Induced Amorphization

Generally, the amorphous state of a solid is considered to be a frozen liquid due to the lack of the long range order. The ion bombardment is a frequently used technique to induce the crystalline-amorphous phase transformation in thin films and near surface regions of solids. Experimental studies have demonstrated that the formation of an amorphous phase for a given material depends on the ion species and its energy, the fluence and ion current density as well as the irradiation temperature (see, e.g., [55, 56]). Based on these studies, it can be concluded that the crystalline-to-amorphous phase transition (phase transition of the first order) is the result of the continuous defect accumulation process or the ion bombardment induced rapid quenching.

4.9.1 Damage Buildup

It can be supposed that isolated Frenkel pairs are generated at the beginning of the ion irradiation, where the defect generation rate will be proportional to the ion flux. A fraction of these interstitial-vacancy pairs immediately recombine after the defect formation process in dependence on the temperature. With increase concentration of ion beam-induced defects, the probability of interactions of the point defects among themselves and with other defects rises (e.g., dislocations, grain boundaries etc. and also the free surface). In particular, processes like recombination, annihilation, clustering and diffusion of defects must be involved in the description of the temporal evolution of the defect concentration at higher temperatures during or after ion irradiation.

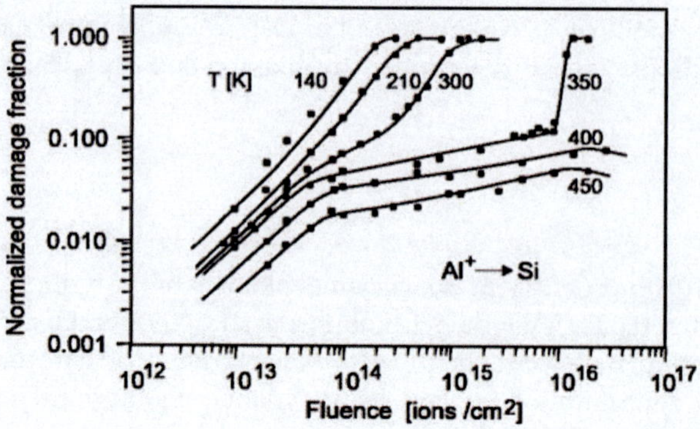

Fig. 4.18 Normalized defect fraction versus ion fluence for Si implanted with Al ions at different temperatures. The damage profiles are measured by optical reflectivity depth profiling [57]

A typical measurement of this damage buildup after high-energy Al ion implantation in silicon is shown in Fig. 4.18. The Al ion fluence dependence of damage in Si at the depth of the damage maximum is plotted for different target temperatures [57]. For low fluences damage increases linearly with fluence for all temperatures as long as each ion produces the same amount of damage. For higher target temperatures a more intense point annihilation/recombination within single collision cascades leads to decreasing damage buildup efficiency. For temperatures > 300 K, the defect clustering and increase of the amorphous volume fraction results in a sublinear fluence dependence of the damage buildup. This multi-step damage buildup has been observed for several compound semiconductors [58] and oxide crystals [59]. Finally, the irradiated target material is amorphized (normalized damage fraction ≈1, see Fig. 4.18).

In the idealized case, every incident ion produces the same average number of displaced atoms (for example described by the modified Kinchin–Pease relationship) independent of the subsequent ions. Gradually, the concentration of defects is increased by ion fluence (time of ion implantation). The kinetic of the defect production can be described under the assumptions that (i) the temperature is low, so that the generated defect can be considered as immobile, (ii) no annihilation of defects takes place and (iii) no defect agglomerations significantly influence the concentration of defects. According to (4.2), the first ion (n = 1) generates $N_d(E)$ defects (displaced atoms) on average. Then, the total number of defects produced after the second incident ion is

$$N_d(n = 2) = \langle N_d(E)\rangle + \langle N_d(E)\rangle\left[1 - \frac{\langle N_d(E)\rangle}{N}\right] = 2\langle N_d(E)\rangle - \frac{\langle N_d(E)\rangle^2}{N}, \tag{4.68}$$

where N is the atomic number density of the target material. After an incidence of n ions, the total number of defects is

$$N_d(n) = -N\sum_{i=1}^{n}\binom{n}{i}\left[-\frac{\langle N_d(E)\rangle}{N}\right] = N - N\left[-\frac{\langle N_d(E)\rangle}{N}\right]^n. \tag{4.69}$$

If $N_d \ll N$, then $\left[-\frac{\langle N_d(E)\rangle}{N}\right]^n \approx \exp\left[-n\frac{\langle N_d(E)\rangle}{N}\right]$ and the last equation can be rewritten as

$$N_d(n) = N - N\exp\left[-n\frac{\langle N_d(E)\rangle}{N}\right] \quad \text{or} \quad \frac{N_d(n)}{N} \equiv f_D = 1 - \exp\left[-n\frac{\langle N_d(E)\rangle}{N}\right], \tag{4.70}$$

where f_D represents the normalized damage fraction.

This description is referred to as model of direct ion impact and is based on the idea that each ion generates a damage volume in the irradiated material and each subsequent ion also hits an undamaged region. Apart from this model, Gibbons [60] has proposed an overlap damage model for the damage buildup after ion irradiation. This model can be considered to be a straightforward extension of the direct ion impact model, because an overlap of damaged regions is involved, i.e. damage is the result of the accumulation of damages caused by a larger number of single ion impacts. A permanent damage is only produced after a certain number of damaged regions have been created and these regions overlap. In this model, the normalized disorder is proportional to $1 - \frac{A\Phi^k}{k!}\exp(-A\Phi)$, where A is the cross-section of the disordered region and k is the number of ion impacts necessary to create permanent damage.

This model is frequently applied to describe the ion beam-induced amorphization versus ion fluence [see (4.79)]. Experimental observations have verified that the damage buildup after ion irradiation is significantly more complicated than described by the direct impact or overlap models. As a consequence, several other models have been suggested. For example, Hecking et al. [61] have assumed a gradual growth of the damage regions up to amorphization and involved the temperature dependence of this process, while Weber et al. [62] have suggested a description of the damage buildup by combination of different mechanisms (direct impact, accumulation of defects and stimulated defect formation). Figure 4.19 schematically illustrates the dependence of the disordering after ion irradiation events on the normalized fluence for different models. Jagielski and Thomé [59] studied all these models in detail and have drawn the following conclusions: First, the assumption that the damage buildup should involve more than one step (c.f. simple shape of the curves in Fig. 4.19) and second that damage buildup (amorphization) driven by chemical effects must be considered. Thus, they have proposed a multi-step model based on the assumption that the damage accumulation can be described as a discontinuous process composed of several steps of structural transformations from perfectly crystalline to heavily disordered or amorphous structure. Among others, this model could be successfully applied to explain the multi-step damage accumulation of oxide crystals after ion bombardment.

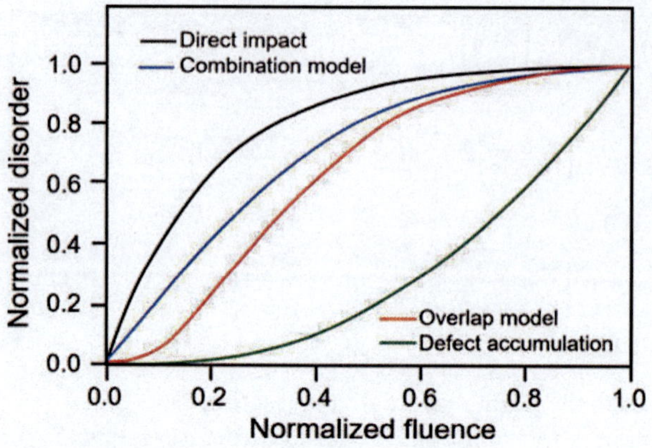

Fig. 4.19 Normalized disorder versus normalized fluence of different damage buildup models (figure is taken from [59] and modified)

4.9.2 *Kinetics of Ion Beam Amorphization*

The passage of an energetic ion through a solid lattice initiates a sequence of displacement events that leads to structural or thermodynamic instabilities of the crystalline structure [63]. If the concentration of defects in a crystalline material exceeds a critical value during implantation, a phase transformation occurs, changing the crystalline solid into an amorphous solid. An equivalent statement is that a threshold damage density is necessary for the amorphization to take place. From the thermodynamic point of view, the amorphization during particle bombardment starts, when the free energies of the defected state and that of amorphous state are equalized. Figure 4.20 shows a schematic of the damage and the concentration distributions after ion implantation as function of the depth. The peak position of the damage profile lies closer to the surface than that of the concentration distribution. By applying an amorphization criterion in which amorphization is assumed to occur wherever a critical damage level is exceeded, it is possible to evaluate the thickness of the amorphous layer. This damage level corresponds to an ion fluence, referred to as amophization fluence, Φ_{am}. On the other hand, if the implanted concentration exceeds a critical concentration, N_c, an amorphous phase is thermodynamically more stable than a crystalline phase. Besides, either a new amorphous phase (alloy, solid solution, etc.,) is formed by the implanted material, or the structural implantation-induced damages are stabilized by the implanted particles. Depending on the condition of ion implantation, different layers can be distinguished as a result of the superposition of the concentration of implanted particles (which are capable to stabilize the amorphous phase) and damage distribution (see Fig. 4.20).

It is widely accepted, that ion bombardment induced defects (point defects, Frenkel pairs, extended defects, etc.,) cause chemical disorder, which creates displacements in addition to the displacements resulting from thermal vibration.

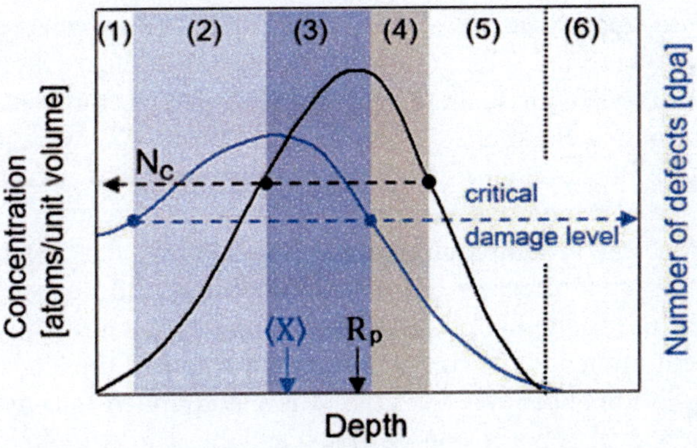

Fig. 4.20 Schematic representation of the damage (blue) and concentration distribution (black) after ion implantation with a fluence higher than the amorphization threshold fluence, where (1) is a crystalline, defect-rich region, (2) is the amorphous layer, (3) is an amorphous compound layer, with the implanted ion species is capable to stabilize the amorphous state (implanted concentration $N_c \geq$ the concentration to stabilize the amorphous phase or form a compound), (4) defect-rich crystalline layer, (5) defect-poor implanted layer and (6) non-implanted substrate material

Then, the mean square static displacement, δ^2 (see, e.g., Sect. 8.2.1.3), should influence the melting temperature, T_M^d, at which a damaged crystal becomes instable [64]. Under the assumption, that the δ^2 is a measure of the topological and chemical disorder, Lam et al. [65], and Okamoto et al. [66] have derived a generalized Lindemann phenomenological melting criterion [67] for the description of the crystalline-to-amorphous transition in ion beam defect-induced crystalline solids. According to these authors, the melting temperature of the damaged crystal is given by

$$T_M^d = \frac{4\pi^2}{9}\frac{Mk_B\theta_d^2}{h^2}\langle\delta_{cri}^2\rangle \quad \text{with } \theta_d^2 = \theta_o^2\left(1 - \frac{\langle\delta^2\rangle}{\langle\delta_{cri}^2\rangle}\right), \tag{4.71}$$

where M is the average atomic mass, k_B is the Boltzmann constant, h is the Planck constant, θ_d and θ_o are the Debye temperatures of the damaged and the perfect crystal, respectively. It should be noted, that the disorder is also accompanied by a strong decrease in the average shear modulus, i.e.

$$\frac{T_M^d}{T_M^o} = \frac{\theta_d^2}{\theta_o^2} = \frac{G_d}{G_o} = \left(1 - \frac{\langle\delta^2\rangle}{\langle\delta_{cri}^2\rangle}\right), \tag{4.72}$$

where G_d and G_o are the shear moduli of the damaged and perfect crystals, respectively. Hence, $\langle\delta_{cri}^2\rangle$ is a critical value of the mean-square static displacement at which the damaged crystal becomes instable (amorphous). In contrast to the Lindemann relationship, the melting temperature (and also the Debye temperature and the shear modulus) is no longer constant. The melting temperature decreases linearly with the increase of the mean square static displacement, $\langle\delta^2\rangle$. Consequently, the ion beam-induced amorphization can be expected when mean square static displacement, $\langle \delta^2 \rangle$, attains a critical value, $\langle \delta^2_{cri} \rangle$. Experimental studies, e.g., by Linker [68], have confirmed this behavior. In these investigations, the structural changes of the ion bombarded materials were studied by X-ray diffraction. The average static atomic displacement, $\langle \delta^2 \rangle$, were deduced from the static Debye–Waller factor, which was given by the slope of a modified Wilson plot (measure of the average displacement of the atoms of damaged sample in relation to the perfect sample) and the volume fraction of the amorphous phase (e.g., determined by RBS channeling) [69]. For example, Linker [69] found in the case of boron ion implantation of Nb at LNT, that the boron concentration and the corresponding distortion of the Nb lattice increases with increasing ion fluence. When 5 at.% B is incorporated into niobium interstitially, a saturation level of boron concentration and thus a maximum of the distortion is attained and the amorphization process starts. The onset of amorphization occurs when $\langle \delta^2 \rangle$ achieves a critical value of about 1.4 nm.

Thus, the mean square atomic displacement is a useful measure of the enthalpy stored in the lattice of the ion bombarded materials [63]. As is well-known, the necessary energy (i.e. enthalpy change $\Delta H_{c-a} = H_{cryst.} - H_{amorp.}$) to transform a stable or metastable phase in an amorphous phase of an ion bombarded material can be obtained (i) by the energy, which is deposited in the crystal lattice during ion bombardment or (ii) by formation of damages (point defects, dislocations, etc.) [60].

As many experimental studies, for example, by means of transmission electron microscopy, X-ray diffraction, low-energy electron diffraction, Rutherford backscattering, Raman and infrared spectroscopy and also electron spin resonance have demonstrated that the ion beam-induced crystalline-to-amorphous transition is strongly dependent on the parameters of the ion irradiation [44]. The process of the damage accumulation as preliminary stage of the amorphization is determined by a competition between the ion beam-induced defect generation and recovery. The amorphization process is controlled by different irradiation parameters:

- *Ion fluence*

  The damage accumulation versus the ion fluence is characterized by three distinct regions (Fig. 4.21). Initially, in region I, the damage volume in the implanted solid grows slowly with a sublinear dependence on ion fluence Φ (damaged volume $\propto \Phi^x$, $x < 1$) until a critical (threshold) fluence is reached. Then, in region II, an extreme superlinear growth can be expected (with $x > 1$). This rapid growth of the defect concentration is attributed to the reduction of the threshold displacement energy in a pre-damaged layer and therefore a larger number of defects can be produced. Finally, a saturation of the defect concentration and a collapse-like amorphization occurs (region III). The implanted fluence, which corresponds to the first appearance of the closed amorphous layer is called the amorphization fluence. In summary, a sigmoidal dependence of the amorphous fraction on the fluence can be observed for a large number of materials.

- *Ion mass*

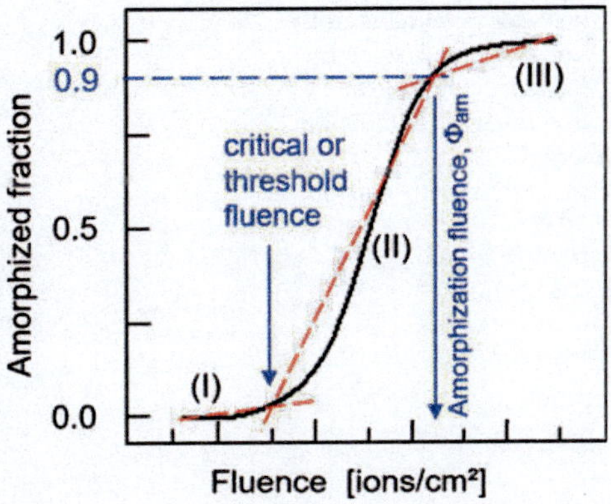

Fig. 4.21 Schematic of the amorphous fraction in dependence on the ion fluence

Generally, with increase of the ion mass the amorphization process begins with lower fluences, because a higher ion mass is causing more atomic displacements for the same incident ion energy. Therefore, it can be anticipated that heavy ions create directly amorphous regions whereas light ions produce isolated point defects or small defect clusters. Consequently, a higher fluence in the case of low-mass ions is required to initialize the crystalline-to-amorphous phase transition.

- *Ion current density*

Also influence of the ion flux can be discussed as interaction between the damage production and the ion beam-induced defect annihilation. At low temperatures, the impact of the ion current density is insignificantly small because the contribution of recovery is small due to the limited irradiation time. It was observed that a critical ion current density is necessary to generate defects at constant ion fluence and at medium and higher irradiation temperatures. Above the critical ion current density an amorphous film can be formed, where this current density threshold for medium ion masses is some orders of magnitude higher than for heavy ions. The significantly higher nuclear stopping of heavy ions in contrast to ions with a medium mass results in a more effective ion beam-induced recovery. The amorphization fluence is also dependent on the ion current density (schematically shown in Fig. 4.22). Increasing ion current density leads to an increase of the amorphization threshold fluence.

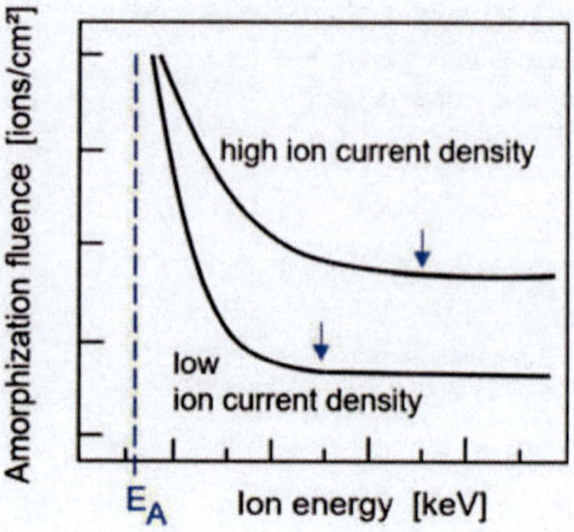

Fig. 4.22 Schematic of the amorphization fluence in dependence on the ion current density and the ion energy

- *Ion energy*

  The influence of the ion energy on the amorphization or in particular on the threshold amorphization fluence is only slight. In general, an amorphization is only possible above a critical ion energy (E_A). With an increase of the ion energy, the amorphization fluence for heavy ions is reduced, while a partial increase of this threshold fluence was registered for light ions. Under the condition that only the nuclear energy loss is responsible for the defect formation, this contradictory behavior of light and heavy ions can be explained by the fact that the contribution to the nuclear energy loss rises compared to the electronic energy loss with rising energy for heavy ions. Above a critical ion energy (represented by arrows in Fig. 4.22) the threshold ion fluence is independent of the ion energy (see also model of critical energy density in Sect. 4.9.3).

- *Irradiation temperature*

  The temperature during ion bombardment strongly influences the amorphization of ion bombarded crystals, especially the magnitude of the amorphization threshold fluence. It is well-known that amorphization is only possible below a critical temperature, T_{am}, for a given ion fluence (Fig. 4.23). With an increase

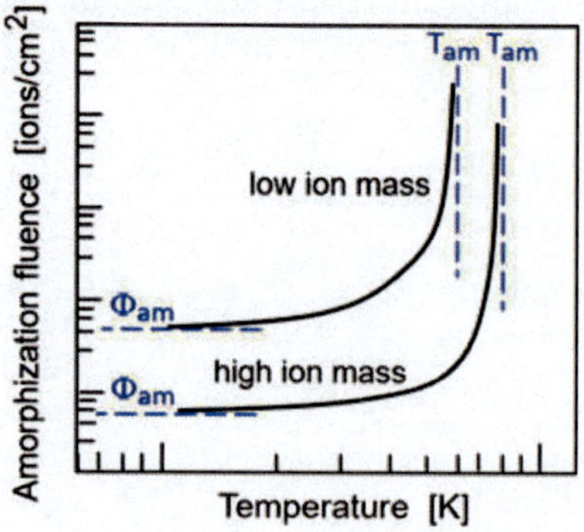

Fig. 4.23 Schematic of the amorphization fluence in dependence on the temperature during ion implantationion implantation

of the atomic number, the critical temperature rises. This means that particularly for light ions a significant influence of the temperature on the amorphization can be expected. For low implantation temperatures the recovery of implantation induced defects is small. Nearly, all defects are preserved. Higher implantation temperatures, where the generation and annihilation of the defect is balanced, a small increase of the temperature results in the aforementioned strongly non-linear behavior of the defect accumulation (region III in Fig. 4.20). Therefore, amorphization fluence, Φ_{am}, increases with the temperature and decreases with atomic number of the ion species (Fig. 4.23).

4.9.3 Amorphization Models

Based on the extensive experimental investigations, several models have been proposed to explain the results of the ion beam-induced amorphization process. In the following, the most common models are shortly summarized.

Models of a Critical Energy Density

The nuclear stopping of incident ions due to the interaction with lattice atoms leads to an increase of the deposited energy in the crystalline solid. If the deposited energy is higher than the critical energy, referred to as critical energy density E_c, it is assumed that a transformation of the lattice into an amorphous state is energetically favorable. For example, this critical energy density for silicon could be determined in order of $5 \ldots 6 \times 10^{23}$ eV/cm^3 for a variety of different ions after low-temperature bombardment [70–72]. The value of the critical energy density is dependent on the ion mass and the temperature of the target during ion bombardment. By increasing the temperature, the diffusion of defects and the thermal lattice recovery become increasingly more significant and the stored energy density is reduced. Consequently, it can be expected that the ion fluence to reach the total amorphization at higher temperatures is necessarily larger than at low temperatures, because the mobility of the implanted ions and the generated point defects is reduced at lower temperatures. According to Dennis and Hale [70] the critical energy density is given by

$$E_c = \frac{\Phi}{x_m} \nu(E), \tag{4.73}$$

where Φ is the ion fluence, x_m the depth corresponding to 90% damage accumulation (see Fig. 4.21) and $\nu(E)$ is energy deposited in nuclear collision processes [see also (4.11)].

Models of Overlapped Damage Regions

In contrast to the models which have assumed a crystalline-to-amorphous transition at a critical stored energy density, Morehead and Crowder [73, 74] as well as Gibbons [60] have proposed models which are characterized by continuous amorphization through an overlap process of locally amorphized regions induced by ion

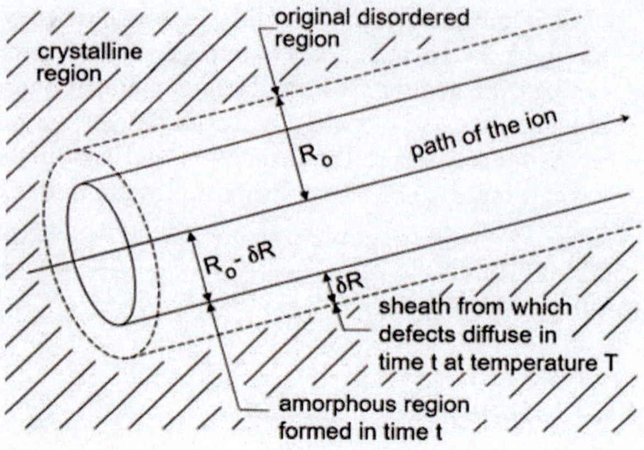

Fig. 4.24 The damage regions surrounding the path of an incident ion. The inner core (solid line) is assumed to be amorphous (figure adapted from [73])

bombardment. Each incident ion generates a cylindrical disordered region with the radius R_o around the ion track (Fig. 4.24).

The defects (vacancies) within the circular ring δR are able to diffuse during the ion bombardment. This process results in a partial recovery of the structure. But, the core around the ion track with the radius $R_o - \delta R$ remains amorphous. By increasing ion fluence, these amorphous regions overlap and form a continuous amorphous layer. The amorphous area per ion, $\pi(R_o - \delta E)^2$, grows with increasing ion mass, while the area of the circular ring grows with raising temperatures. According to this model, every cascade consists of a strongly damaged core and a lightly damaged annular region, where fluence creating an amorphous layer for the case that the temperature is very low ($T \approx 0$ K), i.e. $\delta R = 0$. The corresponding amorphization fluence is given by

$$\Phi_{am}(T = 0) = E_d N/(dE/dx)_n. \tag{4.74}$$

Morehead and Crowder [73, 74] have also shown that the fluence of amorphization at an implantation temperature $T > 0$ K is inverse proportional to square of the radius of the damage cascade. Consequently, as temperatures increase, the fraction of the damaged cascade region is reduced with a recovery rate proportional to $\exp(E_D/k_B T)$, where E_D is the activation energy for the diffusion in the circular ring δR (Fig. 4.24). Under the assumption that the damage region has a spherical shape, the amorphization fluence for temperatures $T > 0$ K, i.e. $\delta R > 0$, can be expressed by

$$\Phi_{am}(T > 0) = \Phi_{am}(T = 0)/\left[1 - K\left(\frac{dE}{dx}\right)_n^{-1/2} \exp\left(-\frac{E_D}{k_B T}\right)\right]^2, \tag{4.75}$$

where K is a material-dependent constant (e.g., for Ge is $K = 200$ (keV/μm)$^{1/2}$) and E_D the activation energy for diffusion in the region around the core (e.g., for Ge, $E_D = 0.09$ eV) [70]). Then, the temperature above which an ion beam-induced amorphization cannot be realized is given by (c.f. Fig. 4.23)

$$T_{am} = \frac{-E_D}{k_B \cdot \ln\left[\left(\frac{dE}{dx}\right)_n^{1/2} / K\right]}. \tag{4.76}$$

Above this critical temperature, the recovery leads to a complete recrystallized (polycrystalline) solid, because of higher atomic mobility.

Based on this consideration, Gibbons [60] assumed, that the amorphous region is also a circular cylinder with fixed length and an area projected on the surface of the crystal. Then, the growth of the amorphous region can be written as

$$\frac{dA_A}{d\Phi} = A_i\left(1 - \frac{A_A}{A_o}\right) \tag{4.77}$$

with the solution

$$A_A = A_o\left[1 - \exp(-A_i\Phi)\right], \tag{4.78}$$

where Φ is the ion fluence, A_A the total surface area covered by amorphous regions, A_i the cross-sectional area of an individual amorphous region and A_o the total area that is being implanted. For light ions, it is accepted that an individual ion only creates a defect-rich, no-amorphous region. Then, an overlap of these defect-rich regions is needed to generate amorphous regions. A straightforward extension of the direct-impact model (see also Sect. 4.9.1), called overlap damage model, yields to the fractional amorphous area A_A/A_o in the case of the n-fold overlap of the defect regions given by

$$A_A = A_o\left\{1 - \left[\sum_{k=0}^{n} \frac{(A_i\Phi)^k}{k!}\right]\exp(-A_i\Phi)\right\}, \tag{4.79}$$

where the summation limiting integer n is the overlap number. For example, $n = 0$ corresponds to direct-impact amorphization. Consequently, n corresponds to the number of ions necessary to damage a defect region before it becomes amorphous. Figure 4.25 (left) shows the fraction of the amorphous material A_A/A_o as function of the ion fluence according to the direct-impact and the overlap damage model. The total amorphization is assumed for $A_A/A_o \geq 0.9$. The influence of the temperature is not considered in these models.

Model Based on Rate Equations

In the model by Hecking et al. [61], it is supposed that after (high-energy) ion implantation, amorphous and crystalline regions simultaneously exist, where the crystalline region contains point defects. Then, the changing of the relative volume content of the amorphous phase and point damages as function of the ion fluence can be described by a coupled differential equation system, expressed by

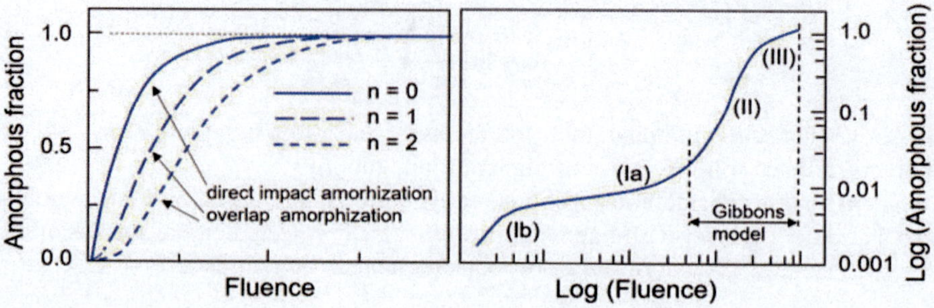

Fig. 4.25 Schematic representations of the amorphous fraction as function of ion fluence according to direct-impact (solid line) and overlap-cascade (dashed lines) models by Gibbons [60] (left) and according to the damage and amorphization model by Hecking et al. [61] (right)

$$\frac{dn_d}{d\Phi} = P_d\left(\exp - R^2\Phi^2\right)(1 - n_a) + Cn_d\left(1 - \frac{n_d}{n_d^*(1 - n_a)}\right) - \frac{dn_a}{d\Phi}\frac{n_d}{(1 - n_a)} \tag{4.80}$$

$$\frac{dn_a}{d\Phi} = (P_a + A_s n_a)(1 - n_a). \tag{4.81}$$

Thus, the first term in (4.80) characterizes the production of point defects, where P_d is the proportionality constant of the point defect production. The exponential expression in the first term describes the probability of the spontaneous point defect recombination for high fluences (rescaled with constant R), and the factor $(1 - n_a)$ considers that point defects can be created in the non-amorphous region, only. The second term in (4.80) characterized the agglomeration of point defects. This contribution is proportional to the concentration of these defects, where C is the proportionality constant and $n_d{}^*$ is the point defect saturation concentration. When $n_d \ll n_d{}^*$, this contribution is insignificant. The last term in (4.80) describes the growth of the amorphous fraction in contrast to the crystalline fraction. With (4.81) the growth of the amorphous phase by ion beam-induced generation of amorphous regions (rescaled with constant P_a) and by stimulated amorphization at the crystalline-amorphous interfaces (rescaled with constant A_s) is described. This system of differential equations can be numerically solved and the free parameters can be fitted. According to Hecking et al. [61], the development of the defects versus ion fluence can be subdivided in four fluence regions (see Fig. 4.25, right). In the region of weak damage production (region I in Fig. 4.25, right), the damage concentration rises proportionally to the ion fluence. In dependence on mass of the applied ion species and the temperature a more or less distinctive plateau is formed (region II) which demonstrates the balance between the generation and the annihilation of defects. In region III, more and more defect clusters are formed which cannot be recombined. Above a defined fluence a collapse-like amorphization (region IV) can be observed by further accumulation of irradiation induced defects. The results of the defect evolution in semiconductor materials after ion implantation can be explained consistently with this model [75].

In addition to the above briefly described amorphization model, numerous other variations of these models were suggested to interpret the experimental results of the crystalline-to-amorphous transition under ion bombardment. Comprehensive overviews have been published on the field of the ion beam-induced amorphization of selected materials (e.g., for semiconductors [58, 76], or for ceramics [77]). But, no single model seems to able to explain all experimental observation, completely.

4.10 List of Symbols

Symbol	Notation
A	Cross-section of disordered region
A_A	Total surface area covered by amorphous regions
A_{BM}, a_{BM}	Coefficients of the Born–Mayer potential
A_i	Cross-sectional area of an individual amorphous region
A_o	Total implanted area
C_i, C_v	Concentration of interstitials, vacancies
C_s	Concentration of sinks
D_{cas}	Effective diffusion coefficient in the cascade
D_{th}	Thermal diffusivity
D_i, D_v	Diffusion coefficients of interstitials, vacancies
E	Energy of incident ion
E' (x)	Energy of incident ion in the depth x
ϵ	Energy of knocked-on lattice atom
E_B	Atomic binding energy
E_d	Threshold displacement energy
E_D	Activation energy of diffusion
$E^v{}_f$	Activation energy of vacancy formation
$E^v{}_m$	Activation energy of the vacancy migration
E_s	Activation energy (threshold energy) for a jump in the thermal spike
E_{cut}	Cut off energy
E' (x)	Energy of the incident ion in the depth x
F_D	Deposited energy
G_d	Shear modulus of the damaged crystal
G_o	Shear modulus of the perfect crystal
ΔH_{sub}	Heat of sublimation
J	Ion current density (ion flux)
K_o	Defect production (generation) rate (defects per time unit)
K_{iv}, K_{vs}, K_{is}	Rate constants for reactions indicated by the suffix combination

(continued)

(continued)

Symbol	Notation
M	Average atomic mass
M_1	Particle (ion) mass
N	Atomic number density
N_d	Average number of displaced atoms
Q	Fraction of incident energy converted into thermal energy (lattice vibrations)
R(E)	Total path-length
R_D	Displacement generation rate
$S_n(E)$	Nuclear stopping power
T	Transferred energy or temperature
T_H	Homologous temperature
$T_M{}^d$	Melting temperature of damaged crystal
T_M	Melting temperature
$T_M{}^o$	Melting temperature of perfect crystal
T_{am}	Amorphization threshold temperature
T_{cyl}	Temperature in radial distance for a cylindrical thermal spike
T_e	Temperature of the electronic system
T_n	Temperature of the lattice system
T_{sph}	Temperature in radial distance for a spherical thermal spike
U	Binding energy
V_{cas}	Volume of cascade
$\langle X \rangle$	Average damage depth
$\langle \Delta X \rangle$	Average longitudinal damage straggling (average damage width)
$\langle Y \rangle$	Average transversal straggling
a	Lattice plane spacing or ion range
b_{arc}, c_{arc}	Constants of the arc-dpa model
c	Specific heat capacity
c_e	Specific heat capacity of the electronic system
c_n	Specific heat capacity of the lattice system
g	Electron–phonon coupling constant
m	Power law varialble (=1/s)
n_υ	Vacancy density
s	Power variable in the inverse power potential
t_{cas}	Life time of cascade
v	Particle (ion) velocity
Φ	Ion fluence
Φ_{am}	Amorphization threshold fluence
γ	Transfer energy efficiency factor

(continued)

(continued)

Symbol	Notation
$\langle \delta^2 \rangle$	Mean squared range of the displaced atoms (square static displacement)
ε	Reduced energy
$\eta(T)$	Inelastic energy loss by electronic excitation
η	Total number of rearrangement processes in the thermal spike
η_{cyl}	Total number of rearrangement processes in a cylindrical thermal spike
η_{sph}	Total number of rearrangement processes in a spherical thermal spike
θ_d	Debye temperatures of damaged crystal
θ_o	Debye temperatures of perfect crystal
$\langle \theta \rangle$	Energy density of the cascade
κ_o	Thermal conductivity
μ	Transversal straggling of the damage distribution
$\nu(r, t)$	Jump rate
$\nu(T)$	Damage energy
ξ	Displacement efficiency
ξ_{arc}	Relative damage efficiency in arc-dpa model
ξ_{vs}, ξ_{is}	Interaction radii for the reaction between the sink and vacancy, interstitial
ρ	Mass density
σ	Longitudinal straggling of the damage distribution or total scattering cross-section
σ_d	Damage energy cross-section
σ_{dam}	Total damage energy cross-section
σ_x, σ_y	Contraction parameters

References

1. M.W. Thompson, *Defects and Radiation Damage in Metals* (Cambridge University Press, Cambridge, 1969)
2. K.B. Winterbon, *Ion Implantation Range and Energy Deposition Distributions*, vol. 2 (Plenum Press, New York, 1975)
3. J.A. Davies, Collision cascades and spike effects, in *Surface Modification and Alloying by Laser, Ion, and Electron Beams*, ed. by J.M. Poate, G. Foti, D.C. Jacobson (Plenum Press, New York, 1983), pp. 189–219
4. M. Nastasi, J.W. Mayer, J.K. Hirvonen, *Ion-Solid Interactions* (Cambridge University Press, Cambridge, 1996)
5. R.S. Averback, T.D. de la Rubia, Displacement damage in irradiated metals and semiconductors, in *Solid State Physics*, vol. 51, ed. by H. Ehrenfest, F. Spaepen (Academic Press, New York, 1998), pp. 281–402
6. G.S. Was, *Fundamentals of Radiation Materials Science* (Springer, New York, 2007)
7. R.E. Stoller, Primary radiation damage formation, in *Comprehensive Nuclear Materials*, vol. 1, ed. by R.J.M. Konings (Elsevier, Amsterdam 2012), pp. 293–332

8. D.R. Olander, A.T. Motta, in *Light Water Reactor Material, Volume I: Fundamentals* (American Nuclear Society, LaGrange Park, 2017)
9. A. Mutzke, R. Schneider, W. Eckstein, R. Dohmen, K. Schmidt, U. von Toussaint, G. Badelow, *SD TrimSP Version 6.00* (IPP Report 2019-02, Munich 2019)
10. H.H. Andersen, The depth resolution of sputter profiling. Appl. Phys. **18**, 131–140 (1979)
11. W.E. King, K.L. Merkle, M. Meshii, Threshold energy surface and Frenkel pair resistivity for Cu. J. Nucl. Mater **117**, 12–25 (1983)
12. T.E. Mitchell et al., in *Fundamental Aspects of Radiation Damage in Metals*, vol. 1, ed. by M.T. Robinson, F.W. Young (US GPO, Washington, DC, 1976), p. 73
13. P. Olsson, S.C. Bequart, C. Domain, Ab initio threshold displacement energies in iron. Mat. Res. Lett. **4**, 219–225 (2016)
14. D.R. Olander, in *Fundamental Aspects of Nuclear Reactor Fuel Elements* (National Technical Information Service, Springfield, 1976), pp. 373–417 (Chapter 17)
15. G.H. Kinchin, R.S. Pease, The displacement of atoms in solids. Rep. Progr. Phys. **18**, 1–51 (1955)
16. M.T. Robinson, On energy dependence of neutron radiation damage in solids, in *Nuclear Fusion Reactors* (British Nuclear Energy Society, London, 1970), pp. 346–378
17. J. Lindhard, V. Nielsen, M. Scharff, P.V. Thomsen, Integral equation governing radiation effects—notes on atomic collision III. Det. Kgl. Danske Vid. Selskab. Mat.-Fys. Medd. **33**(10) (1963)
18. M.J. Norgett, M.T. Robinson, I.M. Torrens, A proposed method for calculating displacement dose rates. Nucl. Eng. Des. **33**, 50–54 (1975)
19. M.T. Robinson, The influence of the scattering low on the radiation damage displacement cascade. Phil. Mag. **12**, 741–765 (1965)
20. M.T. Robinson, I.M. Torrens, Computer simulation of atomic-displacement cascades in solids in binary collision approximation. Phys. Rev. B **9**, 5008–5024 (1974)
21. P. Sigmund, A note on integral equations of Kinchin-Pease type. Rad. Eff. **1**, 15–18 (1969)
22. P. Sigmund, On the number of atoms displaced by implanted ions or energetic recoil atoms. Appl. Phys. Lett. **14**, 114–116 (1969)
23. R.S. Averback, R. Benedek, K.L. Merkle, Ion-irradiation studies of the damage function of copper and silver. Phys. Rev. B **18**, 4156–4171 (1978). Efficiency of defect production in cascades. J. Nucl. Mater. **69** & **70**, 786–789 (1978)
24. S.J. Zinkle, B.N. Singh, Analysis of displacement damage and defect production under cascade damage conditions. J. Nucl. Mater. **199**, 173–191 (1993)
25. S.J. Kim, M.A. Nicolet, R.S. Averback, D. Peak, Low-temperature ion beam mixing in metal. Phys. Rev. B **3**, 38–49 (1988)
26. K. Nordlund, A.E. Sand, F. Granberg, S.J. Zinkle, R. Stoller, R.S. Averback, T. Suzudo, L. Malerba, F. Banhart, W.J. Weber, F. Willaime, S. Dudarev, D. Simeone, Primary radiation damage in materials—review of current understanding and proposed new standard displacement damage model to incorporate in cascade defect production efficiency and mixing effects. Report NEA/NSC/DOC (2015) 9 (Nuclear Energy Agency, OECD 2015)
27. K. Nordlund, S.J. Zinkle, A.E. Sand, F. Granberg, R.S. Averback, R. Stoller, T. Suzudo, L. Malerba, F. Banhart, W.J. Weber, F. Willaime, S. Dudarev, D. Simeone, Improving atomic displacement and replacement calculations with physically realistic damage models. Nat. Comm. **9**, 1084 (2018)
28. Y. Konobeyev, U. Fischer, Y.A. Korovin, S.P. Simakov, Evaluation of effective threshold displacement energies and other data required for the calculation of advanced atomic displacement cross-sections, Nucl. Energy Technol. **3**, 169–175 (2017)
29. K.B. Winterbon, P. Sigmund, J.B. Sanders, Spatial distribution of energy deposited by atomic particles in elastic collisions. Det. Kgl. Danske Vid. Selskab. Mat.-Fys. Medd. **37**(14) (1970)
30. D.K. Brice, Spatial distribution of energy deposited into atomic processes in ion-implanted silicon. Rad. Eff. **6**, 77–87 (1970)
31. K.B. Winterbon, Heavy-ion range profiles and associated damage distributions. Rad. Eff. **13**, 215–226 (1972)

32. J. Lindhard, V. Nielsen, M. Scharff, Approximation method in classical scattering by screened coulomb fields—notes on atomic collision I. Det. Kgl. Danske Vid. Selskab. Mat.-Fys. Medd. **36**(10) (1968)
33. P. Sigmund, G.P. Scheidler, G. Roth, Spatial distribution of defects in cascades. Report BNL 50083, C-52 (1968), p. 375
34. P. Sigmund, Energy density and time constant of heavy-ion-induced elastic-collision spikes in solids. Appl. Phys. Lett. **25**, 169–171 (1974). Theory of sputtering. I. Sputtering yield of amorphous and polycrystalline targets. Phys. Rev. **184**, 383–416 (1969)
35. R.M. Bradley, Exact linear dispersion relation for the Sigmund model of ion sputtering. Phys. Rev. B **84**, 075413 (2011)
36. M.Z. Hossain, J.B. Freund, H.T. Johnson, Ion impact energy distribution and sputtering of Si and Ge. J. Appl. Phys. **111**, 103513 (2012)
37. G. Hobler, R.M. Bradley, H.M. Urbassek, Probing the limitations of Sigmund's model of spatially resolved sputtering using Monte Carlo simulations. Phys. Rev. B **93**, 205443 (2016)
38. M. Feix, A.K. Hartmann, R. Kree, J. Moñoz-Garcia, R. Cuerno, Influence of collision cascade statistics on pattern formation of ion-sputtered surfaces. Phys. Rev. B **71**, 125407 (2005)
39. W. Bolse, Ion-beam-induced atomic transport through bi-layer interfaces of low- and medium-Z metals and their nitrides. Mater. Sci. Eng. R. Rep. **12**, 53–121 (1994)
40. R.S. Averback, Atomic displacement processes in irradiated metals. J. Nucl. Mater. **216**, 49–62 (1994)
41. F. Seitz, S. Koehler, Displacement of atoms during irradiation, in *Solid State Physics*, vol. 2, ed. by F. Seitz, D. Turnbull (Academic Press, New York, 1956), pp. 307–442
42. J.A. Brinkman, On nature of radiation damage in metals. J. Appl. Phys. **25**, 961–970 (1954)
43. A Seeger, On the theory of radiation damage and radiation hardening, in *Proceedings of the Second UN International Conference on the Peaceful Uses of Atomic Energy*, vol. 8 (Geneva 1958), p. 250
44. M. Nastasi, J.W. Mayer, Ion beam mixing in metallic and semiconductor materials. Mater. Sci. Eng. R **12**, 1–52 (1994)
45. G.H. Vineyard, Thermal spikes and activated processes. Rad. Eff. Def. Solids **29**, 245–248 (1976)
46. H. Hofsäss, H. Feldermann, R. Merk, M. Sebastian, C. Ronning, Cylindrical spike model for the formation of diamond like thin films by ion deposition. Appl. Phys. A **66**, 153–181 (1998)
47. H.M. Naguib, R. Kelly, Criteria for bombardment-induced structural changes in non-metallic solids. Rad. Effects **25**, 1–12 (1975)
48. L.T. Chadderton, On the anatomy of a fission fragment track. Int. J. Rad. Appl. Instrum. Part D, Nucl. Tracks Rad. Measur. **15**, 11–29 (1988)
49. I.M. Lifshitz, M.I. Kaganov, L.V. Tanatarov, On the theory of radiation-induced changes in metals. J. Nucl. Energy A **12**, 69–78 (1960)
50. Z.G. Wang, C. Dufour, E. Paumier, M. Toulemonde, The Se sensitivity of metals under swift-heavy-ion irradiation: a transient thermal process. J. Phys. Condens. Matter **6**, 6733–6750 (1995)
51. R. Sizmann, Effect of radiation upon diffusion in metals. J. Nucl. Mater. **69–70**, 386–412 (1978)
52. W.M. Lomer, *Diffusion Coefficients in Copper Under Fast Neutron Irradiation* (UKAEA Report AERE-T, 1954)
53. D. Brailsford, R. Bullough, The rate theory of swelling due to void growth in irradiated metals. J. Nucl. Mater. **44**, 121–135 (1972)
54. K.C. Russel, Phase stability under irradiation. Progr. Mater Sci. **28**, 229–434 (1984)
55. B. Rauschenbach, V. Heera, Amorphisation of metals by ion implantation and ion beam mixing. Defect Diffus. Forum **57–58**, 143–188 (1988)
56. B. Rauschenbach, G. Otto, K. Hohmuth, V. Herra, Structural investigations of amorphised iron and nickel by high-fluence metalloid ion implantation. J. Phys. F Met. Phys. **17**, 2207–2216 (1987)
57. B. Pfeifer, J.K.N. Lindner, B. Rauschenbach, B. Stritzker, 2 MeV aluminum implantation into silicolr: radiation damage. Nucl. Instr. Meth. Phys. Res. B **96**, 150–154 (1995)

58. E. Wendler, Mechanisms of damage formation in semiconductors. Nucl. Instr. Meth. Phys. Res. B **267**, 2680–2689 (2009)
59. J. Jagielski, L. Thomé, Multi-step damage accumulation in irradiated crystal. Appl. Phys. A **97**, 147–155 (2009)
60. J.F. Gibbons, Ion implantation in semiconductors—part II: damage production and annealing. Proc. IEEE **60**, 1062–1096 (1972)
61. N. Hecking, K.F. Heidemann, E. te Kaat, Model of temperature dependent defect interaction and amorphization in crystalline silicon during irradiation. Nucl. Instr. Meth. Phys. Res. B **15**, 760–764 (1986)
62. W.J. Weber, R.C. Ewing, C.R.A. Catlow, T. Diaz de la Rubia, L.W. Hobbs, C. Kinoshita, Hj. Matzke, A.T. Motta, M. Nastasi, E.K.H. Salje, E.R. Vance, S.J. Zinkl, Radiation effects in crystalline ceramics for the immobilization of high-level nuclear waste and plutonium. J. Mater. Res. **13**, 1434–1484 (1998)
63. H. Trinkaus, Ion beam-induced amorphization of crystalline solids: mechanism and modeling. Mater. Sci. Forum **248**(249), 3–12 (1997)
64. S. Voronel, A. Rabinovich, V. Kisliuk, T. Steinberg, Sverbilova, Universality of physical properties of disorderd alloys. Phys. Rev. Lett. **23**, 2402–2405 (1988)
65. N.Q. Lam, P.R. Okamoto, M. Li, Disorder-induced amorphization. J. Nucl. Mater. **251**, 89–97 (1997)
66. P.R. Okamoto, N.Q. Lam, L.E. Rehn, Physics of crystal-to-glass transformation. Solid State Phys. **52**, 1–135 (1999)
67. F.A. Lindemann, Über die Berechnung der molekularen Eigenfrequenzen. Phys. Z. **14**, 609–612 (1910)
68. G. Linker, Amorphization of Niobium films by boron ion implantation. Mater. Sci. Eng. **69**, 105–110 (1985)
69. G. Linker, X-ray study of the defect structure in ion implanted niobium and molybdenum superconductors. Nucl. Instr. Meth. **182–183**, 501–508 (1981)
70. J.R. Dennis, E.B. Hale, Crystalline to amorphous transformation in ion-implanted silicon: a composite model. J. Appl. Phys. **49**, 1119–1127 (1978)
71. S.T. Picraux, F.L. Vook, Ionization, thermal, and flux dependences of implantation disorder in silicon. Rad. Eff. **11**, 179–192 (1971). S.T. Picraux, W. Weisenberger, F.L. Vook, Low temperature channeling measurements of ion implantation lattice disorder in single crystal silicon. Rad. Eff. **7**, 101–107 (1971)
72. S. Prussin, D.I. Margolese, R.N. Tauber, Formation of amorphous layers by ion implantation. J. Appl. Phys. **57**, 180–185 (1985)
73. F.F. Morehead, B.L. Crowder, A model for the formation of amorphous Si by ion bombardment. Rad. Eff. **6**, 27–32 (1970)
74. F.F. Morehead, Jr. B.L. Crowder, A model for the formation of amorphous Si by ion bombardment, in *Ion Implantation*, ed. by F.H. Eisen, L.T. Chadderton (Gordon and Breach Science Publishers, London, 1971), pp. 25–30
75. E. Wendler, W. Wesch, III–V semiconductors under ion bombardment—studied by RBS in situ at 15 K, in *Advances in Solid State Physics*, vol. 44 ed. by B. Kramer (Springer, Berlin, Heidelberg, 2004), pp. 339–350
76. L. Pelaz, L.A. Marqués, J. Barbolla, Ion-beam-induced amorphization and recrystallization in silicon. J. Appl. Phys. **96**, 5947–5976 (2004)
77. W.J. Weber, Models and mechanisms of irradiation-induced amorphization in ceramics, Nucl. Instr. Meth. Phys. Res. B 166–167, 98–106 (2000)

Chapter 5
Sputtering

Abstract Sputtering techniques are of great importance for both academic research and commercial applications. This chapter summarizes the fundamentals of the sputtering process, especially after low-energy ion bombardment. The theoretical description of the total sputtering yield, and the energy and angular distribution of the sputtered atoms is based on the assumption of a linear cascade regime. The description of the sputtering process in the linear regime according to Sigmund for amorphous materials by means of the Boltzmann transport equation is briefly presented, the energy dependence of the sputtering yield is demonstrated, and the influence of critical parameters, such as surface binding energy, threshold energy and material correction factor is discussed. The dependence of the sputtering yield on the ion incidence angle and also the energy and angular distributions of the sputtered particles are considered in detail. The particular case of preferential sputtering in low-energy ion bombardment of binary materials is addressed, considering both the mass effect and the bonding effect. An important aspect of this chapter is the measurement of the sputtering yield and energy and angular distributions of the sputtered particles. Finally, the reflection of low-energy ions from surfaces is briefly presented.

An academically interesting and technological significant result of the bombardment of solid surfaces with energetic ions is the erosion of surface by sputtering. This process is characterized quantitatively by the sputtering yield. The sputtering yield describes how many particles of the surface are removed after collision with an ion. This dimensionless value can vary about many magnitudes depending on experimental parameters and the ion-target combination. For low and middle ion energies (up to some keV) and perpendicular ion bombardment, the sputtering yield can be expected to be between 0.1 and 20, approximately.

The multitude of options for the use of technologies which are based on the sputtering phenomenon have triggered widespread experimental and theoretical studies of the sputtering effects. Precisely controlled removal of material from the surface on the atomic scale as well as the intensive flux of sputtered particles are the determining

B. Rauschenbach, *Low-Energy Ion Irradiation of Materials*, Springer Series in Materials Science 324, https://doi.org/10.1007/978-3-030-97277-6_5

advantages of sputter technologies, compared to other methods. Traditionally, sputtering techniques with low ion energies between about 0.1 keV and 10 keV are at the center of many commercial applications. Typical areas of applications are:

(1) Deposition of thin films

Sputter technologies which are used to prepare thin films are characterized by a process wherein a noble gas is introduced into a vacuum chamber and then a cathode (target) of the material which should be deposited is electrically energized to produce a self-sustaining plasma. The gas atoms become positively charged gas ions in the plasma and accelerated towards a target surface with the objective to sputter atoms, ions and molecules from the target. The sputtered particles pass through the vacuum chamber and are deposited on the substrate surface as a thin film. Frequently, strong electric and magnetic fields are utilized in a so-called magnetron to confine charged plasma particles to the vicinity of the surface of the sputter target in order to generate a higher ionization efficiency.

(2) Depth profiling for thin film analysis

The most widely used technique to obtain sectioning for the in-depth profiling analysis is the sputtering of films. This technique is universally applicable (e.g., for secondary ion mass spectrometry, Auger electron mass spectroscopy, photoelectron spectroscopy, etc.). The sputter ion sources are usually operated with noble gas ions and at energies between some hundred eV to some keV.

(3) Cleaning of surface under vacuum conditions

Ion beam induced desorption is a key technology to clean substrates, for example prior to the deposition of thin films or thin film analysis. Cleaning by ion bombardment involves desorption of adsorbates and the removal of the topmost layer (e.g., oxidized surface layers) of the substrates. The energy of the noble gas ions used here is higher than a few tens of eV.

(4) Ion beam etching

Ion beam etching is an isotropic etching technology that faithfully reproduces the mask pattern on the surface of the product. In this dry etching process an ion beam is applied to sputter material exposed by a mask (e.g., photo resist) to get the desired pattern. Ion beam etching is an enabling process technology for precision microelectronic devices, sensors, and optical as well microwave components which require long operational lifetimes and precise performance specifications.

(5) Surface smoothing and figuring

Smoothing and figuring procedures based on low-energy ion beam sputtering become more and more established in high precision surface and thin film processing. Especially in fields like nanotechnology, extreme ultraviolet lithography, high-precision optics, satellite communication optics and large wafer planarization, ultra-smooth surfaces with nanometer and sub-nanometer depth accuracy over the entire spatial wavelength range are requested.

(6) Formation of nano/microstructures

Due to the interaction of low-energy ions with the substrate, the surface topography is modified and under certain conditions regular nanostructures such as ripples, or dots, etc. can evolve by self-organization processes. It has been found that both ion beam parameters and substrate parameters are responsible for the features of such nanopatterns. The diversity of nanostructures which can be formed in only one step in a wide variety of materials (e.g., elemental and compound semiconductors, mono- and polycrystalline metals, dielectrics) makes this technique an attractive alternative route for the production of nanopatterned surfaces. A second route is the deposition of sputtered material under glancing angle condition. Nano- and micropatterns can be produced for numerous applications.

A number of extensive overview articles and textbooks have been published to give comprehensive reviews about the fundamental processes, the theoretical description of sputtering phenomena and the experimentally measured sputtering data (e.g., see [1–7]).

5.1 Sputtering Yield

The sputtering yield Y is defined as the average number of sputtered particles (ions, atoms, etc.) per incident particle and can be determined experimentally. It is assumed that in measurable quantities the sputtering yield is given by

$$Y = \frac{average\,no.\,of\ sputtered\ particles}{incident\ particle}. \tag{5.1}$$

Then, the erosion velocity during ion bombardment normal to the surface is given by

$$v_o \equiv \frac{dh}{dt} = M_2 \cdot J \cdot \frac{Y}{\rho} = \frac{JY}{N}, \tag{5.2}$$

where h is the eroded depth, ρ is the mass density of the bombarded material, M_2 is the mass of the target atoms (mass number), N the atomic number density of the target material and J is the ion beam current density (ion flux). A calculation of the eroded layer thickness can be realized when the sputtering yield and the ion current density are known. Erosion rates for low-energy ion irradiation vary, typically between 10^{-3} and 10^2 nm/s.

In Fig. 5.1, the erosion process is schematically shown. The impingement of energetic ions on surfaces induces a number of effects. For energies higher than a threshold energy (see Sect. 5.4.1), the incident ion can transfer sufficient energy to displace target atoms from its sites in the solid or at the surface so that these sputtered atoms are removed into the vacuum. As a result, the surface is eroded. The incident particle primarily dissipates its energy by elastic collisions with target

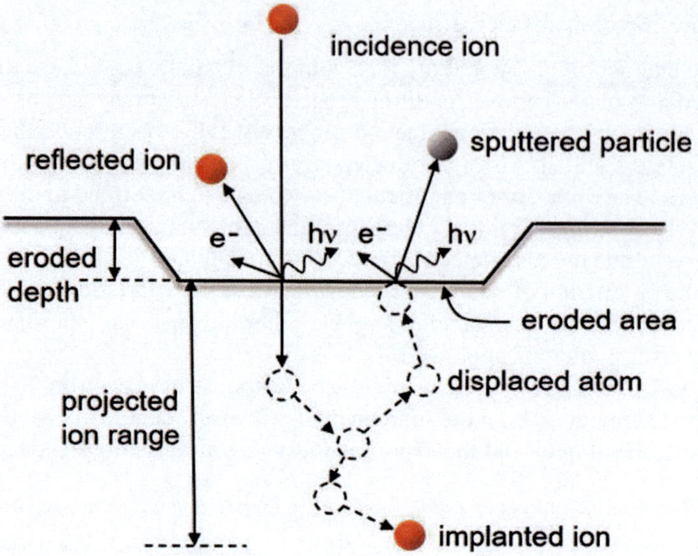

Fig. 5.1 Schematic picture of the surface erosion process by ion bombardment

atoms. This process generates recoiled target atoms, which have enough energy to collide with other target atoms (c.f. Fig. 4.1). It should be noted that reflected ions and photons are also created by the ion bombardment. Experimental studies have shown that the sputtering yield is dependent on the properties of the bombarding particles (kinetic energy, atomic mass and angle of incidence of the incoming particle) and the target parameters (surface binding energy, atomic mass, density of the target and also crystallographic orientation of lattice planes with respect to the direction of ion incidence).

5.2 Theoretical Aspects of Sputtering

At primary ion energies higher than the threshold energy E_{th} and lower than a few kilo-electron-volts, the projectile hits a first target atom and transfers most of the energy in binary elastic collisions. Such an interaction is referred to as near-threshold or knock-on regime (Fig. 5.2, left). This regime cannot be comprehensively described by the conventional solution of the Boltzmann transport equation [8, 9]. For higher energies, from a few keV up to about 100 keV, the incident ion is capable to produce a cascade of recoil atoms along the ion path. In this so-called linear cascade regime (Fig. 5.2, middle), i.e. for a sufficiently small number and isotropic distribution of binary collisions within the cascade, the energy transferred in the binary collisions between the incident particle and the target atoms is capable to displace atoms from its places in the solid. In general, the energy of the displaced atoms is high enough to displace

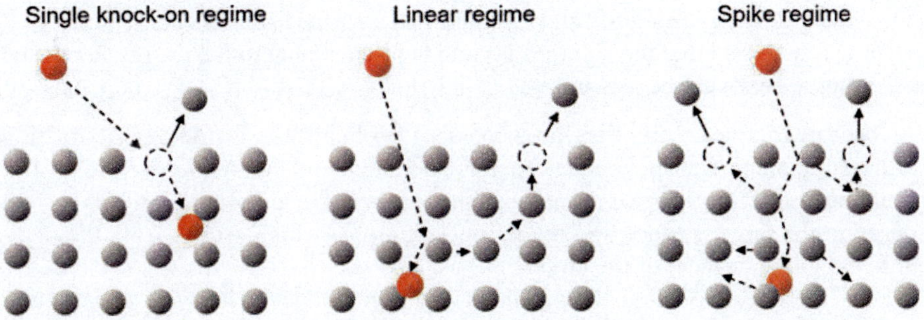

Fig. 5.2 Schematic presentations of the three regimes of sputtering according to Sigmund [9]

further atoms. The resulting collision cascade is characterized by a distribution of interstitial atoms and vacancies along the ion track. The number of the recoil atoms as result of every binary collisions process is linearly proportional to the amount of the energy loss and can be described by using the linearized Boltzmann transport equation [8]. A collision sequence can induce an ejection of surface atoms if the transferred energy overcomes the surface binding energy U_s (Fig. 5.2). In summary, in the linear collision cascade regime, the incident particle collides with the target atoms, with an energy sufficient to displace them (called primary recoils). If the transmitted energy is high enough, these primary recoils are capable to displace secondary recoils. Such collision processes result in a linear collision cascade. Recoil atoms which reach the surface and have an energy higher than the surface binding energy escape the surface and represent the flux of the sputtered atoms.

For ion energies higher than some keV and particularly for heavier particles, including molecules, the assumption of separated binary collisions must be abandoned and a collision processes, in which many atoms are involved simultaneously, can be assumed (Fig. 5.2, right). The deposited energy overcomes the binding energy of all atoms in the vicinity of the ion track. These so-called spikes are characterized by an abrupt, nearly discontinuous change in pressure, temperature and density in the spike region. The number of sputtered atoms increases nonlinearly with the ion energy (see Sect. 4.6).

According to Sigmund [8, 9], the sputtering process in the linear regime can be described in the framework of linear collision cascades for amorphous materials by solving the Boltzmann transport equation using appropriate cross-sections for elastic collisions.

The linear cascade theory by Sigmund is based on some fundamental assumptions (see e.g., reviews [8, 9]):

(i) The collisions between the incident particle and the target atoms as well as between the target atoms themselves are binary collisions,
(ii) the target material is homogeneous and isotropic,
(iii) inelastic energy losses are neglected,
(iv) an isotropic distribution of the particle velocities is given,

(v) the ion energy is significantly higher than a surface barrier energy and
(vi) it is assumed that the sputtering yield is proportional to the energy deposited in the near-surface region.

These restrictive assumptions are necessary to obtain a first-order approximation of the Boltzmann transport equation for calculation of the sputtering yield. The particular aim of this theory is to calculate the number of recoiled atoms, the deposited energy in the target surface and the number of atoms which escape from the near-surface region (surface) of the target.

Thus, the linear collision cascade theory by Sigmund [8, 9] answers in particular the following questions:

(i) How many atoms are recoiled?

Under the assumption that a primary ion or a recoil with the energy E formed a collision cascade, the average number of atoms with an energy E_o within this induced collision cascade is $n(E, E_o)$. Sigmund found an expression for this average number by solving the Boltzmann transport equation to

$$n(E, E_o) \approx \Gamma_m \frac{E}{E_o} \quad \text{for}\, \gamma E \gg E_o, \tag{5.3}$$

where $\Gamma_m = m/(\psi(1) - \psi(1-m))$ and $\psi(x) = \frac{d}{dx}\log[\Gamma(x)]$. The gamma function, Γ_m, is depending on m, where m is a parameter which characterizes the scattering cross-section (see Sect. 2.2.5). The function Γ_m for characteristic values of m is given in Table 5.1.

It is known that the energy deposited in the target is dissipated in elastic nuclear and inelastic electron collisions (see Lindhard's energy partition, Sect. 4.2.2)

$$E = \nu(E) + \eta(E), \tag{5.4}$$

where $\eta(E)$ is the average amount of energy that is transferred in inelastic processes and $\nu(E)$ the average amount of energy that is transferred into elastic collision processes. For the calculation of the average number of recoils the energy dissipation to electrons can be neglected, because the energy transferred to the electron cloud will immediately be shared with all other electrons. Consequently, E can be replaced by $\nu(E)$ and the number of atoms $F(E, E_o)\,dE$ recoiling into a given energy interval (E_o, dE_o) is given by the recoil density

$$F(E, E_o) = -\frac{dn(E, E_o)}{dE_o} = \Gamma_m \frac{\nu(E)}{E_o^2} \quad \text{for} \quad \gamma E \gg E_o. \tag{5.5}$$

Table 5.1 Summary of the values for m and Γ_m

m	0	0.25	0.333	0.5
Γ_m	0.608	0.491	0.452	0.361

With the help (5.3), it can be assumed that the recoil density is isotropic because recoils of higher order dominate the collision cascade. Then, the recoil density after normalization is

$$F(E, E_o, \boldsymbol{\Omega}_o)dE_o d^2\Omega_o = \Gamma_m \frac{\nu(E)}{E_o^2} dE_o \frac{d^2\Omega_o}{4\pi}, \tag{5.6}$$

and describes the average number of recoils with initial E and direction Ω in the energy interval (E_o, dE_o) and direction interval ($\boldsymbol{\Omega}_0$, $d^2\Omega_o$).

(ii) Which fraction of the recoiled atoms are capable to escape from the surface?

It is also assumed by Sigmund [8, 9], that the total number G (E, E_o) of recoils which are moved per unit time in the interval of energy dE_o is given by incident particles flux per unit time J and the number of recoils n(E, E_o) generated by this flux in the interval of the time dt_o, i.e. the total number of recoils can be expressed by

$$G(E, E_o)dE_o = J n(E, E_o)dt_o, \tag{5.7}$$

where the average time is given by

$$dt_o = \frac{dE_o}{|dE_o/dt_o|} = \frac{dE_o}{v|dE_o/dx|}. \tag{5.8}$$

This is the time that is needed by a recoil atom to slow down from $E_o + dE_o$ to E_o. The energy E_o is the energy of the target atom with velocity v and n(E, E_o) is given in (5.2). Substituting (5.5) and (5.8) in (5.7) gives the number of all atoms within the cascade

$$G(E, E_o)dE_o = J \frac{E_o dE_o}{v\left|\frac{dE_o}{dx}\right|} F(E, E_o). \tag{5.9}$$

This is the total number of all moved atoms. But only the target atoms which are capable to leave the surface are of interest. Consequently, only these moved target atoms are considered. These particles are capable to reach the target surface under a specific direction Ω, where it is assumed that the directions of the moving atoms are isotropically distributed.

Thus, the flux of the recoils in the direction of the surface is given by the product of the number of all moved target atoms (recoils), G (E, E_o), and the particle velocity v. According to Sigmund [8, 9], the mean number of atoms moving with an energy (E_o, dE_o) in the direction $\left(\boldsymbol{\Omega}_o, d^2\Omega\right)$, provides an ion flux J per unit time with energy E and direction $\boldsymbol{\Omega}$, which can be expressed by

$$G(E, \boldsymbol{\Omega}, E_o, \boldsymbol{\Omega}_o, \mathbf{r})dE_o d^2\Omega_o d^3r$$
$$= J F_D(E, \boldsymbol{\Omega}, , r)\frac{\Gamma_m}{E_o v|dE_o/dx|} dE \frac{d^2\Omega_o}{4\pi} d^3r, \tag{5.10}$$

if a source at $\mathbf{r} = 0$ is assumed. When taking into account that only particles which moved in the direction of the surface should be involved, this equation must be multiplied with $\mathbf{v} = \mathrm{v}\,\boldsymbol{\Omega}_o$. After multiplication and integration of (5.10) over a plane (surface) $\mathrm{x} = 0$, the flux of recoils at the surface is given by

$$J_s(\mathrm{E}_o, \boldsymbol{\Omega}_o)dE_o, d^2\Omega_o = JF_D(E, \theta, r = 0)\frac{\cos\theta_o}{4\pi}\frac{\Gamma_m}{\mathrm{E}_o|dE_o/dx|}dEd^2\Omega_o. \quad (5.11)$$

This flux of moved recoils reaches the surface, i.e. these particles can be sputtered, if an energetic surface barrier, $\mathrm{U_s}$, can be overcome (see Sects. 5.3 and 5.4).

5.3 Energy Dependence of the Sputtering Yield

According to the linear cascade theory by Sigmund [8], sputtering yield for mono-element materials in the linear cascade regime is strongly dependent on the ion incident energy. When target atoms escape the target surface with a probability P(E,θ), the energy of these particles must be higher than the surface binding energy $\mathrm{U_S}$. Then, the sputtering yield can be determined by integrating (5.11) over θ and E and dividing the sputter current by the incidence particle current J [9]. This yields to an expression of the sputtering yield as function of two factors

$$Y = \Lambda \cdot F_D(E, \theta, r). \quad (5.12)$$

The material constant (first factor) is specified by

$$\Lambda = \frac{\Gamma_m}{2}\int\frac{dE_o}{E_o|dE_o/dx|}\int d(\cos\theta_o)|\cos\theta_o|P(E_o, \theta_o), \quad (5.13)$$

where the escape probability is given by

$$P(E_o, \theta_o) = \begin{cases} 1\ for\ E_o\cos^2\theta_o > U_s \\ 0\ for\ E_o\cos^2\theta_o < U_s \end{cases} \quad (5.14)$$

for a planar surface barrier. The material factor Λ can be expressed by

$$\Lambda = \frac{1-m}{2NC_m}\Gamma_m\int_o^{\pi/2}\cos\theta_o\sin\theta_o d\theta_o\int_{U_s\cos^{-2}\theta_o}^{\infty}\frac{dE_o}{E_o^{2-2m}} \quad (5.15)$$

by using the nuclear stopping power of the target material as introduced by Lindhard, Nielsen and Scharff [10] (see Sect. 3.1.1), where N is the atomic number density of the target material and $\mathrm{C_m}$ is a function of the masses of the incident particle and the target atom, (see 2.79). The integration of (5.15) leads to

$$\Lambda = \frac{\Gamma_m}{8(1-2m)} \frac{1}{NC_m U_s^{1-2m}}, \tag{5.16}$$

for low-energy collisions m = 0 and $\Gamma_{m=o} = 0.608$ (see Table 5.1). Thereby $\Gamma_m/[8(1-2\,m)] = 3/4\pi^2 = 0.067$. Therefore, the (5.16) can be simplified to

$$\Lambda = \frac{3}{4\pi^2} \frac{1}{NC_o U_s}. \tag{5.17}$$

The cross-section constant for m = 0 [see (2.81)] can be expressed by

$$C_o = \frac{\pi}{2} \lambda_0 a_{BM}^2 \cong 0.0181 \text{ nm}^2, \tag{5.18}$$

because a Born–Mayer low-energy interatomic potential (see Sect. 2.1.1.2) is assumed. Then the material factor can be given by

$$\Lambda = \frac{3}{4\pi^2} \frac{1}{C_0 N U_s} = \frac{4.2 \text{ nm}^{-2}}{N U_s}, \tag{5.19}$$

where $3/(4C_oN) \approx 41.5 \text{ nm}^{-2}/N$ is the mean escape depth of the sputtered atoms [8].

The second factor in (5.12), F_D, is the elastic energy deposited at the surface (see 4.33), which can be represented as

$$F_D(E) = \alpha N S_n(E), \tag{5.20}$$

where the material correction factor α is an energy independent function of the mass ratio between target (M_2) and the projectile (M_1) and can be approximately determined (see Sect. 5.4.2). $S_n(E)$ is the nuclear stopping cross-section at the energy E (see Sect. 3.1). Combining (5.19) and (5.20), the sputtering yield after ion bombardment can be expressed as

$$Y = \Lambda F_D(E) = 4.2 \text{ nm}^{-2} \alpha \frac{S_n(E)}{U_s}. \tag{5.21}$$

It is obvious that the sputtering yield is indeed linearly related to the first power of the energy E. As an example, Fig. 5.3 (left) shows the calculated energy dependent-sputtering yields of Si after perpendicular ion bombardment with ions of the carbon group elements (left) and the sputtering yield of different elements after low-energy Ar ion sputtering (right).

If the energy of the incident ion is smaller than 1 keV, Sigmund [8] suggested a modified relationship of sputtering yield at perpendicular incidence, where T_{max} is the maximum transferred energy [see (2.36)]. Using (5.12), it follows that the sputtering yield can be given by

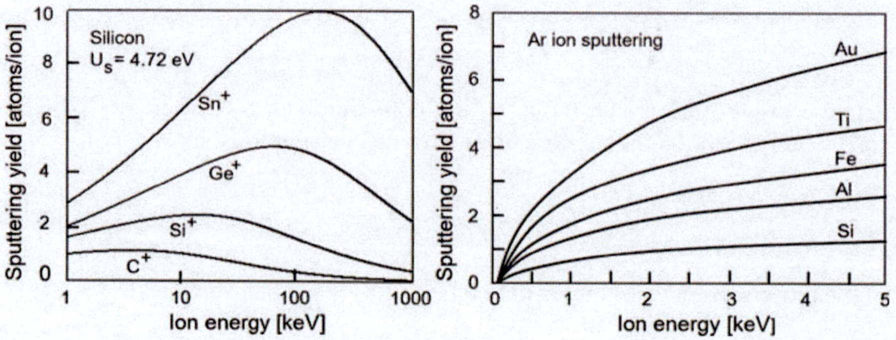

Fig. 5.3 Left: Energy dependence of the sputtering yield of Si for bombardment with ions of the carbon group elements and at normal incidence. Right: Energy dependence on the sputtering yield for different metals and Si after Ar ion sputtering at normal ion incidence. The nuclear cross section $S_n(E_0)$ given by Lindhard et al. [10] for a Thomas–Fermi interaction and the reduced nuclear stopping cross section $s_n(\varepsilon)$ given in Matsunami et al. [11] are used for the analytical calculation

$$Y = \Lambda\alpha N S_n(E) = \frac{3}{4\pi^2}\alpha\frac{T_{max}}{U_s} \cong 0.3\alpha\left(\frac{M_1 M_2}{M_1 + M_2}\right)\frac{E}{U_s}. \tag{5.22}$$

According to this equation the sputtering yield is again linearly depending on the incident ion energy in the low-energy range.

Sigmund's theory provides the basis for other models that determine the energy dependence of the sputtering yield. The dependence of the sputtering yield on the energy of the incident ions is reproduced acceptably in this theory for most ion-target combinations in a wide range of energies. Many refinements and corrections have been suggested with the aim to reduce the discrepancies between theoretical results and the experimental measurements. For example:

(i) In comparison to measured sputtering yields, the application of the Thomas–Fermi screening function and the selected power law fitting variable λ_o (see Sect. 2.2.5 and (2.81) leads to systematically higher sputtering yields for ion energies of more than some 10 keV. The Thomas–Fermi function overestimates the nuclear stopping power. The agreement for a broad energy range can be improved by using a Kr–C potential (see e.g., [12]).

(ii) For $Z_1 > 50$, it was observed that the measured sputtering yields are significantly larger in comparison to the values predicted by the linear cascade theory by Sigmund. The contribution of non-linear or spike effects on sputtering yield is meaningful (see e.g., [13]).

(iii) The inelastic energy loss must be considered when calculating the sputtering yield (e.g., see [14]).

(iv) For ion energies near the threshold energy, Sigmund's assumption of a linear cascade regime (e.g., binary collisions) is probably no longer valid, because a solution of the Boltzmann transport equation required the condition $E \gg U_s$. Most likely, the description of the low energetic ion–solid-interaction by the application of a Thomas–Fermi potential is too inaccurate or/and the velocity distribution of the recoils is not completely isotropic.

(v) Another problem is that exact values for the surface binding energy U_s and the so-called material correction factor α are often unknown.

At low energies, typically below some hundred electron-volts, it is expected that the linear cascade theory of physical sputtering fails, i.e. the root-square dependence of the sputtering yield from the ion energy is no longer fulfilled. Minor deviations from the $\sqrt{E}$ – dependence can be verified. Several proposals in literature provide modified sputtering yield expressions for low ion energies [15].

It should be noted that in contrast to the expected sputtering yields according to Sigmund's linear cascade theory [8, 9] much higher sputtering yields could be observed for incident ion energy range (> some 10 keV) after bombardment with heavy, high-energy ions or molecules [16–19]. The spike sputtering or non-linear sputtering is given, if the sputtering yields are a lot larger than those predicted by the linear cascade theory. A heavy incident particle is particularly capable of generating a collisional spike (see Fig. 5.2), i.e. the majority of the atoms within this limited target volume are simultaneously in motion. An important consequence of this highly excited state is the intensive emission (sputtering) of target atoms and molecules from local regions around the point of ion impact. Several models have been developed to explain sputtering in this spike regime. These models are frequently based on the thermal evaporation from the spike (e.g., [20–22]), or characterized by shock wave propagation induced sputtering (see e.g., [23, 24]) as well as crater formation (see e.g., [25]).

5.4 Sputtering Parameters

5.4.1 Surface Binding Energy and Threshold Energy

Within the linear cascade theory, the surface binding energy is of particular importance for the calculation of the sputtering yield, because the sputtering yield is inversely proportional to its surface binding energy [see (5.21)]. Without better knowledge, the surface binding energy, U_s, is assumed to be correlated to the sublimation energy (or cohesive energy).

Consequently, adequate knowledge of the cohesive energy is of substantial importance for the determination of the sputtering yield. Cohesive energy (sublimation energy) is defined, on the one hand, as the energy per atom required to completely dissolve a solid into free atoms. On the other hand, it is defined as the energy required to form a vacancy in a solid [26, 27].

Here, the cohesive energy is determined by the energy which is necessary to remove surface atoms far away from the solid, i.e. to generate a permanent surface vacancy. In the case of two atoms interacting over a distance (pair potential), the cohesive energy is given by $\Delta H_{coh} = E_B \cdot Z_B/2$, where E_B is the binding energy and Z_B is the bulk coordination number. Then, the surface binding energy can be expressed by

$$U_s = Z_S E_B = 2\frac{Z_S}{Z_B}\Delta H_{coh}, \quad (5.23)$$

where Z_S is the surface coordination number (number of nearest neighbor atoms in the surface). For example, for a fcc (100) surface, $Z_B = 12$ and $Z_s = 8$ and thus $U_s = 1.33\Delta H_{coh}$. The values for the sublimation energies are tabulated and vary between 2 and 9 eV/atom. Sometimes, it is assumed that the energy to sputter an atom from the surface has to be 30–40% greater than the surface binding energy or enthalpy of formation of the bulk material [28].

This usually ignores the fact that the surface binding energy is dependent on the orientation of the crystalline surface structure and the topography. Since the sputtering yield is inversely proportional to the surface energy [see (5.19)], a dependence of the sputtering yield of single-crystalline surfaces on the crystallographic orientation in the order Y(111) > Y(100) > Y(110) for fcc materials and Y(110) > Y(100) > Y(111) for bcc materials can be expected. These sequential arrangements correlate with the planar atom density at the surface of the different crystalline structures (see also Table 5.2).

If the target material is composed of more than one element, e.g., A_xB_y, the heat of sublimation (cohesive energy) of all participating atoms must be considered. According to Eckstein and Biersack [29] the surface energy is given by

$$U_s \approx \frac{x\Delta H^A_{coh} + y\Delta H^B_{coh}}{x+y} - \Delta H_f^{A_xB_y} \quad (5.24)$$

and can be approximately determined by using the enthalpy of formation (heat of formation), ΔH_f, of components A and B as well as of the compound A_xB_y. Note that the heat of formation is small in comparison to the heat of sublimation.

The primary recoils after collision must have an energy which is sufficient to overcome the surface potential. This energy, the so-called threshold energy E_{th} can be interpreted as the kinetic energy for incident ions to generate an observable sputtering effect. It is known that the sputtering yield is proportional to the energy deposited in the near-surface region. The transferred energy by elastic collisions and with that also the sputtering yield are reduced with decreasing ion incident energy. For the simplest case, where the mass of the projectile M_1 is smaller than the mass of the target atom M_2 (Fig. 5.4), an atom at the surface or in the near-surface region is sputtered, if the minimum transferable energy after central elastic collision is identical or greater than the surface binding energy (surface escape barrier), i.e. $\gamma E \geq U_s$, where $\gamma = 4M_1M_2/(M_1 + M_2)^2 = 4\,M/(M+1)^2$ and $M = M_1/M_2$. Consequently, the threshold energy can be defined by [30]

$$E_{th} = U_s/\gamma, \quad (5.25)$$

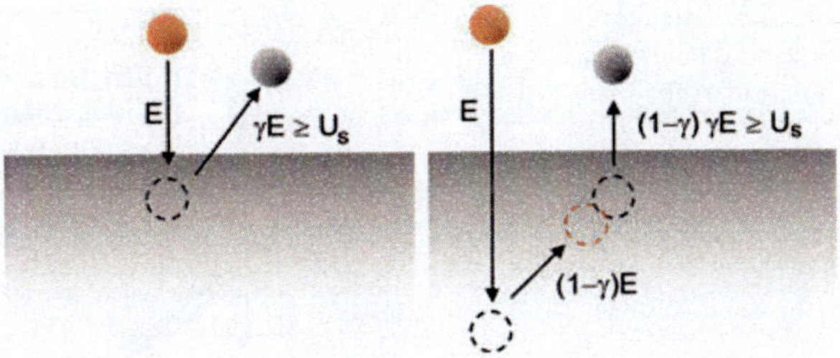

Fig. 5.4 Collision kinetics of energetic projectile impacts leading to ejection of surface atoms

which means that below the threshold energy no sputtering occurs. More commonly, an ion with energy E is elastically backscattered at greater target depths (see Fig. 5.4 on the right side). This backscattered atom with an energy of (1 − γ)E induces a sputter process with a surface atom by elastic collision. In this case, a sputtering is possible if

$$(1-)\gamma E \geq U_s \tag{5.26}$$

and consequently, the threshold energy is given by

$$E_{th} = U_s/(1-\gamma)\gamma. \tag{5.27}$$

It should be noted that under realistic conditions, on the one hand, collision sequences can occur due to variation of the scattering angles between 90° and 180° or small-angle scattering and, on the other hand, the electronic energy loss can no longer be neglected.

Numerous experimental studies have demonstrated that the simple relations of (5.25) and (5.26) are not capable to comprehensively determine the correlation between the relative threshold energy E_{th}/U_s and the ion-to-target mass ratio M_1/M_2 [31].

Therefore, analytical expressions for the relative threshold energy as function of the mass ratio were proposed, based on approximations of experimental or theoretical data. Some of the most frequently applied expressions have been proposed by Bohdansky et al. [31]

$$\frac{E_{th}}{U_s} = \begin{cases} \frac{1}{(1-\gamma)\gamma} & for\ M \leq 0.2 \\ 8M^{2/5} & for\ M > 0.2 \end{cases}, \tag{5.28}$$

by Matsunami et al. [32]

$$\frac{E_{th}}{U_s} = 1.9 + M + 0.134M^{-1.24}, \tag{5.29}$$

by Yamamura and Tawara [33]

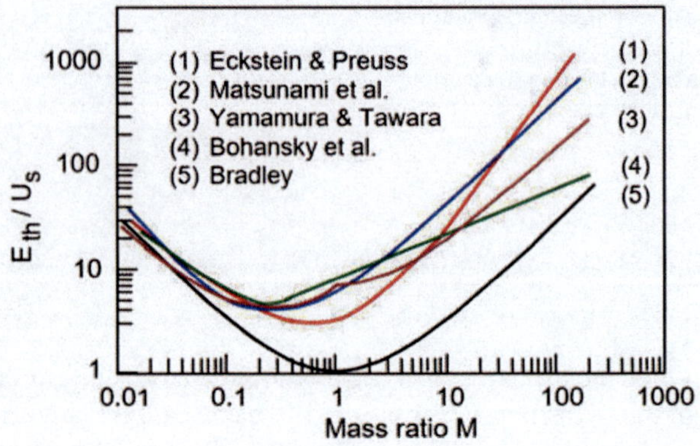

Fig. 5.5 The relative threshold energy E_{th}/U_s as function of the mass ratio M = M_1/M_2

$$\frac{E_{th}}{U_s} = \begin{cases} \frac{6.7}{\gamma} & for\ M \geq 1 \\ \frac{1+5.7M}{\gamma} & for\ M \leq 1 \end{cases}, \tag{5.30}$$

and by Eckstein and Preuss [34]

$$\frac{E_{th}}{U_s} = \frac{1}{\gamma}\left[\left(\frac{0.3198}{M}\right)^{-0.5279} + 1\right]. \tag{5.31}$$

The relative threshold energies, E_{th}/U_s, calculated by (5.28)–(5.31) are shown in Fig. 5.5 in dependence on the mass ratio M. For comparison, the relationship of the relative threshold energy by Bradley [30] is also presented. For low ion energies (< 1000 eV), Zalm [15] has proposed a simple relationship, $E_{th}/U_s \cong 8$, independent of ion species.

In the experiment, the threshold energy can be determined by extrapolating the sputtering yield for low energies, (see e.g., [35]). Typically, the threshold energy ranges between ten and hundred eV. For example, according to (5.27)–(5.30), the threshold energy E_{th} for the Ar ion bombardment of Si can be determined to be between 15.6 eV and 43.6 eV. It should be noted that the threshold energy is also dependent on the angle of incidence [36].

5.4.2 Material Correction Factor

The material correction factor was introduced by Sigmund [8, 9] (see 5.21) to describe the amount of energy, relative to the nuclear energy loss, which is deposited in the near-surface region into elastic collisions. This factor depends on the mass ratio M and the power law variable m for the power potential (see Sect. 2.1). Zalm [37] pointed out that the dependence of the factor α on the mass ratio M (= M_1/M_2) given by Sigmund can be well fitted by

$$\alpha = 0.15 + \frac{0.13}{M}. \tag{5.32}$$

Similar approximations of the form

$$\alpha = 0.15\left(1 + \frac{1}{M}\right)^{q}, \tag{5.33}$$

were suggested, where q is 0.7 [38] and 0.85 [39], respectively. Bohdansky [40] used a formula given by Chen [41] for the mass ratio M in the range between 0.5 and 10

$$\alpha = 0.3M^{-2/3}. \tag{5.34}$$

Later, Yamamura und Tawara [33] have proposed a semi-analytical expression for the material correction factor on the basis of approximations of experimental data

$$\alpha = \begin{cases} 0.249M^{-0.56} + 0.0035M^{-1.5} & for\ M \leq 1 \\ 0.0875M^{0.15} + 0.165M^{-1} & for\ M \geq 1 \end{cases}. \tag{5.35}$$

In Fig. 5.6 the material correction factor is shown as function of the mass ratio. It is obvious that the material correction factor is independent of the energy and the escape depth of the sputtered particles.

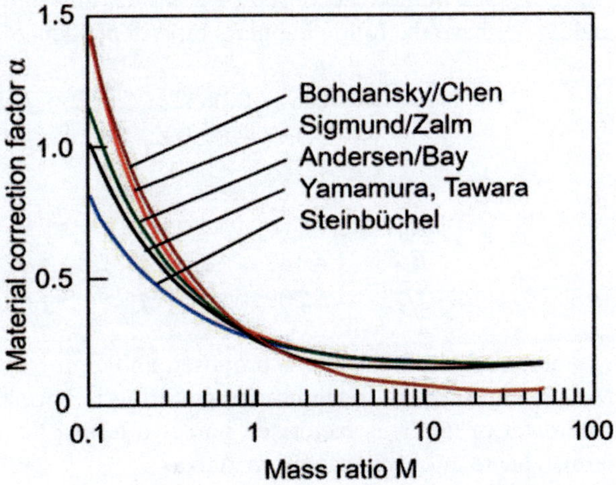

Fig. 5.6 Material correction factor α as function of the mass ratio $M = M_1/M_2$

5.5 Semi-empirical Approaches to Calculate the Energy-Depend Sputtering Yield

Besides Sigmund's analytical linear cascade theory, many semi-empirical formulae have been proposed in order to calculate the sputtering yield at normal incidence as a function of ion energy. Semi-empirical formulas, based on the expansion of (5.22), were created to fit the experimental data for a wide range of ion-target combinations. The aim is to find a universal relationship which precisely describes the energy dependence of sputtering yield.

The first empirical relations for estimating the sputtering yield at normal incidence have been published by Bohdansky et al. [42] and Matsunami et al. [43]. Bohdansky et al. [31, 42] distinguish between two sputtering regimes: the near-threshold regime and the linear cascade regime. The proposed analytic expressions are only applicable for any ion-target combinations. A simpler empirical formula was predicted by Matsunami et al. [11, 43]. This formula for high-energy light-ion sputtering, the so-called first Matsunami formula, is taking into account the threshold energy E_{th} (see Sect. 5.4.1) in the original Sigmund relationship,

$$Y(E) = 4.2\frac{F_D(E)}{NU_s}\left[1-\sqrt{\frac{E_{th}}{E}}\right] = 4.2\frac{\alpha}{U_s}S_n(E)\left[1-\sqrt{\frac{E_{th}}{E}}\right], \tag{5.36}$$

where N is the atomic number density, $S_n(E)$ is the nuclear stopping cross-section (see 3.23) and numerical constant 4.2 has the unit [nm^{-2}] [see (5.21)]. The values α and E_{th}/U_s have the following mass ratio dependence

$$\alpha = \begin{cases} 0.1019 + 0.0842M^{-0.9805} & for\ M > 2.163 \\ -0.4137 + 0.6092M^{-0.1708} & for\ M < 2.163 \end{cases} \tag{5.37}$$

and

$$\frac{E_{th}}{U_s} = \begin{cases} 4.143 + 11.46M^{0.5004} & for\ M > 3.115 \\ 5.809 + 2.791M^{-0.4816} & for\ M < 3.115. \end{cases} \tag{5.38}$$

Yamamura et al. [44] have proposed an empirical relation, the so-called second Matsunami formula, for light-ion sputtering at normal incidence, where the fitting parameter of (5.35) is factorized into two terms, the inelastic term and the elastic term. This relationship can be written as

$$Y(E) = 4.2\frac{\alpha}{U_s}\frac{S_n(E)}{1+0.35s_e(\varepsilon)}\left[1-\sqrt{\frac{E_{th}}{E}}\right]^2, \tag{5.39}$$

where $s_e(\varepsilon)$ is the reduced electronic stopping power [(3.28)]. The best-fit values for α and E_{th}/U_s are

$$\alpha = \begin{cases} 0.10 + 0.155M^{-0.73} & for\ M > 50 \\ 0.321 + 0.0332M^{-1.1} & for\ M < 50 \end{cases} \tag{5.40}$$

and

$$\frac{E_{th}}{U_s} = 1.5\frac{\left(1.5 + 1.38M^h\right)^2}{\gamma}, \tag{5.41}$$

where h = 0.834 for $M_2 > M_1$ and h = 0.18 for $M_2 < M_1$.

Based on the sputtering formula given by Sigmund, (5.22), Strydom and Gries [45] have also developed an analytic expression for the sputtering yield. They determined empirically the values for the reduced nuclear power stopping, $s_n(\varepsilon)$, and α.

Another semi-empirical expression for the sputtering yield after normal ion bombardment has been given by Bohdansky [40] for low ion energy sputtering, because there the calculated low-energy sputtering yields are usually overestimated due a threshold effect at the surface. By replacing the Thomas–Fermi nuclear stopping cross-section in this original formula with the Kr–C stopping cross-section $s_n^{KrC}(\varepsilon)$, a new fit formula, the so-called revised Bohdansky formula,

$$Y(E) = Qs_n^{KrC}(\varepsilon)\left[1 - \left(\sqrt{\frac{E_{th}}{E}}\right)^{2/3}\right]\left(1 - \sqrt{\frac{E_{th}}{E}}\right)^2 \tag{5.42}$$

has been obtained [40, 46, 47], where the relation (5.29) is used for the threshold energy and Q is a scaling factor. On the basis of this relationship, the sputtering yield can be calculated in dependence on the mass and atomic numbers of the involved ions and target atoms, respectively (see table in [32]).

Yamamura and Tawara [33] have predicted an improved formula, called the third Matsunami formula, for the ion-induced sputtering yields from monatomic solids at normal incidence with the objective of describing the energy dependence of sputtering yields as precisely as possible. In this empirical relationship, the Lindhard approximation for the reduced nuclear stopping power, $s_n(\varepsilon)$, and the reduced electronic stopping power, $s_e(\varepsilon)$, are used, and m = 0.4 is assumed. Then, this empirical formula is given as

$$Y(E) = 4.2\frac{Q\alpha}{U_s}\frac{S_n(E)}{1 + \Gamma k_{LS}\varepsilon^{0.3}}\left[1 - \sqrt{\frac{E_{th}}{E}}\right]^s, \tag{5.43}$$

where E is the ion incident energy, $\Gamma = W/\left[1 + (M_1/7)^3\right]$, k_{LS} is the dimensionless Lindhard electronic stopping coefficient (see 3.28) and $S_n(E)$ is the nuclear stopping cross section (see 3.21). Q, W and s are functions of the target atomic number and are tabulated by Yamamura and Tawara [33]. E_{th} and α are given by relations (5.30) and (5.35), respectively.

The third Matsunami formula as well as the revised Bohdansky formula represent experimental results within an error range of 30% [47]. Especially in cases of high incident energies and energies in the threshold energy region, deviations between the calculated and experimental values of the sputtering yields can be observed. Consequently, Eckstein and Preuss [34] have proposed a fit formula which gives a better description of the available data of the sputtering yield

$$Y(E) = q s_n^{KrC}(\varepsilon)\left(\frac{E_{th}}{E} - 1\right)^{\mu} \Big/ \left[\frac{\lambda}{w(\varepsilon)} + \left(\frac{E_{th}}{E} - 1\right)^{\mu}\right], \tag{5.44}$$

where ε is the Lindhard reduced energy (see 3.13) and $w(\varepsilon) = \varepsilon + 0.1728\sqrt{\varepsilon} + 0.008^{0.1504}$. The three parameters q, μ and λ are tabulated [36]. The assigned accuracy of the experimental values is about 10%.

The comparison with experimental results in Fig. 5.7 demonstrates that as well as providing similar values for the sputtering yield under normal incidence, the proposed semi-empirical equations reproduce the experimental results very well with regard to measuring precision.

For the calculation of the sputtering yield at low ion energies, Falcone [50] has also used the concept of the threshold energy. In contrast to Sigmund [8], it is assumed that atoms are only sputtered from the first monolayer. Then, the sputtering yield for low ion energies is given by

$$Y_F = Y\left(1 - \frac{E_{th}}{E}\right)^2, \tag{5.45}$$

where Y is the sputtering yield defined by (5.22). The following figure, Fig. 5.8, shows a comparison of the sputtering yield calculated by the Sigmund standard

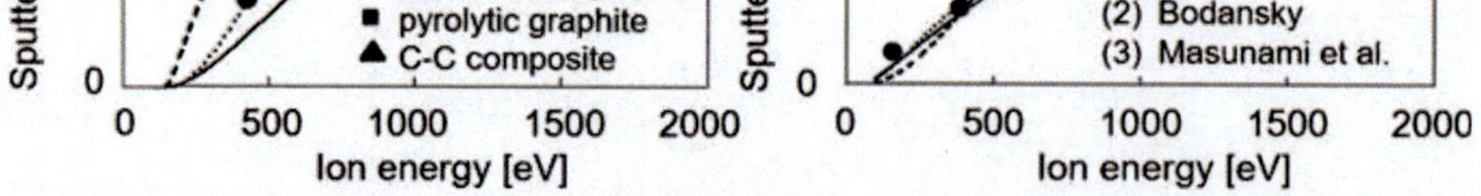

Fig. 5.7 Energy dependence of the sputtering yield of several carbon modifications and molybdenum after low-energy Xe ion bombardment under normal incidence condition at room temperature. The sputtering yields are calculated by applying the semi-empirical formulae (dashed line—Eckstein and Preuss [34], dotted line—revised Bohdansky formula [40] and solid line—third Matsunami formula [43]) and compared with experimental results (black points, taken from [48, 49])

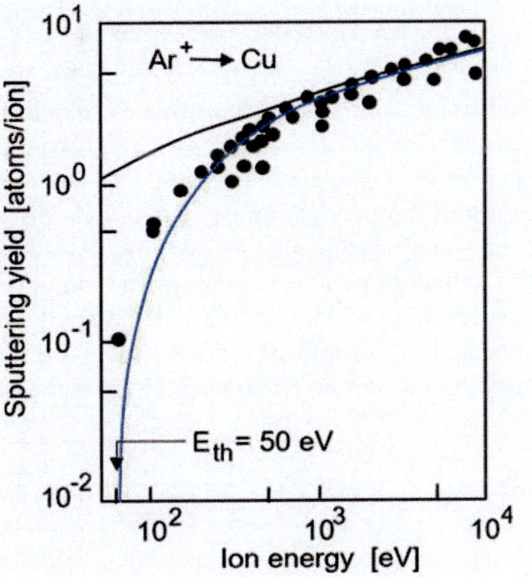

Fig. 5.8 Sputtering yield of Cu bombarded by Ar ions as function of low ion energies. The black line and the blue line give sputtering yield according to the Sigmund theory and the Falcone correction, respectively. The points represent measured results given in a report by Matsunami et al. [43] (adapted from [50])

equation, (5.21) and with the correction given by Falcone, (5.45), for the low ion energy region. It has to be said that this correction significantly improved the standard Sigmund relationship for low-energy region.

It should be noted that the semi-empirical relationships discussed above also reflect the sputter yield at low incidence energies very well.

In addition to semi-empirical approaches to determine the sputtering yield, binary collision Monte Carlo or molecular dynamics codes are available and widely used to simulate the sputtering process and thereby calculate the sputtering yield for numerous ion-target combinations (details see Sect. 3.6).

5.6 Dependence of the Sputtering Yield on the Angle of Ion Incidence

As is generally known, the sputtering yield for amorphous materials is a function of both the angle of ion incidence θ and the azimuthal angle φ (see e.g., [36]). Since a two-dimensional representation of $Y = Y(\theta)$ is usually preferred, the erosion rate is given by

$$v = \frac{JY}{N} \cos\theta = v_o \cos\theta \tag{5.46}$$

(c.f. 5.2), i.e. v is the local erosion rate of the surface along the normal direction so the surface height reduction is $dh/dt = v/\cos\theta = v_o$. Experimental studies have shown a typical dependence of the sputtering yield on the angle of ion incidence (Fig. 5.9). With increasing ion incident angle ($\theta > 0$, off-normal sputtering) or the gradient $dY/d\theta$, the ion-induced collision cascades are developing ever closer to the surface. As a result of collisions between the projectile ion or recoiling target atoms with surface atoms, more near-surface atoms will be sputtered until a maximum incidence angle is reached ($\theta_p \approx 60°$-$80°$). This maximum angle was derived by Lindhard [51] and used by Stewart and Thompson [52] in the following form

$$\theta_p = \frac{\pi}{2} - \sqrt{5\pi a_o^2 N^{2/3} \frac{Z_1 Z_2}{\left(Z_1^{2/3} + Z_2^{2/3}\right)} \frac{E_R}{E}}, \tag{5.47}$$

where a_o is the Bohr radius (= 0.053 nm), Z_1 and Z_2 are atomic numbers of the ion and the target atom, respectively, $E_R = 13.6$ eV is the Rydberg energy, N is the atomic number density of the bombarded substrate and E is the ion energy.

After this maximum, the sputtering yield decreases when increasing the angle of ion incidence because screening effects of neighboring near-surface atoms prevent incident atoms to enter the surface. At extremely large incident angles, almost all incident ions are reflected without generating sputtered atoms in the near-surface region. This means that for large incidence angles the sputtering yield quickly drops to zero at $\theta = 90°$.

For crystalline materials, the sputtering yield can have more than one maximum value, because the faces of the crystalline material are oriented at different angles to the incident ion beam.

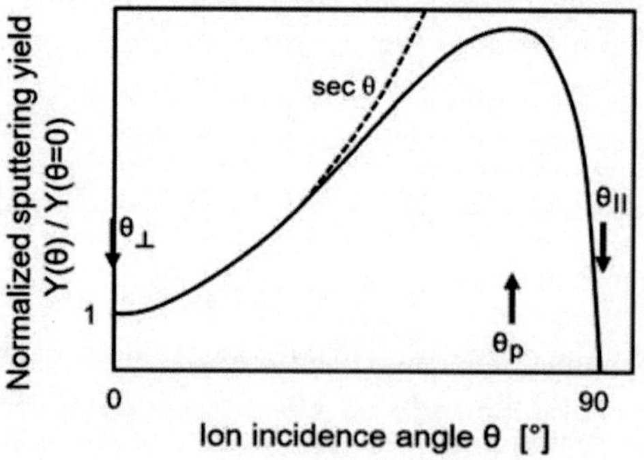

Fig. 5.9 Schematic picture of the normalized sputtering yield versus the angle of ion incidence. θ_p is the incidence angle, where the sputtering yield is a maximum, $\theta_\perp$ is the incidence angle at normal ion bombardment and θ_{II} is the angle of glancing ion bombardment

Stewart and Thompson [52] have also shown that the sputtering yield is approximately proportional to the secant of the angle of ion incident (dashed line in Fig. 5.9), i.e.

$$Y(\theta) \propto \sec(\theta) \equiv \frac{1}{\cos(\theta)} \tag{5.48}$$

for incident angles up to about 60°– 65° and ion energies up to about 100 keV.

Another possibility to describe the dependence of the sputtering yield on the angle of ion incidence was demonstrated by Ducommun et al. [53]. These authors fit this dependence by a sixth-power polynomial to

$$Y(\theta) = \sum_{n=1}^{n=6} a_n \cos^n \theta, \tag{5.49}$$

where the parameters a_n must be determined by experimental measurements of the sputtering yield versus the incident angle. Then, the erosion rate can be given by (c.f. 5.2)

$$v = \frac{J}{N} \sum_{n=1}^{n=6} a^n \cos^n \theta. \tag{5.50}$$

According to Sigmund [8], the angle dependence of the sputtering yield for non-grazing angles ($0 < \theta < 70°$) can be described by

$$Y(E, \theta) = Y(E, 0) \cos^{-f}(\theta), \tag{5.51}$$

where Y(E, 0) is the sputtering yield at normal ion incidence and exponent f is a parameter which should be 5/3 for $M_2/M_1 \leq 3$ and 1 for $M_2/M_1 > 5$. The parameter f can be experimentally estimated by drawing $\ln[Y(E,\theta)/Y(E,0)]$ versus $\cos^{-f}(\theta)$ and taking the slope which is equal to – f.

Yamamura [54] has extended the Sigmund formula for the angle dependence of the sputtering yield over the whole angular range and has proposed a formula for the angular dependence of light and heavy ion sputtering to

$$Y(E, \theta) = Y(E, 0) \cos^{-f}(\theta) \exp\left[-\Sigma\left(\cos^{-1} - 1\right)\right], \tag{5.52}$$

where the parameters f and Σ are determined by fitting the available experimental data. A maximum sputtering yield is given for the angle of ion incidence

$$\theta_p = \cos\left(\frac{f}{\Sigma}\right), \tag{5.53}$$

so that the (5.52) can be assigned to

$$Y(E, \theta) = Y(E, 0) \cos^{-f}(\theta) \exp\left\{f\left[1 - (\cos\theta)^{-1}\right]\cos\theta_p\right\}. \tag{5.54}$$

The best agreement with (5.54) for incidence angle dependent sputtering yields obtained in experiments with light ion bombardment can be found by

$$f = U_s^{1/2}\left(0.94 - 1.33 \cdot 10^{-3} M^{-1}\right). \tag{5.55}$$

The incidence angle θ_p for the maximum of the sputtering yield is given by

$$\theta_p = \frac{\pi}{2} - \frac{a_L N^{1/3}}{\sqrt{2\varepsilon (U_s/\gamma E)^{1/2}}}, \tag{5.56}$$

where U_s is the surface binding energy, γ is the transfer energy efficiency coefficient, a_L is the Lindhard screening length [see Table 2.1], ε is the reduced energy and N is the atomic number density of the target material. This formula also includes the sharp decreasing of the sputtering at large incident angles [c.f. (5.47)]. A particularly good agreement with the published experimental results was achieved for light ion bombardment.

For heavy ion bombardment, the formation of a collision cascade must be considered. For this purpose, Yamamura [40] used an extended version of the general (5.52) for angular sputtering,

$$Y(E,\theta) = Y(E,0)\cos^{-f}(\theta)\exp\left[-\Sigma(\cos\theta)^{-1} - 1)\right]\left[\frac{1 - \sqrt{E_{th}/E}\cos\theta}{1 - \sqrt{E_{th}/E}}\right], \tag{5.57}$$

where the threshold energy is given by

$$\frac{E_{th}}{U_s} = \frac{1.5}{\gamma}\left(1 + 0.138 M^x\right)^2 \text{ with x} = 0.834 \text{ for M} < 1 \text{ and x} = 0.18 \text{ for M} > 1. \tag{5.58}$$

Another empirical formula was proposed by Eckstein and Preuss [34], which leads to a more detailed analysis of the angular sputtering yield especially for low mass ratios and energies near the threshold energy. The fit formula

$$Y(E,\theta) = Y(E,0)\left\{\cos\left[\left(\frac{\theta}{\theta_o}\frac{\pi}{2}\right)^c\right]\right\}^{-f}\exp\left\{b\left(1 - 1\Big/\cos\left[\left(\frac{\theta}{\theta_o}\frac{\pi}{2}\right)^c\right]\right)\right\} \tag{5.59}$$

is based on three parameters f, b and c, which are tabulated for different ion-target combinations in [32], where

$$\theta_0 = \pi - \arccos\sqrt{\frac{1}{1 - \sqrt{E_{th}/U_s}}} \geq \frac{\pi}{2}. \tag{5.60}$$

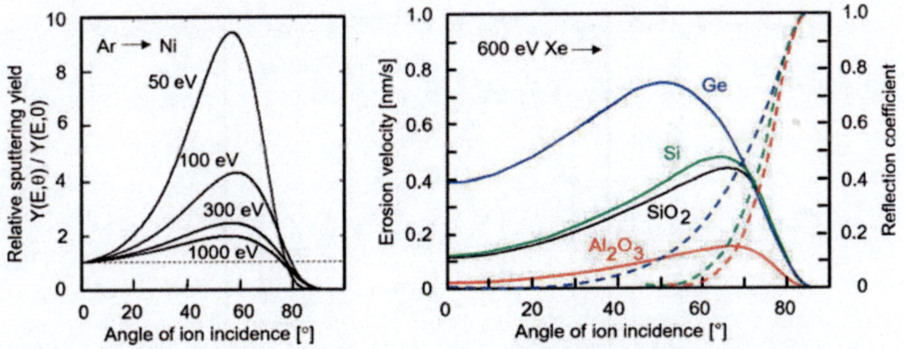

Fig. 5.10 Left: The incident angle dependence of the sputtering yield for nickel after low-energy Ar ion bombardment with energies between 50 and 1000 eV. The relative sputtering yields were calculated by the Eckstein-Preuss empirical formula. Right: The calculated erosion velocity (solid lines) and reflection coefficients (dashed lines) for different materials as function of the angle of ion incidence after 600 eV Xe ion bombardment and an ion current density of 300 $\mu A/cm^2$ [55]

According to (5.59), the relative sputtering yield for nickel after Ar ion bombardment for four low ion energies between 50 eV and 1000 eV in dependence on the angle of incidence is calculated using (5.59) and shown in Fig. 5.10 (left). In general, the relative sputtering yields decrease with increasing ion energy. Consequently, the erosion velocity given by (5.2) is also depending on the ion incidence angle θ and is given by (5.46). As an example, the erosion velocity of different materials as function of the angle of ion incidence after 600 eV Xe ion bombardment is shown in Fig. 5.10 (right). Initially, the erosion rate increases to a maximum erosion rate between 50° and 70°. For larger angles, the erosional velocity decreases and the reflection probability increases significantly.

It should be noted that for crystalline material, the sputtering yield is further influenced by orientation of the ion beam with respect to the surface normal of a single crystal surface or with respect to several oriented grains of polycrystalline materials. Thus, a fraction of ions is capable of channeling in deeper regions of the crystal (channeling effect) and cannot contribute to the sputtering yield. Consequently, it can be observed for ion energies higher than some kilo-electron-volts that (i) the sputtering yield varies depending on the crystalline orientation [3] and (ii) differently oriented faces of polycrystalline materials become eroded with different erosion velocities [56]. For example, Southern et al. [57] have determined the sputtering yield of differently oriented Cu single crystals (fcc) as function of the Ar ion energy between 1.5 keV and 5 keV. They found the lowest sputtering yield for the [110] oriented surface and the highest sputtering yield for the [111] oriented surface, independent of the ion energy. The sputtering yield of GaAs (110) single crystals after low-energy Ar ion bombardment is shown in Fig. 5.11, where the Ar beam was directed along three different directions [110], [111] and [100] [58]. The result of $Y_{[110]} < Y_{[111]} < Y_{[100]}$ can be interpreted on the basis of two geometric arguments. First, at off-normal incidence (ion bombardment along [111] and [100]) more

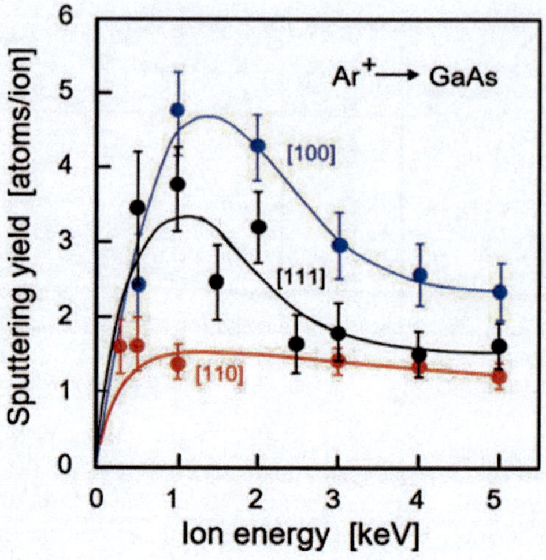

Fig. 5.11 Sputtering yield of GaAs as function of the Ar ion energy in dependence on the orientation of the crystal surface. The figure is taken from [58] and modified

ion-target atom interactions take place near the surface. Second, the lowest and the highest density of atomic columns are given for ions moving along [110] and [100] directions, respectively.

According to the channeling model (details see Sect. 10.5.1.5) by van Wyk and Smith [59] and by Dobrev [60], it can be expected that crystal grains with the lowest area packing density perpendicular to the ion flux direction are preferentially conserved because of their low sputtering yield, i.e. the most open lattice direction aligned with the ion beam has the lowest nuclear energy loss. Consequently, the sputtering yield is proportional to the packing density (and inverse proportional to the surface binding energy, see 5.21). Table 5.2 demonstrates the correlation between the expected sputtering yield and the packing density for the most common crystal structures. In many experimental studies, the correlation $Y_{(111)} > Y_{(100)} > Y_{(110)}$ for fcc crystal structures and $Y_{(110)} > Y_{(100)} > Y_{(211)}$ for bcc structures could be confirmed. It must be mentioned that the application of this correlation to polycrystalline surfaces ultimately leads to the well-known orientation-dependent sputter erosion of particular grains of polycrystalline surfaces (see Sect. 6.3.2).

The models proposed by Odintsov [61], Martynenko [62] and Onderdelinden [63] provide further detailed explanations for the dependence of the sputtering yield on the ion energy and the crystallographic orientation.

Table 5.2 Relative density of crystal planes of the most common crystal structures

fcc-structure (e.g., Al, Ni, Pd, Pt, Cu, Ag, Au)									
Plane	(111)	(100)	(110)	(311)	(210)	(211)	(221)	(310)	(320)
Rel. density to (111)	1.000	0.866	0.612	0.522	0.387	0.354	0.289	0.274	0.240
Expected Sp.-yield	$Y_{(111)} > Y_{(100)} > Y_{(110)} > Y_{(311)} > Y_{(210)} > Y_{(211)} > Y_{(221)} \approx Y_{(310)} > Y_{(320)}$								
bcc-structure (e.g., Fe, W, Mo, Cr, Ta, Nb)									
Plane	(110)	(100)	(211)	(310)	(211)	(210)	(221)	(311)	(320)
Rel. density to (110)	1.000	0.707	0.578	0.447	0.409	0.316	0.236	0.213	0.196
Expected Sp.-yield	$Y_{(110)} > Y_{(100)} > Y_{(211)} > Y_{(310)} > Y_{(211)} > Y_{(210)} > Y_{(221)} \approx Y_{(310)} \approx Y_{(320)}$								
hcp-structure (e.g., Zr, Hf, Os, Co, Cd, Zn), r = c/a, where c and a are lattice constants									
Plane	(0001)	$(10\bar{1}0)$	$(10\bar{1}1)$	$(10\bar{1}2)$	$(11\bar{2}2)$				
Rel. density to (0001)	1.000	$3/2r$	$\sqrt{3/(4r^2+3)}$	$\sqrt{3/(4r^2+12)}$	$1/2\sqrt{r^2+1}$				
Rel. density for r = 1.6	1.000	0.938	0.476	0.367	0.265				
Expected Sp.-yield	$Y_{(0001)} > Y_{(10\bar{1}0)} > Y_{(10\bar{1}1)} > Y_{(10\bar{1}2)} > Y_{(10\bar{2}2)}$								

5.7 Energy and Angular Distributions of the Sputtered Particles

Apart from using total sputtering yield to characterize sputtering processes, the differential sputtering yield is an important parameter for many applications. For example, the description of the re-deposition or reflection of sputtered particles is usually based on a detailed analysis of their angular and energy distributions. For an extensive overview about the energy and angular distributions see [64] and [65].

Based on the linear cascade model by Sigmund [9], the flux of emitted target atoms from the surface, J_s, can be described by (5.12). Only when the energy of this target atom is larger than the surface binding energy, U_s, these atoms are capable to leave the target surface. If the conservation of momentum is valid, it can be expected that the normal component of the momentum is reduced while the tangential component is unchanged (Fig. 5.12). Therefore, the following relationships can be obtained

$$E_{sp} \cos^2 \theta_{sp} = E \cos^2 \theta_o - U_s \tag{5.61}$$

and

$$E_{sp} \sin^2 \theta_{sp} = E \sin^2 \theta_o. \tag{5.62}$$

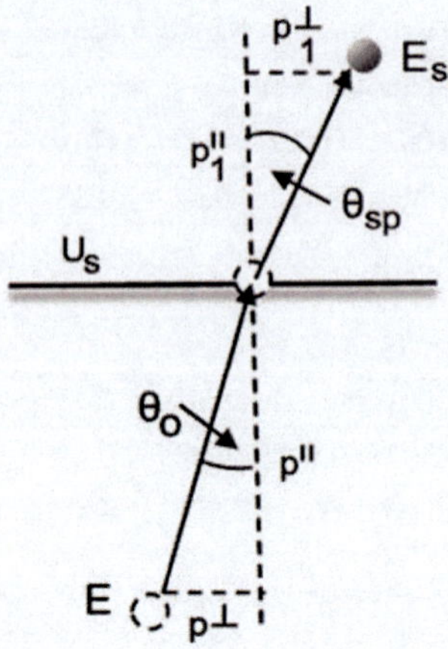

Fig. 5.12 Schematic diagram of the changes of the momentum for sputtered atoms (after Sigmund [9]). The picture demonstrates the effect of surface potential barrier. The normal and the tangential components of momentum p are shown (θ_o is the in-side incidence angle in contrast to the out-side incident angle θ)

Therewith the energy and the direction (emission angle) of the sputtered atom can be expressed by

$$E_{sp} = E - U_s \tag{5.63}$$

and

$$\cos\theta_{sp} = \sqrt{\frac{E\cos^2\theta_0 - U_s}{E_{sp}}}. \tag{5.64}$$

From (5.64) follows that $E\cos\theta_o d^2\Omega_o = E_{sp}\cos\theta_{sp} d^2\Omega_{sp}$, Based on this result and equation (5.14), the differential sputtering yield according to Sigmund is given by

$$d^3Y = F_D(E,\theta,0)\frac{\Gamma_m}{4\pi}\frac{E_{sp}dE_{sp}}{\left(E_{sp}+U_s\right)^2|dE/dx|_{E=E_{sp}+U_s}}\cos\theta_{sp}d^2\Omega_{sp}. \tag{5.65}$$

It should be noted that the ratio of the flux of recoils at the surface to the incident ion flux, (J_s/J), is equivalent to sputtering yield Y.

By inserting the relation for the nuclear stopping cross-section, given by (3.4), for $|dE_o/dx|$ in the nominator of (5.65), the differential sputtering yield can be expressed by

$$\frac{d^3Y}{dE_{sp}d^2\Omega_o} = F_D(E,\theta,0)\frac{\Gamma_m}{4\pi}\frac{1-m}{NC_m}\frac{E_{sp}}{\left(E_{sp}+U_s\right)^{3-2m}}\cos\theta_{sp}$$

$$\approx \frac{E_{sp}}{\left(E_{sp}+U_s\right)^{3-2m}}\cos\theta_{sp}, \tag{5.66}$$

i.e. the angular distribution has a cosine-like shape. The energy distribution of the sputtered particles is given by

$$N(E) \propto \frac{E_{sp}}{\left(E_{sp}+U_s\right)^{3-2m}}, \tag{5.67}$$

for a fixed incidence angle. Then, for low-energy ion sputtering (m = 0), the energy distribution of the sputtered particles is proportional to $E_{sp}/(E_{sp}+U_s)^3$.

Equation (5.66) gives the differential energy and angular distribution of sputtered particles with emission energy E_{sp} into the solid angle Ω around the polar angle θ of the emission. This relation is valid as long as the assumption of an isotropic cascade is justified [66]. Based on this equation, the differential angular and energy distribution of the sputtered particles can be looked at separately (see Sect. 5.2).

Thompson [67] has achieved a similar expression for the energy distribution of the sputtered particles. He has found a relation for the energy distribution of the sputtered particles on the assumption that isotropic recoil cascades of low-energy recoil atoms are created by primary recoil particles. The distribution of the kinetic energy of sputtered particles into $d\Omega$ at the polar (emission) angle of these particles is given by

$$F(E)dEd\Omega \propto \frac{E_{sp}}{\left(U_s+E_{sp}\right)^{1-2m}}\left[1-\left(\frac{U_s+E_{sp}}{\gamma E}\right)^{1-m}\right]\cos\theta_{sp}dEd\Omega, \tag{5.68}$$

where γ is given by Eq. (2.28), E and E_{sp} are the incident ion and sputtered atom energy, respectively. For m = 0 (hard sphere collision) and the maximum transferable energy $\gamma E = T_{max}$ being significantly higher than the energy of the sputtered particles ($\gamma E \gg U_s$), the so-called Thompson energy spectrum can be obtained to

$$F(E,\theta)dEd\Omega \propto \frac{E_{sp}}{\left(U_s+E_{sp}\right)^3}\cos\theta_{sp}dEd\Omega. \tag{5.69}$$

This result was also derived by Sigmund [c.f. (5.66)].

5.7.1 Numerical Analysis of the Energy Distribution

Meyer et al. [68] have calculated the energy distributions of sputtered Nb and Cu atoms ejected from amorphous targets under low-energy Ar bombardment. Their formula is based on the interaction of elastic (hard) spheres and can be used to calculate the subsequent energy loss of the ejected atoms due to collision in the sputtering gas. The energy of the sputtered particles is determined to

$$E_{sp} = \left(E_{sp,0} - k_B T_g\right) \exp(n\xi) + k_B T_g, \tag{5.70}$$

where $E_{sp,0}$ is the energy of the sputtered particle as it leaves the sputter target (i.e. before the first collision with a sputter gas atom), T_g is the temperature of the sputter gas, n is the collision number related to the distance Λ traveled by the sputtered particle and given by $n = \Lambda/\lambda$ and $\xi = \ln\left(E_{sp}^{after}/E_{sp}^{before}\right)$, the ratio of the energies before and after a collision (details see Sect. 10.1.2.2). The mean free path $= k_B T_g/p\sigma$ is given by I.3 (see Appendix I), where p is the total gas pressure of the sputter gas and collision cross-section $\sigma = \pi(r_g + r_{sp})^2$ and r_g and r_{sp} are the Lennard–Jones radii of the sputter gas and the sputtered target atoms, respectively. The ratio of the energies before and after a collision is given by (10.12).

Numerical procedures to calculate the energy distribution of sputtered particles are based on the Monte Carlo simulation method. According to the Thompson theory of atomic collision cascade, the kinetic energy of the sputtered atoms, E_{sp}, can be described by a normalized probability density function [69]

$$N\left(E_{sp}\right) = \begin{cases} 2\left(1 + \frac{U_s}{\gamma E}\right)^2 \frac{U_s E_{sp}}{\left(U_s + E_{sp}\right)^3} & if\, 0 \le E_{sp} \le \gamma E \\ 0 & if\, E_{sp} > \gamma E \end{cases}, \tag{5.71}$$

where E is the energy of the incident ion, $\gamma = 4M_I M_{sp}/\left(M_I + M_{sp}\right)^2$ is the transfer energy efficiency factor and M_I and M_{sp} are the masses of the ion and target (sputtered) species, respectively.

Various authors have determined the mean energy of the sputtered atoms (details see Appendix J) and found very similar expressions given by

$$\left\langle E_{sp}\right\rangle = 2U_s\left[\ln\left(\frac{E}{E_{th}}\right) - 3/2\right] \text{ or } \left\langle E_{sp}\right\rangle = 2U_s\left[\ln\left(\frac{T_{max}}{U_s}\right) - 3/2\right], \tag{5.72}$$

where in the latter case a simplified energy distribution with $F\left(E_{sp}\right) = 2U_s E_{sp}/\left(U_s + E_{sp}\right)^3$ [c.f. (5.71)], was assumed [T_{max} is the maximum transferred energy, see (2.26)].

The energy distribution of sputtered particles according to (5.69) shows a peak at $U_s/2$ and a characteristic falloff towards high energies as E^{-2}. Figure 5.13 (left) shows the normalized energy distribution of sputtered Ag atoms after perpendicular low-energy Ar ion bombardment. This distribution has a characteristic maximum at $U_s/2$ in its low-energy part, which is independent of the ion species, incident angle and ejection angle. The high-energy part of this distribution asymptotically decreases at higher emission energies ($E_{sp}{}^{-2}$ tail). In Fig. 5.13 (right), the calculated energy distribution of aluminum atoms for normal incidence as

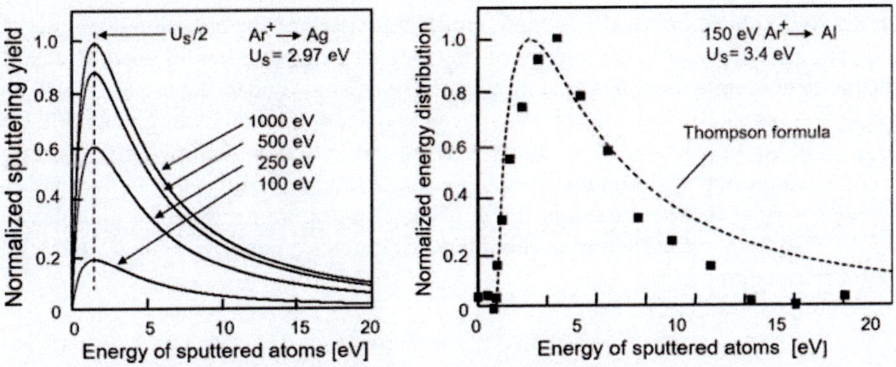

Fig. 5.13 Left: Normalized sputtering yield of Ag after perpendicular Ar ion bombardment at different ion energies calculated with the Thompson formula. Right: Measured energy distribution of sputtered Al atoms after 150 eV Ar ion bombardment (black squares). For comparison the distribution calculated by the Thompson formula is shown (figure is adapted from [70])

well as emission measured by laser induced fluorescence spectroscopy is depicted [70]. For comparison, the Thompson distribution is also shown (dashed line). On the one hand, the breakdown of the high energy fall-off is clearly observable. But, on the other hand, the Thompson formula only approximately describes the high-energy tail of the distribution. Especially for light ions and sub-keV incident energies these deviations from the Thompson formula, which are caused by the dominance of single knock-on processes (Fig. 5.2), are known. Therefore, different modifications of the Thompson formula were proposed. Falcone [50] has developed the following relationship

$$Y(E)dE \propto \frac{E_{sp}dE}{(E_{sp}+U_s)^3} \ln\left(\frac{\gamma E}{E_{sp}+U_s}\right), \tag{5.73}$$

which characterizes the energy distribution of the sputtered atoms for low-energy light ion bombardment of heavy target materials. The derivation of this relation is based on the assumption that the emitted atoms are primary recoils as a result of elastic collision processes. Similarly, Kenmotsu et al. [71] have derived the relation

$$Y(E)dE \propto \frac{E_{sp}dE}{(E_{sp}+U_s)^{8/5}} \left[\ln\left(\frac{\gamma(1-\gamma)E}{E_{sp}+U_s}\right)\right]^2 \tag{5.74}$$

as well, for the case of the light-ion bombardment. It should be noted that in contrast to the Thompson formula, (5.69), the last two relations [(5.73) and (5.74)] are functions of the incident ion energy.

According to the relations by Sigmund [(5.65)] and Thompson [(5.68)], the energy of the sputtered particles is dependent on the primary ion energy as well as the incidence angle. Based on the considerations by Sigmund and Thompson, Goehlich et al. [70, 72] have proposed a simple model to determine the emission angle θ_{sp} of the sputtered atoms and the energy of the sputtered recoil atoms in dependence of the incidence angle (here θ is the out-side emission angle and θ_o the in-side emission angle (c.f. Fig. 5.12). The energy transferred to the target atom T was given in (2.27) to $T = \gamma E\cos^2\varphi_o$, where φ_o is the recoil angle of the target atom. This recoiling angle is connected with the incident angle θ and the angle θ_o by the angle relation $\varphi_o = \pi - \theta_o - \theta$, where the possible emission angles are restricted to the interval $\pi/2 - \theta < \theta_o < \pi/2$. Therefore, the relationship by Sigmund, given by (5.62), can be transcribed to

$$(T - U_s)\cos^2\theta_{sp} = T\cos^2\theta_0 - U_s. \tag{5.75}$$

Then, the emission angle of the sputtered atoms and the energy of the direct recoil atom E_1 (see Chap. 2) are given by

$$\cos\theta_{sp} = \sqrt{\frac{\cos(\theta_o + \theta)^2(cos\theta_o)^2 - U_s/\gamma E}{\cos^2(\theta_o + \theta)^2 - U_s/\gamma E}} \tag{5.76}$$

and

$$E_{sp} = \gamma E\cos(\theta_o + \theta)^2 - U_s, \tag{5.77}$$

respectively. This equation allows to represent the energy of the sputtered particles as function of the emission angle θ_{sp}.

5.7.2 *Spatial Differential Sputtering Yield*

The spatial differential sputter yield $dY/d\Omega$ provides the number of sputtered atoms per incident particle in a particular direction (solid angle Ω) in dependence of the polar angle α and the azimuthal angle φ. According to Sigmund's theory [(5.66)], the spatial or angular distribution of the sputtered particles is proportional to the cosine of the incident angle, i.e. $dY/d\Omega \propto \cos\theta$ (see Appendix K). The cosine law results from an isotropic angular distribution of recoil atoms in the formed cascade (Fig. 5.14a). For low incident ion energy, a collision cascade is generated, which however not completely developed. This causes a non-isotropic angular distribution of sputtered particles with an under-cosine ($n < 1$) or heart-shaped distribution ($n \ll 1$). For high incident energies, the angular distribution of the sputtered particles tends to be a cosine-like distribution ($n > 1$). For even higher incident energies an over-cosine distribution was observed. It is assumed that recoils from deeper regions or recoils generated by multiple collisions (higher-order recoil atoms) in the cascade contribute to the sputtering yield. Several fitting methods have been applied to reproduce these distributions. Based on the cosine law by Knudsen (see Appendix K), it can be

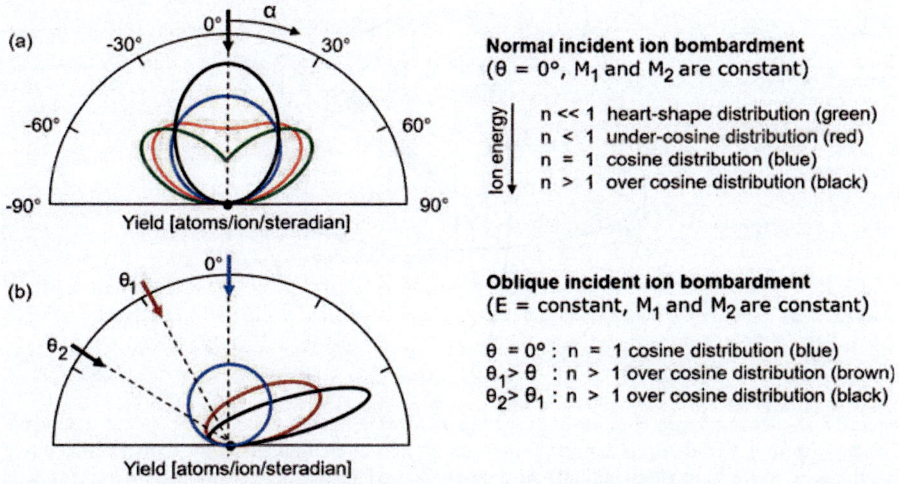

Fig. 5.14 Schematic illustration of different angular distributions of sputtered atoms. **a** Shape of the angular distribution of sputtered atoms as function of the ion incidence energy at normal ion bombardment. For energies slightly higher than the threshold energy a heart-shaped distribution can be expected. α and θ are the polar emission angle and the incident angle, respectively. **b** Shape of the angular distribution of sputtered atoms as function of the ion incidence angle

assumed that the probability of the sputtered particle emission is proportional to $\cos^n(\theta)$, where n is a real number and can be adapted to fit the measured distribution [73]. But, with this fit-procedure, the heart-shape distribution cannot be matched.

Therefore, Yamamura et al. [68] have proposed an extended expression which gives the probability of the sputtering of particles in the solid angle $d\Omega$ proportional to $\cos\theta\,(1 + B\cos^2\theta)$. In this relation, the parameter B is used to describe the shape of the cosine distribution ($B > 0$ and $-0.5 < B < 0$ give the cosine distribution, the over-cosine distribution and the under-cosine distribution, respectively).

At oblique incident bombardment (Fig. 5.14b), the maximum of the angular distribution is located in the forward direction. This direction is not significantly changed by variations of the ion incidence angle and can be explained on the basis of the direct knock-out model [52]. In particular, for the irradiation of surfaces with light ions (e.g., H or He ions) at larger angles of incidence, a superposition of distributions of the sputtered atoms can be observed, which is the result of different interaction processes. These superimposed distributions may be composed of contributions from (i) recoil atoms in the cascade, (ii) the non-isotropic distribution of recoils, and (iii) backscattered atoms generated inside the material. The result is a very complex shape of the angular distribution of sputtered atoms.

The number of sputtered particles being emitted in a direction with the differential solid angle $d\Omega$ is determined by the polar angle α, where the differential solid angle element $d\Omega$ is given by (see Fig. 5.15)

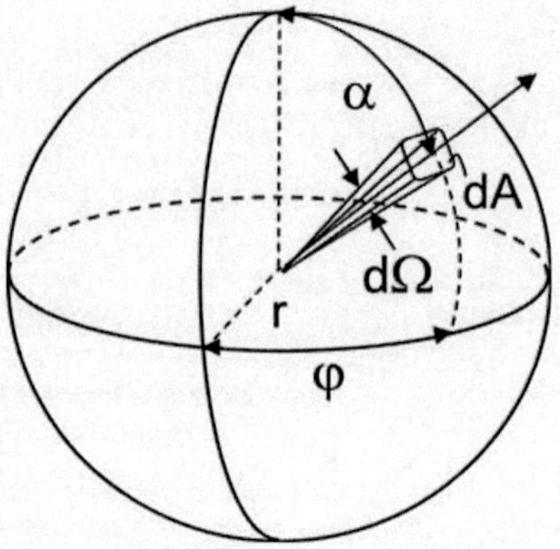

Fig. 5.15 Schematic representation of the geometrical arrangement for angle dependent sputtering. The angle α (unit is [radian]) is defined as distance along a circle divided by the radius of that circle and the solid angle (unit is [steradian]) the area on the surface of a sphere divided by the radius

$$d\Omega = \frac{r \cdot \cos\alpha \cdot dA}{r^3}. \tag{5.78}$$

Then, the local flux J_Ω of the sputtered particles through the surface element dA can be expressed by

$$J_\Omega = J_s \frac{\cos\alpha}{r^2}, \tag{5.79}$$

where J_s is the total flux of the sputtered particles.

Therefore, it is most commonly assumed that the angular distribution of the sputtered particles exhibits a diffuse (cosine-like) distribution. The description of sputtering typically applies the differential sputter yield $dY/d\Omega$, which gives the number of sputtered particles per ion and per steradian. Therefore, the differential sputtering yield is capable to quantify the sputtering yield as function of the polar angle α (varying between $0 \leq \alpha \leq \pi/2$) and the azimuthal angle φ (varying between $0 \leq \varphi \leq 2\pi$) relative to the surface normal (see Fig. 5.15). Not unexpectedly, the differential sputtering yield is also dependent on incidence angle θ, the ion energy E and the masses of incident and target atoms. The total sputtering yield can thus be determined by integrating the differential sputtering yield over the solid angle hemisphere.

In a first approximation, it is assumed that the ions are focused on the target surface to provide a sputtering point source. Then, the number of sputtered particles which are being emitted within differential solid angle $d\Omega$ ($= \sin\theta\, d\theta\, d\varphi = d(\cos\theta)\, d\varphi$) is given by

$$Y(\alpha, \varphi) = \int_\Omega y(\alpha, \varphi) d\Omega = \int_0^{2\pi} \int_0^{\pi/2} y(\alpha, \varphi) \sin\alpha \; d\alpha d\varphi, \tag{5.80}$$

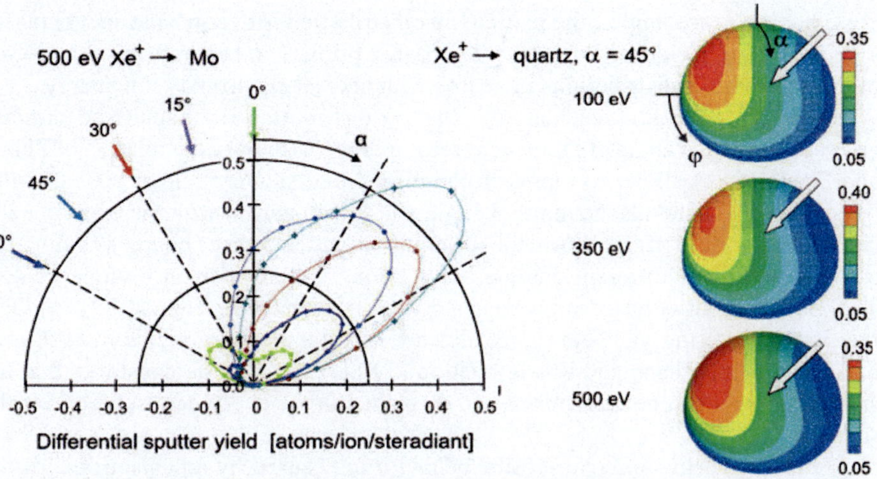

Fig. 5.16 Left: Experimental (symbols) flux distribution of sputtered Mo particles after 500 eV Xe-ion bombardment as function of the primary Xe ion incident angle. α is the polar emission angle. The Monte Carlo code TRIM. SP was used to simulate the Ge flux distribution. Right: The best-fit Zhang relation of measured differential sputter yields of quartz after Xe ion bombardment as function of the incident ion energy is plotted in dependence of both polar angle α and azimuthal angle φ. The differential sputter is normalized with respect to the total sputtering yield. The colors correspond to the normalized yield in the given direction (figures are taken from [76] and modified)

where α is the polar angle and φ is the azimuthal angle (Fig. 5.15).

Zhang and Zhang [74] have derived an expression for the differential sputtering yield y(α,φ) which is based on modifications of previous formula by Yamamura [75]

$$y(E,\theta,\alpha,\varphi) = Y(E,\theta)\cdot\cos\alpha\cdot\left[1-\frac{1}{2}\sqrt{\frac{E_{th}}{E}}\left(\cos\varphi\cdot z(\alpha)+\frac{3\pi}{4}\sin\theta\sin\alpha\cos\varphi\right)\right], \tag{5.81}$$

where

$$z(\alpha) = \frac{3\sin^2\alpha-1}{\sin^2\alpha}+\frac{\cos^2\alpha\left(3\sin^2\alpha-1\right)}{2\sin^3\alpha}\cdot\ln\left(\frac{1+\sin\alpha}{1-\sin\alpha}\right) \tag{5.82}$$

and Y(E,θ) is an expression for the sputtering yield, e.g., see (5.22). Note that for high ion energies the Zhang formula for the differential angular sputter yield distribution is reduced to the simple cosine sputtering yield distribution y = Y cos (α/π). With (5.80) the total sputtering yield can be determined by integration over the hemisphere. As an example, Fig. 5.16 shows the measured and simulated particle flux distribution of sputtered Mo atoms in dependence of the primary Xe ion energy and the incident angle. The particle flux increases with increasing primary ion energy. The particle flux distribution is under-cosine for perpendicular Xe ion bombardment.

For other incidence angles, the particle flux distribution is over-cosine and increases with increasing incidence angle and decreasing primary ion energy. Also the tilting of the particle flux distributions increases with decreasing primary ion energy.

Beyond the simple representation of the flux distribution of the sputtered particles as a function of the angle of incidence (see e.g., Fig. 5.16, left), the relation of Zhang and Zhang (5.81) allows to represent the differential sputter yield as a function of both polar angle and also azimuthal angle and to compare it with the experimental results. Figure 5.16, right, shows the differential sputter yield of quartz as a function of both angles for different energies of Xe ions and an incidence angle of 45°. The measurement is based on a quartz crystal microbalance method [10.76]. The measured sputtering yields are normalized by the total sputter yields in each case und fitted by the Zhang and Zhang relation (5.81). The figure demonstrates that the sputtering yield can be determined and displayed in the entire hemisphere over the sample.

The under-cosine and over-cosine behavior are caused by an anisotropic distribution of the recoil flux in the cascade. For the lower primary ion energy and the heavier primary ions, this anisotropy is closer to the target surface and therefore, the influence on the angular distribution of the sputtered particles is reinforced. Based on a momentum consideration, Stepanova and Drew [77] have demonstrated, that the anisotropy of the energy distribution of sputtered atoms scales with a factor $\left[M_1\left(E_{sp}+U_s\right)/M_2E\right]^{1/2}$, where E_{sp} is the energy of the sputtered atoms and E is the energy of the incident ion. If this factor is significantly smaller than 1, the anisotropy can be neglected. For most low-energy ion sputter experiments, this is not given for both normal and oblique ion incidences.

The distinction between energy and angular distribution as described above is an approximation. In reality, the angular distribution is also dependent on the energy of the sputtered atoms, i.e. the angular and energy distribution of the sputtered particles are coupled [78].

It should be noted that the physics underlying sputtering with monoatomic projectiles is understood reasonably well. The sputtering dynamic is generally described as a linear collision cascade modelled by the analytical transport theory. However, deviations were observed when the incident projectiles contain more than one particle (molecules) or/and the sputtered particles contain two or more identical atoms [74].

5.8 Sputtering of Compounds—Preferential Sputtering

Extensive experimental studies have shown that the sputtering yields of the individual components of a multicomponent material after sputtering are not proportional to its relative concentrations due to preferential sputtering [80, 81]. Preferential sputtering results in surface enrichment of one component in binary or ternary alloys, oxides and other compounds. For example, sputtering is a widely used method for depth analysis in combination with different analytical techniques like secondary ion mass spectrometry (SIMS), Auger electron (AES) and photoelectron spectroscopy (XPS). One of the reasons for this is that changes to the near-surface composition of studied solids induced by preferential sputtering must be considered for quantitative depth analysis.

Several reviews have been published on the preferential sputtering with a focus on the sputtering of alloys, oxides and compound semiconductors [78, 81–83].

For preferential sputtering, the meaning of sputtering yield (given in Sect. 5.1) must be redefined. Now, a partial sputtering yield Y_i of i-th component of a multi-component material is given by the average number of sputtered particles of the i-the component per incident ion, expressed by

$$Y_i = \frac{av.no.of\ sputtered\ particles\ of\ the\ i{-}th\ component\ of\ a\ mulitic omp.material}{incident\ particle}.$$

The correlation between the total sputtering yield given in (5.1) and the partial sputtering yield Y_i is given by [1]

$$Y = \sum_i Y_i C_i^s, \tag{5.83}$$

where Y_i is the sputtering yield of the component i and C_i^s is the concentration of the component i at the surface. During ion bombardment of surfaces, the sputtering yield of a multicomponent system is modified when the surface concentration is changed. This so-called preferential sputtering is a well-known phenomenon and can be observed measuring the concentration distributions by means of ion beam analysis or ion beam sputtering methods. The near-surface atoms are ejected into the gas phase by energetic ion bombardment, leading to an erosion of the multi-component material surface. Since the sputtering yields of several elements in this multicomponent material are likely to differ from each other, atoms of different elements are sputtered off with different probabilities, which gives rise to the preferential sputtering phenomenon. As result of this process, an equilibrium will develop between the sputtering yields of the different components and the bulk as well as surface concentrations.

As an example [80], the dependence of the sputtering yield ratio Y_{Ta}/Y_{Si} on the Ar ion energy for $TaSi_2$ is illustrated in Fig. 5.17. The ion bombardment induced surface composition changes with ion energy. At an ion energy < 2–3 keV the ratio Y_{Ta}/Y_{Si} is changed significantly, while for ion energies > 3 keV the ratio Y_{Ta}/Y_{Si} shows only a slight energy dependence. Such a behavior is typical for preferential sputtering of metal silicide.

In the simplest case of sputter erosion of a two-component system, $A_{1-x}B_x$ with x = 0 – 1, the number of incident ions per unit area and unit time, J (current density or ion flux), should be proportional to the flux of the sputtered species and the sputtering yields of the individual components, i.e

$$J_A = JY_A C_A^s \text{ and } J_B = JY_B C_B^s = JY_B(1 - C_A^s), \tag{5.84}$$

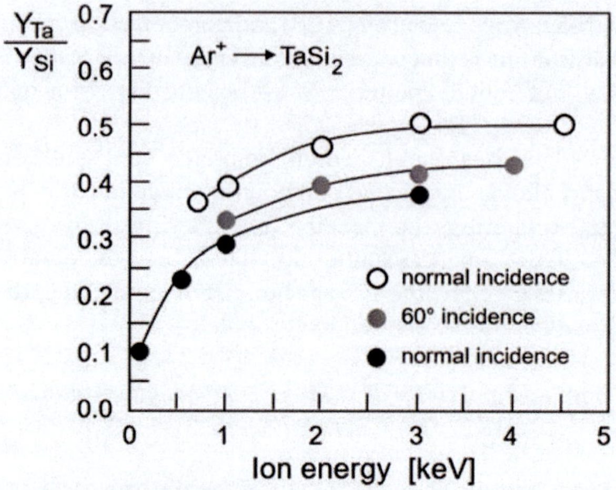

Fig. 5.17 Summary of AES measurements of the ratio Y_{Ta}/Y_{Si} as function of the Ar ion energy for $TaSi_2$ bombarded by Ar ions. The results according to normal ion incidence (open and black points) were taken from two different publications (figure is adapted from Ref. [80] and modified)

where C_A^s and C_B^s are the fraction of lattice places at the surface occupied by A atoms and B atoms, respectively. Assuming steady-state conditions between the surface compositions for the sputtering of binary materials, the partial sputtering yields can be obtained [78, 84], because these fluxes should be also proportional to their bulk compositions C_A and C_B given by

$$\frac{J_A}{J_B} = \frac{Y_B C_A}{1 - C_A}. \tag{5.85}$$

When the component sputtering yields are unequal, i.e. $Y_A \neq Y_B$, the surface concentration of the components A and B under steady-state conditions can be derived at by using (5.84) and (5.83) to

$$C_A^s = \frac{Y_B C_A}{Y_A(1 - C_A) + Y_B C_A} \text{ and } C_B^s = 1 - C_A^s, \tag{5.86}$$

respectively. Equation (5.86) can be transformed into the more general sputter yield ratio expressed by

$$\frac{Y_A}{Y_B} = \left(\frac{C_A}{C_B}\right)\left(\frac{C_B^s}{C_A^s}\right). \tag{5.87}$$

Consequently, the surface is enriched by component A when $Y_B > Y_A$ and is enriched by B species when $Y_B < Y_A$. Hence, the partial sputtering yields can be determined only if the surface concentrations as well as the bulk concentrations are known (or can be measured).

In literature, various models are proposed to explain the concentration changes in binary compounds under ion bombardment. According to Betz and Wehner [78], either mass effects or binding effects are responsible for enrichment or depletion effects. Therefore, preferential sputtering has been studied for isotopic samples, i.e. monoatomic samples of two or more isotopes [85, 86]. For such systems the preferential sputtering is only determined by mass differences, because the influence of the binding energy can be neglected.

(1) *Mass effect*

The sputtering yield in the framework of the linear collision theory by Sigmund was given by (5.21). Following this equation, the partial sputtering yield Y_i of the i-th component for low energies can be formulated as

$$Y_i = \Lambda_i F_D(E, \theta, 0) \text{ with } i = A, B, \tag{5.88}$$

where F_D is the elastic energy deposited at the surface and given by (5.20). Sigmund has shown [87] that the ratio of mutual nuclear stopping cross-sections for the components A and B can be described by

$$\frac{S_{AB}(E)}{S_{BA}(E)} = \left(\frac{M_A}{M_B}\right)^{2m}, \tag{5.89}$$

where S_{AB} and S_{BA} are the mutual stopping cross-sections for an A atom hitting a B atom and for a B atom hitting an A atom, respectively. M_A and M_B are the masses of the sputtered components, respectively. The parameter m is a power law variable, which characterizes the scattering cross-section (see e.g., Table 2.2). For low ion energies (< 1 keV), this parameter $m \leq 0.25$. According to (5.19), the ratio of the material factors Λ_A and Λ_B is proportional to the ratio of surface binding energies $\left(U_s^B / U_s^A\right)^{1-2m}$. Combining these two expressions, the ratio of the partial sputtering yields for two-component material can be obtained to

$$\frac{Y_A}{Y_B} = \left(\frac{M_B}{M_A}\right)^{2m} \left(\frac{U_s^A}{U_s^B}\right)^{1-2m}. \tag{5.90}$$

Therefore, an enrichment of elements with higher masses can be expected in conformity with numerous experimental results. If the mass difference effect is small, the changed concentration in the near-surface region can be governed by the difference in surface binding energies. Substituting (5.90) into (5.87) yields to a relationship,

$$\frac{C_s^A}{C_s^B} = \left(\frac{C_A}{C_B}\right)\left(\frac{M_A}{M_B}\right)^{2m} \left(\frac{U_s^A}{U_s^B}\right)^{1-2m}, \tag{5.91}$$

which characterizes the enrichment (depletion) on the surface under low-energy ion bombardment.

For higher ion energies and heavy incident ions, the concept of the linear cascade theory must be replaced with the concept of the spike regime [9]. Then, according to Sigmund, the ratio of the partial sputtering yields is given by

$$\frac{Y_A}{Y_B} = \left(\frac{C_A}{C_B}\right)\left(\frac{M_B}{M_A}\right)^{1/2} \exp\left(\frac{U_s^B - U_s^A}{k_B T_{spike}}\right), \tag{5.92}$$

where T_{spike} is the spike temperature which is dependent on the material properties and the parameters of the ion bombardment (see Sect. 4.6.2).

(2) *Binding effect*

The influence of the surface binding energy on the sputtering yield was discussed in Sect. 5.4.1. However, the surface binding energy is also dependent on its lattice position and the chemical nature of neighboring atoms.

Malherbe et al. [88] calculated the binding energies by using a model based on a modification of the Pauling formalism [89] for bond energies in a covalent bond. There, it is assumed, that the average binding energy of a two-component molecule is the arithmetic average of the binding energies of the respective elements. Then, the average binding energy $U_{A_{1-x}B_x}$ of a chemical compound $A_{1-x}B_x$ is given by

$$U_{(A_{1-x}B_x)} = (1-x)U_{AA} + xU_{BB} + (1-x)U_{AB} + \frac{1}{2}(\varepsilon_A + \varepsilon_B)^2, \tag{5.93}$$

where U_{AA}, U_{BB} and U_{AB} are the binding energies between components A and B, and ε_A and ε_B are the electronegativities of the components A and B, respectively. Consequently, the ratio of the surface binding energies can be expressed as

$$\frac{U_s^B}{U_s^A} = \frac{C_B U_{BB} + C_A U_{AB}}{C_B U_{AB} + C_A U_{AA}}. \tag{5.94}$$

By using this binding energy relation and introducing it to (5.88), the concentration ratio C_s^A/C_s^B at the surface (enrichment or depletion) can be obtained and compared with the measured results.

Kelly [90] has introduced an expression for the surface binding energy that considers the influence of the nearest neighbor binding strengths U_s^{AA}, U_s^{BB} and U_s^{AB}. Then, for example, the surface binding energy of the component A can be given by

$$U_s^A = -C_s^A Z_s U_s^{AA} - C_s^B Z_s U_s^{AB}, \tag{5.95}$$

where Z_s is the surface coordination number. The surface binding energies can be expressed in a first approximation by the heats of sublimation ΔH_A and ΔH_B of the pure components by means of

$$\frac{U_s^A}{U_s^B} \cong \frac{(1+C_s^B)\Delta H_B + (1+C_s^B)\Delta H_A}{(1+C_s^A)\Delta H_A + (1+C_s^A)\Delta H_B}. \tag{5.96}$$

Thus, the ratio of the partial sputtering yields after Kelly [81] is obtained to

$$\frac{Y_A}{Y_B} = \frac{(C_A+\gamma C_B)}{(C_B+\gamma C_A)} \frac{(1+C_s^B)\Delta H_B + (1+C_s^B)\Delta H_A}{(1+C_S^A)\Delta H_A + (1+C_s^A)\Delta H_B}, \tag{5.97}$$

where the transfer energy efficiency factor, γ, is defined by (2.28). In many studies the enrichment of one component was qualitatively determined in good agreement with (5.97). However, in some cases the proposed enrichment was smaller than that which was observed during experiments.

The effect of ion bombardment induced compositional redistribution is dominant for low temperatures (room temperature and less). As the temperature is increased, other processes influence the composition in the near-surface region with thickness in order of the range of the incident ions, because the ion irradiation induced point defects become mobile (see Sect. 4.6). In addition to preferential sputtering, processes like radiation enhanced diffusion, radiation induced segregation and Gibbsian adsorption must also be considered [81].

In numerous experimental for binary compounds $A_{1\text{-}x}B_x$, a linear dependence of the sputtering yield on composition x has been demonstrated [76, 80, 82, 88, 91]. But, a nonlinear behavior of the sputtering yield for binary $Si_{1-x}Ge_x$ was also observed [92] and simulated [93]. Figure 5.18 shows the sputtering yield as function

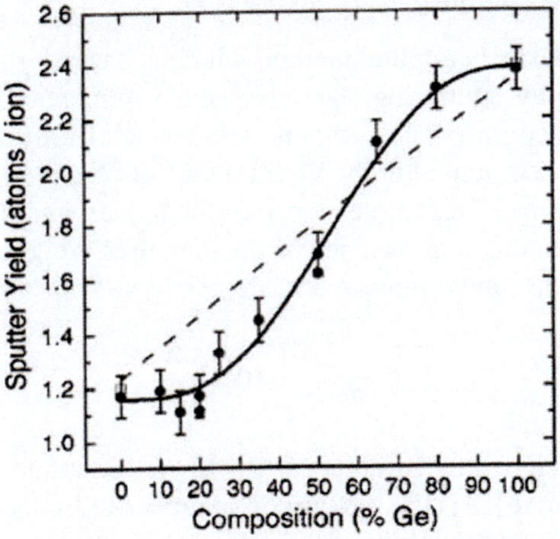

Fig. 5.18 Composition dependence of the sputter yield for $Si_{1-x}Ge_x$ bombarded with 3 keV Ar ions. The measured sputter yields (points and squares) are compared with theoretically calculated values (solid line). The dashed line is calculated by the linear sputtering theory, (5.21), where the surface binding energy is calculated by a weighted average of the constituents (figure adapted from [93])

of the composition after 3 keV Ar ion bombardment. The nonlinear dependence of the experimentally observed $Si_{1-x}Ge_x$ sputtering yield on composition could be calculated using linear cascade theory, with an expression for the surface binding energy which takes into account the alloy effect.

5.9 Measurement of the Sputtering Yields

Ever since 1852, results of sputter experiments have been published [94]. Nowadays, a variety of methods are used to very exactly measure the sputtering yield in dependence of primary ion energy, the ion and target atom mass, the ion fluence, the ion incident angle, as well as the temperature. These methods are able to detect average sputter yields $< 10^{-4}$ atoms/ion and to thus measure very near the sputtering threshold. However, it must be remarked that each of these methods has its advantages and disadvantages, which should be carefully reviewed before application.

5.9.1 Measurement of the Total Sputtering Yield

The total sputtering yield measurements are based on the principles of either measurement of the target mass or areal density loss, or measurement of the target thickness reduction. The most frequently used methods are:

- Measurement of weight loss

This is the most extensively applied method which is based on measuring mass loss in the target caused by sputtering. For these measurements, the samples have to be outgassed in a vacuum oven and then weighed with a precision microbalance before the samples are mounted in the vacuum sputter chamber. After sputtering for a defined period of time, the samples are weighted again and the total erosion rate can be calculated from the total fluence and the measured weight loss. The sputtering yield for perpendicular ion bombardment is given by [95].

$$Y(\theta = 0) = 10^5 \frac{\Delta W}{ItM_2}, \tag{5.98}$$

where ΔW is the weight loss in units of [g], M_2 is the atomic mass of the target material in units of [amu], I is the ion current in units of [A] and t the exposure time in units of [s]. The erosion depth h can be calculated by

$$h = 10^5 \frac{\Delta W}{NM_2A}, \tag{5.99}$$

where the A is the total sputtered area and N the atomic number density of the sputtered target.

Uncertainties in the measurement of the total sputtering yield result from finite accuracy of mass, current and time measurements. The main disadvantages of this method are re-deposition of sputtered material onto the sputtered surface as well as possible incorporation of incident ions, which can lead to irrepressible weight increase. Consequently, the weight loss data must be corrected [88]. The limits for the detection of mass change and the measurement of sputtering yields are about 5 μg and 10^{-3} atoms/ion, respectively.

- Quartz crystal microbalance method

The quartz crystal or oscillator microbalance method proposed by Sauerbrey [96] is based on monitoring the reference frequency of a piezoelectric quartz crystal disc. With this technique, the total sputtering yield can be determined in-situ by collecting sputtered material on one side of the oscillated quartz disc, without calibration. The oscillation frequency shifts to lower resonant frequencies, in proportion to the mass of the sputter deposited material. The change in resonant frequency correlates with the mass changes described by the Sauerbrey relationship [96]

$$\Delta f = -C_c \cdot \Delta M = -\frac{2nf^2}{\sqrt{\rho G}}\,\Delta \mathrm{M}, \tag{5.100}$$

where Δf is the observed frequency change in units of [Hz], ΔM the change of the mass per unit area in units of [g/cm^2], C_c is the sensitivity factor for the crystal used (e.g., 56.6 Hz $\mu g^{-1} cm^2$ for a 5 MHz AT-cut quartz crystal at room temperature), f is the resonant frequency of the fundamental mode of the crystal in units of [Hz], n is the number of the harmonic at which the crystal is driven, ρ is the mass density of quartz (= 2.648 g/cm^3) and G is the shear modulus of quartz (= 2.947×10^{11} g cm^{-1} s^{-2}). Mass changes in the order of 1 ng/cm^2 can be detected that way. Thereafter, the sputtering yield can be determined using (5.100) with a sensibility of up to 10^{-5} atoms/ion.

- Sputter crater measurement [97]

The total sputtering yield can be also determined by measuring the sputtered volume, i.e. the sputtered area A times sputtered depth h. Tactile and optical profilometry, atomic force microscopy (AFM) and scanning electron microscopy (SEM) are the most frequently applied methods to reconstruct the sputter crater and to measure the amount of the material lost after the sputtering process.

(a) A tactile profilometer is based on a tip which probes the sputtered area in a straight line across the zenith and records the deflection of the contact point perpendicular to it track. A more extensive variation is the 3D coordinate tactile measuring technique, where the position of the probe tip is controlled by a three-axis measuring arrangement. Thus, the material lost as well as the sputtering yield can be determined by comparison of the depth profile before and after the sputtering procedure. A resolution of about 1 nm can be obtained.

(b) Optical profilometry methods [98] including white light interferometric (WLI) or confocal techniques can be used. WLI applies the so-called Mirau technique where a mirror inside the optical system is used to create interference between the light reflected from the sputtered surface and the reference light signal. These techniques are capable to give a depth resolution of about 0.1 nm. In a similar way, a laser interferometer can be applied to probe the depth of a sputtered area in comparison to the non-sputtered area.

(c) AFMs operate by measuring the force between a needle and the sputtered surface. A tip mounted on a cantilever scans the surface. A laser beam which is reflected from the cantilever is detected by a position-sensitive detector and provides a topographical image of the surface with atomic resolution.

(d) In the scanning electron microscopy, a focused electron beam is scanned over the surface and excites the emission of secondary and backscattered electrons. These are then used to generate a topographical image.

The ion bombardment can lead to changes of the surface topography at the bottom of the crater as well as the sidewalls, thus causing problems in the quantitative determination of the sputtered volume.

5.9.2 *Measurement of Energy and Angular Distributions*

A frequently used technique to determine the energy distribution of sputtered particles is the laser induced (excited) fluorescence (LIF) [70, 99]. In this process, a laser beam is tuned to a fixed frequency so that the sputtered atoms are put in an excited state. After some nanoseconds the excited level drops down to a non-excited state and emits a photon of well-known frequency. This fluorescence spectrum is measured by an optical spectrometer placed over the bombarded target. From this spectrum, the particle velocity (energy) and the atomic density (sputtering yield) can be in-situ determined, while the sputter process running.

Another important method to determine the energy distribution is the time-of-flight secondary ion mass spectrometry (SIMS), especially the secondary neutral mass spectrometry (SNMS). The majority of sputtered particles are neutral. These neutral gas phase atoms are ionized before they are collected in a time-of-flight mass (ToF) spectrometer. In the ToF spectrometer, the ions are accelerated to the same potential. Then, the flight time to the detector is related to the mass of the ions and the intensity of the measured signal is proportional to the sputtering yield. SIMS has the advantage of having a lower detection limit than SNMS, while SNMS is better quantifiable.

Two major methods are usually applied to measure the angular distribution:

(a) Samples are placed in the half-space over the target surface at well-defined positions relative to the target. During the sputtering process, these samples are coated with the sputtered target material. After the experiment, the thickness of the deposited film can be determined by most commonly known thin film analysis methods like Rutherford backscattering, secondary ion mass spectrometry, optical profilometry or X-ray and optical reflectance measurements.

(b) A second principle is based on moving a sensor into the half-space over the sample, with the aim to detect sputtered particles at well-defined positions (with respect to the target surface or vice versa) in-situ. Typical detectors are quartz crystal microbalance plates or energy and mass selective spectrometers.

More detailed descriptions of the measuring techniques for the determination of energy and angular distributions of sputtered particles have been reported (see e.g., [64, 65, 100]). An extensive review about the methods to measure the sputtering yield is given by Andersen and Bay [101].

5.10 Reflection

A fraction of the incoming ions is backscattered and doesn't contribute to the sputtering of the surface. Reflection (backscattering) of ions with energies up to some keV is of significant importance for thin film deposition research and ion etching processes. Much data on backscattering has been accumulated both from simulations and experiments (for reviews see e.g., [102–104]).

By ion bombardment, the scattering of these energetic particles on surface atoms in rest (substrate atoms) can occurs in a forward direction when the mass of the ion is larger than the mass of the substrate atom and in forward and backward directions when the mass of the ion is lower than the substrate atom (see Chap. 2). The reflected particles are neutrals. The energy of these particles ranges from nearly zero up to the initial ion bombardment energy.

Reflection is described by the particle or total reflection coefficient R_N (ratio of reflected to incident particles) and the energy reflection coefficient R_E (ratio of R_N to incident energy). The reflection coefficients are given by [105]

$$R_N = \int_0^E \int_0^{\pi/2} \int_0^{2\pi} f(E_R, \theta, \varphi) d\varphi d\theta dE \tag{5.101}$$

and

$$R_E = \frac{R_N}{E} = \frac{1}{E} \int_0^E \int_0^{\pi/2} \int_0^{2\pi} f(E_R, \theta, \varphi) d\varphi d\theta dE, \tag{5.102}$$

where E is the energy of the incidence ion, $f(E_R,\theta,\varphi)$ the energy-angular distribution of the reflected particles, which is dependent on the energy of the reflected particles E_R, the polar angle, θ, and the azimuthal angle, φ, of the emitted particles. With decreasing of ion energy, the direction of the preferential reflection is continuously shifted from the surface normal of the bombarded material to a region of specular reflection of the incoming ions (similar to the angular distribution of sputtered particle, c.f. Fig. 5.14).

Early calculations of reflection by Bøttiger et al. [106] are based on random slowing-down caused by elastic and inelastic collisions in an infinite medium. A first model, the one-collision model, was proposed [107, 108], which explains the reflection by one large angle collision where the particle move on a straight trajectory. Later, Eckstein and Biersack [109] have demonstrated that under some approximation, analytical expressions can be obtained for energy and angular distributions of reflected particles as well as particle and energy reflection coefficients. Parilis et al. [110], in particular, have proposed a two-collision model for the intermediate energy range.

The particle reflection coefficient R_N can be measured by the radioactive tracer method, by nuclear reaction techniques or by measurement of the energy distribution of reflected particles at a fixed emission angle. The last method as well as calorimetric methods are used to measure the energy reflection coefficient R_E (see e.g., overview about the various measurement methods by Mashkova [111]). Frequently, computer simulations are used to determine reflection coefficients which are based on binary collision models, where the KrC-potential is mostly applied. Sometimes, the universal potential (ZBL potential) and the Moliere potential are used too [102, 103]. Advanced Monte Carlo simulation programs, like TRIM and MARLOW and their versions (see Sect. 3.6), provide high-precision data of the reflection coefficients. An example is shown in Fig. 5.19. The contour plot of the reflection coefficient R_N of 1 keV Ar ions on a Ni surface for the ion incident angle of 60° is shown as result of the TRIM simulation [112]. The backscattered particles have the largest coefficient in the forward direction.

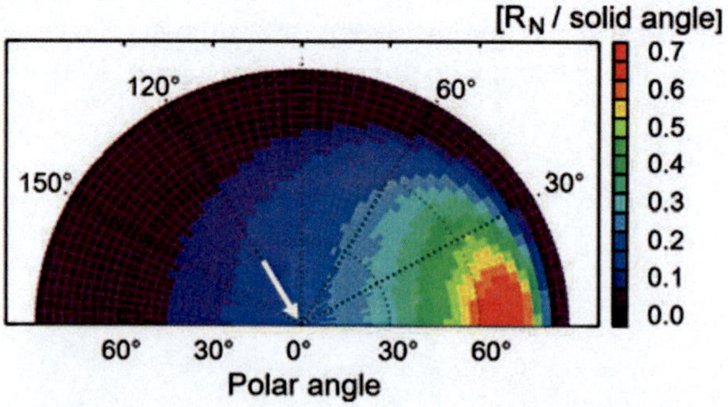

Fig. 5.19 Contour plot of the angular distribution of the reflection coefficient R_N per solid angle. Ni target is bombarded with 10^8 Ar ions at an angle of 60° (white arrow). The energy of the Ar ions is 1 keV (figure is adapted from [112])

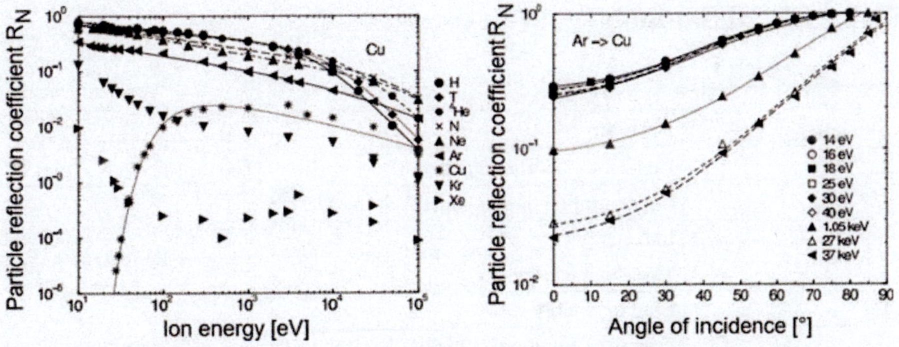

Fig. 5.20 Calculated particle reflection coefficient R_N at normal incidence of Cu as function of ion energy for different ion species (left) and as a function of the incidence angle of Ar ions for different ion energies (right). Figures adapted from [105]

Calculations of the reflection coefficients based on the Boltzmann transport theory are also known (see e.g., [113]). Experimental as well as calculated data of the reflection coefficients are mostly available for light ion bombardment, whereas values for the reflection coefficient after heavy ion bombardment have rarely been published. In Fig. 5.20 two examples of calculated total reflection coefficients as function of the incident energy (left) and the incidence angle (right) are presented. In general, following trends can be observed [102–105, 109, 111, 114]:

(i) The reflection coefficients decrease with increasing ion energy for given ion-substrate combinations.

(ii) The reflection coefficients increase by raising the mass of the incident ion and with increasing mass ratio M_{target}/M_{ion} under normal ion incidence.

(iii) For self-ion bombardment, the reflection coefficients decrease for low energies, because the surface binding potential accelerates the incoming atom and decelerates the backscattered atom.

(iv) R_N and R_E exhibit equal dependence on the incident energy and angle for the same ion-target combinations.

(v) In contrast to sputtered atoms, the energy distributions of backscattered particles are extended to higher energies. The mean energy of reflected atoms is usually higher than this energy for sputtered atoms.

(vi) The reflection coefficients increase with incidence angle, whereby the rate of increase arises with the increase of the dimensionless reduced energy.

In literature some analytic approximations for the calculation of the reflection coefficient have been published [105, 112, 115–119].

5.11 List of Symbols

Symbol	Notation
C_i	Concentration of component i
C_i^s	Concentration of component i at the surface
E	Ion energy
E_{th}	Threshold energy
E_B	Binding energy
E_{sp}	Energy of sputtered particle
$\langle E_{sp} \rangle$	Mean energy of sputtered particles
$F_D(E)$	Deposited energy
ΔH_{coh}	Cohesive energy
ΔH_f	Heat of formation
J	Ion current density (ion flux)
J_i	Particle flux of component i
J_s	Flux of sputtered particles
M	Mass ratio ($= M_1/M_2$)
M_i	Mass of particle i
ΔM	The change of the mass per unit area
N	Atomic number density
P	Probability
R_E	Energy reflection coefficient
R_N	Total reflection coefficient
$S_n(E)$	Nuclear stopping cross-section
T_g	Temperature of sputter gas
T_{max}	Maximum transferred energy
U_s	Surface binding energy
V_s	Sputtered volume
Z_B, Z_S	Bulk and surface coordination numbers
Y	Sputtering yield
Y_i	Sputtering yield of the component i
a_L	Lindhard screening length
d	Thickness
f	Resonant frequency of the fundamental mode of the crystal
h	Height, depth
i	Ion current
k_{LS}	Dimensionless Lindhard electronic stopping coefficient
p	Pressure

(continued)

(continued)

Symbol	Notation
$s_e(\varepsilon)$	Reduced electronic stopping power
$s^{KrC}{}_n(\varepsilon)$	Reduced Kr–C nuclear stopping power
m	Power law variable
$n(E,E_o)$	Number of recoils
v	Erosion rate (velocity)
v_i	Velocity of particle i
v_o	Erosion rate (velocity) normal to the surface
Λ	Material contant
Φ	Ion fluence
Ω	Solid angle
α	Material correction factor, polar emission angle
γ	Transfer energy efficiency coefficient
ε	Lindhard reduced energy
ε_i	Electronegativity of component i
$\eta(E)$	Inelastic energy loss to electrons
θ	Angle of ion incidence, polar angle
θ_p	Maximum sputtering angle
θ_{sp}	Emission angle of the sputtered atoms
λ	Mean free path
μ	Shear modulus
ρ	Mass density
$\nu(E)$	Elastic energy loss to nuclei (damage energy)
φ	Azimuthal angle

References

1. P. Sigmund, Collison theory of displacement damage, ion ranges, and sputtering. Phys. Roum. Rev. **17** 823–870, 969–1000, and 1079–1106 (1972)
2. H. Oechsner, Sputtering – a review of some recent experimental and theoretical aspects. Appl. Phys. **8**, 185–198 (1975)
3. R. Behrisch (Ed.), *Sputtering by particle bombardment I*, Topics in Appl. Physics, Vol. 47, (Springer-Verlag Berlin Heidelberg, 1981)
4. R. Behrisch (Ed.), *Sputtering by particle bombardment II*, Topics in Appl. Physics, Vol. 52, (Springer-Verlag Berlin Heidelberg, 1983)
5. R. Behrisch (Ed.), *Sputtering by Particle Bombardment III*, Topics in Appl. Physics, Vol. 64, (Springer-Verlag Berlin Heidelberg, 1991)
6. H. Gnaser, *Low-energy ion irradiation of solid surfaces* (Springer-Verlag, Berlin Heidelberg, 1999)

7. R. Behrisch, W. Eckstein (ed.), *Sputtering by Particle Bombardment IV*, Topics in Appl. Physics, Vol. 110, (Springer-Verlag Berlin Heidelberg, 2007)
8. P. Sigmund, Theory of sputtering I. Sputtering yield of amorphous and polycrystalline targets, Phys. Rev. **184**, 383–416 (1969)
9. P. Sigmund, Sputtering by ion bombardment: theoretical concepts, in Ref. [3], pp. 9–71
10. J. Lindhard, V. Nielsen, M. Scharff, Approximation method in classical scattering by ccreened Coulomb fields – notes on atomic collision I, Det. Kgl. Danske Vid. Selskab., Mat.-Fys. Medd. **36**, no. 10 (1968)
11. N. Matsunami, Y. Yamammura, Y. Itikawa, N. Itoh, Y. Kazumata, S. Miyagawa, K. Morita, R. Shimizu, A semiempirical formula for the energy dependence of sputtering yield. Rad. Effects Lett. **57**, 15–216 (1980)
12. W.D. Wilson, L.G. Haggmark, J.P. Biersack, Calculation of nuclear stopping, ranges, and straggling in low-energy range. Phys. Rev. B **15**, 2458–2468 (1977)
13. H.H. Andersen, H.L. Bay, Sputtering-yield studies on silicon and silver targets. Rad. Effects **19**, 139–146 (1973)
14. J.P. Biersack, W. Eckstein, Sputtering studies with the Monte Carlo program TRIM.SP, Appl. Phys. **174**, 257–269 (1984)
15. P.C. Zalm, Some useful yield estimates for ion beam sputtering and ion plating at low bombarding energies. J. Vac. Sci. Technol. B **2**, 151–152 (1984)
16. G. Dupp, A. Scharmann, Die Zerstäubung von Kupfer durch Ne^+-, Ar^+- Kr^+- und Xe^+-Ionen im Energiebereich von 75 keV bis 1 MeV. Z. Physik **192**, 284–298 (1966)
17. H.L. Bay, H.H. Andersen, W.O. Hofer, O. Nielsen, The energy dependence of gold selfsputtering. Nucl. Instr. Meth. **132**, 301–305 (1976)
18. D.A. Thompson, High density cascade effects, Rad. Eff. **56**, 105–150 (1981) and S.S. Johar, D.A. Thompson, Spike effects in heavy-ion sputtering of Ag, Au and Pt thin films, Surf. Sci. **90**, 319–330 (1979)
19. K.L. Merkle, W. Jäger, Direct observation of spike effects in heavy-ion sputtering. Phil. Mag A **44**, 741–762 (1981)
20. P. Sigmund, C. Claussen, Sputtering from elastic collision spikes in heavy-ion-bombarded metals. J. Appl. Phys. **52**, 990–993 (1981)
21. R. Kelly, The mechanisms of sputtering. Part I: Prompt and slow collisional sputtering, Rad. Effects **80**, 273–317 (1984) and The varieties of surface alteration: structural, topographical, and compositional, *in Ion Bombardment Modification of Surfaces: Fundamentals and Applications*, ed. by O. Auciello, R. Kelly, (Elsevier, Amsterdam, 1984) pp 79–126
22. P. Sigmund, M. Szymonski, Temperature-dependent sputtering of metals and insulators. Appl. Phys. A **33**, 141–152 (1984)
23. G. Carter, Spike and shock processes in high energy deposition density atomic collision events in solids, Rad. Effects Lett. **43**, 193–199 (1979) and A semi quantitative approach to ion impact induced shock processes in solids, Rad. Effects Lett. **50**, 105–109 (1980)
24. B. Rauschenbach, K. Hohmuth, Analysis of heavy-ion sputtering based on the shock wave model. Phys. Stat. Sol. (a) **79**, 395–400 (1983)
25. K.L. Merkle, W. Jäger, Direct observation of spike effects in heavy-ion sputtering. Phil. Mag. A **44**, 741–762 (1981)
26. H. Gades, H.M. Urbassek, Surface binding energies of alloys: a many-body approach. Nucl. Instr. Meth in Phys. Res. B **88**, 218–228 (1994)
27. H. Gnaser, Energy and angular distributions of sputtered species, in Ref. [7], pp. 231–329
28. B.J. Garrison, N. Winograd, D. Lo, T.A. Tombrello, M.H. Shapiro, Energy cost to sputter an atom from a surface in keV ion bombardment processes. Surf. Sci. **180**, L129–L133 (1987)
29. W. Eckstein, J.P. Biersack, Computer simulation of two-component target sputtering. Appl. Phys. A **37**, 95–108 (1985)
30. R.C. Bradley, Sputtering of alkali atoms by inert gas ions of low-energy. Phys. Rev. **93**, 1421–1430 (1954)
31. J. Bohdansky, J. Roth, H.L. Bay, An analytical formula and important parameters for low-energy ion sputtering. J. Appl. Phys. **51**, 2861–2865 (1980)

32. N. Matsunami, Y. Yamamura, Y. Itikawa, N. Itoh, Y. Kazumata, S. Miyagawa, K. Morita, R. Shimizu, H. Tawara., Energy-dependence of ion-induced sputtering yields of monoatomic solids, Atomic Data and Nuclear Table **31**,1–80 (1984)
33. Y. Yamamura, H. Tawara, Energy dependence of ion-induced sputtering yields from monoatomic solids at normal incidences, Atomic Data and Nuclear Tables, **62**, 149–253 (1996), [Erratum in Atomic Data and Nuclear Tables, **63**, 353 (1996)]
34. W. Eckstein, R. Preuss, New fit formulae for the sputtering yield. J. of Nucl. Mater. **320**, 209–213 (2003)
35. R.V. Stuardt, G.K. Wehner, Sputtering yield at very low bombarding ion energies. J. Appl. Phys. **33**, 2345–2352 (1962)
36. W. Eckstein, Sputtering yields, in Ref. [7], pp. 33–187
37. P.C. Zalm, Energy dependence of the sputtering yield of silicon bombarded with neon, argon, krypton, and xenon ions. J. Appl. Phys. **54**, 2660–2666 (1983)
38. H.H. Andersen, H.L. Bay, Heavy-ion sputtering yields of gold: Further evidence of nonlinear effects. J. Appl. Phys. **46**, 2416–2422 (1975)
39. Ch. Steinbrüchel, A simple formula for low-energy sputtering yields, Appl. Phys. A **36**, 37–42 (1985)
40. J. Bohdansky, A universal relation for the sputtering yield of monoatomic solids at normal ion incidence. Nucl. Instr. Meth. in Phys. Res. B **2**, 587–591 (1984)
41. G.L. Chen, unpublished, quoted in Ref. [22]
42. J. Bohdansky, Important sputtering yield data for Tokamaks: A comparison of measurements and estimates. J. Nucl. Mater. **93**(94), 44–60 (1980)
43. Y. Matsunami, Y. Yamamura, Y. Itikawa, N. Irtoh, Y. Kazumata, S. Miyagawa, K. Morita, R. Shimizu, Energy dependence of sputtering yields of monatomic solids, Report JPPJ-Am-14 (1980)
44. Y. Yamamura, N. Matsunami, N. Itoh, Theoretical studies on an empirical formula for sputtering yield at normal incidence. Rad. Eff. **71**, 65–86 (1983)
45. H.J. Strydom, W.H. Gries, A comparison of three versions of Sigmund's model of sputtering using experimental results, Rad. Eff. Lett. **86** (1984) 145–151 and W.H. Gries, H.J. Strydom, A table of normalized sputtering yields for mono-elemental polycrystalline targets, Report SMAT, 1984
46. C. Gracía-Rosales, W. Eckstein, J. Roth, Revised formulae for sputtering data. J. Nucl. Mater. **218**, 8–17 (1994)
47. W. Eckstein, C. Garcís-Rosales, J. Roth, W. Ottenberger, Sputtering Data, IPP – Report 9/82, 1993
48. M. Tartz, T. Heyn, C. Bundesmann, C. Zimmermann, H. Neumann, Sputter yields of Mo, Ti, W, Al, Ag under xenon ion incidence. Eur. Phys. J. D **61**, 587–592 (2011)
49. M. Tartz, H. Neumann, H. Leiter, J. Esch, Pyrolytic graphite and carbon-carbon sputter behaviour under xenon ion incidence, Proceed. 29th Intern. Electric Propulsion Conference, IEPC 2005-143, Princeton 2005
50. G. Falcone, Sputtering theory. Rivista del Nuovo Cimento **13**(1), 1–52 (1990)
51. J. Lindhard, Influence of crystal lattice on motion of energetic charged particles, Det. Kgl. Danske Vid. Selskab., Mat.-Fys. Medd. **34** no. 14 (1965)
52. A.D.G. Stewart, M.W. Thompson, Microtopography of surfaces eroded by ion bombardment. J. Mater. Sci. **4**, 56–60 (1969)
53. J.P. Ducommun, M. Cantagel, M. Moulin, Evolution of well-defined surface contour submitted to ion bombardment: computer simulation and experimental investigation. J. Mater. Sci. **10**, 52–62 (1975)
54. Y. Yamamura, An empirical formula for angular dependence of sputtering yields Rad. Effects **80**, 57–72 (1984)
55. M. Teichmann, J. Lorbeer, F. Frost, B. Rauschenbach, Ripple coarsening on ion beam-eroded surfaces, Nanoscale Res. Lett. **4** (1914) 439
56. D. Manova, M. Schreck, S. Mändl, B. Stritzker, B. Rauschenbach, Orientation dependent sputter yield of aluminium. Surf. Coat. Technol. **151–152**, 72–75 (2002)

57. A.L. Southern, W.R. Willis, M.T. Robinson, Sputtering experiments with 1- to 5-keV *Ar+* ions. J. Appl. Phys. **34**, 153–163 (1963)
58. X.-S. Wang, R.J. Pechman, J.H. Weaver, Ion sputtering of GaAs(110): From individual bombardment events to multilayer removal. J. Vac. Sci. Technol. B **13**, 2031–2040 (1995)
59. G.N. van Wyk, N.J. Smith, Ion bombardment induced preferential orientation in polycrystalline Cu Targets. Rad. Eff. **38**, 245–247 (1987)
60. D. Dobrev, Ion-beam-induced texture formation in vacuum-condensed thin metal films. Thin Solid Films **92**, 41–53 (1982)
61. D.D. Odintsov, Dependence of single-crystal sputtering on the direction of incidence of particles. Soviet Physics-Solid State **5**, 813–815 (1963)
62. Yu.V. Martynenko, Theory of single crystal sputtering. Soviet Physics-Solid State **6**, 1581–1585 (1965)
63. D. Onderdelinden, *The influence of channeling on Cu single-crystal sputtering* Appl. Phys. Lett. **8**, 189–191 (1966)
64. H. Gnaser, Energy and angular distributions of sputtered species, in Ref. [7], pp. 231–328
65. W.O. Hofer, Angular, energy and mass distribution of sputtered particles, in Ref. [5], pp. 15–90
66. T. Ono, T. Kenmotsu, T. Muramoto, Simulation of sputtering process, *in Reactive Sputter Deposition*, ed. by D. Depla, S. Mahieu, (Springer-Verlag, Berlin, Heidelberg 2008) pp 1–42
67. M.W. Thompson, II. The energy spectrum of ejected atoms during the high energy sputtering of gold, Phil. Mag. **18**, 377–411 (1968) and Physical mechanisms of sputtering, Phys. Rep. **69**, 335–371 (1981)
68. K. Meyer, I.K. Schuller, C.M. Falco, Thermalization of sputtered atoms. J. Appl. Phys. **52**, 5803–5805 (1981)
69. V.V. Serikov, K. Nanbu, Monte Carlo numerical analysis of target erosion and film growth in three-dimensional sputter chamber. J. Vac. Sci. Technol. A **14**, 3108–3123 (1996)
70. A. Goehlich, D. Gillmann, H.F. Döbele, Angular resolved energy distributions of sputtered atoms at low bombarding energy. Nucl. Instr. Meth in Phys. Res. B **164–165**, 843–839 (2000)
71. T. Kenmotsu, Y. Yamamura, T. Ono, T. Kawamura, A new formula for energy spectrum of sputtered atoms due to low-energy. J. Plasma and Fusion Res. **80**, 406–409 (2004)
72. A. Goehlich, N. Niemöller, H.F. Döbele, Anisotopiy effects in physical sputtering investigated by laser-induced fluorescence spectroscopy. Phys. Rev. **62**, 9349–9358 (2000)
73. Y. Yamamura, T. Takiguchi, M. Ishida, Energy and angular distributions of sputtered atoms at normal incidence. Rad. Eff. Def. Solids **118**, 237–261 (1991)
74. Z.L. Zhang, L. Zhang, Anisotropic angular distribution of sputtered atoms. Rad. Eff. & Defects in Solids **159**, 301–307 (2004)
75. Y. Yamamura, Contribution of anisotropic velocity distribution of recoil atoms to sputtering and angular distributions of sputtered atoms. Rad. Eff. **55**, 49–55 (1981)
76. K. Zoerb, J.D. Williams, D.D. Williams, A.P. Yalin, Differential sputtering yields of refractory metals by xenon, krypton, and argon ion bombardment at normal and oblique incidences, Proceed. 29th Intern. Electr. Propulsion Conf., Princeton Univ. 2005, paper IEPC-2005-293 and A.P. Yalin, B. Rubin, S.R. Domingue, Z. Glueckert, J.D. Williams, Differential sputter yields of boron nitride, quartz, and kapton due to low-energy Xe+ bombardment, Proceed. 43rd AIAA Joint Propulsion Conf., Cincinnati 2007, paper 2007–5314
77. M. Stepanova, S.K. Dew, Estimates of differential sputtering yields for deposition application. J. Vac. Sci. Technol. A **19**, 2805–2816 (2001)
78. G. Betz, G.K. Wehner, Sputtering of multicomponent materials, in Ref. [4], pp. 11–90
79. H.H. Anderson, Nonlinear effects in collisional sputtering under cluster impact, Det. Kgl. Danske Vid. Selskab., Mat.-Fys. Medd. **43**, 127–154 (1993)
80. V.I. Zaporozchenko, M.G. Stepanova, Preferential sputtering in binary targets. Progr. Surf. Sci. **49**, 155–196 (1995)
81. R. Kelly, Surface compositional changes by particle bombardment, in *Chemistry and Physics of Solid Surfaces V*, ed. by. R. Vanselow, R. Howe, (Springer Verlag Berlin, 1984) pp. 159–182
82. R. Shimizu, Preferential sputtering. Nucl. Instr. Meth in Phys. Res. B **18**, 486–495 (1987)

83. J.B. Malherbe, Sputtering of compound semiconductor surfaces. *II.* Compositional changes and radiation-induced topography and damage, Crit. Rev. Solid Stat. Mater. Sci. **19**, 129–195 (1994)
84. R. Shimizu, N. Saeki, Study of preferential sputtering on binary alloy by in-situ Auger measurement of sputtered and sputter-deposited surfaces. Surf. Sci. **62**, 751–755 (1977)
85. R.R. Olsen, M.E. King, G.K. Wehner, Mass effects on angular distribution of sputtered atoms. J. Appl. Phys. **50**, 3677–3683 (1979)
86. H. Gnaser, H. Oechsner, Mass effects on angular distribution of sputtered atoms. Phys. Rev. Lett. **63**, 2673–2676 (1989)
87. P. Sigmund, Sputtering of single and multiple component materials. J. Vac. Sci. Technol. **17**, 396–399 (1980)
88. J.B. Malherbe, S. Hofmann, J.M. Sam, Preferential sputtering of oxides: A comparison of model predictions with experimental data. Appl. Surf. Sci. **27**, 355–365 (1986)
89. L. Pauling, *The Nature of the Chemical Bond* (Cornell University Press, Ithaca, New York, 1960)
90. R. Kelly, An attempt to understand preferential sputtering, Nucl. Instr. Meth. **149**, 553–558 (1978) and On the problem of whether mass or chemical bonding is more important to bombardment-induced compositional changes in alloys and oxides Surf. Sci. **100**, 85–107 (1980)
91. N.Q. Lam, H. Wiedersich, Bombardment-induced segregation and redistribution. Nucl. Instr. Meth in Phys. Res. B **18**, 471–485 (1987)
92. V. Tuboltsev, P. Kalkanen, M. Kolodyazhnaya, J. Räisänen, Composition dependence of $Si_{1-x}Ge_x$ sputter yield, Phys. Rev. B **72**, 205434 (2005)
93. M.Z. Hossain, J.B. Freund, H.T. Johnson, Differential sputter yields in $Si_{1-x}Ge_x$, J. Appl. Phys. **103**, 073508 (2008)
94. W.R. Grove, On the electrochemical polarity of gases, Phil. Trans. Royal Soc. (London) B **142**, 87–102 (1852)
95. N. Laegreid, G.K. Wehner, Sputtering yields of metals for low-energy Ar^+-, and Ne^+- ions with energies from 50 to 600 eV. J. of Appl. Phys. **32**, 365–368 (1961)
96. G. Sauerbrey, Verwendung von Schwingquarzen zur Wägung dünner Schichten und zur Mikrowägung. Z. für Physik **155**, 206–222 (1959)
97. A.W. Czanderna, T.E. Madey, C.J. Powell [Eds.), *Beam Effects, Surface Topography, and Depth Profiling in Surface Analysis*, (Kluwer Academic Publ., New York 2002)
98. F. Gao, R.K. Leach, J. Petzing, J.M. Coupland, Surface measurement errors using commercial scanning white light interferometers, Meas. Sci. Technol. **19**, 015303 (2008)
99. T. Whitaker, A. Li, P. Jones, R. Watts, Angular distributions of sputtered zirconium atoms. J. Chem. Phys **98**, 5887–5898 (1993)
100. G. Betz, K. Wien, Energy and angular distributions of sputtered particles. Intern. J. Mass Spectr. Ion Processes **140**, 1–110 (1994)
101. H.H. Andersen, H.I. Bay, Sputtering yield measurements, in Ref. [3], pp.145–218
102. E.S. Mashkova, V.A. Molchanov, *Medium-Energy Ion Reflection from Solids* (North-Holland, Amsterdam, 1985)
103. T. Tabata, R. Ito, Y. Itikawa, N. Itoh, K. Morita, Backscattering coefficients of H, D, and He ions from solids. At. Data Nucl. Data Tables **28**, 493–530 (1983)
104. W. Eckstein, *Computer Simulation of Ion-Solid Interactions*, Springer Series in Material Science, Vol. 10, (Springer, Berlin 1991)
105. W. Eckstein, Reflection (backscattering), (IPP report 17/12, 2009)
106. J. BøSttiger, J.A Davies, P. Sigmund, K.B. Winterbon, On the reflection coefficient of keV heavy-ion beams from solid targets, Rad. Eff. **11**, 69–78 (1971)
107. G.M. McCracken, N.J. Freeman, Back-scattering of keV hydrogen ions in solids. J. Phys. B **2**, 661–668 (1969)
108. J. Vucanic, P. Sigmund, Total backscattering of keV light ions from solid targets in single-collision approximation. Appl. Phys. **11**, 265–272 (1976)

109. W. Eckstein, J.P. Biersack, The reflection of light swift particles from heavy solid targets. Z.Physik A **310**, 1–8 (1983)
110. E.S. Parilis, V.K. Verleger, Energy spectra and charge states of light atoms scattered by solid surface, J. Nucl. Mater. **93** & **94**, 512–517 (1980)
111. E.S. Mashkova, Particle and energy reflection from solid surfaces. Rad. Eff. **54**, 1–28 (1981)
112. A. Mutzke, R. Schneider, W. Eckstein, R. Dohmen, K. Schmidt, U. von Toussaint, G. Badelow, SD TrimSP Version 6.00,(IPP Report 2019-02, 2019)
113. U. Littmark, A. Gras-Marti, Energy spectra of light ions backscattered from random solids. Appl. Phys. **16**, 247–253 (1978)
114. C.A. Ordonez, R.E. Peterkin, Numerical determination of plasma ion reflection coefficients at a sheath-surface interface. J. Nucl. Mater. **228**, 201–206 (1991)
115. V.S. Remizovich, M.I. Ryazanov, I.S. Tilinin, Energy and angular distributions of particles reflected in glancing incidence of a beam of ions on the surface of a material. Sov. Phys. JETP **52**, 225–230 (1981)
116. Y. Xia, X. Xu, C. Tan, L. Mei, H. Yang, X. Sun, Reflection of light ion from solid surfaces studied by Monte Carlo simulation and transport theory. J. Appl. Phys. **69**, 439–446 (1991)
117. E.W. Thomas, R.K. Janev, J.J. Smith, Scaling of particle reflection coefficients. Nucl. Instr. Meth. B **69**, 427–436 (1992)
118. R. Wedell, J.-P. Kuska, Theoretical description of sputtering at grazing incidence. Nucl. Instr. Meth. in Phys. Res. B **78**, 204–211 (1993)
119. W. Eckstein, Physical sputtering and reflection processes in plasma-wall interactions. J. Nucl. Mater. **248**, 1–8 (1997)

Part II
Applications

Chapter 6
Evolution of Topography Under Low-Energy Ion Bombardment

Abstract The sputtering of the surface by low-energy ion irradiation presented in the last chapter leads directly to roughening of the surface. The behavior can be described by relations of dynamic scaling theory and stochastic growth equations. The surface after ion irradiation with ion energies above the displacement energy is initially characterized by the formation of vacancies, interstitial sites, and adatoms, some of which coalesce into clusters. The kinetics of growth of surface defects in the early stages of ion irradiation is influenced primarily by two factors, the ion energy and the temperature during bombardment. In the specific case of a high nuclear energy loss, carters are created. As ion irradiation progresses, increasingly extended defects form on the surface, whereby a distinction must be made between intra- and Inter-crystalline surface defects. The evolution of individual defects such as cones, pyramids, etch pits, facets, etc. is discussed in detail as a function of the irradiation parameters. The theoretical description of ion beam-induced surface evolution is based on the assumption that the sputtering yield is exclusively a function of the local surface curvature and higher spatial derivatives of the local surface height. This assumption allows the surface evolution to be considered spatially and temporally as a moving wavefront and to apply geometrical methods developed in optics. Finally, secondary ion beam-induced mechanisms, such as grazing incidence ion reflection, re-deposition, shadowing, surface diffusion, non-uniform bombardment, viscous flow, and swelling are presented, which can have an additional significant effect on the topography evolution.

One academically interesting and technologically significant result of the bombardment of solid surfaces with low-energy ions is the erosion of surfaces. The roughening of a surface as a result of the erosion (sputtering) process is the first observable change following ion bombardment. If the energy of the ions is higher than the displacement energy (see Chap. 5), the erosion on the surface begins with the sputtering of target atoms and the formation of adatoms (atoms adsorbed on the surface), and adatom clusters on the surface, along with the formation of vacancies and interstitials in the subsurface region. The development of sophisticated experimental techniques and comprehensive molecular dynamical simulations have made it possible to study the formation of interstitials and vacancies in the near-surface regions and

B. Rauschenbach, *Low-Energy Ion Irradiation of Materials*, Springer Series in Materials Science 324, https://doi.org/10.1007/978-3-030-97277-6_6

adatoms on the surface following low-energy ion bombardment. For low ion energies (up to a few thousand electron-volts), surface effects (surface relaxation, free surface-governed defect agglomeration, diffusion on the surface, viscous flow, etc.) significantly influence the surface morphology.

This chapter will initially address the kinetic roughening of surfaces following low-energy ion bombardment and the description of this within the framework of dynamic scaling theory (Sect. 6.1). Following this, there will be a presentation of some aspects of the early defect evolution on surfaces or in the near-surface region following ion bombardment with single or a small number of ions as well as the formation of extended defects in crystalline materials, based on the presence of an additional diffusion barrier (Sect. 6.2), classification of several surface defects as a result of low-energy ion sputtering (Sect. 6.3), kinematic description of the evolution of the bombarded surfaces (Sect. 6.4), and the contribution of secondary processes to surface evolution, including ion reflection, re-deposition, shadowing, ion beam-induced surface diffusion viscous flow, and swelling (Sect. 6.5).

6.1 Ion Beam-Induced Roughening

For over a century now, it has been known that ion bombardment can result in surface roughening on the microscopic scale [1]. Because of the statistical nature of the sputtering process, this process of surface roughening is inevitable [2]. Individual ions randomly strike the bombarded surface and statistically erode the surface. Thus, a layer-by-layer removal cannot be expected. The surface tends to become rougher and rougher with time of bombardment. Experimental studies (here, a ceramic surface after 2 keV Xe ion bombardment, Fig. 6.1) confirm that on an initially flat surface, large-scale mountain-like structures evolves during longer duration of ion bombardment, i.e., these structures are produced as a result of the ion beam-induced erosion in the presence of noise fluctuation. This phenomenon, known as white noise, can be modelled by adding a Gaussian random variable, η, to the stochastic rate equation of the evolution of the topography in the result of low-energy ion bombardment (see e.g., Sect. 6.1.3). The effect of this statistic fluctuation on the evolution of the surface is profound. Figure 6.1 demonstrates the evolution in presence of white noise. Bales et al. [3] simulated the bombardment of an initially smooth surface in the absence and in the presence of noise fluctuation. They were able to show that roughening is possible only under the assumption of white noise, i.e. the statistical fluctuation of the incoming ions.

In contrast to the statistic fluctuation on the surface caused by ion bombardment, a tendency toward layer-by-layer erosion of single crystalline solids after ion bombardment can be observed if the sputtering yield is smaller than one, i.e. for ion energies lower than any hundred electron-volts. In this case, the collision cascade is not completely formed and the sputtering is limited to surface atoms. For example,

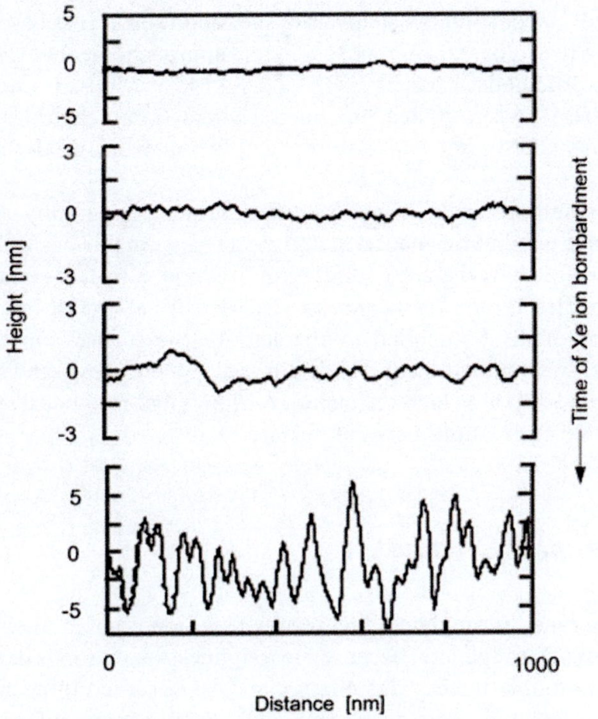

Fig. 6.1 Roughness evolution as function of the time of ion bombardment (ion fluence) of a ceramic surface after perpendicular 2 keV Xe ion bombardment at room temperature

a layer-by-layer erosion could be observed for 200 eV and 250 eV Xe ion bombardment of Si(111) [4] and 500 eV Ar ion bombardment of Si(111) surfaces at 900 K [5, 6], as could also be demonstrated using computer simulations [7, 8].

Roughness characterises the topography of a solid surface (see, e.g., [9, 10] or Appendix L). The roughness of a two-dimensional surface can be described by the interface width w, which gives the root-mean-square (rms) roughness fluctuation at the height h(x, y, t) to

$$w(L, t) = \sqrt{\left\langle \left[h(x, y, t) - \overline{h}(x, y, t) \right]^2 \right\rangle}, \tag{6.1}$$

where the brackets $\langle \ldots \rangle$ denote the ensemble average and L the lateral sample length. The mean height is given by

$$\overline{h} = \frac{1}{L^2} \sum_{x=1}^{L} \sum_{y=1}^{L} h(x, y) \tag{6.2}$$

(see Appendix L). A quantitative analysis of the rough surfaces is based on different, non-destructive metrological approaches. In one approach, mechanical profilometry scanning probe methods, such as STM and AFM, as well as scanning electron microscopy (SEM), were applied. In another, diffraction techniques, such as light scattering, X-ray scattering, and electron diffraction were used (for details, see Appendix L).

In general, methods used to characterize surface topography are capable of providing surface height information in digitized form. On this basis, different statistical parameters, such as the auto-correlation function, R(r, t), height-height correlation function, H(r, t), and the power spectral density after fast Fourier transform (FFT) can additionally be applied to characterize the surface topography and its dynamics under ion bombardment (for details, see Appendixes L and M). In contrast to the interface width, w, which characterizes the global roughness, the mentioned functions provide correlations between surface features.

6.1.1 Roughness Evolution

The increase in random roughness that results from ion bombardment is frequently referred to as kinetic roughness, because the roughness evolution is dependent on the duration of ion bombardment. This effect could be observed in nearly all material classes (metal, semiconductors, insulators, etc.). As an example, Fig. 6.2 shows the evolution of the roughness of silicon, deduced from AFM images after noble gas ion bombardment, as a function of the time of bombardment [11]. For short erosion times of up to 5 min (corresponding to ion fluence of 5.6×10^{17} ions/cm^2), the roughness

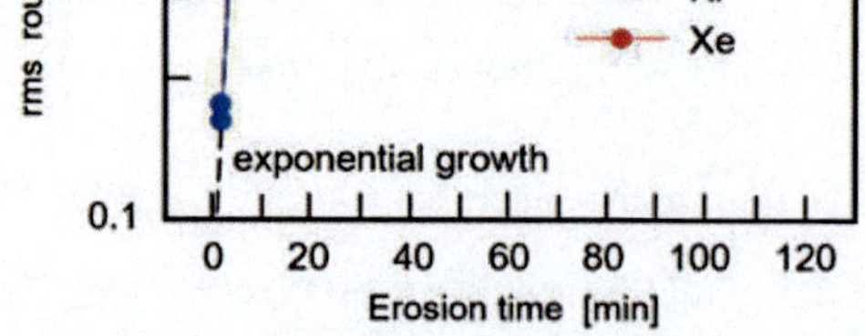

Fig. 6.2 Evolution of the rms roughness of Si with time of ion bombardment (ion fluence) for Ar, Kr, or Xe ion (ion energy is 1200 eV, angle of ion incidence is $\theta = 15°$). The dashed line corresponds to an exponential growth of the surface roughness for the initial stage of sputtering [11]

seems to grow exponentially (see dashed line in Fig. 6.2). For a time of ion bombardment of approximately 9 min (corresponding to ion fluence of 1×10^{18} ions/cm^2), the roughness saturates and remains constant during further sputtering. This behavior has been confirmed several times, e.g., for ZnO [12], for glass [13], for InP [14], and for silicon [15] after low-energy rare gas ion bombardment. For shorter times of ion bombardment (or lower ion fluences, $< 10^{17}$ ions/cm^2, approximately), i.e. in the event that roughness saturation has not yet been achieved (c.f. Fig. 6.2), the evolution of the rms roughness versus ion fluence (time of ion irradiation) has been interpreted either as an approximately linear increase [16, 17] or an increase proportional to the square of the time of ion bombardment [18].

6.1.2 Dynamic Scaling of the Roughness Evolution

The evolution of the surface roughness under ion bombardment can be described by the concept of dynamic scaling, i.e. that the evolution of the surface height contour can be characterized by appropriate parameters [19–24]. Within this concept, it is assumed that the roughness evolution is time- and scale-invariant, where this evolution is characterized by scaling parameters such as the roughness exponent α, the growth exponent β, and the coarsening exponent or dynamic exponent z (for details, see Appendix M).

A suitable possibility for studying the problem of the time-dependent growth of interface width (rms roughness) as a function of the parameters of ion bombardment (ion fluence, temperature during bombardment, ion energy, etc.) within the framework of dynamic scaling is based on the application of stochastic (continuum) growth equations [19–24]. The most relevant information regarding the dynamic behavior of the roughness evolution can be obtained from the temporal dependence of the roughening process. For self-affine surfaces (for a definition, see Appendix M), it is known that the roughness grows with time as a power law at an exponent β (growth exponent) until the roughness reaches a stationary regime, where the roughness saturates. The stationary roughness and the saturation time are also dependent on the system size, characterized by the roughness exponent α.

It is assumed that a surface is initially eroded by ion irradiation over time and its roughness then increases as a result of the stochastic ion bombardment process. If a section of the surface at size L is perpendicular to the direction of the global ion etch direction, then the surface can be described by a simple function h(x, t). Because of an increase in random fluctuations in the heights, the average width of the interface (rms roughness), as defined by (6.1), grows over time. Initially, the interface width, w, grows over the time of ion bombardment (fluence) for $t \leq \tau_s$, where τ_s is the saturation time. Beyond this time τ_s, the roughness is independent of the time of ion bombardment. According to Family and Vicsek [25], this behavior can be described by a dynamic scaling relation (Appendix M), expressed by

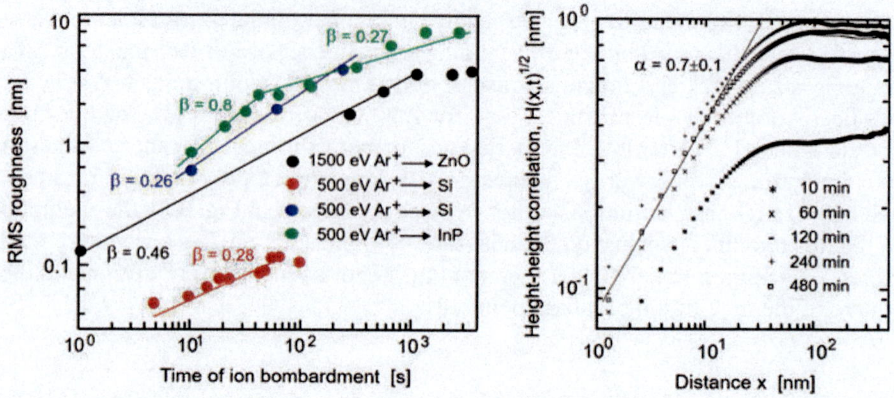

Fig. 6.3 Left: log–log plot of the interface width w(t) as function of the time of ion bombardment for different sputter materials (● [12], ● [31], ● [15], ● [14]) Right: log–log plot of the square root of the one-dimensional height-height correlation function as a function of distance for Si(111) after 500 eV Ar ion bombardment at 610 K (figure taken from [15])

$$w(L,t) \propto \begin{cases} t^{\beta} & \text{for } \tau_o \ll t \ll \tau_s \\ L^{\alpha} & \text{for } t \gg \tau_s \end{cases}. \tag{6.3}$$

The crossover between both roughness regimes (see Fig. M.1 in Appendix M) occurs at ξ, the lateral correlation length, which signifies the distance at which the surface features are no longer correlated. The correlation length scales as $\xi \propto t^{1/z}$, where the dynamic exponent $z = \alpha/\beta$.

The three exponents and the correlation length can be experimentally determined (for details see Appendix M). The growth exponent β is extracted from the linear fit to the log–log plot of the interface width as a function of the time of bombardment, as shown in Fig. 6.3 (left). It is notable that the experimentally determined growth exponents possesses values between 0.2 and 0.5. However, significantly larger β-values ≥ 0.8 were also identified. Frequently, the roughness exponent α from the log–log plot of height-height correlation function versus scan length for different sputtering times was simultaneously extracted for $t > \tau_s$ (see Appendix M). An example is shown in Fig. 6.3 (right) [15]. In this case, a roughness exponent α = 0.7 ± 0.1 was evaluated. Several authors have demonstrated that the roughness exponent is nearly invariant with regard to time (ion fluence). For example, weighted averages of α = 0.77 ± 0.04 [26] and α = 0.7–0.8 [14] were experimentally determined, and α = 0.7–0.8 [27] and α = 0.75–0.8 [28] were calculated. The values of the experimentally determined scaling exponents α and β are listed in Table 6.1. It is obvious that the values of the growth exponent β vary between approximately 0.1 and 0.3 and between 0.8 and 1. Within the framework of the dynamic scaling theory, four different mechanisms contribute to the roughening/smoothing of surfaces by ion bombardment: (i) random sputtering, (ii) erosion instability, as proposed by Bradley and Harper [29] (see Sect. 8.2.1.1), (iii) surface diffusion (see Sect. 6.5.4),

Table 6.1 Measured scaling exponents for different materials after low-energy ion bombardment (θ is the incident angle, f-SiO_2 is fused silica, ML is monolayer)

Material	Ion energy [eV]	Scaling exponents		Remarks	Refs.
		Roughness exponent α	Growth exponent β		
Pt	9700	0.77 ± 0.04	0.36 ± 0.05	$\theta = 45°$	1
Fe	5000	0.53 ± 0.02		$\theta = 25°$	2
Ni	1000	0.49 ± 0.05			3
Graphite	5000	~0.2–0.4		For lower fluences, only	4, 5
Si(111)	300, 1000	0.23 ± 0.08 at 1 keV 0.3 ± 0.05 at 300 eV	0.07 ± 0.01 at 1 keV 0.14 ± 0.002 at 300 eV	$\theta = 55°$	6
	500	1.15 ± 0.08		T = 300–600 °C	7
	500	0.7 ± 0.1	0.25 ± 0.05	T = 610 K	8
Si (001)	200–3000	1.03	0.28		9
Ge(001)	1000	~0.4 for T < 488 K ~0.1 for T > 488 K		Abrupt transition between two dynamic regimes as function of temperature	10
	30	0.34 ± 0.05 for T = 20–200 °C 0.95 ± 0.02 for T = 225–325 °C 0.52 ± 0.07 for T = 350–400 °C		Square mounds and pit for T = 20–200 °C Mounds disappear for T = 350–400 °C	11
	1000		0.4 for T < 488 K 0.1 for T > 488 K		12
GaAs(110)	2000		0.3 at 725 K	$\beta > 0.3$ at 625 K for erosion < 1 ML $\beta < 0.3$ at 625 K for erosion > 1 ML	13
	300–5000		0.26–0.31	T = 626–775 K	14
	3000		0.4–0.5	T = 626–775 K	15
ZnO(0001)	1.5	0.75	0.46		16

(continued)

Table 6.1 (continued)

Material	Ion energy [eV]	Scaling exponents		Remarks	Refs.
		Roughness exponent α	Growth exponent β		
InP	500	0.7–0.8	0.80 ± 0.10 0.27 ± 0.06	$\theta = 40°$, similar scaling exponents for $\theta = 0°$	17
f-SiO_2	600–1200		0.46 ± 0.03 at 800 eV	β increases from 0.25 ± 0.02 at 600 eV to 0.93 $\pm$ 0.05 at 1200 eV, ripple formation	18
Glass	800	0.99 ± 0.03	1.02 ± 0.20	$\theta = 35°$–$75°$; ripple formation	19

[1] P. Karmakar D. Ghose, Nucl. Instr. Meth. Phys. Res. B 222 (2004) 477
[2] J. Krim et al., Phys. Rev. Lett. 70 (1993) 57
[3] Z. Csakok, et al., Surf. Sci. 364 (1996) L600
[4] E.A. Eklund et al., Phys. Rev. Lett. 67 (1991) 1759
[5] E.A. Eklund et al., Surf. Sci. 285 (1993) 157
[6] V.I.T.A. de Rooij-Lohmann et al., Appl. Surf. Sci. 256 (2010) 5011
[7] H.-N. Yang, et al., Phys. Rev. B 50 (1994) 7635
[8] A. C.-T. Chan, G.-C. Wang, Surf. Sci. 414 (1998) 17
[9] S.A. Pahloy, et al., Nucl. Instr. Meth. in Phys. Res. B 272 (2012) 206
[10] D.-M. Smilgies et al., Europhys Lett. 38 (1997) 447
[11] D. Chowdhury, et al., Vacuum 107 (2014) 23
[12] D.M. Smilgries, et al., Surf. Sci. 377–379 (1997) 1038
[13] X.S. Wang et al., Surf. Sci. 364 (1996) L 511
[14] X.S. Wang et al. J. Vac. Sci. Technol. B 13 (1995) 2031
[15] R.J. Pechman et al., Phys. Rev. B 51 (1965) 10929
[16] V. Solanki et al., Nucl. Instr. Meth. in Phys. Res. B 434 (2018) 56
[17] F. Frost, et al. Phys. Rev. Lett. 85 (2000) 4116
[18] D. Flamm et al., Appl. Surf. Sci. 179 (2001) 95
[19] A. Toma, et al., Nucl. Instr. Meth. in Phys. Res. B 230 (2005) 551

and (iv) surface curvature-dependent sputtering (see Sect. 6.4.1). Cuerno et al. [27, 30] interpreted the experimental observations based on a stochastic model involving these mechanisms. This simulation provides different roughness regimes over time. In the first regime, the roughness scales at $\beta = 0.38 \pm 0.02$, the second refines this at $\beta > 0.5$, followed by a region where $\beta = 0.23 \pm 0.03$ and a final region where $\beta = 0.28 \pm 0.03$. Because the first regime cannot be observed within most of the experiments and it is difficult, from an experimental point of view, to distinguish the last regime from the third regime, the results of the growth exponent β presented in Table 6.1 satisfy this prediction model by Cuerno et al. within a satisfactory deviation.

Table 6.1 shows also that the values of the experimentally determined roughness exponent usually vary from $0 < \alpha < 1$. This indicates that the studied surfaces can be termed self-affine surfaces, because the rms roughness changes with the lateral sample length L as $w \propto L^{\alpha}$. The linear slope (log w versus log of distance, see Fig. 6.3, right) indicates that the self-affine character of the ion bombarded surface is a result of both roughening and smoothing mechanisms. Consequently, the roughness exponent quantifies roughness changes with the length scale rather than the rms roughness.

It must be noted that knowledge of the dynamic scaling exponent provides only a limited contribution to the understanding of the physical mechanisms that cause roughness evolution on surfaces under ion bombardment. Consequently, stochastic equations (next subchapter) are applied to describe the evolution of the topography under low-energy ion bombardment. The benefit of stochastic equations lies in their ability to deduce the underlying mechanisms on the basis of adopted roughness and smoothing processes as well as their easy applicability.

6.1.3 Stochastic Growth Equations for Surface Erosion by Ion Bombardment

Another possibility for obtaining analytical information about the roughness evolution is based on the construction of a stochastic growth equation to describe the roughness evolution. When the erosion of surfaces by ion bombardment can be interpreted as 'negative surface growth', then this continuum growth equations also models the ion erosion process wherein this equations is anisotropic due to the directionality of the ion beam [32].

The time-dependent evolution of the roughness under ion beam bombardment is oriented on the tool of the theoretical description of the growth of thin films on a mesoscopic scale (see e.g., [19–24]). In general, a temporal variation of the height profile satisfies a stochastic continuum equation of the form

$$\frac{\partial}{\partial t} h(\boldsymbol{r}, t) = G[h(\boldsymbol{r}, t)] + noise, \tag{6.4}$$

where $h(\boldsymbol{r}, t)$ denotes the local height of the surface profile in $\boldsymbol{r} = (x, y)$. Thereby, $G[h, (\boldsymbol{r}, t)]$ is a deterministic functional that includes dynamic surface processes to

Table 6.2 Stochastic growth equations and scaling exponents for various local growth models in 2 + 1 dimensions (η is the white-noise simulating random fluctuation during sputtering, ν and λ are parameters of the model)

Stochastic growth equation		Scaling exponents			Refs.
		α	β	z	
Random erosion	$\frac{\partial h}{\partial t} = \eta$		0.5		[28]
Edwards-Wilkinson eq.	$\frac{\partial h}{\partial t} = \nu\nabla^2 h + \eta$	0	0	2	[20, 33]
Kardar-Parisi-Zhang eq.	$\frac{\partial h}{\partial t} = \nu\nabla^2 h + \frac{\lambda}{2}\|\nabla h\|^2 + \eta$	0.38	0.24	1.58	[20, 34]
Mullins diffusion eq.	$\frac{\partial h}{\partial t} = -K\nabla^4 h + \eta$	1	0.25	4	[35]
Kuramoto–Sivashinsky eq.	$\frac{\partial h}{\partial t} = \nu\nabla^2 h - K\nabla^4 h + \frac{\lambda}{2}\nabla^2\|\nabla h\|^2 + \eta$	0.75–0.80	0.22–0.25	3–4	[28][a]
		0.25–0.28	0.16–0.21		[28][b]

[a] Early time values of exponents, [b] Long time values of exponents

be taken into consideration [24] and the noise, denoted by $\eta = \eta(\mathbf{r}, t)$, is the random noise that exist during the ion bombardment due to the stochastic nature of this process (for details, see (M.6) of Appendix M).

With the help of (6.4), the topography can be described by a function that is dependent on the local coordinates, $\mathbf{r} = (x, y)$, and the time t. Every height over the substrate can then be assigned a height h at every point $\mathbf{r}$ and every time. In the simplest case, the dynamic at the surface is determined by three processes: (i) deposition/implantation of incidence particles, (ii) surface diffusion, and (iii) sputtering of surface atoms. Assuming conservation of matter on the surface, the functional must be the divergence of the height change.

Over recent decades, numerous equations have been suggested that aim to confirm a concrete growth model and to obtain analytic values for the exponents α, β and $z = \alpha/\beta$ (Table 6.2).

The simplest equation of this type was suggested by Edwards and Wilkinson [33]. This field equation has a functional $G(\nabla h)$ which can be written in form of a divergence, $G(\nabla h) = -\upsilon\nabla^2 h$, where the Laplacian term, $\upsilon\nabla^2 h$, in this equation is referred to as a surface relaxation term, because the effect of this term is to smooth the surface by diffusion [24]. According to Mullins [35], changes in the surface topography (smoothing) can be described by $\partial h/\partial t = -K\nabla^4 h$ (see Table 6.2), where the coefficient K is governed by the Arrhenius law $K \propto \exp(-E_{sd}/k_B T)$ and E_{sd} is the activation energy for nearest-neighbor hopping on the surface. Based on the consideration of symmetry, Kardar et al. [34] proposed a non-linear continuum equation, which contains the term $\frac{\lambda}{2}|\nabla h|^2$ with the parameter λ. When the local angle of ion incidence increases (e.g., an incident ion hits the flank of a surface elevation), then the sputtering yield rises (curvature-dependent sputtering yield, see Sect. 6.4.1). Thus, these regions of the surface are subjected to additional sputtering, i.e. the contribution of this effect to the height function h can be approximated by $\frac{\lambda}{2}|\nabla h|^2$ with $\lambda > 0$ (see Table 6.2). The non-deterministic Kuramoto–Sivashinsky equation in Table 6.2 (see Sect. 8.2.1.2) combines the discussed terms [36]. The first term of this equation in Table 6.2 again describes the effect of erosion, the second term is the surface diffusion, and the third term is a non-linear, non-conservative term that again decreases the height h by sputtering normal to the local surface (see also Kardar-Parisi-Zhang equation). With help of the both the $\nabla^2 h$ and $|\nabla h|^2$ terms, the effect of the knock-out of surface atoms by incoming ions is described [28], where the derivation of these terms is based on the well-known model by Sigmund (see Chap. 5).

The theoretical results of the scaling exponents predicted by different models are summarized in Table 6.2. Numerous experimental investigations and computer simulations have shown that the evolution under ion bombardment can be described well within the framework of the kinetic roughening theory. Frequently, these studies are based on a comparison of the roughness and dynamic exponents as a result of adopted roughening and smoothing mechanisms (see Table 6.2) and the experimentally or theoretically determined scaling exponents. In the following, some representative results will be introduced:

- Koponen et al. [37] studied the roughening of carbon surfaces following low-energy Ar ion bombardment by means of Monte Carlo simulations with variation of the ion incidence angle and temperature. They found that the roughness exponent decreases from $\alpha \approx 0.37$–0.45 after normal ion incidence up to $\alpha \approx 0.25$ after 60° ion incidence, while the growth exponent is virtually independent of the ion incidence angle θ. The dynamic exponent $z = \alpha/\beta$ was about 3.2–4.0, typical for a diffusion-controlled relaxation process (see Table 6.2). The dependence of the roughness exponent on the ion energy up to 10 keV was also studied [38]. They identified a linear relation, $\alpha = (0.011\ \mathrm{nm}^{-1}) \cdot a \cdot \cos\theta$, where a is the average depth of the deposited energy (see also Fig. 4.10).
- Cuerno et al. [27, 30] studied the development of the growth exponent by numerical integration of the noisy Kuramoto–Sivashinsky equation in (1 + 1) dimensions. Initially, values of growth exponent were obtained that were consistent with the values of the diffusion equation by Mullins [35]. With increasing time, the growth exponent increases up to a maximal value and then decreases to values that are consistent with values predicted according to the Edwards-Wilkinson equation. Roughness exponents expected by the Kardar-Parisi-Zhang equation were obtained only for times smaller the time of saturation τ_s.
- Drotar et al. [28] performed numerical simulations of the noisy Kuramoto-Sivashinsky equation in 2 + 1 dimensions and found the scaling exponents $\alpha = 0.75$–0.8, $\beta = 0.22$–0.25, and $z = 3$–4. These exponents, derived at the early-time region, are in good agreement with those obtained after ion bombardment at low energies (< 1 keV). For longer times, the scaling exponents are significantly lower than the exponents obtained when using the Kardar-Parisi-Zhang equation.
- Scaling exponents $\alpha = 0.2$–0.4 and $z = 1.6$–1.8 were found after Ar ion bombardment of graphite [17], which are approximately consistent with the predicted values of the Kardar-Parisi-Zhang equation. But, as the authors point out, the

topographical changes are strongly dependent on the conditions of the ion bombardment (temperature, ion fluence, and flux).

- The effect of temperature on roughness due surface diffusion is recognizable in Table 6.1. Yang et al. [39] studied the roughness evolution in time on Si(111) under Ar ion bombardment as a function of the temperature between room temperature and 650 °C. They observed anomalous behavior of the height-height correlation function on the short range scale for t > 450 °C, which can be interpreted as a dynamic phase transition between smooth phase at high temperatures and rough phase at low temperatures.
- Wang et al. [40] found that after Ar ion bombardment at 625 K, GaAs(110) surfaces were rougher on a small scale than those bombarded at 725 K. On a large scale, surfaces bombarded at 625 K were smoother than those bombarded at 725 K. Consequently, on a large scale, surfaces sputtered at a higher temperature become rougher than those at a lower temperature.
- A sharp transition between two roughening regimes, characterized by growth exponent $\beta = 0.4$ for $T < 488$ K and $\beta = 0.1$ for $T > 488$ K, was detected by Smilgies et al. [41] after 1 keV Xe ion bombardment of Ge. Chan and Wang [14] could not identify the determined values of the scaling exponents after 500 eV Ar ion bombardment of Si(111) at 610 K with a diffusion bias roughening process. The roughness morphology was consistent with the early-time behavior of the Kuramoto–Sivashinsky equation, i.e. it corresponds to the initial stage of the dynamic scaling theory.
- Pahlovy et al. [31] and Chowdhury et al. [42] estimated roughness exponents close to one after low-energy Ar ion bombardment of single crystalline Si and Ge, respectively. Consequently, the surface morphology should be determined by the dynamics of a diffusion-driven process (see Table 6.2).

Due to the complex nature of the problem, it has not been possible to date to develop a comprehensive and accurate description of the evolution of roughness under low-energy ion bombardment. However, numerous experimental investigations and theoretical studies on the roughness of ion beam eroded surfaces have been performed, which allowed prediction of morphological evolution within a limited framework.

6.2 Formation of Surface Defects by Low-Energy Ion Irradiation

The bombardment of solids with energetic ions results in atomic displacements in the bulk and at the surface and, consequently, in changes to the surface topography. If sufficient energy is transferred in a single collision with the target atoms to surpass the local binding energy, the atoms in the bulk can be displaced (see Chap. 4) or surface atoms can be sputtered (see Chap. 5). The displaced atoms can create a higher generation of displacements, which ultimately leads to the formation of a collision cascade (spike). When the ion energy is reduced, the processes of defect formation (generation of interstitials, vacancies, and clusters of these defects) shift increasingly to the near-surface region or to the surface. It is significant that a notable number of atoms (known as adatoms) remain on the surface following the collision process.

6.2.1 Defect Distribution and Defect Evolution After Ion Bombardment

The implantation process of low-energy ions results in the generation of ion beam-induced defects (interstitials, vacancies). It should be taken into account that the distribution of the interstitials and vacancies represents strictly net-distributions of defects that have survived recombination and annihilation processes immediately following the ion beam-induced defect generation process. These temperature-dependent annealing processes are capable of significantly reducing the number of vacancies and interstitials.

Adatoms can also be produced at the surface, using ion bombardment. Harrison and Webb [43] predicted this defect type using MD simulations of 5 keV Ar ion bombardment of Cu surfaces. These simulations indicate that atoms from the Cu target are pushed onto the surface. Sputtering of target atoms and its re-deposition onto the surface is a second option to form adatoms.

Thus, three types of surface point defects (interstitials, vacancies, and adatoms, as well as their clusters and cluster islands) contribute to surface topography. For instance, Mäkinen et al. [44] studied the defect distribution in Al by positron annihilation in dependence on Ar ion energy, fluence, and incident angle as well as temperature and the recovery regime. The shape of the defect profiles varies only slightly with the sputtering energy and incident angle. Defect production at less than 1 keV Ar ion energies is typical for production of 1–5 vacancies per incident ion. The temperature (150 K and 300 K) during ion bombardment has little effect on vacancy profiles. Defects anneal out gradually between 100 °C and 400 °C. Molecular dynamic simulations have been used to obtain a detailed understanding of the phenomena occurring during and after low-energy atom bombardment of surfaces. Karetta and Urbassek [45] studied, among others, the vacancy and interstitial distribution, the spontaneous defect recombination, and the number of surface vacancies and adatoms produced at low energy ion bombardment (30 eV and 100 eV) of Cu at two different incidence angles (0° and 60°). In this case as well, vacancies only formed close to the surface and interstitials were only produced further inside the target by long-range replacement sequences in the crystalline target. In addition, a considerable number of adatoms were created on top of the target. Kornich and Betz [46] illustrated through MD simulations that the creation of stable interstitials decreases with temperature due to the recombination of interstitials and vacancies at higher ambient temperatures. Production of vacancies (at the surface) increases due to the increasing number of adatoms generated.

MD simulations have been applied in studies of defect production in semiconductors, particularly in silicon, after low-energy ion bombardment. Murty and Atwater [47] studied defect generation in the near-surface region of Si(001) after non-normal Ar ion bombardment and found that for ion energies < 20 eV, it is exclusively surface atoms that are displaced. MD simulations by Hensel and Urbassek [48] proved that the damage distribution depends on the crystal orientation of the bombarded Si targets. Tarus et al. [49] studied the vacancy and interstitial distributions in Si(001) as a function of depth for ion energies < 1 keV after self-ion bombardment (Fig. 6.4). At low ion energies (< 200 eV), the interstitial distribution is much wider than the vacancy distribution, whereas at higher energies (200 eV and 800 eV), the vacancy distribution approaches the interstitial distribution. The vacancy number peaks strongly at the surface layer, but the percentage of vacancies in the surface layer relative to all vacancies drops as the ion energy rises.

On the basis of experimental studies and MD simulations, a schematic picture of the vacancy and interstitial distribution after ion bombardment with low ion energies follows the results presented above, where this schematic is only assumed to be valid at low temperatures (for temperatures where no notable diffusion and temperature-dependent recombination processes are to be expected). The vacancy distribution is high in the first monolayers and has a tail that extends through many monolayers into the bombarded material. The interstitial profile lies deeper and is the result of collision processes. It should be noted that the distribution of the implanted ion species is slightly deeper than the interstitial profile. At slightly higher ion energy,

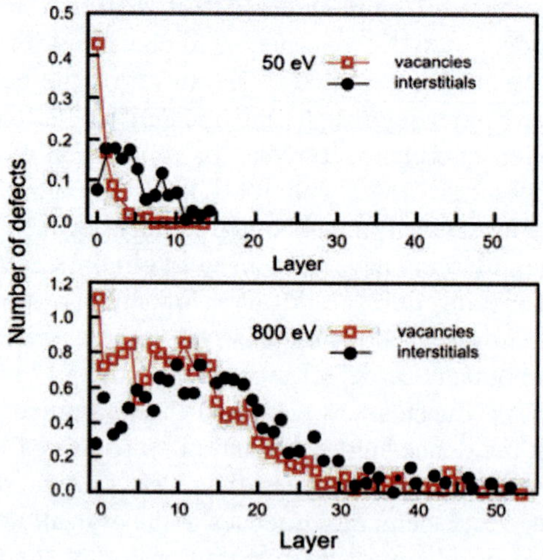

Fig. 6.4 MD simulations of the defect distribution after self-ion bombardment of Si(001) with 50 eV and 800 eV as a function of the depth (in monolayers). Layer 0 represents the surface. The lines are drawn as visual guides (figures adapted from [49] and modified)

vacancy distribution is similar to the interstitial distribution outside the near-surface region. Finally, the concrete concentration of the interstitials and vacancies as a function of the depth is dependent on the ion mass, the mass of the target atoms, the angle of incidence, the impact energy, and the crystal orientation with respect to the direction of the incident ions.

A number of investigations have been carried out to study the temperature dependence of the formation and distribution of defects after low-energy ion bombardment. For example, using MD simulations and variable temperature STM studies, Seki et al. [50] examined the formation and annihilation of defects at higher temperatures on silicon surfaces after Xe single ion impact with ion energies between 1 keV and 5 keV. The average single ion impact traces of a few nanometer diameters are independent of the ion energy. After annealing at higher temperatures (> 400 °C), vacancies diffuse toward the surface (interstitials are immobile at approximately this temperature). A further increase of the temperature (600 °C) leads to the formation of vacancy clusters. At 650 °C, the interstitials diffuse and recombine with surface vacancies.

Of special interest for the evolution of the topography after low-energy ion bombardment is the formation of adatoms on the surface. Adatoms (also called inverse vacancies), the counterparts to the surface vacancies, are isolated atoms bound to the surface, which are capable of moving on the surface by way of atomic jumps. Experimental studies and MD simulations have illustrated a substantial generation of adatoms on the surface after ion bombardment with energies < 10 keV (see next subchapter). On the one hand, the formation of adatoms after ion irradiation is the result of the evolution of a displacement cascade, where the material in the high-energetic core of the cascade expands towards the surface and transports some atoms to the surface. On the other hand, the sputtering process produces isolated atoms, which are also deposited onto the surface (re-deposition).

Because of the high complexity of the processes involved, it is difficult to make generalized predictions about the evolution of defects on the surface and in the subsurface region in dependence on the ion species, ion energy, target material, and temperature. Based on qualitative conceptions developed by Seki et al. [50] and Chan and Chason [51], Fig. 6.5 provides a schematic of the different processes, such as defect generation, diffusion, recombination, and annihilation that takes place after single ion impact at elevated temperatures and low energies and contribute to modifying the topography of the surface. After ion bombardment with energies higher than the displacement energy, vacancies and interstitial atoms are created in the subsurface region (some monolayers, ML) and also in deeper layers. The concentration of vacancies in the near-surface region is higher than the concentration of interstitials. In the period of recovery immediately following ion impact, vacancies can diffuse toward the surface and create vacancy clusters on the surface or, at elevated temperatures, recombine with interstitial atoms in bulk. This is followed by a reduction in the total number of surface defects per ion impact. The size of the vacancy clusters is dependent on the number of vacancies created in the near-surface layer.

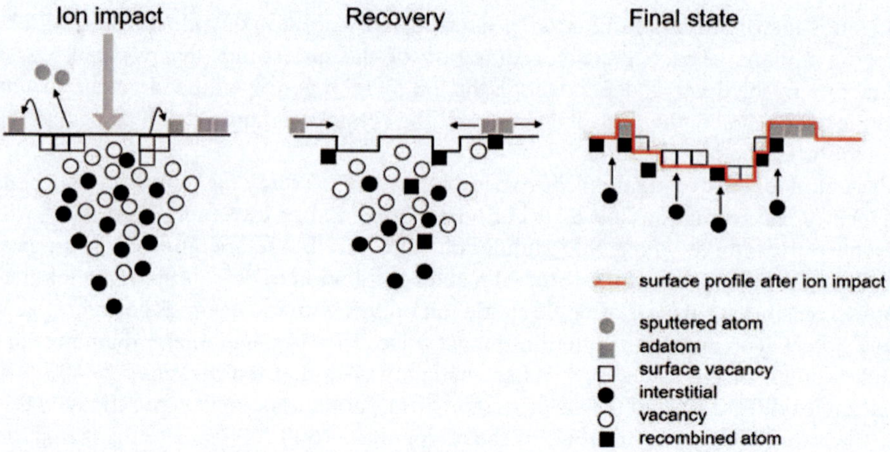

Fig. 6.5 Schematic representation of defect evolution after low-energy single ion impact and subsequent recovery at elevated temperatures. The arrows indicate the possible movement of atoms and the red line indicates the surface following sputtering and recovery (final surface)

Consequently, the size of the surface defects depends only on number of those vacancies that were created in the subsurface layer. At higher temperatures, interstitial atoms can also diffuse and recombine with the surface vacancies. The diffusion and enrichment of vacancies on the surface as well as the annihilation of adatoms by vacancies and the attachment of adatoms on the surface ultimately lead to topographical modifications to the surface (red line in Fig. 6.5).

6.2.2 *Early Stages of the Formation of Surface Defects by Ion Impact*

According to the findings in the previous chapter, it can be expected that in the low-energy ion bombardment of crystalline solids, particularly in metals, vacancies and adatoms are generated at very low ion fluences on the surface and in the near-surface region. At elevated temperatures, these surface point defects are mobile and can form clusters or islands. The early stages of defect formation on surfaces following low-energy ion bombardment has been studied on crystalline metals (Al [52], Pt [53–56], Cu [57, 58], Au [59]) and semiconducting surfaces (GaAs [60], Si [61–66], Ge [67, 68]) using scanning tunneling microscopy (STM), scanning electron microscopy (SEM) or reflection high energy electron diffraction (RHEED). Sometimes, ion fluences are applied in units of the atomic area density of the bombarded crystal face, i.e. one ion per surface atom is referred to as 1 ML (monolayer). Fluences up to about 1 ML are used to study the early stages of the formation of surface defects. For example, Fig. 6.6 (left) shows the topography of GaAs(110)

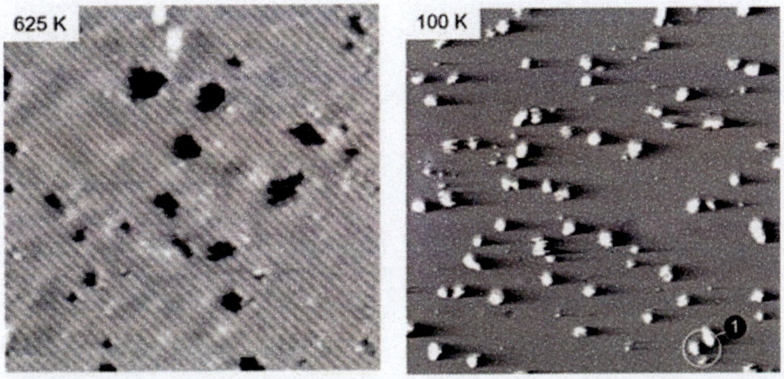

Fig. 6.6 Left: STM image of GaAs(110) after 300 eV Ar ion bombardment with an ion current density of 3 × 10^{11} Ar-ions/(s cm^2) at 625 K. The scan area is 25 × 25 nm^2 (figure adapted from [60]). Right: STM image 1 keV Xe single ion impacts on Al(111) at 100 K and at fluence of 5.3 × 10^{-4} ML. Roughly 70 single adatom islands as well as several adsorbates (white dots) can be distinguished. The circle (1) marks a possible grouping of three islands to represent one impact event. The scan area is 58 × 58 nm^2 (figure adapted from [52])

after 300 eV Ar ion bombardment at higher temperatures. Vacancies (black) and adatom islands (bright) up to a depth of about one monolayer are visible. With an increase in the temperature (up to 775 K), the lateral dimension of these islands increases and their density decreases due to adatom diffusion and adatom-vacancy annihilation processes [59]. STM studies by Zandvliet et al. [62] have confirmed that vacancy clusters are the dominant defects on reconstructed Si surfaces after 3 keV Ar low fluence ion bombardment. Busse et al. [52] have detected adatoms and adatom clusters on crystalline Al surfaces after Xe ion irradiation at low fluences and temperatures. Figure 6.6 (right) shows about 70 adatom clusters after Xe ion irradiation, but no vacancies are visible. The authors estimate the adatom yield to be between 12 and 50 per incident ion, where the bulk vacancy can further reduce by annealing at higher temperatures.

The kinetics of the growth of surface defects in the early stages of ion irradiation are primarily influenced by two factors: the ion energy and the temperature during bombardment, including subsequent annealing.

6.2.2.1 Influencing Factor: Ion Energy

At low energies (< some keV) and low fluences (< 0.5 ions per surface atom), it can be assumed that vacancies and adatoms, including their clusters, are generated only by ion bombardment (no additional thermal generation of point defects on the surface). With increasing the ion energy, one-dimensional defects (dislocations) are increasingly created. This preliminary stage of surface erosion has been studied by de la Fuente et al. [69]. Pt(001) and Au(001) surfaces were bombarded with 600 eV Ar ions at room temperature. For very low fluences, only elongated rectangular depressions (A in Fig. 6.7, left) and individual dislocations (B) could be observed. The

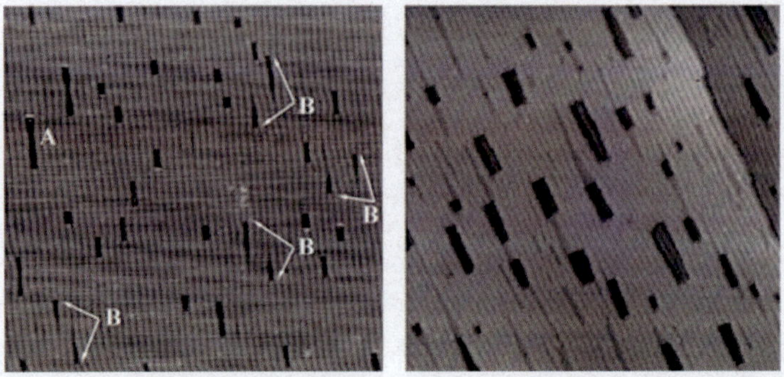

Fig. 6.7 Left: STM image of Au(001) surface after 600 eV Ar ions with a fluence of 6×10^{13} ions/cm^2 ≙ 0.05 ML at room temperature. **a** Shows depressions and **b** individual dislocation. The scan area is 83×83 nm^2. Right: STM image of Au(001) surface after 600 eV Ar ion bombardment with a fluence of 0.25 ML at room temperature. The scan area is 89×89 nm^2 (figures adapted from [69])

depressions are interpreted as dislocation dipoles of opposite Burgers vectors [69]. With increasing ion fluence, vacancy islands become the dominant feature. Figure 6.7 (right) shows large vacancy islands together with elongated troughs with a depth of 0.06 nm. For fluences of about 10^{15} ions/cm^2, other defects (e.g., perpendicular reconstruction domains and unreconstructed patches) become detectable [69].

The number of generated surface defects (adatoms, vacancies, interstitials) is strongly dependent on the ion energy. By monitoring the intensity of the out-of-phase specular RHEED spots, Floro et al. [68] investigated surface defect production from bombardment of the Ge(001) surface by He, Ar, and Xe ions (energy between 70 eV and 500 eV). It was found that the defect generation rate increases with increased ion energy (Fig. 6.8, left) and the point defect diffusion can occur even at −100 °C.

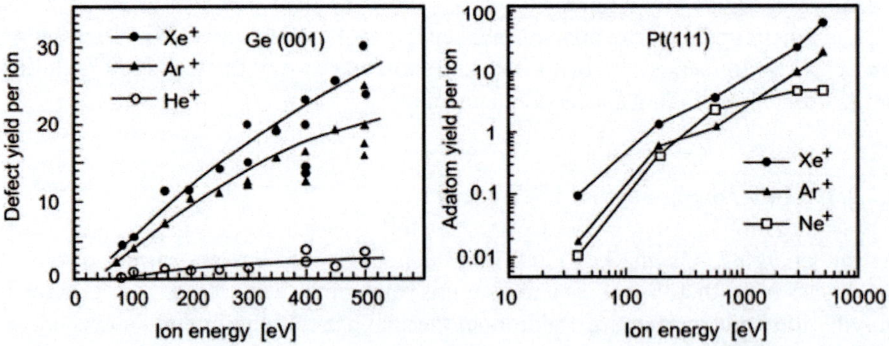

Fig. 6.8 Left: surface defect yield as a function of ion energy for He, Ar and Xe ion bombardment of Ge(001) at −100 °C. The lines are only intended as a visual guide (figure adapted from [68]). Right: experimentally determined adatom yield of Pt(111) after Ne, Ar und Xe ion bombardment at 150 K as a function of ion energy (figure adapted from [53])

As shown in Fig. 6.8, left, at temperatures of –100 °C, Ar and Xe ion bombardment significantly enhances surface defect generation (up to 25 defects/ion for 500 eV Xe ion bombardment). Due to its much lower mass, He ion irradiation produces nearly an order of magnitude fewer surface point defects. At elevated substrate temperatures, the surface defect generation is reduced for Ar and Xe ion bombardment, presumably due to defect annealing.

Michely and Teichert [53] were able to demonstrate experimentally (Fig. 6.8, right) that the adatom yield increases steeply in the low-energy region (up to about 200 eV). However, for higher ion energies (keV energy range), adatom production is significantly increased. In particular, the ratio of adatoms to the sputtering atoms, expected to be 1 [70], is larger (up to 3) in the keV energy range. It is assumed that non-linear spike effects cause this increase. Chey et al. [67] studied the generation of vacancies and adatoms, including the formation of adatom islands, on Ge surfaces after Xe ion irradiation as a function of ion energy between 20 eV and 240 eV at 165 °C. At this temperature, Ge remains crystalline and the measured adatom and vacancy production rates include annihilation of adatom vacancy pairs due to finite point defect mobility. In the energy range studied, the number of vacancies increased from about 0.01 vacancies/ion to about 1 vacancy/ion, and the number of adatoms increased from about 0.008 adatoms/ion to about 0.1 adatoms/ion. STM studies allowed the observation that the vacancy island density increases with increasing ion energy and that the ratio of adatoms to vacancies is approximately constant for ion energies > 40 eV.

The interpretation of the obtained experimental results is frequently based on the support of MD simulations. For example, Gades and Urbassek [71] investigated bombardment of Pt(111) surfaces with noble gas ions with energies between 100 eV and 3 keV, and, using STM, Michely and Teichert [53] determined the adatom formation yield after single noble gas ion impact on Pt(111) with energies between 40 eV and 5 keV at a temperature of 150 K. The adatom yield increases with increasing energy, where the simulations predict higher yields than the STM studies. In these MD simulations, it was also shown that the ratio of adatoms to sputtered particles is between 2 and 4 for not-too-low ion energies, because of the lower formation energy of adatoms. With increasing bombardment energy, the number of adatoms and sputtered particles increases more strongly than the number of surface vacancies. This is due to the increasing number of adatoms and sputtered particles originating from deeper target layers. Gades and Urbassek [71] were also able to confirm the experimentally observed trend [53] that for rare low ion energies ($\leq$ 100 eV), the Pt adatom yields are significantly higher than sputtering yields and that above about 1 keV, the sputtering yield dominates the adatom yield. Ghaly et al. [72] carried out MD simulations to compare the damage production mechanisms at solid surfaces during bombardment with kilo-electron-volt ions. Light ions tend to induce linear collision cascades that create individual defects on the surface. Bombardment with heavy ions produces damage concentrated at the point of impact. These MD simulations also demonstrate that the high energy density deposited near the surface can cause processes such as local melting and micro-explosions, leading to the formation of craters and large adatom clusters near the point of impact (see Sect. 6.2.2.3).

6.2.2.2 Influencing Factor: Temperature

Of particular interest is the influence of applied temperature on the modification of the surface topography of single crystalline metals after erosion of the first tenth of a monolayer (i.e. during the initial stage of topography evolution). In principle, two borderline situations can be distinguished with respect to the influence of temperature during ion irradiation: (i) ion bombardment at temperatures of zero Kelvin, then, no thermal healing is to be expected, but only the effects of ion irradiation, including irradiation-induced relaxation. And (ii) ion irradiation at temperatures > 0 K, then, ion irradiation-induced healing must be included in the discussion. For example, the evolution of Pt(111) topography has been studied for fluences < 1 ML after 600 eV Ar ion beam erosion between 350 K and 650 K [72] and between 625 K and 910 K [73] as well as after 1 keV Xe ion bombardment between 600 K and 800 K [74]. These experiments have shown that adatoms are produced in large numbers, where, above a temperature of about 550 K, the adatoms form islands [54]. The topography of Pt(111) surfaces after 1 keV ion beam erosion at normal ion incidence at these high temperatures is very similar. It seems to be independent of the temperature, and is characterized by the formation of prolonged vacancy islands, which are nearly hexagonal, with steps oriented along the densely packed [110] directions (Fig. 6.9). Some of the small vacancy islands exhibit a triangular shape. It is evident that the vacancy islands reflect the hexagonal symmetry of the Pt(111) surface. The further evolution of the topography as a function of the erosion time (fluence > 1 ML) is strictly dependent on the temperature (see Sect. 6.2.3).

Experimental studies by Busse et al. [52] have shown that a very efficient adatom production and the formation of stable vacancy clusters beneath the surface occur

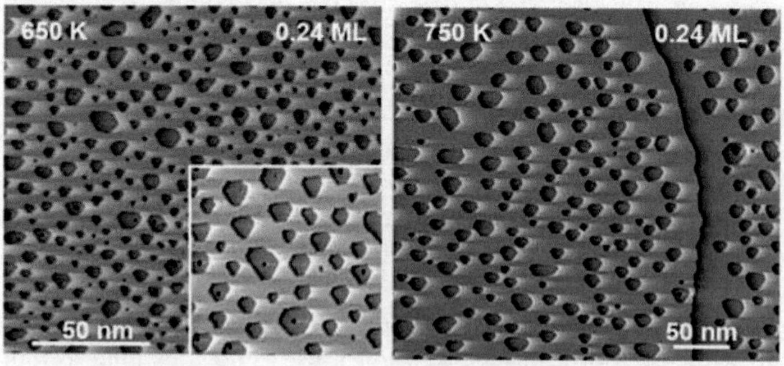

Fig. 6.9 STM images of Pt(111) after 1 keV Xe ion bombardment at 650 K (left) and 750 K (right) after sputter erosion of one tenth of a monolayer. One monolayer (ML) corresponds to the number of atoms needed to create an atomic layer (for Pt(111), 1 ML = 1.504×10^{15} atoms/cm^2). Figures adapted from [74]

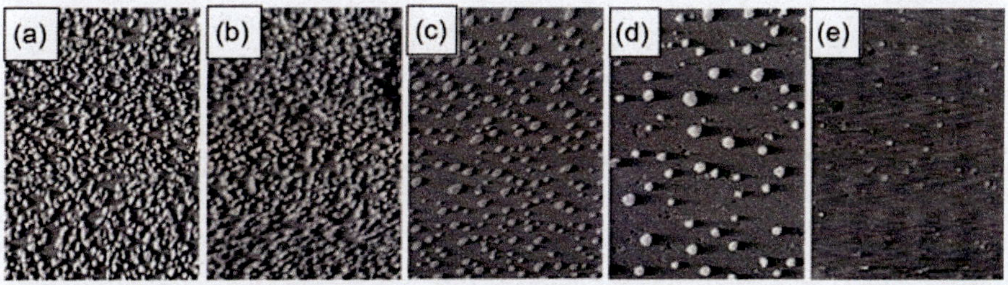

Fig. 6.10 STM images of an isochronal annealing sequence on Al(111) after 1 keV Xe ion bombardment at 100 K and a fluence of 6×10^{-3} ML [1 ML $\triangleq 1.41 \times 10^{15}$ atoms/cm^2 for Al(111)]. **a** After bombardment at 100 K, **b** annealed for 20 s at 150 K, **c** annealed for 20 s at 250 K, **d** annealed for 20 s at 300 K and **e** annealed at for 20 s at 450 K. The scan area is 120×80 nm^2 (figures adapted from [52])

during the initial stages of noble gas ion bombardment of Al(111) at lower temperatures (between 100 K and 400 K). Figure 6.10 illustrates the isochronal annealing on the adatom island density after 1 keV Xe ion bombardment of Al(111) surfaces at 100 K and a fluence of 6×10^{-3} ML. It could be observed that the adatom coverage is reduced in two steps. The first step in the reduction of adatom coverage takes place between 150 and 200 K through vacancy migration and the second step takes place through vacancy cluster dissociation. In contrast, island size increases. At 450 K, all adatoms on the surface have been removed. Supporting MD simulations explain the underlying mechanism on the basis of the thermal spike model. The extreme energy density in the spike causes a material transport onto the surface. No pits or carters could be observed. According to these authors [52], the ion bombardment induced growth (formation of adatom layers on the surface of the substrate) should also observable for other lighter elements at low ion energies and temperatures.

As is known, the recovery of energetic ion-induced bulk defects as a function of temperature can be described by the widely accepted five-stage process (see, e.g., [75]). Stage I (for most metals at homologous temperatures ≤ 0.15) is characterized by interstitials that are mobile and can recombine with vacancies or form interstitial clusters. In stage II, these interstitial clusters grow by coalescence. This process continues up to stage V. Vacancies become mobile in stage III and can form vacancy clusters or annihilate at interstitials. In stage IV, the vacancy clusters dissociate thermally. In stage V, the final stage (for most metals at a homologous temperature of about 0.5), all ion beam-induced defects disappear. How far this five-stage process can be applied to describe the temperature-dependent behavior of ion beam-induced surface defects is unknown, particularly with regard to the role of adatoms. Albeit, Busse et al. [52] determined the evolution of adatom coverage of Al(111) surfaces during annealing and were able to identify the recovery stages III and V.

The experiments performed point to the fact that with increasing temperature, the following processes of defect recovery at the surface are released: At the lowest temperature (some ten K), the onset of adatom diffusion can be anticipated and at slightly higher temperatures, the onset of the interstitial diffusion can be anticipated.

In a next stage, the vacancies become mobile (T > 150 K) and the adatoms and surface atoms at the step edges can then evaporate. At higher temperatures (T > room temperature), vacancy and interstitial clusters dissociate.

6.2.2.3 Crater Formation After Low-Energy Ion Impact

It is known that the impact of ionized clusters or high-energy heavy ions creates carters with dimensions ranging from nanometers up to micrometers, where the dimensions of the carters correlate with the deposited nuclear energy loss [76]. The cratering effect for high-energy single ion and ionized cluster bombardment has been computationally and experimentally studied in detail (see, e.g., [76–78]). In contrast, studies on crater formation after single ion bombardment with low-energies (< some keV) are rare, although the local surface height change for this technologically significant ion energy range can be of great importance for several applications.

Harrison and Webb [43] used computer simulations to illustrate the formation of carters on Cu surfaces after 1 keV Ar ion impacts. The carters are formed by atoms displaced laterally or downward, initiating replacement collision sequences. The distribution of pit sizes is determined primarily by the type of target material and, to a lesser extent, the ion energy. Using molecular dynamic simulations, Kalyanasundaram et al. [79, 80] and later, Yang et al. [81] have studied the surface height change, the so-called crater function (an average of the results of numerous MD simulation events), after 500 eV single Ar ion bombardment of Si surfaces in dependence on the ion angle of incidence. As an example, Fig. 6.11 illustrates the calculated shape of carters after Ar ion bombardment for an incident angle between 0° (normal incidence) and 67°. The impact of an ion causes a deep crater, surrounded by a crater rim. With an increased angle of ion incidence, this rim becomes increasingly asymmetrical in shape. Kalyanasundaram et al. [79, 80] and Yang et al. [81] showed that the formation of carters at the point of ion impact induces a significant mass redistribution, i.e. the crater shape is determined by local mass rearrangement

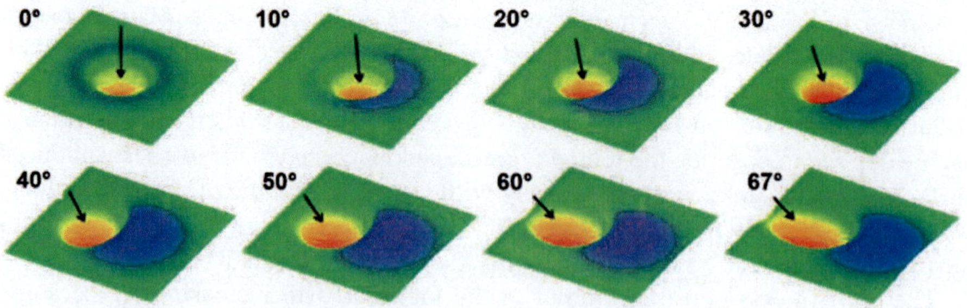

Fig. 6.11 Left: calculated crater function for 500 eV single Ar ion impacts on silicon in dependence on an incidence angle between 0° and 67°. The direction of ion arrival is marked by black arrows (figures adapted from [81])

and not only by the sputtering, i.e. the crater volume is the sum of the sputtered atoms and the atoms relocated into the rim (indicated in blue in Fig. 6.11). The shape of the crater is the result of the ion impact induced formation of vacancies and interstitials, the diffusion of these defects, and the subsequent recombination among themselves and with the surface [68], where the shape of a crater is strongly dependent on the ion energy, the angle of incidence, and the mass ratio of ion mass to the target mass. The crater rim is characterized by an excess of interstitials, while the crater floor shows an enrichment of vacancies. The change of height (shape) due to atom rearrangement is described by an empirical function, the so-called crater function. For example, Kalyanasundaram et al. [79, 80] applied a discrete function consisting of the difference of two Gaussian functions with eight free parameters. Note that the crater function is an important input for explaining ripple formation based on the continuum theory (see Chap. 8).

6.2.3 Formation of Extended Surface Defects

In the previous subchapter, the early steps of surface evolution under low-energy ion bombardment were discussed. Following this low-fluence ion bombardment (fluences $\leq$ 1 ML, approximately) the topography is characterized by leaving of surface particles due to sputtering and the formation of isolated vacancies and adatoms as well as vacancy and adatom clusters. At moderate temperatures, these surface defects are immobile. It can be expected that with an increase of the ion fluence > 1 ML, i.e. fluences > 5×10^{15} ions/cm^2, approximately, a more developed topography should be observable.

The topography evolution of crystalline materials following low-energy ion beam erosion and fluences up to some hundred MLs has been comprehensively studied in detail, e.g., for Au(111) [59, 82–84], Ag(001) [85, 86], Pt(111) [54–56, 73, 74, 87], Cu(001) [57, 88, 89], Cu(111) [58, 90], Si(100] [47, 63, 64], Ge(001) [67], GaAs(110) [40, 60] and graphite [91]. For instance, Fig. 6.12 shows the topography of Pt(111) surfaces after Xe ion bombardment in dependence on ion fluence (in MLs) for two different temperatures (600 K and 750 K). The STM images illustrate the development of topography after bombardment at 600 K [74]. Initially, small monolayer deep vacancy islands are formed (c.f. Fig. 6.10), where the morphology is still close to a layer-to-layer removal. New vacancy islands are formed, particularly on the bottom terraces of the vacancy islands. These islands develop into regular hexagonal pits consisting of stacked vacancy islands and remainder pyramids connected by ridges. This preference for nucleation of islands on the bottom, as compared to the other terraces, ultimately causes the pit formation (Fig. 6.13, left). Similar hillock-like surface morphology with a diameter of about 5 nm was observed by Feenstra and Oehrlein [92] following 500 eV Ar ion bombardment of Si(100) surfaces. For surfaces with a squared surface atom arrangement (e.g., Ag(001) [64]), squared pits replace the hexagonal ones. Experimental studies by Kalff et al. [74] and Michely et al. [93] also indicate that with increasing ion fluence at erosion temperatures

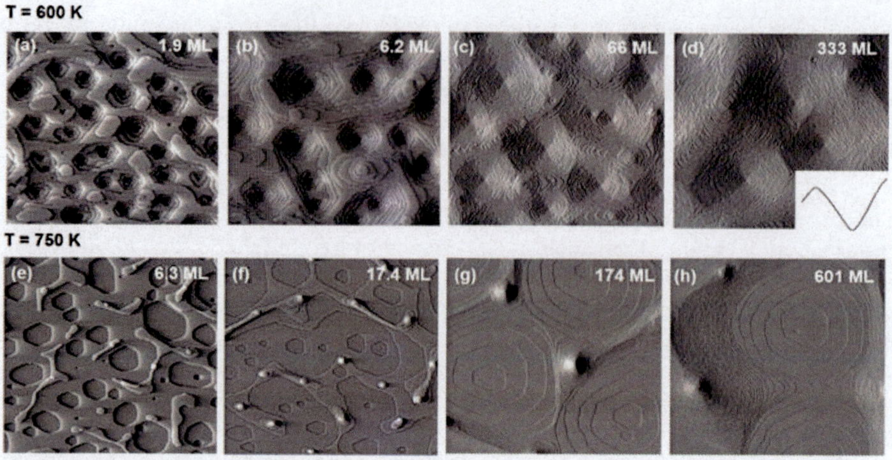

Fig. 6.12 STM images after 1 keV Xe ion beam erosion at 600 K (above) and 750 K (bottom). The inset in (**d**) represents a line scan through a pit. The scan area in **a** and **b** is 81 × 81 nm^2, in **c** and **d** 156 × 156 nm^2, in **e** and **f** 161 × 161 nm^2 and in **g** and **h** 364 × 364 nm^2 (figures adapted from [74])

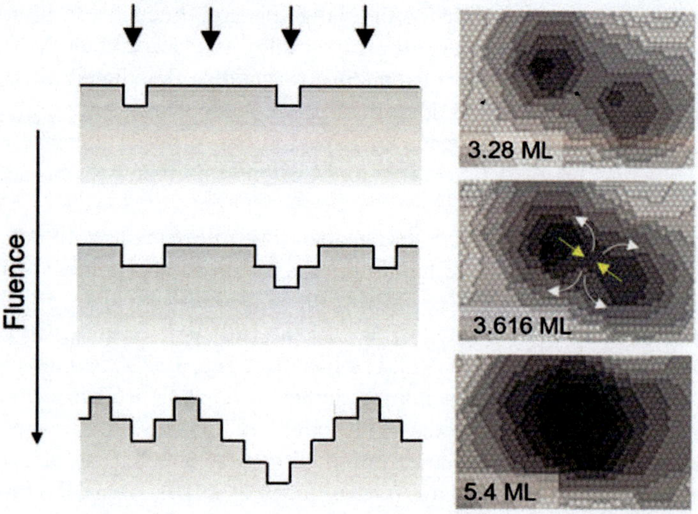

Fig. 6.13 Left: schematic diagram of the formation of pits on crystalline Pt surfaces at temperatures <700 K. Right: sequence of snapshots of evolution of two pits obtained by MC simulation (figures adapted from [94])

< 700 K, pit coalescence becomes a dominant process. Strobel et al. [94] studied the coarsening of pits by way of 3D kinetic lattice Monte Carlo simulations. If two adjoining pits are separated only by a thin ridge (Fig. 6.13, right), the coalescence of both pits begins with a strong edge diffusion (diffusion of atoms along steps), which results in the movement of material away from the locations of high curvature (white arrows in Fig. 6.13, right). The driving force behind this process is the line stress of the new common steps of the joint islands. In response to this stress, a vacancy diffusion is triggered, which results in pit-shaped rearrangement (yellow arrows in Fig. 6.13, right). The roughness increases with the time of ion bombardment in accordance with a power law, where the growth exponent $\beta = 0.58$.

A drastic change in the topographic evolution after 1 keV Xe ion bombardment of Pt(111) surfaces can be observed for temperatures > 700 K. A transition from three-dimensional ion impact induced pits to two-dimensional, i.e. layer-by-layer removal with increasing ion flux, could be proved. Again, the initial stage (0.24 ML) is characterized by a flat surface with small vacancy islands (c.f. Fig. 6.9). In contrast to the ion bombardment at lower temperatures, the surface remains flat with increasing fluence. The vacancy islands are elongated adatom islands that have formed and are left over (Fig. 6.12f). With increasing surface removal, pit walls are built up between remainder pyramids (Fig. 6.12f, g). When the expanding vacancy layers strike the small reminder walls, vacancy coalescence is hindered, i.e. the surface remains flat (Fig. 6.12h).

Kalff et al. [74] explained the development of such micro-pyramids on the basis of the effect of a step-edge barrier on the incorporation of ion beam-induced vacancies at ascending steps (in analogy to the Ehrlich-Schwoebel barrier (see Appendix N) for the incorporation of adatoms). For temperatures < 650 K…700 K, adatoms are not able to descend the crystal steps due to the Ehrlich-Schwoebel barrier along that direction. At higher temperatures, however, the onset of step atom detachment can be expected (thermally activated adatom formation). Kinetic MC simulations [94] have shown that at higher temperatures, this smoothing process considerably slows down pit coarsening and impedes pit shape relaxation.

Consequently, the formation of extended defects in crystalline materials is based on the presence of an additional diffusion barrier and follows the symmetry of the crystallographic directions, i.e. in addition to the curvature-dependent sputtering yield (see Sect. 6.4), other effects can be responsible for (or can contribute to) the destabilization of crystalline surfaces during ion bombardment. The pit formation can be explained on the basis of the step edge barrier for vacancies, i.e. they can barely anneal at ascending steps, but are more facile at descending steps.

At temperatures of around 700 K, adatom detachment sets in and pit formation ceases. This leaves only remainder pyramids, which need not be nucleated.

6.3 Low-Energy Ion Beam-Induced Surface Defects

According to Carter [95], the topography evolution on crystalline material surfaces under ion bombardment can be attributed to two basic processes: on the one hand, intrinsic processes that result from the variation of the sputtering yield with the crystal orientation and on the other hand, extrinsic processes caused by perturbations of the sputtering process (impurities, surface inhomogeneity, inclusions, etc.). Accordingly, ion beam-induced defects on crystalline and amorphous surfaces are typically introduced as the result of the variation in the irradiation conditions (fluence, energy, incidence angle, etc.). The generation and development of topographical defects under low-energy ion bombardment have been preferentially studied on crystalline surfaces, while there have been only sporadic studies on ion beam-induced defects on amorphous surfaces.

It is advantageous to distinguish between the topographical defects on amorphous and crystalline surfaces following ion bombardment, where for crystalline surfaces, a distinction is drawn between intra-crystalline and inter-crystalline topography [95]. In the first case, the topography evolution *on the surface* of single crystals is examined, while in the second case, the topographical evolution *between* differently oriented crystallites is observed.

6.3.1 *Intra-crystalline Surface Defects*

The ion beam-induced surface defects that occur on the surface of a single crystal can be divided into the following classes:

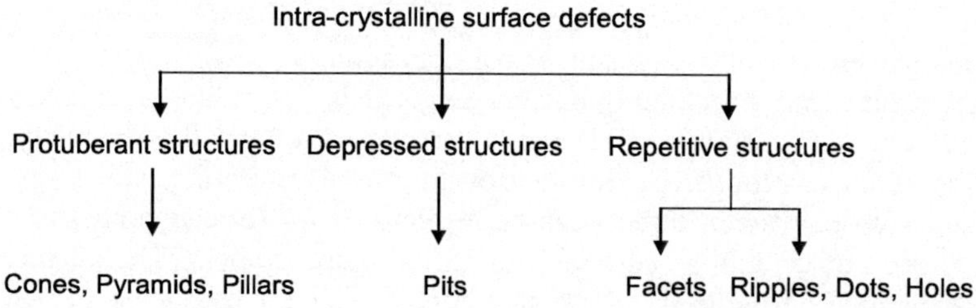

The depressed structures result from a local enhanced sputtering effect, while the protuberant structures are caused by enhanced sputtering around the protuberance. In contrast to these isolated defects, periodic structures result from the overlap of isolated structures or of highly correlated nonlinear effects. Due to the high interest that ripple structures, in particular, hold for various technological applications, these will be discussed separately (see Chap. 8). It must be noted that the surface defects are subject to continuous transformation in dependence on the ion fluence and temperature. It thus follows that the shape of surface defects can change over the course of progressive ion bombardment. Carter and Navinsek [95] and Navinsek [96] have provided excellent overviews of the formation of topographic defects on surfaces under low-energy ion bombardment. In the following, the most frequently occurring intra-crystalline surface defects will be introduced (ripple formation as well as dots and holes will be discussed separately in Chap. 8).

6.3.1.1 Cones and Pyramids

Cones, pillars, and pyramids formed on solid surfaces under low-energy ion bombardment have a very similar shape. The term 'pyramid' is used instead 'cone' to emphasize that pyramids have a defined crystallographic base (square, rhombohedral, hexagonal, etc.), i.e. a geometric shape. In contrast to pyramids and cones, the diameter of pillars changes only slightly with increasing height.

The first report on the formation of cones under low-energy ion bombardment was published by Güntherschulze and Tollmien [97]. They studied cone formation on selected metal surfaces and determined a relationship between the erosion velocity and the cone angle, already taking into account the re-deposition of the sputtered material. With increased sputtering time, increasing cone dimensions were observed. The researchers assumed that impurities with a sputtering yield lower than that of the bombarded material caused cone formation. Stewart and Thompson [98] attached an ion source to an early model of a scanning electron microscope. This allowed them to directly study the evolution of cones on different metal surfaces under Ar ion bombardment. They proposed an initial sputter-erosion model to explain the evolution of the cones and reasoned that the sputter-induced cones were the result of non-uniformities in the sputtering yield over the surface caused by foreign particles (impurities), either as perturbations on the surface (contamination) or as inclusions in the bulk target.

Over the past several decades, cone/pyramid formation by ion bombardment has been studied on metals by Wehner and Hajicek [99], Wilson and Kidd [100], Gvosdover et al. [101], Whitton et al. [102], Wehner [103, 104], Rossnagel and Robinson [105, 106], Goto and Suzuki [107], Sen and Ghose [108]; on different semiconducting materials by Wilson [109], Barber [110], Gvosdover et al. [101], Nozu et al. [111], Meng et al. [112]; on stainless steel by Witcomb [113]; on carbon by Yusop et al. [114]; on graphite by Floro et al. [115]; on bioglass by Popa et al. [116]; and on countless other materials. Most of these studies assume that the controlled fabrication of cones is based on the simultaneous resourcing of seeding materials by co-deposition, concurrent sputtering, or coating of the sample surface with a thin film (thickness in order of some nanometers) of the seed material on the surface of the sputtered target before ion bombardment. As an example, an SEM image of silicon and copper cones is shown in Fig. 6.14. The cones possess a uniform shape. In all cases, impurity sources consisted of Ni (left) and titanium or carbon films (right) were used. In general, the seeding material targets are placed at an angle to the ion beam such that some of the sputtered impurity atoms impinge on the bombarded substrate (for an experimental arrangement see, e.g., Fig. 8.17). It is assumed that the presence

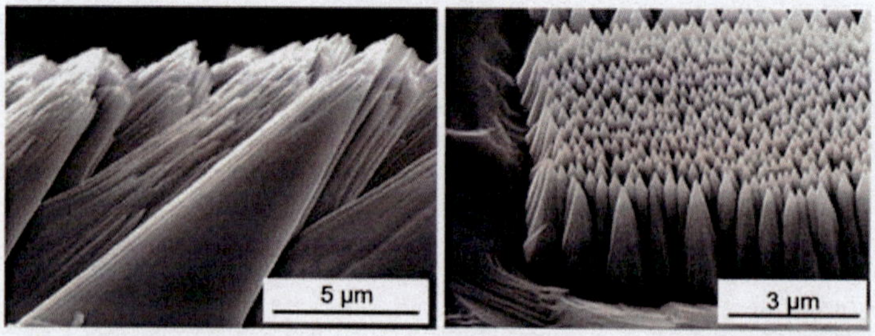

Fig. 6.14 Left: cross-sectional SEM image of Si cones on Si at an ion energy of 900 eV, an incident angle of 40° and substrate temperature of 300 °C. Ni was used as seeding material (figure adopted from [112]). Right: SEM image of Cu(13 3 2) after bombardment with 5×10^{18} Kr ions/cm^2 at normal incidence. Ti or carbon was sputter-deposited onto Cu before Kr ion bombardment (figure adapted from [120])

of surface impurities, initial surface asperities, and/or subsurface defects (inclusions) are responsible for cone evolution. In Fig. 6.15, the evolution of cones is schematically illustrated in the presence of a surface impurity (above) and a subsurface inclusion (below) under normal low-energy ion bombardment. During ion bombardment, the surface is eroded at the erosion rate given by (5.2). If the target is covered by impurities characterized by a sputtering yield lower than the surrounding material, the underlying material is then protected from sputtering. Wehner and Hajicek [99] and Kaufman and Robinson [117, 118] and later Begrambekov et al. [119] assumed that impurity atoms move about the substrate surface by thermally activated surface diffusion and gather into localized clusters. Because these clusters act as shields and prevent the sputtering of the substrate material, they initiate the formation of conical

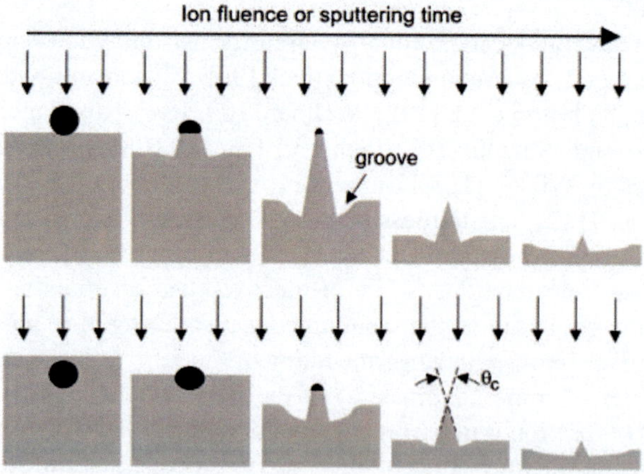

Fig. 6.15 Schematic picture of the evolution of cones as function of the sputtering time (ion fluence). Above: particles (impurities) occur as perturbations on the surface. Below: inclusion occurs in the subsurface region, where θ_c is the final angle of the cone faces

nanostructures. When the shielding particles are eroded, the cones on the target have smooth sides and an equilibrium configuration characterized by an apex angle θ_c (Fig. 6.15). According to Stewart and Thompson [98], the apex angle correlates to the ion incidence angle θ_p, i.e. the ion incidence angle at which the sputtering yield is at a maximum (c.f. Fig. 5.9). This apex angle is given by

$$\theta_c = \pi - 2\theta_p, \tag{6.5}$$

where its axis is directed towards the ion beam direction (Fig. 6.15). When the ion bombardment is continued, the cones decrease in size and eventually disappear.

Based on experimental studies, Wehner [104] reasoned that the described prerequisite for the formation of cones or pyramids (seed atoms with a lower sputtering yield on substrate materials with a higher sputtering yield) is applicable only to a limited degree. Instead, a lower sputtering yield and a higher melting point of the seed materials in comparison to the substrate material are the conditions required to create cones.

Whitton et al. [102] and Carter et al. [95] have opined another hypothesis for the formation of cones or pyramids under low-energy ion bombardment. They have asserted that pyramids and cones develop due to crystallographic reasons in the absence of impurities, where these authors have concluded that the formation of pyramids is associated with defects (dislocations). These structural perturbations in the subsurface region give rise to a variable sputtering yield. The combination of this effect with the additional influence of the crystallographic orientation on the sputtering yield determines the final topography.

Sigmund [121] has given an alternative explanation of the formation of cones, including the grooves, by application of the sputtering theory (see Sect. 5.2). Assuming a non-plane surface, i.e. the surface possesses spatial inhomogeneity (e.g., a ridge located on a plane surface, Fig. 6.16), then it must be taken into account that as a result of the ion impact, target atoms are not only sputtered from the point of ion impact, P, but also from points at distances in order of magnitude of the collision cascade (Fig. 6.16, right side of the ridge). It is evident that the maximum of the sputtering yield curve is further down the slope than the point of ion impact (point of ion impact P), i.e. this point is not identical with the point of the maximum sputtering yield, Q (point of the maximum energy deposited at the surface). Consequently, the local sputtering yield varies as a function of the slope of small irregularities on a surface. On this basis, both the drastic enhancement of the roughness surfaces and the stability of cones and pyramids can be explained. In the case of ion bombardment of the ridge (Fig. 6.16, left), the sputtering velocity is greatest at point C, slightly lower at point B (slope), significantly lower on the plane (point D) and lowest near the top (point A) [121]. Consequently, the evolution of small cones is enhanced and the formation of grooves around the cones can be also explained (see also Fig. 6.32, right). It should be noted that mechanism of surface roughening under ion bombardment proposed by Sigmund has been used by Bradley and Harper [29] to describe the evolution of ripple patterns (see Sect. 8.2.1.1).

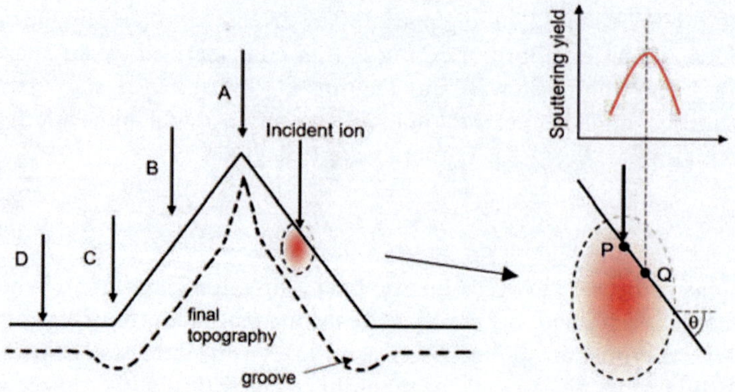

Fig. 6.16 Schematic picture of a ridge before and after sputtering. As an example, a contour plot of deposited energy on a flank is shown. P and Q are the point of ion impact and the maximum of the sputtering yield, respectively. θ is the angle of ion incidence (the left figure is taken from [121] and modified)

In ion sputtering studies, the formation of grooves has been observed around the base of the cones. Bayly [122] has suggested that the majority of ions are reflected on the steeply angled conical flanks of the cones or pyramids (particularly if the ion incidence angle $\theta > \theta_c/2$). An enhanced local flux of ions is then generated, which leads to the creation of grooves by sputtering around the cones as well as concave depressions at the bottom of troughs (c.f. Fig. 6.15 or Fig. 6.16).

In addition, cones can be also obtained on compound materials (e.g., III-V semiconductors [123, 124]) after low-energy ion bombardment. Trynkiewicz et al. [125, 126] conducted a detailed study on the formation of cones (pillars, nanorods) on InSb(001), InP(001), InAs(001) and GaSb(001) surfaces after 3 keV Ar ion bombardment at temperatures between −129 °C and 150 °C. With increasing temperature, a change from a small-dot pattern to well-developed pillar or column array was observed. Figure 6.17 (left), illustrates two examples of the formation of a regular

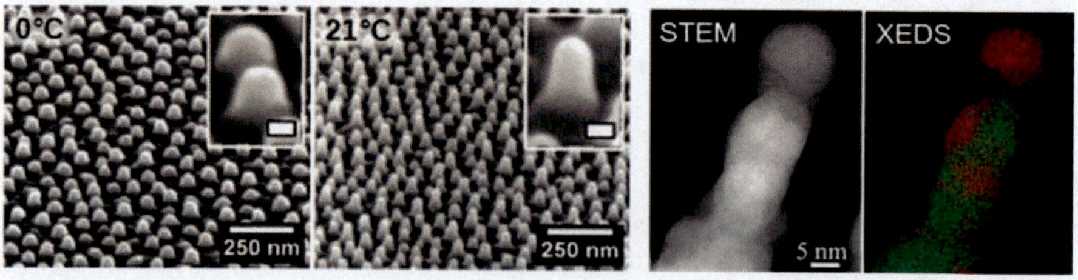

Fig. 6.17 Left SEM images of GaSb surfaces after 3 keV Ar ion bombardment at two different temperatures. The insets represent a corresponding single structure with scale bar of 25 nm for each image (figures are taken from [125, 126]). Right: dark-field STEM and the corresponding XEDS images of single InSb pillars produced by 30 keV Ga ion bombardment at normal incidence. The color red represents the element In and the color green the element Sb (figures are taken from [127])

network of cone-like structures on GaSb surfaces. Independent of the III-V semiconducting material, a transition from small islands via high-aspect ratio pillars to featureless surface structures via increase in the temperature was observed. By means of chemical analysis, it could be proven that the spherical cap at the top of these nanostructures is enriched by the metallic component of the sputtered material (Fig. 6.17, right) [127] due to the difference in sputtering yields between Ga or In and Sb, P or As within these binary compounds (see Sect. 5.8). Based on this non-stoichiometric sputtering, Le Roy et al. [128, 129] proposed a phenomenological model to explain the formation of nanostructures as the result of the ion sputtering of binary materials. The preferential sputtering of the non-metal component leads to enrichment of the metallic component and thus the formation of metal clusters, which act as local sputter masks. This masking effect and the diffusion of metal components in the surface layer cause vertical elongation of the nanowire.

The re-deposition of atoms sputtered from the surrounding materials also significantly influences the formation of cones. As is well known, there are differences in the flux of the re-deposited material between upstream and downstream movement on the inclined cone (see Sect. 6.3.1.3). Based on this finding, Wu and Goldman [130] have extended the Le Roy model of the cone formation to include comprehension of the re-deposition of sputtered atoms.

Finally, it should be noted that due to their fast formation rate, the ability to achieve high aspect ratios, and the possibility of fabricating dense patterns over large surface areas, cones formed by low-energy ion sputtering are of great importance to applications. For example, spatially ordered nanocones with a diameter of any ten nanometers and an average height > 100 nm could be obtained following Ar ion bombardment of III-V semiconducting materials with ion energies < 1 keV and fluences of about 10^{18} ions/cm^2 [123, 124]. Interestingly, the growth rate is enhanced by increasing temperature.

6.3.1.2 Etch Pits

Etch pits are the most frequently observed surface defects following ion bombardment. Under perpendicular ion impact, individual crystallites are damaged by the sputtering of substrate material from the surface. For example, the formation of pits has been studied in detail following low-energy ion beam erosion for metals, such as Cu(111) [89, 131], Cu(001) [57, 88], Pt(111) [54, 80, 87], Ag(001) [85, 86], Au(111) [81, 83], W [96]; for highly oriented pyrolytic graphite (HOPG) [132]; and for semiconductors, such as Si [133–135], Ge(001) [67, 135], and GaAs(110) [90].

As an example, Fig. 6.18 shows etch pit formation on a single Cu crystal surface after intensive Ar ion bombardment. The shape of the pits is quadratic to rectangular, i.e. the pits show the tendency to form regular symmetric habits according to the crystallographic orientation of the substrate material with respect to the direction of the incidence ions. Thus, these surface features are directly correlated to the crystallographic structure type and the orientation of the target material as well as the ion irradiation conditions (incident angle, ion fluence, and ion current density).

In general, with increasing ion fluence, the pit dimension increases linearly and the pit density increases with cube-like power law. For low fluences ($\approx 10^{17}$ ions/cm^2), the fraction of the pitted surface area is between 10^{-4} % and 10^{-1} %, while for fluences $>10^{18} \ldots 10^{19}$ ions/cm^2, the etch pits begin to overlap [136]. Etch pits on single crystalline surfaces are frequently characterized by two-, three-, or four-fold symmetry and are shaped like a square, hexagonal trapezoidal, or trigonal inverted pyramid (c.f. Fig. 6.18), where the faces of these inverted pyramids correspond to low-index planes [137]. Etch pits independent of the crystal orientation were partially observed (e.g., as result of Ar ion bombardment of (100), (110) or (111) oriented Si) [135]. This result was interpreted under the assumption that the pits form in a predominantly amorphous layer on the surface, generated by ion bombardment.

The shape of the etch pits is strongly dependent on the angle of ion incidence. Studies have indicated that when materials are bombarded under non-normal ion incidence conditions, the shape of the etch pits differs considerably from those on the same material under normal ion incidence. Under oblique ion irradiation, an increased sputtering yield (higher erosion rates) and sloped etch pits can be expected. Additionally, it is assumed that reflected ions contribute substantially to the formation of etch pits [138]. At oblique ion sputtering, the reflection of ions from steep slopes of a rough surface can be expected. These reflected ions lead to an enhanced ion flux, i.e. to an additional contribution by the reflected ions to the sputtering process at other fractions of the surface. The enhancement of the erosion rate increases with increasing angle of ion incidence and decreasing ion-to-substrate mass ratio (see Sects. 5.3 and 5.6). Cunningham et al. [139] and Magnusson et al. [140] were the first to show that differentially shaped etch pits can be obtained when the ion bombardment occurs at oblique angle ion incidence. Nelson and Mazey [141], and later Carter et al. [133], observed that large triangular depressions emerge on the Si surface. Figure 6.19 illustrates these triangular depressions following Xe ion bombardment at different ion energies [142, 143]. A peculiarity of these Si etch pit patterns is that they are independent of the crystal orientation. Motohiro and Taga [143] have demonstrated that the shape dependence of etch pits is also valid for amorphous materials (fused silica). The shape of the etch pits changed systematically from isolated circular pits

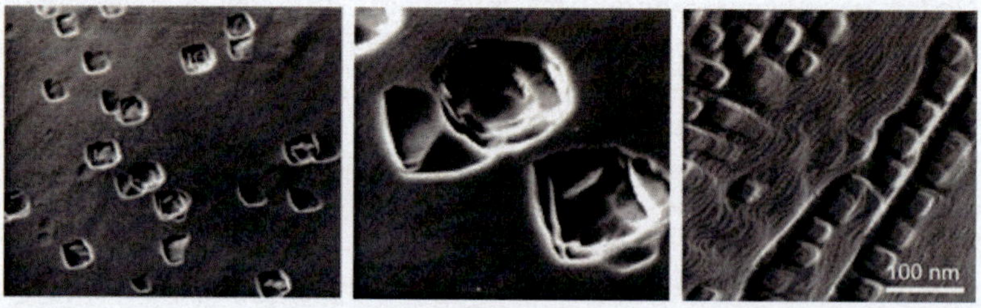

Fig. 6.18 Left and middle: SEM images of pits upon (11 31) orientated Cu surface formed by normal incidence 1×10^{19} Ar ions/cm^2 bombardment at different magnifications (figures adapted from [131]). Right: STM image of pit on a high-stepped Ag(001) surface after 1 keV Ar ion bombardment with a fluence of 1×10^{13} ions/cm^2 at 440 K (figure adapted from [86])

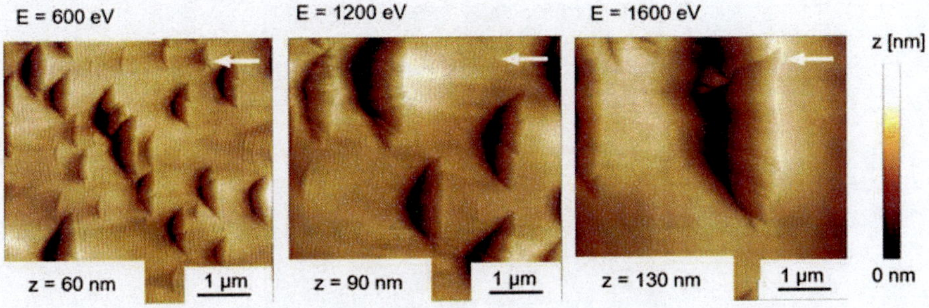

Fig. 6.19 AFM images of Si(100) surfaces after bombardment of 6.74×10^{18} Xe/cm^2 at room temperature and at different ion energies. The ion incidence angle is 75°. The white arrows indicate the projection of the ion incidence direction (figures adapted from [142])

at 0° to crescent etch pits at 22.5° and to a ripple structure at 45°. The origin of etch pits under ion bombardment has not been fully elucidated [138, 144]. On the one hand, it is assumed that dislocations are initially present in the near-surface region before ion bombardment, which act as seed points for the formation of ion beam-induced etch points. On the other hand, it is suggested that ion radiation induced defects agglomerate to seed points (defect clusters). In both cases, it can be expected that the binding energy around a dislocation or within a defect cluster is significantly reduced and hence the sputtering yield is reinforced. Based on experimental studies on Si pit evolution following 7 keV Ar ion or Xe ion bombardment as a function of the ion fluence, Carter et al. [133, 134] have assumed that before pits were elaborated at the surface, there were extended defects below the surface, located at the interface between the amorphized Si surface layer and the crystalline substrate, which acted as seed points for pit generation. This structure then transformed into surficial pits.

6.3.1.3 Facets

Experimental studies have demonstrated that the process of coarsening under oblique low-energy ion bombardment can result in the formation of facets. Ogilvie [145] and Balarin and Hilbert [146] were first able to prove the ion bombardment induced formation of facets. Ogilvie [145] observed preferred oriented crystallites on the Ag surface after Ar ion bombardment with energies up to 4 keV. Balarin and Hilbert [146] bombarded Cu surfaces with 40 keV Mg ions at an angle of 70° with respect to the surface normal. After 20 h, ion beam-induced facets on the Cu surface could be observed. Later, in many further studies, facet formation was observed on single and polycrystalline metal surfaces (Cu, Nb, Ag, Au, etc., see, e.g., [147]) as well as glass and NaCl (see references in [95]), Si [102, 148, 149], Ge [150], and amorphous SiO_2 [143, 151]. As an example, Fig. 6.20 shows the formation of a faceted Ge surface after Xe ion bombardment under an ion incidence angle of 75°. Following ion bombardment, these sawtooth-like profiles with rms roughness values of about

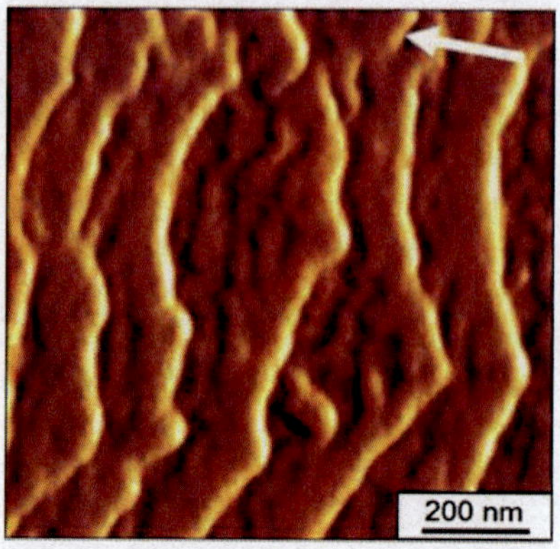

Fig. 6.20 AFM images of a Ge surface (amplitude error image) after Xe ion bombardment with 6.7×10^{18} ions/cm^2, an ion energy of 2 keV and an angle of ion incidence of 75°. The rms roughness before ion bombardment was 0.5 nm. The arrow shows the direction of ion incidence (adapted from [150])

7 nm are formed. The topography is characterized by extended facets on the downstream side and an angle of 5° of the local surface normal with respect to the global surface normal. With increased ion energy, the period and the amplitude become more pronounced [150].

The evolution of ion beam-induced facets or terraces on the surface is investigated by measuring the height profiles. Figure 6.21 (left) shows the evolution height profiles of the germanium surfaces following 1.2 keV Xe ion bombardment at an ion incidence angle of $\theta = 75°$ as function of the ion fluence [142]. Following a fluence of 1.35×10^{19} ions/cm^2, a faceted surface contour is formed. From these data, the local ion incidence angles on the upstream side, φ_{up}, and on the downstream side, φ_{down}, are determined (slope distribution functions), where the gradient of the surface, $\partial h/\partial x$, is given by $\tan \varphi = \partial h/\partial x$ (Fig. 6.21, right). An example of the evolution of facets on fused silica surfaces as a function of the time of bombardment (ion fluence) is shown in Fig. 6.22, as a function of the local incidence angle. In these experiments, fused silica surfaces with a well-defined surface topography (pre-patterned surfaces with ripples perpendicular to the direction of the incident ion beam) were used.

Figure 6.22, demonstrates the expected distribution of the local ion incidence angles with a maximum at 45° before ion bombardment. After about 20 min ($\equiv 2.25 \times 10^{18}$ ions/cm^2), facets appear, indicated by the formation of two maxima (Fig. 6.22). These are more pronounced after a bombardment time of 240 min ($\equiv 2.70 \times 10^{19}$ ions/cm^2). A stable facet topography is then obtained by distinct maxima represented by the facet angles of the upstream side, $\varphi_{up} = 3°$ and downstream side, $\varphi_{down} = 58°$. The process of coarsening at the transition from rippled to faceted surfaces, including the increase of the faceted structure (amplitude of facets), has been confirmed in numerous studies (see, e.g., [148, 152, 153]).

Based on the development of a model to describe the erosion of the sinusoidally rippled surface in dependence on the sputtering yield and the spatial derivatives of

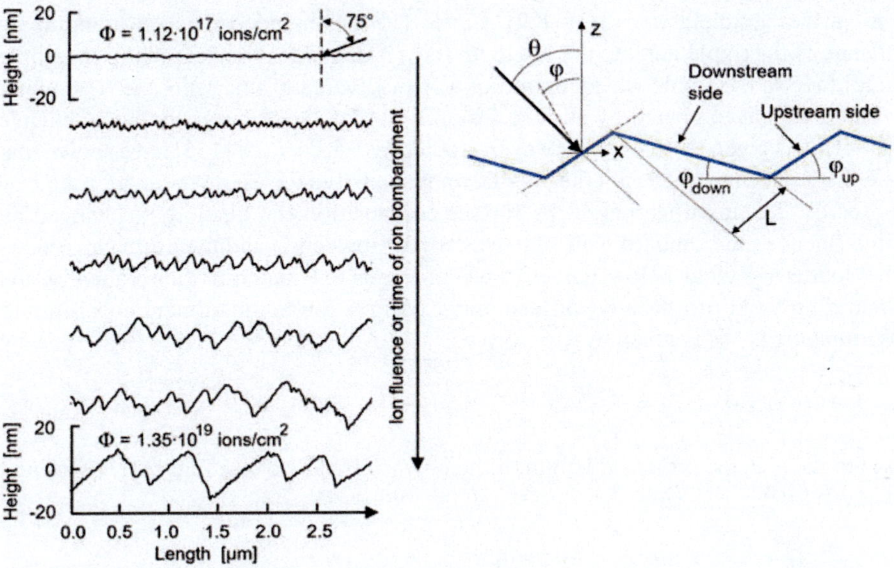

Fig. 6.21 Left: measured height profiles on Ge surfaces after 1.2 keV Xe ion bombardment at an ion incidence angle of 75°. The ion fluence ranged between 1.12×10^{17} cm^{-2} (above) 1.35×10^{19} cm^{-2} (below). Right: schematic picture of a faceted surface under oblique ion bombardment

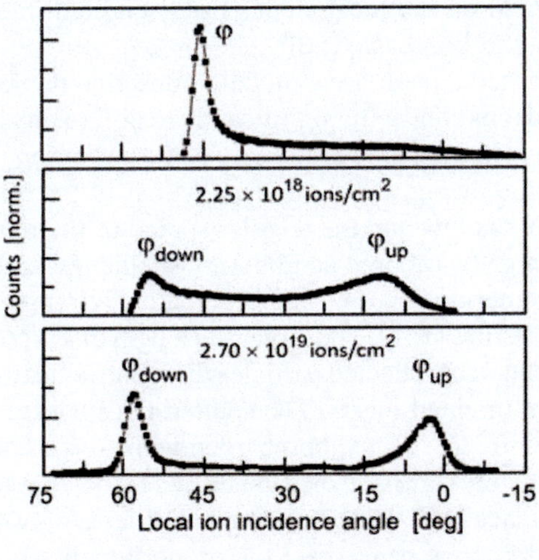

Fig. 6.22 Local ion incidence angle distributions on fused silica surfaces after 2 keV Ar ion bombardment at an angle of ion incidence of 30° as function of the ions fluence (time of ion bombardment). Above: initial situation characterized by orthogonal configuration of ripples with respect to the ion beam before ion bombardment. Middle: facets appears. Below: facets are fully formed

the surface gradient (see Sect. 8.2), Carter [154] concluded that the topographic change of the ripple habit to the sawtooth form is dependent on the ratio of the ripple amplitude to the ripple wavelength. For a critical value of this ratio, the shadowing of the incident ion beam by surface features can lead to this topographical change [criterion is given by (6.43)]. However, studies by Wittmaarck [155] have shown that the shadowing process is not the only factor responsible for the sawtooth morphology observed. The modification of the surface composition can likewise be changed by high fluence ion bombardment (fluences > 10^{18} ions/cm^2) and therefore can reduce the sputtering yield below that of the non-implanted material. Consequently, the sputtering yield of a faceted surface under oblique ion bombardment according to Wittmaarck [155] is given by

$$Y(\theta, \varphi_{up}, \varphi_{down}) = L_{down} Y(\theta + \varphi_{up}) + [1 - L_{down}] Y(\theta - \varphi_{down}), \tag{6.6}$$

where L_{down} is the projected length of the downstream side (see Fig. 7.21, right) and is given by

$$L_{down} = 1 \Big/ \left\{ 1 + \left[\frac{\sin \varphi_{down}}{\sin \varphi_{up}} \right] \left[\frac{\cos(\theta - \varphi_{up})}{\cos(\theta - \varphi_{down})} \right] \right\}. \tag{6.7}$$

Wittmaarck's assumption that the change in the sputtering yield is based on the dependence of the yield on the local slopes has been confirmed for small surface gradients by Makeev and Barabasi [156].

Hauffe [157] proposed a qualitative model to describe the facet development. It assumes that reflected ions contribute significantly to the erosion of surfaces. Because the reflectivity of the incoming ions is strongly enhanced at ion incidence angles $\theta >$ 65°…70° (c.f. Fig. 5.20 or Fig. 6.36), the energy loss of the reflected particles due to the collision process is low and the angular spread of the reflected ions is small. Ions scattered from large facets onto neighboring smaller facets enhance the erosion of these facets by sputtering. Figure 6.23 schematically represents the coarsening process according to Hauffe [157]. The larger the length of a facet on the downstream side, the larger the fraction of reflected particles that contribute to the sputtering of the neighboring upstream oriented facets. The sputtering of the facets on the upstream side thus becomes reinforced. For Si bombarded with Ar ions, the facet lengths are known to be approximately Gaussian distributed and to increase linearly with erosion depth, i.e. with ion fluence [149, 158]. In a simplistic view, for two neighboring facets of the lengths L_a and L_b, one can expect the proportionality $L_a/L_b \approx v_a/v_b$, where v_a and v_b are the corresponding local erosion velocities (Fig. 6.23). The coarsening process can be verified by comparison of the feature heights h(0) and h(t). According to this model, the additional backscattered ion flux is responsible for the coarsening of the topography. It should be noted that the faceted topography formed is independent of the crystal structure of the bombarded material.

It has also been suggested that surface imperfections, e.g., localized impurities, contaminants, etc., protrusions, or ion bombardment-induced defects (e.g., ion

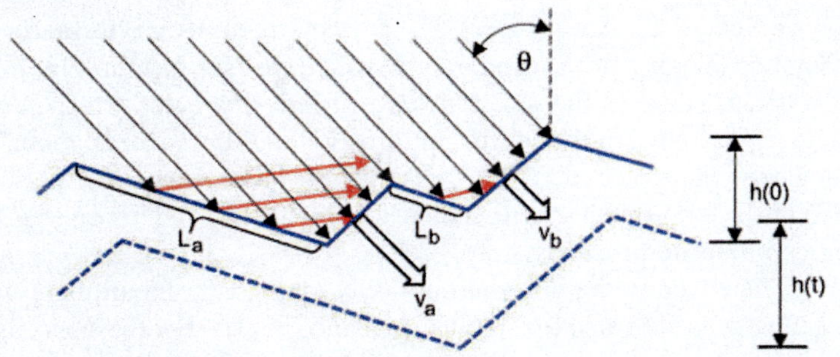

Fig. 6.23 Schematic presentation of the coarsening process under oblique ion bombardment (v is erosion velocity, θ is the angle of ion incidence, L_a, L_b are face lengths on the downstream side and h(0) and h(t) represents the feature heightsat the time t = 0 (solid line) and a subsequent moment (dotted line) of the ion bombardment, respectively) (figure adapted from [157] and modified)

bombardment-induced defects such as etch pits) are starting points for the development of repetitive features (facets) [95]. With increased ion fluence, these features increasingly grow and overlap. The growth associated with a coarsening of topographical areas and a significant increase in roughness. This process also leads to the formation of faceted structures on both crystalline surfaces and amorphous materials, among others. The orientation of the facets is typical for this prismatic topography. These are aligned approximately parallel and normal to the direction of the incoming ions. As an example, Fig. 6.24 shows the roughness of Si and Ge after Kr ion bombardment at room temperature as a function of the ion fluence. For low ion angles of incidence up to about θ = 60°, a slow rise in the rms roughness can be measured (Fig. 6.24). With an increase of the incidence angle up to 60°, a moderate increase of the rms roughness can be observed, which corresponds to the formation

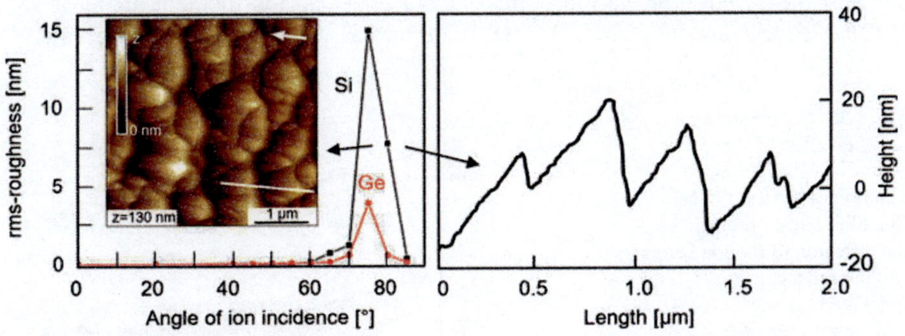

Fig. 6.24 Left: root mean square roughness of Si and Ge in dependence on the angle of Kr ion incidence. The ion energy was 1200 eV, the ion current density was 300 $\mu A/cm^2$ and the ion fluence 6.75×10^{18} ions/cm^2. The inset shows an AFM images of Si surfaces after Kr ion bombardment with a fluence of 6.75×10^{18} ions/cm at an angle of ion incidence of 82°. The white arrow indicates the direction of the Kr ion beam. Right: scan of the surface profile along the white line in the AFM image (left). Figures adapted from [142, 159]

of ripples (see Sect. 8.1). Above $\theta = 75°$, faceted structures are formed and there is a significant coarsening of the topographic structures (see inset in Fig. 6.24, left). Figure 6.24 (right) shows a measured surface profile, characterized by a sawtooth contour. The angle between the ion beam direction and the facet is about 84°. For glancing angle ion incidence ($\theta > 85°$), the roughness drops dramatically (rms roughness $w < 0.5$ nm). Consequently, with increasing ion fluence and ion energy, the facet coarsening continues to proceeds.

In general, facets could be observed not only after ion bombardment under ion incidence angles > 65° but also for smaller incidence angles. For these, no significant contribution to the process of coarsening by backscattered ions can be expected. For example, Engler et al. [160] determined that for an ion incidence angle of $\theta = 63°$, the additional contribution of reflected ions to the total ion flux is less than 0.01%, i.e. it can be expected that not only the reflected ions contribute to evolution of the topography (facet formation).

The Hauffe mechanism cannot explain the development of the facets as a function of the incident ion fluence. Nobes et al. [161] and later Johnson [162] predicted that gradient-dependent sputtering of a curved surface with increasing time (fluence) ultimately tends to equilibrium slope of the bombarded protuberant structure. Based on the dependence of the sputtering yield on the angle of ion incidence, they predicted that the local slope of such curved surfaces evolves towards structure angles for which the temporal change of the surface curvature is zero, i.e. $dY(\theta)/d\theta = 0$. This situation is given for the ion incidence angle $\theta = 0°$, 90°, and the angle for maximum sputtering yield θ_p. In Fig. 6.25, this evolution process in the development of facets is schematically represented. Beginning with a rough surface, a profile with rising fluence develops, characterized by two non-equivalent slopes. In ideal circumstances, the equilibrium profile of the sawtooth-like facets has a short upstream side with a local ion incidence angle of 0° and an elongated downstream side with a local ion incidence angle of 90°. Exemplary, Fig. 6.24 (right) shows surface profiles of Ge surfaces measured after Xe ion bombardment under an ion incident angle of 75° [142]. A strong tendency to form facets under increased ion fluence is observable.

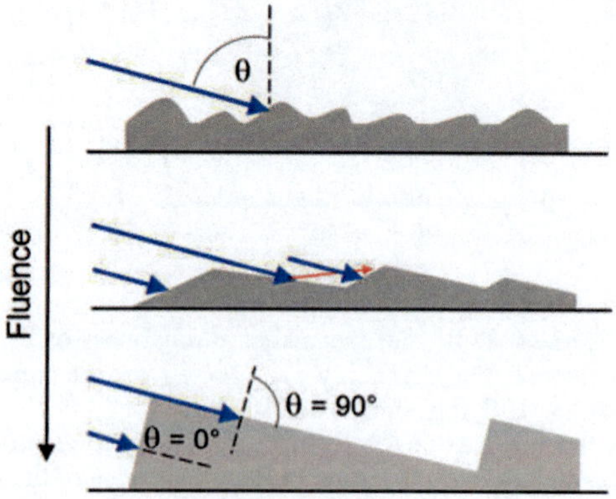

Fig. 6.25 Schematic representation of the evolution of facets as a function of the ion fluence. The blue arrows represent the direction of the ion beam and the red arrow the reflected ions

This dynamics of the surface evolution with increasing ion fluence was studied by Chowdhury and Ghose [163]. They determined angles of 85° and 10° for the equilibrium state of facets on Si(100) surfaces after 500 eV Ar ion sputtering at an incidence angle of 75° and fluences up to 1.7×10^{19} ions/cm^2.

6.3.2 *Inter-crystalline Surface Defects*

The difference of the sputtering yield on crystal orientation between neighboring crystallites is reflected in different erosion rates. Consequently, the grains eroded after sputtering are either depressed or protuberant relative to their immediate vicinity. As result of this process, the surface becomes increasingly rough and different elevations are visible across the surface. Grain-orientation-dependent sputter yield has been reported in polycrystalline materials (e.g., Cu [164], Fe [165], Ti and Cu [166], Al [167], Fe-Si-Al alloy [168]). Figure 6.26 demonstrates the dependence of the sputtering yield [168]. The EBSD mapping in Fig. 6.26, left, shows the orientation of the grains by means of a false color representation. The grains of the polycrystalline alloy are oriented randomly to each other. The AFM image of three grains (Fig. 6.26, right) illustrates the significant height difference between the grains (20 nm…300 nm) after 6 keV Ar ion bombardment with a fluence of 3×10^{18} ions/cm^2.

The variation in height between different crystallites is dependent on (i) the angle of incidence, because sputtering yield is a function of this angle (see Sect. 5.6) and (ii) the surface binding energy, U_s, since the sputtering yield is also dependent on the crystallite orientation. For example, for fcc metals (e.g., Cu), $U_{s,111} > U_{s,100} > U_{s,110}$ and therefore $Y_{110} > Y_{100} > Y_{111}$ (see Sect. 5.4.1).

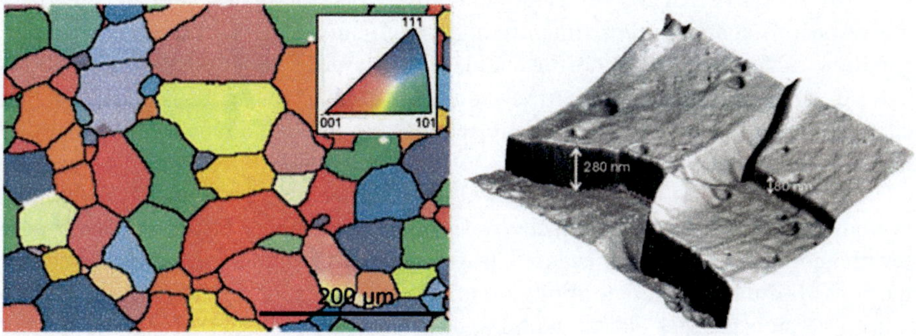

Fig. 6.26 Left: electron backscatter diffraction (EBSD) analysis of the grain orientation on the surface of an Fe-Si-Al alloy sample. Right: AFM image (scan size is 10 µm × 10 µm) of a sample area with three grains after Ar ion bombardment (figures adapted from [168])

Polycrystalline materials consist of grains (crystallites), separated by grain boundaries. Grain boundaries between the crystallites play an important role in the evolution of the topography under ion bombardment. Grain boundaries are planar (two-dimensional) lattice defects where two crystal lattices of different orientation and/or different crystallographic structure meet. When the misfit orientation between the grains is not too large, these grain boundaries contain line defects and dislocations. In the simplest case of a tilt between two grains, a sequence of edge dislocations (in the twist case, a network of screw dislocations) occurs. When these dislocations are considered as regions of lower binding energies because of the mismatch of the crystal lattices, then an increasing sputtering yield can be expected in the region around the dislocations (around the grain boundary). Where the grain boundary intersects the bombarded surface, increased ion beam etching is evident and leads to the formation of pits or grooves between the grains (see Fig. 6.26, right).

It is also known that in some materials, impurities segregate to the dislocations [169]. The segregated impurities are capable of agglomerating and dramatically reducing/strengthening the erosion rate. Under this condition, additional increased sputter etching around the grain boundary can be expected.

6.4 Analysis of Surface Evolution Under Ion Bombardment

Topography changes under conditions of ion beam-induced surface erosion can be analyzed using geometric methods. The sputtering parameters, ion current density, atomic number density, and sputtering yield that determine the rate of erosion are dependent upon both spatial and time coordinates. As a consequence, an analysis of sputter erosion induced topography changes are only possible if the number of these parameters is reduced or it can be assumed that the parameters are invariant with respect to spatial and/or time coordinates. It should be further noted that for crystalline materials, the sputtering yield, which is dependent on the ion incident angle, θ, between the surface normal and the ion beam direction, is additionally dependent on the azimuthal angle (i.e. orientation of the individual crystal planes to the ion beam direction).

Historically, the first observation of well-defined surface structures as result of an ion bombardment process was made by Güntherschulze and Tollmien [97]. Later, Stewart and Thompson [98] presented the first sputter-erosion model and opened the field for two-dimensional determination of topography resulting from an ion beam-induced erosion process. These authors, as well as Nobes et al. [161] and Carter et al. [170], were able to explain the formation of hillocks or cones on surfaces under ion bombardment (see Sect. 6.3.1.1). Under the condition that the sputtering coefficient is a function of the angle between the direction of ion incidence to the surface and the normal to the surface θ (see Fig. 5.9), it could be concluded that cones will form on a surface when the semi-vertical angle of the cone tends to the angle of the ions to the normal at which the sputtering yield is a maximum (i.e. $\theta \rightarrow \pi/2 - \theta_p$). Nobes et al. [161] and Carter et al. [170] derived analytical equations of the motion of the changing surface topography and demonstrated that a steady-state can be obtained only in conditions where the surface topography consists of planes aligned either parallel or perpendicular to the direction of the ion incidence. Barber et al. [110] applied the kinematic theorems of crystal dissolution theory by Frank [171] to predict topography evolution as result of ion bombardment where ion reflection and re-deposition of sputtered material have not been taken into account. Ducommun et al. [172] and Carter et al. [170] studied in detail the surface evolution of two-dimensional surfaces from a theoretical point of view.

Many experimental and theoretical studies of surface evolution following ion beam erosion (see reviews [173–176]) have led to a general concept based on an eikonal equation or Huygens principle of wave front propagation [177, 178]. The eroded surface is considered to be a propagation wavefront, where the Huygens construction in each point on a surface is assumed to be an effective Huygens wavelet oscillator. This is equivalent to the propagation of the light in geometrical optics, where all points of a wave front may be regarded as new sources of wavelets that expand in every direction. This method, known as the *kinematic erosion method*, allows the modeling of surface evolution under ion bombardment. A second approach is the *method of characteristics* (this can be considered equivalent to 'ray tracing' in geometrical optics) developed by Smith et al. [179, 180]. They recognized that the propagation of a wave can be described via the method of characteristic trajectories (for details, see [174]). The method of characteristics was used to trace the development of three-dimensional topographic structures (edges, facets, cones, etc.) due to multiple stationary points given by the angular dependence of the sputtering yield. This approach is capable of describing the surface evolution under ion bombardment in inhomogeneous and time-dependent systems and can be applied when the surface remains continuously differentiable during the process of evolution (see, e.g., [181]).

On the basis of different methods, it is possible to make quantitative predictions of topography formed by low-energy ion bombardment when this technique is used to shape surface reliefs and topographies in the semiconductor and optical industries (see Chap. 7) as well as to explain the process of surface roughening and uncertainness of ion depth profiling in the field of surface analysis.

6.4.1 *Kinematic Erosion Theory–Gradient-Dependent Sputtering*

The primary condition of the description of the surface evolution under ion bombardment is based on the assumption, that (i) a spatially uniform ion beam with an averaged mean flux density (ion flux), J, is incident onto the surface under an incident angle θ, (ii) the atomic number density N of the bombarded target material is invariant and (iii) the sputtering yield Y is only a function of the ion angle of incidence θ. It must be recalled (c.f. Fig. 5.9) that the sputtering yield is approximately proportional to the secant of the angle of ion incident (dashed line in Fig. 6.27) up to about

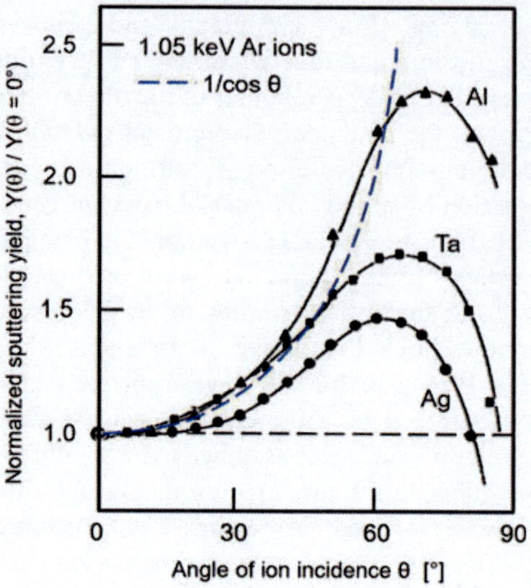

Fig. 6.27 Experimentally measured sputtering yield Y(θ)/Y(θ = 0) as function of the angle of ion incidence θ for different metals. The energy of the Ar ions was 1.05 keV. The dashed line represents the 1/cosθ dependence of the normalized sputtering yield (figure adapted from [182])

60°. The description of the temporal evolution of surfaces under ion bombardment is based exclusively on the fundamental premise that sputtering alone is responsible for the erosion of the surface. Secondary processes contributing to the evolution of the surface under ion bombardment (see Sect. 6.5) are ignored. Fig. 6.28 schematically shows the situation in which a curved surface, h(x, t), is subjected to a uniform ion flux with an average ion current density, J, at an angle of incidence θ . For simplification, the surface height, h, is assumed be independent of y-surface coordinate (two-dimensional case). Then, the local erosion rate along the normal direction in the –hP direction is given by

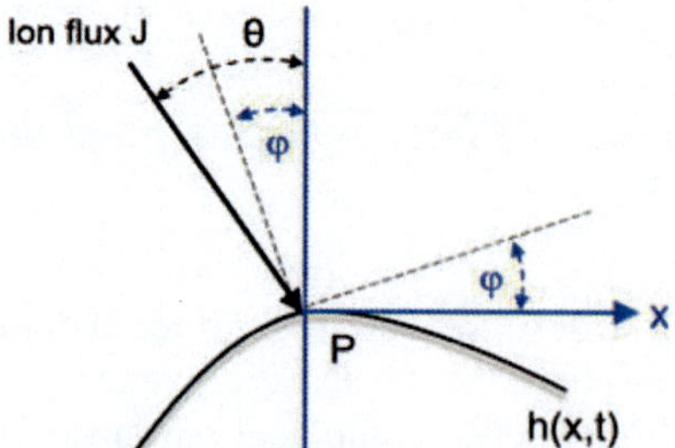

Fig. 6.28 Schematic picture of the gradient-dependent sputtering, where θ is the global incidence angle and φ is the local incidence angle

$$\frac{\partial h}{\partial t} = \frac{J}{N} Y \cos\theta = v_o \cos\theta \equiv v, \tag{6.8}$$

(compare this equation with the Eqs. (5.2) and (5.49). Note that a negative sign, i.e. $-\partial h/\partial t$, is often used in the literature to demonstrate the erosion of the surface. Each point, x, on a curved surface (e.g., the point of ion impact P) can be described at time t (erosion time) by the graph h(x,t). The gradient (slope), ∂h/∂x, at this point is given by tanφ (φ is the local slope angle, see Fig. 6.28) and the associated surface arc element is given by $\sqrt{1 + (\partial h/\partial x)^2}$. Consequently, this term must be included in Eq. (6.8) because the surface recedes in the direction of the local surface normal as a result of ion beam induced erosion, i.e.

$$\frac{\partial h}{\partial t} = v\sqrt{1 + \left(\frac{\partial h}{\partial x}\right)^2} = v\sqrt{1 + \tan^2\varphi} \tag{6.9}$$

or

$$v = \frac{1}{\sqrt{1 + \tan^2\varphi}} \frac{\partial h}{\partial t}. \tag{6.10}$$

Thus, the regression of the surface caused by the ion irradiation can be projected onto the h-axis (see Fig. 6.28). Both equations, Eqs. (6.9) and (6.10), clearly show that the erosion of a surface under ion bombardment can be assumed to be exclusively a function of the local surface gradient (contributions of secondary processes are not considered).

Nobes et al. [161] and Carter et al. [147, 170] proposed an initial two-dimensional model to determine the spatiotemporal evolution of the surface topography of an amorphous, isotropic solid. This model was derived in the light of the kinematic wave treatment by Frank [171] and based on the conclusion, that the sputtering yield is dependent on the gradient of the surface with respect to the incident ion flux direction.

A two-dimensional surface element AB (blue solid line in Fig. 6.29) at the time t is considered. The tangent to AB at A has an angle of θ with the horizontal coordinate axis 0x. This is also the angle between the normal to AB and the vertical coordinate axis 0h (angle of ion incidence).

According to Fig. 6.29, the tangential angle at A is given by θ (angle of ion incidence and at B by θ + (∂θ/∂x), i.e. the erosion rate (c.f. 6.10) at A can be expressed by (J/N)Y(θ)cosθ and at B by (J/N)Y(θ + dθ)cos(θ + dθ).

The sputtering induced erosion of this element in the time dt results in a change of the topography from AB to A′B′ due only to the variation of the sputtering yield with the angle of ion incidence. Because the sputtering yield, Y(θ), increases with the angle of ion incidence, θ, it follows that BB′ exceeds AA′ by the distance

$$CB' = \frac{J}{N}\frac{\partial}{\partial\theta}(Y(\theta)\cos\theta)dt \quad \text{and} \quad A'C = R\frac{d\theta}{dx}dx. \tag{6.11}$$

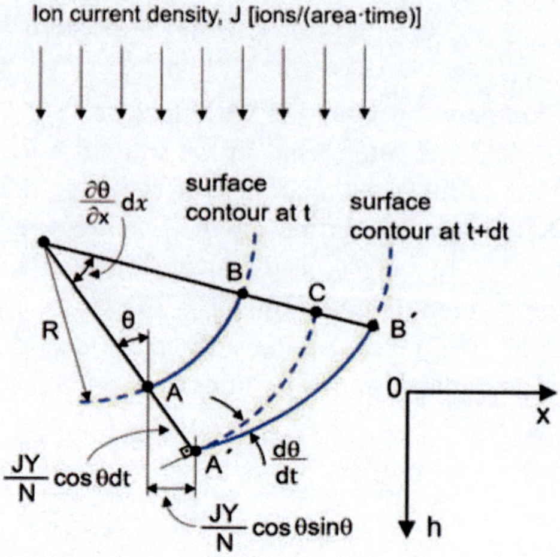

Fig. 6.29 Schematic of sputter induced erosion of a line contour by vertical ion bombardment (figure is taken from [170] and modified)

On the base of geometrical arguments, Carter et al. [170, 183], have shown that the radius of curvature of AB in Fig. 6.29 is given by

$$R = -\frac{CB'}{A'C} = -\frac{J}{N}\frac{\partial}{\partial\theta}(Y(\theta)\cos\theta)\frac{dt}{d\theta} \quad \text{or}$$
$$\frac{d\theta}{dt} = -\frac{J}{NR}\frac{\partial}{\partial\theta}(Y(\theta)\cos\theta). \tag{6.12}$$

Apparently, the incidence angle θ is dependent on x and t. It is then essential that

$$d\theta = \left(\frac{\partial\theta}{\partial x}\right)_t dx + \left(\frac{\partial\theta}{\partial t}\right)_x dt \quad \text{or rather} \quad \frac{d\theta}{dt} = \left(\frac{\partial\theta}{\partial x}\right)_t dx + \left(\frac{\partial\theta}{\partial t}\right)_x. \tag{6.13}$$

From (6.8), it follows that the rate of change of point A is given by

$$\frac{dx}{dt} = \frac{J}{N}Y(\theta)\cos\theta\sin\theta. \tag{6.14}$$

Analytic expression for the temporal development of surface points is then given by [170]

$$\left(\frac{\partial\theta}{\partial t}\right)_x = \frac{J}{NR}\frac{\partial Y(\theta)}{\partial\theta}\cos\theta - \frac{J}{NR}Y(\theta)\sin\theta = \frac{J}{NR}\cos\theta\frac{\partial Y(\theta)}{\partial\theta}. \tag{6.15}$$

By substitution of the radius of curvature of AB, (6.12), it follows that

$$\left(\frac{\partial\theta}{\partial t}\right)_x = \frac{J}{NR}\frac{\partial Y(\theta)}{\partial\theta}\cos\theta = \frac{J}{N}\frac{\partial Y(\theta)}{\partial\theta}\cos^2\theta\left(\frac{\partial\theta}{\partial x}\right)_t \quad \text{or rather}$$

$$\frac{\left(\frac{\partial \theta}{\partial t}\right)_x}{\left(\frac{\partial \theta}{\partial x}\right)_t} = \left(\frac{dx}{dt}\right)_\theta = \frac{J}{N}\frac{\partial Y(\theta)}{\partial \theta}\cos^2\theta \equiv v_x. \tag{6.16}$$

This last equation indicates the variation of the tangential angle with both spatial coordinate x and time t. In principle, if the Y-θ function (see Fig. 5.9) is known, (6.16) can be solved for θ as function of x and t.

The steady-state (equilibrium) will be reached if the time rate of topography changes is zero, i.e. is given by

$$\frac{J}{N}\frac{\partial Y(\theta)}{\partial \theta}\cos^2\theta = 0. \tag{6.17}$$

Consequently, equilibrium is achieved when either $\cos\theta = 0$ (for $\theta = \pm\pi/2$) or $dY/d\theta = 0$ (for $\theta = 0$ or $\pm\theta_P$) [170]. The equilibrium contour is thus given only for planar surfaces at $\theta = \pi/2$ (vertical surface) or $\theta = 0$ (horizontal surface). For arbitrary values of θ, equilibrium (stable surface contours) cannot be achieved.

Carter et al. [183] also derived an analytic expression for the temporal development of surface points as a function of the spatial coordinate h and the time t which are

$$\left(\frac{\partial \theta}{\partial t}\right)_h = \frac{J}{N}\left[\frac{\partial Y(\theta)}{\partial \theta}\sin\theta\cos\theta - Y(\theta)\right]\left(\frac{\partial \theta}{\partial h}\right)_t \tag{6.18}$$

and

$$\frac{\left(\frac{\partial \theta}{\partial t}\right)_h}{\left(\frac{\partial \theta}{\partial h}\right)_t} = \left(\frac{dh}{dt}\right)_\theta = \frac{J}{N}\left[\frac{\partial Y(\theta)}{\partial \theta}\sin\theta\cos\theta - Y(\theta)\right] \equiv v_h. \tag{6.19}$$

These are identical to the kinematic wave equations [184]. Equations (6.16) and (6.19) describe the movement of a point surface in space and time at a given orientation if the surface is sputtered. Carter et al. [176] referred to this circumstance as surface gradient-dependent sputtering.

The erosion velocities parallel, v_x (6.16), and perpendicular, v_h (6.19), to the surface are thus obtained for a one-dimensional surface h(x, t). The direction of the motion of points of constant orientation θ (slope of the trajectory) is given by forming the ratio of the (6.16) and (6.19) as

$$\tan\varphi = \frac{v_h}{v_x} = \frac{\left(\frac{dh}{dt}\right)_\theta}{\left(\frac{dx}{dt}\right)_\theta} = \frac{dh}{dx} = \left[\frac{\partial Y(\theta)}{\partial \theta}\sin\theta\cos\theta - Y(\theta)\right] / \frac{\partial Y(\theta)}{\partial \theta}\cos^2\theta \tag{6.20}$$

and the velocity of such point motion can also be determined from (6.16) and (6.19) [185, 186] to be

$$v = \sqrt{\left(\frac{dx}{dt}\right)_\theta + \left(\frac{dh}{dt}\right)_\theta} = \frac{J}{N}\sqrt{\left[\frac{dY(\theta)}{d\theta}\cos^2\theta\right]^2 + \left[\frac{dY(\theta)}{d\theta}\sin\theta\cos\theta - Y(\theta)\right]^2}$$
$$= \frac{J}{N}\sqrt{\left[\frac{dY(\theta)}{d\theta}\cos\theta\right]^2 + [Y(\theta)\cos\theta]^2}. \quad (6.21)$$

The spatial velocities of points on the surface are given by [154]

$$\left(\frac{dx}{dt}\right)_h = \frac{J}{N}\frac{Y(\theta)}{\tan\theta} \quad \text{and} \quad \left(\frac{dh}{dt}\right)_x = \frac{J}{N}Y(\theta). \quad (6.22)$$

In general, graphical (geometric) or numerical methods have been applied with the aim to find the time dependent topography evolution.

Geometrical methods for reconstructing erosion profiles are based on relationships between the initial surface contour and the surface profile after a finite period of ion sputtering. It has been demonstrated above that such relationships exist between small surface elements of identical orientation. It was shown that for any surface element, the erosion velocity and direction of movement depend on the orientation as well as on the orientation-dependent sputtering yield [see (6.16) and (6.19)].

If the infinitesimal changes along the x- and the h-coordinates are known as a function of time for different values of the orientation θ, obtained from the Y-θ function (see Fig. 5.9), the erosion contour corresponding to any sputter time can be determined. A geometrical approach is important if the time-dependency of motion of a surface cannot be solved analytically and if the motion of discontinuities in a surface profile must be described. Barber [110] proposed a graphic technique, known as erosion slowness curves, to construct such erosion profiles. This approach is also based upon the kinematic theory of the orientation-dependent dissolution of crystals by Frank [171] and provides a polar diagram of the reciprocal of the surface erosion ($1/Y\cos\theta$ measured in the surface normal direction) as a function of the surface orientation θ. The geometrical construction technique and its application, which allow the topographical development to be predicted following ion bombardment, have been discussed in detail (see e.g., [185, 187]).

Numerical computational methods have also been applied to study the development of the surface contour under ion bombardment [185]. These methods trace the changes in the topography for short time periods. At first, an initial topography is specified and the ion beam distribution as well as the sputtering yield as a function of the ion incidence angle (Y-θ function) are defined. Both the Y-θ function and the ion flux distribution can be measured experimentally and then input into analytic functions. In general, the normalized function $Y(\theta)/Y(\theta = 0)$ is given by a polynomial and the normalized function by $J(x)/J(x = 0) = \exp(-x^2/2\sigma^2)$, where σ^2 is the variance of a Gaussian distribution. The time steps between calculations for each change of the topography are set to be so small that topographical changes in order of a fraction of the variance can be expected. According to Carter et al. [185, 187], such a program consists of the following steps:

- the changes dx_n and dh_n in the h direction for the first time step are calculated (details see, e.g., Ref. [186]),
- with these values dx_n and dh_n, the new surface contour is reconstructed,
- after reconstruction, the values of θ_n are evaluated by $\tan\theta_n = (h_{n+1} - h)/(x_{n+1} - x)$,
- then, the erosion process is calculated for each x_n, where the erosion rate at $\theta = 0$ is multiplied by value of $Y(\theta)/Y(\theta = 0)$ and
- the newly calculated points of the surface contour replace the former surface points.

By following this procedure sequentially, the surface contour can be calculated following a given number of time steps.

Application Examples

Assuming a uniform ion flux, a number of different authors [172, 188, 189] have solved (6.10) by iterative approaches, with the aim of determining the change of a sinusoid wave form. In an initial attempt, using computer simulation, Catana et al. [188] studied the development of steady-state topography by uniform ion bombardment of an initially sinusoidal contour. The sputtering yield as a function of the incident angle was approximated by a fourth-order polynomial. The rapid changing of the surface contour within the initial time stages is noteworthy. The sine-contour developed into a triangular form of a half-angle θ_n, where the sputtering yield at θ_n is equal to that at $\theta = 0$. In contrast, the downward contour became straight. This corresponds to the experimentally proven formation of cones and other protuberant structures following ion bombardment of a three-dimensional topography (see Sect. 6.3).

Ishitani et al. [189] discussed in detail the development of concave downward and upward surfaces and tabulated the slopes of equilibrium surface contours for various initial conditions. For example, the erosion of a concave downward surface contour by ion sputtering in the negative h direction (erosion) is schematically shown in Fig. 6.30, left. The influence of the ion bombardment and the initial incidence ion angle θ is discussed at the points A, B and C, where $x_A < x_B < x_C$. Between A and C on the surface, θ_B is initially always $\theta_p \geq \theta_A > \theta_B > \theta_C$, where θ_p is the angle at which $Y(\theta)$ reaches its maximum (Y_p, see Fig. 5.9). Hence, $(\partial Y/\partial\theta) > 0$ and $(\partial\theta/\partial x) < 0$, $(\partial\theta/\partial t)$ is positive according to (6.16), and $Y_A > Y_B > Y_C$. Consequently, the surface contour tends towards a straight line at certain angles θ in the range between θ_A and θ_p. The time-dependent development of the concave downward (middle) and upward (right) surface contours by ion sputtering are schematically shown in Fig. 6.30 as a function of the time of ion bombardment.

Ducommun et al. [172] also described the evolution of a sine-type surface under ion bombardment using a non-iterative computer simulation. The original contour is thereby taken to be the envelope of tangential lines, and the progressive motion of each tangential segment at a given slope due the ion beam-induced erosion was calculated.

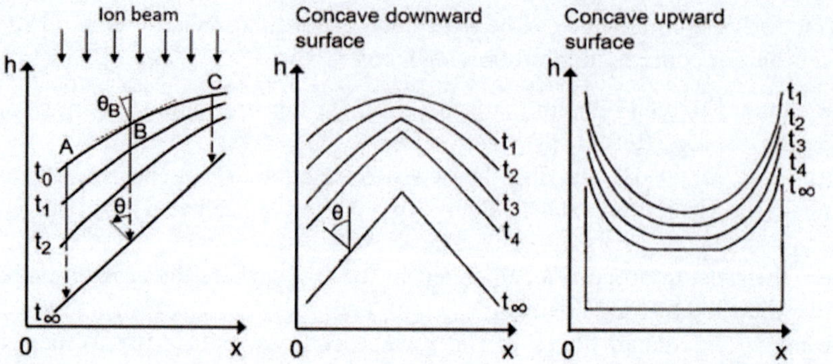

Fig. 6.30 Left: schematic depiction of the time-dependent erosion by sputtering of a concave downward surface. Middle: development of a concave downward surface as a function of the erosion time. Right: development of a concave upward surface as a function of the erosion time. Figures adapted from [189]

The recalculated tangents then configure the envelope of the eroded topography. The authors used this approach to solve (6.16) and (6.19) under the condition of a sinusoidal surface contour defined by the function $f(x) = k \cdot \sin(x)$. The general tendency is to transform the sinusoidal contour into a horizontal straight line for three different values of k. The stages of this evolution are different, depending on the maximum slope θ_p (incidence angle, where the sputtering yield is a maximum) of the initial profile. It could be asserted that (i) the horizontal planes are steady-state contours, (ii) vertical planes remain unchanged by ion beam erosion, and (iii) planes with a slope $\pm\theta_p$ initially appear before these vanish. The studies by Ducommun et al. [172] confirmed the models by Carter et al. [183] and Barber et al. [110].

Ducommun et al. [190] also applied their approach to analyzing the evolution of initial step profiles as a function of the erosion time. Figure 6.31, left, shows the development of the surface contour under uniform ion bombardment. The graphic representation exhibits a rounded surface at the end of the sputtering surface. The simulated results are compared with results of ion etching experiments. Steps are etched into silicon substrates, where the surface of silicon was partially covered by a thick vanadium mask. Figure 6.31, left, shows cross-section SEM images of a Si step before ($t = 0$ min) and after ($t = 45$ min) sputtering. The size and angle measurements

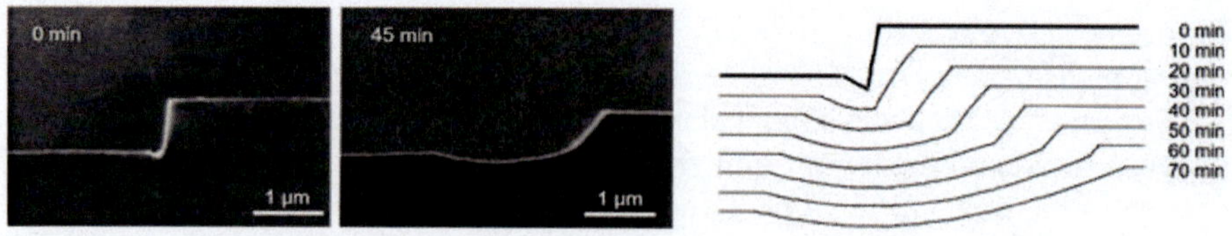

Fig. 6.31 Left: SEM images of a Si step profiles before (0 min) and after 45 min of ion bombardment. Right: calculated development of a step contour by sputtering as function of the sputtering time. Figures adapted from [190]

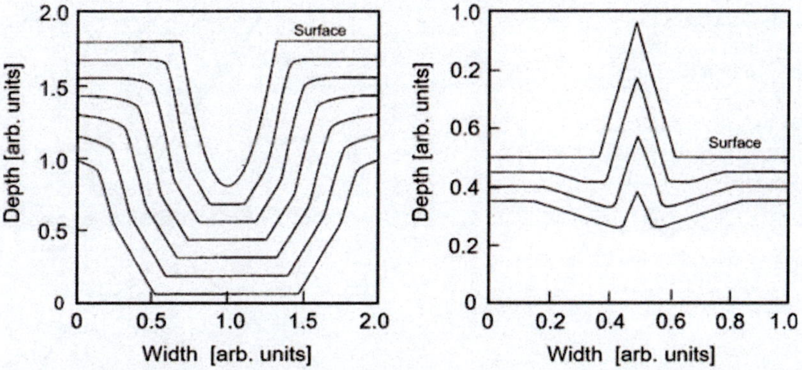

Fig. 6.32 Left: evolution of a locally perturbed planar surface contour by ion beam sputtering (figure adapted from [177]). Right: simulation of evolution of a triangle-plane section in which reflection from the triangle section enhances the rate of the basal region locally, i.e. a groove is formed (figure adapted from [181])

of the rounded corner are in good agreement with the values provided by computer simulations.

The evolution of a locally perturbed planar surface under ion beam erosion has also been studied [177]. Figure 6.32, left, illustrates the development the surface contour of a pit by ion sputtering in dependence on the sputter time. It becomes apparent that the actual form of the perturbation should not affect the final shape of the sputter etch pit, which has also been experimentally observed. According to Katardjiev et al. [177], an overlap of neighboring etch pits results in the formation of pyramids or cones, which could not be observed in the simulation, because only the evolution of a single perturbation was simulated. Figure 6.32, right, illustrates the evolution of a cone on an Si surface (c.f. Figs. 6.14 and 6.15) after 1 keV Ar ion bombardment, where ion reflection is taken into account [181]. The reflection coefficient was chosen to be 0 for angles of up to 65° then rises for angles larger than 75°.

6.4.2 *Method of Characteristics*

Alongside the kinematic method, Smith et al. [79, 80] and Carter [47] provided another approach, which treats the moving (bombarded) surface as a propagating wave front which can be described by partial differential equation. This partial differential equation of first order can be solved by the method of characteristics. This method is based on the conversion of a partial differential equation into an ordinary differential equation on defined areas, the so-called characteristics, by means of a suitable coordinate transformation (details see e.g. [84, 174]).

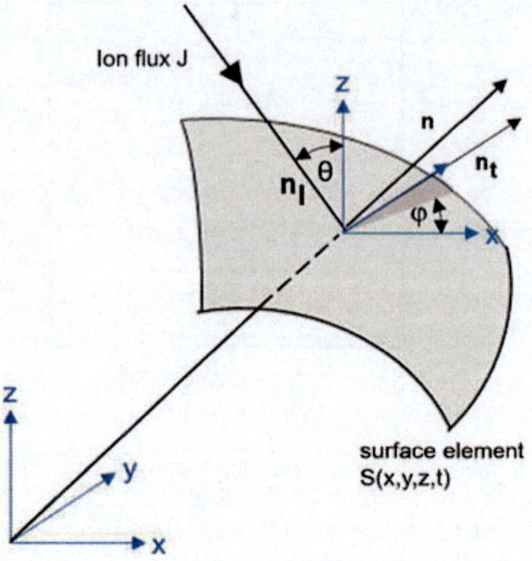

Fig. 6.33 Schematic diagram illustrating the geometry of bombardment of a surface element (to distinguish between the kinematic erosion theory (sect. 6.4.1) and the method of characteristic (Sect. 6.4.2), the height coordinates were denoted by h and z, respectively)

In general, the height of a surface can be described by a function z = u(x,y,t), where x,y,z the three spatial coordinates and t is the time of ion irradiation. If a surface function S of the type

$$S = z - u(x, y, t) = 0, \tag{6.23}$$

is introduced, the evolution of the surface as a function of time can be given by

$$\frac{dS}{dt} = \frac{\partial S}{\partial z}\frac{dz}{dt} + \frac{\partial S}{\partial x}\frac{dz}{dt} + \frac{\partial S}{\partial y}\frac{dz}{dt} + \frac{\partial S}{\partial t} = 0. \tag{6.24}$$

where x, y and z are the spatial variables in Cartesian coordinates and t is the time of the ion bombardment (Fig. 6.33).

The ion beam is incident in the direction of the unit vector n_I under the incident angle θ with respect to the unit vector of the surface normal n, where the surface normal varies spatially and temporally with surface evolution. According to Hamaguchi et al. [191], the velocity of a given surface point as a function of time can be expressed by

$$\left(\frac{dx}{dt}, \frac{dy}{dt}, \frac{dz}{dt}\right) = v\boldsymbol{n} + v_t\boldsymbol{n}_t, \tag{6.25}$$

where **n** and $\mathbf{n_t}$ are the surface normal and tangential vectors, respectively, and v (given by 6.8) and v_t are surface velocities (or local etching rates) in the normal and tangential directions, respectively (see Fig. 6.33). Based on geometric considerations, these vectors can be expressed by

$$\boldsymbol{n} = \frac{\frac{\partial S}{\partial x}\boldsymbol{i} + \frac{\partial S}{\partial y}\boldsymbol{j} + \frac{\partial S}{\partial z}\boldsymbol{k}}{\sqrt{\left(\frac{\partial S}{\partial x}\right)^2 + \left(\frac{\partial S}{\partial y}\right)^2 + \left(\frac{\partial S}{\partial z}\right)^2}} \quad \text{and}$$

$$\boldsymbol{n_t} = \frac{\left(-\frac{\partial S}{\partial y} - \frac{\partial S}{\partial z}\right)\boldsymbol{i} + \left(\frac{\partial S}{\partial x} - \frac{\partial S}{\partial z}\right)\boldsymbol{j} + \left(\frac{\partial S}{\partial x} + \frac{\partial S}{\partial y}\right)\boldsymbol{k}}{\sqrt{\left(\frac{\partial S}{\partial x}\right)^2 + \left(\frac{\partial S}{\partial y}\right)^2 + \left(\frac{\partial S}{\partial z}\right)^2}}, \tag{6.26}$$

where $\mathbf{n_t}$ is perpendicular to $\mathbf{n}$ and $\mathbf{i}$, $\mathbf{j}$ and $\mathbf{k}$ are the unit vectors in the x, y and z directions, respectively. The substituting of these equations in (6.25) leads to

$$\begin{pmatrix} \frac{dx}{dt} \\ \frac{dy}{dt} \\ \frac{dz}{dt} \end{pmatrix} = \frac{1}{\sqrt{\left(\frac{\partial S}{\partial x}\right)^2 + \left(\frac{\partial S}{\partial y}\right)^2 + \left(\frac{\partial S}{\partial z}\right)^2}} \left[v \begin{pmatrix} \frac{\partial S}{\partial x} \\ \frac{\partial S}{\partial y} \\ \frac{\partial S}{\partial z} \end{pmatrix} + v_t \begin{pmatrix} -\frac{\partial S}{\partial y} - \frac{\partial S}{\partial z} \\ \frac{\partial S}{\partial x} - \frac{\partial S}{\partial z} \\ \frac{\partial S}{\partial x} + \frac{\partial S}{\partial z} \end{pmatrix} \right]. \tag{6.27}$$

Inserting into equation (6.24) results in

$$\frac{dS}{dt} = v\sqrt{\left(\frac{\partial S}{\partial x}\right)^2 + \left(\frac{\partial S}{\partial y}\right)^2 + \left(\frac{\partial S}{\partial z}\right)^2} + \frac{\partial S}{\partial t} = 0. \tag{6.28}$$

This approach leaves an equation that is dependent only on the local etching rate v. Comparing (6.24) and (6.28), it follows that

$$\frac{\partial S}{\partial x} = -\frac{\partial u}{\partial x}, \frac{\partial S}{\partial y} = -\frac{\partial u}{\partial y}, \frac{\partial S}{\partial z} = 1, \frac{\partial S}{\partial t} = -\frac{\partial u}{\partial t}, \tag{6.29}$$

so that (6.28) can be rewritten by

$$\frac{dS}{dt} = v\sqrt{\left(\frac{\partial u}{\partial x}\right)^2 + \left(\frac{\partial u}{\partial y}\right)^2 + 1} + \frac{\partial u}{\partial t} = 0 \quad \text{or}$$

$$\frac{\partial u}{\partial t} = v\sqrt{\left(\frac{\partial u}{\partial x}\right)^2 + \left(\frac{\partial u}{\partial y}\right)^2 + 1}. \tag{6.30}$$

With $\frac{\partial u}{\partial x} = p$, $\frac{\partial u}{\partial y} = q$ and $\frac{\partial u}{\partial t} = r$, the final surface evolution equation can be obtained to be

$$r = v\sqrt{p^2 + q^2 + 1}. \tag{6.31}$$

This first-order hyperbolic partial differential equation of a class, known as the Hamilton–Jacobi equation, can be integrated by transformation into a set of ordinary differential equations using the method of characteristics (for details, see [174]).

Based on a geometrical consideration, the integration is carried out along the trajectories or characteristic curves. Smith et al. [145, 146] first introduced this method to predict the spatial and temporal evolution of the topography under ion bombardment. According to Hamaguchi et al. [191] and Arnold et al. [192] for the two-dimensional case, and Klein and Ramirez [193] for the three-dimensional case, the following set of ordinary differential equations has been obtained after application of the method of characteristics. For three-dimensional surface evolution, the set of equations is given by

$$\begin{aligned}
\frac{dx}{dt} &= -\frac{v\tan\psi_x}{\sqrt{1+\tan^2\psi_x+\tan^2\psi_y}} - \frac{v_{\psi_x}\sqrt{1+\tan^2\psi_x+\tan^2\psi_y}}{\tan^2\psi_x+1},\\
\frac{dy}{dt} &= -\frac{v\tan\psi_y}{\sqrt{1+\tan^2\psi_x+\tan^2\psi_y}} - \frac{v_{\psi_y}\sqrt{1+\tan^2\psi_x+\tan^2\psi_y}}{\tan^2\psi_y+1},\\
\frac{dz}{dt} &= -\frac{v\left(\tan^2\psi_x+\tan^2\psi_y\right)}{\sqrt{1+\tan^2\psi_x+\tan^2\psi_y}}\\
&\quad -\left(\frac{v_{\psi_x}\tan\psi_x}{\tan^2\psi_x+1}+\frac{v_{\psi_y}\tan\psi_y}{\tan^2\psi_y+1}-v\right)\sqrt{1+\tan^2\psi_x+\tan^2\psi_y},
\end{aligned} \tag{6.32}$$

where ψ_x and ψ_y are the elevation angles that the surface forms with the positive x-axis and the positive y-axis, respectively. These angles are given by

$$\tan\psi_x = p = \frac{dz}{dx} \quad \text{and} \quad \tan\psi_y = q = \frac{dz}{dy}. \tag{6.33}$$

The erosion velocity or erosion rate defined by (6.8) is both spatially and temporally dependent on the coordinates x, y, z, the time t, the orientation of the normal vector to the local surface (unit vector **n**), and the vector in the direction of the incoming ion beam (unit vector $\mathbf{n_I}$), i.e. v = v (x, y, z, t, **n**, $\mathbf{n_I}$). The unit vector in the direction of the ion beam can be described by $\boldsymbol{n_I} = A\boldsymbol{i} + B\boldsymbol{j} + C\boldsymbol{k}$, where A, B, and C are the magnitude of this vector in the x, y and h directions, respectively, and **i**, **j**, and **k** are again the unit vectors. An expression for the local surface unit vector, **n**, is given in (6.26). Based on a geometric consideration, the ion beam incidence, θ, can be expressed in a universally valid form by

$$\cos\theta = \boldsymbol{n}\cdot\boldsymbol{n_I} = \frac{A\frac{\partial S}{\partial x}+B\frac{\partial S}{\partial y}+C\frac{\partial S}{\partial z}}{\sqrt{\left(\frac{\partial S}{\partial x}\right)^2+\left(\frac{\partial S}{\partial y}\right)^2+\left(\frac{\partial S}{\partial z}\right)^2}} = \frac{A\frac{\partial S}{\partial x}/\frac{\partial S}{\partial h}+B\frac{\partial S}{\partial y}/\frac{\partial S}{\partial z}+C}{\sqrt{\left(\frac{\partial S}{\partial x}/\frac{\partial S}{\partial z}\right)^2+\left(\frac{\partial S}{\partial y}/\frac{\partial S}{\partial z}\right)^2+1}}. \tag{6.34}$$

Substituting (6.32) in (6.34), it follows that

$$\cos\theta = \frac{A\tan\psi_x + B\tan\psi_y + C}{\sqrt{1+\tan^2\psi_x + \tan^2\psi_y}}. \tag{6.35}$$

Recalling from (6.8) that the erosion rate behaves as $v = \frac{Y(\theta)}{N} J \cos\theta$, this expression can be written as

$$v = \frac{Y(\theta)}{N} J \frac{A\tan\psi_x + B\tan\psi_y + C}{\sqrt{1+\tan^2\psi_x + \tan^2\psi_y}}, \tag{6.36}$$

where A, B, and C must be determined according to the experimental arrangement. For example, Ducommun et al. [190] have fitted the sputtering yield in dependence on the angle of incidence by sixth order polynomial, $Y(\theta) = \sum_{n=1}^{6} a_n \cos^n\theta$. Equation (6.36) is then given by

$$v = \frac{J}{N}\sum_{n=1}^{n=6} a^n \frac{A\tan\psi_x + B\tan\psi_y + C}{\sqrt{1+\tan^2\psi_x + \tan^2\psi_y}}, \tag{6.37}$$

where coefficients a_n are determined by regressing experimental data of sputtering yield versus incident angle. Based on such relationships the development of the surface topography of amorphous and crystalline solid surfaces can be quantitatively described. This approach has been extended to include the effects of ion reflection and re-deposition [194, 195].

Application Examples

The application of the methods of characteristics is demonstrated in Fig. 6.34 for two-dimensional cases. It is assumed that the inert gas ion etching rate v is dependent only on the angle of incidence. On the left side of Fig. 6.34, a trench whose cross-section is a semicircle at time $t = 0$ and whose mask cannot be eroded was vertically irradiated. On the left side of this figure, some selected characteristics are shown to illustrate the formation of facets. On the right side of Fig. 6.34, the formation of a rectangular trench under uniform inert gas ion bombardment is shown. For the etching rate, a Gaussian-like distribution is assumed. In this case, selected characteristics are also visualized for $x < 0$. The bottom of the trench tends to form a tapered profile.

6.5 Secondary Processes Contributing to Surface Evolution

Experimental studies have demonstrated that not all ion beam-induced topographical features can be adequately explained solely through consideration of the variations in the sputtering yield as a function of the ion incidence angle (surface gradient-dependent sputtering). Moreover, several secondary ion beam-induced mechanisms have been discussed, which may be capable of influencing topographical development.

Bayly [122] was the first to suggest that the mechanisms of ion reflection at grazing incidence and the re-deposition of material onto closely adjacent planes are also involved. In addition, a number of other ion beam-induced effects can also contribute to changes in the topography during ion bombardment. Sigmund [121] discussed a local variation of the sputtering yield due to non-uniform energy deposition. Based on this mechanism, additional sputtered atoms can be generated if the size of surface feature is comparable to or smaller than the region of the deposited energy. Carter [196] introduced surface diffusion as another mechanism. It is assumed that topographical development can be influenced by both sputtering and surface diffusion during ion bombardment. Carter et al. [186] also considered the influence of spatially non-uniform ion bombardment on topography evolution. It should be further pointed out that numerous ion beam-induced phenomena at the evolution of the surface topography can be reasonably interpreted by assumption of a radiation-induced viscous flow [16, 197]. Finally, irradiation with high ion fluences leads to surface topography changes, called swelling, which must be taken into account.

In the following chapters, secondary processes, such as ion reflection, re-deposition, shadowing, surface diffusion, non-uniform ion bombardment, viscous flow and swelling will be introduced, all of which additionally contribute to the evolution of ion bombarded surfaces.

6.5.1 Reflection of Ions Under Grazing Incidence

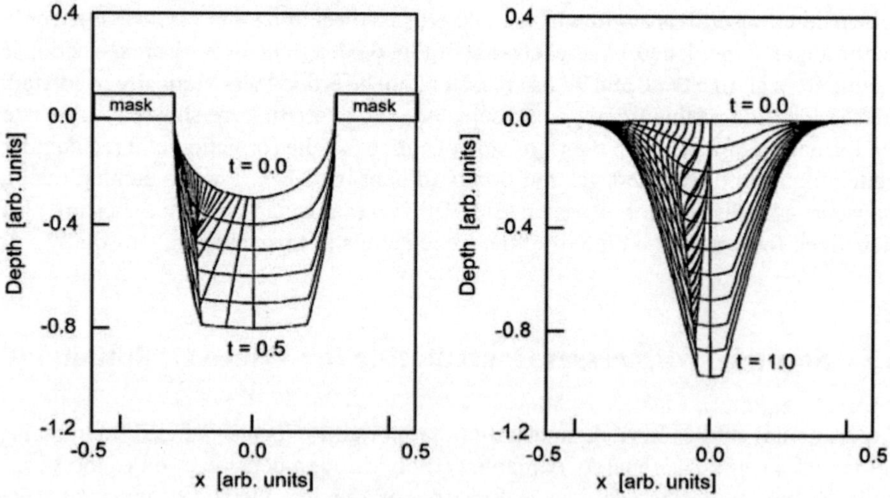

Fig. 6.34 Evolution of semicircular trench (left) and an initially flat surfaces (right) under uniform inert gas ion bombardment. For $x < 0$, some selected characteristics are shown. Right: the distribution of the erosion rate is $v(x, \theta) = v(\theta) \exp[-(x/\sigma)]$ with $\sigma = 0.2$ (figures adapted from [191])

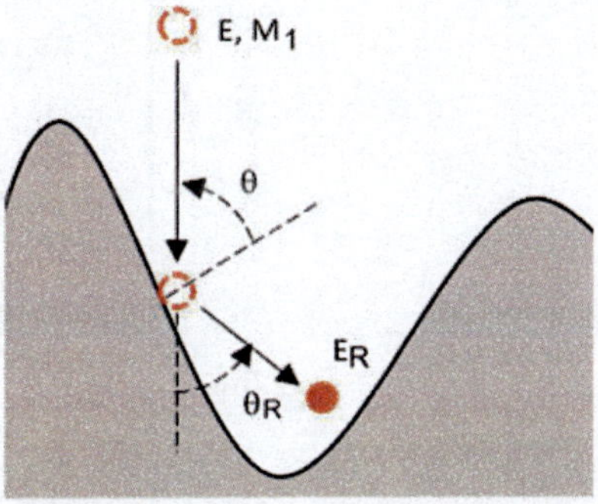

Fig. 6.35 Schematic presentation of sputtering by reflected ions

In Sect. 5.10, the fundamental processes involved in the reflection of ions were presented. It should be remembered that the reflection probability increases significantly with increasing angle of incidence θ (see Sect. 5.10 and Figs. 5.10 and 5.20). The influence of this reflection on the evolution of the topography must be considered primarily in regard to near-to-surface features with steep sides. Experimental and computational studies have demonstrated the significant impact on topographic evolution produced by the combined bombardment of a primary ion beam and secondary particles generated by backscattered collision processes.

Under the condition that collisions between incident particles are disregarded and the ions travel in a straight line, these ions can be reflected upon hitting the surface (Fig. 6.35). The backscattering process of the incident ions with energies > 100 eV can be described by an elastic two-body scattering event. Assuming a hard-sphere collision, the energy of the reflected ion, E_R, with known mass, M_1, and initial energy, E, after elastic collision with a surface atom at rest with mass, M_2, is given by

$$E_R = E\left[\left(M\cos\theta_R \pm \sqrt{1 - M^2\sin^2\theta_R}\right)/(1+M)\right]^2 \tag{6.38}$$

where θ_R is the final scattering angle for an incident ion after collision (see kinematic factor, (2.26)) and $M = M_1/M_2$. For $M_2/M_1 > 1$ only the plus sign applies while for $M_2/M_1 < 1$ both plus and minus are valid. It is evident that these reflected ions can possess sufficient energy to significantly modify the neighboring surface. In particular, a significant impact of reflected ions on topographic evolution is expected at grazing ion incidence, low ion energies (up to a few keV), and when the mass of the primary ion is larger than the mass of the target atom. Under these conditions, the backscattered ions are also capable of sputtering material from the surface of the substrate, and these sputtered particles either disintegrate or can be re-deposited onto surfaces. The re-deposition onto adjacent surfaces (sidewall of a trench, etc.) leads to an accumulation of the sputtered material. Consequently, when a surface is simultaneously subjected to direct ion beam erosion and re-deposition of the sputtered material, the re-deposition process decelerates the erosion.

Two properties of the reflection process, in particular, must be taken into account:

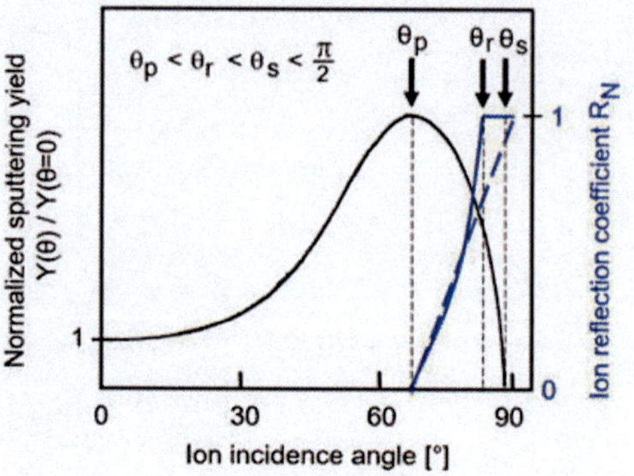

Fig. 6.36 Schematic representation of the dependence of the normalized sputtering yield (black solid line), the ion reflection coefficient (ratio of reflected to incident particles, blue solid line) and a linear approximation of the normalized reflection flux (dashed blue line). θ_p is the angle of the maximum sputtering yield, θ_r is the angle above which reflection is total and θ_s is the angle at which the sputtering yield is zero

1. Dependence of the reflection yield on the angle of ion incidence

   In general, an ion is reflected when the angle of ion incidence θ is larger than the critical angle θ_p, which is assumed to be the value of the angle of ion incidence θ corresponding to the maximum sputtering yield (see Sect. 5.6 and Figs. 5.9 and 6.36). For $\theta > \theta_P$, the sputtering yield decreases with increasing θ to $\theta_s \leq 90°$, due to the increased probability of ions being reflected from the surface without significant energy loss. The drop-off in sputtering yield at grazing angles is associated with ion reflection. According to Stewart and Thompson [98], this critical angle θ_p is given by (5.50). Consequently, for $\theta > \theta_p$, the probability of the reflection of the ions significantly increases. Only a small fraction of the ion energy is transferred to the target atoms [198]. Based on a proposal by Smith and Tagg [195], the flux of reflected ions, $J_R(\theta)$, can be approximately assumed to be a linear function of the angle of incidence (dashed blue line in Fig. 6.36) and is given by

$$J_R(\theta) \approx \begin{cases} 0 & \text{for}\, 0 \leq \theta < \theta_p \\ \frac{J}{N}\left(\frac{\theta-\theta_p}{\pi/2-\theta_p}\right) & \text{for}\, \theta_p \leq \theta \end{cases}, \tag{6.39}$$

   where J is the primary ion flux.

2. Energy distribution of reflected ions

   The energy and angular distributions of the reflected ions following low-energy bombardment of surfaces have been comprehensively investigated (see, e.g., [199, 200]). Fig. 6.37 shows a schematic depiction of the normalized yield of the reflected ions as a function of the normalized energy of the reflected ions and the ion incidence angle. It is evident that the energy distributions of reflected ions under grazing ion incidence bombardment ($\theta > 70°$) reaches a maximum

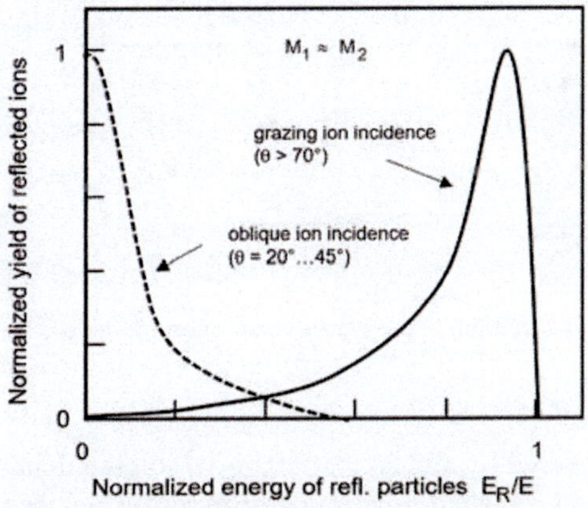

Fig. 6.37 Schematic representation of the normalized energy distribution, E_R/E, of backscattered ions for two ranges of the primary ion incidence angle: (0°...45°), oblique ion incidence (25°...45°) and grazing ion incidence (>70°). It should be noted, that the angular ranges are strongly dependent on the ion energy E (here, E is in the range of a few keV) and the mass ratio of the target atom to the projectile (here $M_2/M_1 \approx 1$). E_R is the energy of the reflected particle

near the incident ion energy E, whereas the energy distribution for reflected ions after approximately vertical ion bombardment ($\theta < 45°$) is characterized by a maximum at low energies. It should be taken into consideration that the high-energy peak of the distribution after grazing ion bombardment can be observed when the ratio of the target mass to the projectile mass is sufficiently high and that this peak disappears with increasing ion energy. The calculation of the reflection coefficients as a function of ion energy and the angle of incidence for an arbitrary number of mass ratios of target mass to the ion mass has been implemented in many simulation programs (see Sect. 3.5.2).

Smith et al. [194] applied two different approaches for determining the evolution of topography under simultaneous bombardment of the surface by primary and reflected ions. On the one hand, the method of characteristics (see Sect. 6.4.2) was used and on the other hand, an alternative method for computing the sputtering of ions reflected from two-dimensional surfaces was developed [201]. Several modifications of the topography can be traced back exclusively to the simultaneous impact of the primary ion bombardment and the reflection of grazing incident ions.

Example: Trenching

To provide an example, the development of trenches through ion beam etching has been explained by the reflection of ions on the sidewall, which leads to an increased ion flux near the base of the sidewall and thus results in increased erosion rates. The formation of trenches around the base of edges of a mask by ion beam erosion has been frequently observed (see, e.g., [202, 203]). In particular, this effect is verifiable if a mask is applied that is more resistant than the substrate material under ion beam erosion. Figure 6.38 schematically illustrates the trench formation in a

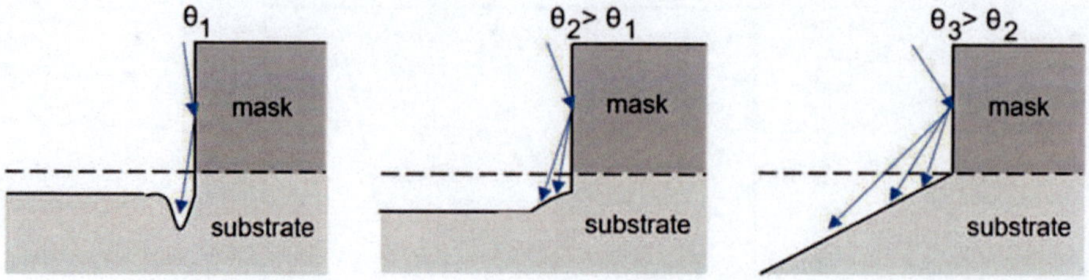

Fig. 6.38 Schematic representation of the influence of reflected ions on the topography at different ion incident angles

substrate partially covered by a mask. At a normal ion incident angle, the increased flux (primary ions and reflected ions from the sidewall) causes trenching. With an increased angle of the ion incident, the erosion rate is increasingly reduced and sloped planes are formed.

6.5.2 Re-deposition

On the basis of topographical studies of glass following Ar ion bombardment, Bayly [122] was the first to recognize that the concurrent process of the re-deposition of the sputtered material has a significant influence on the final topography. This process is schematically illustrated in Fig. 6.39. Under uniform ion bombardment, erosion of the surface with local erosion velocities, which are dependent on the local ion incidence angle, can be expected. For rough surfaces, the re-deposition of sputtered particles must be taken into account. In contrast to sputtered particles from elevations of the surface, particles from the sidewall and the bottom of valleys are re-deposited onto

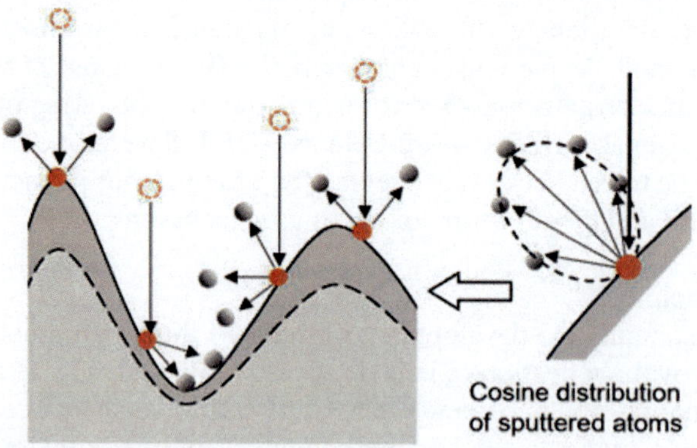

Fig. 6.39 Schematic presentation of the non-conform surface evolution if re-sputtering must be taken into account. Solid and dashed lines represent the surface contour at the time t and the time t + dt, respectively

closely adjacent planes. This leads to a reduced net erosion velocity at the sidewalls and the valleys. It can be assumed that the sputtered atoms leave the surface in approximately the form of a cosine distribution (see Sect. 5.7.2).

The re-deposited material can be deleterious to the performance of microelectronic device structures produced by ion beam etching. The influence of re-deposition on the topography of surfaces after ion beam etching was recognized very early (see, e.g., [162, 202, 204–206]). For example, Johnson [162] studied in detail the impact of re-deposition on the ion beam-induced erosion profiles of grating reliefs. In contrast to the significant deposition of sputtered material after oblique (grazing incidence) ion etching, a lower deposition of re-sputtered material was observed after perpendicular ion bombardment. This illustrates that a significant buildup of re-sputtered material cannot occur if erosion and re-deposition act simultaneously on locations of the bombarded substrate [162, 203].

Initial attempts to analytically describe the influence of re-deposition on the development of selected topographies were carried out by Belson and Wilson [207]. They calculated the flux of re-sputtered particles onto different two-dimensional asperities of the surface. In the simplest case, a linear asperity was assumed, as shown in Fig. 6.40, left. The ion etching of the asperity is disregarded and the trajectories of all sputtered particles are straight lines. It is also assumed that each point of the line with length l is capable of sputtering particles at a constant rate per unit length and per unit time. As an example, Fig. 6.40, left, schematically shows the sputtering of particles per time unit from the surface element dx into the angle element dφ. These re-sputtered particles are deposited on the surface element du at the height h of the asperity.

According to Belson and Wilson [207], the normalized flux density of the re-deposited particles, J_{resp}, toward a linear asperity is given by

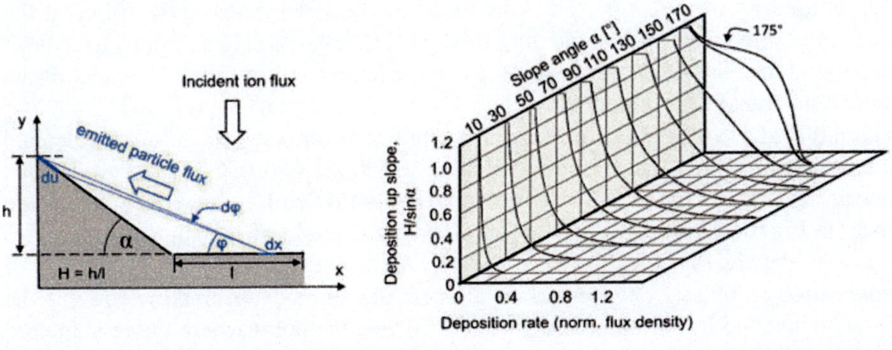

Fig. 6.40 Left: schematic image of the re-deposition of a linear asperity on a surface. Right: Variation of the normalized re-deposited atom flux density with the density up-slope and the slope angle (figure adapted from [207])

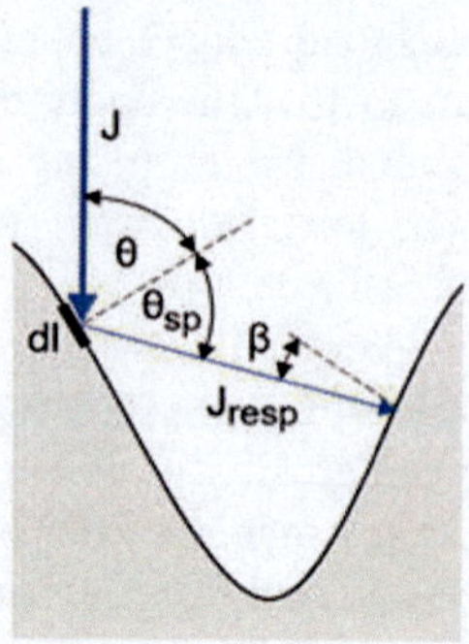

Fig. 6.41 Schematic representation of the re-sputtering model by Smith et al. [174], where J is the incidence ion flux, J_{resp} is the flux of the re-sputtered particles, θ is the ion incidence angle, θ_{sp} is the emission angle of the sputtered particles, β is the incidence angle of the sputtered particles. The dashed black lines represent the local surface normal

$$J_{resp} = \frac{1}{2\pi}\sin\alpha \cdot \ln\left[\frac{(1 + H\cot\alpha)^2 + H^2}{H^2 \mathrm{cosec}^2\alpha}\right] - \frac{\cos\alpha}{\pi}\left[\tan^{-1}\left(\frac{1}{H} + \cot\alpha\right) - \left(\frac{\pi}{2} - \alpha\right)\right], \tag{6.40}$$

where the formula symbols are shown in the drawing (Fig. 6.40, left). J_{resp}, represents the ratio of incident ion flux density, J, to the emitted particle flux per unit length of source. It is obvious that flux density is dependent only on the angle α between the rising edge of the asperity and the flat surface. Figure 6.40, right, also shows the normalized flux density or deposition rate, J_{resp}, in dependence on the distance up the slope and the slope angle α. This figure illustrates that the deposition rate is very high at the foot of the slope. The amount of re-sputtered material deposited at the foot of the slope rises with angle α up to about 130°. Analytical solutions have been also found for deposition onto asperities with a sinusoidal profile. The effect of the sputter etching was not included.

Smith et al. [194] applied the method of characteristics (see Sect. 6.4.2) to calculate redeposition profiles. They assumed that an ion flux J with a Gaussian profile hits the surface and generates a flux of re-deposited particles J_{resp}, where θ_{sp} and β are given in Fig. 6.41 (see also Sect. 5.7.1). The distribution of the sputtered particles is energy dependent (see also Sect. 5.7). In the two-dimensional case, the flux of redeposited particles from the surface element dl can be determined where d is the distance traveled by the sputtered particle between the point where the re-sputtered particle is generated and the point of re-deposition, and a two-dimensional sputtered particle distribution function is assumed. In the simplest case [208], the normalized flux densities of the redeposited material from the bottom of a rectangular trench (H is the depth and D is the width of the trench) to the sidewalls is given by

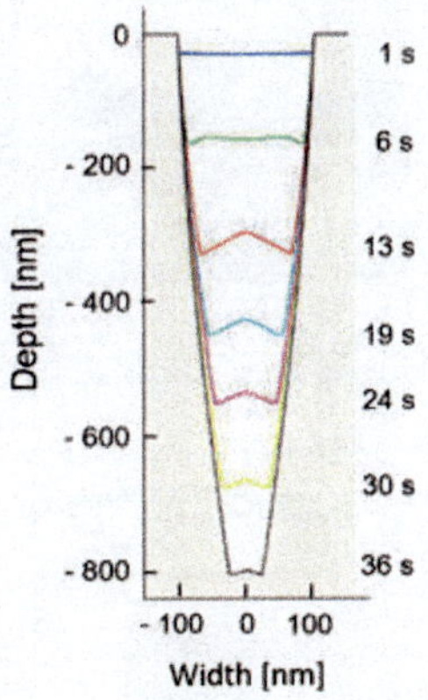

Fig. 6.42 Sputtering-simulated trench with re-deposition and secondary sputtering produced by a 1.5 nm wide ion beam (ion current density 5 mA/cm^2) over a scan width of 200 nm as function of sputter time (figure adapted from [214])

$$J_{resp} = \frac{1}{2}\int_0^D \frac{\cos\beta\cos\theta_{sp}}{d}dx = \frac{H}{2}\int_0^D \frac{x}{d^3}dx, \tag{6.41}$$

where x is the spatial coordinate of the trench width and varies between 0 and D, $d = \sqrt{x^2 + H^2}$, $\cos\beta = x/d$ and $\cos\theta = H/d$ (see also Fig. 6.41). The assumption here is that the sputtered atoms have a cosine distribution and that the sticking coefficient is equal to one (see Appendix K).

For example, Smith et al. [194] investigated the problem of re-deposition in initially rectangular grooves or trenches as a function of beam energy and groove width. They have shown that the buildup of re-deposited material is greatest at lower energies, but that the sidewalls of the grooves become less steep and the originally vertical sidewalls become sloped due to re-deposition.

Meanwhile, re-deposition is implemented in many ion surface simulator and ion–solid simulation programs (see e.g., [208–213]). For instance, Lindsey and Hobler [214] used a segment-based model in two dimensions combined with Monte Carlo simulations to study the influence of re-deposition and the sputtering by backscattered ions on the shape of trenches. Figure 6.42 demonstrates this influence on both the trench shape and the depth of the trench. Re-deposition leads to inclined sidewalls and a bending up of the bottom toward the sidewalls, while sputtering by backscattered ion causes the formation of a pair of micro-trenches that increase in size until they meet in the center of the trench (w-shaped trench). Bizyukov et al. [215] prepared Si pitch grating specimens (250 nm for pits and 250 nm for grates) on Ta layers and bombarded these with 6 keV Ar ions at an angle of 42°. The evolution of the pitch grating structure was studied experimentally and by means of numerical simulation using the SDTrimSP-2D code [213]. Figure 6.43a–d shows a series of SEM cross-section images. The simulated surface contours are indicated by red lines. With increasing ion fluence, the rectangular surface structure initially becomes rounded, then scalene triangles are formed. In addition, Fig. 6.43e–h also displays the calculated cross-sectional areas, where the red squares in Fig. 6.43a–d correspond to the surface

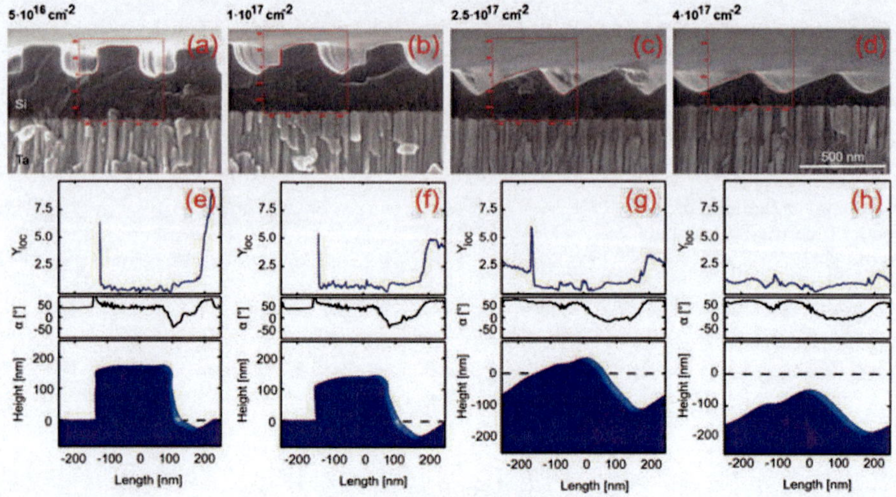

Fig. 6.43 **a–d** SEM cross-section images of Si pitch grating after Ar ion bombardment as a function of the ion fluence. The angle of ion incidence is $\theta = 48°$. Red colored graphs are the surface contour simulated by SDTrimSP-2D code. All scales are given in nanometers. **e–h**: calculated re-deposition yields Y_{loc} (above, blue lines), the calculated local angle of ion incidence α (middle, black lines) and the calculated cross-sectional area (below) of the bombarded structure (figures adapted from [215] and modified)

contours in Fig. 6.43e–h. Initially (ion fluence < 10^{17} ions/cm^2), the re-deposition is small, because only Si atoms from the right plane of the feature can be removed, while other areas of the structure cannot be reached from the incident ion beam. For higher fluences (1×10^{17} ions/cm^2), the sputtered Si atoms are deposited on the initially vertical structure planes. A small fraction of Si atoms is deposited in the horizontal region. When the shadowed regions disappear completely (2.5×10^{17} ions/cm^2), the initially rectangular shape of the structure tends to a scalene triangle, with the angle determined by the ion beam incidence. The simulation of this system for higher fluence (4×10^{17} ions/cm^2) shows that the triangular shape becomes increasingly smooth [215].

With regard to the stability of surface patterns, Diddens and Linz [216] studied the temporal evolution of an initially flat surface using ion beam erosion and re-deposition with a generated numerical algorithm. For this purpose, a standard-model equation for pure erosion processes under perpendicular ion incidence, the (isotrope) Kuramoto–Sivashinsky equation (see Chap. 8), is coupled with a functional of the re-deposition. Without term of re-deposition, the topographic evolution shows no regular surface patterns, only unsteady rough surfaces. The spatiotemporal development of these surfaces is chaotic. If the re-deposition is included, an initially irregular topography develops to an array of nanodots with short-range hexagonal order. It is clearly evident that re-deposition acts as a stabilizing mechanism for patterning on surfaces.

6.5.3 Shadowing

The evolution of the topography under low-energy ion bombardment is determined by a further nonlocal effect. The shadow effect is characterized by taller features on the surface blocking the incoming ion flux under oblique in bombardment (Fig. 6.44). Consequently, the crest of the hill is hit by more incident ions than the valley. In contrast to re-deposition, shadowing therefore tends to enhance the surface roughness [24]. The extent of shading of the valley in Fig. 6.44 depends on the angle of incidence of the ion beam θ. Where the ion incidence is perpendicular ($\theta = 0°$), no shadowing can be observed, while for ($\theta > 0°$) the effect of shadowing on the geometry depicted will result in partial shadowing. According to Fig. 6.44, the tangent of the incident angle, tan θ, and the differences in distance between two points, $\Delta x = x_o - x$, and in the height, $\Delta y = h_o - h$, determine whether a point on the adjacent feature is shadowed from the incident flux (details to exposing height, see Sect. 11.5). The condition that shadowing occurs on the adjacent feature can then be expressed by

$$h > h_o - \frac{\Delta x}{\tan\theta}, \tag{6.42}$$

where $\tan\theta = \Delta h/\Delta x$. Based on this condition, the point at which the ion collides with the surface can be determined. In dependence on the ion energy, the incident angle, and the sticking coefficient, an ion can be deposited, reflected, or sputtered. The exposure height, Λ, hit by the incident ions (see Fig. 6.44) is given by $\Lambda = \Delta x/\tan\theta$ for the case that the distance between the adjacent elevations is smaller than so-called shadowing length (details see Sect. 11.5).

For the special case of non-normal incidence ion bombardment of a sinusoidally rippled surface [154], this limiting condition for non-shadowing, (6.42), can be rewritten to

$$\frac{2\pi h_o}{\Delta x} \le \tan\left(\frac{\pi}{2} - \theta\right), \tag{6.43}$$

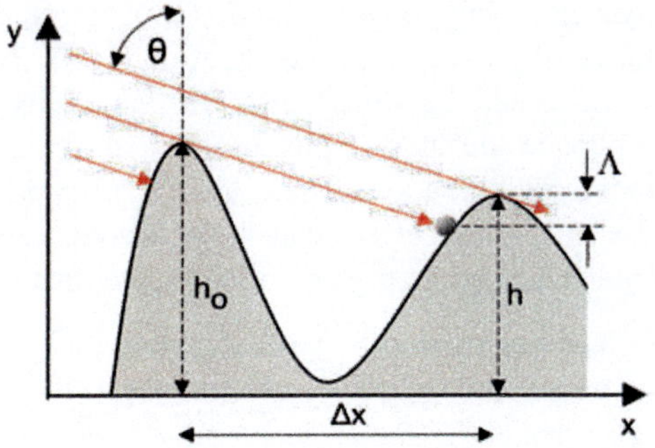

Fig. 6.44 Schematic of the shadowing effect. Δx is the distance between two adjacent elevations and Λ is the exposure height hit by the incident ions

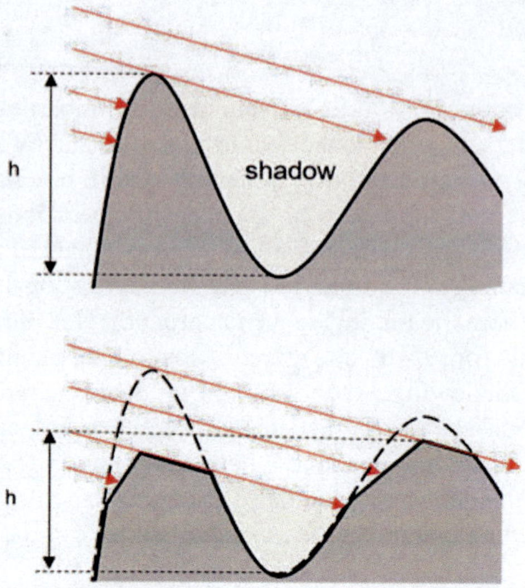

Fig. 6.45 Schematic of the local shadowing effect on the topography of the surface by ion bombardment. h represents the average feature heights (in accordance with [154]). Above: initial situation. Bottom: following ion bombardment

where h_o is here the ripple amplitude (height of the sinusoidal elevation) and Δx is here the ripple wave length (distance between two sinusoidal elevations). Consequently, the ratio $h_o/\Delta x$ determines the condition at which shadowing occurs for a given ion incidence angle θ. Shadowing increases in importance as the incidence angle is increased, i.e. shadowing plays a minor role in near-normal ion bombardment.

One problem with the shadowing effect is that only the crest of the neighboring features can be hit by the incoming ions. Consequently, shadowing is combined with a preferred erosion of the peaks, whereas the valleys will be protected from the erosion process. As shown schematically in Fig. 6.44, the time of ion bombardment can significantly influence the changes to the topography. Shadowing is thus accompanied by a smoothing effect (reduced roughness, compare the average feature heights h in Fig. 6.45) and can initiate faceting [154], (see Sect. 6.3.1.3).

It should be noted that (i) shadowing and re-deposition of sputtered atoms influence the topography of surface in a contradictory manner. Shadowing tends to increase the roughness, while re-deposition reduces the roughness, and (ii) the shadowing effect can be consequently used to create nanostructures through sputtering-induced glancing angle deposition (see Chap. 11).

6.5.4 *Surface Diffusion*

In this subchapter, the development of the topography will be discussed when both ion beam-induced sputtering and surface diffusion can occur during ion bombardment. Experimental studies have shown that temperature can significantly influence changes to the topography (see, e.g., [73, 82, 88, 217]). According to Herring [218] and Mullins [35], it is possible to describe the diffusion on isotropic surfaces in terms of a driving force for the material transport without detailed consideration of the atomistic processes of the diffusion.

It is assumed that the change in the topography is caused only by the movement of atoms on the surface through thermal diffusion. All other possibilities of material transport on the surface are disregarded. For a convex (concave) surface, the curvature is positive (negative) and thus the chemical potential of an atom on such surface is higher (lower) than that on a flat surface (Fig. 6.46). Consequently, the transport from the flat surface to a convex (concave) surface results in an increase (decrease) in the surface chemical potential. Chemical potential, μ, is a thermodynamic quantity that expresses the incremental energy content of a system per mole of the chemical substance. From a thermodynamic point of view, an atom on a convex surface possesses the highest chemical potential, whereas an atom on a concave surface has the lowest chemical potential. Consequently, the transport (diffusion) of surface atoms (adatoms) is expected from the peaks to the valleys. This consideration is only valid when the surface slope is small. The surface curvature $\partial^2 h/\partial x^2$ or $\partial^2 h/\partial y^2$ can then be assumed to be proportional to the chemical potential.

It is assumed that the topography of a surface can be described by its height h as the function h(x, y, t). The surface is characterized by small perturbations from the plane h = 0. For a perturbation, the principal local curvatures are given by

$$\frac{1}{R_x} \approx -\partial^2 h/\partial x^2 \quad \text{and} \quad \frac{1}{R_y} \approx -\partial^2 h/\partial y^2, \tag{6.44}$$

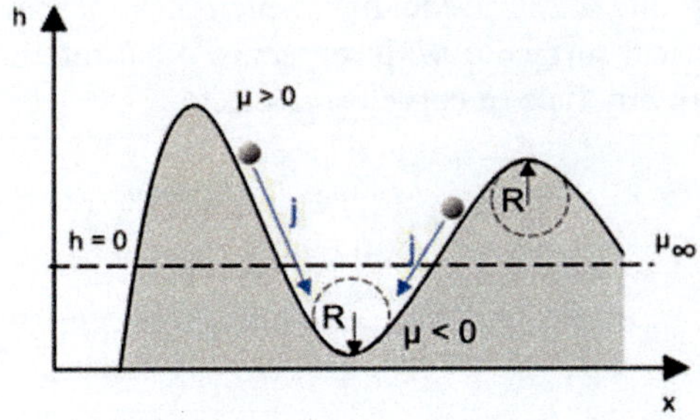

Fig. 6.46 Schematic of the surface diffusion on a curved surface. μ is the chemical surface potential. The dashed line represents the chemical surface potential of the flat surface μ_∞. R is the local radius of curvature

where R_x and R_y are the radii of the curvature. The curvature is taken to be positive if the perturbation is convex and negative if the perturbation is concave.

The diffusion of an atom on a surface with convex and concave perturbations is subject to local changes of the chemical surface potential $\Delta\mu = \mu_c - \mu_\infty$, where μ_c is the chemical surface potential of curved surface dependent on the radius of the curvature, and μ_∞ is the chemical surface potential of the flat surface (Fig. 6.46). Thus, a convection current of particles is generated by diffusion in a potential gradient. The driving force for this convection current (a vector tangential to the surface) is given by

$$\boldsymbol{F} = -\nabla\mu. \tag{6.45}$$

The surface particle flux (number of atoms per unit length and unit time) caused by thermal surface self-diffusion is linearly proportional to the driving force and can be expressed by

$$\boldsymbol{j} = -M\boldsymbol{F} = M\nabla\mu, \tag{6.46}$$

where M is the mobility given by the Einstein relation, $M = N_s/k_B T$, D_{sd} is the coefficient of the self-diffusion on the surface and N_s is the number of atoms per unit area.

The surface extends when an additional atom is added and recedes when an atom leaves the surface. The divergence of the convection current, $\nabla \boldsymbol{j}$, describes the number of atoms diffusing in or out of the surface element per unit area and unit time. The temporal change of surface topography can then be expressed by the continuity equation of mass

$$\frac{\partial h}{\partial t} = -\Omega\nabla \cdot \boldsymbol{j}, \tag{6.47}$$

where Ω is the atomic volume of the mobile species.

The well-known Young–Laplace equation, which describes the chemical potential of an atom in a spherical surface μ with respect to a flat reference surface, can be readily generalized for any type of curved surfaces to

$$\mu = \gamma\Omega\left(\frac{1}{R_x} + \frac{1}{R_y}\right), \tag{6.48}$$

where $\gamma = \gamma(\theta)$ is the surface free energy per unit area. Combining (6.46), (6.47) and (6.48) leads to

$$\frac{\partial h}{\partial t} = \Omega^2\gamma M\nabla^2\left(\frac{1}{R_x} + \frac{1}{R_y}\right) = \Omega^2\gamma\frac{N_s D_{sd}}{k_B T}\nabla^2\left(\frac{1}{R_x} + \frac{1}{R_y}\right) = K\nabla^2\left(\frac{1}{R_x} + \frac{1}{R_y}\right) \tag{6.49}$$

with the parameter

$$K = \Omega^2 \gamma \frac{N_s}{k_B T} D_{sd} = \Omega^2 \gamma \frac{N_s}{k_B T} D_o \exp\left(-\frac{E_{sd}}{k_B T}\right), \tag{6.50}$$

where D_o is the pre-exponential factor of the diffusion coefficient and E_{sd} is the activation energy of the surface diffusion.

Using (6.44) the change of the topography through diffusion on the surface driven by local differences of the chemical surface potential is given by

$$\frac{\partial h}{\partial t} = -\mathrm{K}\left(\frac{\partial^4 h}{\partial x^4} + \frac{\partial^4 h}{\partial x^2 \partial y^2} + \frac{\partial^4 h}{\partial y^4}\right). \tag{6.51}$$

In the special case that the perturbation possesses a spherical shape, (6.51) can be reduced to

$$\frac{\partial h}{\partial t} = -\mathrm{K}\left(\frac{\partial^4 h}{\partial x^4}\right) = -\frac{\Omega^2 \gamma D_{sd} N_s}{k_B T} \nabla^4 h. \tag{6.52}$$

Equation (6.52) can be interpreted as a contribution of surface diffusion to the surface evolution. Under the precondition that the initial contour is known, the partial differential equations (6.51) or (6.52) describe the evolution of the topography caused by thermal surface diffusion within the period of time t. This process is known to be temperature-dependent. For numerous metals, the self-diffusion coefficient on surfaces can be assumed by $D_{sd}/D_o \approx \exp(-9.5/T_H)$, T_H is the homologous temperature, which is calculated by normalizing the actual substrate temperature to the melting-point temperature of the material (both in units of [K]) [219].

It has been demonstrated [133, 143] that the erosion rate for both convex and concave features on the surface under ion bombardment along the normal direction is given by (6.8). According to Carter [196], contributions to the erosion rate (evolution of the topography) caused by ion beam sputtering, (6.8) and by surface diffusion, (6.52) can be summarized, because the processes are uncorrelated (negating the fact that ion beam generated defects are capable to influencing of surface diffusion). The total rate of erosion (erosion velocity) for the simplest case is then given by

$$\frac{\partial h}{\partial t} = -\left[\frac{Y(\theta)}{N} J \cos\theta + \mathrm{K}\nabla^4 h\right]. \tag{6.53}$$

The effective sputtering yield, Y_{eff}, which contains both contributions, can be found by

$$-\frac{Y_{eff}}{N} J \cos\theta = -\left[\frac{Y(\theta)}{N} J \cos\theta + \mathrm{K}\nabla^4 h\right] \tag{6.54}$$

or

$$Y_{eff} = Y(\theta) + \frac{N}{J\cos\theta} K\nabla^4 h = Y(\theta) + \frac{N}{J\cos\theta} \frac{\Omega^2 \gamma N_2 D_{sd}}{k_B T} \nabla^4 h. \quad (6.55)$$

Based on this equation, both the velocity and the motion of atoms on the surface can be determined according to the previously demonstrated method (see Sect. 6.4), which Carter et al. [196] and Ducommun et al. [136] used to successfully derive the rate of change of the radius of curvature.

It should be noted that the Herring-Mullins consideration is misemployed in the case of crystalline surfaces, because the surface energy is dependent on the orientation of the different crystal faces. It can be expected that relaxation coefficient K, in particular, must be chosen in such a manner that this coefficient is dependent on temperature and ion flux [51].

6.5.5 Non-uniform Ion Bombardment

Nearly all theoretical studies have assumed that ion flux is spatially uniform. However, for many applications such as depth analysis of composition profiles, self-organized pattern formation or smoothing of relief patterns by local ion bombardment, the ion beam used is characterized by a spatially variable ion flux distribution. Consequently, this beam divergence (ions incident under a defined angular range) is expected to make an additional contribution to topography evolution by means of the angle-dependent secondary processes (e.g., reflection, re-sputtering, swelling, etc.). The ion beam density distribution can be approximately described by a Gaussian-like distribution

$$J(x, y) = J_o \exp\left\{-\left[\frac{(x - x_o)^2}{2\sigma_x^2} + \frac{(y - y_o)^2}{2\sigma_y^2}\right]\right\}, \quad (6.56)$$

where J_o is the ion current density at the center of the beam (x_o, y_o) and σ_x, σ_y are the standard derivations of the Gaussian distribution in x and y direction, respectively. In general, it can be assumed that the distribution of the ion current density is symmetrical, i.e. $\sigma_x = \sigma_y = \sigma$.

Assuming rectilinear ion trajectories, the beam divergence (beam broadening), φ_{div}, for circular beams can be estimated either on the basis of a geometrical consideration or on the basis of a Gaussian-like distribution of the ion flux to be

$$\varphi_{div} = \arctan\left(\frac{r_B - r_A}{d}\right) = \arctan\left(\frac{\frac{FWHM}{2} - r_A}{d}\right), \quad (6.57)$$

where r_B is the radius of the beam cross-section at distance d from the extraction aperture (e.g., on the surface of the irradiated substrate), r_A is the radius of the extraction aperture (e.g., radius of the ion beam at the source exit), and FWHM is the full width at half maximum of the ion beam.

In many broad beam ion sources, the first grid determines the kinetic energy of the ions and the second grid, the accelerating grid, determines the divergence of the extracted ions. Tartz et al. [220] discussed in detail the impact of the hole diameter and the distance between the grids in the ion source on the beam divergence. For example, it has been shown that the optical parameters of the ion beam, especially the influence of accelerator voltage on the divergence of the ion beam and the angular distribution of the ions within the ion beam, play a role in pattern formation [221].

The additionally introduced parameter of angular distribution provides an extra degree of freedom for controlling the pattern formation process and for producing large-scale ordered nanostructures of different forms [222]. Johnson [162] studied the influence of the ion beam divergence angle on the facet formation (angle between coincident facets) or blaze angle of GaAs after 2 keV Ar ion bombardment, where a Gaussian angular distribution centered about the nominal angle of incidence with an angular spread between points of half-maximum intensity was assumed. It could be observed that a considerable influence of the beam divergence on the facet angle for beam tilt angles > 40°. Using continuum theory and Monte Carlo simulations, Kree et al. [223] were able to successfully describe the importance of the ion beam divergence on the formation of self-organized pattern formation.

Carter et al. [186] studied the influence of a stationary ion beam with a non-uniform flux, $J(x) = J_o\exp[-x^2/2\sigma_x^2]$, on the development of a time-dependent surface profile of an initially flat surface. As described above (see Sect. 6.4), the motion of each surface point over small time increments could be derived from the normal of the effective erosion curves. Computer simulations have demonstrated that for short erosion times, the erosion profile shows an unchanged Gaussian-like profile. For erosion depths ≥ standard deviation of the ion beam σ, the erosion profiles reflect an increasingly abnormal Gaussian profile. Figure 6.47 illustrates a computer-simulated erosion profile for different time steps generated by an ion beam with a standard deviation of 3.5σ. The initial Gaussian profile degenerates toward a profile

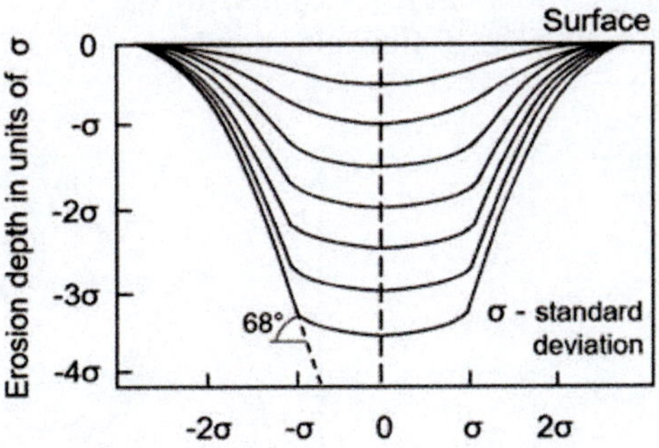

Fig. 6.47 Computer-simulated erosion profiles after ion bombardment with a Gaussian ion flux profile, where a standard deviation of the ion beam of 3.5σ is assumed (figure adapted from [186])

with a flattish bottom and steep sides. The rising angle of the sides tends to 68°. Consequently, the effect is unimportant at small erosion depths, but is significant if the depth of the erosion crater is comparable to the width of the ion beam.

6.5.6 Viscous Flow

The plastic deformation of different materials during ion bombardment has been studied in detail (see, e.g., [224]). It could be observed that the damage caused by irradiation leads to density changes, where the density change in crystalline materials is achieved by transformation to the amorphous phase (see Sect. 4.10). The ion beam-induced density change in both crystalline and amorphous materials is strongly linked to a plastic deformation that relieves the stress in the ion-irradiated region.

A viscous flow is a cooperative rearrangement of particles (atoms, molecules) in a solid (or fluid) in which the viscosity is significant. Viscosity is a measure of the resistance of a fluid to a change in shape, or the movement of neighboring portions of the fluid relative to one another. Viscosity may be thought of as internal friction between the particles, as one part of a fluid system is forced to move in opposite directions to neighboring parts. If the shear rate between the adjacent parts is linearly dependent on the shear stress (shear velocity), then this fluid can be described as a Newtonian fluid, where the proportionality constant between the shear stress and the shear rate is given by the viscosity.

The roughness of a solid surface has a tendency to minimize its free surface energy by such means as viscous flowing. The local curvature of a rough surface generates stress gradients, caused by surface tension, where this tension is proportional to the surface curvature. Consequently, the viscous flow in the near-surface region results in a relaxation of the surface, i.e. mounds on the surface deliquesce and valleys fill up. This is referred to as surface leveling (smoothing, see Chap. 7) of viscous, horizontal films of finite thickness and infinite extent (schematic in Fig. 6.48). The flow behavior of linear viscous Newtonian liquids can be described by the Navier–Stokes equation [225]. Assuming that the frictional force caused by the viscosity is the dominant driving force for the flow (i.e. small Reynolds number), the Navier–Stokes equation

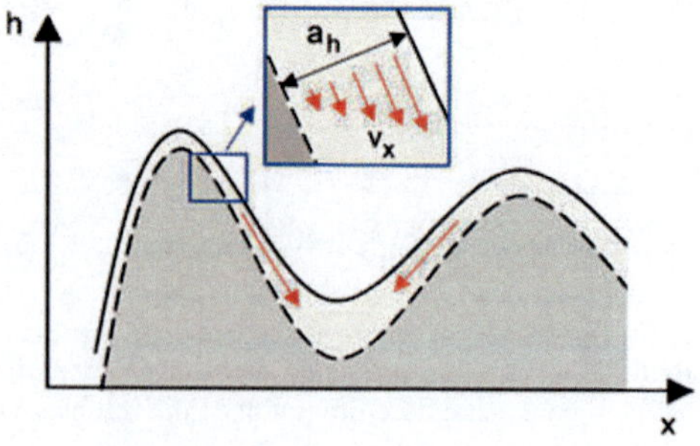

Fig. 6.48 Schematic picture of the stress-induced viscous flow

can be simplified by the Stokes equation. Orchard [226] was the first to theoretically treat surface-tension-driven leveling in Newtonian films in a two-dimensional case. Under the conditions, that (i) no shear forces occurs on the surface of the liquid and (ii) the flow velocity at the solid–liquid interface is zero (Fig. 6.48), the Stokes equation was solved in the Fourier space and leads to an expression relating the decay rate of each individual Fourier mode to the wavenumber, q, the equilibrium thickness of the film, a_h, the surface tension (free surface energy), γ, and the fluid viscosity, η_o. According to Orchard [226], the relaxation rate (height change) is given by

$$\frac{\partial \tilde{h}(q,t)}{\partial t} = -\frac{\gamma a_h^3}{3\eta_o} q^4 \tilde{h}(q,t), \tag{6.58}$$

where $\tilde{h}$ is the Fourier transform of the height coordinate h, γ is the free surface energy per unit area (surface tension), η_o is the viscosity (internal friction), a_h is the thickness of the liquid-like film on the surface, and q is the wavenumber in the Fourier space. This equation is valid for a viscous flow in a confined surface layer, i.e. if the structure dimension is comparable to the film thickness or $a_h q \ll 1$ (Orchard limit). Otherwise, (6.58) is given by $\partial \tilde{h}(q,t)/\partial t = -\frac{\gamma}{2\eta_o} q \tilde{h}(q,t)$ for thicker films.

Stillwagon and Larsen [227] later dealt with the solution of the Stokes equation in the space for the two-dimensional case and, analogous to the approach of these authors, Vauth [228] addressed for the three-dimensional case. According to Landau and Lifshitz [225], the Stokes equation is given by

$$\nabla p = \eta_o \Delta \boldsymbol{v}, \tag{6.59}$$

where **v** is the vector of the velocity and, p is pressure. When the liquid film thickness slowly varies in the position space, then the vertical velocity will be approximately that of a uniform film thickness. This so-called lubrication approximation [227] can be expressed as

$$\left|\frac{\partial^2 v_i}{\partial x^2}\right| \ll \left|\frac{\partial^2 v_i}{\partial h^2}\right| \quad \text{and} \quad \left|\frac{\partial^2 v_i}{\partial y^2}\right| \ll \left|\frac{\partial^2 v_i}{\partial h^2}\right| \tag{6.60}$$

with i = x, y. Thus, the Stokes equation can be redrafted as

$$\frac{\partial p}{\partial x} = \eta_o \frac{\partial^2 v_x}{\partial h^2} \quad \text{and} \quad \frac{\partial p}{\partial y} = \eta_o \frac{\partial^2 v_y}{\partial h^2}. \tag{6.61}$$

According to Orchard [226], identical boundary conditions on the velocity are assumed for the integration of (6.60). This means that no slip occurs at the substrate surface (i.e. $v_i = 0$ for the thickness of the viscous film = a_h), and no shear forces at the film-air surface (i.e. $\partial v_i/\partial h = 0$, if the thickness of the viscous film = 0) [207]. Then, the integration yields

$$v_x = \frac{1}{\eta_o}\frac{\partial p}{\partial x}\left(\frac{1}{2}h^2 - \frac{1}{2}a_h^2\right) \quad \text{and} \quad v_y = \frac{1}{\eta_o}\frac{\partial p}{\partial y}\left(\frac{1}{2}h^2 - \frac{1}{2}a_h^2\right). \tag{6.62}$$

These equations describe the thickness changes in time under the influence of a driving force. Assuming conservation of mass and the mass continuity equation in the form $\partial v_x/\partial x + \partial v_x/\partial x = 0 = \nabla \cdot \boldsymbol{v}$ for an incompressible flow, the rate of topographic height change can be determined [228] to be

$$\frac{\partial h}{\partial t} = -\nabla\left[p\frac{1}{\eta_o}\int_{-a_h}^{0}\left(\frac{1}{2}h^2 - \frac{1}{2}a_h^2\right)dh\right] = \frac{a_h^3}{3\eta_o}\nabla(\nabla p). \tag{6.63}$$

According to Landau and Lifshitz [225], the approximation $p = -\gamma\nabla^2 h$ is valid in the absence of shear stress. The time-dependent change in the topography (relaxation rate) in the real space is then given by the Orchard formula [226] as

$$\frac{\partial h}{\partial t} = -\frac{\gamma a_h^3}{3\eta_o}\nabla^4 h = -B\nabla^4 h, \tag{6.64}$$

i.e. the relaxation rate, $\partial h/\partial t$, is proportional to $\nabla^4 h$ in the real space or proportional to the wavenumber q^4 in the reciprocal space (see Chap. 7). It is obvious that the near-surface mass transport by viscous flow influences the topography in the same manner as does surface diffusion (c.f. 6.65 and 6.50).

In general, viscous flow is thermally generated. However, the energetic ion incidence in a solid is also capable of inducing a viscous-like state [229, 230]. Within the collision cascade, a super-saturation of interstitial-like or/and vacancy-like defects are generated [231, 232]. Above a threshold ion fluence, the near-surface region is amorphized (for semiconductors, in order of 10^{14} ions/cm^2). The thickness of this amorphized film beneath the surface is maximal in order of the penetration depth of the ions, i.e. a few nanometers for low-energy ion bombardment. During the duration of the ion bombardment (in order of picoseconds), volumetric expansion around the implanted ion and the generated defects creates a stress state [233]. In a significantly longer period of relaxation, generated residual stresses confined in the thin amorphous layer lead to material redistribution by means of a viscous flow process [234]. It can thus be expected that the ion beam-induced fluidity contributes to relaxing the free surface energy on non-planar surfaces.

For example, Trushin [235] has explained the plastic flow of the material as a result of spatial redistribution of ion irradiation induced vacancies and interstitial atoms. Interstitial atoms and vacancies generated during ion irradiation introduce a distortion of volume proportional to their atomic volume. In contrast to vacancies, interstitial atoms are more diffusive and interact strongly with a field of sinks (dislocations, dislocation loops, etc., see Sect. 4.8). Consequently, the volume of the material increases, because (i) sinks grow through a contributing stream of interstitials to the sinks and (ii) interstitial atoms precipitate out on surfaces (surface acts as a sink). The uncompensated vacancies in the volume form clusters that can be transformed into vacancy dislocation loops or vacancy voids (see, e.g., Wang and Birtcher [236]).

Particularly in the case of *high-energy ion bombardment* of solids, ranging from some tens of keV up to MeV, it is known that the surface can be significantly modified, where these topographic changes are exclusively interpreted on the basis of the ion beam-induced (enhanced) viscous flow [199, 210, 237–239].

While viscous flow has been identified as playing a significant role in the high-energy regime, the atomic-scaled mechanism of the surface-confined viscous flow during *low-energy ion bombardment* (energies lower than a few keV) has only been investigated sporadically. In contrast to the high-energy radiation induced viscous flow in the bulk, it can be expected that low-energy ion bombardment generates a viscous flow limited to the near-surface region of a stiff bulk material.

Mayer et al. [197] observed that ion bombardment of SiO_2 with H, He, and Xe ions at energies $\leq$1 keV caused viscous flow that led to surface smoothing, and evaluated an ion-enhanced viscosity. The viscous relaxation per ion scales as the square root of the ion beam-induced displacements in the film over the range of the ion damaged depth and is strongly dependent on the ion energy. The magnitude of the radiation-induced viscosity of SiO_2 for light ion bombardment (1–20 $\times$ 10^{12} Pa s) is comparable to the thermal viscosity of this material at about 1000 °C. In contrast, Ziberi [240] determined values of viscosity per ion in Si and Ge of about 2.1 $\times$ 10^9 Pa s and 1.1 $\times$ 10^9 Pa s, respectively, after heavy ion (Xe ion) bombardment at room temperature and with an ion current density of 300 μA/cm^2. This is significantly less than the values reported by Mayer et al. [197]. Mayr et al. [239] and Castro and Cuerno [232] reported similar values (10^9 Pa s and 6 $\times$ 10^8 Pa s, respectively) of the viscosity of Si after low-energy ion bombardment. Norris et al. [238] determined the viscosity by means of the equation

$$\eta = 1.25\frac{E_d a_h}{H\Omega J E}, \tag{6.65}$$

where E_d is the displacement energy (see Sect. 4.1 and Appendix H), J is the ion flux, E the ion energy, and Ω is the atomic volume. H is ion beam-induced fluidity (reciprocal of the viscosity $\times$ ion flux) and can be determined by MD simulations. For example, Vauth and Mayr [241] calculated H = 1.04 $\times$ 10^{-9} (Pa dpa)$^{-1}$ for Si after 1 keV ion bombardment at room temperature. Here, it must be pointed out that Mayr et al. [239] introduced a scaling factor (in unities of [(Pa dpa)$^{-1}$]) to compare the ion beam-induced viscosity for different ion energies and angles of ion incidence.

Umbach et al. studied the relaxation process caused by viscous flow within the near-surface range in the order of the ion range after Ar ion bombardment of SiO_2 with energies ranging between 0.5 keV and 2 keV. They found a significant viscous flow in the low and medium temperature range (between room temperature and 200 °C) caused by atomic rearrangements and dominated by the energetic ion beam-induced collision cascade, where the wave number, q, is only weakly dependent on temperature. Above this temperature, Arrhenius-like behavior has been observed, which is provoked by thermally induced relaxation.

The two relaxation mechanisms (surface diffusion and surface-confined viscous flow, which show the same wave-vector dependence) are distinguishable by the dependence of their relaxation rate on ion energy. The dominance of the surface viscous flow at ambient temperature has been confirmed by theoretical investigations for keV-ion bombarded amorphous materials [241].

Zhou et al. [242] and Babonneau et al. [243] compared the influence of ion beam-induced viscous relaxation and surface diffusion on the smoothing of the topography of sapphire and amorphous alumina surfaces, respectively. These surfaces were bombarded by low-energy Ar ions at temperatures between room temperature and significantly higher temperatures. Zhou et al. [242] concluded that both factors play an important role in changing surface topography, because at low temperatures, the viscous flow determines the smoothing of the surface, while at higher temperatures, it is suggested that surface diffusion increases gains influence. Conversely, Babonneau et al. [243] indicated that the surface-confined ion-induced viscous flow dominates over thermal diffusion and ion-induced effective surface diffusion.

The experimental studies have shown that fluidity (the reciprocal of viscosity) is proportional to the ion flux, proportional to the ion energy, and weakly dependent on temperature. In general, it can be assumed that the relaxation rates after low-energy ion bombardment are significantly increased (i.e. viscosity is decreased) in comparison to thermally induced viscous flow.

6.5.7 Swelling

Simultaneously with sputtering, swelling of the near-surface areas of the target can also take place. It is well-known that ion implantation creates defects and finally leads to amorphization (see Chap. 4) associated with the change in volume or density of the implanted material. This process of material change leads to expansion in surface area, referred to as swelling.

In particular, for low ion energies, the volume expansion due to sputtering is reduced, i.e., the expansion of the implanted material versus the irradiation time can be described by the competition between the rates volume swelling and the erosion [see Eq. (5.46)]. The change in height, taking into account that the sputtering counteracts the expansion of surface, is given in a first approximation by

$$\frac{\partial h(x,t)}{dt} = \frac{1}{N \cdot cos\theta}[1 - Y(\theta)] \tag{6.66}$$

where J is the ion current density, N is the atomic number density, Y(θ) is the sputtering yield and θ is the angle of ion incidence.

The swelling effect has been observed in nearly all classes of materials (semiconductors, metals, alloy, ceramics, diamonds, etc.). In order to study swelling as a function of ion beam parameters, the procedure of measuring step heights is performed. A mask is used on selected areas of the sample, which is impermeable to the energetic ions. This creates a surface step between swelling and non-swelling areas after ion bombardment. This allows a direct measurement of the swelling integrated along the ion path. It could be that the surface elevation (swelling) by ion implantation continuously increases with the ion fluence, where step heights the implanted and non-implanted regions of some ten nanometers after irradiation with ion energies up to some ten kilo-electron-volts and step height up to hundred nanometers after high-energy ion irradiation were reached. For example, Giri et al. [244] studied the swelling in low-energy self-ion implanted silicon. The swelling height perpendicular to the surface is approximately cubic root-dependent on the Si ion fluence and the density of the irradiated amorphous Si is reduced by about 3.1% to the non-irradiated crystalline Si.

It is generally accepted that the ion radiation induced elevation of the materials, particularly with regard to covalently bound materials, including crystalline silicon, is caused by plastic flow [224]. According to Trushin [235], viscous flow of the material can be explained as being a result of the spatial redistribution of ion-irradiation induced vacancies and interstitial atoms (see Sect. 6.5.6). Tamulevicius et al. [245] have taken this idea and proposed a model explaining the swelling of silicon as a diffusion-like process, where the flux of interstitials out of the surface is driven by the gradient of the concentration of interstitials and the gradient of mechanical stresses in the ion-implanted region of the solid. The interstitials form a new crystallographic plane on the top of the solid.

Swelling is associated with the build-up of compressive stress in the implanted area. During swelling, the implanted layer can freely expand only along the normal to the surface (visible in the step at the boundary of the implanted region). The implantation-induced expansion parallel to the surface is restricted by the surrounding non-implanted regions, which induces the compressive stress. Note also that the uniform increase in swelling height at low fluences can be accompanied by the formation of blisters on the nanometer or micrometer scale at higher fluences [246].

6.6 List of Symbols

Symbol	Notation
D_o	Pre-exponential factor of the diffusion coefficient
D_{sd}	Coefficient of self-diffusion
E_d	Displacement energy

(continued)

(continued)

Symbol	Notation
E_R	Energy of reflected ions
E_{sd}	Activation energy of surface diffusion
H	Ion beam-induced fluidity (reciprocal of the viscosity × ion flux)
H(r, t)	Height-height correlation function
J	Ion current density (ion flux)
J_R	Flux of reflected ions
L	Lateral sample length
N_s	Number of atoms per unit area
M	Mobility
M_i	Mass of particle i
R	Radius of curvature
J_{resp}	Ratio of the incident flux density to emitted flux per unit length
T	Temperature
T_H	Homologous temperature
U_s	Surface binding energy
Y	Sputtering yield
Y_{eff}	Effective sputtering yield
a	Depth of energy distribution
a_h	Thickness of liquid-like film on the surface
h	Height
$\overline{h}$	Mean height
j	Surface particle flux
p	Pressure
q	Wavenumber
r_s	Beam radius
s	Sticking coefficient
v	Erosion velocity (erosion rate)
w	Interface width or roughness
z	Coarsening exponent or dynamic exponent (= α/β)
Λ	Exposure height
Φ	Ion fluence
Ω	Atomic volume
α	Roughness exponent
β	Growth exponent
γ	Surface free energy per unit area
θ	Angle of ion incidence
θ_c	Apex angle of a cone

(continued)

(continued)

Symbol	Notation
θ_p	Angle of ion incidence, where sputtering yield is a maximum
φ	Local ion incidence angle
μ	Chemical potential
ξ	Correlation length
η	Gaussian random variable
η_o	Viscosity
σ	Standard deviation
τ_s	Saturation time

References

1. J. Stark, G. Wendt, Über das Eindringen von Kanalstrahlen in feste Körper. Ann. Physik **38**, 921–940 (1912)
2. G. Carter, J.S. Colligan, *Ion Bombardment of Solids* (Heinemann, London, 1968)
3. G.S. Bales, R. Bruinsma, E.A. Eklund, R.P.U. Karunasiri, J. Rudnick, A. Zangwill, Growth and erosion of thin solid films. Science **249**, 264–268 (1990)
4. P. Bedrossian, J.E. Houston, J.Y. Tsao, E. Chason, S.T. Picraux, Layer-by-layer sputtering and epitaxy of Si(100). Phys. Rev. Lett. **67**, 124–127 (1991)
5. H. Watanabe, M. Ichikawa, Initial stage of layer-by-layer sputtering of Si(111) surfaces studied by scanning reflection electron microscopy. Appl. Phys. Lett. **68**, 2514–2516 (1996)
6. H. Watanabe, M. Ichikawa, Layer-by-layer sputtering of Si(111) and (001) surfaces mediated by surface vacancy diffusion: surface physics and application for nanofabrication. J. Vac. Sci. Technol. B **15**, 2666–2671 (1997)
7. E. Chason, P. Bedrossian, J.E. Houston, J.Y. Tsao, B.W. Dodson, S.T. Picraurx, Simulations of layer-by-layer sputtering during epitaxy. Appl. Phys. Lett. **59**, 3533–3535 (1991)
8. P. Smilauer, M.R. Wilby, D.D. Vvedensky, Morphology of singular and vicinal metal surfaces sputtered at different temperatures. Surf. Sci. Lett. **291**, L733–L738 (1993) and Shape of the surface-step-density oscillations during sputtering of singular and vicinal surfaces. Phys. Rev. B **48**, 4968–4971 (1993)
9. Y. Zhao, G.-C. Wang, T.-M. Lu, *Characterization of Amorphous and Crystalline Rough Surface: Principles and Application* (Academic Press, San Diego, 2001)
10. J.M. Bennett, L. Mattsson, *Introduction to Surface Roughness and Scattering* (Optical Society of America, Washington, 1989)
11. B. Ziberi, F. Frost, Th. Höche, B. Rauschenbach, Ripple pattern formation on silicon surfaces by low-energy ion beam erosion: experiment and theory. Phys. Rev. B **72**, 235310 (2005)
12. V. Solanki, D. Kabiraj, D.K. Avasthi, S. Varma, Topographic evolution and scaling study of ZnO (0001) single crystal after low-energy atom beam irradiation. Nucl. Instr. Meth. Phys. Res. B **343**, 56–60 (2018)
13. A. Toma, F. Buatier de Mongeot, R. Buzio, G. Firpo, S.R. Bhattacharyya, C. Boragno, U. Valbusa, Ion beam erosion of amorphous materials: evolution of surface morphology. Nucl. Instr. Meth. Phys. Res. B **230**, 551–554 (2005)
14. F. Frost, A. Schindler, F. Bigl, Roughness evolution of ion sputtered rotating InP surfaces: pattern formation and scaling laws. Phys. Rev. Lett. **85**, 4116–4119 (2000)
15. A.C.-T. Chan, G.-C. Wang, Roughness evolution of Si(111) by low-energy ion bombardment. Surf. Sci. **414**, 17–22 (1998)

16. T.M. Mayer, E. Chason, A.J. Howard, Roughening instability and ion-induced viscous relaxation of SiO_2 surfaces. J. Appl. Phys. **76**, 1633–1643 (1994)
17. E.A. Eklund, E.J. Snyder, R.S. Williams, Correlation from randomness: quantitative analysis of ion-etched graphite surfaces using the scanning tunneling microscope. Surf. Sci. **285**, 157–180 (1993)
18. E. Chason, T.M. Mayer, A. Payne, D. Wu, In situ energy dispersive x-ray reflectivity measurements of H ion bombardment on SiO_2/Si and Si. Appl. Phys. Lett. **60**, 2353–2355 (1992)
19. F. Family, T. Viscek, *Dynamics of Fractal Surfaces* (World Scientific, Singapore, 1991)
20. A.-L. Barabási, H.E. Stanley, *Fractal Concepts in Surface Growth* (Cambridge University Press, Cambridge, 1995)
21. T. Halpin-Healy, Y.-C. Zhang, Kinetic roughening phenomena, stochastic growth, directed polymers and all that. Phys. Rep. **254**, 215–414 (1995)
22. J. Krug, Origins of scale invariance in growth processes. Adv. Phys. **46**, 139–282 (1997)
23. P. Meakin, *Fractals, Scaling and Growth far from Equilibrium* (Cambridge University Press, Cambridge, 1998)
24. M. Pelliccione, T.-M. Lu, *Evolution of Thin Film Morphology* (Springer, Berlin, 2008)
25. F. Family, T. Vicsek, Scaling of the active zone in the Eden process on percolation networks and the ballistic deposition model. J. Phys. A Math. Gen. **18**, L75–L81 (1985)
26. P. Karmakar, D. Ghose, Low-energy Ar^+ ion beam-induced kinetic roughening of thin Pt films on a Si substrate. Nucl. Instr. Meth. Phys. Res. B **222**, 477–483 (2004)
27. R. Cuerno, H.A. Makse, S. Tomassone, S.T. Harrington, H.E. Stanley, Stochastic model for surface erosion via ion-sputtering: dynamical evolution from ripple morphology to rough morphology. Phys. Rev. Lett. **75**, 4464–4467 (1995)
28. J.T. Drotar, Y.-P. Zhao, T.-M. Lu, G.-C. Wang, Numerical analysis of the noisy Kuramoto-Sivashinsky equation in 2+1 dimensions. Phys. Rev. E **59**, 177–185 (1999) and Surface roughening in shadowing growth and etching in 2+1 dimensions. Phys. Rev. B **62**, 2118–2125 (2000)
29. R.M. Bradley, J.M.E. Harper, Theory of ripple topography induced by ion bombardment. J. Vac. Sci. Technol. A **6**, 2390–2396 (1988)
30. R. Cuerno, A.-L. Barabási, Dynamic scaling of ion-sputtered surfaces. Phys. Rev. Lett. **74**, 4746–4749 (1995)
31. S.A. Pahlovy, S.F. Mahmud, K. Yanagimoto, N. Aikawa, I. Miyamoto, Roughening and smoothing behavior of single crystal Si by low energy Ar^+ ion bombardment. Nucl. Instr. Meth. Phys. Res. B **272**, 206–209 (2012)
32. A. Keller, S. Facsko, R. Cuerno, Numerical integrator for continuum equations of surface growth and erosion, in *Computational Nanotechnology: Modeling and Application with MATLAB*, ed. by S.M. Musa (Taylor & Francis Group, Boca Raton, 2012), pp. 189–215
33. S.F. Edwards, D.R. Wilkinson, The surface statistics of a granular aggregate. Proc. Roy. Soc. Lond. Ser. A **381**, 17–31 (1982)
34. M. Kadar, G. Parisi, Y.-C. Zhang, Dynamic scaling of growing interfaces. Phys. Rev. Lett. **56**, 889–892 (1986)
35. W.W. Mullins, Theory of thermal grooving. J. Appl. Phys. **28**, 333–339 (1957) and Flattening of a nearly plane solid surface due to capillarity. J. Appl. Phys. **30**, 77–83 (1959)
36. D. Michelson, Steady solutions of the Kuramoto-Sivashisky equation. Physica D **19**, 89–111 (1986)
37. I. Koponen, M. Hautala, O.-P. Sievänen, Simulations of submicrometer-scale roughening on ion-bombarded solid surfaces. Phys. Rev. B **54**, 13502–13505 (1996)
38. I. Koponen, M. Hautala, O.-P. Sievänen, Simulations of self-affine roughening and ripple formation on ion bombarded amorphous carbon surfaces. Nucl. Instr. Meth. B **129**, 349–355 (1997)
39. H.-N. Yang, G.-C. Wang, T.-M. Lu, Anomalous dynamic scaling on the ion-sputtered Si(111) surface. Phys. Rev. B **50**, 7635–7639 (1994)

40. X.-S. Wang, R.J. Pechman, J.H. Weaver, Trends in surface roughening: analysis of ion-sputtered GaAs(110). Surf. Sci. **364**, L511–L518 (1996)
41. D.-M. Smilgies, P.J. Eng, E. Landemark, M. Nielsen, Sputtering of Ge (001): transition between dynamic scaling regimes. Surf. Sci. **377**, 1038–1041 (1997)
42. D. Chowdhury, D. Ghose, S.A. Mollick, Homoepitaxy of germanium by hyperthermal ion irradiation. Vacuum **107**, 23–27 (2014)
43. D.E. Harrison, R.P. Webb, The creation of surface damages by ion beam bombardment. Nucl. Instr. Meth. Phys. Res. **218**, 727–736 (1983) and R.P. Webb, D.E. Harrison, Computer simulation of pit formation in metals by ion bombardment. Phys. Rev. Lett. **50**, 1478–1481 (1983)
44. J. Mäkinen, A. Vehanen, P. Hautojärvi, H. Huomo, J. Lahtinen, R.M. Nieminen, S. Valkealahti, Vacancy-type defect distribution near argon sputtered Al(110) surface studied by variable-energy positrons and molecular dynamics simulations. Surf. Sci. **175**, 385–414 (1986)
45. F. Karetta, H.M. Urbassek, Moleculardynamics simulations of bulk and surface damage production in low energy Cu→Cu bombardment. J. Appl. Phys. **71**, 5410–5418 (1992)
46. G. Kornich, G. Betz, Two step simulation of low energy ion beam mixing at different temperatures. Nucl. Instr. Meth. Phys. Res. B **117**, 81–89 (1996)
47. M.V.R. Murty, H.A. Atwater, Defect generation and morphology of (001) Si surfaces during low-energy Ar-ion bombardment. Phys. Rev. B **45**, 1507–1510 (1992)
48. H. Hensel, H.M. Urbassek, Implantation and damage under low-energy Si self-bombardment. Phys. Rev. B **57**, 4756–4763 (1998) and Disordering and annealing of a Si surface under low-energy Si bombardment. Rad. Eff. Defects Solids **143**, 297–315 (1997)
49. J. Tarus, K. Nordlund, A. Kuronen, J. Keinonen, Effect of surface on defect creation by self-ion bombardment. Phys. Rev. B **58**, 9907–9915 (1998)
50. T. Seki, T. Aoki, J. Matsuo, I. Yamada, STM observation of surface vacancies created by ion impact. Nucl. Instrum. Meth. Phys. Res. B **164–165**, 650–655 (2000)
51. W.L. Chan, E. Chason, Making waves: kinetic processes controlling surface evolution during low energy ion sputtering. J. Appl. Phys. **101**, 121301 (2007)
52. C. Busse, C. Engin, H. Hansen, U. Linke, T. Michely, H.M. Urbassek, Adatom formation and atomic layer growth on Al(111) by ion bombardment: experiments and molecular dynamics simulations. Surf. Sci. **488**, 346–366 (2001)
53. T. Michely, C. Teichert, Adatom yields, sputtering yields, and damage patterns of single-ion impacts on Pt(111). Phys. Rev. B **50**, 11156–11166 (1994)
54. T. Michely, G. Comsa, Generation and nucleation of adatoms during ion bombardment of Pt(111). Phys. Rev. B **44**, 8411–8414 (1991)
55. C. Teichert, M. Hohage, T. Michely, G. Comsa, Nuclei of the Pt(111) network reconstruction created by single ion impacts. Phys. Rev. Lett. **72**, 1682–1685 (1994)
56. M. Morgenstern, T. Michely, G. Comsa, Collective effects in adatom production by 4.5 keV rare gas impacts on Pt(111)—a low temperature scanning tunneling microscopy analysis. Philos. Mag. A **79**, 775–794 (1999)
57. J.C. Girard, Y. Samson, S. Gauthier, S. Rousset, J. Klein, STM study of the nucleation and annealing of ion bombardment induced defects on Cu(100). Surf. Sci. **302**, 73–80 (1994)
58. J. Naumann, J. Osing, A.J. Quinn, I.V. Shvets, Morphology of sputtering damage on Cu(111) studied by scanning tunneling microscopy. Surf. Sci. **388**, 212–219 (1997)
59. C.A. Lang, C.F. Quate, J. Nogami, Initial stages of sputtering on Au(111) as seen by scanning tunneling microscopy. Appl. Phys. Lett. **59**, 1696–1698 (1991)
60. X.-S. Wang, R.J. Pechman, J.H. Weaver, Ion sputtering of GaAs(110): from individual bombardment events to multilayer removal. J. Vac. Sci. Technol. B **51**, 10931–10940 (1995) and R.J. Pechman, X.-S. Wang, J.H. Weaver, Vacancy kinetics and sputtering of GaAs(110). Phys. Rev. B **51**, 10929–10936 (1995)
61. K. Kyuno, D.G. Cahill, R.S. Averback, J. Tarus, K. Nordlund, Surface defects and bulk defect migration produced by ion bombardment of Si(001). Phys. Rev. Lett. **83**, 4788–4791 (1999)
62. H.J.W. Zandvliet, H.B. Elswijk, E.J. van Loenen, I.S.T. Tsong, Scanning tunneling microscopy and spectroscopy of ion-bombarded Si(111) and Si(100) surfaces. Phys. Rev. B **46**, 7581–7587 (1992)

63. H. Feil, H.J.W. Zandvliet, M.-H. Tsai, J.D. Dow, I.S.T. Tsong, Random and ordered defects on ion-bombarded Si (100)-(2× 1) surfaces. Phys. Rev. Lett. **69**, 3076–3079 (1992)
64. T. Bedrosssian, Klitsner, anisotropic vacancy kinetics and single-domain stabilization on Si(100) -2 x 1. Phys. Rev. Lett. **68**, 646–649 (1992)
65. P. Bedrossian, Generation and healing of low-energy ion-induced defects on Si(100) -2 × 1. Surf. Sci. **301**, 223–232 (1994)
66. P.J. Bedrossian, T.D. de la Rubia, Surface segregation of low-energy ion-induced defects in Si. J. Vac. Sci. Technol. A **16**, 1043–1046 (1998)
67. S.J. Chey, J.E. van Nostrand, D.G. Cahill, Surface morphology of Ge(001) during etching by law energy ions. Phys. Rev. B **52**, 16696–16701 (1995)
68. J.A. Floro, B.K. Kellerman, E. Chason, S.T. Picraux, D.K. Brice, K.M. Horn, Surface defect production on Ge(001) during low-energy ion bombardment. J. Appl. Phys. **77**, 2351–2357 (1995)
69. O.R. de la Fuente, M.A. Gonzalez, J.M. Rojo, Ion bombardment of reconstructed metal surfaces: from two-dimensional dislocation dipoles to vacancy pits. Phys. Rev. B **63**, 085420 (2001)
70. R.P. Webb, D.E. Harrison Jr., The formation of target adatoms during the quenching of an energetic ion induced cascade in a crystalline metal surface. Rad. Eff. Defects Solids **86**, 15–18 (1984)
71. H. Gades, H.M. Urbassek, Molecular-dynamics simulation of adatom formation under keV-ion bombardment of Pt(111). Phys. Rev. B **50**, 11167–11174 (1994)
72. M. Ghaly, K. Nordlund, R.S. Averback, Molecular dynamics investigations of surface damage produced by kiloelectronvolt self-bombardment of solids. Philos. Mag. A **79**, 795–820 (1999)
73. T. Michely, G. Comsa, Temperature dependence of the sputtering morphology of Pt(111). Surf. Sci. **256**, 217–226 (1991)
74. M. Kalff, G. Comsa, T. Michely, Temperature dependent morphological evolution of Pt(111) by ion erosion: destabilization, phase coexistence and coursing. Surf. Sci. **486**, 103–135 (2001)
75. H. Ullmaier, P. Ehrhart, P. Jung, H. Schultz, Atomic defects in metals, in *Landolt-Bornstein, Numerical Data and Functional Relationships in Science and Technology*, vol. III, no. 25 (Springer, Berlin, 1991)
76. K.L. Merkle, W. Jäger, Direct observation of spike effects in heavy-ion sputtering. Phil. Mag. A **44**, 741–762 (1981)
77. J. Nordlund, M. Keinonen, R.S. Ghaly, Averback, recoils, flows and explosions: surface damage mechanisms in metals and semiconductors during 50 eV-50 keV ion bombardment. Nucl. Instrum. Methods Phys. Res. B **148**, 74–82 (1999)
78. E.M. Bringa, J. Keinonen, K. Nordlund, Cratering-energy regimes: from linear collison cascades to heat spikes to macroscopic impacts. Phys. Rev. B **64**, 235426 (2001)
79. N. Kalyanasundaram, M. Ghazisaeidi, J.B. Freund, H.T. Johnson, Single impact crater function for ion bombardment of silicon. Appl. Phys. Lett. **92**, 131909 (2008)
80. N. Kalyanasundaram, J.B. Freund, H.T. Johnson, A multiscale crater function model for ion-induced pattern formation in silicon. J. Phys. Condens. Matter **21**, 224018 (2009)
81. Z. Yang, M. Lively, J.P. Allain, Atomistic simulation of ion beam patterning with crater functions. Nucl. Instr. Meth. Phys. Res. B **307**, 189–193 (2013)
82. M.V.R. Murty, T. Curcic, A. Judy, B.H. Cooper, A.R. Woll, J.D. Brock, S. Kycia, R.L. Headrick, X-ray scattering study of the surface morphology of Au(111) during Ar+ ion irradiation. Phys. Rev. Lett. **80**, 4713–4716 (1998)
83. M.V.R. Murty, T. Curcic, A. Judy, B.H. Cooper, A.R. Woll, J.D. Brock, S. Kycia, R.L. Headrick, Real-time x-ray scattering study of surface morphology evolution during ion erosion and epitaxial growth of Au(111). Phys. Rev. B **60**, 16956–16964 (1999)
84. T. Michely, K.H. Besocke, G. Comsa, Observation of sputtering damage on Au(111). Surf. Sci. Lett. **230**, L135–L139 (1990)
85. G. Costantini, F. Buatier de Mongeot, C. Boragno, U. Valbusa, Is ion sputtering always a "negative homoepitaxial deposition"? Phys. Rev. Lett. **86**, 838–841 (2001)

86. G. Costantini, S. Rusponi, R. Gianotti, C. Boragno, U. Valbusa, Temperature evolution of nanostructures induced by Ar+ sputtering on Ag(001). Surf. Sci. **416**, 245–254 (1998)
87. B. Poelsema, R. Kunkel, L.K. Verheij, G. Comsa, Mechanisms for annealing of ion-bombardment-induced defects on Pt(111). Phys. Rev. B **41**, 11609–11611 (1990)
88. H.J. Ernst, The pattern formation during ion bombardment of Cu(001) investigated with helium atom beam scattering. Surf. Sci. **383**, L755–L759 (1997)
89. M. Ritter, M. Stindtmann, M. Farle, K. Baberschke, Nanostructuring of the Cu(001) surface by ion bombardment: a STM study. Surf. Sci. **348**, 243–252 (1996)
90. J. Naumann, J. Osing, A.J. Osing, I.V. Shvets, Morphology of sputtering damage on Cu(111) studied by scanning tunneling microscopy. Surf. Sci. **388**, 212–219 (1997)
91. J.R. Hahn, H. Kang, STM study of Ar+-induced defects produced by near-threshold energy collision. Surf. Sci. **357–358**, 165–169 (1996)
92. R.M. Feenstra, G.S. Oehrlein, Surface morphology of oxidized and ion-etched silicon by scanning tunneling microscopy. App. Phys. Lett. **47**, 97–99 (1985)
93. T. Michely, M. Kalff, G. Comsa, M. Strobel, K.-H. Heinig, Coarsening mechanisms in surface morphological evolution, in *Atomic Aspects of Epitaxial Growth.* ed. by M. Kotrla, N.I. Papanicolaou, D.D. Vvedensky, L.T. Wille (Springer, Dordrecht, 2002), pp. 185–196
94. M. Strobel, K.-H. Heinig, T. Michely, Mechanisms of pit coarsening in ion erosion of fcc(111) surfaces: a kinetic 3D lattice Monte-Carlo study. Surf. Sci. **486**, 136–156 (2001)
95. G. Carter, B. Navinsek, J.L. Whitton, Heavy ion sputtering induced surface topography development, in *Sputtering by Particle Bombardment II*, ed. by R. Behrisch. Topics in Applied Physics, vol. 52 (Springer, Berlin, 1983), pp. 231–269
96. B. Navinsek, Sputtering—surface changes induced by ion bombardment. Prog. Surf. Sci. **7**, 49–70 (1976)
97. A. Güntherschulze, W. Tollmien, Neue Untersuchungen über Kathodenzerstäubung der Glimmentladung. V. Die Oberflächenbeschaffenheit der Kathode, Z. Physik **119**, 685–695 (1942)
98. A.D.G. Stewart, M.W. Thompson, Microtopography of surfaces eroded by ion bombardment. J. Mater. Sci. **4**, 56–60 (1969)
99. G.K. Wehner, D.J. Hajicek, Cone formation on metal targets during sputtering. J. Appl. Phys. **42**, 1145–1149 (1971)
100. I.H. Wilson, M.W. Kidd, A study of cones developed by ion-bombardment of gold. J. Mat. Sci. **6**, 1362–1366 (1971)
101. R.S. Gvosdover, V.M. Efremenkova, L.B. Shelyakin, V.E. Wrasova, Formation of cones during sputtering. Rad. Eff. **27**, 237–244 (1976)
102. J.L. Whitton, L. Tanovic, J.S. Williams, The production of regular pyramids on argon ion bombarded surfaces of cupper crystals. Appl. Surf. Sci. **1**, 408–413 (1978)
103. G.K. Wehner, Whiskers, cones and pyramids created in sputtering by ion bombardment. Report NASA CR-159549 (1979)
104. G.K. Wehner, Cone formation as a result of whisker growth on ion bombarded metal surfaces. J.Vac. Sci. Technol. A **3**, 1821–1835 (1985)
105. S.M. Rossnagel, R.S. Robinson, The time development of impurity-generated sputter cones on copper. Rad. Eff. **58**, 11–16 (1981)
106. S.M. Rossnagel, R.S. Robinson, Surface diffusion activation energy determination using ion beam microtexturing. J. Vac. Sci. Technol. **20**, 195–198 (1982)
107. K. Goto, K. Suzuki, Dynamic observation of cone formation on Cu by ion sputtering. Nucl. Instr. Meth. Phys. Res. B **33**, 569–573 (1988)
108. A.K. Sen, D. Ghose, Ar+-ion sputtering-induced surface topography development on aluminium. J. Mater. Sci. **28**, 1525–1531 (1993)
109. I.H. Wilson, The topography of sputtered semiconductors. Rad. Eff. **18**, 95–103 (1973)
110. D.J. Barber, F.C. Frank, M. Moss, J.-W. Steeds, I.S.T. Tsong, Prediction of ion bombarded surface topographies using Frank's kinematic theory of crystal dissolution. J. Mater. Sci. **8**, 1030–1040 (1973)

111. M. Nozu, M. Tanemura, F. Okuyama, Direct evidence for In-crystallite growth on sputter-induced InP cones. Surf. Sci. Lett. **304**, L468–L474 (1994)
112. F.Y. Meng, W.K. Wong, N.G. Shang, Q. Li, I. Bello, Multi-tip cones induced by ion-bombardment. Vacuum **66**, 71–76 (2002)
113. M.J. Witcomb, The development of ion-bombardment surface structures on stainless steel. J. Mat. Sci. **9**, 551–563 (1974)
114. M.Z. Yusop, K. Yamaguchi, T. Suzuki, P. Ghosh, A. Hayashi, Y. Hayashi, M. Tanemura, Morphology and size of ion induced carbon nanofibers: effect of ion incidence angle, sputtering rate, and temperature. Jap. J. Appl. Phys. **50**, 01AF1 (2011)
115. J.A. Floro, S.M. Rossnagel, R.S. Robinson, Ion-bombardment-induced whisker formation on graphite. J. Vac. Sci. Technol. A **1**, 1398–1402 (1983)
116. A.C. Popa, G.E. Stan, C. Besleaga, L. Ion, V.A. Maraloiu, D.U. Tulyaganov, J.M.F. Ferreira, Submicrometer hollow bioglass cones deposited by radio frequency magnetron sputtering: formation mechanism, properties, and prospective biomedical applications. ACS Appl. Mater. Interfaces **8**, 4357–4367 (2018)
117. H.R. Kaufman, R.S. Robinson, Ion beam texturing of surfaces. J. Vac. Sci. Technol. **16**, 175–178 (1979)
118. S.M. Rossnagel, R.S. Robinson, Ion beam-induced topography and surface diffusion. J. Vac. Sci. Technol. **21**, 790–797 (1982)
119. L.B. Begrambekov, A.M. Zakharov, V.G. Telkovsky, Peculiarities and mechanism of the cone growth under ion bombardment. Nucl. Instr. Meth. Phys. Res. B **115**, 456–460 (1996)
120. R. Reiche, W. Hauffe, Pyramid formation on a high index copper bicrystal during bombardment with 10 keV argon and krypton ions. Appl. Surf. Sci. **165**, 279–287 (2000)
121. P. Sigmund, A mechanism of surface micro-roughening by ion bombardment. J. Mater. Sci. **8**, 1545–1553 (1973)
122. A.R. Bayly, Secondary processes in the evolution of sputter-topographies. J. Mater. Sci. **7**, 404–412 (1972)
123. Y. Yuba, S. Hazama, K. Gamo, Nanostructure fabrication of InP by low energy ion beams. Nucl. Instr. Meth. B **206**, 648–652 (2003)
124. I.S. Nerbø, M. Kildemo, S. Le Roy, I. Simonsen, E. Søndergård, L. Holt, J.C. Walmsley, Characterization of nanostructured GaSb: comparison between large-area optical and local direct microscopic techniques. Appl. Opt. **47**, 5130–5139 (2008)
125. E. Trynkiewicz, B.R. Jany, D. Wrana, F. Krok, Thermally controlled growth of surface nanostructures on ion-modified AIII-BV semiconductor crystals. Appl. Surf. Sci. **427**, 349–356 (2018)
126. E. Trynkiewicz, B.R. Jany, A. Janas, F. Krok, Recent developments in ion beam-induced nanostructures on AIII-BV compound semiconductors. J. Condens. Matter **30**, 304005 (2018)
127. M. Kang, J.H. Wu, W. Ye, Y. Jiang, E.A. Robb, C. Chen, R.S. Goldman, Formation and evolution of ripples on ion irradiated semiconductor surfaces. Appl. Phys. Lett. **104**, 052103 (2014)
128. S. Le Roy, E. Barthel, N. Brun, A. Lelarge, E. Søndergård, Self-sustained etch masking: a general concept to initiate the formation of nanopatterns during ion erosion. J. Appl. Phys. **106**, 094308 (2009)
129. S. Le Roy, E. Søndergård, M. Kildemo, I.S. Nerbø, M. Plapp, Diffuse-interface model for nanopatterning induced by self-sustained ion-etch masking. Phys. Rev. B **81**, 161401 (2010)
130. J.H. Wu, R.S. Goldman, Mechanisms of nanorod growth on focused ion beam-irradiated semiconductor surfaces: role of redeposition. Appl. Phys. Lett. **100**, 053103 (2012)
131. L. Tanovic, N. Tanovic, G. Carter, M.J. Nobes, C. Cave, N. Al-Quadi, Further studies of topography evolution on ion bombardment 1131 copper. Vacuum **45**, 929–935 (1994)
132. Y.-J. Zhu, A. Schnieders, J.D. Alexander, T.P. Beebe, Properties of gold nanostructures on highly oriented pyrolytic graphite. Langmuir **18**, 5728–5733 (2002)
133. G. Carter, M.J. Nobes, F. Paton, J.S. Williams, J.L. Whitton, Ion bombardment induced ripple topography on amorphous solids. Rad. Eff. **33**, 65–73 (1977)

134. G. Carter, G.W. Lewis, M.J. Nobes, J. Cox, W. Begemann, The effect of ion species on bombardment induced topography during ion etching of silicon. Vacuum **34**, 445–450 (1984)
135. S.J. Chey, D.G. Cahill, Surface defects created by low energy ($20<E<240$ eV) ion bombardment of Ge(001). Surf. Sci. **380**, 377–384 (1997)
136. G. Carter, I.V. Katardjiev, M.J. Nobes, J.L. Whitton, Sputtering-induced surface topography on F.C.C. metals. Mater. Sci. Eng. **90**, 21–32 (1987)
137. L. Tanovic, N. Perovic, The surface microtopography induced by ion bombardment on Cu single crystals. Nucl. Instr. Meth. **132**, 393–3396 (1976)
138. G. Carter, I.V. Katardijev, M.J. Nobes, J.L. Whitton, Topography evolution in atomic scale growth and erosion continuum. Scan. Microsc. **2**, 1293–1327 (1988)
139. R.L. Cunningham, P. Haymann, C. Lecomte, W.J. Moore, J.T. Trillat, Etching of surfaces with 8-keV argon ions. J. Appl. Phys. **31**, 839–842 (1960)
140. G.D. Magnuson, B.B. Meckel, P.A. Harkins, Etch effects from oblique-incidence ion bombardment. J. Appl. Phys. **32**, 369–374 (1991)
141. R.S. Nelson, D.J. Mazey, Surface damage and topography changes produced during sputtering, in *Ion Surface Interaction, Sputtering, and Related Phenomena*, ed. by R. Behrisch, W. Heiland, W. Poschenrieder, P. Staib, H. Verbeek (Gordon and Breach, London, 1973), pp. 199–206
142. M. Teichmann, Musterbildung auf Si- und Ge-Oberflächen durch niederenegetische Ionenstrahlerosion, Dissertation, University Leipzig (2014)
143. T. Mohohiro, Y. Taga, Characteristic erosion of silica by oblique argon ion beam bombardment. Thin Solid Films **147**, 153–165 (1987)
144. I.H. Wilson, The topography of ion-bombarded surfaces, in *Materials Modification by High-Fluence Ion Beams*. ed. by R. Kelly, M. Fernanda da Silva (Kluwer, Dordrecht, 1989), pp. 421–466
145. G.J. Ogilvie, Surface structure of silver crystals after argon-ion bombardment. J. Phys. Chem. Solids **10**, 222–228 (1959)
146. M. Balarin, F. Hilbert, Die Einwirkung energiereicher Ionen auf Metalloberflächen. J. Phys. Chem. Solids **20**, 138–145 (1961)
147. G. Carter, The physics and applications of ion beam erosion. J. Phys. D Appl. Phys. **34**, R1–R22 (2001)
148. G.W. Lewis, M.J. Nobes, G. Carter, J.L. Whitton, The mechanisms of etch pit and ripple structure formation on ion bombarded Si and other amorphous solids. Nucl. Instr. Meth. **170**, 363–369 (1980)
149. G. Carter, M.J. Nobes, G.W. Lewis, C.R. Brown, The effect of surface topography evolution on sputter profiling depth resolution in Si. Surf. Interf. Anal. **7**, 35–40 (1985)
150. M. Teichmann, J. Lorbeer, B. Ziberi, F. Frost, B. Rauschenbach, Pattern formation on Ge by low energy ion beam erosion. New J. Phys. **15**, 103029 (2013)
151. J. Völlner, B. Ziberi, F. Frost, B. Rauschenbach, Topography evolution mechanism on fused silica during low-energy ion beam sputtering. J. Appl. Phys. **109**, 043501 (2011)
152. T. Basu, D.P. Datta, Z. Som, Transition from ripples to faceted structures under low-energy argon ion bombardment of silicon: understanding the role of shadowing and sputtering. Nanoscale Res. Lett. **8**, 289 (2013)
153. G. Carter, V. Vishnyakov, Ne^+ and Ar^+ bombardment-induced topography on Si. Surf. Interface Anal. **23**, 514–520 (1995)
154. G. Carter, The effects of surface ripples on sputtering erosion rates and secondary ion emission yields. J. Appl. Phys. **85**, 455–459 (1999)
155. K. Wittmaarck, Effect of surface roughening on secondary ion yields and erosion rates of silicon subject to oblique oxygen bombardment. J. Vac. Sci. Technol. A **8**, 2246–2250 (1990)
156. M.A. Makeev, A.-L. Barabasi, Secondary ion yield changes on rippled interfaces. Appl. Phys. Lett. **72**, 906–908 (1998)
157. W. Hauffe, Faceting mechanism in the sputtering process. Phys. Stat. Sol. (A) **35**, K93–K96 (1976)

158. G. Carter, M.J. Nobes, I. Abril, R. Garcia-Molina, Facet development and its influence on depth resolution during sputtering of Si. Surf. Interface Anal. **7**, 41–48 (1985)
159. M. Teichmann, J. Lorbeer, F. Frost, B. Rauschenbach, Ripple coarsening on ion beam-eroded surfaces. Nanoscale Res. Lett. **9**, 439–446 (2014)
160. M. Engler, S. Macko, F. Frost, T. Michely, Evolution of ion beam-induced patterns on Si(001). Phys. Rev. B **89**, 245412 (2014)
161. M.J. Nobes, J.S. Colligon, G. Carter, The equilibrium topography of sputtered amorphous solids. J. Mater. Sci. **4**, 730–733 (1969)
162. L.F. Johnson, Evolution of grating profiles under ion beam erosion. Appl. Opt. **18**, 2559–2574 (1979)
163. D. Chowdhury, D. Ghose, Super-roughening scaling behaviour of Si surface morphology at grazing incidence low energy ion beam sputtering. Appl. Surf. Sci. **324**, 517–524 (2015)
164. R.S. Nelson, D.J. Mazey, Surface damage and topography changes produced by sputtering. Rad. Eff. **18**, 127–134 (1973)
165. F. Vasiliu, I.A. Teodorescu, F. Glodeanu, SEM investigations of iron surface ion erosion as a function of specimen temperature and incidence angle. J. Mater. Sci. **10**, 399–405 (1975)
166. A. Barna, P.B. Barna, Analysis of the development of large area surface topography during ion etching. Vacuum **40**, 115–120 (1990)
167. D. Manova, M. Schreck, S. Mändl, B. Stritzker, B. Rauschenbach, Orientation dependent sputter yield of aluminium. Surf. Coat. Technol. **151–152**, 72–75 (2002)
168. B.Š Batič, M. Jenko, Orientation-dependent ion beam sputtering at normal incidence conditions in FeSiAl alloy. J. Vac. Sci. Technol. A **28**, 741–744 (2010)
169. P. Lejcek, *Grain Boundary Segregation in Metals* (Springer, Berlin, 2010)
170. G. Carter, J.S. Colligon, M.J. Nobes, The equilibrium topography of sputtered amorphous solids II. J. Mater. Sci. **6**, 115–117 (1971)
171. F.C. Frank, *Growth and Perfection of Crystals* (Wiley, New York, 1958) and F.C. Frank, On the kinematic theory of crystal growth and dissolution processes II. Z. Phys. Chem. **77**, 84–92 (1972)
172. J.P. Ducommun, M. Cantagel, M. Marchal, Development of a general surface contour by ion erosion. Theory and computer simulation. J. Mater. Sci. **9**, 725–736 (1974)
173. G.M. McCracken, The behaviour of surfaces under ion bombardment. Rep. Prog. Phys. **38**, 241–327 (1975)
174. E. Zauderer, *Partial Differential Equations of Applied Mathematics* (Wiley-Interscience, New York, 1989)
175. R. Smith, G. Carter, M.J. Nobes, The theory of surface erosion by ion bombardment. Proc. Roy. Soc. Lond. A **407**, 405–433 (1986)
176. G. Carter, M.J. Nobes, I.V. Katardjiev, J.L. Whitton, The role of bombardment induced defects in initiation of topography on ion bombarded f.c.c. metals. Defect Diff. Forum **57–58**, 97–126 (1988)
177. I.V. Katardjiev, G. Garter, N.J. Nobes, A fundamental approach to surface evolution during growth and erosion. Vacuum **39**, 1069–1075 (1989)
178. I.V. Katardjiev, G. Garter, N.J. Nobes, The application of the Huygens principle to surface evolution in inhomogeneous, anisotropic and time dependent systems. J. Phys. D Appl. Phys. **22**, 1813–1824 (1989)
179. R. Smith, J.M. Walls, The development of a general three-dimensional surface under ion bombardment. Phil. Mag. A **42**, 235–248 (1980)
180. R. Smith, T.P. Valkering, J.M. Walls, The erosion of amorphous and crystalline surfaces by ion bombardment. Phil. Mag. A **44**, 879–893 (1981)
181. I.V. Katardjiev, G. Garter, N.J. Nobes, Modelling of surface evolution during growth and erosion. Int. J. Numer. Model. Electr. Netw. Devices Fields **3**, 137–155 (1990)
182. H. Oechsner, Untersuchungen zur Festkörperzerstäubung bei schiefwinkligem Ionenbeschuß polykristalliner Metalloberflächen im Energiebereich um 1 keV. Z. Physik **261**, 37–58 (1973)
183. G. Carter, J.S. Colligon, M.J. Nobes, The growth of topography during sputtering of amorphous solids. J. Mater. Sci. **8**, 1473–1481 (1973)

184. G.B. Whitham, *Linear and Non-linear Waves* (Wiley, New York, 1974)
185. G. Carter, J.S. Colligon, M.J. Nobes, Analytical modelling of sputter induced surface morphology. Rad. Eff. **31**, 65–87 (1977)
186. G. Carter, M.J. Nobes, K.I. Arshak, R.P. Webb, D. Evanson, B.D.L. Eghawary, J.H. Williamson, The influence of non-uniform incident flux upon surface erosion processes. J. Mater. Sci. **14**, 728–736 (1979)
187. G. Carter, M.J. Nobes, K.I. Arshak, The application of erosion slowness theory to prediction of surface contour generation during particulate ablation of solids. Wear **53**, 245–261 (1979)
188. C. Catana, J.S. Colligon, G. Carter, The equilibrium topography of sputtered amorphous solids, III. Computer simulation. J. Mater. Sci. **7**, 467–471 (1972)
189. T. Ishitani, M. Kato, R. Shimizu, Comments on "The equilibrium topography of sputtered amorphous solids." J. Mater. Sci. **9**, 505–508 (1974)
190. J.P. Ducommun, M. Cantagel, M. Moulin, Evolution of well-defined surface contour submitted to ion bombardment: computer simulation and experimental investigation. J. Mater. Sci. **10**, 52–62 (1975)
191. S. Hamaguchi, M. Dalvie, R.T. Farouki, S. Sethuraman, A shock-tracking algorithm for surface evolution under reactive-ion etching. J. Appl. Phys. **74**, 5172–5184 (1993)
192. J.C. Arnold, H.H. Sawin, M. Dalvie, S. Hamaguchi, Simulation of surface topography evolution during plasma etching by the method of characteristics. J. Vac. Sci. Technol. A **12**, 620–634 (1994)
193. E.J. Klein, W.F. Ramirez, Consideration of local shadowing and ion beam voltage effects in the prediction of a surface evolving under ion milling. J. Vac. Sci. Technol. A **18**, 166–175 (2000)
194. R. Smith, D.D. Makh, J.M. Walls, Surface morphology during ion etching, the influence of redeposition. Phil. Mag. A **47**, 453–481 (1983)
195. R. Smith, M.A. Tagg, J.M. Walls, Deterministic models of ion erosion, reflection and redeposition. Vacuum **34**, 175–180 (1984)
196. G. Carter, The influence of surface diffusion on topography development of an amorphous solid during sputtering. J. Mater. Sci. **11**, 1091–1098 (1976)
197. C.C. Umbach, R.L. Headrick, K.-C. Chang, Spontaneous nanoscale corrugation of ion-eroded SiO_2: the role of ion-irradiation-enhanced viscous flow. Phys. Rev. Lett. **87**, 246104 (2001)
198. M. Hou, M.T. Robinson, The conditions for total reflection of low-energy atoms from crystal surfaces. Appl. Phys. **17**, 371–375 (1978)
199. W. Eckstein, *Computer Simulation of Ion-Solid Interactions*. Springer Series in Material Science, vol. 10 (Springer, Berlin, 1991)
200. W. Eckstein, Reflection (backscattering). IPP Report 17/12 (2009)
201. R. Smith, M.A. Tagg, An algorithm to calculate secondary sputtering by the reflection of ions in two dimensions. Vacuum **36**, 285–288 (1986)
202. R.E. Lee, Microfabrication of ion beam etching. J. Vac. Sci. Technol. **16**, 164–170 (1979)
203. D.W. Younner, C.M. Haynes, Modeling ion beam milling. J. Vac. Sci. Technol. **21**, 677–680 (1982)
204. F. Johnson, K.A. Ingersoll, Asymmetric triangular grating profiles with 900 groove angles produced by ion beam erosion. Appl. Opt. **20**, 2951–2961 (1981)
205. P.G. Glöersen, Ion beam etching. J. Vac. Sci. Technol. **12**, 28–35 (1975) and Masking for ion beam etching. Solid State Technol. **19**, 68–73 (1976)
206. R.E. Chapman, Redeposition: a factor in ion beam etching topography. J. Mater. Sci. **12**, 1125–1133 (1977)
207. J. Belson, I.H. Wilson, Flux density equations for topographical evolution of features on ion bombarded surfaces. Rad. Eff. **51**, 27–34 (1980) and Theory of redeposition of sputtered flux on to surface asperities. Nucl. Instr. Meth. **182/183**, 275–281 (1981)
208. M. Utlaut, Focused Ion Beam, in *Handbook of Charged Particle Optics*, ed. by J. Orloff (CRC Press, Boca Raton 2009), pp. 523–600
209. N. Yamamuchi, T. Yachi, T. Wada, A pattern edge profile simulation for oblique ion milling. J. Vac. Sci. Technol. A **2**, 1552–1557 (1984)

210. T. Ishitani, T. Ohnishi, Modeling of sputtering and redeposition in focused-ion beam trench milling. J. Vac. Sci. Technol. A **9**, 3084–3089 (1991)
211. E. Platzgummer, A. Biedermann, H. Langfischer, S. Eder-Kapl, M. Kuemmel, S. Cernusca, H. Loeschner, C. Lehrer, L. Frey, A. Lugstein, E. Bertagnolli, Simulation of ion beam direct structuring for 3D nanoimprint template fabrication. Microelectr. Eng. **83**, 936–939 (2006)
212. H.-B. Kim, G. Hobler, A. Steiger, A. Lugstein, E. Bertagnolli, Simulation of ion beam-induced micro/nano fabrication. J. Micromech. Microeng. **17**, 1178–1183 (2007)
213. A. Mutzke, R. Schneider, SDTrimSP-2D: simulation of particles bombarding on a two dimensional target version 1.0. IPP Report 12/4, MPI Garching (2009)
214. S. Lindsey, G. Hobler, The significance of redeposition and backscattering in nanostructure formation by focused ion beams. Nucl. Instr. Meth. Phys. Res. B **282**, 12–16 (2012)
215. I. Bizyukov, A. Mutzke, R. Schneider, J. Davies, Evolution of the 2D surface structure of a silicon pitch grating under argon ion bombardment: experiment and modeling. Nucl. Instr. Meth. Phys. Res. B **268**, 2631–2638 (2010)
216. C. Diddens, S.J. Linz, Redeposition during ion beam erosion can stabilize well-ordered nanostructures. Europhys. Lett. **104**, 17010 (2013)
217. S. Rusponi, G. Costantini, C. Boragno, U. Valbusa, Temperature and time evolution of ripple structure induced by ion sputtering on Cu(110). Phys. Low-Dim. Struct. **5**(6), 96–99 (1999)
218. C. Herring, Surface tension as a motivation for sintering, in *The Physics of Powder Metallurgy*, ed. by W.E. Kingston (McGraw-Hill Book Company, New York, 1951), pp. 143–179 and The use of classical macroscopic concepts in surface-energy problems, in *Sintering and Properties of Solid Surfaces*, ed. by R. Gomer, C.S. Smith (Chicago University Press, Chicago, 1953), pp. 5–72
219. S. Liedtke-Grüner, C. Grüner, A. Lotnyk, J.W. Gerlach, B. Rauschenbach, Biaxially textured titanium thin films by oblique angle deposition: conditions and growth mechanisms. Phys. Stat. Sol. (A) **217**, 1900636 (2019)
220. M. Tartz, E. Hartmann, F. Scholze, H. Neumann, Modeling of broad beam ion sources. Rev. Sci. Instrum. **69**, 1147–1149 (1998)
221. B. Ziberi, F. Frost, M. Tartz, H. Neumann, B. Rauschenbach, Importance of ion beam parameters on self-organized pattern formation on semiconductor surfaces by ion beam erosion. Thin Solid Film **459**, 106–110 (2004)
222. B. Ziberi, F. Frost, M. Tartz, H. Neumann, B. Rauschenbach, Ripple rotation, pattern transitions and long range ordered dots on silicon by ion beam erosion. Appl. Phys. Lett. **92**, 063102 (2008)
223. R. Kree, T. Yasseri, A.K. Hartmann, The influence of beam divergence on ion beam-induced surface patterns. Nucl. Instr. Meth. Phys. Res. B **267**, 1407–1411 (2009)
224. C.A. Volkert, Stress and plastic flow in silicon during amorphization by ion bombardment. J. Appl. Phys. **70**, 3521–3527 (1991) and C.A. Volkert, A. Polman, Radiation-enhanced plastic flow of covalent materials during ion irradiation. Mater. Res. Soc. Symp. Proc. **235**, 3–14 (1992)
225. L.D. Landau, E.M. Lifshitz, Fluid Mechanics, vol. IV (Institute of Physical Problems, Pergamon Press, Oxford, 1966)
226. S.E. Orchard, On surface levelling in viscous liquids and gels. Appl. Sci. Res. Section A, **11**, 451–464 (1963)
227. L.E. Stillwagon, R.G. Larson, Fundamentals of topographic substrate leveling. J. Appl. Phys. **63**, 5251–5258 (1988)
228. S. Vauth, Glättungsmechanismen beim Ionenbeschuss rauer amorpher Oberflächen, Dissertation, University Göttingen (2007)
229. A.S. Rudy, V.K. Smirnov, Hydrodynamic model of wave-ordered structures formed by ion bombardment of solids. Nucl. Instr. Meth. Phys. Res. B **159**, 52–59 (1999)
230. M. Castro, R. Cuerno, Ion induced solid flow. arXiv:1007.2144v1 (2010)
231. S. Roorda, W.C. Sinke, J.M. Poate, D.C. Jacobson, S. Dierker, B.S. Dennis, D.J. Eaglesham, F. Spaepen, P. Fuoss, Structural relaxation and defect annihilation in pure amorphous silicon. Phys. Rev. B **44**, 3702–3725 (1991)

232. S.T. Pantelides, Defects in amorphous silicon: a new perspective. Phys. Rev. Lett. **57**, 2979–2983 (1986)
233. N.V. Medhekar, W.L. Chan, V.B. Shenoy, E. Chason, Stress-enhanced pattern formation on surfaces during low energy ion bombardment. J. Phys. Condens. Matter **21**, 224021 (2009)
234. A. Witvrouw, F. Spaepen, Viscosity and elastic constants of amorphous Si and Ge. J. Appl. Phys. **74**, 7154–7161 (1993)
235. Yu.V. Trushin, Theoretical concepts of radiation swelling of materials and the characteristics of sinks. J. Tech. Phys. **39**, 564–568 (1994)
236. L.M. Wang, R.C. Birtcher, Radiation-induced formation of cavities in amorphous germanium. Appl. Phys. Lett. **55**, 2494–2497 (1989)
237. S.G. Mayr, The role of ion irradiation induced viscous flow in thin film structuring: hard matter going soft. Adv. Solid State Phys. **45**, 327–338 (2005)
238. S.A. Norris, J. Samela, L. Bukonte, M. Backman, D. Flyura, K. Nordlund, C.S. Madi, M.P. Brenner, M.J. Aziz, Molecular dynamics of single-particle impacts predicts phase diagrams for large scale pattern formation. Nat. Commun. **2**, 276–281 (2011)
239. S.G Mayr, Y. Ashkenazy, K. Albe, R.S. Averback, Mechanisms of radiation-induced viscous flow: role of point defects. Phys. Rev. Lett. **90**, 055505 (2003)
240. B. Ziberi, Ion beam-induced pattern formation on Si and Ge surfaces, Dissertation, University Leipzig (2008)
241. S. Vauth, S.G. Mayr, Relevance of surface viscous flow, surface diffusion, and ballistic effects in keV ion smoothing of amorphous surfaces. Phys. Rev. B **75**, 224107 (2007)
242. H. Zhou, Y. Wang, L. Zhou, R.L. Headrick, A.S. Özcan, Y. Wang, G. Özaydin, K.F. Ludwig, Jr., D.P. Siddons, Wavelength tunability of ion-bombardment-induced ripples on sapphire. Phys. Rev. B **75**, 155416 (2017)
243. D. Babonneau, E. Vandenhecke, S. Camelio, Formation of nanoripples on amorphous alumina thin films during low-energy ion beam sputtering: experiments and simulations. Phys. Rev. B **95**, 085412 (2017)
244. P.K. Giri, V. Raineri, G. Franzo, E. Rimini, Mechanism of swelling in low-energy ion-irradiated silicon. Phys. Rev. B **65**, 012110 (2001) and P.K. Giri studies on the surface swelling of ion-irradiated silicon: role of defects. Mater. Sci. Eng. B **121**, 238–243 (2005)
245. S. Tamulevicius, I. Pozéla, M. Andrulevicius, A simple model of radiation swelling of silicon. Mater. Sci. Eng. B **40**, 141–146 (1996)
246. G.M. McCracken, The behaviour of surfaces under ion bombardment, Reports on Progress in Physics, **38** 241-327 (1975)

Chapter 7
Ion Beam Figuring and Smoothing

Abstract The technologies, ion beam figuring (IBF) and ion beam-induced smoothing (IBS), are used to precisely remove imperfections or correct surface shape in a predetermined and controlled manner. The IBF method uses the computer-aided codes to realize the desired surface topography, where a spatially and temporally stable ion beam is passed vertically over the surface at a fixed distance in a high-vacuum environment. The main input variables of this approach, the ion beam removal function, the surface error function and the dwell time procedure, are presented. The various algorithms (Fourier transform algorithm, iterative dwell algorithm, matrix based algorithm, Bayesian algorithm) for determining the dwell time of the ion beam over each object point to be machined and the effect of temperature during the figuring procedure are discussed. The application of IBF technology for the correction of shape errors on surfaces to achieve depth accuracies in the nanometer and sub-nanometer range over the entire spectrum of the spatial surface wavelength is demonstrated with selected examples. Ion beam-induced smoothing (IBS) focuses on feature processing (spatial wavelength < a few microns and height amplitudes on the order of nanometers) with the aim of producing ultra-smooth surfaces. In addition to direct smoothing by low-energy ions, the technologies of smoothing with a planarization layer and smoothing by means of ions incident at a very oblique angle have also become established. Atomistic surface relaxation processes such as ion beam enhanced viscous flow, thermally activated surface diffusion, effective ion-induced diffusion, and ballistic mass redistribution can contribute to the smoothing process.

The controlled modification of the topography is of primary importance in numerous contemporary industrial applications, particularly in microelectronics, metrology, mechanical engineering, and the optical industry. In these fields, the requirements for precision and on the geometry become increasingly challenging. Industrial applications based on ion-solid interaction, such as ion beam etching, reactive ion beam etching, chemically activated ion beam etching, ion beam assisted deposition, ion beam sputtering, focused ion beam writing, and shaping, etc. are increasingly being applied to smooth out surfaces or to shape surface profiles. This makes high-precision

B. Rauschenbach, *Low-Energy Ion Irradiation of Materials*, Springer Series in Materials Science 324, https://doi.org/10.1007/978-3-030-97277-6_7

predictions about topographical evolution under low-energy ion bombardment absolutely essential. Not least, the depth resolution of destructive analytical surface and thin film techniques, such as AES, SIMS, XPS etc., are strongly limited by the variation of the erosion profile during ion sputtering. Having broad knowledge with regard to the change of topography due to ion beam erosion, including the interpretation of the results measured, would be extremely advantageous.

Each method for manufacturing precision surfaces has its performance limitations, which unavoidably lead to either to deviations from the ideal shape or to unacceptable roughness. Conventional grinding, turning, and milling techniques are thus applied consecutively to correct these fabrication limitations. Moreover, a finishing process is frequently required to correct the shape and the flatness so that the desired accuracy can be attained.

Ion beam figuring and ion beam-induced smoothing are technologies that have been developed to polish surfaces so that imperfections can be removed or the surface shape can be corrected in a predetermined and controlled fashion. These methods use accelerated ions for the precision removal of atoms on the surface. This allows a deterministic and controlled final processing of surfaces, which avoids problems associated with traditional polishing and finishing technologies.

The *ion beam figuring* (IBF) process is characterized by physical sputtering of materials primarily specializing in correction of surface errors and contour with nanometer precision (Sect. 7.1). IBF is a highly deterministic (leads to the expected results), non-contact, extremely stable and controllable method with a high process convergence (> 90% compared to the theoretical input requirement).

The *ion beam smoothing* (IBS) of surfaces utilizes relaxation mechanisms induced by ion bombardment in the near-surface region to decrease surface roughness and remove defects, i.e. the task of this process is surface polishing (Sect. 7.2). Because fundamental relaxation processes occur on atomic scale, based on this method, spatial distances up to about a few 100 nm can be efficiently smoothed. When a planarization layer is previously deposited on the surface and the layer, together with the rough surface, is then removed by ion bombardment, surfaces can be also smoothed.

7.1 Ion Beam Figuring

The ion beam figuring technique is based on the fundamental finding that local topographical gradients on the surface are the driving force behind the evolution of the surface topography under low-energy ion bombardment (see Chap. 6). In the presence of protrusions and/or depressions on the surface, this phenomenon of gradient-dependent sputtering leads to different erosion velocities if no further surface transport and evolution processes capable of affecting the evolution of the topography are involved.

As is generally known, the erosion rate or velocity, v, parallel to the surface normal can be expressed by $v = (JY(\theta)/N)\cos\theta$ (see 6.9), where J is the ion current density, Y is the sputtering yield, θ is the angle of ion incidence and N is the atomic number density. In Sect. 6.4.1, findings from Nobes et al. [1] and Carter et al. [2, 3] were presented, showing that the erosion velocities perpendicular and parallel the surface (6.16) and (6.19) in the light of the kinematic wave treatment by Frank [4] can be given by

$$v_x = \frac{J}{N}\frac{dY(\theta)}{d\theta}\cos^2\theta \text{ and } v_h = \frac{J}{N}\left[\frac{dY(\theta)}{d\theta}\sin\theta\cos\theta - Y(\theta)\right], \tag{7.1}$$

where it was assumed that the surface lay in the x0h plane and was bombarded by ions with uniform ion current density J in the 0h direction (see Fig. 6.29).

A consistent application of these equations on the evolution of topography of initially rough surfaces under ion irradiation results in a reduction of the roughness (see, e.g., [5] and examples in Sect. 6.4.1). In principle, if the lateral dimensions are comparable, depressions erode more slowly than protrusions.

7.1.1 The Principle of the IBF Method

The ion beam figuring method is a specific variant of ion beam sputtering in which the material removal is temporally and spatially controlled. Figure 7.1 (left) is a schematic illustration of the IBF principle.

This method is a noncontact machining technique in which an ion beam is directed toward a workpiece to remove material by sputtering. In contrast to other polishing methods, such as mechanical grinding or magnetorheological finishing, the material removal rate is very low. The removal areas vary between some mm^2 and several cm^2. Typically, in this method, a focused beam is guided over the surface of the workpiece on a predetermined scanning path (Fig. 7.1, right). The scanning velocity of the ion beam on the surface is determined by the dwell time of the beam on each

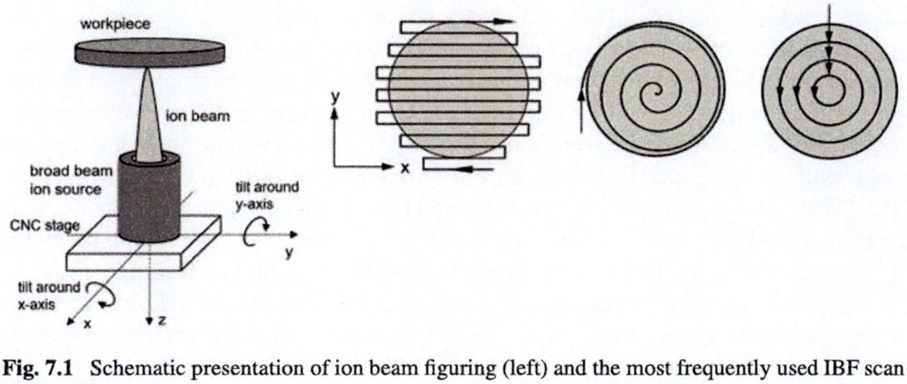

Fig. 7.1 Schematic presentation of ion beam figuring (left) and the most frequently used IBF scan paths (right)

point on the surface. This time is proportional to the desired material removal (see Sect. 7.1.2.3).

Preferentially, hot-filament Kaufman and 13.56 MHz RF Ar ion sources with a multi-grid ion optical multi-aperture beam extraction system are applied in an energy range between some hundred eV and few keV. The generated ions form an axial-symmetrical ion beam with a Gaussian-like distribution of the accelerated ions. A neutralizer outside the source supplies electrons to neutralize the ion beam to prevent charging isolating target components. Faraday cup arrays are installed to measure the ion current distribution of the ion source and to control the ion beam position.

In dependence on the elongation of the spatial error features or the roughness of surface, ion beams with spot sizes between a few centimeter and some tenth-millimeter FWHM are used. In general, for correction of long spatial wavelength errors > 1 mm (for details see Sect. 7.1.4, Table 7.1), ion sources providing an ion beam with a Gaussian ion current distribution and FWHM $\geq$ 10 mm are used, while for the correction of spatial wavelength errors < 1 mm, the ion beam possesses a FWHM < 10 mm, e.g., achieved through the use of diaphragms of different sizes in front of the ion source [6–8].

Inside the vacuum chamber, the broad ion beam source is installed on a computer-controlled multi-axis CNC stage. For plane surfaces of the workpiece, the temporally and spatially stable ion beam can then be scanned along the x- and y-axes (Fig. 7.1, right). For ion beam machining of 3-dimensional components (non-plane components), the fact that the working distance is not constant must be taken into account. Further axes must be implicated, with the objective of achieving a constant working distance at each working point on the surface. Consequently, the z-axis is varied to guarantee the same distance between the ion source and the surface. Further, two additional tilt axes are used to orient the ion source perpendicular to the bombarded surface. Consequently, a 5-axis CNC system is necessary to ensure a constant ion beam removal function (Sect. 7.1.2.1) at each surface point. Reviews of IBF technology may be found in [9–11].

The first application of the ion milling technique was reported by Castaing and Laborie [12] for the preparation of TEM specimens from hardening alloys. About ten years later, Meinel et al. [13], and later Schroeder et al. [14], discovered that high-energy positive ions effect a uniform removal of a surface of fused silica and other insulators. It was anticipated that a significant feature of polishing optical materials with a particle beam would be the opportunity to fabricate highly precise optical surfaces. In the 1970s, there were attempts to apply this new figuring technology to surfaces of various materials (fused silica, different glasses, ultra-low expansion glass (ULE), ceramics, magnetic orthoferrites, etc.) [15–17]. Typically, in these studies, ion energies of between some ten and some hundred keV were used (see bibliography in [18]). By the end of the 1970s, low-energy ion sources (in particular Kaufman sources [19]) could be used for ion beam-based surface modification. After Kaufman low-energy ion sources were made available, Wilson et al. [20] successfully smoothed fused silica, copper, and Zerodur surfaces. Since the 1980s, several research and development IBF facilities have been constructed and expanded. For example, the Eastman Kodak company in Rochester, New York, has established a 2.5 $\times$ 2.5 $\times$

$0.6\ m^3$ IBF system and has been capable to polish an axial-symmetric aspheric fused silica optics with a diameter up to 80 cm for space applications [21]. Over the past two decades, numerous institutes have developed their own IBF machines and several companies are offering commercial IBF facilities equipped with computer-controlled 3- or 5-axis motion systems. Consequently, a rapidly increasing commercialization of the IBF technology can be established. One example is the commercial IBF plant developed by NTGL GmbH and Leibniz Institute of Surface Modification Leipzig [11]. This IBF plant has a $1.3 \times 1.5\ m^2$ cylindrical vacuum chamber (base pressure 10^{-4} Pa) and is fitted with a load-lock chamber. 25 kg optics up to 70 cm diameter can be handled. The system is equipped with a 40-mm RF broad-beam ion source and a 5-axes positioning system.

Presently, research and development in the field of ion beam figuring is focused on (i) the development of IBF technologies to correct middle- and high-frequency errors on spherical and aspherical surfaces, (ii) realization of sub-nanometer rms roughness over the entire spectrum of spatial surface wavelengths, (iii) the search for simple, fast and high-precision methods for calculating dwell time, and (iv) application of IBF on metals and temperature-sensitive materials.

7.1.2 *Ion Beam Figuring Procedure*

In general, ion beam figuring was established for the final correction of surface figure errors. Figure 7.2 shows a schematic representation of the several steps involved in this procedure. The IBF procedure starts with measurements of the topography of the

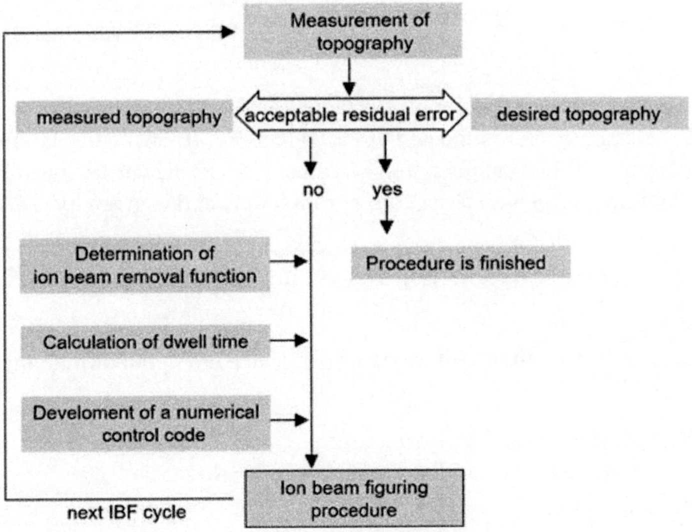

Fig. 7.2 Schematic representation of the flow of the IBF procedure

workpiece and the determination of the removal profile. The topography measurements (x-y array height map) are compared with the desired topography. On this basis, the desired removal topography can be determined by comparison of the measured results with the desired surface topography. The ion beam removal function (BRF, see Sect. 7.1.2.1) is determined by measurement of the depth distribution of ion beam etched profile as a function of the radial distance from the center of the ion beam. Both magnitudes, the BRF function and the removal error map, are necessary for calculation of the dwell time using a deconvolution procedure (see Sect. 7.1.2.5). A software code is then developed to automate the ion beam figuring process. Each discretized point on the surface is thereby assigned a dwell time, the scanning path is pinpointed, and the scanning velocity is determined. The ion beam figuring machining system uses the computer-aided codes to achieve the desired surface topography, a process in which a spatially and temporally stable ion beam is guided perpendicularly over the surface at a fixed distance in a high-vacuum environment. In some cases, according to the dependence of the angle of incidence on the sputtering yield, a tilted ion beam is used to enhance the removal velocity (see Sect. 5.6). This is especially important for enhancing production efficiency, for instance, for thin film thickness trimming in mass production.

7.1.2.1 Ion Beam Removal Function

The diameter and the intensity (ion current density, J) of the ion beam determine the capability of the ion beam as a tool. The radial ion intensity distribution of a circular Gaussian beam can be expressed by [22]

$$J(r) = \frac{j}{2\pi\sigma^2}\exp\left[-\left(\frac{r^2}{2\sigma^2}\right)\right], \tag{7.2}$$

where σ is the standard derivation of the Gaussian distribution, r is the radial coordinate and the beam center is located at $r = 0$ (c.f. Eq. (6.56)) and j is ion current. The total current within a radius r_s (e.g., radius of a circular orifice) can be obtain from (7.2) by integration with respect to r from 0 to r_s and is given by

$$j(r_s) = j\left\{1 - \exp\left[-\left(\frac{r_s^2}{2\sigma^2}\right)\right]\right\}. \tag{7.3}$$

Additional different other definitions of beam size are applied (note that e is the Euler number):

- width at 1/e of the Gaussian distribution: $w_{1/e} = 2\sigma\sqrt{2} \approx 2.83\sigma$
- width at 1/e^2 at the Gaussian distribution: $w_{1/e^2} = 4\sigma$
- full width at half maximum: $FWHM = 2\sigma\sqrt{2\ln 2} \approx 2.355\sigma$.

The erosion of the surface using an ion beam is characterized by the ion beam removal function (BRF), which describes the rate of material removal (in the unit [length per time unit]) from a workpiece at (x, y), when the beam is directed at (x′, y′), i.e. the BRF describes the material removal profile (the 'footprint') and determines the volume of the material removal. With the objective of determining of the BRF, for example, the ion beam is applied for static bombardment of the material for a given time. Figure 7.3a shows a schematic image of the footprint in a material following by an ion bombardment with a rotationally symmetrical Gaussian shaped ion beam for different times of bombardment. Interferometric measurements of footprints in fused silica are shown in Fig. 7.3c, d. The BRF is thus frequently approximated by an axial symmetrical Gaussian distribution. For example, Fig. 7.3b demonstrates the conformity of the measured footprint in quartz after 1000 eV Ar bombardment with a two-dimensional Gaussian fit [8].

Then, the volumetric removal or beam removal function (see Sect. 7.1.4), is given by

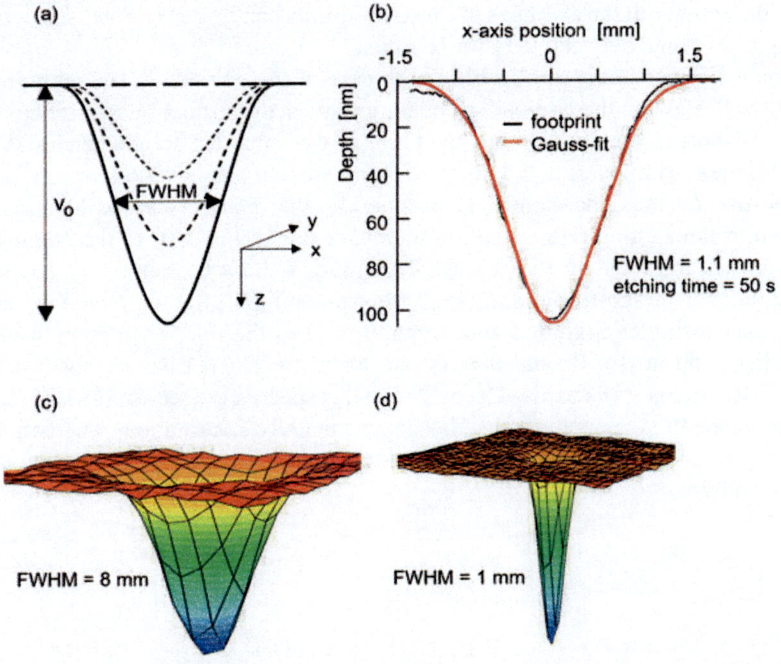

Fig. 7.3 **a** Schematic representation of a cross-section footprint after bombardment with a Gaussian-shaped ion beam (the dashed lines indicate distribution to earlier times of ion bombardment), **b** comparison of the measured footprint (black profile) in quartz after Ar ion etching for 50 s and the fitted Gauss distribution (red profile), **c** example of an ion beam footprint measured in fused silica using interferometry (RF ion source without diaphragm), **d** example of an ion beam footprint measured in fused silica using interferometry (RF ion source with diaphragm (diameter of the diaphragm in front of the RF ion source was 1 mm). Figures (**b**–**d**) are taken from [8]

$$v(x, y) = v_o \exp\left[-\frac{1}{2}\left(\frac{x^2}{\sigma_x^2} + \frac{y^2}{\sigma_y^2}\right)\right]$$
$$= \frac{v_r}{2\pi\sigma_x^2\sigma_y^2} \exp\left[-\frac{1}{2}\left(\frac{x^2}{\sigma_x^2} + \frac{y^2}{\sigma_y^2}\right)\right], \qquad (7.4)$$

where v_o is the peak removal rate and v_r is the volume removal rate. In the axial symmetrical case, x = y, consequently, the integration above the removal (etching) rate leads to the volume removal rate (removed volume per time unit) by

$$R(x, y) = v_r = \int_{-\infty}^{\infty} dx \int_{-\infty}^{\infty} v(x, y)dy = 2\pi\sigma^2 v_o \approx 1.133 v_o (FWHM)^2. \qquad (7.5)$$

From interferometric measurement of numerous footprints, the BRF can be evaluated and the etching rates R(x, y) as well as the standard derivation or the diameter of the ion beam with the averages of these footprints can be determined. Typical IBF etching rates range between 0.1 and 10 nm/s.

Equation (7.5) is only applicable in the case of the perpendicular ion bombardment ($\theta = 0°$). When the circular ion beam impinges the surface at an incident angle $\theta > 0°$ (off-normal condition), then the circularly bombarded area degenerates to an elliptical area. Moreover, it is known that the erosion rate is also strongly dependent on the ion incidence angle. Consequently, the removal rate is dependent on the ratio of the etching rate at the ion incidence angle $v_r(\theta > 0)$ to the etching rate under normal ion incidence $v_r(\theta = 0)$. This ratio, $v_r(\theta)/v_r(0)$, can be approximated by the ratio of the corresponding sputtering rates $Y(\theta)/Y(0)$ for incidence angles $\theta < \theta_p$. According to Sigmund and Yamamura [23], the relationships between the sputtering yield at $\theta > 0°$ and $\theta = 0°$ are given by $Y(\theta)/Y(0) = \cos^{-f}(\theta)$ and $Y(\theta)/Y(0) = \cos^{-f}(\theta)\exp\left[-\Sigma\left(\cos^{-1} - 1\right)\right]$, respectively [see (5.54) and (5.55)]. The fit parameters Σ and f must be experimentally determined. The ion beam removal rate under off-normal ion bombardment and tilting along the x-direction is then given by

$$R(x, y) \approx v_o \frac{v_r(\theta)}{v_r(0)} \exp\left\{-\frac{\left[(x\cos\theta)^2 + y^2\right]}{2\sigma^2}\right\}$$
$$\approx v_o \frac{Y(\theta)}{Y(0)} \exp\left\{-\frac{\left[(x\cos\theta)^2 + y^2\right]}{2\sigma^2}\right\}. \qquad (7.6)$$

Consequently, the deformation of the footprint of the ion beam removal function by oblique angle ion bombardment must be taken into consideration and must be corrected in the figuring of spherical and aspherical surfaces. In praxis, a 5-axis movement mechanism in an IBF plant guaranteed that the ion beam could be moved normal to the surface at a constant distance.

7.1.2.2 Surface Error Function

The difference between measured and desired heights at each sample point (x, y) is described by surface error function given by

$$Z(x, y) = z_{mes}(x, y) - z_{des}(x, y), \tag{7.7}$$

where z_{mes} and z_{des} are the measured and the desired heights at the point (x, y) on the surface of the workpiece (target), respectively. Consequently, Z(x, y) quantifies the material to be removed. Based on interferometric measurements or AFM/STM studies, a surface error contour map can be established.

7.1.2.3 Dwell Time Procedure

In ion beam figuring technology, one of the most important problems is determination of the dwell time. According to Allen and Romig [21], the development of dwell time algorithms is based on assumptions that the removal of material (sputtering yield) is isotropic and proportional to dwell time and that the ion beam is temporally and spatially invariant. Input parameters for these algorithms are (i) the ion beam removal function, R(x, y), (ii) the surface error contour map, Z(x, y), (iii) the shape of the surface (plane, spherical, aspherical, etc.), and (iv) the scan path (meander, spiral, circular, etc. see Fig. 7.1).

Wilson and McNeil [24] and Drueding et al. [25] proposed a two-dimensional description of the IBF process by a convolution operation of the ion beam removal function R (x, y) at point (x, y) and the process time $\tau(x, y)$ at this point. The material removed (surface error function) at this point is thus given by

$$Z(x, y) = R(x, y) \otimes \tau(x, y) = \int_{-\infty}^{\infty} \int_{-\infty}^{\infty} R(x - x', y - y') \tau(x', y') dx'dy', \tag{7.8}$$

where $\tau(x', y')$ is the time that the ion beam is directed at point (x', y'), i.e. the dwell time, and the symbol $\otimes$ represents the convolution operation. It must also be taken into account that the ion source dwell time at each point (x′, y′) for the time τ(x′, y′) also removes material at the point in the vicinity of the ion beam center. During the movement of the ion source over the surface, a uniform amount of material removal can be achieved if the beam characteristic is spatially invariant and the velocity between the dwell points is constant.

7.1.2.4 Experimental Routines for Determining Dwell Time

IBF standard technology uses a constant and stable ion beam that moves via computer control across the entire workpiece with variable scan line velocity according to the local dwell time required for the specified removal of material. Typically, the ion beam scans the surface in a meander-, spiral- or cycloid-shaped path (c.f. Fig. 7.1, right). The beam tool size is adjusted according to the spatial surface error size processing. The scanning velocity of the ion beam on the surface is determined by the dwell time of the beam on each point on the surface. This time is proportional to the desired material removal. Consequently, the dwell time can be equalized with the amount of the removed material at this point. On the basis of this consideration, the integral removed material function (7.6) can be discretized. For this purpose, the surface of the workpiece is subdivided into square grids (Fig. 7.4). It is assumed that the ion beam is positioned at the point (x_i, y_j) and the dwell time at this point is given by $\tau(x_i, y_j)$, where the dwell time is considered to be the time that the ion beam remains on a surface point (x_i, y_j) before moving to the next surface point. The material removed per unit time in space domain is then given by $R(x - x_i, y - y_j)$. Therefore, the material removal at the point (x, y) is given by $R(x - x_i, y - y_j)\tau(x_i, y_j)$. With scanning of the ion beam over all dwell points (c.f. Fig. 7.4), the effective value of the material removal is determined by the discretized equation.

$$Z(x_k, y_k) = \sum_{i=1}^{N} R(x_k - x_i, y_k - y_j)\tau(x_i, y_j), \tag{7.9}$$

where N is the total number of dwell points on the surface and $R(x_k - x_i, y_k - y_j)$ gives the value of material removal per unit time at the point (x_k, y_k), under the condition that the ion beam dwells at the point (x_i, y_j) for the time $\tau(x_i, y_j)$.

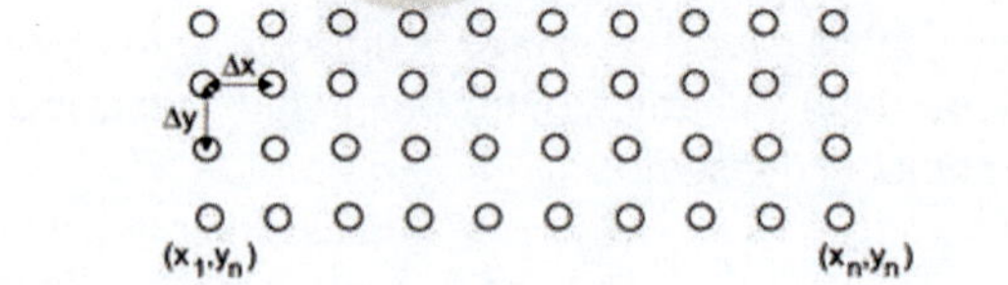

Fig. 7.4 Schematic image of the removal process. The grey area represents the ion beam at the dwell point (x_i, y_i)

From an experimental point of view, the x-y square grid array should be so chosen that the discretization using measurements of the surface topography by interferometry is identical to the discretization for the deconvolution.

7.1.2.5 Deconvolution Procedures

To determine the dwell time for a given surface error function Z(x, y) and beam removal function R(x, y), the inverse problem must be solved, i.e. a mathematical deconvolution procedure must be carried out. This ambitious approach is difficult, because the iterative approaches used to solve the deconvolution problem (i.e. the determination of the dwell time) under the auspices of the following four requirements must be numerically stable [26, 27]:

- the solution must be non-negative, because IBF does not have the ability to deposit material,
- the dwell time map should closely duplicate the desired removal map,
- the entire dwell time should be short as possible and
- the calculation time for the dwell time should be acceptable.

Several algorithms have been proposed to resolve the deconvolution problem under the named restrictions. MATLAB codes for numerous dwell time algorithms can be uploaded to a public GitHub repository [28].

(1) Fourier transform based dwell time algorithm

According to Wilson and McNeil [24, 29], the convolution operation, (7.8), can be transcribed in the frequency domain, which enables multiplication of Fourier transformed (FT) functions, given by

$$FT[Z(x, y)] = FT[R(x, y)] \cdot FT[\tau(x, y)]. \tag{7.10}$$

The dwell time can be then obtained by

$$\tau(x, y) = FT^{-1}\left\{\frac{FT[Z(x, y)]}{FT[R(x, y)]}\right\} = FT^{-1}\left\{\frac{FT[removal\ function]}{FT[ion\ beam\ removal\ funtion]}\right\}, \tag{7.11}$$

where FT^{-1} is the inverse Fourier transform. Because the fast Fourier transform can be used, this calculation is very efficient. However, the Fourier transform of the ion beam removal function in (7.10) can be zero or close to zero. Thus, the noise in the function Z(x, y) at selected frequencies is strongly amplified, i.e. the deconvolution operation becomes instable. Consequently, Wilson and McNeil [24] proposed applying an inverse threshold filter to overcome this problem, where the threshold value introduced, which filters the small frequencies (the denominator R(x, y) in (7.11), can be used only if R(x, y) > value of the threshold filter value). The determination of the optimal threshold value is complicated. Wang et al. [26] have recently proposed a robust iterative Fourier-transform-based algorithm, involving an optimized threshold filter, to determine an optimized dwell time map.

(2) Iterative dwell time algorithm

The discretized equation, (7.6), can be rewritten in matrix form. The solution of this linear equation system, under the condition that Z(x, y) and R(x, y) are known, is then a problem of deconvolution. Based on their positive performance, different iterative algorithms of deconvolution are available in the field of ion beam figuring. Specifically, the van Cittert [30] and Richardson-Lucy iteration methods [31] are used to determine the final topography and dwell time in IBF of planar and spherical surfaces, while the Gold algorithm [32] is applied for the calculation of these magnitudes for aspherical surfaces. In principle, the iterative algorithms are based on (i) finding an approximation for deconvolution, (ii) realizing the convolution of this conjecture, (iii) determination of the difference between the topography after convolution and the measured topography (determination of the residual) and (iv) adjusting the conjecture to improve the approximation of the topography. For example, the van Cittert method is an iterative time-domain transform that generates successive approximations of the desired topography. The effect of the smoothing is attenuated using the following iterative process [25]:

$$\tau_{n+1} = \tau_n + \varepsilon(Z - R \otimes \tau_n), \tag{7.12}$$

where τ_{n+1} is the new estimated dwell time from the previous one τ_n, n is the number of the steps in the iteration and ε is a scaling factor that ensures the convergence of (7.12). For a discrete linear system, (7.12) becomes

$$\tau_{n+1} = \tau_n + \varepsilon([Z] - [R]\tau_n) \tag{7.13}$$

where [ ] represents a matrix. The convergence condition of this equation is satisfied if the diagonal of the matrix $[R]_{ii} > \sum_{j \neq i} [R]_{ij}$. On this basis, the van Cittert algorithm of deconvolution can be used (for a detailed description, see, e.g., [33]). This algorithm converges very quickly. Convergence is achieved as the difference between the original topography and the smoothed topography at n-th iteration approaches approximately zero. A critical factor in the quality of the reconstructed topography is the prior mitigation of map noise through Gaussian smoothing, which simultaneously attenuates both noise and topography signals. The residual spatial structures are then amplified. In contrast, the Richardson-Lucy algorithm can be applied for a variety of reasons, such as that it does take the type of noise affecting the topography into consideration. The advantage is that this algorithm is simple to implement, however, it can become unstable [34]. The iterative dwell time algorithms were successfully applied to control ion beam figuring processes with ultra-high precision [35–37].

(3) Matrix-based dwell time algorithm

An alternative approach to determining the dwell time is based on the solution of a set of linear equations, where the dwell time is found in the solution of this over-estimated linear equation system. This method was initially proposed by Carnal et al. [38]. They described the convolution operation in (7.10) as a product of a matrix containing beam characteristics and a vector of the dwell time values. For this purpose, the discretized equation, (7.9) can be expressed as vector equation

$$\mathbf{Z} = [\mathbf{R}] \cdot \tau \text{ or} \begin{pmatrix} Z_o \\ Z_1 \\ \vdots \\ Z_{N-1} \end{pmatrix} = \begin{pmatrix} R_{1,1} & R_{1,2} & \ldots R_{1,n-1} \\ R_{1,1} & R_{2,2} & \ldots R_{2,n-1} \\ \vdots & \vdots & \vdots\vdots \\ R_{N-1,1} & R_{N-1,2} & \ldots R_{N-1,n-1} \end{pmatrix} \begin{pmatrix} \tau_o \\ \tau_1 \\ \vdots \\ \tau_{n-1} \end{pmatrix} \tag{7.14}$$

where $\mathbf{Z}$ and $\boldsymbol{\tau}$ are column vectors of the dimensions N and n, respectively, where N is the total number of removal points on the surface and n is the total number of the dwell time points, and [**R**] is a N × n matrix.

On the basis of this transformation, the dwell time can be determined by solving this system of linear equations. The main problem with this approach is that Gaussian elimination to solve a system of linear equations is only applicable to a limited degree because the matrix is ill-conditioned and rank-deficient. Consequently, Carnal et al. [38] and Wu et al. [34] have used so-called 'single value decomposition' to find a unique solution. However, for large-scale matrix, the computational operating expense is so high that Carnal et al. [38], Zhou et al. [39] and Wang et al. [26, 27] used a special algorithm, the LSQR algorithm, proposed by Paige and Saunders [40] to numerically calculate large, sparse, rank-deficient system of equations.

(4) Bayesian based dwell time algorithm

All dwell time algorithms presented are characterized by the fact that the expected non-negativity of the dwell time cannot be guaranteed. Jiao et al. [41] proposed an iterative dwell time algorithm based on the Bayesian statistics. Assuming that the material removal function (surface error function), Z(x, y), and dwell time $\tau(x, y)$ are both random, then the Bayesian theory predicts that

$$P(\tau|Z) = P(Z|\tau)\frac{P(\tau)}{P(Z)}, \tag{7.15}$$

where the posterior, $P(\tau|Z)$, describes the probability of the dwell time τ being true given that the material removal function, Z, is true, the likelihood, $P(Z|\tau)$, describes the probability of Z being true given that τ is true, the prior, $P(\tau)$, is the probability of τ being true and the P(Z) the probability of Z is true. Thus, a distinction is to be made between the real topography, i.e. without noise, measuring effects, blurring, etc., and the measured topography. The aim of this algorithm is then to reconstruct the dwell time, τ(x,y), from the material error function Z(x,y) at each point (x,y).

Within the framework of the Bayesian method (details see e.g., [42]), an initial probability distribution P(τ) of the τ(x,y) is assumed, which contains information on the expected material removal function. If τ(x,y) is assumed to be the correct dwell time, it is also necessary to construct a probability function of the measured removal function or ion beam density $P(Z|\tau)$.

Assuming the ion beam density probability distribution $P(\tau|Z)$ is a Poisson distribution, the dwell time can be found by maximum a posteriori as $\min_\tau J(\tau)$ [41], where $J(\tau) = \int_{-\infty}^{\infty}\int_{-\infty}^{\infty}\left[R \otimes \tau - Z \cdot \log(R \otimes \tau)\right]dxdy$. The optimization of this condition can be obtained using calculus of variation to

$$\frac{R(-x,-y)}{\int_{-\infty}^{\infty}\int_{-\infty}^{\infty} Rdxdy} \otimes \frac{Z}{R \otimes \tau} = 1. \tag{7.16}$$

The Richardson–Lucy multiplicative algorithm [31] is then applied to determine dwell time at the (n+1)-th iterative step to

$$\tau_{n+1} = \tau_n \left[\frac{R(-x,-y)}{\int_{-\infty}^{\infty}\int_{-\infty}^{\infty} Rdxdy} \otimes \frac{Z}{R \otimes \tau_n}\right]. \tag{7.17}$$

Jiao et al. [41] have demonstrated that the ion beam figuring technology based on the Bayesian principle is a high-efficiency deterministic process and can be used to correct surface errors rapidly and precisely.

7.1.3 Temperature During Ion Beam Figuring Process

It is known that the collision of incident ions with target atoms generates measurable heat (temperature rise on the surface of the workpiece). Allen and Romig [21] measured the increase in temperature on a 3.4 mm fused silica plate during an ion figuring uniform removal scan. When power equivalent to 70 W was transferred by the ion beam, maximum temperatures of 125 °C and 65 °C were detected on the front and the back sides of the plate, respectively. Later, Ghigo et al. [43] and Gailly et al. [44] measured the temperature increase on a hexagonal Zerodur mirror (1.45 m corner-to-corner distance) during IBF runs because thermal hysteresis of this material is affected at temperatures above 120–130 °C. For example, if the power of the ion beam was 75 W, a maximal temperature of 160 °C was achieved on the front side of the mirror. Xie et al. [45] measured a temperature increase on a BK7 glass surface after 700 eV Ar ion bombardment (ion current was 60 mA, beam diameter was 48 mm) of up to about 100° during a time of ion bombardment of 60 s. They were also able to simulate the distribution of thermal-induced stress on the surface. Figure 7.5 shows an example of the temperature increase on the surface of fused silica at steady-state Ar ion bombardment [46]. The peak of temperature appears at

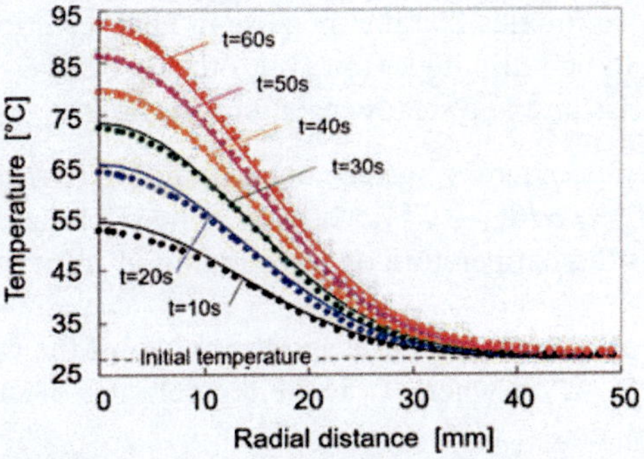

Fig. 7.5 Temperature distribution on a fused silica surface along radial direction after Ar ion bombardment for different times of bombardment. Solid lines represent the results after finite element simulation and the dots represent the measured temperatures by infrared thermal imager. The ion energy is 600 eV, the beam diameter is about 80 mm and the angle of ion incidence is 45° (Figure is adapted from [46])

the center of an ion beam on the surface and the temperature decreases along the radial direction of the surface.

It can be established that the temperature created during the IBF process can significantly influences the temperature-sensitive properties of the bombarded materials (stress, thermal hysteresis, expansion, index of refraction, etc.).

Assuming uniform beam incidence on the solid surface, the total input power is then given by $P = \frac{E}{e}JA$, where E is the acceleration energy, J is the ion current density, A is the bombarded area ($= 2\pi r_s^2$, where r_s is the radius of the ion beam) and e is the elementary electric charge.

Under the condition that energy loss processes remain disregarded, the ion beam power P deposited in the sample during the time of bombardment leads to an increase of the temperature given by

$$\Delta T \cong \frac{P}{V\rho C}\Delta t \approx \frac{P}{2\pi r_s^2 2R_p \rho c}\Delta t, \tag{7.18}$$

where V is the volume in which the energy of the ion beam is deposited (where $V = 2\pi r_s^2 \cdot 2R_p$, approximately, and R_p is the mean projected ion range), ρ is the mass density, and c is specific heat of the applied material. Based on (7.18), approximate information about the increase in temperature at a point on the surface can be obtained. For example, an Ar ion beam with a radius of 8 mm hits the surfaces of Cu and fused silica and transfers a power equivalent of 50 W. After a dwell time of 15 μs, the bombarded regions on the surfaces of Cu and fused silica have warmed about 40° and 25°, respectively. This implies that the sample heating in most IBF processes can be negligible if the dwell time is short and the input power is not too high.

Equation (7.18) also implies that the increase in heat resulting from the increase in dwell time seems to be unlimited. However, other processes ultimately lead to a limited temperature rise and to a steady-state temperature:

(i) radiation loss of bombarded surface to surrounding in front of the workpiece, described by $\alpha \Delta T_1^4 = \alpha\left(T_w^4 - T_s^4\right)$, where T_w is mass of material being irradiated by ion beam, T_s is the temperature of surrounding in front of the workpiece and α is a constant,
(ii) radiation loss of bombarded surface to the holder of the workpiece, described by $\beta \Delta T_2^4 = \alpha\left(T_w^4 - T_s^4\right)$, where T_h is the temperature of the workpiece holder and β is a constant, and
(iii) loss by heat conduction from the bombarded material to the workpiece holder, described by $\kappa \Delta T_2 = \kappa (T_w - T_h)$ where κ is the heat conduction constant.

According to a detailed analysis of the temperature increase under ion bombardment, taking into account the mentioned processes that can influence the temperature balance, Parry [47] was able to derive the following relationship

$$\Delta T = \frac{\Delta t}{V\rho C}\left(P - \alpha \Delta T_1^4 - \beta \Delta T_2^4 - \kappa \Delta T_2\right) \tag{7.19}$$

with

$$\alpha = \sigma_s\, A\, \varepsilon_w\, \varepsilon_s/[\varepsilon_s + \varepsilon_w(1-\varepsilon_s] \quad \text{and} \quad \beta = \sigma_s\, A\, \varepsilon_w\, \varepsilon_h/[\varepsilon_h + \varepsilon_w(1-\varepsilon_h] \tag{7.20}$$

where A is the bombarded area, σ_s is the Stephan-Boltzmann constant (= 5.67 $\cdot 10^{-12}$ $Wcm^{-2}K^{-4}$), and ε_w, ε_s, and ε_h are emissivities of workpiece, surrounding and holder, respectively.

It remains unconsidered that atoms with the average energy $\langle E_{sp}\rangle$ removed during the sputter process reduce the introduced power (the mean energy of sputtered atoms is given in Appendix J). This contribution can be determined by the number of sputtered atoms given by the volume removed.

During the IPF process, the heat source (ion beam) is guided over the surface along a precisely defined path (see Fig. 7.4). The ion beam dwells for a time interval τ (dwell time) at each point of the surface (x_i, y_j). This result in an increase in temperature, ΔT, at the surface. During the dwell time at the next point, e.g., (x_{i+1}, y_j), the ion beam has heated this point on the surface to a temperature, while the temperature at the previously heated point (x_i, y_j) drops. Thus, in general, numerous dwell points on the surface are involved in this heating-cooling process induced by the ion beam (see Fig. 7.4). For example, Li et al. [46] developed a finite element method to calculate the surface temperature of nonlinear optical KDP crystals (KH_2PO_4) during the IBF process.

7.1.4 Ion Beam Figuring Applications

Low-energy ion beam technologies are well introduced for shape correction or figuring of surfaces to achieve nanometer and sub-nanometer depth accuracies over the entire spectrum of spatial surface wavelength down to a micro-roughness level of only micrometer and sub-micrometer lateral feature sizes [48]. Currently, ion beam figuring is well established in high-precision surface and thin film processing for advanced optical and mechanical products. The examples that follow illustrate the application of IBF technology for shape error correction on selected surfaces.

Fused Silica (Fused Quartz) for Optical Components

Glasses and fused silica or fused quartz are indispensable amorphous materials in the optical industry. Due to their unique properties, both materials are used in the manufacture of mirrors and lenses as well as UV and IR transmitting optics. Figure errors at the surface or in the near-surface region limit the field of application. These errors are defined according to their spatial wavelength or spatial frequency, where a distinction is made between three spatial frequency/spatial wavelength intervals to categorize the defects according to their optical influence (Table 7.1). For example, HSFR causes high-angle scattering and a loss mechanism in the EUV spectral range, while MSFR creates small-angle scattering of light that reduces the image contrast. However, this categorization is arbitrary and is not used uniformly.

Wilson and McNeil [20] and Weiser [49] introduced low-energy ion beam figuring to correct the surface of fused silica. Ar ions generated in Kaufman ion sources with an energy up to 1.5 keV were applied to demonstrate that ion beam figuring is a suitable tool for correcting surfaces in the sub-micrometer range. The ability of IBF to shape fused silica surfaces is illustrated in Fig. 7.6. The IBF surface figure of a plane fused silica plate with a diameter of 100 mm could be significantly improved. The pv-values and the rms roughness values (LSFR) were reduced from about pv = 68 nm and rms = 12 nm to pv = 7.75 nm and rms = 1.55 nm, respectively. On the right side of Fig. 7.6, the successful surface shape error correction of a small convex lens is depicted. A 1.1 mm FWHM Gaussian ion beam is used to reduce the pv-height of 22.6 nm and the rms of 4.5 nm to pv = 6.50 nm and rms = 0.6 nm.

Silicon Substrates for Synchrotron Beam Line Optics

X-ray mirrors are standard components for collimating and focusing at synchrotron beam lines. These grazing incidence mirrors provide an achromatic component to focus the diverging beam radiated by the source into a small spot for sample illumination. X-ray mirrors work at low grazing angles (typically a few milliradians) and

Table 7.1 Classification of the surface topography according to the International Organization of Standardization (ISO), (HSFR—high-spatial frequency roughness, MFSR—mid-spatial frequency roughness, LSFR—low-spatial frequency roughness)

	Mirco-roughness (HSFR)	Waviness (MFSR)	Shape (LSFR)
Spatial wavelength	≤20 μm	20 μm–1 mm	≥1 mm

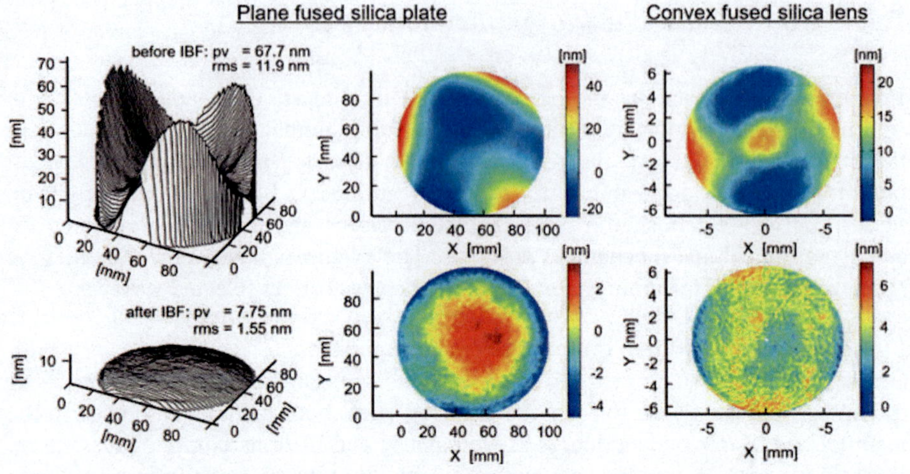

Fig. 7.6 Left: 3D surface topography of a plane fused silica plate with a diameter of 100 mm before (above) and after IBF (below). Middle: Topography (false color representation) of the plane fused silica plate measured by interferometry before (above) and after IBF (below). Figures are adapted from [50]. Right: Interferometer measurement (false color representation) of the surface error topography of a convex fused silica lens (diameter is 13.5 mm, radius of curvature is −10.44 mm). Figures are adapted from [51]

must be large enough (typically a few decimeter) to collect a sizable fraction of the entire beam. Accordingly, this application requires mirrors with a nanometer to sub-nanometer rms figure height error for surface error wavelengths ranging from a few millimeters to 1 m for hard X-ray optics. To achieve this performance, the surface of different Si beam line optics has been corrected by IBF. Two examples are shown in Fig. 7.7. For stable and reliable long-term operation with a rotationally symmetrical

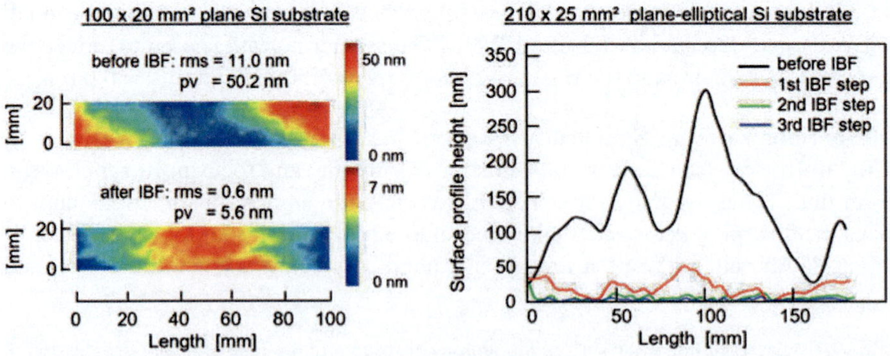

Fig. 7.7 Left: Interferometer surface topography measurements (false color representation) of Si substrate for synchrotron beam line application before (at the top) and after four IBF cycles. Right: Surface height error profiles along the central line of a plane-elliptical single crystalline Si substrate for synchrotron beam line application before and after three IBF cycles [52]

Gaussian beam shape. A 13.56 MHz RF ion source was used, which operates with an 8-mm FWHM beam. This beam is reduced by diaphragms of different sizes to 2 mm, 1 mm and 0.5 mm FWHM, respectively. IBF is performed using controlled scanning lines across the surface with a dwell time distribution calculated by a deconvolution procedure proportional to the desired material removal depth distribution. The initial surface shape error of plane Si substrate (Fig. 7.7, left) is characterized by a pv-value of 50.2 nm and a rms roughness of 11 nm. The surface was machined several times using IBF technology (four IBF cycles), twice with a 4 mm FWHM beam and then twice again with a 2 mm FWHM beam. According to interferometer measurements, a final height error of pv = 5.6 nm and rms = 0.6 nm was achieved [50]. The second example (Fig. 7.7, right) displays the surface height error profiles along the central line of a plane-elliptical Si X-ray beam optics before and after three IBF cycles to correct the surface figure error [52]. The initial shape error of 47.3 nm rms and 191.1 nm pv could be reduced to 1.4 nm and 8.8 nm, respectively.

Ultra-low Thermal Expansion Materials for EUV and DUV Lithography
The dimensional stability of mechanical components for optical applications can be realized by glass-ceramics with a near-zero thermal expansion coefficient, such as Zerodur or ULE. These materials are characterized by excellent thermal sensitivity, extreme hardness, and brittleness. The manufacturing of ultra-precise surfaces of these materials can be optimally performed through the use of non-contact IBF as a finishing step.

Allen et al. [53] reported on surface error figuring correction of a 60-degree concave off-axis outer petal section of a mirror fabricated from ULE material. The inside radius was 68 cm and the outside radius 131 cm. The Ar ion beam figuring was carried out in three runs with three different beam removal functions between 15 and 5 cm and beam powers between 126 and 6 W. The total correction time was 41 h. The three-dimensional contour plot before IBF is shown in Fig. 7.8 (left). The initial surface error before the first IBF cycle was rms $= 0.62 \cdot \lambda$ and pv $= 5.04 \cdot \lambda$ ($\lambda = 632.8$ nm). After the final (fourth) figuring cycle, the rms roughness was rms $= 0.015 \cdot \lambda$ and pv $= 0.301 \cdot \lambda$. An improvement of the surface figure was consistently achieved at each cycle and convergence ratios on average greater than a factor of 2.5 were achieved. In comparison, conventional methods generally yield convergence values < 1.3 per correction cycle [53].

The second example in this subchapter demonstrates the manufacturing of a rectangular concave spherical Zerodur mirror with extremely high surface accuracy ($< \lambda/600$ rms). An efficient and precise correction of low and mid-spatial frequency surface errors was based on the scanning of the ion source with a variable velocity along a defined path, where the ion beam impinges the surface under normal ion incidence. In Sect. 7.1.2.1, it was reported that findings showed that ion beam figuring of curved surfaces must be performed using a 5-axis motion system to fulfill all conditions of the IBF technology.

Figure 7.9 illustrates the surface of an 80 cm $\times$ 150 cm rectangular Zerodur mirror for the next generation of lithography, extreme ultraviolet lithography, before and after ion polishing. A surface of this EUV blank possesses a concave curvature to

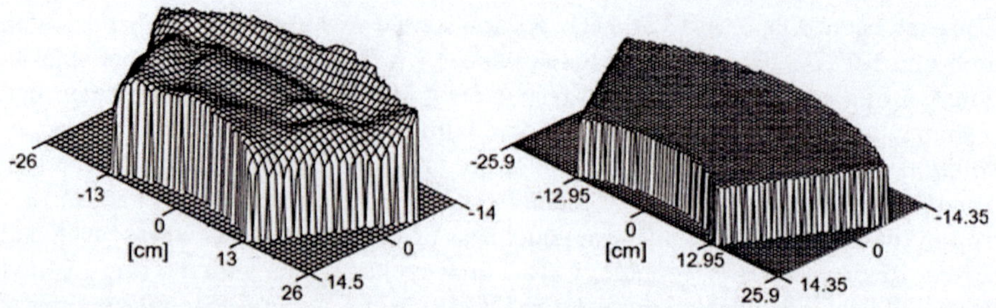

Fig. 7.8 Three-dimensional contour plots of an off-axis petal before the first (left) and after the third ion beam figuring cycle (right). Figure was adapted from [53]

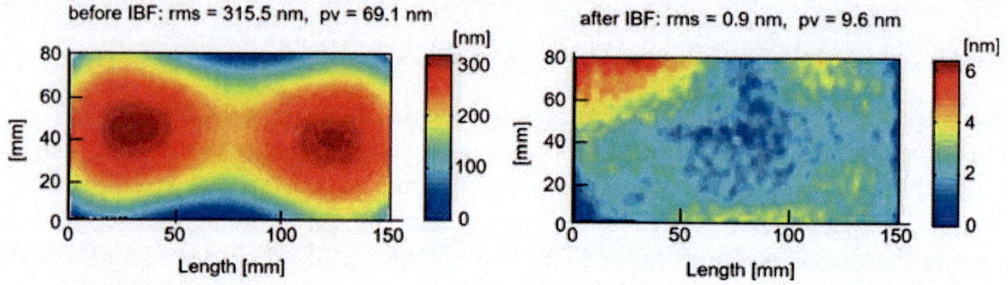

Fig. 7.9 Interferometric measurement of the roughness of a spherical Zerodur rectangular mirror with extreme accuracy in shape (< $\lambda/600$ rms) [11, 54]

serve as X-ray optical component. The rms roughness of the shape could be reduced from 69.1 nm (pv = 315.5 nm) to 900 pm (pv = 6.9 nm) after IBF [11, 54].

In summary, it can be ascertained that IBF offers several advantages compared to conventional abrasive figuring methods in terms of achieving extremely precisely figured surfaces over the entire spatial wavelength range as the final step in shaping surfaces. These are, in detail:

- the high nanometer or sub-nanometer precision of large- and mid-spatial frequency roughness surface figure errors, because the IBF procedures are accurately controlled,
- IBF is a non-contact method, i.e. problems with abrasive techniques, such as edge roll-off effects, damages in the near-surface region caused by the tool, etc., are irrelevant,
- long-term process stability, because the process of the ion–solid interaction can be precisely predicted,
- no restrictions with regard to material class (glasses, metals, semiconductors, ceramics), the state of the materials (amorphous, polycrystalline, or mono-crystalline), and the shape (plane, spherical, aspherical, free-shaped surfaces),
- a simple material removal function (Gaussian distribution of the ion current), which is stable during processing,
- the IBF process quickly converges to the desired topography.

These advantages are, however, accompanied by drawbacks. IBF is a vacuum technique, the removal rate is small, and the investment is large. Additionally, the maximum velocity and acceleration/deceleration of the mechanical multi-axis motion system is limited (mass inertia of the ion source or/and the workpiece) and thermal expansion of large workpieces due to the ion beam power can be a problem.

Based on recent developments (see, e.g., [54]) in the field of surface processing using pulsed ion beam irradiation and the associated pulse width modulation technique, the disadvantage of acceleration limitation due to the mass inertia of the ion source could be partially compensated. This leads to a sometimes major reduction in the machining time (at the absolute minimum of the topology, no more material removal takes place) and thus to increased machining accuracy (the lower volume removed leads to a lower re-deposition, which is essentially unavoidable in the ion sputtering process).

7.2 Ion Beam Smoothing

In contrast to the ion beam figuring technique, ion beam-induced smoothing (IBS) is focused on the processing of features at the micrometer and nanometer scale (spatial wavelength < some micrometer and height amplitudes in order of nanometers) to produce ultra-smooth surfaces. The creation of differently shaped workpieces with surfaces characterized by atomistic precision is absolutely essential for modern applications in optics, lithography, and microelectronics.

Following the discovery by Meinel et al. [13] and Schroeder et al. [14] that high-energy ions affect the shape of surfaces (IBF) (see Sect. 7.1.1), initial studies by Wilson [55], Tarasevich [56] and Perveyev et al. [57] in the 1970s examined the application of IBS for glasses and fused silica. Later, following the development of low-energy broad-beam ion sources, ion energies lower than 10 keV were increasingly applied to smooth surfaces.

For example, Fig. 7.10 demonstrates the potential of ion beam-induced smoothing to produce ultra-smooth surfaces on a silicon substrate used for synchrotron optics [58]. In this figure, the power spectral density (PSD) curves of the Si surface before and after 500 eV Ar ion beam smoothing are compared (the associated dependence of the roughness on the smoothing time is shown in Fig. 7.12). A decrease in the spatial wavelength surface features can be observed for all spatial frequencies. In the steady-state (after 180 min), the PSD graph is proportional to q^{-2} for small frequencies and proportional to q^{-4} for higher frequencies, where $f = q/2\pi$ and q is the wavenumber. The q^{-2} and q^{-4} behavior can be interpreted as ballistic mass drift and viscous flow mechanism induced by ion bombardment, respectively (for details, see Sect. 7.2.1). The transition from q^{-2} to q^{-4} behavior after the smoothing process was observed at spatial wavelengths of about 30 nm. Among different alternative ion beam smoothing methods, direct ion beam smoothing (Sect. 7.2.2), ion beam smoothing with planarization layer (Sect. 7.2.3), and glancing angle ion beam smoothing (Sect. 7.2.4) can be distinguished.

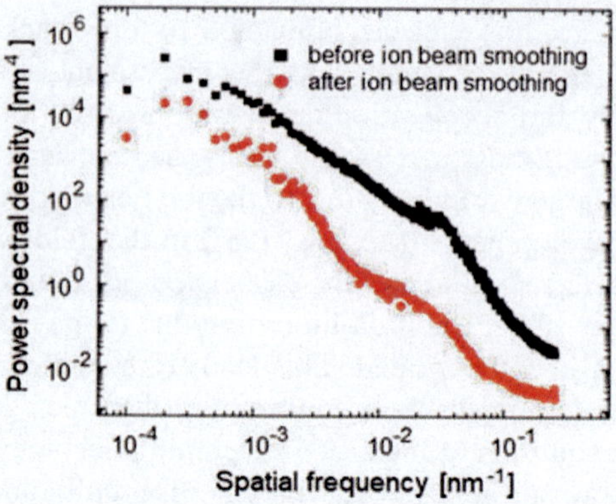

Fig. 7.10 The PSD graphs of Si before ion beam smoothing (i.e. after magnetorheological smoothing) and after ion beam smoothing (processing conditions: 500 eV Ar ion bombardment ion incident angle was 45° and the ion current density was 300 μA/cm^2). Figure is taken from [58]

7.2.1 Atomistic Processes of Ion Beam Smoothing

It should be noted that within the framework of ion beam figuring, the evolution of the topography at large length scales can be exclusively described by the variation of the sputtering yield Y(θ) in dependence on the local incident angle θ (for details, see Sect. 5.6). However, in smoothing processes focused on smaller length scales, down to the micrometer and nanometer scale, atomistic processes become meaningful.

As previously stated (Chap. 6), the sputtering due to ion impact results in sputtering and roughening of surfaces, where the sputtering yield is dependent on the surface gradient. The roughening can be compensated for through relaxation processes, such as surface diffusion, viscous flow, and/or surface atomic drift processes. Consequently, a steady-state value of the roughness/smoothing can be achieved.

Based on the theory of crystal dissolution, Barber et al. [59], Smith et al. [60] and Carter et al. [61] derived geometrical methods including numerical simulation to describe the evolution of topography under ion bombardment, where the erosion rate normal to the surface is dependent on the surface gradient (for details see Sect. 6.4.1). Later, Bradley and Harper [62] proposed a stochastic partial differential equation to describe the evolution of the surface height, where the roughness is slight, i.e. for small surface gradients (details see Sect. 8.2.1.1). They assumed a deterministic system, i.e., ion beam without divergence and sputtering continuous throughout space and time. The equation is stochastic because a random noise term must be added due to the statistical arrival of the ions. This equation has been studied by numerous researchers (e.g., [63–66]) for different assumptions with regard to the nature of relaxation mechanisms.

For one-dimensional surfaces, it is assumed that an ion hits the surface at an angle θ. The surface is then sputtered, where the sputtering yield, Y, is a function of (i) the angle between local surface normal and the ion incident direction and (ii) the local curvature of the bombarded surface. Where ion sputtering is exclusively responsible for the change of topography [62], the time-dependent changing of the height can be expressed by [c.f. (8.4)]

$$\frac{\partial h}{\partial t} = -\frac{J}{N}Y(\theta) + \frac{J}{N}\frac{\partial}{\partial \theta}[Y(\theta)\cos\theta]\frac{\partial h}{\partial x} - \frac{Ja}{N}Y(\theta)\Gamma_x(\theta)\frac{\partial^2 h}{\partial x^2}, \tag{7.21}$$

where J is the ion flux perpendicular to the surface, N is the atomic number density, a is the mean depth of the energy deposition, and Γ_x is a function of the incident angle θ (see Appendix O). If the distribution of the energy damage function is assumed to be approximately spherical (c.f. Sect. 4.6), then $\Gamma_x(\theta) = \sin^2 -(\cos^2\theta/2)(1+\sin^2\theta)$.

In Chap. 8, it will be demonstrated that (7.21) can be extended when diverse relaxation processes on the surface under low-energy ion bombardment are involved. The (7.21) then becomes

$$\begin{aligned}\frac{\partial h}{\partial t} = &-\frac{J}{N}Y(\theta) + \frac{J}{N}\frac{\partial}{\partial \theta}[Y(\theta)\cos\theta]\frac{\partial h}{\partial x} - \frac{Ja}{N}Y(\theta)\Gamma_x(\theta)\frac{\partial^2 h}{\partial x^2}\\ &-\left[K\frac{\partial^4 h}{\partial x^4} + B\frac{\partial^4 h}{\partial x^4} + D_{xx}\frac{\partial^4 h}{\partial x^4} + N_x\frac{\partial^2 h}{\partial x^2}\right],\end{aligned} \tag{7.22}$$

where viscous flow, thermally activated and effective ion-induced diffusion, and ballistic drift parallel to the surface are taken into consideration:

$K\frac{\partial^4 h}{\partial x^4}$ → thermally activated surface diffusion (see Sect. 6.5.4)

$B\frac{\partial^4 h}{\partial x^4}$ → viscous flow (see Sect. 6.5.6)

$D_{xx}\frac{\partial^4 h}{\partial x^4}$ → effective ion-induced diffusion (see Sect. 8.2.1.2)

$N_x\frac{\partial^2 h}{\partial x^2}$ → ballistic mass redistribution (see Sect. 8.2.1.3)

It was demonstrated in Chap. 6 that this type of stochastic equation is capable of yielding numerical results for scaling exponents. However, the underlying physical processes that result in the roughening or smoothing of surfaces after ion beam erosion remain unidentified.

Herring [67] was the first to identify that the four different time-dependent sintering processes obey scaling laws. Later, Tong and Williams [68] applied this scaling approach in analyzing the kinetic mechanisms of the evolution of surfaces by a stochastic rate equation in Fourier space (see Appendix M) expressed by

$$\frac{\partial \tilde{h}(\boldsymbol{q},t)}{\partial t} = -\tilde{h}(|\boldsymbol{q}|,t)C(q) + \eta(\boldsymbol{q},t) \tag{7.23}$$

where $\tilde{h}$ is the Fourier transform of h, **q** is the wave vector ($q = 2\pi/\lambda = 2\pi f$) and η is the spatially and temporally uncorrelated Gaussian noise. Assuming four different roughening and smoothing processes effect the surface evolution, then the function C(q) is given by

$$C(q) = \sum_{n=1}^{4} a_n q^n, \tag{7.24}$$

where $a_n \geq 0$ are constants (in units of [length^n/time]), which characterizes the mass transport mechanism on the surface. The power n of the polynomial q^n in (7.24) represents different evolution mechanisms (see Appendix M).

According to Herring [67], each C(q)-term in Eq. (7.24) is correlated to a specific roughening or smoothing process. Equation (7.24) can be solved by applying the radially averaged power spectral density function, PSD (see Appendix L). For initially flat surfaces the solution is given by

$$PSD(q,t) = \Omega(q)\frac{1 - \exp(-2tC(q))}{C(q)}, \tag{7.25}$$

where Ω is the fluctuation or noise strength ($\propto$ ion fluence). For initially rough surfaces the solution is given by

$$PSD(q,t) = PSD(q,0) \cdot \exp(-2tC(q)) + \Omega(q)\frac{1 - \exp(-2tC(q))}{C(q)}. \tag{7.26}$$

It is evident that for frequencies q, the surface is smoothed when $C(q) < 0$. For short times of ion bombardment, the PSD is determined by the white noise to $PSD(q, t \rightarrow 0) = \Omega(q)t$. For longer times, the steady-state solution is

$$PSD(q, t \rightarrow \infty) = -\Omega(q)/C(q), \tag{7.27}$$

i.e. this equation indicates that the PSD of the topography is strongly related to the power n in (7.24). An important consequence of this relationship is that the minimum roughness is determined on the one hand by the sputter noise and, on the other, by the relaxation processes involved (see below).

By integration of (7.25) or (7.26), the interface width (or rms roughness) can be obtained to [69]

$$w^2 = \frac{1}{2\pi}\int_0^{\infty} PSD(q,t)q\,dq, \tag{7.28}$$

where it is assumed that the roughness is isotopic in the surface plane.

In accordance with the theoretical considerations of Herring [67] and Mullins [70], the exponent n of q is an integer between 1 and 4 for different surface relaxation mechanisms. Consequently, the slope of PSD(q,t) versus q in a log-log diagram provides the dominant mechanisms of surface relaxation [70].

In particular, Herring [67] and Mullins [70] have identified four surface relaxation processes, where two processes, viscous flow and surface diffusion, are significant for relaxation following low-energy ion bombardment of surfaces. In contrast, the surface relaxation processes of bulk diffusion and evaporation/condensation have no relevance for relaxation following ion bombardment.

7.2.2 *Relevant Smoothing Mechanisms*

Ion Beam Enhanced Viscous Flow

In Sect. 6.5.3, the viscous flow is defined as a cooperative rearrangement of atoms in the near-surface region under the influence of the viscosity (resistivity against topographic changes caused by driving force, e.g., ion bombardment). According to the Orchard limit [72], a distinction must be made between two limiting cases. For thin viscous films on the surface of a substrate, the rate of height change in the Fourier space, $\partial\tilde{h}/\partial t$, is proportional to the wavenumber q^{-4} [see (6.58)] under the condition that $a_h q \ll 1$, where a_h is the thickness of the viscous film. Usually, this condition is only fulfilled in the low-energy region because the ion range is small. For high-energy ion bombardment, i.e. larger viscous film thicknesses ($a_h q > 1$), the height change is proportional to q^{-1} [71].

Thermally Activated Surface Diffusion

As is generally known, the chemical potential of the species on a surface is a function of the local curvature of the surface, where regions of negative curvature (e.g., crests, hills) have a higher chemical potential than do those with a positive curvature (e.g., troughs, valleys). This potential drives the mass transport from crests to valleys. Consequently, the surface tends to flatten. In Sect. 6.5.4, it is shown that the rate of height change is proportional to $\nabla^4 h$ [e.g., see (6.52)].

Complementary to this, two further relaxation processes were suggested to explain the ion beam-induced surface phenomena:

Effective Ion-Induced Diffusion

Makeev and Barabasi [73] introduced a new smoothing mechanism: effective ion-induced diffusion (see Sect. 8.2.1.2). This newly created diffusion mechanism does not describe a true mass transport, but is a consequence of the higher-order derivatives in the expansion of the erosion rate [c.f. (7.4)]. This expansion results in the change in average height with time given by

$$\frac{\partial h}{\partial t} = -\left(D_{xx}\frac{\partial^4 h}{\partial x^4} + D_{yy}\frac{\partial^4 h}{\partial y^4}\right), \tag{7.29}$$

where the constants D_{xx} and D_{yy} are given in Appendix O. According to these authors, the process of effective ion-induced diffusion is included in the description of the topography change under ion bombardment by expanding (7.21) with the relationship [see also (8.16)]. Analogous to the thermally activated surface diffusion is the process $\propto q^{-4}$. In detail, the perpendicular low-energy ion bombardment enhances the effective ion-induced surface diffusion, thus, for large angles of ion incidence this diffusion process can be suppressed. According to Makeev and Barabasi [73], (7.29) provides a contribution to the reorganization of the surface, corresponding to a diffusion-like mechanism but without real atomic diffusion.

Ballistic Mass Redistribution

Carter and Vishnyakov [74] developed a mass redistribution model in which collision cascade mass transport and redistribution are considered to be counteracted by Bradley-Harper instability (for details, see Chap. 8). They have shown that the ion-atom and atom-atom collision processes in ion bombarded near-surface regions of solids provide a ballistic atomic drift parallel to the surface (in addition to the isotropic ballistic diffusivity). The gradient of this particle flux can compensate for the curvature-dependent sputtering process and leads to net smoothing, particularly for normal and near-normal incidences. The change in average height with time is then proportional to $\partial^2 h/\partial x^2$ (c.f. 8.2.1.3). The smoothing rate is thus proportional to the second-order spatial derivative (i.e. $\propto q^{-2}$).

Table 7.2 provides a summary of the relaxation processes discussed, along with their wavenumber dependence.

In general, it can be expected that some of the relaxation processes are simultaneously active under low-energy ion bombardment of surfaces. Figure 7.11 illustrates this situation. Using normal incidence Ar ion bombardment (ion energy between 400 eV and 1 keV and the ion current between 10 and 30 mA), Liao et al. [75] smoothed three different materials: fused silica (amorphous), quartz (crystalline) and Zerodur (glass ceramic). The roughness of both the fused silica and quartz could be reduced to 0.1 nm rms, while an increase was observed in the roughness of Zerodur surfaces. It should be noted that the rms roughness of Zerodur could be reduced to a level of 0.15 nm using the so-called 'ion beam material-adding technology' proposed by Liao et al. [76]. The power spectral density functions of the materials studied after ion beam smoothing are proportional to q^{-2} at low frequencies and q^{-4} at high

Table 7.2 Surface relaxation processes (a_h is the thickness of the viscous film)

Relaxation mechanism	Wavenumber dependence (for steady-state, i.e. $t \rightarrow \infty$)	References
Viscous flow	$\propto q^{-4}$ for $a_h q \ll 1$ $\propto q^{-1}$ for $a_h q > 1$	[71]
Thermally activated surface diffusion	$\propto q^{-4}$	[69]
Effective ion-induced diffusion	$\propto q^{-4}$	[72]
Ballistic mass redistribution (ballistic drift)	$\propto q^{-2}$	[73]

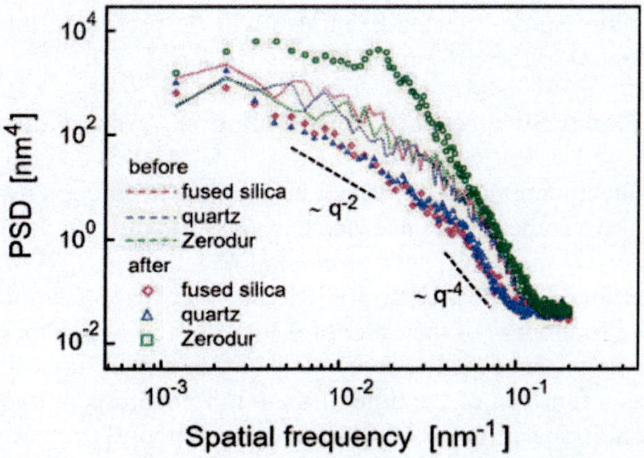

Fig. 7.11 Comparison of the PSD functions of fused silica, quartz and Zerodur before and after ion beam-induced smoothing. Peak at the spatial frequency 0.018 nm^{-1} indicates the formation of nanopatterns on Zerodur surface. (figure adapted from [75] and modified)

frequencies, where the slopes −2 and −4 correspond to the ballistic smoothing and the smoothing caused by ion-induced viscous flow (Fig. 7.11). The transition from the ballistic drift dominant smoothing to the viscous for smaller spatial wavelengths was determined to be 30 nm. This is in agreement with the results of simulations by Vauth and Mayr [77] for 1 keV ion bombardment of the amorphous materials Si and CuTi and confirms the results for 500 eV Ar ion bombardment of silicon [78].

It can be reasoned that ion beam-induced smoothing is dominated by the mechanisms of the ion beam-induced ballistic drift at larger spatial wavelengths (larger surface features) and the viscous surface flow at smaller spatial wavelengths (smaller surface features) while the contributions to smoothing caused by surface diffusion processes are insignificant.

7.2.3 *Direct Ion Beam Smoothing of Material Surfaces*

Experiments aimed at smoothing surfaces are performed in conventional high- or ultrahigh vacuum facilities (base pressure between 10^{-3} and 10^{-8} Pa). Usually, broad-beam Kaufman-, RF- or ECR-type ion sources are applied, which are capable of producing rare gas ions with energies between 100 eV and some keV and ion fluxes up to 50 mA/cm^2. Samples are mounted on a heated or cooled sample holder. In general, the sample stage can be rotated and tilted with respect to the surface normal. On this basis, smoothing experiments can be performed through variation of the ion energy, ion current density, angle of ion incidence, and temperature during ion bombardment. The ion beam smoothing of larger substrates requires equipment comparable to that for the IBF technique.

In the following, some selected examples will be presented that illustrate the tremendous potential of the IBS procedure.

Smoothing of Planar Surface/Roughness Evolution Versus Time of Ion Bombardment

The ion beam direct smoothing of planar surfaces has been demonstrated from the atomistic scale up to some tens of microns in spatial wavelength for numerous technologically important materials, such as crystalline silicon [58, 78, 79], amorphous silicon [80], Zerodur [75, 78], ULE [48, 81, 82] and fused silica [75, 83–85], to achieve a surface roughness of the order of 0.1–0.2 nm rms roughness.

Of particular importance for the application of this technology is the evolution of the roughness as a function of the time of the ion bombardment under the adopted experimental conditions. Figure 7.12 Illustrates the temporal evolution of the roughness on Si surfaces during Ar ion beam smoothing (ion energy is 500 eV, ion incidence angle 45°, ion current density 0.3 mA/cm^2, and simultaneous sample rotation during ion bombardment). Additionally, two AFM images for the initial surface (before smoothing, $t = 0$ min) and after smoothing ($t = 180$ min) are shown [57]. The rms roughness was reduced from 2.25 nm to < 0.2 nm. The experimental results of the temporal evolution of the surface roughness (black dots in Fig. 7.11) can be explained by assumption of an ion beam-induced ballistic drift mechanism proposed by Carter and Vishnyakov [74].

The smoothing of plane surfaces of fused silica under near-normal ion bombardment is demonstrated in Fig. 7.13 [48]. The figure shows AFM images and the corresponding circularly averaged power spectral (PSD) curves before and after direct ion beam smoothing. Initially, the roughness of the polished fused silica wafer was 0.31 nm rms. Under optimized condition (ion energy is 800 eV, ion incidence angle is 20° and simultaneous sample rotation during ion bombardment) the roughness could be reduced to 0.08 nm rms after Ar ion bombardment of 20 min. From Fig. 7.13, it is also evident, that smoothing occurs over all spatial frequencies measured by AFM.

Fig. 7.12 Time evolution of rms surface roughness of Si surface during ion-beam smoothing. The two AFM images were taken from the initial surface and after smoothing. The black dots are experimental results and the red curve was obtained by assumption of a ballistic drift mechanism [58]

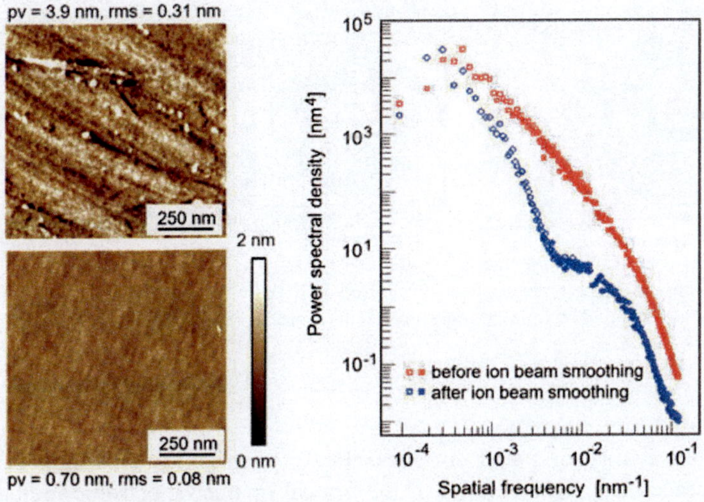

Fig. 7.13 AFM images and the corresponding PSD curves of polished fused silica surfaces before and after Ar ion beam-induced smoothing. The ion energy was 800 eV, the ion current density was 400 μA/cm^2, the ion incidence angle was 20° and the time of bombardment was 20 min [48]

Smoothing of Spheres

The potential of ion beam-induced smoothing technology to polish curved surfaces was demonstrated by the example of the polishing of Si spheres. Isotope-pure Si spheres were produced with the aim of redefining of the Avogadro constant by 'counting' the atoms in 1 kg perfect single crystal Si spheres (for details, see [86]). Silicon was used because it is one of the best known materials and can be grown as highly enriched ^{28}Si single crystals (99.99 %). In order to achieve a relative uncertainty at the determination of the Avogadro constant $< 2 \times 10^{-8}$, the deviation from the ideal roundness of the Si sphere should be < 10 nm peak-to-valley (pv).

Isotopically enriched ^{28}Si spheres with a diameter of 93.7 mm were mounted in a 1 m^3 high-vacuum IBF machine equipped with three linear stages and three rotary axes for computer-controlled coordinated motions, a 40 mm diameter RF ion source with a multi-grid system for ion beam extraction and ring-shaped hot filament neutralizer for the beam neutralization [87].

Figure 7.14, left, shows the topography of a chemo-mechanically pre-finished Si sphere in sinusoidal projection (pseudo-cylindrical equal-area map projection) before IBF. The surface is characterized by a roughness error of 7.95 nm rms and a peak-to-valley value of 43.29 nm. The dwell time calculation, Fig. 7.14, middle, indicates a similar picture, unsurprisingly. For IBF a tool path was chosen that follows an Archimedean spiral with constant distance between successive turns. Figure 7.14, right, shows the computer-simulated result with a residual error topology after an Ar ion smoothing time of 397 min with 0.06 nm rms and a peak-to-valley value of 3.87 nm [87].

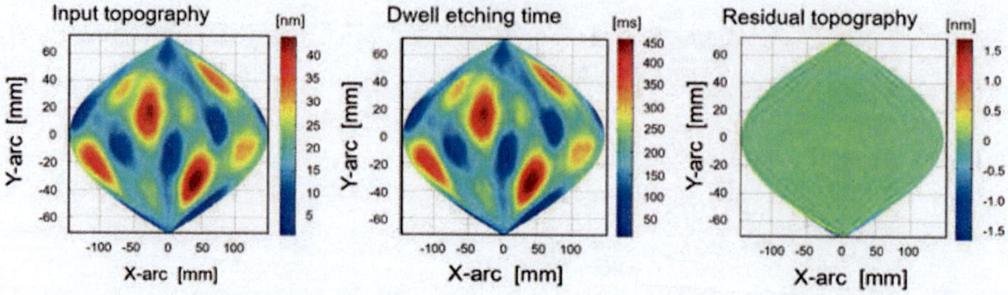

Fig. 7.14 Surface figure error topography before IBF (left), calculated dwell time distribution (middle) and determined residual topography (right) of a Si sphere in sinusoidal projection (figures taken from [87])

Smoothing of Metal Surfaces

Because the quality of metal surfaces after conventional smoothing procedures does not satisfy the requirements of the optical or microelectronic industries, ion beam-induced smoothing of polycrystalline metal surfaces is applied. For example, polycrystalline copper is widely employed as a wiring material for large integrated circuits and microelectrical-mechanical systems. In general, wet etching processes, such as chemical-mechanical polishing, are used as etching methods for Cu wiring. Following this etching process, the surface is characterized by protuberances of approximately 200 nm in width and 30 nm in height. Hino et al. [88] and Kobayashi et al. [89] have used IBS to smooth the rough surface of polycrystalline copper using Ar ion bombardment under off-normal ion incident conditions and at different temperatures. It was observed (Fig. 7.15) that with increase of the ion fluence, the temperature (up to 300 °C), and the angle of incidence (up to 70°), the surface became significantly smoother (height of protuberances was only 10–15 nm [88]). The authors suggest that an increase in the substrate temperature enhances the surface transport mechanisms.

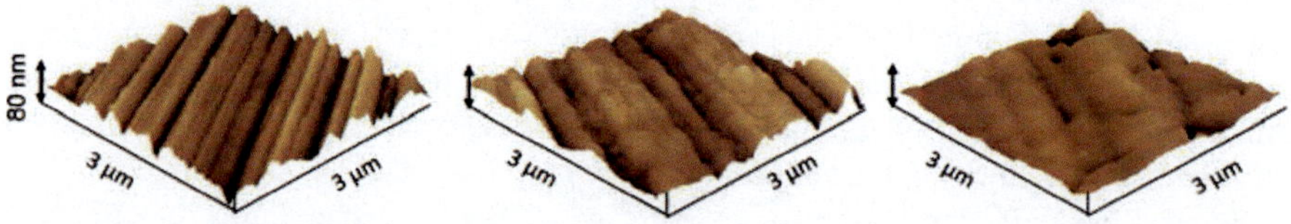

Fig. 7.15 Topography measured using AFM of polycrystalline Cu before (left) after irradiation with 1 keV Ar ions and fluences of 0.6×10^{16} ions/cm^2 (middle) and 5.3×10^{16} ions/cm^2 (right) at 573 K. The angle of ion incidence was 70°. The figures are taken from [89]

7.2.4 Ion Beam Smoothing with Planarization Layer

More than four decades ago, Johnson et al. [90, 91] proposed an innovative polishing technique, known as planarization. As shown schematically in Fig. 7.16, left, a rough surface is covered with a planarization layer (deposited film), which acts as sacrificial film under ion bombardment. Typically, spin coating is used to deposit a thin planar film onto the surface, which is then annealed at higher temperatures. In general, photoresists are an ideal choice, because this deposition material creates smooth films and the thickness of the films can be controlled by the spin speed. The thickness of this film should be larger than the roughness of the deposited surface, i.e. ranging between about 100 nm and up to any micrometer. Johnson et al. [90, 91] recommend a thickness equivalent to 2–3 times the maximum amplitude of the surface roughness. The erosion of the rough material with the sacrificial layer using low-energy noble ion bombardment is carried out at an appropriate off-normal ion incidence angle. According to Johnson et al. [91], this angle is given for the case where the erosion rates (see Sect. 5.1) of the deposited film and the material are identical. For example, Fig. 7.16, right, shows the erosion rate for fused silica (SiO_2) and the photoresist AZ-1350 as a function of the ion incidence angle. In this case, a planarization angle of 59° could be determined [91]. It can then be expected that the ion bombardment causes a uniform removal of the deposited layer and the underlying material, leading to a smooth surface. Understandably, the planarization angle is strongly dependent on the material-film combination used, the ion species, and the ion energy. For example, with a 1.3 μm thick film of the photoresist AZ-1350, Yamauchi et al. [92] found a planarization angle > 60° for 600 eV Ar bombardment of Si surfaces.

Later, Stognij and Novitskii [93] proposed an alternative planarization technique, characterized by a sacrificial film that is continuously deposited during the

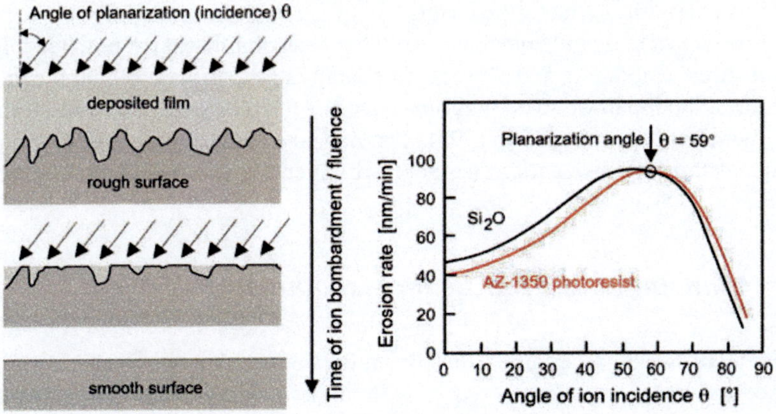

Fig. 7.16 Left: Schematic of the planarization technique. Right: Angular dependence of the erosion rate of fused silica (SiO_2) and the photoresist AZ-1350 under 2 keV Ar ion bombardment (figure on the right side is adapted from [92] and modified)

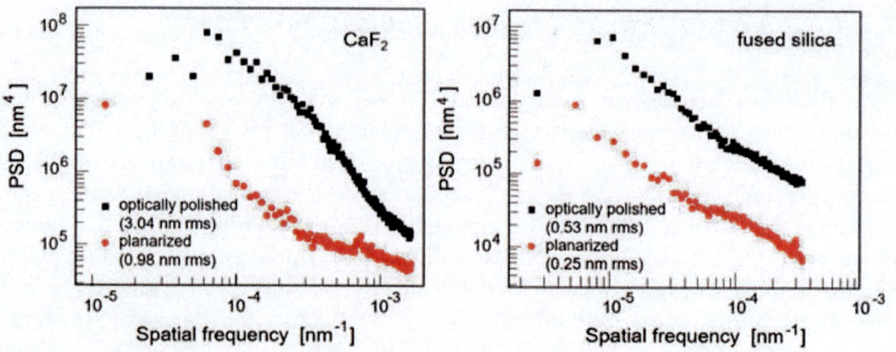

Fig. 7.17 Calculated power spectral density functions (PSD) of CaF_2 and fused silica surfaces before and after ion beam planarization [11]

ion bombardment instead of the deposition of the planarization layer before ion bombardment.

Numerous studies have demonstrated the feasibility of ion beam polishing of different materials through the application of a planarization layer. Substantial progress could be made in reducing roughness, in particular, for the polishing of long wavelength structures (low spatial frequencies). The planarization technique has been successfully applied for the smoothing of diamond [94], diamond-turned NiP [95], or metal surfaces [96, 97]. Figure 7.17 illustrates the application of this technique for polishing CaF_2 and fused silica [11]. The power spectral density distributions derived from AFM measurements indicate that the roughness reduction takes place over all measured spatial frequencies between about 10^{-5} nm^{-1} and about 10^{-3} nm^{-1}. The rms roughness in both materials could be significantly reduced.

Ion beam-induced planarization technique can be assessed to be a final processing step that successfully reduces the roughness of surfaces from the micrometer up to the nanometer scale. It could be demonstrated that a combination of both techniques, ion beam direct smoothing and subsequent planarization, offers sub-nanometer scale roughness reduction down to 0.1 nm rms, covering a large range of spatial frequency surface wavelength structures [52, 98]. The primary drawback of the planarization technique is the extensive experimental effort (layer deposition and ion irradiation).

7.2.5 Glancing Angle Ion Beam Smoothing

In Sect. 5.10, it was ascertained that the reflection coefficient (i) decreases with increased ion energy and (ii) increases with angle of ion incidence. Consequently, low-energy ion bombardment under glancing angle conditions has been successfully applied to produce contamination-free surfaces with no increase in ion beam-induced defects in the near-surface region (see e.g., [99] and Sect. 9.5), where the low damage density is explained by increase in ion reflection and decrease of projected ion range.

It is also known that near-specular ion scattering from the flat surface can be expected when the reflection coefficient is unity. Small features on the surface, such as terraces or steps, then cause a changed angular and energy-loss distribution of scattered ions. The temporal evolution of these distributions was frequently interpreted as a reduction/enhancement the surface roughness (see, e.g., [100]).

Based on these low-energy ion beam studies, a technique for the preparation of extremely smooth surfaces of crystalline solids using glancing angle ion bombardment has been proposed [101]. This technique is predicated on the difference in the interaction of incident ions with smooth and rough crystalline surfaces. Figure 7.18 shows schematically the evolution of the surface topography at glancing angle ion bombardment as a function of the ion fluence. If a grazing incidence ion hits an atomically smooth surface at an angle of φ_S, it can then be expected that this ion will be reflected, because only a small part of the kinetic energy and momentum is transferred to the surface (c.f. Fig. 6.36). In contrast, if the ion impinges on a protuberant structure (islands, steps, kinks, etc., see Fig. 7.18), the incidence angle is larger ($\varphi_r > \varphi_S$). Thus, significant energy and momentum transfer to the substrate occur and lead to sputtering of this protuberant structure. It can be expected that the average sputtering yield dependent on the angle of ion incidence (see Chap. 5 or Fig. 6.27) is no longer constant at glancing ion incidence. With increasing time of

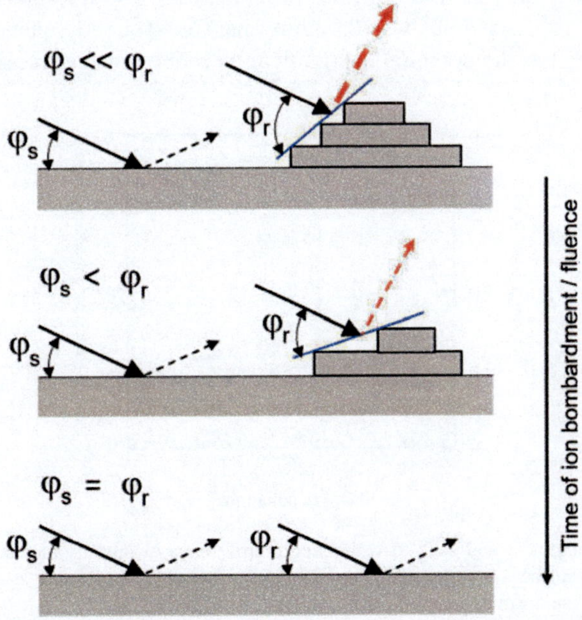

Fig. 7.18 Schematic of glancing angle ion beam smoothing in dependence on the time of ion bombardment, where φ_S is the angle between the smooth surface and the direction of the incident ions and φ_r is the effective angle between the slope of the protuberant structure (blue line) and the direction of the incident ion. The red arrows indicate the flux of the sputtered substrate atoms

ion bombardment (fluence), the effective ion incidence angle, φ_r, decreases to the incidence angle of the smooth surface, ($\varphi_r = \varphi_s$), due to sputtering. Ultimately, a smoother surface can be obtained.

The glancing angle experiments are performed in an ultrahigh vacuum chamber containing Kaufman, RF, or ECR ion sources. The divergence of the exit ion beam (see Sect. 6.5.5) should be as small as possible. The surface is exposed to noble gas ions incident at an angle $\varphi \leq 3\ldots10°$ relative to the surface plane at various temperatures and for various times of ion bombardment. The ion energy ranges between some hundred electron-volts and about 5 keV. Frequently, the vacuum chamber is equipped with surface analysis techniques (RHEED, LEED, STM, quadrupole mass spectrometer, etc.) to control the surface in-situ during ion bombardment.

In the literature, the smoothing of different materials was studied using glancing angle ion bombardment. For example, the evolution of the topography of thin 60 nm Cr films on BK7 glass substrates [101], the smoothing of Ar ion bombarded (001) GaAs [102], (100) diamond [103], and (001) Si surfaces [104] was studied in detail. The following example demonstrates the capability of this technique for smoothing both air-cleaved and mechanically polished (111)-oriented CaF_2 surfaces using 4.5 keV Ar ion bombardment at a glancing angle of 10° to the surface [105, 106]. Figure 7.19 illustrates the dependence of the rms roughness and the peak-to-valley height difference on the ion fluence. The initial roughness of the non-irradiated surface was 0.64 nm rms and the final roughness after ion bombardment with a fluence of 3×10^{17} ions/mm^2 was 0.12 nm rms. Consequently, glancing angle ion beam smoothing has the potential to significantly reduce surface roughness.

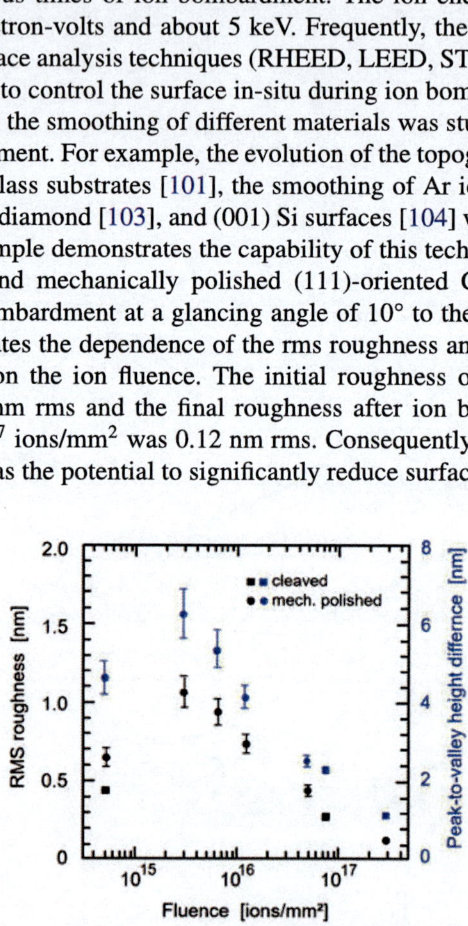

Fig. 7.19 Rms roughness and peak-to-valley height difference of initially mechanically polished and cleaved CaF_2 surfaces in dependence on the Ar ion fluence. The glancing angle was 10° with respect to the surface normal (adapted from [106] and modified)

7.3 List of Symbols

Symbol	Notation
A	Area
C	Specific heat
D	Constant of thermal diffusion
J	Ion current density (flux)
N	Atomic number density
P	Ion beam power
R_p	Projected range of ions
R(x, y)	Volume removal rate (removed volume per unit time)
T	Temperature
T_o	Ambient temperature
U	Beam voltage
V	Volume
Y	Sputtering yield
Z(x, y)	Surface error function
a_h	Thickness of the viscous film
e	Elementary electric charge
j	Ion current
r_s	Radius of the circular ion beam
q	Wavenumber
v	Erosion velocity (rate), removal rate
Ω	Fluctuation or noise strength
ε	Emissivity of sample
η	Spatially and temporally uncorrelated Gaussian noise
θ	Angle of ion incidence
λ	Wavelength
ρ	Mass density
σ	Standard derivation of the Gaussian distribution
σ_s	Stephan-Boltzmann constant
τ	Dwell time

References

1. M.J. Nobes, J.S. Colligon, G. Carter, The equilibrium topography of sputtered amorphous solids. J. Mater Sci. **4**, 730–733 (1969)
2. G. Carter, J.S. Colligon, M.J. Nobes, The equilibrium topography of sputtered amorphous solids II. J. Mater Sci. **6**, 115–117 (1971)
3. G. Carter, The physics and applications of ion beam erosion. J. Phys. D: Appl. Phys. **34**, R1–R22 (2001)
4. F.C. Frank, On the kinematic theory of crystal growth and dissolution processes II. Z. Phys. Chem. **77**, 84–92 (1972)
5. G. Carter, Theory of surface erosion and growth, in *Erosion and Growth of Solids Stimulated by Atom and Ion Beams*, ed. by G. Kiriakidis, G. Carter, J.L. Whitton, (Proceed. NATO ASI Series E 112, Martinus Nijhoff', Dordrecht, 1986) pp. 70–97
6. A. Schindler, G. Boehm, T. Haensel, W. Frank, A. Nickel, B. Rauschenbach, F. Bigl, Precision optical asphere fabrication by plasma jet chemical etching (PJCE) and ion beam figuring. Proc. SPIE **4451**, 242–248 (2001)
7. T. Hänsel, P. Seidel, A. Nickel, A. Schindler, B. Rauschenbach, Deterministic ion beam figuring of surface errors in sub-millimeter spatial wavelength range, in *Proceeding of 6th EUSPAN International Conferences Baden/Wien* (2006)
8. A. Schindler, T. Hänsel, A. Nickel, F. Frost, H.-J. Thomas, H. Neumann, G. Seidenkranz, R. Schwabe, S. Gürtler, S. Görsch, A. Bogatz, B. Rauschenbach, Ion beam figuring (IBF) solutions for high performance optics surface finishing from meter to millimeter spatial wavelength range, in *Proceeding of 3rd International Conferences on Leading Edge Manufacturing in 21st century*, Nagoya (2005)
9. N. Savvides, A. Knittel, Ion beam figuring of optics, CSIRO report CIP2152 (2004).
10. X. Xuhui, L. Shengyi, Ion beam figuring technology, in *Handbook of Manufacturing Engineering and Technology*, ed. by A.Y.C. Nee, (Springer, London 2015), pp. 1343–1390
11. A. Schindler, T. Hänsel, D. Flamm, G. Boehm, F. Frost, R. Fechner, B. Rauschenbach, Ion beam and plasma jet etching for optical component fabrication. Proc SPIE **4440**, 217–227 (2001)
12. R. Castaing, P. Laborie, Aspects partieuliers de l'etude des métaux en coupes mines. Compt. Rend. **238**, 1885–1887 (1954)
13. A.B. Meinel, S. Bashkin, D.A. Loomis, Controlled figuring of optical surfaces by energetic ionic beams. Appl. Optics **4**, 1647–1647 (1965)
14. J.B. Schroeder, S. Bashkin, J.F. Nester, Ionic polishing of optical surfaces. Appl. Optics **5**, 1031–1034 (1966)
15. P.H. Schmidt, E.G. Spencer, E.M. Walters, Ion milling of magnetic oxide platelets for the removal of surface and near-surface imperfections and defects. J. Appl. Phys. **41**, 4740–4742 (1970)
16. B. Schroeder, H.D. Dieselman, J.W. Douglass, Technical feasibility of figuring optical surfaces by ion polishing. Appl. Optics **10**, 295–299 (1970)
17. R.A. House II., J.R. Bettis, A.H. Guenther, Appl. Optics **16**, 1486–1488 (1977)
18. D.T. Hawkins, Ion milling (ion beam etching), 1954–1975: a bibliography. J. Vac. Sci. Technol. **12**, 1389–1397 (1975)
19. H.R. Kaufman, P.D. Reader, G.C. Isaacson, Ion sources for ion machining applications. AIAA J. **15**, 843–847 (1977)
20. S.R. Wilson, J.R. McNeil, Surface figuring using neutral ion beams. Proc. SPIE **966**, 74–81 (1989)
21. L.N. Allen, H.W. Romig, Demonstration of an ion figuring process. Proc. SPIE **1333**, 22–33 (1990)
22. R.L. Seliger, W.P. Fleming, Focused ion beams in microfabrication. J. Appl. Phys. **45**, 1416–1422 (1974)

23. P. Sigmund, Theory of sputtering I. Sputtering yield of amorphous and polycrystalline targets, Phys. Rev. **184**, 383–416 (1969) and Y. Yamamura, An empirical formula for angular dependence of sputtering yields. Rad. Effects **80**, 57–72 (1984)
24. S. Wilson, J. McNeil, Neutral ion beam figuring of large optical surfaces. Curr. Dev. Optical Eng. **II**(818), 320–325 (1987)
25. T.W. Drueding, T.G. Bifano, S.C. Fawcett, Contouring algorithm for ion figuring. Precis. Eng. **17**, 10–21 (1995)
26. T. Wang, L. Huang, H. Kang, H. Choi, D.W. Kim, K. Tayabaly, M. Idir, Transform-based dwell time algorithm for ultra-precision ion beam figuring of synchrotron mirrors. Sci. Rep **10**, 813 (2020)
27. T. Wang, L. Huang, M. Vescovi, D. Kuhne, K. Tayabaly, N. Bouet, M. Idir, Study on an effective one-dimensional ion-beam figuring method. Opt. Express **27**, 15368–15381 (2019)
28. https://zenodo.org/record/3834856
29. S. Wilson, D. Reicher, C. Kranenberg, J. McNeil, P. White, P. Martin, D. McCready, Ion beam milling of fused silica for window fabrication, in *Laser-Induced Damage in Optical Materials*, ed. by H. Bennett, L. Chase, A. Guenther, B. Newnam, M. Soileau, (West Conshohocken), ASTM International **1441**, 82–86 (1991)
30. P.H. van Cittert, Zum Einfluss der Spaltbreite auf die Intensitätsverteilung in Spektrallinien II. Z. Phys. **69**, 298–308 (1931)
31. W.H. Richardson, Bayesian-based iterative method of image restoration, J. Opt. Soc. Am. **62**, 55–59 (1972) and L.B. Lucy, An iterative technique for the rectification of observed distributions. Astronom. J. **79**, 745–754 (1974)
32. R. Gold, An iterative unfolding method for response matrices, (Argonne National Laboratory Report ANL-6984, 1964)
33. Ch. Xu, I, Aissaoui, S. Jacquey, Algebraic analysis of the Van Cittert iterative method of deconvolution with a general relaxation factor. J. Opt. Soc. Am. A **11** (1994) 2804-2808
34. J.F. Wu, Z.W. Lu, H.X. Zhang, T.S. Wang, Dwell time algorithm in ion beam figuring. Appl. Opt. **48**, 3930–3937 (2009)
35. T. Hänsel, A. Nickel, A. Schindler, Ion beam figuring of strongly curved surfaces with a (x, y, z) linear three-axes system, OAS Tech. Digest (Opt. Soc. Am., 2008) paper JWD6
36. Y.S. Ghim, S.-J. Shin, H.-G. Rhee, H.-S. Yang, Y.-M. Lee, Ultra-precision surface polishing using ion beam figuring. Proc. SPIE **8416** (2012)
37. A. Haberl, R. Rascher, Yet one more dwell time algorithm. Proc. SPIE **103026** (2017)
38. C.L. Carnal, C.M. Egert, K.W. Hylton, Advanced matrix-based algorithm for ion-beam milling of optical components. Proc. SPIE **1752**, 54–63 (1992)
39. L. Zhou, Y. Dai, X. Xie, C. Jiao, S. Li, Model and method to determine dwell time in ion beam figuring. Nanotechnol. Precis. Eng. **5**, 107–112 (2007)
40. C.C. Paige, M.A. Saunders, LSQR: An algorithm for sparse linear equations and sparse least squares. ACM Trans. Math Softw **8**, 43–71 (1982)
41. C. Jiao, S. Li, X. Xie, Algorithm for ion beam figuring of low-gradient mirrors. Appl. Opt. **48**, 4090–4096 (2009)
42. J.M. Bernardo, A.F.M. Smith, *Bayesian theory* (Wiley, Chichester, 2000)
43. M. Ghigo, G. Vecchi, S. Basso, O. Citterio, M. Civitani, G. Pareschi, G. Sironi, Ion beam figuring technique used as final step in the manufacturing of the optics for the E-ELT. Mem. Soc. Astro. **86**, 412–415 (2015)
44. P. Gailly, J.P. Collette, L. Renson, J.-P. Tock, Ion beam figuring of small BK7 and zerodur optics: thermal effects. Proc. SPIE **3739** (1999)
45. X. Xie, Y. Hao, L. Zhou, Y. Dai, S. Li, High thermal expansion optical component machined by ion beam figuring. Opt. Eng. **51**, 013401 (2012)
46. F. Li, X. Xie, G. Tie, H. Hu, L. Zhou, Research on temperature field of KDP crystal under ion beam cleaning. Appl. Opt. **56**, 4888–5489 (2016)
47. P.D. Parry, Target heating during ion implantation, J. Vac. Sci. Technol. **13**, 622–629 (1976) and Localized substrate heating during ion implantation, **15**, 111–115 (1978)

48. A. Schindler, T. Hänsel, F. Frost, G. Böhm, W. Frank, A. Nickel, T. Arnold, R. Schwabe, S. Gürtler, S. Görsch, B. Rauschenbach, Modern methods of highly precise figuring and polishing, in *Proceed of 3rd DGG Symposium, Glass Science and Technology* (vol. 78 C, 2005), pp. 111–114
49. M. Weiser, Quantitative investigations of the removal of glass material by low energy ion beams with the use of optical interferometry. Nucl. Instr. Meth. Phys. Res. B **80**(81), 1174–1177 (1993)
50. T. Hänsel, F. Frost, A. Nickel, A. Schindler, Ultra-precision surface finishing by ion beam techniques. Vak. Forsch. Praxis **19**, 24–30 (2007)
51. T. Hänsel, A. Nickel, A. Schindler, H.-J. Thomas, Ion beam figuring surface finishing of x-ray and synchrotron beam line optics using stitching interferometry for the surface topography measurement, in *Conference on Optical Society America* (OSA, Rochester 2004), Paper OMD 5
52. A. Schindler, T. Hänsel, F. Frost, A. Nickel, R. Fechner, B. Rauschenbach, Recent achievements on ion techniques for optics fabrication, in *Conference on Optical Society America* (OSA, Rochester 2004), Paper OMC 3
53. L.N. Allen, J.J. Hannon, R.W. Wambach, Final surface error correction of an off-axis aspheric petal by ion figuring. Proc. SPIE **1543** (1992)
54. A. Schindler, F. Frost, A. Nickel, T. Hänsel, B. Rauschenbach, Ion beam assisted smoothing of surfaces, in *Proceeding of 1st International Conference on Micro- and Nano-Technology Vienna* (2005), pp. 367–374 and A. Schindler, Tutorial on recent advances in ion beam and plasma jet processing, (Optical Fabrication and Testing (OAS), Paper OW4D.1, 2012) pp. 43–45
55. R.E. Wilson, Ionic polishing of fused silica and glass. Opt. Technol. **2**, 19–26 (1970)
56. M. Tarasevich, Ion beam erosion of rough glass surfaces. Appl. Opt. **9**, 173–176 (1970
57. A.F. Perveyev, V.V. IL'in, A.V. Mikhaylov, Ion polishing of Glass. Sov. J. Opt. Technol. **39**, 622–624 (1972)
58. F. Frost, B. Ziberi, A. Schindler, B. Rauschenbach, Surface engineering with ion beams: from self-organized nanostructures to ultra-smooth surfaces. Appl. Phys. A **91**, 551–559 (2008)
59. D.J. Barber, F.C. Frank, M. Moss, J.-W. Steeds, I.S.T. Tsong, Prediction of ion bombarded surface topographies using Frank's kinematic theory of crystal dissolution. J. Mater Sci. **8**, 1030–1040 (1973)
60. R. Smith, M.A. Tagg, J.M. Walls, Deterministic models of ion erosion, reflection and redeposition. Vacuum **34**, 175–180 (1984)
61. G. Carter, J.S. Colligon, M.J. Nobes, Analytical modelling of sputter induced surface morphology. Rad. Eff. **31**, 65–87 (1977)
62. R.M. Bradley, J.M.E. Harper, Theory of ripple topography induced by ion bombardment. J. Vac. Sci. Technol. A **6**, 2390–2396 (1988)
63. G. Carter, M.J. Nobes, I.V. Katardjiev, The production of repetitive surface features by oblique incidence ion bombardment. Phil. Mag. B **68**, 231–236 (1993)
64. E. Chason, T.M. Mayer, B.K. Kellerman, D.T. McIlroy, A.J. Howard, Roughening instability and evolution of the Ge(001) surface during lon sputtering. Phys. Rev. Lett. **72**, 3040–3043 (1994)
65. R. Cuerno, A.-L. Barabási, Dynamic scaling of ion–sputtered surfaces. Phys. Rev. **74**, 4746–4749 (1995)
66. R. Cuerno, H.A. Makse, S. Tomassone, S.T. Harrington, H.E. Stanley, Stochastic model for surface erosion via ion-sputtering: dynamical evolution from ripple morphology to rough morphology. Phys. Rev. Lett. **75**, 4464–4467 (1995)
67. C. Herring, Effect of change of scale on sintering phenomena. J. Appl. Phys. **21**, 301–303 (1950)
68. W.M. Tong, R.S. Williams, Kinetics of surface growth: Phenomenology, scaling, and mechanisms of smoothing and roughening. Ann. Rev. Phys. Chem. **45**, 401–438 (1994)
69. D.G. Stearns, Stochastic model for thin film growth and erosion. Appl. Phys. Lett. **62**, 1745–1747 (1993)

70. W.W. Mullins, Theory of thermal grooving, J. Appl. Phys. **28**, 333–339 (1957). W.W. Mullins, Flattening of a nearly plane solid surface due to capillarity. J. Appl. Phys. **30**, 77–83 (1959)
71. A.-L. Barabási, H.E. Stanley, *Fractal Concepts in Surface Growth* (Cambridge University Press, Cambridge, 1995)
72. S.E. Orchard, On surface levelling in viscous liquids and gels. Appl. Sci. Res., Section A. **11**, 451–464 (1963)
73. M.A. Makeev, A.-L. Barabási, Ion-induced effective surface diffusion in ion sputtering. Appl. Phys. Lett. **71**, 2800–2802 (1997)
74. G. Carter, V. Vishnyakov, Roughening and ripple instabilities on ion-bombarded Si. Phys. Rev. B **54**, 17647–17653 (1996)
75. W. Liao, Y. Dai, X. Xie, L. Zhou, Microscopic morphology evolution during ion beam smoothing of Zerodur surfaces. Opt. Express **22**, 377–386 (2014)
76. W. Liao, Y. Dai, X. Xie, L. Zhou, Deterministic ion beam material adding technology for high-precision optical surfaces. Appl. Optics **52**, 1302–1309 (2013)
77. S. Vauth, S.G. Mayr, Relevance of surface viscous flow, surface diffusion, and ballistic effects in keV ion smoothing of amorphous surfaces. Phys. Rev. B **75**, 224107 (2007)
78. F. Frost, R. Fechner, B. Ziberi, J. Völlner, D. Flamm, A Schindler, Large area smoothing of surfaces by ion bombardment: fundamentals and applications. J. Phys.: Condens. Matter. **21**, 224026 (2009)
79. E. Ziegler, L. Peverini, N. Vaxelaire, A. Cordon-Rodriguez, A. Rommeveaux, I.V. Kozhevnikov, J. Susini, Evolution of surface roughness in silicon X-ray mirrors exposed to a low-energy ion beam. Nucl. Instr. Meth. in Phys. Res. A **616**, 188–192 (2010)
80. A.J.R. van den Boogaard, E. Louis, E. Zoethout, S. Müllender, F. Bijkerk, Surface morphology of Kr+-polished amorphous Si layers. J. Vac. Sci Technol. A **28**, 552–558 (2010)
81. K. Morijiri, H. Endo, K. Morikaawa, S.A. Pahlovy, I. Miyamoto, 0.5 keV Xe+ ion beam nano smoothing of ULE substrate after processing with 3.0–10.0 keV Xe+ ion beam, Microelectr. Eng. **88**, 2694–2696 (2011)
82. H. Endo, T. Inaba, S.A. Pahlovy, I. Miyamoto, Low energy Xe+ ion beam machining of ULE substrates for EUVL projection optics—evaluation of high-spatial frequency roughness. Microelectr. Engng. **87**, 982–984 (2010)
83. M. Xu, Y. Dai, L. Zhou, X. Peng, S. Chen, W. Liao, Evolution mechanism of surface roughness during ion beam sputtering of fused silica. Appl. Optics **57**, 5566–5573 (2018)
84. N.I. Chkhalo, S.A. Churin, M.S. Mikhaylenko, A.E. Pestov, V.N. Polkovnikov, N.N. Salashchenko, M.V. Zorina, Ion-beam polishing of fused silica substrates for imaging soft X-ray and extreme ultraviolet optics. Appl. Optics **55**, 1249–1256 (2016)
85. N.I. Chkhalo, S.A. Churin, A.E. Pestov, N.N. Salashchenko, Yu.A. Vainer, M.V. Zorina, Roughness measurement and ion-beam polishing of super-smooth optical surfaces of fused quartz and optical ceramics. Opt. Express **22**, 20094–20106 (2014)
86. P. Becker, H. Friedrich, K. Fujii, W. Giardini, G. Mana, A. Picard, H.-J. Pohl, H. Riemann, S. Valkiers, The Avogadro constant determination via enriched silicon-28. Meas. Sci. Technol. **20**, 092002 (2009)
87. T. Arnold, F. Pietag, Ion beam figuring machine for ultra-precision silicon spheres correction. Prec. Eng. **41**, 119–125 (2015)
88. T. Hino, T. Taguchi, Y. Yamauchi, Y. Hirohata, M. Nishikawa, Surface flatness of polycrystalline copper after argon ion etching followed by annealing. J. Vac. Sci. Technol. B **22**, 2632–2634 (2004)
89. T. Kobayashi, Y. Nobuta, Y. Yamauchi, T. Hino, Oblique argon ion etching for copper at elevated temperature. J. Plasma Fusion Res. Series **8**, 1358–1360 (2009)
90. L.F. Johnson, K.A. Ingersoll, Ion polishing with the aid of a planarizing film. Appl. Opt. **22**, 1165–1167 (1983)
91. L.F. Johnson, K.A. Ingersoll, D. Kahng, Planarization of patterned surfaces by ion beam erosion. Appl. Phys. Lett. **40**, 636–638 (1982)
92. N. Yamauchi, T. Yachi, T. Wada, A pattern edge profile simulation for oblique ion milling. J. Vac. Sci. Technol. A**2**, 1552–1557 (1984)

93. A.I. Stognij, N.N. Novitskii, An ion-beam apparatus for the surface planarization of oxide materials. Instr. Exp. Tech. **45**, 141–151 (2002)
94. D.F. Grogan, T. Zhao, B.G. Bovard, H.A. Macleod, Planarizing technique for ion-beam polishing of diamond films. Appl. Optics **31**, 1483–1487 (1992)
95. Y. Li, H. Takino, F. Frost, Ion beam planarization of diamond turned surfaces with various roughness profiles. Opt. Express **25**, 7828–7838 (2017)
96. F. Frost, H. Takino, R. Fechner, A. Schindler, N. Ohi, K. Nomura, Smoothing of diamond-turned copper surfaces using ion beams with aid of planarizing film Jap. J. Appl. Phys. **46**, 6071–6073 (2007)
97. M. Ulitschka, J. Bauer, F. Frost, T. Arnold, Ion beam planarization of optical aluminum surfaces. J. Astron. Telesc. Instrum. Syst. **6**, 014001 (2020)
98. F. Frost, R. Fechner, D. Flamm, B. Zberi, W. Frank, A. Schindler, Ion beam assisted smoothing of optical surfaces. Appl. Phys. A **78**, 651–654 (2004)
99. P. Oelhafen, J.L. Freeouf, G.D. Pettit, J.M. Woodall, Elevated temperature low energy ion cleaning of GaAs. J. Vac. Sci. Technol. B **1**, 787–790 (1983)
100. U. von Gemmingen, R. Sizmann, Charge states of slow hydrogen ions reflected at single crystal surfaces. Surf. Sci. **114**, 445–458 (1982). K.J. Snowden, D.J. O'Conner, R.J. MacDonald, Observation of skipping motion in small-angle ion-surface scattering. Phys. Rev. Lett. **61**, 1760–176 (1988)
101. M. Holzwarth, M. Wissing, D.S. Simeonova, S. Tzanev, K.J. Snowdon, O.I. Yordanov, Preparation of atomically smooth surfaces via sputtering under glancing incidence conditions. Surf. Sci. **331–333**, 1093–1098 (1995)
102. J.G.C. Labanda, S.A. Barnett, L. Hultman, Sputter cleaning and smoothening of GaAs(001) using glancing-angle ion bombardment. Appl. Phys. Lett. **66**, 3114–3116 (1995) and Effects of glancing-angle ion bombardment on GaAs(001). J. Vac. Sci. Technol. B **13**, 2260–2268 (1995)
103. B. Koslowski, S. Strobel, P. Ziemann, Ion polishing of a diamond (100) surface artificially roughened on the nanoscale. Diamond Rel. Mater. **9**, 1159–1163 (2000)
104. J.G.C. Labanda, S.A. Barnett, L. Hultman, Damage-free cleaning of Si (001) using glancing-angle ion bombardment. J. Vac. Sci. Technol. B **16**, 1885–1890 (1998)
105. M. Wißing, M. Holtzwarth, D.S. Simeonova, K.J. Snowdon, An apparatus for glancing incidence ion beam polishing and characterization of surfaces to angstrom-scale root-mean square roughness. Rev. Sci. Instr. **67**, 4314–4320 (1996)
106. M. Wißing, M. Batzill, K.J. Snowdon, Preparation by glancing incidence ion irradiation of CaF_2 surfaces with angstrom-scale rms roughness. Nanotechnology **8**, 40–45 (1997)

Chapter 8
Low-Energy Ion Beam Bombardment-Induced Nanostructures

Abstract An attractive ion beam method is the possibility of spontaneous formation of ordered surface patterns in the form of nanodots/nanoholes or sinusoidal modulations of the surface (ripples) in the nanometer range. This method, based on self-organization, is characterized by the interplay of two low-energy processes induced by ion beams. The ion bombardment roughens the surface, while relaxation processes such as surface diffusion or/and beam-induced viscous flow smooth the surface. In this chapter, the formation of nanoripples with and without metallic contaminants is presented and the dependence of ripple formation on temperature, ion incidence angle, ion energy and co-deposited metal concentration is discussed. Bradley and Harper have proposed a continuum theory to describe the topography evolution and pattern formation. This theory is based on curvature-dependent sputtering, which is proportional to the locally deposited energy. In the following, it will be shown that, on the one hand, this theoretical concept can be extended by introducing nonlinear terms and, on the other hand, that the formation of surface patterns can be also explained by a directional redistribution of mass. Finally, the great application potential of this technology for effective, low-cost and scalable patterning of large areas of all materials is demonstrated.

In principle, nanostructures are produced on surfaces on the basis of two strategies. On one hand, the top-down strategy, characterized by the application of lithography techniques, and, on the other hand, the utilization of self-organization processes. An attractive method of the latter strategy is the possibility of spontaneous formation of surface patterns as nanodots/nanoholes or sinusoidal modulations of the surface (ripples) in the nanometer scale, as the result of low-energy ion bombardment. This method based on self-organization could open a new route to fabricating ordered patterns on the nanometer scale over large areas, with short preparation times. This method seems to be capable of overcoming the limitations of lithography. Comprehensive reviews on this topic are published in [1–9].

In Fig. 8.1, the method is schematically displayed. This method, a destructive route to preparing nanostructures [10], is characterized by the interplay between two low-energy ion beam-induced processes. The ion bombardment roughens the surface, whereas relaxation processes such as surface diffusion or/and radiation-induced

B. Rauschenbach, *Low-Energy Ion Irradiation of Materials*, Springer Series in Materials Science 324, https://doi.org/10.1007/978-3-030-97277-6_8

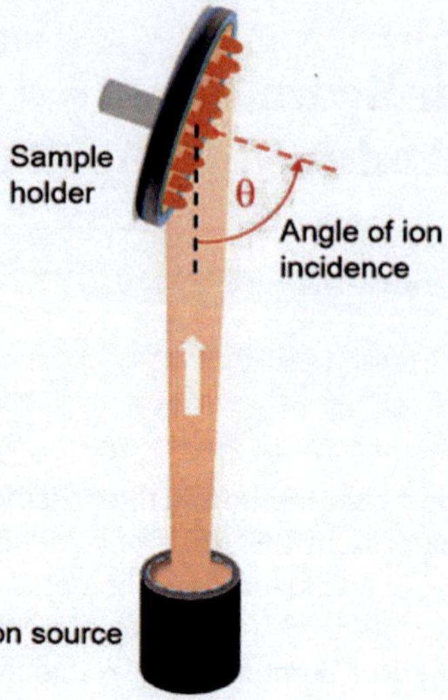

Fig. 8.1 Schematic configuration of the formation of nanostructures by low-energy ion beam sputtering. The angles θ characterize the tilt angle

viscous flow smooth the surface. Both processes can be substantially controlled through the choice of the experimental conditions (ion species, ion energy, angle of ion incidence, temperature, etc.). Depending on the incident ion species, the substrate (target), and the experimental condition of low-energy ion bombardment, a variety of different surface topographies can be generated. Of particular importance are self-organized ion beam patterns, such as ripples and dots with nano- and micrometer periodicity.

The following chapter focuses on this technique, selected experimental results, and theoretical concepts. It provides an overview of the experimental results in the field of nanostructuring of surfaces induced by low-energy ion beam sputtering (Sect. 8.1). This is followed by a presentation of the most frequently discussed theoretical concepts of nanopatterning after low-energy ion bombardment (Sect. 8.2), a brief report about the formation of ripples on polycrystalline surfaces (Sect. 8.3) and some aspects regarding the potential application of nanoripples and nanodots (Sect. 8.4).

8.1 Nanoripples Produced by Low-Energy Ion Bombardment

The first observation of the formation of ripple-like structures by low-energy ion bombardment was made in the early 1960s. Following Ar ion bombardment with energies between 4 keV and 10 keV at glancing angles, Haymann and Trillat [11] observed a characteristic pattern of grooves on electropolished uranium surfaces, oriented parallel to the direction of the ion beam. Cunningham et al. [12], from the same lab, have shown that such grooves (ripples) can be reproduced on Au surfaces after 8 keV Ar ion bombardment for 120 min (ion current density ranged between 50 and 200 $\mu A/cm^2$) at 70° ion incidence. Navez et al. [13] have demonstrated the formation of ripple-like patterns on glass surfaces after 4 keV air ion bombardment at incident angles ranging from 30° to 80° (Fig. 8.2). When the ion incident angle was smaller than a define incidence angle (Fig. 8.2, right), they were able to observe wave-like periodic ripples with a wavelength between 30 nm and 120 nm, where the wave-vector was oriented parallel to the direction of the ion beam (parallel mode ripples). For incident angles larger than this defined ion incidence angle, the ripples were rotated by 90° in the direction of incident ions (perpendicular mode ripples, Fig. 8.2, right). Over the course of the following years, periodic surface structures were repeatedly observed. For example, Nelson and Mazey [14] found ripple-like structures on Cu surfaces after Xe ion bombardment and Whitton et al. [15] observed a fine-scale ripple structure on Cu following 40 keV Ar ion bombardment with 10^{19} ions/cm^2. Wave-like structures on Si after 40 keV Ar ion bombardment at incidence angles of 45° and 60° were observed by Carter et al. [16]. Later, Zalar [17] was able to correlate the onset of the ripple evolution with changes in the sputtering yield that take place during surface depth profiling.

In subsequent decades, low-energy ion beam erosion-induced ripple nanostructures were observed on the surfaces of crystalline and amorphous materials as well as on the surfaces of all types of materials, including semiconductors, metals, polymers, and isolators. On the basis of this fact, it can be assumed that the phenomenon of nanopatterning under ion bombardment is observable on all materials, i.e. this phenomenon seems to be fundamental in nature.

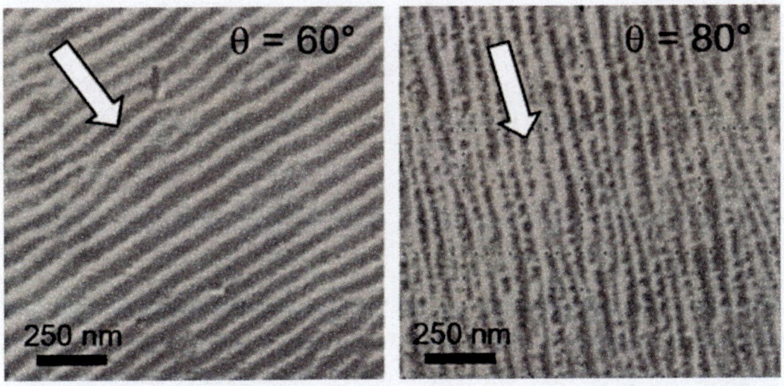

Fig. 8.2 Surface topography of glass after 4 keV air ion bombardment, dependent on the angle of ion incidence. Arrows indicate the projected ion beam direction. Figure adapted from [13]

The preferred methods for the characterization/observation of ion beam-induced topographical changes during and after ion bombardment include scanning probe microscopy (STM, AFM), scanning and transmission electron microscopy (SEM, TEM), as well as light and X-ray scattering techniques.

In recent decades, an enormous number of papers have been published that focus on the ion bombardment-induced formation of periodic ripples and dots. To illustrate, Fig. 8.3 shows an AFM image and corresponding FFT spectra as well as the calculated PSD function of a Si surface after non-normal Xe ion bombardment [2]. The AFM image gives an impression of the surface topography (Fig. 8.3, left). For the height fluctuations analysis of the resulting topography, the most significant statistical parameter is the root mean square (rms) roughness (for details, see Appendix L), which can be calculated from the height profiles of the AFM images. The evolution of the surface roughness is characterized by two exponents, the growth exponent β and the roughness exponent α. Exponent β can be extracted from a linear fit of the log–log plot of the rms roughness (e.g., measured using AFM) versus the time of ion bombardment, while exponent α can be established from a fit to the linear part of the log–log plot of the height-height correlation function (see Appendix L).

By performing a two-dimensional fast Fourier transform (2D-FFT) of the height profiles from AFM or STM images, information about the presence of dominant frequencies or periodic elements in the topography can be detected. The FFT image (Fig. 8.3, middle) shows a strong anisotropy. The spots are aligned in the direction of the wave-vector of the ripples. The main spot indicates the characteristic frequency of the ripples, i.e. the inverse of the separation of the features in real space. In this regard, the separation of the features will be considered to be the wavelength, λ, for the ripples and the mean size for the dots. Higher lateral ordering is indicated by additional spots (ripples) or rings (dots), while the spread of the spots is related to

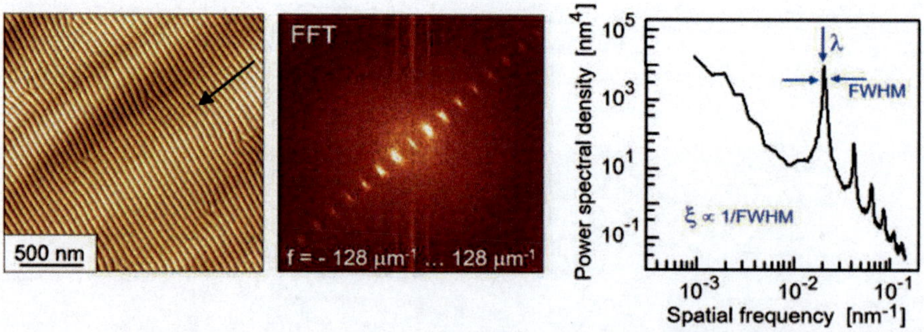

Fig. 8.3 Typical AFM image (left), corresponding FFT image (middle), and angular averaged PSD function (right) of a Si surface after 1200 eV Xe ion beam erosion under non-normal ion incidence at room temperature (black arrow indicates the projected ion beam direction, f is the spatial frequency, λ is the wavelength of the ripples, ξ is the correlation length, FWHM is the full width of the half maximum)

the homogeneity and spatial correlation of the features. The information from the FFT images can be quantified by calculating the power spectral density function (PSD, for details see Appendix L). The PSD function (Fig. 8.3, right) is a second-order statistics quantity that describes the relationship of two points on the surface and provides quantitative information about height and lateral distributions of these topographic points. If one or more characteristic peaks occur in the PSD function, then regular features are present on the surface. The peak with the highest intensity is the first-order peak. Its position indicates the characteristic spatial frequency of the ripples, while additional peaks demonstrate a high lateral ordering. This lateral ordering, given by the system correlation length, ξ, can be deduced from the PSD function by determining the full width of the half maximum (FWHM) of the first-order peak (Fig. 8.3). The system correlation length is then inversely proportional to FWHM ($\xi \propto 1/\text{FWHM}$) [18].

In the past, prior to the finding that impurities can significantly alter the formation of nanopatterns [8, 19], the effect of impurities had not received any attention. The low-energy ion bombardment induced formation of nanoripples on the surfaces of diverse materials has been studied exclusively on the basis of the parameter of ion irradiation, including ion species, ion energy, angle of ion incidence, temperature during bombardment, ion fluence, and ion flux. For example, Fig. 8.4 shows that after Kr ion bombardment, ordered nanoripples already form on Si surfaces based

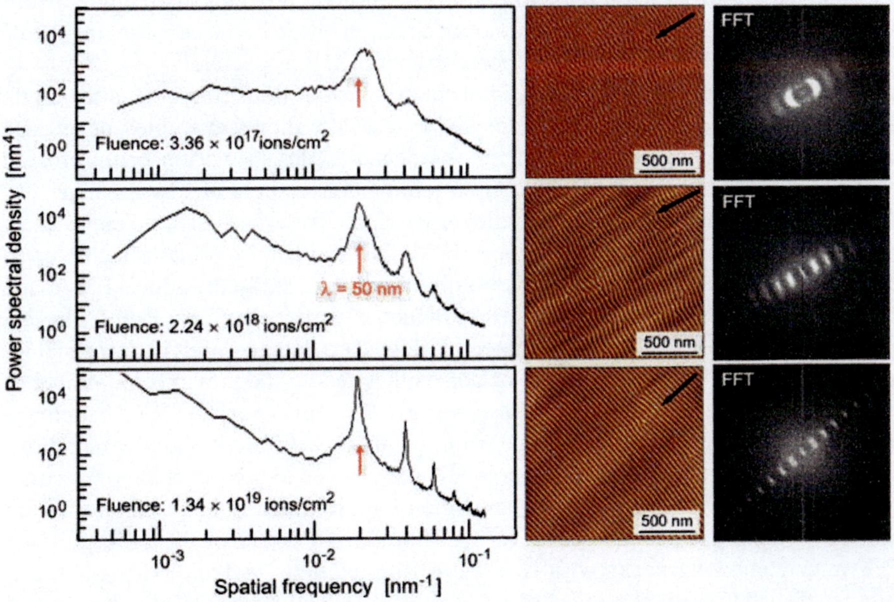

Fig. 8.4 Angular averaged PSD functions and AFM images and the corresponding Fourier spectrum (FFT image range $\pm 127.5\ \mu m^{-1}$) of self-organized Si ripple patterns produced by Kr^+ ion beam erosion at an ion energy of 1200 eV and an ion incidence angle of 15° for different ion fluences. The black arrows give the projected ion beam direction

on ion fluence at an ion incidence angle of 15° with respect to the normal surface [2, 20, 21]. These experimental observations reveal that ripples begin to form immediately following commencement of the bombardment. The corresponding FFT image reveals that the wave-vector of ripples is parallel to the projection of the ion beam on the surface. The position of the first peak in the PSD functions specifies the characteristic wave-vector of ripples, i.e. the wavelength of ripples (here about 50 nm). The radial width of the peak shows how ordered the ripples are, i.e. the less wide the peak, the more ordered are the ripples. Also, the angular width of the Fourier peak (which determines the angular distribution of the ripples) decreases with erosion time. This means that the homogeneity and ordering (alignment) of the ripples improves. The appearance of multiple peaks is an additional indication of the improved order of ripples. In prolonged sputtering (ion fluence of 1.35×10^{19} cm^{-2}), the ripples are almost perfectly aligned.

A second example is the dot-ripple and ripple-dot transition, which seems to be triggered by concurrent metal atom contamination during the ion bombardment. Neither theoretical considerations nor experimental models have predicted the ripple-dot transitions in an impurity-free case. Based on the known fact that low metal concentrations act as seeds for the evolution of cones on Si surfaces (see Sect. 6.3.1.1 and [22]), Allmers et al. [23] studied the dot-ripple transition on GaSb surfaces after 3 keV Ar ion bombardment of samples mounted on either tantalum or molybdenum plates with molybdenum clamping strips. They observed that dots are formed up to incidence angles of at least 10° and that ripple formation begins at an angle of 20° and higher. The dot-ripple transition was also proven on Ge surfaces following 2 keV [24] and 1 keV Xe ion near-normal bombardment [25], respectively.

Ziberi et al. [26] investigated in detail the ripple-dot transition versus the angle of ion incidence on ion beam eroded Si. Figure 8.5 shows examples of surface topographies evolving for different ion incidence angles in Xe ion beam erosion with no sample rotation [2, 26, 27]. At an ion incidence angle of 5° with respect to the surface, normal well-ordered parallel mode ripple patterns evolve on the surface. By further increasing the ion incidence angle, this topography remains stable up to 23°, where a mixture of dots and ripples is observed on the surface. As the AFM image shows, dot and ripple structures form simultaneously on the surface. Ripples have a slightly curved form and are interrupted by dots. The dots themselves form mainly along the ripples, i.e. the alignment of dots is dictated by the previous alignment of ripples. By increasing the ion incidence angle to 25°, this coexistence of patterns is retained. This AFM image reveals that ripples with three different spatial orientations are formed on the surface (ripples perpendicular aligned to the ion beam projection, ripples with a wave-vector forming an angle of approximately 60° to the ion beam projection, and ripples with a curved form). This is also reflected on the corresponding FFT image, showing peaks with two distinct orientations. Additionally, the broad angular distribution of the first-order spots in the FFT, those that are shaped like semi-circles, is due to the influence of the curved ripples.

With a further increase of $\theta = 26°$, the AFM image shows nanodot structures aligned primarily in two directions, crossing each other, creating a 90° angle between both directions. It is interesting that the direction of both predominant structures is

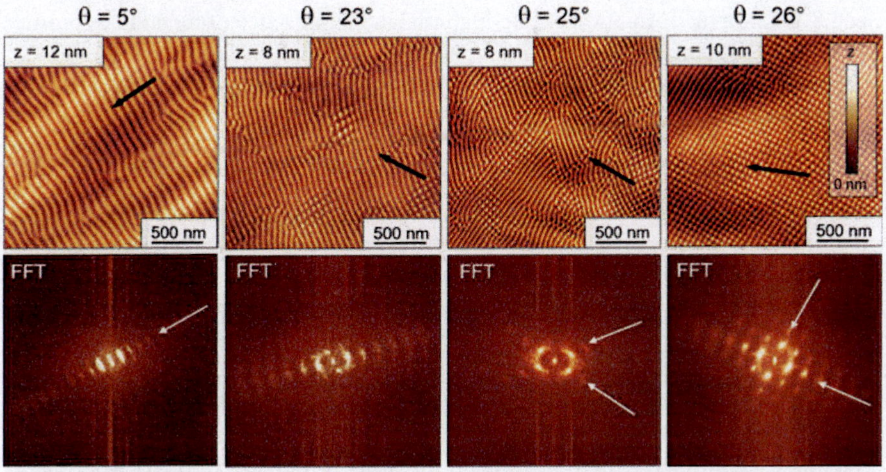

Fig. 8.5 AFM images and the corresponding FFT images of the topography of Si after bombardment with 2000 eV Xe ions at room temperature without sample rotation in dependence on the angle of ion incidence. The ion fluence was fixed at 6.7×10^{18} ions/cm^2. The black arrows indicate the projected ion beam direction, and the white arrows point out the two distinct wave-vectors

rotated with respect to the ion beam projection, i.e. these are neither perpendicular nor parallel to the ion beam projection. Along these structures, dots with a chain-like shape dominate the surface. The dots show an almost perfect lateral ordering. This is reflected in the corresponding FFT image, with the equidistant first-order spots implying the same periodicity of the dots in the two structures and wave-vectors perpendicular to each other.

A typical feature of all the experiments presented up to this time is that the concentration of impurities was not studied and any correlation between nanopatterning and metal contamination was not addressed.

Since it has been discovered that low concentrations of metallic impurities play an important role in the ion beam-induced nanopatterning of surfaces, it is important to distinguish between patterning that occurs with and without impurities [8]. This distinction is not unproblematic, because up until now, the influence of metallic contamination on the nanopatterning has been almost exclusively studied only for silicon substrates in detail. For other substrate materials (non-Si semiconductors, metals, polymers, isolators, etc.) upon which the formation of ripples and dots on the surface could be verified, very little is known about the influence of metallic impurities.

But, it is evident that the simultaneous contamination by metal impurities during low-energy ion bombardment provides an additional degree of freedom for controlling the pattern formation process, for improving the nanopatterning, and for producing large-scale ordered nanostructures of different shapes. Consequently, this chapter will focus on nanopatterning in the absence of metallic contamination

(Sect. 8.1.1), on the formation of nanopatterns in the presence of metallic impurities (Sect. 8.1.2), and on the formation of nanodots and nanoholes (Sect. 8.1.3).

8.1.1 Formation of Ripples Without Metallic Contamination

Experiments conducted under the condition that no metallic contamination be allowed to influence the nanopatterning allow a deep insight into the basic mechanisms of the ion beam-induced formation of ripples. In this chapter, the only measurements and results introduced and discussed are those for which metallic contaminations could be explicitly excluded.

Since the important studies by Ozaydin et al. [19, 28] on the dramatic influence of small concentrations of metal impurities on ion beam-induced nanopatterning of semiconductor surfaces, there have been attempts to conduct sputter experiments under contamination-free conditions or to quantify the impurity concentration. Madi et al. [29–31] carried out such experiments in a carefully shielded vacuum chamber to eliminate the possibility of metallic contamination and have studied the energy and angle dependence of Si ripple formation after Ar ion bombardment with energies between 250 eV and 1000 eV at ion incidence angles, θ, up to 85° with respect to the normal surface of the substrate. The results of these experiments are summarized in a phase diagram shown in Fig. 8.6 [31]. For small ion incidence angles ($\theta < 50°$), a smoothing of the surface has been observed, while for incidence angles $\theta > 50°$ the surface becomes instable and ripples with a wave-vector form parallel to the ion beam. For very large angles of ion incidence (grazing ion bombardment), the wave-vector of ripples changes and is oriented parallel to the direction of the ion beam. Consequently, ripples appear only when a critical angle of ion incidence, θ_c, is

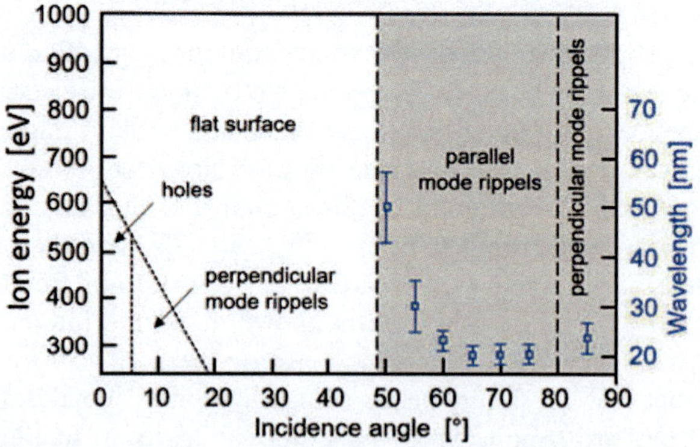

Fig. 8.6 Phase diagram of pattern formation on Si(001) surface after Ar ion beam sputtering at room temperature and a fluence of 3.8×10^{18} Ar ions/cm^2. The light gray region indicates pattern formation without the influence of multiple scattered Si substrate atoms. Figure is adapted from [30, 31] and modified

exceeded. Taking into account that in these experiments, scattered Ar and redeposited Si particles sputtered from the Si shield reach the Si substrate surface, ripples and holes are generated (Fig. 8.6, dotted lines) for small incidence angles ($\theta < 20°$) and Ar ion energies up to about 650 eV [30]. The wavelength of the ripples decreases rapidly with increasing ion incidence angle (Fig. 8.6, blue points). Perkinson et al. [32] have developed a phase diagram of nanopatterns on Ge surfaces in dependence on Kr ion energy (250 and 500 eV) and ion incidence angles in the impurity-free case. This phase diagram is qualitatively similar to the diagram developed by Madi et al. (see Fig. 8.4), i.e. a smooth region, a region with parallel mode ripples, and a region with perpendicular mode ripples could be observed, independent of the ion energy. Both phase diagrams differ significantly with regard to magnitude of the transition angle. The stability-instability transition angle, θ_c, after Kr ion bombardment of Ge surfaces occurs at about 15° higher than this angle for Ar ion bombardment of Si surfaces.

It is known that the existence of a flat surface under ion bombardment is inconsistent with the Bradley–Harper theory (see Sect. 8.2.1.1). Therefore, the model of ion impact induced mass redistribution as well as a crater function theory (see Sect. 8.2.2) are applied to explain the evolution of the ion beam-induced topography of silicon in particular [33].

8.1.1.1 Angle Dependence of Transition from the Stable to the Instable Surface

The transition from the stable smooth surface to the instable rippled surface was studied by Madi et al. [33] (see Fig. 8.6) and Anzenberg et al. [34], where this stability-instability transition of the Si surface under 1 keV Ar ion bombardment at room temperature and different incidence angles was measured in real-time using GISAXS. On the basis of these measurements, the exponential growth [c.f. (8.8)] can be analyzed at early time points and low fluences. The sign used for the maximum growth rate indicates stability (negative sign) or instability (positive sign) of the surface (see also Fig. 8.12). The linear dispersion relations, R(q), leads to a critical angle $\theta_c = 47° \pm 2.5°$ [33].

Figure 8.7 illustrates the stability-instability transition of the topography on a Si surface after 1200 eV Kr ion bombardment with a fluence of 6.74×10^{18} ions/cm^2 [35]. The AFM image sequence indicates that the surface was flat up to about an incidence angle of $\theta = 60°$, where the roughness was still lower than the initial roughness of the Si substrate. Beyond this angle, the roughness increases slightly. At $\theta = 65°$, ripples with a wave-vector parallel to the ion beam are formed. The analysis of the PSD spectra has shown that the stability-instability transition can be expected at an angle $\theta_c = 63°$. In contrast to the measurements by Madi et al., a transition of the oriented ripples into an orthogonal arrangement (perpendicular mode ripples) could not be observed. Instead, faceted Si surfaces were found for angles $\theta \geq 80°$. Macko et al. [36] and Engler [37] were also able to observe ripples with a small

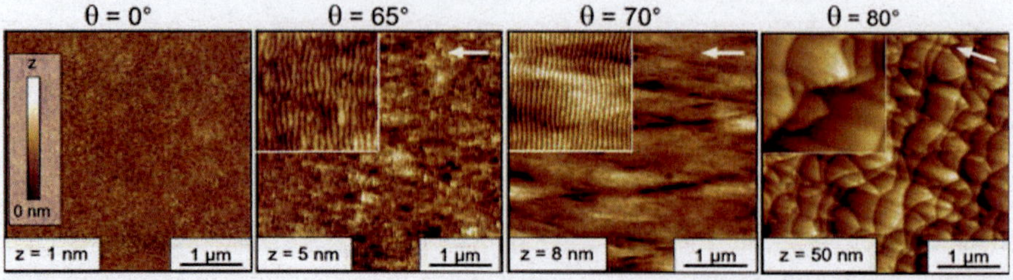

Fig. 8.7 AFM images of Si after Kr ion bombardment at different angles of ion incidence. The white arrows indicate the projection of the ion beam direction. The ion energy was 1200 eV, the ion current density 300 μA/cm^2, and the ion fluence 6.47×10^{18} ions/cm^2. The different height scales of the images are specified in each image

amplitude and a wave-vector parallel to the ion beam azimuth after 2 keV Kr ion bombardment of Si(001) surfaces at room temperature at an incidence angle $\theta > 55°$.

Basu et al. [38] studied the topography evolution of silicon after 500 eV Ar ion bombardment at room temperature and found (i) a transition angle $\theta_c = 67°$ for the appearance of parallel mode ripples and (ii) a transition to perpendicular mode ripples at 80°. At identical conditions of ion bombardment, a critical angle of about $\theta_c = 55°$ and the transition to the perpendicular mode ripple formation for $\theta \geq 80°$ were determined by Chowdhury and Ghose [39]. Si(001) was bombarded by Ar ions within the energy range of 300 and 1100 eV under different incidence angles by Castro et al. [40]. In conformance with the results from Madi et al. [31], they observed the critical angle $\theta_c = 50°$ for the formation of parallel mode ripples, and a decrease of the wavelength with increasing incidence angle.

In similar impurity-free experiments, the transition angle was also determined on germanium surfaces after low-energy noble gas ion bombardment. Perkinson et al. [32] found that a critical transition angle between stable and rippled Ge surfaces occurs at 57.5° for two different ion energies (250 eV and 500 eV). Other studies using GISAXS have determined a critical angle of about 62° after 1 keV Kr ion bombardment [41] and AFM studies determined a transition angle of about 70° following 1.2 keV Kr ion bombardment [42].

The origin of the instability of the ion-bombarded surfaces (silicon) for ion incidence angles $\theta \geq \theta_c$ is explained by applying the Carter–Vishnyakov formalism [43–45], in which the ion-impact-induced mass redistribution dominates the process of ion beam-induced erosion (see Sect. 8.2.1.3). In the model by Castro and Cuerno [46], a driving force in the direction of the ion beam proportional to the cosine of incidence angle is assumed to induce the local mass redistribution. Independent of the ion species and ion mass, a critical angle of $\theta_c = 45°$ was predicted. On the basis of the crater function model (see Sect. 8.2.2) by Norris et al. [9, 47, 48], a critical angle can be determined by MD simulation based on the ion-substrate combination and the ion energy. For example, a transition angle $\theta_c = 48°$ was obtained after 250 eV Ar ion bombardment of Si [48].

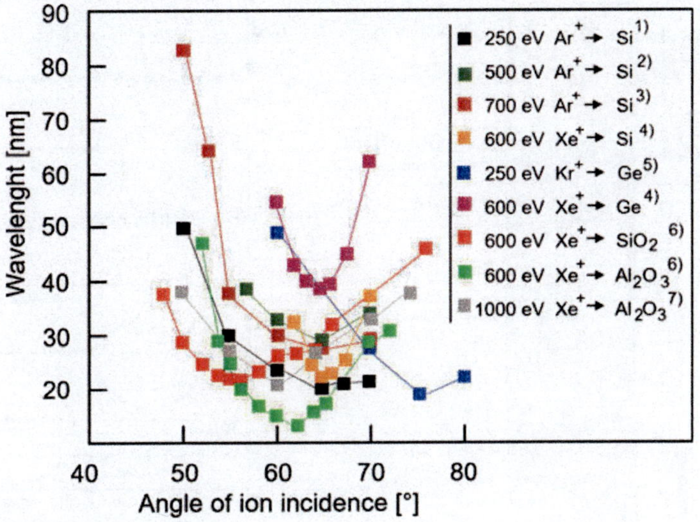

Fig. 8.8 Ripple wavelength as function of the angle of ion incidence. References: [1][29], [2][39], [3][40], [4][35, 49], [5][32], [6][50], [7][51]

8.1.1.2 Ripple Wavelength Dependence on Incidence Angle

These studies are focused on the dependence of the ripple wavelength on the incidence angle beyond the critical angle θ_c. For example, the development of the wavelength can be measured in-situ and ex-situ using GISAXS or determined by calculation of the PSD from AFM/STM images. Figure 8.8 shows the wavelength of ripples on Si, Ge, SiO_2 and Al_2O_3 surfaces without metallic contamination as a function of the ion incidence angle. In this figure, the experimentally determined transition angle between the stable and rippled surfaces for Si, $\theta_C \approx 50°$, and for Ge, $\theta_C \approx 58°$, can be extracted. Typically, it can be observed that the wavelength initially decreases when the angle of ion incidence is increased. Thereafter, an increase of the wavelength with an increase to the angle can be observed. This behavior is particularly pronounced if heavy ion species (here, Kr and Xe ions) are applied. The minimum wavelength appears to be correlated with the maximum angle-dependent sputtering yield at θ_p for each material (see Fig. 5.9).

8.1.1.3 Ripple Formation Dependence on Ion Fluence

It must be remembered that the wavelength of ripples is also dependent on bombardment time (ion fluence). Up until now, wavelength dependence on fluence, the amplitude of the ripples, and the roughness have been studied exclusively on silicon, germanium and fused silica. A typical example of the topographical evolution of surface patterns under low-energy ion bombardment at room temperatures as a function of

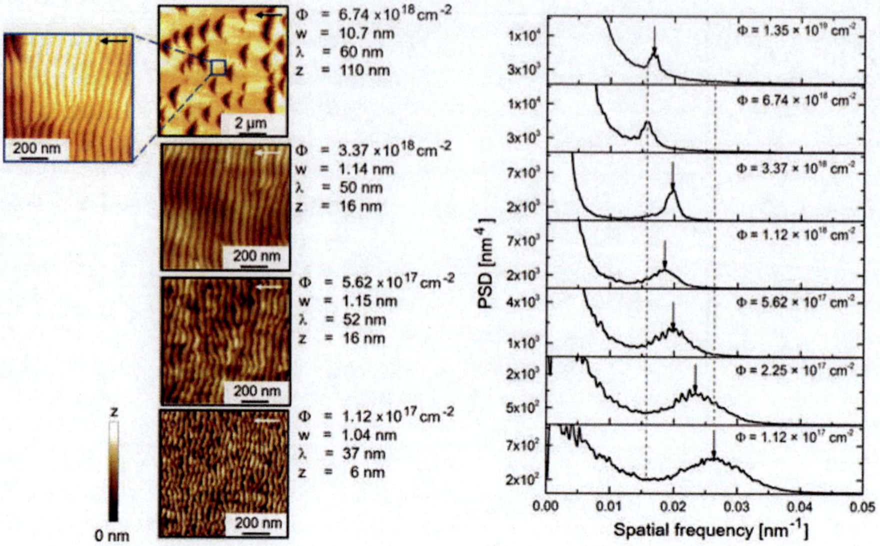

Fig. 8.9 AFM images (left) and calculated PSD spectra (right) of Si following Xe ion bombardment with ion fluence at an angle of ion incidence of 75°. The ion energy was 1200 eV and the ion current density 300 μA/cm^2. The arrows indicate the projection of the ion beam direction. For each AFM image, the rms roughness, w, the ripple wavelength, λ, and the height scale, z, are given. The dotted vertical lines in the calculated PSD mark the shortest and longest ripple wavelengths, respectively, and the vertical arrows highlight the wavelength at the corresponding fluence

the ion fluence is shown in Fig. 8.9 [52]. For these experiments, an incidence angle of $\theta = 75°$ was selected because a distinct ripple formation in the metal contamination-free case is expected. Some characteristic AFM images of the ripple pattern formed on silicon surfaces are shown in Fig. 8.9, with variation of the fluence between 1.12×10^{17} cm^{-2} and 1.35×10^{19} cm^{-2}. At the shortest erosion time, an irregular ripple pattern with a mean wavelength of about 37 nm evolves on the surface. With increased ion fluence, the ripple wavelength increases slightly more, up to 50 ± 2 nm, and the surface roughness also increases. This coarsening process can lead to more significantly distinctive nanopatterns. At high fluences ($> 5 \times 10^{18}$ cm^{-2}), a saturation of the roughness and the mean ripple wavelength can be determined. Similar behavior was also observed in other studies [5, 40].

Between of this highly regular pattern, at higher fluences, large triangular depressions emerge on the surface. These faceted structures have a well-defined orientation toward the global surface normal (for details on facet formation, see Sect. 6.3.1.3). Between the triangular depressions, the surface is characterized by distinct ripple formation (e.g., see the magnified AFM image after Xe ion bombardment with a fluence of 6.74×10^{18} cm^{-2}). The stack of PSD curves in Fig. 8.9 (right) summarizes the development of the Si surface based on fluence. This thereby illustrates the general trend of the wavelength as well as the roughness evolution. Essentially the same behavior was observed for Si after lower ion energy (600 eV) [53], for Ge [49],

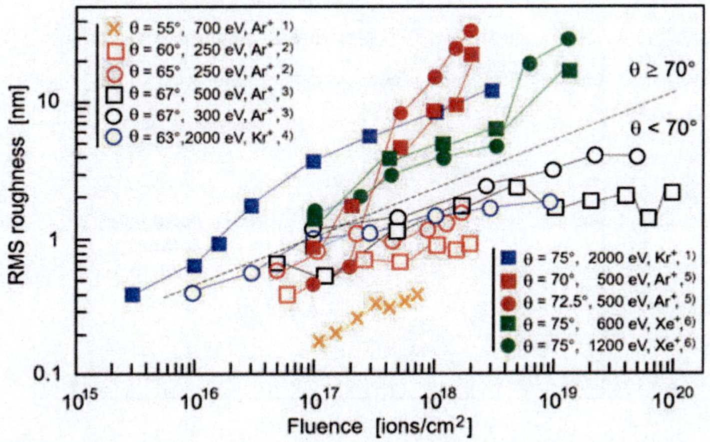

Fig. 8.10 Evolution of the rms roughness of Si versus ion fluence (time of bombardment) for ion energies $\leq$ 2000 eV and different ion species after rare gas ion bombardment at room temperature under different angles of ion incidence. Closed symbols represent results for $\theta \geq 70°$, open symbols represent results for $60° \leq \theta < 70°$ (see dashed line), and crosswise symbols represent results for $\theta < 60°$. References: [1][40], [2][57], [3][5, 58], [4][37], [5][55], [6][35]

and for surfaces of fused silica [52, 54], where, in the last case, the ripple coarsening is much more pronounced at higher incidence angles.

Figure 8.10 provides a summary of some experimental measurements of roughness evolution as a function of ion fluence for silicon bombarded with different rare ion species at room temperature. This figure indicates that the rms roughness increases with ion fluence up to a saturation fluence, independent of the ion species and ion energy. When the angle of ion incidence is increased, the roughness rises significantly (e.g., compare the results following $\theta < 70°$ [open symbols in Fig. 8.54) with results after $\theta \geq 70°$ (closed symbols)]. It should be noted that a pronounced tendency of the transition of parallel mode ripples to transform into a faceted topography can be observed for ion incidence angles $\theta > 70°$ and high ion energies [55, 56]. At high fluences, a tendency to saturation of the roughness in this double-logarithmic diagram is obvious (Fig. 8.10).

Figure 8.11 is a schematic representation of the dependence of roughness on ion fluence after bombardment with low-energy ions in the absence of metallic contamination or simultaneous metal co-deposition. In the linear regime (e.g., in the case of Si for fluences < 10^{16} ions/cm^2), the roughness is unchanged by the fluence. According to the Bradley–Harper theory and also the momentum-based continuum theory, this is to be expected (see Sect. 8.2.1). With higher fluences in the non-linear regime, ripple coarsening begins. The coarsening is caused by the different propagation velocities of the ripples, where the velocity is proportional to $dY/d\theta$ [59]. Due to the different shapes of the ripples, different velocities can be expected. Consequently, the faster moving ripples will overtake the slower ripples, merging together to form larger ripples. In this stage of ripple development,

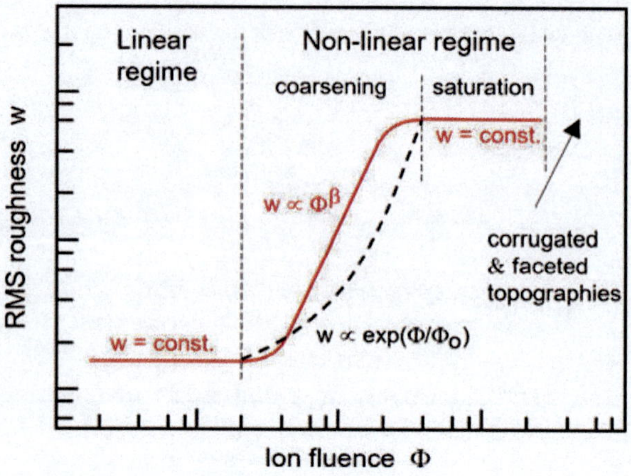

Fig. 8.11 Schematic of roughness as function of the ion fluence (bombardment time)

the rms roughness, w, can be fitted by an exponential relation $w \propto \exp(\Phi/\Phi_o)$, where Φ_o is the threshold fluence for appearance of nanoripples and Φ is the ion fluence used. For example, experimental studies by Castro et al. [40] have confirmed this dependence for time of bombardment up to 25 min ($\Phi \approx 2.8 \times 10^{17}$ ions/cm^2) after 700 eV Ar ion bombardment of Si at an incidence angle of 55° and an ion current density of 30 μA/cm^2. It should be noted that in the majority of experiments focused on the development of ripples under ion bombardment, the linear regime (expected at low fluences) was inaccessible, because the ion fluxes (> 200 μA/cm^2) applied were too high, i.e. the ion bombardment times were too short.

In this non-linear regime, the roughness can be fitted by the proportional relation $w \propto \Phi^{\beta}$, where β is the growth exponent (see Appendix L). For example, Keller et al. [5, 58] found a growth exponent of $\beta = 0.28$ and $\beta = 0.27$ for 500 eV and 300 eV Ar ion bombardment of Si, respectively, and Chowdhury et al. [60] determined a growth exponent of $\beta = 0.42$ for ripples after 30 eV Ar ion bombardment of GaAs. With increased ion fluence, the coarsening of the ripples advances and ultimately leads to a saturation of the roughness. With increased ion energy, this saturation is reached earlier (lower fluences).

In addition to the evolution of the ripple pattern, the formation of corrugations (see e.g., [49, 52, 58, 61] or Fig. 8.9) or triangular hillocks were observed for high ion fluences (> 5×10^{18} cm^{-2}) and then the formation of facets for fluences of 6.74×10^{18} Xe ions/cm^2 (see e.g., [49, 55]). The formation of faceted, sawtooth-like structures is probably the result of geometrical shadowing or/and the impact of reflected particles (see Sect. 6.3.1.3), thus the formation of corrugations is not affected by shadowing. For grazing and near-grazing ion incidence angles (i.e. θ > 75°…80°), the phenomenon of rotation of the ripple pattern from a parallel to a perpendicular orientation with respect to the direction of the ion beam also influences the roughness evolution. Consequently, the rms roughness at high ion

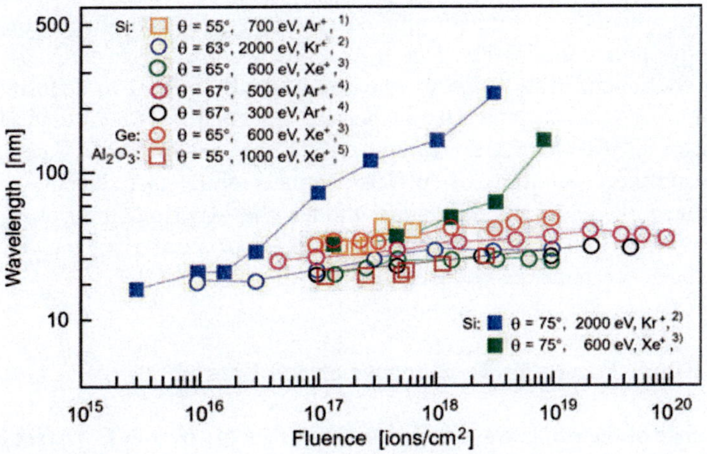

Fig. 8.12 Wavelength of ripples versus ion fluence (time of bombardment) for ion energies $\leq$ 2000 eV and different ion species after rare gas ion bombardment at room temperature under different angles of ion incidence. Closed symbols represent results for $\theta < 70°$, open symbols represent results for $\theta > 70°$. References: [1][40], [2][37], [3][49], [4][5, 58], [5][51]

fluences is no longer determined by the amplitude of the ripple pattern but rather by the corrugation, facets, triangular hillocks, or other depressions.

The time-dependent evolution of the ripple wavelength at room temperature is shown in Fig. 8.12. In this figure, several experimental measurements of the ripple wavelength on different materials (Si, Ge, amorphous alumina) are summarized. Comparable to the fluence-dependent roughness evolution, the ripple evolution can be subdivided into two regions of the ion incidence angle. For ion incidence angles $\theta_c < \theta < 70°$ (open symbols in Fig. 8.10), a coarsening is evident, independent of the ion species, ion energy (< 2 keV), and the substrate material. A few studies (e.g., [62]) have found that the ripple wavelength remains constant and the amplitude of the ripples continuously increases with the ion fluence (time of ion bombardment). In most cases, the ripple wavelength increases according to the power law relation $\lambda \propto \Phi^n$, where the coarsening exponent n ranged between 0.1 and 0.34. Makeev et al. [63] have predicted that the wavelength decreases with an increase in the ion energy, indicating that the ion-induced diffusion is the determining process of surface smoothing. This behavior could not be confirmed in all cases studied. Frequently, saturation of the wavelength can be observed at higher fluences. The linear regime at very low fluences, characterized by constant wavelengths and exponential increase of the ripple amplitudes, was inaccessible (cannot be observed) in the majority of the experiments.

For ion incidence angles $\theta > 70°$ (grazing incidence, closed symbols in Fig. 8.12), a completely different dependence of the ripple wavelength on ion fluence can be assessed. The ripple wavelength increases significantly with increasing fluence. This behavior can be expected if the parallel-mode ripples at low fluences are transformed into coarse structures (facets, hillocks, etc.).

This experimentally observed ripple coarsening could be confirmed on the basis of a hydrodynamical model (see Sect. 8.2.1.2) by Muñoz-García et al. [64, 65]. The nonlinear coefficient in this theory, which described the local re-deposition of the sputtered material and the material transport in the near-surface region, stimulates the coarsening of the wavelength. For example, Babonneau et al. [51] have demonstrated that based on the assumption of surface-confined ion beam-induced viscous flow in the framework of the hydrodynamic model, the experimentally observed time evolution of ripples on amorphous alumina (Al_2O_3) surfaces for ion incidence angles $< 65°$ can be successfully reproduced.

8.1.1.4 Ripple Formation Dependence on Ion Energy

A dependence of the ripple wavelength on the ion energy by $\lambda \propto E^{-1/2}$ (see Sect. 8.2) can be expected if thermally activated surface diffusion is the dominant process for surface smoothing [1]. In the absence of thermal diffusion, the ion-induced effective surface diffusion (see Sect. 8.2) dominates the topography evolution and an energy dependence of the wavelength given by $\lambda \propto E$ can be anticipated [66].

There are only a small number of known studies that have examined ripple wavelength as a function of ion energy under low-energy ion bombardment of impurity-free substrates. Most experimental studies have shown that a power law relation $\lambda(E) \propto E^p$ with $p \in (0.5–1.1)$ successfully describes the wavelength-ion energy dependence. Castro et al. [40] have determined a linear relation, i.e. $p = 1$, between the wavelength of the ripples and the ion energy after Ar ion sputtering of Si with ion energies between 300 eV and 1100 eV. According to these authors, this linear dependence on ion energy indicates that the mechanism of stress-induced viscous flow determines the ripple formation in amorphized near-surface regions of ion-bombarded elemental semiconductors (see Sect. 8.2.1.2). Studies on the ripple formation after 1 keV Xe ion bombardment of amorphous alumina films by Babonneau et al. [51] have confirmed the assumption from the model of ion beam-induced viscous flow. Figure 8.13 shows an example of the ripple wavelength on alumina (measured using GISAXS as a function of the ion energy). A power-law dependence of the ripple wavelength with an exponent of $p = 0.58$ was observed. In Fig. 8.13, to predict the dependence of the wavelength on the ion energy between the investigated energy range (blue curve), the consistency of the viscous flow model by Castro et al. [40] is also displayed.

The influence of ion energy between 100 eV and 600 eV on the ripple wavelength of Si for a fixed ion fluence and ion-current density at room temperature were studied by Chowdhury and Ghose [39]. They observed that the ripple amplitude decreases with ion energy up to 300 eV ($p = -0.26$) and then rises ($p = 0.83$). The ripple formation on Si and Ge surfaces after Kr and Xe ion bombardment in the energy range between 400 eV and 2000 eV at room temperature and at incidence angles of 65° and 75° was studied in detail [35, 49]. A power law exponent of $p = 1.01$ was found for silicon, and for germanium, an exponent of $p = 0.81$.

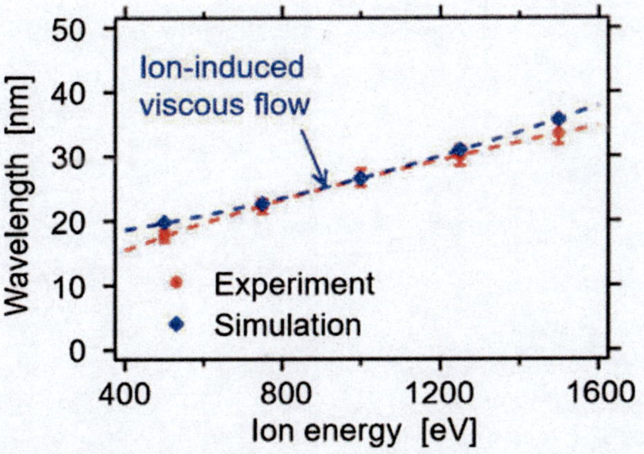

Fig. 8.13 Wavelength dependence of nanoripples produced by Xe ion bombardment of amorphous alumina at room temperature as a function of the ion energy. The samples were bombarded for 340 s (ion flux was 40 mA/cm^2) at an incidence angle of 55°. Figure is adapted from [51]

8.1.2 Formation of Ripples with Simultaneous Metallic Incorporation

The strong influence of impurity materials (mainly metals) on surface evolution during ion bombardment [22, 67–69] has been known for a long time. For example, Rossnagel and Robinson [68] used a simple arrangement for the simultaneous contamination of samples during ion beam erosion (see also Fig. 8.17) and were able to achieve a continuous impurity flux. The concentration of the deposited impurities varied between 0.2% and 3%. The ion beam-induced formation of cones was studied under these conditions (see Sect. 6.3.1.1). According to the authors, above a critical temperature, the surface diffusion of the metallic impurities leads to clusters. These clusters are characterized by a change in the sputtering yield and ultimately to surface patterning.

It has also been known for several decades now that metal contamination of sputter target during the sputtering has a significant influence on the target topography [61, 67]. It has also been partially accepted that metallic impurities generated by ion bombardment of sample holders, components of the ion source, or the wall of the vacuum chamber during the sputtering process have a considerable effect on the topography evolution of the sputtered sample [62, 70].

Ever since the seminal work by Ozaydin et al. [19, 28], it has been known that low concentrations of metallic impurities significantly influence the tendency for nanoripples to be generated. They studied Si(001) surfaces following 1 keV Ar ion bombardment under normal ion incidence conditions with and without simultaneous incorporation of Mo (this metal was sputtered away from the sample holder clamps) and found that in the absence of metallic contamination, the surface was smooth, whereas in the presence of Mo contamination, the surface exhibited nanodots. Figure 8.14

Fig. 8.14 AFM images of Si(001) surfaces after 1 keV Ar ion bombardment under normal incidence without (left) and with (right) Mo seeding. The Mo comes from the target-fixing clamps. Figure is adopted from [19]

displays a Si(001) surface without (left) and with (right) Mo incorporation after a sputtering time of 75 min. The authors concluded that the dot pattern develops only as a result of Mo seeding. They suggest that molybdenum silicides are formed, which reduce the sputtering yield. It was roughly estimated that a metal content of $>10^{15}$ metal atoms\cm^2 is necessary to substantially influence the topographical evolution under ion bombardment [45].

The experiments were primarily carried out in such a manner that the incorporated metallic particles came from the ion source or sample surroundings [6, 30, 71–73]. The experimental results are very sensitively dependent upon the material that is used to attach the sample to the sample holder, the alignment, the accelerating voltage on the second grid of the ion source, and the grid material. These types of experiments unambiguously demonstrate the strong influence of metallic impurities on the topographical evolution of Si surfaces under low-energy ion bombardment. Figure 8.15 provides an example of the Fe concentration profiles, the Fe atom area density, and the topography of Si after Kr ion bombardment with dependence on the incidence angle [6, 73]. At the low-incidence angle (and high Fe concentration), ripples evolve that are oriented perpendicular to the projected ion beam direction, while at the higher-incidence angle (and lower Fe concentration) the surface remains smooth, i.e. the metal atom contamination from the steel plate is directly responsible for the pattern formation. Interestingly, the Fe contamination is distributed non-uniformly. The HR-TEM micrograph (Fig. 8.16) indicates some dark regions in the crests of the ripples [73]. The composition of the dark regions on the crests were compared with the composition of the valleys by EELS analysis (positions I and II). From the comparison of the two EELS spectra, it is clear that the concentration of Fe and Cr is higher at the peak of the ripple than in the valley. Sánchez-García et al. [71, 72] studied the formation of nanoholes and nanodots on Si surfaces after normal

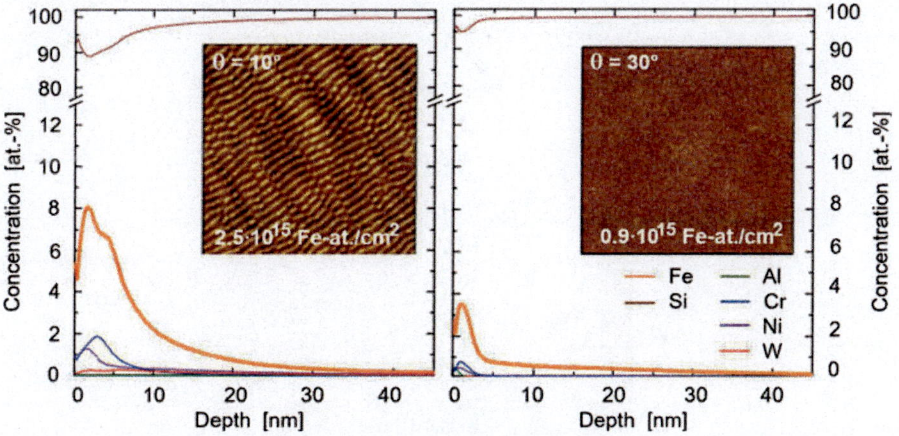

Fig. 8.15 TOF–SIMS concentration profiles after 2 keV Kr ion sputtering of a Si(001) sample at two different angles of ion incidence θ. Behind the grids of the ion source a cylindrical-shaped stainless-steel plate has been installed, which is also sputtered by the Kr ion beam. The ion fluence was 3.8 × 10^{18} ions/cm^2. The SIMS depth profiling was performed using 0.5 keV O_2^+ ions for erosion and 15 keV Ga^+ ions for analysis. The insets show AFM images (2 × 2 μm^2) and the corresponding area densities of Fe measured by RBS

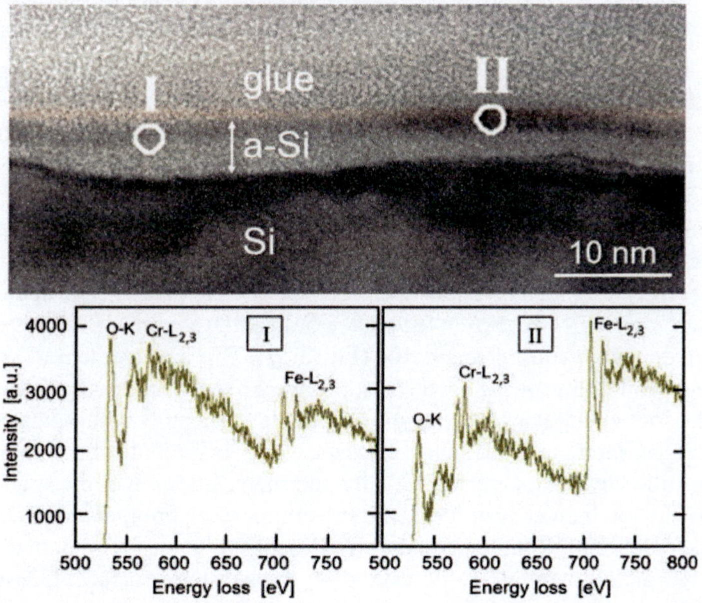

Fig. 8.16 HR-TEM cross-section micrograph and EELS spectra of positions I and II of Si after 2 keV Kr ion sputtering at an angle of θ = 20°. The ion fluence was 3.8 × 10^{18} ions/cm^2 (a-Si is the amorphized Si surface layer)

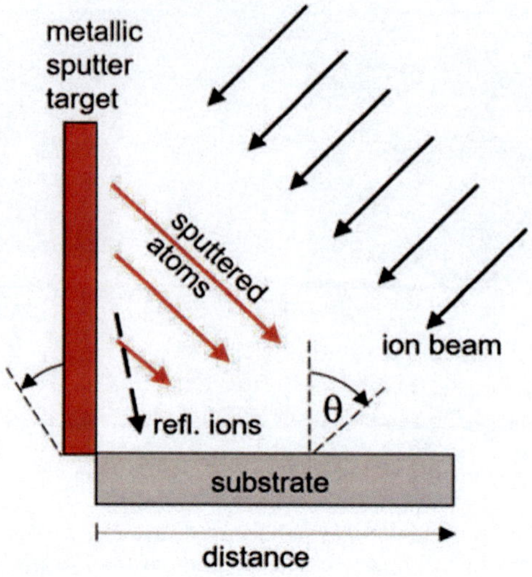

Fig. 8.17 Schematic representation of the experimental setup. θ is the ion incidence angle with respect to surface normal. The sputter target can be also tilted

incidence 1 keV Ar ion bombardment with incorporated Fe and Mo impurities up to an area density of 3.5×10^{15} metal atoms/cm^2. They, too, found that by increasing ion current density or increasing ion fluence at low ion-current density, the topography changes from holes to dots. This change of topography is explained by the decrease in metal contamination. A quantitative analysis of the role of metal contamination on basis of this experimental arrangement is rarely possible. It is assumed in part that the different results of the Si pattern formation in dependence on the ion energy and the ion incidence angle can be explained by an unintentional contamination of the surface during the sputtering experiments.

Consequently, for metal co-deposition, an arrangement of ion beam, sputter target, and substrate was typically chosen [61] that would allow a more quantitative study of metallic co- deposition during sputtering (Fig. 8.17). This arrangement was proposed by Rossnagel and Robinson [68] and variations were used in many other experimental studies. The sputter target is placed at an angle to the ion beam, so that sputtered atoms from its surface impinge on the substrate surface. The magnitude of the impurity flux can be controlled by the ion current density and the positioning of the sputter target. By varying the ion incidence angle θ, the substrate can be bombarded under normal and oblique conditions. Simultaneously, an adjacent sputter target is sputtered, so that the sputtered metallic atoms are deposited on the substrate surface. Consequently, (i) the metal atom flux is proportional to the flux of the primary ion beam and (ii) the surface coverage of metals decreases with the distance from the sputter target [maximum metal concentration on the substrate surface near the contact point with the sputter target (Fig. 8.17)]. This arrangement allows a broad variation of the direction and the flux of the sputtered metallic species. Reflected (scattered) primary ions from the sputter target (for grazing ion incidence and low ion energies in particular, see Sect. 5.10) additionally erode the substrate. In summary, the advantage of this experimental setup is that the ratio of the flux of co-deposited metal atoms to the flux of primary ions can be widely varied during one experiment. Typically, a controlled coverage of the surface with 10^{13}–10^{16} metal atoms/cm^2 is obtained (remain as coverage after sputtering).

Numerous ion beam sputtering experiments with simultaneous metal co-deposition have been carried out, with the aim of clarifying the role of metallic impurities in the formation of nanostructures. It must be acknowledged that all of the studies focus exclusively on silicon, although the phenomenon of pattern formation has been observed in all material classes. The only exceptions are the investigations by Liu et al. [54] of the 400 eV Ar ion beam-induced formation of ripples on fused silica (SiO_2) under concurrent co-deposition of the non-silicide-forming element Al. In Table 8.1, some selected experimental results for silicon are summarized, including experimental conditions.

An experimental, schematically shown in Fig. 8.17, was used to study the formation of patterns on Si [36] following 2 keV Kr ion bombardment at an incidence angle of $\theta = 30°$. They observed, in dependence on the distance, arrays of unordered nanoholes close to the steel plate and then ripples in the metal-rich region, with a wave-vector parallel to the co-deposited metal. At greater distances, weakly ordered arrays of nanodots were found. For the greatest distances, which were not or were only slightly contaminated by Fe, the smooth surface was preserved. Figure 8.18 exhibits the roughness, wavelength, and normalized Fe concentration for the different pattern regions as a function of the distance from the steel plate. The concentration of the co-deposited Fe seems to determine the formation of ripples or dots, along with the fact that the dots contain much less Fe atoms than ripples.

In similar experiments by Zhang et al. [53], the angle of the metal plate was additionally varied between 20° and 60° with respect to the silicon substrate (Fig. 8.17). They found the same hole-ripple-dot-flat-surface sequence of the topography with increasing distance from the metal sputter target for 5 keV Xe ion bombardment of Si. The patterns remain unchanged with increasing ion fluence.

It is obvious that the flux of the sputtered metal atoms is always dependent upon the flux of the primary bombardment ions. With the aim of studying the effect of the impurity flux on topographical evolution, Macko et al. [74] co-evaporated Fe from an e-beam evaporator directly onto the Si surface. These experiments have confirmed the relevance of the Fe atom on Kr ion flux ratio and of the incidence angle between the primary Kr ion flux and the flux of the co-evaporated Fe atoms for the pattern selection. It was further found that pattern formation can be detected for temperatures between 140 K and 440 K.

The experimental results involving co-deposited metal atoms during low-energy noble gas ion bombardment on silicon surfaces have demonstrated that metal contamination is an important parameter in the evolution of the topography. Intentional or incidental metallic contamination can also trigger nanopatterning with ion incidence angles also smaller than the critical angle θ_c and improve the quality of the nanopatterning (increasing system correlation length).

Table 8.1 Main results from selected references on the formation of nanostructures on Si surfaces by low-energy ion beams with simultaneous deposition of metals

Conditions of ion bombardment					Co-deposition	Detected compounds	Remarks/selected results	References
Species	Energy (keV)	Fluence ($\times 10^{18} cm^{-2}$)	Temp.[a] (K)	Angle θ (°)				
Ar	1	~0.3		0	Mo	-	Nanodots formed by Mo seeding	1
	1	~0.4	~350	0	Fe, Mo		Holes and dots formation as function of the ion flux, topography correlates with metal content	2
	0.5, 1		≤975	0	Mo		Impurities do not play a large role during normal-incidence Ion bombardment	3
	1	~4		0	Fe, Mo		Hole and dot production depends on metal content	4
	1.2	1.1	<375		st.St., Si		Ballistic smoothing and ion-enhanced viscous flow dominates the surface relaxation	5
	40	≤6.5		60	Steel	Silicide	Ripple wavelength and pattern dynamics is affected by Fe co-deposition	6
	0.2	≤0.25			Ni, Fe, Cu	Silicide	Silicide formation for Cu, Fe, and	7
	1			0	Fe		Phase separation does not precede the ripple formation; ordering of the ripple pattern increases with increasing ion fluence	8
	2			75	Au		No silicide formation, but the Au impurities induce the pattern formation	9
	1	1.6		0	Fe		Fe pronounces nanopatterns; ripple wave vector parallel to Fe flux direction	10

(continued)

Table 8.1 (continued)

Conditions of ion bombardment					Co-deposition	Detected compounds	Remarks/selected results	References
Species	Energy (keV)	Fluence ($\times 10^{18} cm^{-2}$)	Temp.[a] (K)	Angle θ (°)				
Kr	2	2		30	st.St., C		Sequence of nanopatterns as function of the distance; formation of ripples and dots dependent on metal concentration; no nanopattern after C co-deposition	11
	0.1–2	6.7		20	st.St		Ripples are formed parallel and perpendicular to the ion beam for high Fe concentrations	12
	2	0.59		30	Fe	Fe silicide	The ratio of flux of Fe to flux of Kr and silicide formation are relevant for pattern formation; sharp transition from ripples to a flat surface	13
	2	3.4		0–75	Steel		Fe content depends on angle θ	14
	2	0.59	≤660	30	st.St	Fe-silicide	Formation of sponge, segmented wall and pillar patterns	15
	2			30	Ag, Pd, C, Fe, Pb, Ir	Fe, Ir, Pd, silicide	Formation of silicides is a necessary, but not a sufficient condition for pattern formation	16
Xe	5	0.1		70	Si, Au, Ag, Pt		Ripple formation is not influenced by Si co-deposition, ripple formation in presence of Au, Ag, Pt	17

(continued)

Table 8.1 (continued)

Conditions of ion bombardment					Co-deposition	Detected compounds	Remarks/selected results	References
Species	Energy (keV)	Fluence (× $10^{18}cm^{-2}$)	Temp.[a] (K)	Angle θ (°)				
	5	≤2		≤30	Fe	Fe_xSi	Pronounced pattern formation	18
	5			30	Fe	Fe silicide	Sharp transition from pronounced ripple patterns to a flat surface	19
	5	≤0.5		60–90	Fe, Ni, Cu, Mo, W, Pt, Au	Fe, Mo, Ni, W, and Pt silicide	Pronounced pattern formation for Fe, Mo, Ni, W, Pt, flat surfaces after co-deposition of Cu or Au	20
	5, 10	≤0.5		62–70	Fe		Determination of the ripple propagation velocities	21

st.St. stainless steel

[a]If no temperature data is mentioned, then the ion irradiation experiments were carried out at room temperature

References

1 G. Ozaydin et al., Appl. Phys. Lett. 87 (2005) 163104
2 J.A. Sánchez-García, et al., Nanotechnology 19 (2008) 355306
3 G. Ozaydin-Ince, K.F. Ludwig, J. Phys. Condens. Mater 21 (2009) 224008
4 J.A. Sánchez-García, et al., J. Phys.: Condens. Matter 21 (2009) 224009
5 J. Zhou, et al., J. Appl. Phys. 109 (2011) 104315
6 A. Redondo-Cubero et al., Phys. Rev. B 86 (2012) 085436
7 O. El-Atwani, et al., J. Appl. Phys. 113 (2013) 124305
8 S.K. Vayalil, et al., J. Appl. Phys. 117 (2015) 024309
9 B. Moon et al., Phys. Rev. B. 93 (2016) 115430
10 S.K. Vayalil, et al., Appl. Phys. A 125 (2017) 225
11 S. Macko, et al. Nanotechnology 21 (2010) 085301
12 M. Cornejo et al., Appl. Surf. Sci. 257 (2011) 8659
13 S. Macko, et al., New J. Phys. 13 (2011) 073017
14 M. Cornejo, et al., Appl. Phys. A 102 (2011) 593
15 S. Macko et al., New. J. Phys. 14 (2012) 073003
16 M. Engler et al., Nanotechnology 25 (2014) 115303
17 H. Hofsäss, K. Zhang, Appl. Phys. A 92 (2008) 517
18 K. Zhang et al., New J. Phys. 13 (2011) 013033
19 K. Zhang et al., AIP Advances 2 (2012) 032123
20 H. Hofsäss, et al., Appl. Phys. A 111 (2013) 653
21 H. Hofsäss, et al., Phys. Rev. B 88 (2013) 075426

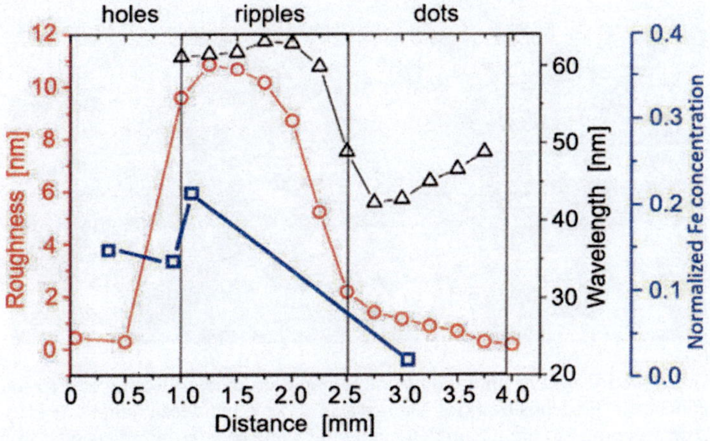

Fig. 8.18 Roughness, wavelength, and maximum Fe concentration normalized to the Si concentration as a function of distance from the steel plate. Figure adapted from [36] and modified

Consequently, the controlled contamination of materials with impurities during ion bombardment could be a selective measure in a variety of applications for tuning the topography, similar to the doping process in semiconductors to modify electrical conductivity.

8.1.2.1 Influence of the Co-deposited Metal Concentration

The concentration of the co-deposited metal atoms is characterized by a lateral gradient along the substrate surface, i.e. it has a maximum close to the metal plate and decreases with distance. The onset of the nanodot formation on Si surfaces was observed for an Fe atom area density between 2×10^{15} atoms/cm^2 and 4×10^{15} atoms/cm^2, whereas ripples appear for contents $> 5 \times 10^{15}$ Fe atoms/cm^2 [36]. With an increase in the primary ion fluence (time of ion bombardment), the metal content becomes saturated, i.e. the steady-state condition is achieved in which deposition and re-sputtering are in balance. For low-energy ion bombardment, a steady-state coverage of $<10^{16}$ co-deposited atoms/cm^2 was measured [36, 53], while for medium ion energy bombardment, the steady-state coverage is about one order of magnitude higher [75].

Figure 8.16 illustrates that the concentration of metallic impurities originating from sputtering from the ion source cathode or the substrate holder under oblique ion bombardment is higher at the crest of the ripple than at the valley. A stronger non-uniform distribution was attained after oblique long-time medium-energy Ar ion bombardment of Si with concurrent Fe deposition [75]. Figure 8.19 shows a cross-sectional TEM micrograph of the ripple structure.

The rear side (far side with respect to the Fe flux) displays a continuous Fe film, while the front side is characterized by discontinuous Fe agglomerations. EDX

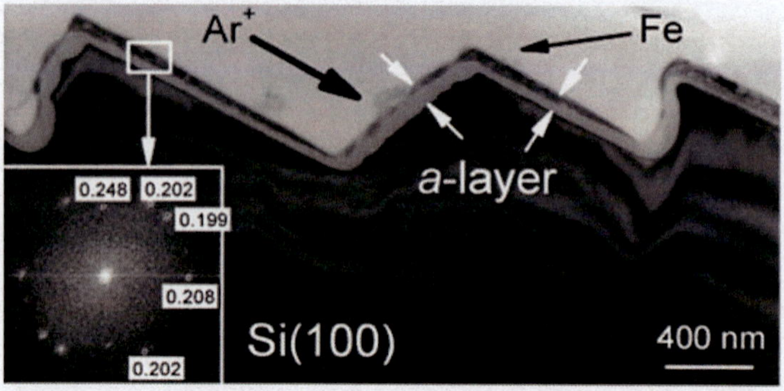

Fig. 8.19 Cross-sectional TEM micrograph of a Si ripple structure following 16 h Ar ion bombardment with concurrent Fe deposition (ion energy was 40 keV, incidence angle was 60°). The inset shows FFT corresponding to the region defined by the white square (figure adapted from [75])

analysis has revealed the Fe content to be four times greater at the rear side than at the front. The main spots in the FFT image (inset in Fig. 8.19) were identified as those from the orthorhombic, semiconducting, low- temperature phase β-$FeSi_2$. For longer ion bombardment times, the topography displays the familiar faceted pattern resulting from asymptotic behavior under shadowing processes (see Sect. 6.3.1.3). The roughness exponent changes from $\alpha = 0.72$ (symmetric ripple topography) to $\alpha = 0.85$ (asymmetric topography of facets) [75]. This transitional behavior is similar for surfaces under ion bombardment with and without metal co-deposition. However, the bombardment time required for this transition is significantly longer in the case of metal-free surfaces, i.e. the incorporated metal atoms promote topographical instability [75, 76].

8.1.2.2 Influence of the Angle of Ion Incidence

Madi et al. [29, 30] studied nanopattern formation after Ar ion sputtering at room temperature in dependence on an ion energy between 250 eV and 1000 eV and an angle of ion incidence θ. In these experiments, any metal contamination was avoided. Following Si sputtering, XPS and EDX studies have proven no evidence of metal contamination. For lower Ar ion energies between 250 eV and about 400 eV, holes were observed for normal and near-normal ion incidence and perpendicularly oriented ripples for θ between 5° and 20°. For higher ion energies (> 500 eV) and angles $\theta > 20°$, the surface remains flat. For incidence angles θ between 50° and 80°, parallel mode ripples were detected, independent of the ion energy. This is in contrast to the Bradley–Harper model, which predicts the formation of patterns for all ion incidence angles. Subsequent to these experiments, it is now assumed that metal contamination is a substantial trigger for the Si surface instabilities observed at ion incidence angles $\theta < 50°$. A number of studies have confirmed this threshold angle of about 50° for Si [74, 77, 78]. Because Si has been the only substrate material used up until now, it is not known whether a threshold angle can also be observed for other materials and incident ion species.

8.1.2.3 Influence of the Flux Ratio

Redondo-Cubero et al. [75] have produced patterns on Si surfaces with concurrent Fe co-deposition using significantly higher Ar ion energy (40 keV) under oblique ion irradiation ($\theta = 60°$). The sequence of the pattern types in dependence on the distance from the Fe plate varies from the ripple pattern to, after longer time of bombardment, facets, to dotted ripples and slight ripples at the greatest distances. They were able to conclude that the pattern dynamics and the range of pattern types are dependent on the ratio of the iron atom flux to the argon ion flux. The larger this ratio, the faster the pattern dynamics develop, i.e. the pattern topography at longer time of bombardment for larger distances from the metal plate is identical to the pattern topography at smaller distances for shorter time of bombardment.

8.1.2.4 Influence of Temperature During Ion Bombardment

For low substrate temperatures (e.g., $T \leq 440$ K [74] and $T \leq 425$ K [79]), i.e. when thermal mobility is less than ion beam-induced mobility, the pattern topography and the topography sequence in dependence on the distance from the metal plate remain consistent. With increased temperatures, the topography becomes increasingly dependent on temperature, because thermal diffusion, silicide formation, and phase separation processes increasingly determine the surface topography. Consequently, a progressive decrease in dot heights for temperatures between 400 K and 500 K was observed [79]. For temperatures above 550 K, the nanopatterns disappear.

It is noteworthy that the crystallization temperature of amorphous thin silicon layers (intrinsic crystallization temperature is about 700 °C) decreases in the presence of metal impurities. For example, the crystallization temperature of amorphous Si contaminated with Ni (silicide former) is about 500 °C and after contamination with Au or Al (form eutectics) is about 180 °C and 150 °C, respectively [80]. It can be ascertained that on the one hand, metal impurities are, to some degree, capable of significantly decreasing the crystallization temperature and, on the other hand, that recrystallization prevents the evolution of nanopatterns. It thus appears that the formation of nanopatterns in the presence of co-deposited metals is possible even at relatively low temperatures.

8.1.2.5 Models of Nanopatterning During Simultaneous Metal Co-deposition

In the literature, numerous explanations have been provided that deal with the description of nanopattern formation in the presence of co-deposited impurities (see [8, 9]). Two primary attempts at explanation will be described:

(1) Variation of the sputtering yield along the surface. In this version, the metallic contamination is non-uniformly distributed due to segregation. If the partial sputtering yield of the metal component differs from the sputtering yield of silicon, then a preferential sputtering (see Sect. 5.8) leads to the enrichment of one component.

(2) In the second explanation, the co-deposited metallic elements form a compound with the Si atoms of the substrate, a metal silicide, because ion bombardment increases the number of free Si bonds, causing a stronger interaction between Si and co-deposited metal atoms. Again, the changed surface chemistry leads to a variation of the sputtering yield.

In early studies of the ion beam sputtering experiments with concurrent metal deposition, it was supposed that the formation of metal silicide effects topographical changes [19, 42]. More detailed analysis using EDX, EELS, and XPS have shown that silicide formation for co-deposited Fe [6, 53, 72, 74, 75, 81–83], Mo [72], Ni [82], Cu [82], Pd [84] can be proven. For example, in-situ XPS measurements by El-Atwami et al. [82] have shown a significant chemical shift in the binding energy of the co-deposited Ni with respect to the pure Ni of about 0.85 eV for N $2p_{3/2}$ and about 1.15 eV for Ni $2p_{1/2}$, indicating the formation of Ni silicide following 200 eV Ar ion bombardment with a fluence of 9×10^{16} ions/cm^2.

The silicide formation of transition metals can be expected when the enthalpy of formation (heat of formation) is significantly negative. In the following table (Table 8.2) the calculated enthalpies of formation for elements used in co-deposition experiments are compared.

The table indicates that for the elements Cu and Au, the formation of silicide is unlikely at room temperature, whereas the substantial negative enthalpies of formation for transition metals Pd and Pt show that the formation of silicide can be expected. The silicide normally exists in an amorphous state after co-deposition at room temperature.

Table 8.2 Calculated enthalpy of formation for silicide of selected transition and noble metals at room temperature. In accordance with the semi-empirical model by Miedema [85], the calculations were carried out for three concentrations

Metal X	Enthalpy of formation (kJ/mol)			Tendency to form silicide
	X_2Si	XSi	XSi_2	
Fe	−25	−30	−13	Strong silicide former
Ni	−34	−36	−19	Strong silicide former
Mo	−25	−32	−17	Strong silicide former
Pd	−51	−59	−43	Very strong silicide former
W	−22	−24	−13	Strong silicide former
Pt	−49	−56	−43	Very strong silicide former
Cu	−3	0	+10	Weak silicide former
Au	−7	−5	0	No stable silicide

Engler et al. [84] studied pattern formation on Si with simultaneous co-deposition of silicide-forming-elements and non-forming elements. For the strong silicide-forming elements Fe and Ir, significant pattern formation could be verified. With non-metal carbon, silicide formation could be expected, but pattern formation following C co-deposition was not observed. For the co-deposited non-silicide-forming metals Ag and Pb, no pattern formation was found. It was concluded that silicide formation is a necessary but not all-sufficient condition for pattern formation. Sánchez-García et al [71] found that the metal content for nanohole patterns is comparatively high and concluded that the generation of different patterns can be understood as a result of the dependence of the ion erosion rate on the local surface morphology and the concentration of metal impurities.

Moon et al. [86] have studied in detail the possibility of generating nanopatterns on Si with and without co-deposition of the non-silicide-forming element gold. The sputter experiments were performed with 2 keV Ar ions under an incidence angle of 75°, since the sputtering yield of Au and Si are equal at this particular angle. The experiments with Au co-deposition exhibited that well-defined nanopatterns with ripples can be observed on the Si substrate, indicating that silicide formation is not a necessary condition for the pattern formation. At very low Au concentrations, the surface remains flat.

As above mentioned, it is assumed that the formation of nanopatterns is based on the difference between the local sputtering yields of silicon and metallic components (pure metal, silicon-metal mixture, or metal silicide component) at the surface or in the near-surface region [19, 36, 71, 87, 88]. It is known [89] that the sputtering yield of metals, Y_M, is greater than the sputtering yield of stable metal silicides, Y_{MSi}, and silicon, Y_{Si}, for noble gas ion energies up to a certain keV, i.e. $Y_{si} < Y_{MSi} < Y_M$. This behavior can be expected, since the surface binding energy of a silicide is increased by the formation of a silicide. Figure 8.20 illustrates the sputtering yield and the ratio of the sputtering yields of iron, silicon, and the stable compound $FeSi_2$ after normal Ar ion incidence. It is interesting that in the energy range under consideration, the sputtering ratios are independent of ion energy beyond 250 eV and vary between about 1.5 and 2.5.

It must be pointed out that the partial sputtering yield of a particular species during the sputtering of composite targets can be enhanced with regard to its pure sputtering yield [90]. This phenomenon, known as sputter yield amplification effect, first observed by Harper et al. [91], results in enhanced sputtering of the substrate species when a high-mass impurity (e.g., metal atoms) on a low-mass substrate (e.g., silicon) is bombarded. The enhancement, with a factor between 2 and 10, was found to increase with the mass and density of the impurity as well as the temperature during ion bombardment, and to decrease with increasing ion energy. As a result of the ion bombardment, the multicomponent material (e.g., metal silicide phase) decomposes increasingly with the ion fluence and one component (here, Si) diffuses to the surface and segregates.

Against the background of (i) higher sputtering yields of Fe and Fe silicide in comparison to Si and (ii) phase separation that can occur under low-energy ion bombardment, Zhang et al. [45] have proposed a model to explain the nanopattern

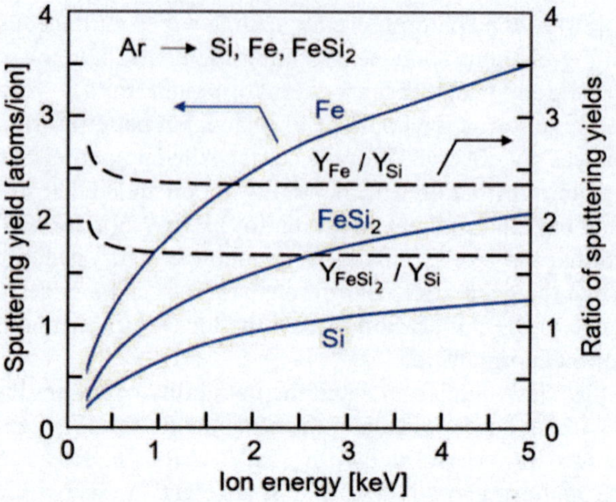

Fig. 8.20 Sputtering yield (solid curves) of Fe, Si, and $FeSi_2$, and the sputtering ratios (dotted lines) of Y_{Fe}/Y_{Si} and Y_{FeSi}/Y_{Si} as a function of Ar ion energy under normal ion incidence. The energy-dependent sputtering yields are calculated using SRIM software

formation. They suggest that a thin silicide film is initially formed. With progressive ion bombardment and concurrent metal co-deposition, silicide phase separation takes places in local surface regions with lower and higher Fe content, driven by ion-induced diffusion. The difference in the sputtering yields between Fe and the amorphous Fe silicide causes the formation of nanodots. With increased sputter time, the directional iron deposition on this rough topography leads to the appearance of shadowing effects that result in ripple development.

The assumption that the ion beam-induced decomposition or phase separation occurs simultaneously with the formation of nanopatterns was studied by Vayalil et al. [83]. They prepared nanoripple patterns on Si using normal incidence of 1 keV Ar ions and simultaneous incorporation of iron. Comparative AFM and GISAXS measurements have shown that phase separation does not precede ripple formation, rather it develops concurrently with the ripple structure.

One of the first predictions of the topography evolution on binary compounds under ion bombardment was made by Shenoy et al. [92], where the surface composition was coupled with surface height. On this basis, Bradley and Shipman [93, 94] developed a theoretical model for a composition-driven patterning mechanism following normal ion incidence. They assumed that (i) surfaces can be unstable without participation of a second species (see Bradley–Harper model, Sect. 8.2.1.1), (ii) the ion beam-induced redistribution of the different components in the near-surface region is substantially different, and (iii) the components diffuse according to Fick's law. It is also assumed that a height field $z = h(x, y, t)$ of the bombarded surface is coupled with two concentration fields, $C_A(x, y, t)$ and $C_B(x, y, t)$, for the two components A and B, where one of these components is preferentially sputtered and the other component is enriched. It is also assumed, that the surface is eroded at a constant velocity under steady-state conditions. As a result of this consideration, Bradley and Shipman [93] proposed linearized equations to describe perturbations to the height and concentration field of one of these two components.

The linear stability analysis on these equations by Shipman and Bradley [94] leads to a range in which linearly unstable wave-vectors exist for appropriated experimental conditions. Later, Bradley [95] expanded this model to include that substrate atoms and co-deposited impurities can react to produce a chemical compound. For example, this model predicts that the more easily sputterable component is enriched at the top of the nanodots (c.f. Figs. 8.16 and 8.19). The quadratic nonlinearities in the equation of the motion, which describes the surface topography changing under ion bombardment, explain the experimentally observed hexagonal pattern formation on the surface of binary compounds.

Recently, Bradley [96] has modified the existing theory of surface evolution with concurrent impurity deposition [92–95] by taking into account that the co-deposited impurities also influence the collision cascade. This provided an explanation for the experimental results obtained by Moon et al. [86].

Norris [97] has extended the Bradley-Shipman model by including the effect of ion-assisted phase separation. The phase separation as a consequence of decomposition is considered to be the driving force behind nanopattern formation. Taking into account the dependence of the sputtering rate on the composition, the topography of the bombarded surface is determined by the changing of the chemical composition. On the basis of this extended model, pattern formation can be explained by a chemically driven finite-wavelength instability even if the erosive instability according to the Bradley-Shipman model is not observed. It is suggested that the stress state in the near-surface region is partially generated by the incorporation of metallic impurities, leading to the formation of nanodots and nanoripples.

Ozaydin et al. [19] have assumed that surface stress evolution during ion sputtering of Mo-seeded Si may be responsible for the formation of nanopatterns. They observed a continuous increase of tensile stresses over time. Stress relief could lead to protuberance formation.

With the objective of integrating surface stress and preferential sputtering into a continuum model to describe surface evolution during ion bombardment and simultaneous metal co-deposition, Zhou and Lu [76] developed a model based on a coupled two-field equation. They have shown that in the presence of metallic species, surface-stress-induced instability, together with curvature-dependent erosion, overcomes ballistic smoothing, leading to dot pattern formation. In addition, preferential sputtering, in conjunction with topographic instability, produces the composite patterning. Based on this model, numerical simulations of normal incidence 1200 eV Ar ion sputtering of Si(100) surfaces with and without Fe incorporation were carried out and compared to the experimental results. An example is shown in Fig. 8.21. Without Fe co-deposition, smooth surfaces were obtained in experiments as well as simulations. In contrast, the PSD functions for Si samples with Fe co-deposition (samples A) indicate that significantly rougher surfaces can be observed. Consequently, comprehension of the of curvature-dependent sputtering and the stress-induced instability induced by metal impurities in an extended continuum description

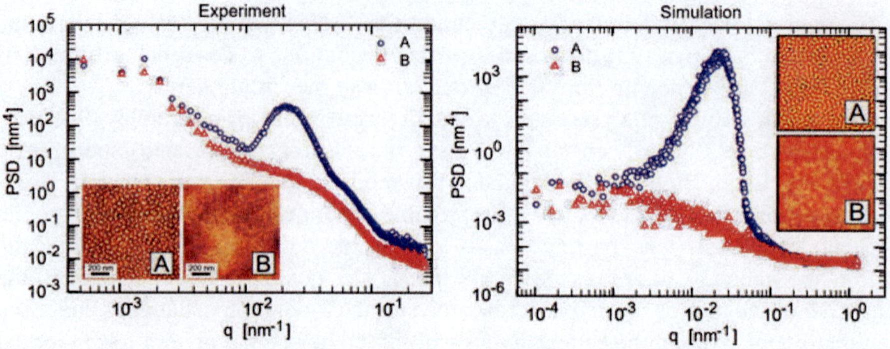

Fig. 8.21 PSD functions from Si(100) surfaces and $1 \times 1\ \mu m^2$ AFM images (insets) after sputtering in presence (**a**) and absence of Fe co-deposition (**b**). Figure adapted from [76]

makes it possible to reproduce results that are in good qualitative agreement with the experimental results.

In summary, it must be acknowledged that the underlying mechanism of pattern formation in the presence of metallic contamination remains virtually unknown. The above discussions indicate that, on the one hand, nanopattern formation produced by ion sputtering at normal and small ion incidence angles can be strongly affected by simultaneous metallic co-deposition. On the other hand, sputtering-induced pattern evolution in the simultaneous presence of metallic impurities cannot be comprehensively explained by either height fluctuations or phase separation processes. Consequently, both experimental studies on the ion beam-induced formation of nanopatterns on non-silicon substrates under concurrent metal co-deposition and theoretical consideration of the involved mechanisms are needed.

8.1.3 *Formation of Nanodot and Nanohole Patterns*

The formation of nanodots by ion beam sputtering is considered important, as this technique has the ability to produce surface patterns characterized by uniform dot size and a near-perfect dot ordering. Undoubtedly, the ion beam-induced formation of nanodots can be also influenced by co-deposited impurities. In fact, few studies are currently known on this topic.

In addition to the formation of periodic ripples after 4 keV air ion bombardment at incident angles ranging from 30° to 80° (cf. Fig. 8.2), Navez et al. [13] also observed, for the first time, randomly distributed dot-like structures on a glass surface under normal ion bombardment, with a diameter of about 40 nm (Fig. 8.22). In the subsequent decades, the mechanisms of spontaneous dot formation during low-energy ion bombardment have only rarely been studied. Nearly 30 years after the discovery of the formation of randomly distributed nanodots as a result of ion beam erosion (Fig. 8.22), the ion beam-induced formation of hexagonal arrays of

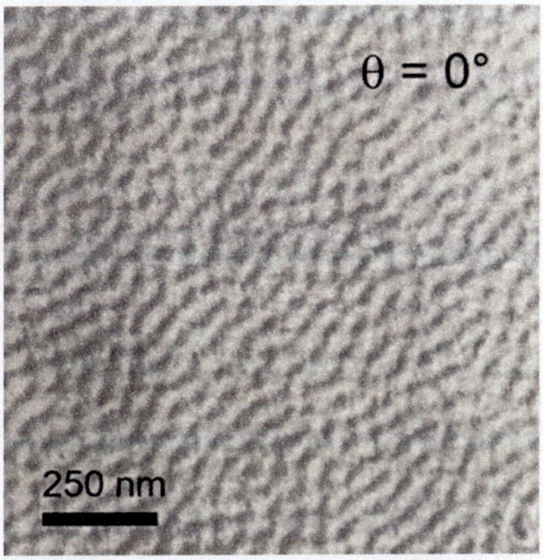

Fig. 8.22 Surface topography of glass after 4 keV air ion bombardment under normal ion incidence. Figure was adapted from [13]

semiconductor nanodots was observed. First, Facsko et al. [98] produced hexagonally ordered nanodots on GaSb surfaces using normal 420 eV Ar ion incidence bombardment (Fig. 8.23, left) and immediately thereafter the same pattern formation [99] was demonstrated on rotating InP surfaces with oblique Ar ion bombardment at energies between 350 eV and 1200 eV (Fig. 8.23, right). In both studies, the dot diameter was < 100 nm. Then, Gago et al. [100] demonstrated the formation of nanodots on silicon surfaces for the first time.

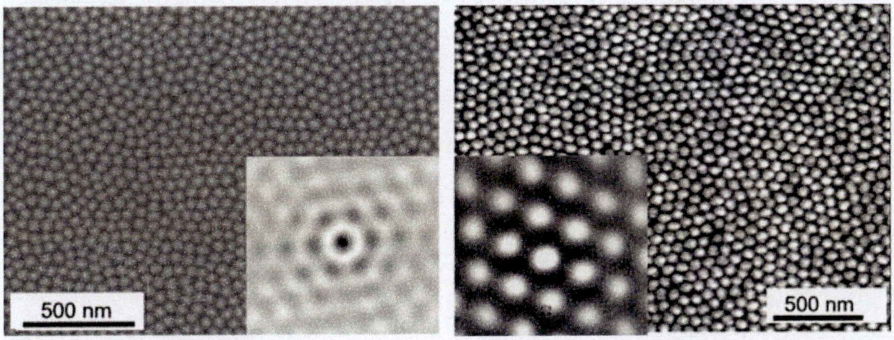

Fig. 8.23 Left: SEM image of hexagonally ordered nanodots on (100) GaSb surface after 4×10^{18} Ar ion/cm^2 bombardment. The ion energy is 420 eV and the angle of ion incidence is 0°. The inset shows the two-dimensional autocorrelation calculated from a magnified area of this SEM image. Figure is adapted from [98]. Right: AFM image of hexagonally ordered nanodots on (100) InP surface after 9×10^{18} Ar ions/cm^2. The ion energy is 500 eV and the angle of ion incidence is 40°. The inset shows the two-dimensional autocorrelation calculated from a magnified area of this AFM image (figure adapted from [99])

In general, experiments that involve producing dots on surfaces using ion beam erosion are performed under UHV conditions. Usually, initially flat surfaces are bombarded with noble gas ions. In the majority of experimental studies, the ion energy (E $\leq$ 2 keV), the angle of ion incidence (θ $\leq$ 90°) and the temperature of the target material during the ion bombardment (T $\leq$ 400 °C) are varied. In principle, three technological methods of low-energy ion bombardment are used: (i) normal incidence ion bombardment, (ii) ion bombardment under oblique incidence, and (iii) ion bombardment under oblique incidence and simultaneous rotation of the sample (the so-called Zalar rotation [101]). The aim of these preparation methods is the preservation of the in-plane surface symmetry. In off-normal sputtering experiments, the anisotropy of the evolution of nanostructures is given by the direction of the incident ions. A continuous rotation of the sample or the ion beam derestricts this limitation and can prevent the evolution of anisotropic nanostructures (ripples). The rotation of the target around its surface normal plays an important role in the evolution of dots in non-normal (off-normal) ion bombardment. Bradley [96] found that the continuous rotation of the bombarded sample enhances the quality of isotropic structures (dots). Muñoz-García et al. [102] derived approximate analytical expressions for the dependence of stationary dot amplitude and lateral size of dots for both the normal and the oblique rotating samples. The rotation frequencies used are in order of 1–30 rotations per minute (rpm). The topography of surfaces characterized by dots and their dependence on the experimental parameters have primarily been investigated by means of scanning probe microscopy (AFM, STM). A representative example of dot structure is provided in Fig. 8.24. The AFM image (left) displays domains of closely packed, hexagonally ordered dot structures (see the 2D autocorrelation function image, Fig. 8.24, left) [103, 104]. Among each other, the individual domains are randomly oriented. This allows a ring-like spectrum in the FFT image to be obtained. A typical nanodot topography on Si surfaces, including the 2D autocorrelation function image after low-energy rare gas ion bombardment, is shown in Fig. 8.24. From the first peak of the angularly averaged PSD function (Fig. 8.24

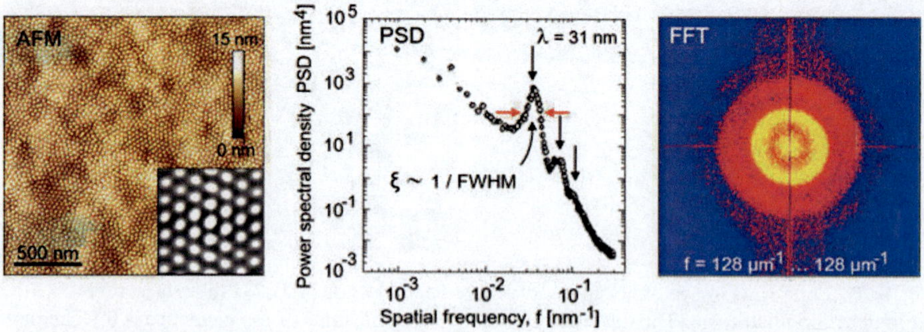

Fig. 8.24 Typical dot structure on Si surfaces after 500 eV Ar ion bombardment at room temperature (figures are adapted from [104]). The angle of ion incidence was 75°, the ion flux 200 μA/cm^2 and the erosion time was 180 min. Left: AFM micrograph (inset shows the 2D autocorrelation function). Middle: angular averaged PSD spectra. Right: the corresponding FFT spectra

middle), a mean dot size of $\lambda = 31$ nm is deduced. The size fluctuation $\Delta\lambda$ of the dots, calculated from the full width at half maximum (FWHM) of the first order PSD peak, is $\pm$ 2.5 nm. In addition, minor peaks are visible (Fig. 8.24, middle), indicating the high lateral ordering. The system correlation length ξ (Appendix L) can be used as quantity to describe the lateral ordering of dots [18], where ξ gives the length scale up to which spatial correlation is present, i.e. the mean domain size. This is deduced from the FWHM of the first order PSD peak and is inversely proportional to the FWHM, $\xi \sim 1/\text{FWHM}$. It should be noted that the lateral correlation length can be also determined using grazing incidence small angle X-ray scattering (GISAXS). For example, lateral correlation lengths for Si nanodots up to 1.9 μm were determined [104]. The bright ring in the Fourier image (Fig. 8.24, right) corresponds to the mean size of the nanodots. Adjoining rings in the Fourier image (up to third order) prove the high degree of uniformity of the dots.

In general, the temporal evolution of nanodots during the ion sputtering process can be described by a power law dependence of the characteristic wavelength $\lambda \propto t^n$ (see Sect. 8.1.2.3), where this wavelength serves as a measure of the lateral dot size. The time-scaling exponent n can be obtained from the linear fit of the log–log plot of the wavelength λ versus the time of bombardment t. The wavelength λ can be extracted from the power spectral density analysis of scanning probe microscopic images and plotted as a function of sputtering time in the log scale.

The evolution of nanodots as result of low-energy ion bombardment was studied preferentially on surfaces of semiconductor materials (Si, Ge) and the III–V compound semiconductors (GaSb, InP, etc.). Figure 8.25 shows some examples for different semiconducting materials. The surface dot density in order of 10^9 up to 10^{12} dots/cm^2 is of great interest for applications in ultra-high-density recording media and optoelectronic devices (see Sect. 8.3). The typical morphology of nanodots is shown in Fig. 8.26. The cross-section TEM micrograph illustrates that the average distance between the dots is approximately equal to their lateral size. The high-resolution transmission electron microscope (HR-TEM) micrograph also shows that the dots have a cone-like shape, covered by an amorphous surface layer [105, 106]. The thickness of this layer, caused by amorphization due ion bombardment, corresponds to the ion range. The dots themselves are crystalline with the same crystalline structure as the non-bombarded material (Fig. 8.26, left), i.e. the crystallinity and orientation of dots and bulk are identical. However, another important difference in the shape of nanodots formed at normal (or near-normal) and grazing ion incidence at simultaneous sample rotation (15 rpm) could be observed [8]. An inspection of the geometrical shape of the different dots using cross-sectional HRTEM (Fig. 8.27) revealed that the nanostructures produced at normal incidence show a conical shape with a rounded apex and sidewall angles from 60° to 70° (Fig. 8.27). Both the width and the height of these dots are in the range of 50 nm. In contrast, the dots formed at glancing angle ion incidence show a sinusoidal shape, with a diameter of about 20 nm and a dot height of about 10 nm (Fig. 8.27). Both types of dots are covered by a thin amorphous layer with a thickness of about 4 nm. With increasing ion energy, the different shapes persist at normal and grazing angle ion incidence. The shapes of

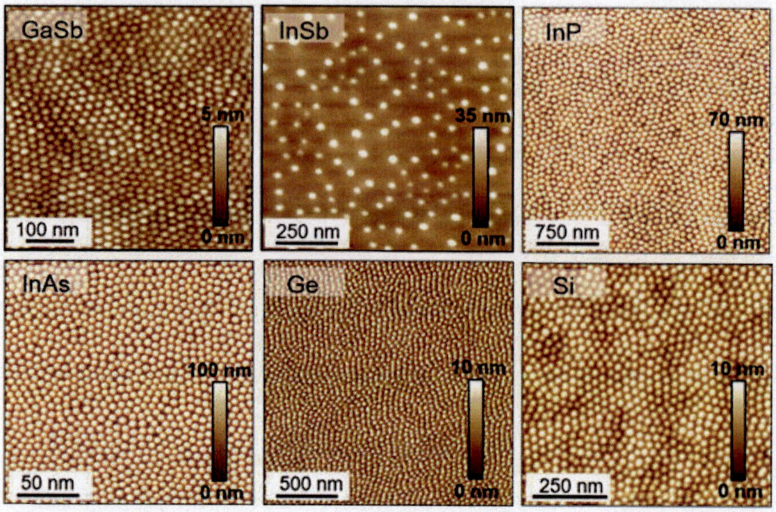

Fig. 8.25 AFM images of semiconductor surfaces after low-energy ion bombardment at room temperature (E—ion energy, J—ion flux, θ—angle of ion incidence, Φ—ion fluence, t_I—duration of ion bombardment). GaSb: Ar ions, E = 500 eV, J = 400 μA/cm^2, θ = 80°, with sample rotation, t_I = 90 min, [107]. InSb: Ar ions, E = 500 eV, J = 400 μA/cm^2, θ = 80°, with rotation, t_I = 10 min, [105]. InP: Ar ions, E = 500 eV, J = 400 μA/cm^2, θ = 50°, with rotation, t_I = 90 min [107]. InAs: Ar ions, E = 1000 eV, J = 270 μA/cm^2, θ = 30°, with rotation, t_I = 60 min, [105]. Ge: Kr ions, E = 2000 eV, Φ = 6.8 × 10^{18} cm^2, θ = 30°, without rotation, t_I = 10 min [104]. Si: Kr ions, E = 1000 eV, Φ = 6.7 × 10^{18} cm^2, θ = 75°, with rotation, [1, 108]

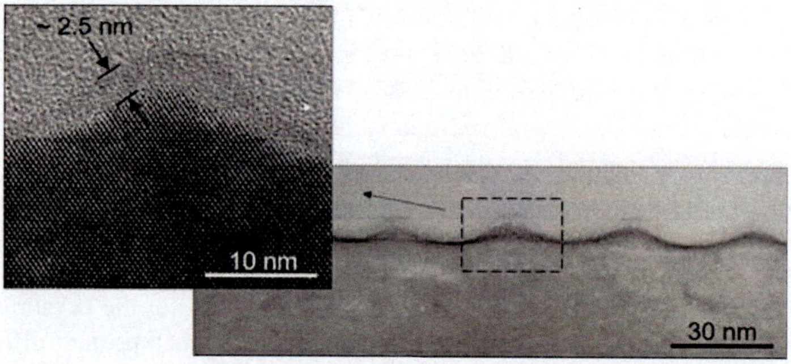

Fig. 8.26 Cross-section TEM micrograph of self-organized Si nanodots produced by Ar ion bombardment at room temperature (ion energy is 500 eV, ion flux is 200 μA/cm^2, angle of ion incidence is 75° and erosion time is 180 min) [103, 108]. HRTEM image (left) shows a single nanodot with an approx. 2.5 nm thick amorphized surface layer

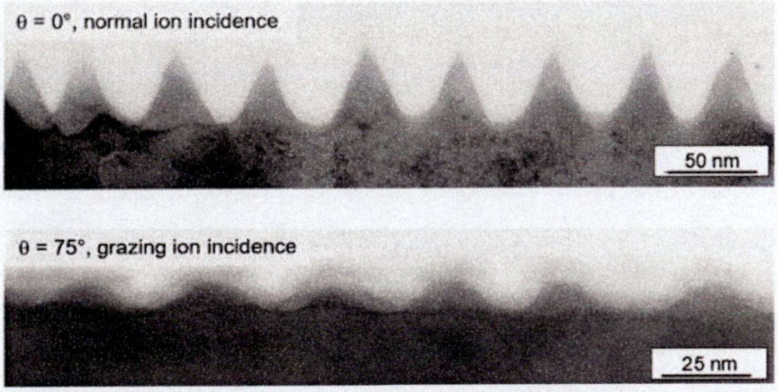

Fig. 8.27 Cross-sectional high-resolution TEM micrographs of GaSb nanodot patterns produced at normal (at the top, 500 eV Ar ions) and grazing ion incidence (below, 1200 eV Ar ions), respectively. The temperature during ion bombardment was 285 K, the ion current density was 300 $\mu A/cm^2$ and the sputter time was 90 min

these nanostructures are in agreement with those obtained by AFM, if tip convolution effects are taken into account.

8.1.3.1 In-Plane Ordering of Nanodots

With the selection of appropriate ion beam conditions, it was observed that self-organized, hexagonally or squarely ordered dot patterns can evolve under low-energy ion sputtering of semiconductor surfaces for normal as well as oblique incidence ion bombardment. These dot patterns are characterized by uniform spacing and size of the dots. Table 8.3 contains a summary of the experimental conditions, the dot diameter, and the lateral dot arrangement.

The ordering of the nanodots is dependent on the ion energy, temperature, angle of ion incidence, and the duration of bombardment (ion fluence). To illustrate, dependence on ion energy and ion fluence is shown in Fig. 8.28 [103, 109]. Silicon surfaces were bombarded with Ar ions at two different ion energies (Fig. 8.28, left and middle). By comparing the FFT and autocorrelation images, it is obvious that the ordering of the dots increases with decreasing ion energy. Following 1800 eV Ar ion bombardment (Fig. 8.28, left), the dot structures are poorly ordered, as can be observed from the Fourier image. Dot patterns with distinct short-range ordering appear on the surface following 500 eV Ar ion bombardment (Fig. 8.28, middle). In the FFT image, a clear ring can be observed. Figure 8.28, right, shows a comparison of PSD functions after 500 eV Ar ion bombardment for different times of ion erosion (fluences). The system correlation length ξ (the average domain size of ordered dots) can be deduced from the FWHM of the first peak. In contrast to the ion energy dependence of the dot ordering, the ordering increases with increasing fluence. At prolonged sputtering, multiple peaks are also visible, indicating the high lateral ordering and size homogeneity of dots.

Table 8.3 Summary of experimentally observed ordered dot patterns

Target	Ion species	Energy (keV)	Temperature (°)	Incidence angle (°)	Dot diameter (nm)	Dot pattern	References
GaSb	Ar	0.42	−60, +60	0	18–50	Hex	1
	Ar	0.075–1.8	−60, +60	0	15–120	Hex	2
	Ar	0.5	<100	0	40–60	Hex	3
	Ar	−2	60	0	20	Hex	4
	Ar	−1.8	<100	0	15–80	Hex	5
	Ar	0.3–2	RT-100	0–90	15–50	Hex	6
	Ar	0.5	<100	0	<55	Hex	7
	Ar	0.3–2	RT-100	0–90	20–110	Hex and square	8
	Ar	0.4	80	0	44	Hex	9
	Ar	3	−129 to 150	0		Hex and square	10
InP	Ar	0.35–1.2	RT	10–80	<85	Hex	11
	Ar	0.075–1.8	−60, +60	0	15–120	Hex	2
	Ar	0.5–2	−170, RT		30–60	Trigonal	12
	Ar	0.2–2	RT-100	0–90	≤85	Hex	6
	Ar	0.3–2	RT-100	0–90		Hex and square	8
	O_2	1	RT	38–62	≤75	Chains	13
	Ar	1	−110, 23, 36	0	73 ± 10	Hex	14
	Ar	3	RT	16	≤90 ± 26	Hex and square	15
	Ar	1.5	RT	0–76	67–114	Square	16
InAs	Ar	1	RT-100	30	60 ± 7	Ordered	8
InSb	Ar	1.2	RT-100	85	55	Ordered	8
	Ar	4	RT	50		Chains	17
Si	Ar	1.2		0	≈50	Hex	18
	Ar	0.1–2	RT	0–75	30	Hex	19
	Ne, Ar, Kr, Xe	0.5–2	RT	75	20–50	Hex	20
	Ar	1–5	RT-800	0	10–70		21
	Ar	1.2		0	40–50	Hex	22
	Ar	1	RT-355	0	30–50	Hex	23

(continued)

Table 8.3 (continued)

Target	Ion species	Energy (keV)	Temperature (°)	Incidence angle (°)	Dot diameter (nm)	Dot pattern	References
	Xe	2	RT	0–45	≤50	Square	24, 25
	Ar	0.1–0.6	RT	75	<50	Hex	26
Ge	Xe	0.8–2	RT	0	30–70	Hex	27
	Xe	2	RT	10–30	0–30	Hex	25, 28

RT room temperature, *hex* hexagonally ordered dot patterns, *square* squarely ordered dot patterns, *trigonal* trigonal ordered dot patterns, *0* ion beam impinging parallel to the surface normal

References
[1] S. Facsko, et al. Science 285 (1999) 1551
[2] S. Facsko et al., Phys. Rev. B 63 (2001) 165329
[3] T. Bobek, et al., Phys. Rev. B 68 (2003) 085324
[4] T. Bobek, et al., Nucl. Instr. Meth in Phys. Res. B 178 (2001) 101
[5] S. Facsko, et al., Appl. Phys. Lett. 80 (2002) 130
[6] F. Frost, B. Rauschenbach, Appl. Phys. A 77 (2003) 1
[7] M. Xu, C. Teichert. J. Appl. Phys. 96 (2004) 2244
[8] F. Frost et al., Nucl. Instr. Meth. in Phys. Res. B 216 (2004) 9
[9] T. Bobek, et al., Adv. Mater. 19 (2007) 4375–4380
[10] E. Trynkiewicz, et al., Condens. Matter 30 (2018) 304005
[11] F. Frost, et al., Phys. Rev. Lett. 85 (2000) 4116
[12] Y. Yuba et al., Nucl. Instr. Meth in Phys. Res. B 206 (2003) 648
[13] S.K. Tan et al. Nucl. Instr. Meth in Phys. Res. B 248 (2006) 83
[14] S.K. Tan, A.T.S. Wee, J. Vac. Sci. Technol. B 24 (2006) 1444
[15] D. Paramanik S. Verma, Nucl. Instr. Meth in Phys. Res. B 266 (2008) 1257
[16] I. Sulania et al., Adv. Mat. Lett. 1 (2010) 118
[17] F. Krok et al., Nucl. Instr. Meth. in Phys. Res. B 212 (2003) 264
[18] R. Gago et al., Appl. Phys. Lett. 78 (2001) 3316
[19] B. Ziberi, et al., Appl. Phys. Lett. 87 (2005) 033113
[20] B. Ziberi et al., Surf. Sci. 600 (2006) 3757
[21] L. Ling, et al. Phys. Rev. B 71 (2005) 155329
[22] R. Gago et al., Appl. Phys. Lett. 89 (2006) 233101
[23] R. Gago et al., Phys. Rev. B 73 (2006) 155414
[24] B. Ziberi, et al., Appl. Phys. Lett. 92 (2008) 063102
[25] B. Rauschenbach, et al., Proceed. 2nd IEEE Intern. Nanoelectr. Conf. Shanghai 2008, p. 93
[26] D. Chowdhury et al., Mat. Sci. Engng. B179 (2015) 1
[27] B. Ziberi et al. J. Phys.: Condens. Matter 21 (2009) 224003
[28] B. Ziberi et al., Appl. Phys. Lett. 88 (2006) 173115

Table 8.3 shows that a hexagonal close-packed two-dimensional arrangement of dots is preferentially formed as a result of the ion beam-induced self-organizing process. Along with square dot patterns, the development of long dot chains was also observed. Notably, the observation of the squared arrangement of nanodots is contrary to theoretical predictions. The regular dot patterns on selected semiconducting materials surfaces appear only within a limited range, dependent on the ion fluence [100], the ion energy [2, 10, 109], the temperature during ion bombardment [105], the angle of ion incidence [105, 107], the ion beam divergence [105], and

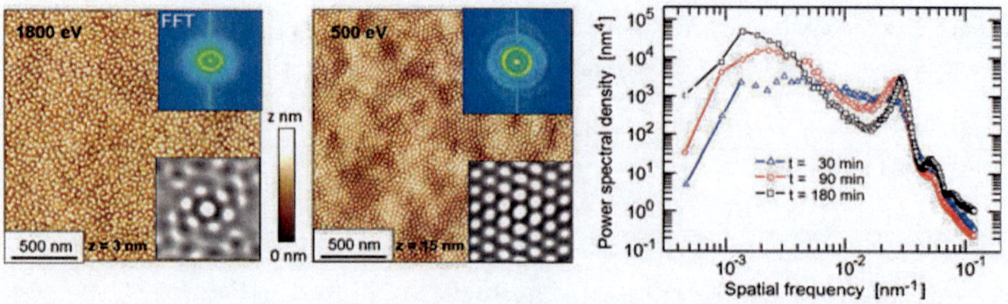

Fig. 8.28 Left: AFM, FFT and autocorrelation function images of Si surfaces eroded by 1800 eV Ar ion beam for 60 min at room temperature. Middle: AFM, FFT and autocorrelation function images of Si surfaces eroded by 500 eV Ar ion beam for 60 min at room temperature. Right: Angular averaged PSD functions of Si eroded by 500 eV Ar ion beam for different erosion times at room temperature. The angle of ion incidence was always 75°

also the concentration of metallic impurities. For example, Fig. 8.29 illustrates the hexagonal and square ordering of the dots as a function of temperature during Ar ion bombardment of InP surfaces [105]. The changing of the symmetry is evident. The corresponding two-dimensional autocorrelation functions highlight the symmetry of the local dot arrangement. Further, the mean separation between the individual dots as well as height of the dots increases with temperature.

The underlying basic mechanism for nanodot evolution is similar to that proposed for periodic nanopattern (ripples) formation by Bradley and Harper [110] (see

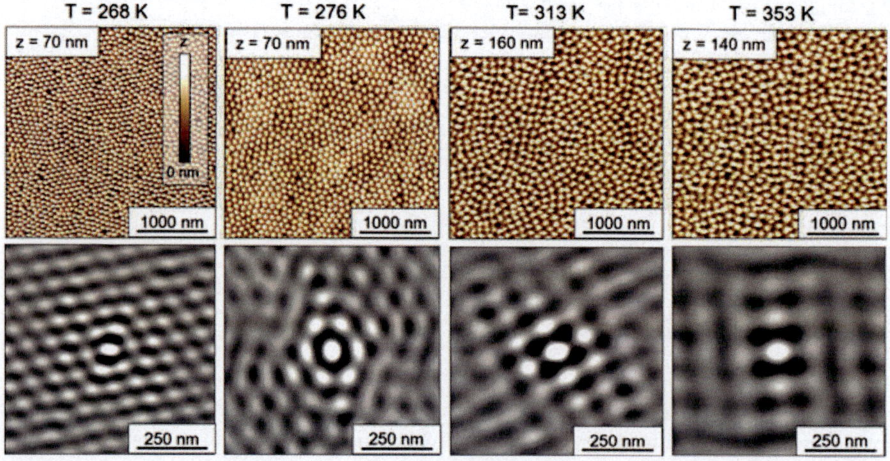

Fig. 8.29 Sequence of AFM images of Ar-ion-sputtered InP surfaces at different temperatures during ion sputtering. The ion energy was 500 eV, the ion current density was 300 μA/cm^2, the angle of ion incidence was 30°, and sputter time was 90 min. In the lower row, the corresponding two-dimensional autocorrelation function images are shown

Sect. 8.2.1.1). However, a general problem in their linear partial differential equation, (8.4), is that it cannot reproduce steady-state ordered patterns. Kahng et al. [111] have explained the formation of spatially ordered uniform nanoscale dots and holes from sputtering under normal incidence conditions using an isotropic Kuramoto–Sivashinsky equation (see Sect. 8.2.1.2), where the sign preceding nonlinear terms differentiates the formation of dots or holes (positive nonlinear terms indicate a hexagonally ordered dot formation while negative nonlinear terms predict hexagonal hole arrangement). The size of these nanostructures is independent of flux and temperature but can be controlled by tuning the ion energy. Unfortunately, the phase diagram by Kahng et al. [111] shows that the technologically important regime of dot formation is very restricted. Analysis indicates that dot formation occurs solely if $\sigma < \mu$ and $a \leq \sigma$, i.e. for anisotropic collision cascades with a large lateral straggling, where a is the ion range and σ and μ are the longitudinal and transversal straggling, respectively (see Sect. 4.6). However, for most ion/substrate combinations, anisotropic energy distributions with $\sigma > \mu$ can be expected [112]. Therefore, based on the theoretical approach using the isotropic Kuramoto–Sivashinsky (KS) equation, the self-organized formation of dots by ion sputtering under normal incidence is limited to a small number of materials. In contrast, the situation is completely different for sputtering with simultaneous sample rotation under oblique ion incidence. The parameter regions fulfilling the conditions for the formation of dots in dependence on the angle of ion incidence could be determined for off-normal cases of ion incidence [107].

The explanation of the dot formation by Facsko et al. [113] is based on the isotropic damped Kuramoto–Sivashinsky equation for normal incidence ion beam erosion (see Sect. 8.2.1.2). They added a damping term to this equation, which yields stable ordered patterns at long-term erosion and were able to demonstrate that the dot patterns can exhibit hexagonal ordering, as experimentally observed on III/V semiconductor surfaces (cf. Fig. 8.42). Chowdhury et al. [114] applied the damped KS equation to oblique angle ion beam erosion of a rapidly rotating sample. They were able to confirm that the damped Kuramoto–Sivashinsky equation (see Sect. 8.2) governs the mechanism of hexagonal dot ordering.

8.1.3.2 Dependence on Ion Energy

The formation of nanodots following noble gas ion bombardment was preferentially examined for ion energy ≤ 2 keV. It appears that a threshold energy of about 100 eV exists, above which the dots are generated [114]. Assuming that the evolution of the topography is determined by an effective ion-induced diffusion process without mass transport [57, 115], it can be expected that the characteristic wavelength increases with energy, in accordance with a power dependence of the type $\lambda \propto E^{2m}$, where the parameter m originates from the interatomic potential determining the stopping cross-section (see Sect. 2.2.4).

Facsko et al. [116] have conducted detailed studies of the dependence of the characteristic wavelength of nanodots on GaSb and InSb after normal Ar ion

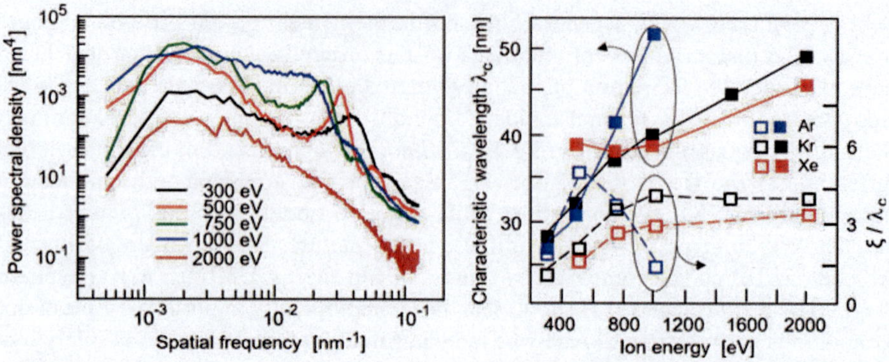

Fig. 8.30 Left: Angular averaged PSD functions deduced from AFM images of silicon following Ar ion bombardment at an incidence angle of 75°, room temperature and a fluence of 1.5×10^{17} ions/cm^2 (erosion time was 180 min) for different ion energies. The first peak corresponds to the characteristic wavelength. Right: Evolution of the characteristic wavelength (left) and the ratio ξ/λ_c (right) on rotating Si surfaces with ion energy for different ion species. The total ion fluence was 6.7×10^{18} cm^{-2} and the angle of ion incidence was 75° (adapted from [108] and modified)

bombardment for ion energies between 75 and 1800 eV. From scaling the wavelength with ion energy, they determined a value of m = 0.25 for GaSb. Wang et al. [117] have shown that the energy dependence of the wavelength also scales with the ion energy. They prepared GaAs nanodots using Ar ion bombardment with energies between 700 and 1000 eV under normal ion beam incidence and found for two different sputtering times m = 0.26 and m = 0.4. This result is supported by Sigmund sputter theory that m is about 1/3 in the low-energy range.

Another study [107] investigated the evolution of the dot sizes on GaSb and InP surface following Ar ion bombardment between 200 and 1200 eV at a fixed ion incidence angle of 40°. It was discovered that this energy range is too small to reliably differentiate between a possible linear dependence of the wavelength on the ion energy or a dependence with a power-law scaling exponent between 0.5 and 1.

Ziberi et al. [2, 108, 109] studied the energy dependence of the nanodot ordering and average dot sizes as a function of ion energy between 300 and 2000 eV after rare gas ion bombardment of silicon at room temperature (Fig. 8.30, left). The PSD functions show that for a specific range of ion energies (300–1000 eV), dot nanostructures are formed on the surface, where the best-ordered dots (dot domains) are observed at an Ar ion energy of 500 eV. By further increasing the ion energy > 500 eV, the mean size of nanostructures increases but the ordering decreases, i.e. multiple peaks in the PSD function disappear (quantitatively shown in Fig. 8.30, right). The increase of the mean dot size corresponding to ion energy can be observed from the shifting of the first-order peak of the PSD functions toward larger wavelengths. At ion energies above 1000 eV, dot nanostructures continuously vanish, until at 2000 eV, no dots can be observed on the surface, i.e. the surface remains smooth. In contrast, Ling et al. [118] have observed nanodots on Si surfaces following Ar ion bombardment for ion energies up to 5 keV.

From Fig. 8.30, right, it is obvious that the nanodots continuously coarsen with the ion energy for Ar ions up to about 500 eV and up to significant higher energies for Kr and Xe ions. The average dot size increases from about 25 nm up to about 50 nm. The power-law scaling exponent, p, seems to be dependent on the ion mass, i.e. $p_{Ar} \approx 0.5 > p_{Kr} \approx 0.3 > p_{Xe} \approx 0.2$.

The increase of dot size with increasing ion energy was confirmed by Chowdhury et al. [114] for Ar ion bombardment of Si surfaces at a grazing incidence angle of 75°. The average dot increases from 26 nm at 200 eV to 46 nm at 600 eV. Simultaneously, the dot amplitude grows and hence, surface roughness increases.

A qualitative measure of the two-dimensional nanodot arrangement is provided by the ratio of the correlation length ξ (equal to the average domain size for dots) to the characteristic wavelength (mean dot sizes) λ_c. Figure 8.30 right, quantitatively illustrates the dependence of this ratio ξ/λ_c on ion energy for Ar, Kr and Xe ion beam erosion of silicon at room temperature [2, 108]. It is evident that for Kr and Xe ions with ion energies > 750 eV, the lateral ordering of dot structures remains constant. For Ar ions, the best lateral ordering of dots is obtained at an energy of 500 eV. Behavior similar to that of Ar ions can be observed for Ne ions under the given sputtering conditions. However, the uniformity and ordering of dots is much less pronounced than for Ar ions.

8.1.3.3 Dependence on Temperature During Ion Bombardment

In some cases, the influence of temperature on the size and ordering of the self-organized dots was studied. According to Bradley and Harper [110] as well as Kahng et al. [111], the characteristic wavelength λ_c of the nanodots (diameter) should be proportional to the temperature T during ion bombardment, because $\lambda_c = 2\pi\sqrt{2K/|v|}$ [see (8.9)] and the factor $K \propto D_o \exp\left(-\frac{E_{sd}}{k_B T}\right)$ [see (6.50)], where D_o is the pre-exponential factor, ν is the largest absolute value of the negative surface tension coefficients, and E_{sd} is the activation energy of the surface diffusion. While Facsko et al. [113] did not ascertain temperature dependence for the characteristic wavelength of GaSb dots prepared by perpendicular 500 eV Ar ion bombardment at two distinct temperatures (+60 °C and −60 °C), Lu et al. [119] found that the diameter of nanodots on MgO surfaces following Ar ion bombardment between 0.5 and 5 keV at temperatures between 700 °C and 1000 °C increases with increasing temperature. They concluded that thermal surface diffusion is the main relaxation mechanism for ion-bombardment-induced surface instability. Ling et al. [118] also found an increase of the average dot diameter with increasing temperatures between room temperature and 800 °C. The effect of sample temperature during ion bombardment on dot diameter and in-plane arrangement of dots was also investigated [105, 107]. Rotating InP surfaces were bombarded by 500 eV Ar ions at an incidence angle of 30° and at temperatures between 265 K and 370 K. Figure 8.31 shows the mean separation between the dots (wavelength) and the normalized dot size fluctuation as a function of the target temperature during the Ar ion bombardment [105]. It is

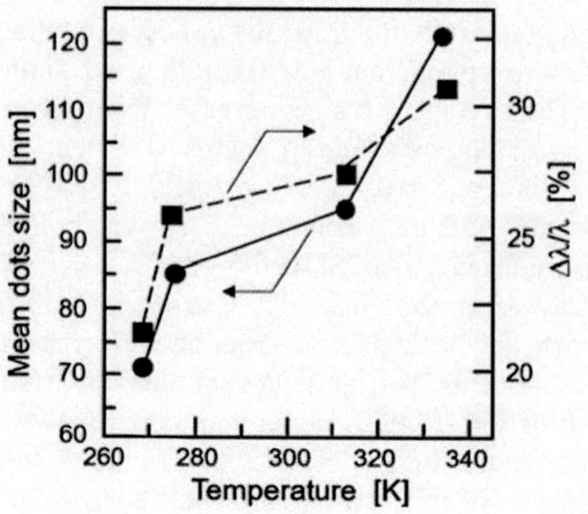

Fig. 8.31 The mean dot sizes (wavelength), λ, and the normalized dot size fluctuation, Δλ/λ, as a function of the target temperature during 500 eV Ar ion bombardment of InP surfaces. The angle of ion incidence was 30° and the ion current density 300 μA/cm² (the solid and dashed lines are merely visual guides)

obvious that both the wavelength and the normalized dot size fluctuation increase with the temperature. Dots usually form hexagonally ordered close-packed arrays, but square arrays have also been observed. The type of ordering was changed by only slight changes in temperature for InP samples that were bombarded by 500 eV Ar ions at an incidence angle 40° with concurrent substrate rotation. Fan et al. [120] also detected an increase in dot size with increasing temperature for low ion current density (20 μA/cm²) following Ar ion bombardment of InP, while a decrease in dot size occurs with high Ar ion fluxes (380 μA/cm²). According to Trynkiewicz et al. [121], an increase in temperature during low-energy noble gas ion bombardment of the formed nanodot patterns on the surface of III–V semiconductor surfaces as InP, InSb, InAs, or GaSb leads to pillar arrays, i.e. the aspect ratio of height to the diameter of the nanostructures increases significantly (see Sect. 6.3.1.1).

Gago et al. [79] have observed three different temperatures regimes in 1 keV Ar ion bombardment of Si at normal incidence. For temperatures up to 425 K, no changes to dot density, inter-dot distances, or dot height occur. For temperatures between 425 K and 500 K, a decrease in dot height and an increase in compressive strain in the crystalline cores of the dots was measured. For temperatures above 500 K, the nanostructures disappear, and the surface becomes flat. The authors had to acknowledge that this behavior cannot be explained on the basis of the existing theories.

8.1.3.4 Dependence on Ion Species

With the exception of only few studies, low-energy Ar ions have generally been used to generate topographical changes on surfaces. The influence of several non-reactive ions (Ar, Ne, Kr, and Xe) on average dot size and the lateral ordering of dots has been compared [108]. It becomes apparent that the evolution of mean dot size and the lateral ordering on Si surfaces have the same tendencies, independent of the ion species. An analogous tendency appears for the evolution of Ge surface topography after Kr or Xe ion bombardment (increase in the lateral ordering of dots and increased mean dot size with ion energy), where the parameters used for the sputtering process are identical to those for Si [2, 63, 104]. Tan et al. [122] bombarded InP surfaces with 1 keV reactive O_2 ions and produced ordered nanodot chains aligned perpendicular to the ion beam direction in ion incidence range between 38° and 44°. This procedure oxidized the nanodots.

8.1.3.5 Dependence on Angle of Ion Incidence

The topography evolution of ion-bombarded GaSb, InP and Si under oblique ion incidence has been intensively studied in terms of nanodot evolution without considering the influence of impurities. For example, Fig. 8.32 displays the topography of an Ar ion-sputtered GaSb surface at different ion-incidence angles [107]. The corresponding two-dimensional autocorrelation functions indicate the local symmetry. At normal ion incidence Ar ion sputtering ($\theta = 0°$) as well as for $\theta = 15°$, the familiar hexagonal arrangement of the dots becomes visible. When the ion incidence angle was increased to 30°, the hexagonal dot array shifted to a square pattern.

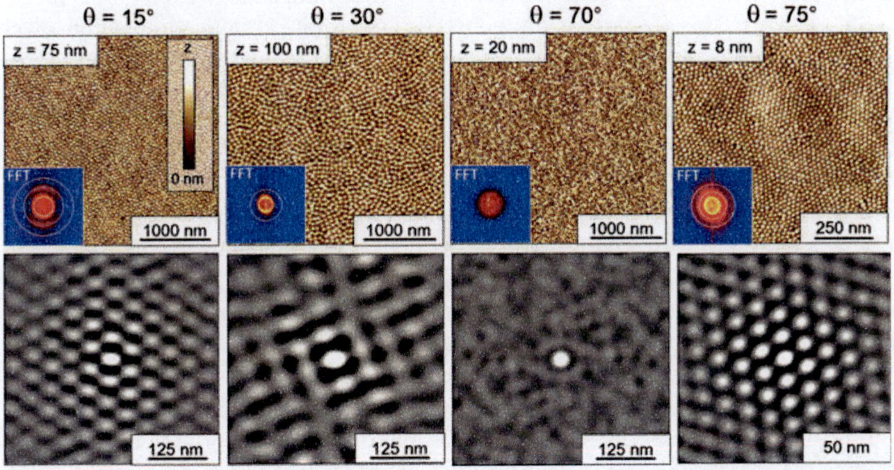

Fig. 8.32 Sequence of AFM images and corresponding FFT images (at the top) of Ar-ion-sputtered GaSb surfaces at different ion incidence angles. The ion energy was 500 eV, the ion current density was 300 $\mu A/cm^2$, the temperature during ion bombardment was 40 °C, and sputter time was 90 min. In the lower row, the corresponding two-dimensional autocorrelation functions images are shown

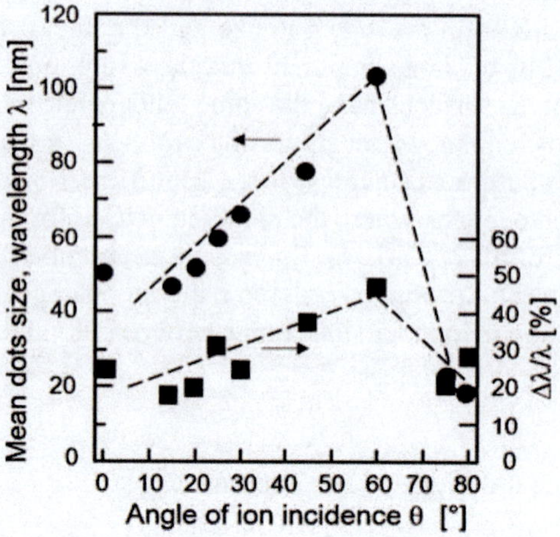

Fig. 8.33 The lateral dot size λ and normalized dot size fluctuation Δλ/λ as a function of ion incidence angle on Ar-ion-bombarded GaSb surfaces. Ion beam processing parameters were the same as in Fig. 8.28 (dashed lines are merely visual guides)

With a further increase of θ to 60°, the hexagonal arrangement emerges again. At θ = 70°, dots with a spherical shape but no local near ordering appear. For more oblique ion incidence (θ = 75°), a hexagonal pattern is once again observed. The high degree of self-organization is clearly visible from the high order satellite peaks in the Fourier spectrum (up to four orders are visible, image range from $-85.3\ \mu m^{-1}$ to $85.3\ \mu m^{-1}$). This scenario is only observed at elevated temperatures. For lower temperatures (e.g., room temperature), only a successive transition from hexagonal to no pattern and back to a hexagonal pattern was found with an increasing ion-incidence angle. Figure 8.32 also shows that dot diameter and height vary with the ion incidence angle (note the different z-scale in Fig. 8.32). Similar behavior could be observed for Ar ion sputtering of InP at room temperature [99], but here, hexagonal rather than square patterns were observed at incidence angles around 30°. In order to give a more quantified description of the dot pattern change with increasing ion incidence, several characteristics have been deduced from the AFM images. These are summarized in Fig. 8.33, which show the mean dot size, wavelength and the dot size fluctuation as a function of the angle of ion incidence [105], determined from the PSD functions for characteristic ion incidence angles. From this figure and consistent with the AFM images in Fig. 8.32, (upper row), it is evident that λ changes with the ion incidence angle and that the ordering is maximal at near-normal (θ = 15°) and grazing ion incidence (θ = 75°), as can be derived from the high order peaks in the PSDs. The lateral dot size reaches a maximum at θ = 60°. The ratio $\Delta\lambda/\lambda$ serves as a measure for ordering of the dot structures and confirms the conclusion that the degree of ordering is maximal at normal/near-normal and grazing ion incidence. Sulania et al. [123] have also studied mean dot size in dependence on the angle of ion incidence. InP(100) surfaces were bombarded with 1.5 keV Ar ions and an ion fluence of $8 \times 10^{16}\ cm^{-2}$ at room temperature. They observed that nanodots with a diameter of between 32 nm and 114 nm form at ion incidence angles up to 23°, where the maximum of dot size was obtained at an angle of 14°. Beyond 23°, an elongation of the nanodots was observed along the direction of the ion beam.

8.1.3.6 Dependence on Ion Fluence (Sputtering Time)

Studies on the influence of ion fluence or the duration of sputtering time on the formation of nanodots are focused on the evolution of roughness and wavelength (dot size). In Sect. 6.1, the kinetics of the roughness evolution during low-energy ion sputtering of solid surfaces was introduced. Here, the functional dependence of the rms roughness or interface width, w, and the spatial wavelength, λ, on the sputtering time (ion fluence) in the evolution of nanodots at low-energy ion bombardment is examined. The rms surface roughness acts as an indicator for the vertical height fluctuations and, hence, for the dot height. Frequently, it is assumed that the characteristic wavelength of the evolving dot patterns can be scaled by $\lambda \propto t^n$, i.e. the time-scaling exponent n can be extracted from a linear fit of the log–log plot of the wavelength versus the irradiation time.

The evolution of the roughness of InP after low-energy Ar ion bombardment in dependence on sputter time exhibits very similar behavior. Initially, roughness increases rapidly over a short period of time as the dots coarsen [99, 124, 125]. For example, a growth exponent of $\beta = 0.8 \pm 0.1$ was determined for rotating InP samples bombarded by oblique 500 eV Ar ions (Fig. 8.34) [99]. A similar value for the growth exponent, $\beta = 0.74 \pm 0.03$, was measured by Tan and Wee [123] following 1 keV normal Ar ion sputtering. According to (6.2), the time-scaling exponent of the nanodot wavelength, λ ($\propto t^n$), has been determined as $n = 0.26 \pm 0.04$ [99] and $n = 0.23 \pm 0.01$ [124] in the coarsening regime, where n is extracted from the linear fit of the log–log plot of the wavelength versus the sputter time. Beyond this regime, in the ordering regime, the growth of the nanodots remains quite low and the ordering of the nanodot arrangement increases with sputtering time. The roughness evolution in this regime rises at a significantly slower slope (Fig. 8.34), saturating [107, 124] or decreasing [125].

In principle, similar behavior in the roughness evolution has been observed for ion-bombarded Si and GaSb at the beginning of ion bombardment. The temporal evolution of roughness of silicon is characterized by an exponential increase across a short timeline. This rapid increase in Si roughness has been observed by Gago et al. [99] at 1 keV normal Ar ion incidence and by Ziberi et al. [108] with oblique ($\theta = 75°$) rare gas ion bombardment. Independent of the ion species, the roughness seems to grow exponentially. For ion fluences $> 1 \times 10^{18}$ cm^{-2} the roughness, i.e. the dot height, saturates and remains constant upon further ion bombardment.

These results are in agreement with studies on the temporal roughness evolution of ion-bombarded GaSb. Several authors [107, 126, 127] have concurrently ascertained that rms roughness increases exponentially in the early regime (coarsening regime). It is interesting to note that the exponential growth occurs within a narrow timeline, independent of the material used (for Ar-ion-bombarded Si, within 300 s, i.e. after a

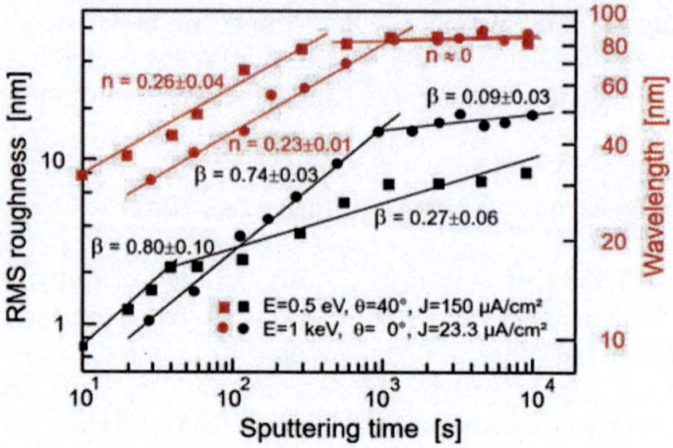

Fig. 8.34 The temporal evolution of the wavelength and rms roughness of InP sputtered by Ar ions. Data adapted from [99] (squares) and [124] (dots)

fluence of 5.6×10^{17} ions/cm^2 [10], and for Ar-ion-bombarded GaSb, within 20 s, i.e. after a fluence of 1×10^{17} ions/cm^2 [126, 127]. Thereafter in the ordering regime, the roughness remains nearly constant [107] or drops slightly [126, 127].

The temporal evolution of the roughness can be interpreted as the evolution of the average dot amplitude with the ion fluence (sputtering time). Consequently, the temporal evolution of the spatial wavelength λ, determined from circularly averaged power spectra, shows similar behavior, i.e. an initially strong increase of λ with the ion fluence within a short sputtering time and thereafter, the saturation of dot growth (but characterized by an improvement in the ordering). However, the slope of the dot dimension in the early stage and the characteristic fluence of transition from the coarsening regime to the ordering regime differs, dependent on the material, temperature, and ion flux. For example, the dependence of the wavelength on the sputtering time for InP bombarded with Ar ions is shown in Fig. 8.30, where the ordering regime was obtained for a fluence of about 1×10^{18} Ar ions/cm^2 [99, 124]. The ranges of the two regimes are distinct and visible. The time scaling of the nanodot wavelength is given by (6.2). The time-scaling exponent n characterizes the dynamics of the dot growth before and after the transition between the two regimes. For InP time-scaling exponents $n = 0.26$ [99, 107] and $n = 0.23$ [124], for GaSb $n = 0.14$ [127], and for Si $n \approx 0.2$ [100] were determined. Wang et al. [117] studied the temporal evolution of the wavelength on GaSb surfaces after normal incidence Ar ion bombardment at temperatures < 80 °C as a function of the ion energy between 700 eV and 1000 eV. They found that the temporal evolution of nanodots is not only dependent on the sputtering time but also the ion energy. The wavelength increases with the ion energy for a given sputtering time (ion fluence), where the exponent of the power law dependence on the wavelength ranged between $n = 0.24$ and $n = 0.21$ for ion energies between 700 eV and 1000 eV.

8.1.3.7 Dependence on Ion Current Density (Ion Flux)

Investigations of the dependence of dot size on ion flux have arrived at contradictory results. On the one hand, for GaSb [116] and InP [107], no dependence of the dot size on the ion current density was reported. On the other hand, the influence of the ion flux on pattern formation and the pattern wavelength of dots could be proven for low-energy Ar ion bombarded InP and Si [114]. Tan and Wee [124] found that the ordering of highly ordered hexagonal nanodots increases with increasing ion current density from 17.4 $\mu A/cm^2$ up to 23.3 $\mu A/cm^2$. Fan et al. [120] found that lateral dot size increases with increasing ion flux up to about 220 $\mu A/cm^2$ for Ar ion sputtering of Si(100) with ion energy of 1.5 keV. Beyond this ion flux, the dot size decreases with the ion current density in good agreement with the predictions of the Bradley–Harper theory (proportional to $J^{-1/2}$). Likewise, Chowdhury et al. [114] found a decrease in the dot wavelength (and the rms roughness) with an increase of the ion current density between 280 $\mu A/cm^2$ and 1.25 mA/cm^2 for silicon following 500 eV Ar ion bombardment.

8.1.3.8 Dependence on Crystallinity and Crystal Orientation

According to the current state of knowledge, nanodots can be produced by low-ion energy bombardment on both crystalline and amorphous surfaces [128]. The pattern formation on crystalline surfaces is only weakly controlled by the crystallographic surface orientation. For example, Gago et al. [129] have observed that the pattern coarsening and ordering dynamics of Si(111) surfaces are faster than those of Si(100) surfaces.

8.1.3.9 Dependence on Ion Beam Divergence

A strong indication of the influence of ion beam divergence on the evolution of surface patterns by ion beam erosion was analyzed by Kree et al. [130]. Based on both continuum theory and Monte-Carlo simulations, they found that the effect of the ion beam profile has a strong influence on the pattern formation. In experimental studies [70], it was demonstrated that ion optical parameters of the ion beam, especially the influence of accelerator voltage on the divergence of the ion beam and the angular distribution of the ions within the ion beam, play an important role in the pattern formation. They investigated the influence of beam divergence (see Chap. 10) on the formation of nanodots on the surface of silicon and III/V semiconductor surfaces at normal and off-normal ion incidence with simultaneous sample rotation after Ar ion bombardment. In order to estimate the influence of the accelerator voltage on the shape of the ion beam, simulations were carried out to determine the shape of the plasma sheath boundary at the grid. The ion trajectories were then calculated by solving the Poisson equation, taking into account the space charge effects. Figure 8.35 (left) shows the distribution of the Ar ion beam, with an ion energy of 500 eV.

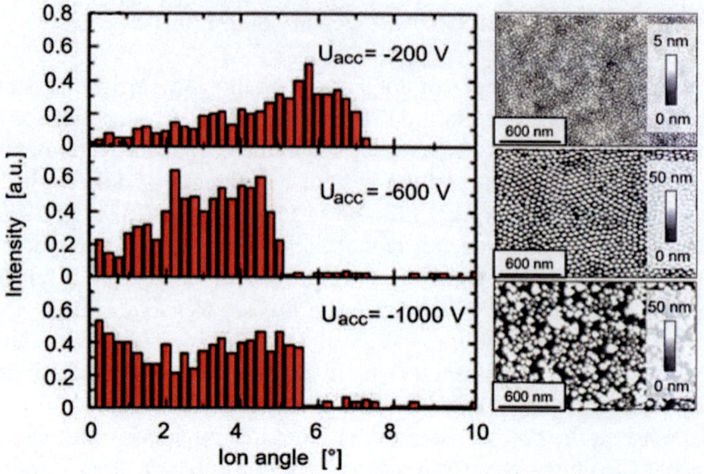

Fig. 8.35 Left: Simulation of ion incidence angle distribution as a function of the accelerator voltage for different acceleration voltages U_{acc}. Right: AFM images of GaSb surfaces after 500 Ar ion sputtering for different acceleration grid voltages (ion beam current density *was* 300 μA/cm^2, the sputter time was 60 min, temperature was 285 K)

Within experimental settings, $U_{acc} = -600$ V yields the ion beam with the smallest divergence and narrowest angle distribution, and therefore the best conditions for the given case of dot formation on GaSb at normal ion incidence. However, for pattern formation on Si surfaces, higher accelerator voltages and therefore greater ion beam divergence and a broader angle distribution seem to be necessary. Figure 8.35 (right) shows corresponding AFM images prepared with accelerator voltages of -200, -600, and -1000 V. From the AFM images, it can be seen that for $U_{acc} = -200$ V and -1000 V, there are no completely developed patterns on the sputtered surface, while at an accelerator voltage of -600 V, well-ordered dots form, with a mean dot diameter of 50 nm. Similarly, it has been found that various accelerator voltages are required for the observation of dot pattern on InAs or Si. Consequently, ion beam divergence has a considerable effect on topography evolution.

It should be noted that published results in the field of ion beam erosion induced nanostructures could sometimes not be reproduced, even when identical experimental conditions were reconstructed. It is likely that the beam divergence and/or the geometrical and ion optical parameters were not taken into account.

It has been shown that metallic impurities incorporated simultaneously with noble gas ion bombardment significantly affect pattern formation (see Sect. 8.1.3). Such impurities can be generated within the ion beam facility during the irradiation process. Varying the ion beam divergence by changing the acceleration voltage then strongly influences the formation of dots, holes, or ripples [6], because a change in beam divergence generally leads to a change in the concentration of metallic contaminants.

8.1.3.10 Nanoholes

In Chap. 6, it was explicated that the formation of ion beam-induced etch pits is intrinsically a phenomenon governed by a competition between the migration of adatoms and vacancies and the conditions of the ion bombardment. In contrast to amorphous or radiation-amorphizable materials such as semiconductors, the higher diffusivity of crystalline metals allows them to recover at lower temperatures following ion bombardment [131]. In addition, the Ehrlich-Schwoebel barrier (see Appendix N) causes an uphill adatom current that leads to the formation of mounds and pyramid-like structures on the surface, which can strengthen the pattern formation. It has been demonstrated (cf. Fig. 6.18), that low-energy ion bombardment at normal and near-normal incidence results in the formation of regular etching pits (nanoholes) on metal surfaces. For example, after normal incidence low-energy noble gas ion bombardment, square etch pits on Cu(001) [132] and Ag(001) [133] and hexagonal pits on Pt(111) [134] and Cu(111) [135] were observed. As an example, Fig. 8.36 shows a periodic pattern of rectangular shaped pits (vacancy islands). The line scans in Fig. 8.36 (right, white lines) demonstrate the periodicity of the pattern. The depth of the crater is 5 ± 1 ML and the side length is about 20 nm at the top. The separation of the pits is on the order of 27 nm along the [110] direction. It is evident that the shape of the etch pits reflects the symmetry of the surface unit cells.

The formation of nanoholes at the surface of amorphous or amorphized materials has rarely been studied. The formation of holes with no regular distribution on Si(001) surfaces sputtered with 500 eV Ar ions at normal incidence with simultaneous sample rotation has been observed. At normal ion incidence, hole coarsening with time of ion bombardment emerges on the Si surface with no preferred orientation or ordering (Fig. 8.37) [106]. By further increasing the incidence angle, the surface roughness decreases and the nanoholes disappear (for example, nanoholes on Si surfaces after

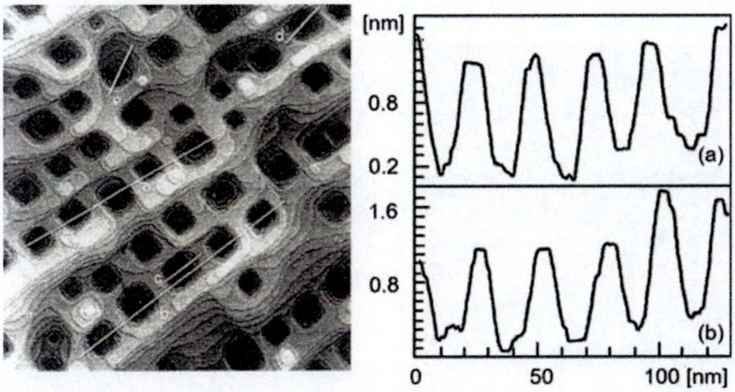

Fig. 8.36 STM image (200 nm × 200 nm) of Cu(001) after 400 eV Ar ion bombardment at 310 K. The angle of ion incidence was 45° and the ion fluence was 5.4×10^{15} ions/cm^2. The line scans **a**, **b** are taken close to the [011l] direction. Line scans across individual vacancy islands are marked by (**c**)–(**e**). Figures adapted from [132]

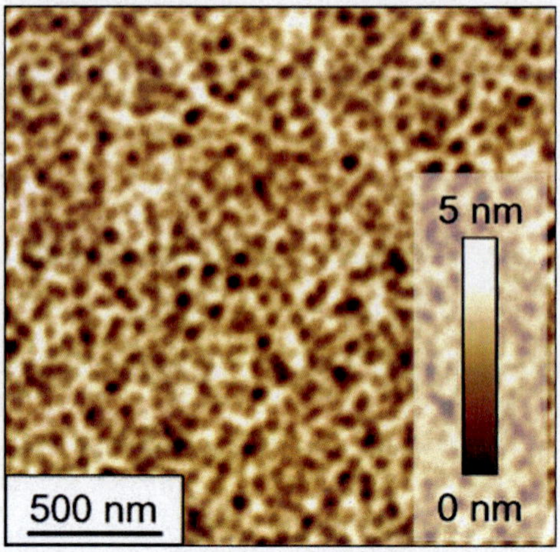

Fig. 8.37 AFM image of 500 eV Ar ion sputtered Si surfaces at normal ion incidence and at room temperature

Xe ion bombardment could only be observed at ion incidence angles < 10° and ion energies between 500 eV and <1 keV). These observations are consistent with the results of hole formation after normal Ar ion bombardment of Si as reported by Madi et al. [30, 31], see also Fig. 8.4)]. They found nanoholes at Ar ion incidence angles < 10° and ion energies between 250 eV and 600 eV.

Ordered arrangements of nanoholes were only found on the surface of Ge after low-energy ion bombardment. Wei et al. [136] have proven that highly ordered hole arrays can be prepared by normal incidence 5 keV Ge ion bombardment of Ge(100) surfaces at room temperature. Figure 8.38 (left), shows a perfect hexagonal arrangement of holes within domains with a size of 500 nm. The mean diameter and the

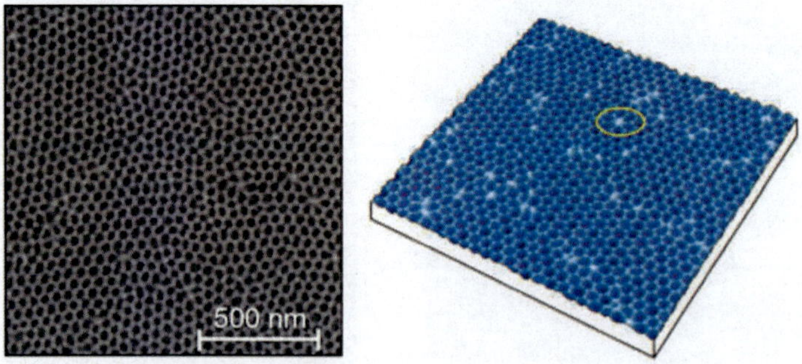

Fig. 8.38 SEM image (left) of hexagonally ordered nanoholes on Ge surfaces after 5 keV Ge ion bombardment at room temperature and simulation of the nanohole structure (right). A point defect (a dot instead of a hole surrounded by the six nearest holes) is shown by the circle (figures adapted from [136])

spacing between the holes were 35 nm and 45 nm, respectively. They found also that the ordering structure is orientation-independent after bombardment of differently oriented Ge samples. It is likely that the amorphous layer induced by ion bombardment leads to this orientation-independent ordering structure. Hole formation starts at a fluence of 6.6×10^{15} ions/cm^2 and forms a network of holes with broad hole distribution. At longer bombardment times, the hole sizes become uniform. Oh et al. [137] also bombarded Ge(100) with 1 keV Ar ions under normal incidence. However, they applied temperatures above the Ge recrystallization temperature of 250 °C and found checkerboard patterns of inverse pyramids with fourfold symmetry along the crystalline $\langle 100 \rangle$ direction, which are attributed to the aggregation of vacancies (Fig. 8.38). The upper temperature was determined by the Ehrlich-Schwoebel barrier (ES barrier, Appendix N) of about 430 °C. Beyond this temperature, the vacancies generated by ion bombardment are capable of overcoming the ES barrier and smoothing the surface.

Wei et al. [136] and Oh et al. [137] applied nonlinear continuum equations (see Sect. 8.2.1.2) to describe the temporal evolution of surface patterning during ion bombardment, in good agreement with the experimental observations. Figure 8.38 (right) demonstrates the high similarity of the topography between the simulation results and the experiment [136]. The perfect hexagonal arrangement of holes, domains, and grain-like boundaries matches very well to the experimental observations. Experimental studies by Chowdhury et al. [138] have confirmed the formation of a checkerboard of nanoholes in conformity with the symmetry of the crystal face after Ar ion bombardment at temperatures between 270 °C and 325 °C. However, these researchers used ion energy of only 30 eV (hyperthermal ions), i.e. ion beam erosion cannot be taken into account to explain the pattern formation. According to this study, the ES barrier is presumed to be highly effective at temperatures above the recrystallization temperature of Ge, so that the total reflection of adatoms at the step edges becomes significant and the observed square symmetric patterns can be obtained.

8.2 Theoretical Concepts of Ion Beam-Induced Pattern Formation

Based on numerous experimental studies, various mechanisms for the development of nanopatterns have been discussed over the past two decades. Within this scope, many different theoretical models have been suggested. These models can be divided into microscopic models and continuum models.

The microscopic description of the topography evolution and pattern formation is based on the atomic structure of bombarded solids where molecular dynamic or

Monte Carlo simulations are applied. A disadvantage here is that it is extremely difficult to describe extensive periodical surface structures using these procedures. Logically, only very few microscopic investigations have been carried out (for example, see [139]).

In contrast, continuum models, which aim at a quantitative analysis of the time evolution of the surface topography under ion bombardment without considering the atomic structure, describe the surface topography by means of adequate partial differential equations [140].

8.2.1 *Continuum Modelling*

Continuum models describe the evolution of the surface, usually given by a two-dimensional height profile (contour) function h(x, y, t), using partial continuum equations. The noise caused by the non-continuous erosion process is considered through a stochastic term $\eta(x, y, t)$. Usually, the coordinate system in which the erosion process is described moves with the erosion velocity v together with the eroded surface.

Essentially, two ion beam-induced bombardment processes contribute to roughening and the creation of spatially periodic surface patterns. Ion beam sputtering leads both to erosion of the surface (erosive process) as well as to mass redistribution as the result of ion beam-induced recoils in the near-surface region (non-erosive process). Continuum models have already been developed to describe surface evolution, including pattern formation for both fields. On the one hand, Bradley and Harper [110] have introduced a linear theory based on curvature-dependent sputtering, which is proportional to the locally deposited energy noted by Sigmund [141]. This theoretical concept was extended by introduction of non-linear terms in the Bradley–Harper equation [56].

On the other hand, Carter and Vishnyakov [142] have suggested a continuum model that explains the formation of surface patterns caused by directed mass redistribution. Recoils generated by ion bombardment momentum transfer near-surface atoms parallel to the substrate surface. The mass redistribution model has been extended by assumption of an anisotropic plastic flow [143] or hydrodynamic behavior [46] of the surface and near-surface region amorphized by ion bombardment.

8.2.1.1 Linear Continuum Model by Bradley and Harper

Based on Sigmund's [141] ellipsoidal distribution of the deposited energy [c.f. (4.33)], Bradley and Harper [110] developed a seminal continuum theory to explain the periodic height modulation on the surface of ion-bombarded amorphous solids (see also Cuerno and Barabási [66]).

They have assumed that the deposited energy on an arbitrary point on the surface is given by the sum of the deposited energy in this point and the contributions of the surrounding points after ion impacts. Consequently, the deposited energy is strongly dependent on the lateral position on a rough surface. In Fig. 8.39, energy deposition is compared after ion impact at points P and Q on surfaces with negative curvature and positive curvature, respectively. The deposited energy after ion impact is characterized by an ellipsoidal cascade with a centered depth (the depth a is given in Fig. 8.39). From this figure, it is obvious that the distance $\overline{Qq}$ is shorter than the distance $\overline{Pp}$. This means that more energy is deposited at point Q than at point P. Thus, the erosion rate at Q is greater than at P, or more generally, the erosion rate is greater in points with positive curvature than in those with negative curvature. Consequently, because the ions that hit the surface at point Q also contribute to point q, a valley erodes faster than a crest due to the deposition of a greater amount of energy at point Q than at point P. Therefore, lateral variation of the energy deposition onto a rough surface depends on the local shape of the surface, i.e. the initial surface roughness is amplified and the surface becomes unstable (so-called micro-roughness instability [141]). The key finding here is that the sputter yield at each point on the surface is proportional to the total amount of energy deposited from all ions.

The surface at time t can be described by the height function z = h(x, y, t). The ion beam in the x–z plane hits the surface at point P at the incidence angle θ, where the deposited energy $F_D(x, y)$, given by (4.23), is deposited into the solid (Fig. 8.40). The erosion in point 0 as result of the collision cascade is of interest. According to Sigmund [141], normal erosion velocity in this point 0(x, y) on the surface can be expressed by

$$v(x, y) = \Lambda \int_x \int_y F_D(x, y) J(x, y, h) dx dy, \tag{8.1}$$

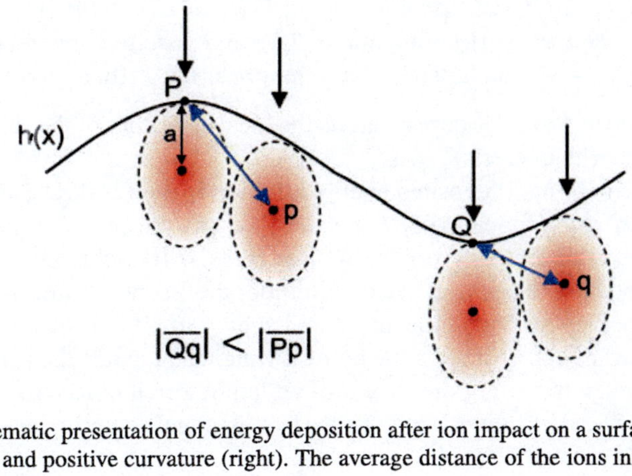

Fig. 8.39 Schematic presentation of energy deposition after ion impact on a surface with negative curvature (left) and positive curvature (right). The average distance of the ions in the target before its stops is given by a

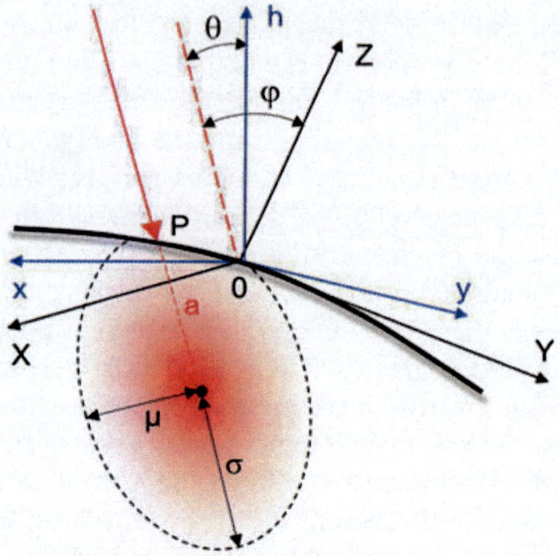

Fig. 8.40 Schematic representation for the calculation of erosion velocity at point 0

where the integration is performed over the region covering all points at which the deposited energy contributes to the erosion at a generic point 0 on the surface [66]. Λ is a material constant that depends on the surface binding energy and the scattering cross-sections [given by (5.20)], and the function J(x, y, h) describes the corrections of the uniform ion flux, $J_{\perp}$,when this is not incident on a planar surface. Makeev et al. [63] have expressed this corrected flux by

$$J(x, y, h) = J_{\perp}\left[\arctan\left(\sqrt{\frac{1}{R_x}} + \sqrt{\frac{1}{R_y}}\right)\right], \tag{8.2}$$

where R_x and R_y are radii of curvature, given by the second deviations of the local surface heights, i.e. $1/R_x = -\left(\partial^2 h/\partial x^2\right)$ and $1/R_y = -\left(\partial^2 h/\partial y^2\right)$.

Bradley and Harper [110] integrated (8.1) and inserted the relation for the deposited energy, (4.33), under several assumptions [66, 110]:

(i) The sputtering yield is dependent on the local curvature of the surface (valleys erode faster than crests).

(ii) The distribution of deposited energy after ion impact is described by a Gaussian distribution function [2], (c.f. (4.33)], where the center of the distribution is located at distance *a* from the surface (Figs. 4.10 and 5.39).

(iii) The deposited energy at each point on the surface after ion impact is determined by adding up the deposited energy of all ions that hit the surface.

(iv) The radius of curvature R at the point at which the ion hits the surface is larger than average travelling distance *a* of an ion before it comes to rest (i.e. R $\gg$ a), where R is positive when the surface is concave and negative when the surface is convex.

(v) The temporal changing of the height is given by [c.f. (6.10)],

$$\frac{\partial h}{\partial t} = -v(\varphi, R_x, R_y)\sqrt{1 + \left(\frac{\partial h}{\partial x}\right)^2}. \tag{8.3}$$

where φ is the is the angle of the beam direction with the local normal to the surface (see Fig. 8.40).

Based on the last assumption, angle φ is dependent on the incident angle θ and the local slopes $\partial h/\partial x$ and $\partial h/\partial y$; Bradley and Harper [110] have demonstrated that the last equation, (8.3), can be expanded in terms of derivatives of the surface height (performing a Taylor series expansion). If the variation of the topography is smooth enough, the products of derivatives of h can be neglected for third or higher orders [66, 110]. Under these assumptions, Bradley and Harper derived the following linear partial differential equation:

$$\begin{aligned}\frac{\partial h}{\partial t} &= -v(\theta) + \frac{\partial v(\theta)}{\partial \theta}\frac{\partial h}{\partial x} + \frac{Ja}{N}Y(\theta)\left\{\left[\Gamma_x(\theta)\frac{\partial^2 h}{\partial x^2}\right] + \left[\Gamma_y(\theta)\frac{\partial^2 h}{\partial y^2}\right]\right\}\\ &= -v + \frac{\partial v}{\partial \theta}\frac{\partial h}{\partial x} + v_x\frac{\partial^2 h}{\partial x^2} + v_y\frac{\partial^2 h}{\partial y^2},\end{aligned} \tag{8.4}$$

where $v = JY\cos\theta/N$ is the erosion rate or erosion velocity of the unperturbed planar surface (c.f. (5.49) and Y is the sputtering yield under normal ion incidence. The so-called effective surface tension coefficients parallel and perpendicular to the projection of the ion beam onto the surface are given by $v_x = \frac{Ja}{N}Y(\theta)\Gamma_x(\theta)$ and $v_y = \frac{Ja}{N}Y(\theta)\Gamma_y(\theta)$, respectively. These functions are dependent not only on the incidence angle θ but also on the parameters of the energy distribution, such as the ion range a, the longitudinal straggling σ, and the lateral straggling μ. The coefficients in (8.4) are provided in Appendix O in the notation introduced by Bradley and Harper [110], (O.1)–(O.6).

The second term, $\frac{\partial v}{\partial \theta}\frac{\partial h}{\partial x} = v'\frac{\partial h}{\partial x}$, is the contribution to the erosion rate that arises when the surface is sloped. The third and fourth terms reflect only the curvature-dependent sputtering. It has been confirmed that the local sputtering yield is proportional to the local curvature of the surface, i.e. surface points with a positive curvature erode faster than surface points with a negative curvature. The surface shows instable behavior under ion bombardment at all incident angles. For the calculation of functions v_x and v_y, the ion range, the longitudinal straggling, and the lateral straggling must be known. This can be evaluated with acceptable accuracy using Monte Carlo simulation programs and the approach proposed by Bolse (Sect. 4.4), i.e. calculating the spatial distribution of the vacancy density produced by primary and secondary collisions.

During the ion bombardment of a surface at finite temperatures, atoms will diffuse on the surface, leading to a surface relaxation (smoothing). The influence of the surface diffusion on the change of the surface height has been studied by Mullins [144, 145]. Bradley and Harper [110] have included this process in their equation of the surface height evolution

$$\frac{\partial h}{\partial t} = -v + v' \frac{\partial h}{\partial x} + v_x \frac{\partial^2 h}{\partial x^2} + v_y \frac{\partial^2 h}{\partial y^2} - K\nabla^4 h + \eta(x, y, t), \tag{8.5}$$

where K is a coefficient dependent on the surface self-diffusion, the free energy per unit area, and the number of atoms per unit area moving across the surface [for details see (6.50)]. The term $K\nabla^4 h$ does not take into account the anisotropic movement of atoms on crystalline surfaces. For single crystalline surfaces, the coefficient K must be replaced by two direction-dependent terms, K_{xx} and K_{yy}.

Since a diffusion of surface atoms is expected from the peaks to the valleys (see Sect. 6.5.4), the diffusion term in (8.5) has a tendency to smooth the surface. The stochastic nature of the arrival of ions on the surface is taken into account by the white noise, η, which is often assumed to be Gaussian [8] (see Appendix M). In summary, it is certain that the periodic modulation of the surface under low-energy ion bombardment is the result of an interplay between curvature-dependent sputtering and surface diffusion.

Linear Stability Analysis

In general, the stability or instability of a flat surface under ion erosion is studied using a linear stability analysis, where a linearized equation of surface motion is assumed for early ion bombardment times. The results of such a stability analysis are growth rates, R, and the propagation velocities of the Fourier modes, ω.

In the Fourier or **q**-space, a periodic perturbation is expressed by amplitude h_a and the wave-vector $\mathbf{q} = (q_x, q_y)$, where $q = 2\pi f = 2\pi/\lambda$, f is the spatial frequency, and λ is the wavelength. Assuming that $h(q_x, q_y, t)$ is the Fourier transform of the height h(x, y, t), (8.5) can be rewritten as

$$\frac{\partial h(q_x, q_y, t)}{\partial t} = -\left[v_x \frac{\partial^2 q_x}{\partial x^2} + v_y \frac{\partial^2 q_y}{\partial y^2} + K\left(q_x^4 + q_y^4\right)\right] h(q_x, q_y, t), \tag{8.6}$$

where the noise term is neglected. Then, a periodic perturbation function (general solution) is introduced in the form of

$$h(q_x, q_y, t) = -vt + h_a \exp\left[i\left(q_x \mathrm{x} + q_y \mathrm{y} - \omega t\right) + Rt\right], \tag{8.7}$$

where the amplitude of the perturbation can be expressed by $h_a(t) = h_a(t = 0)\exp(Rt)$. Equation (8.6) provides the amplification factor of the growth rate of the perturbation, i.e. each component h(q, t) of the surface spatial frequency spectrum grows exponentially with the rate

$$R(q) = -\left[v_x q_x^2 + v_y q_y^2\right] - K\left(q_x^2 + q_y^2\right)^2, \tag{8.8}$$

also called linear dispersion relation, and to the propagation velocity $\omega = -v' q_x$.

In summary, R(q) describes the rate at which the amplitude of a perturbation with wave-vector **q** grows or decays with time, and ω indicates the magnitude and direction of the velocity of the transverse in-plane motion of the disturbances. It should be

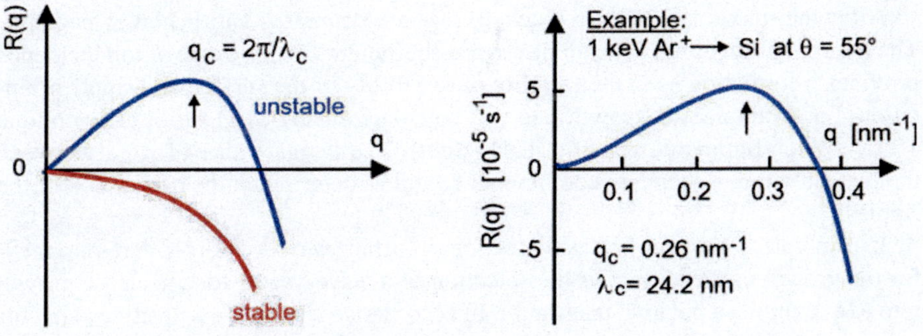

Fig. 8.41 Left: Schematic depiction of the growth rate R(q) as a function of the wave-vector q. Right: Calculated growth in dependence on the wave-vector for silicon bombarded with 1 keV Ar ions at an incidence angle of θ = 55°

noted that in the experiment, the mode with the largest growth rate is observed. It is obvious from (8.8) that R is negative if the surface relaxes by smoothing, while R is positive if roughening is expected. If R > 0, the amplitude grows exponentially and when R < 0 it decays exponentially over time. A schematic depiction of the behavior of R is provided in Fig. 8.41 (left). R(q) increases with increasing q, and R reaches a maximum at $\partial R/\partial q = 0$ and $\partial^2 R/\partial q^2 < 0$. The maximum value of R gives the critical wavelength, $q_c = \sqrt{|v_{x,y}|/2K}$, i.e. indicating the fastest growing spatial frequency, q_c. The wavelength of the surface pattern can then be expected at

$$\lambda_c = \frac{2\pi}{q_c} = 2\pi\sqrt{\frac{2K}{|v_{x,y}|}}. \tag{8.9}$$

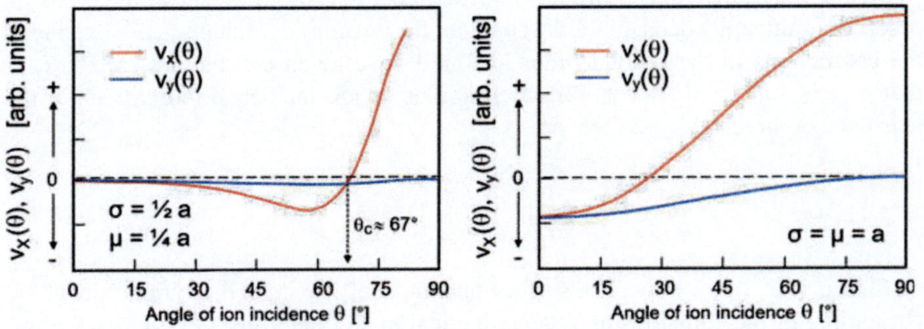

Fig. 8.42 Normalized surface tension coefficients in dependence on the angle of ion incidence for an asymmetrical distribution of the ion energy (σ = a/2 and μ = a/4) and for the symmetrical distribution of deposited energy (σ = μ = a)

After the maximum R(q) is reached, the rate decreases and becomes negative. The interpretation of the dependence of the coefficients on the angle of ion incidence provides information as to the stability or instability of the surface and ripple orientation. For example, the growth rate was calculated using (8.8) for Si after oblique 1 keV Ar ion bombardment (Fig. 8.41, right). The largest value of $|\upsilon_{x,y}|$ between the two curvature-dependent coefficients ν_x and ν_y determined the orientation of the ripples.

In principle, waves can be developed for all surface wave-vectors **q**. A requirement for the growth of the surface is the selection of a wave-vector for which the growth rate R(q) exhibits a positive maximum. In accordance with (8.8), a positive value for R is given only with a negative surface tension coefficient of ν_x and/or ν_y. Surface instability can then be expected because the diffusion coefficient K is always positive. Consequently, the absolute value $|\upsilon_{x,y}|$ in (8.9) is the largest absolute value of the two effective surface tension coefficients ν_x and ν_y.

Figure 8.42 shows the dependence of the coefficients ν_x and ν_y on the angle of ion incidence for an asymmetrical (elliptical) distribution (left) and for the symmetrical (spherical) distribution of the deposited energy (right). In the asymmetrical distribution of the deposited energy (Fig. 8.42, left), the ripples are oriented parallel to the direction of the incident ion beam for incident angles $\theta < \theta_c$ or $\nu_x < \nu_y < 0$, while for $\theta > \theta_c$ or $\nu_x > 0$ and $\nu_y < 0$, the ripples are perpendicular to the direction of the incident ion beam. Consequently, negative values for ν_x and ν_y are indications of topographical instability.

The orientation of ripples from $\nu_x < \nu_y$ to $\nu_x > \nu_y$ changes at the critical angle θ_c [110]. For example, the critical angle is about 67° for $\sigma = a/2$ and $\mu = a/4$ and is about 45° for $\sigma = \mu = a/2$. Feix et al. [146] have compared the numerically calculated values of the surface tension coefficients for different ion energy distributions for 5 keV Cu ion bombardment of Cu. Their results show that the critical angle θ_c is located at about 68° in an exponential energy distribution, as compared to modified Gaussian distribution at about 44° or the Gaussian distribution at 53°.

Within the framework of Bradley–Harper theory [110], the shape of the collision cascade significantly determines the behavior the tension coefficients. They expanded the coefficients to the second order in θ and have given a condition for the case that $-\nu_x$ is smaller than $-\nu_y$ for small angles of ion incidence (condition for the existence of an instable surface) as

$$\left(\frac{a}{\sigma}\right)^2 + \left(\frac{a}{\mu}\right)^2 > 3. \tag{8.10}$$

Figure 8.42, right, shows the surface tension coefficients in the symmetrical case (straggling of the collision cascade is identical in all directions, i.e. $\sigma = \mu$). Consequently, ripples in x-direction cannot be observed, while ripples in y-direction can be expected for small incident angles (not observed in experiments). In contrast, in the asymmetrical case, $\sigma > \sqrt{3}\mu$, ripples should be detected for small angles of incidence

In summary, analysis of the Bradley–Harper equation by determination of the Fourier transform permits some predictions, which could only be partially confirmed experimentally (see Sect. 8.1):

(i) According to the Bradley–Harper equation, the ripple formation can be predicted when either ν_x or ν_y is negative.
(ii) When both ν_x and ν_y are positive, the surface remains stable.
(iii) For $\nu_x < 0$, parallel-mode ripples occur, whereas for $\nu_y < 0$ perpendicular-mode ripples should be expected.
(iv) At the critical angle of ion incidence, θ_c, the ripple orientation changes.
(v) In accordance with (8.7), the amplitude of the ripples should grow exponentially.
(vi) If both magnitudes ν_x and ν_y are negative, then the direction of the ripples is given for where the growth rate R is greatest, i.e. $|\upsilon_x| < |\upsilon_y|$ and $|\upsilon_x| > |\upsilon_y|$ lead to parallel and perpendicular mode ripples, respectively.
(vii) For $\nu_{x,y} = 0$ (normal ion incidence), curvature dependent terms in (8.5) are neglected and pattern formation is impossible.
(viii) From (8.9), the proportionality is [66]

$$\lambda_c = 2\pi\sqrt{\frac{2K}{|\upsilon_{x,y}|}} \propto \sqrt{\frac{1}{Y(E)a(E)}}, \tag{8.11}$$

where the sputtering yield should vary approximately linearly at ion energy E for energies up to any keV (see Sect. 5.3), and the average ion penetration depth, a, should be proportional to E^{2m}, where m ranges between 1/3 and 1/2 for average energies (see Sect. 2.2.5). Consequently, the wavelength is $\lambda \propto 1/\sqrt{E}$ for $m = 1/2$, i.e. the ripple wavelength decreases with an increase of the ion energy.
(ix) In accordance with (8.5), in which $\nu_{x,y}$ is proportional to ion flux, it follows with (8.9) that $\lambda \propto 1/\sqrt{J}$.
(x) The wavelength in (8.11) is independent of ion fluence or sputter (bombardment) time.
(xi) Equation (8.11), together with (6.50), indicate that wavelength is dependent on temperature during ion bombardment, i.e.

$$\lambda_c = 2\pi\sqrt{\frac{2K}{|\upsilon_{x,y}|}} \propto (TJ)^{-1/2}\exp(-E_{sd}/k_B T), \tag{8.12}$$

with $m = 1/2$ (low ion energies).

It can be concluded that the linear Bradley–Harper model provides a fundamental improvement to the understanding of ripple formation at the nanoscale and is a useful model, applicable for explaining experimentally observed processes. However, some findings cannot be explained with this model, such as:

(i) the restricted growth of the ripples in contrast to a theoretically predicted unbounded exponential growth [45, 106, 107],

(ii) in the majority of experiments, ripple wavelength increases with ion energy, while the Bradley–Harper theory predicts decreasing wavelength with ion energy [4, 39]; no ripple coarsening effect of the wavelength, $\lambda \propto t^n$, is predicted,

(iii) in general, no dependence of the wavelength on temperature has been observed, although a $T^{-1/2}$ behavior is expected [62],

(iv) the proportionality between the wavelength and the ion flux, $\lambda \propto J^{-1/2}$, that is expected according to the Bradley-Harper theory, can frequently not be experimentally proven [147].

In the light of the accumulated experimental results, an expansion of the Bradley–Harper theory seems to be advisable.

8.2.1.2 High Order Extensions of the Bradley–Harper Theory

With the objective of explaining the observed phenomena at low-energy ion bombardment of surfaces as instability of ripple amplitude in long-term sputtering or kinetic roughening, a non-linear extension of the linear Bradley–Harper equation and further, its generalization to the Kuramoto–Sivashinsky (KS) equation has been proposed [63, 66, 111, 115, 140, 148–150].

Anisotropic Kuramoto–Sivashinsky Equation

Cuerno et al. [66, 140], in particular, have proposed introducing non-linear terms of the lowest order in the linear Bradley–Harper equation, with the aim of describing the saturation of the ripple amplitude. This non-linear equation, derived following the approach by Bradley and Harper, is given by

$$\frac{\partial h}{\partial t} = -v + v'\frac{\partial h}{\partial x} + v_x\frac{\partial^2 h}{\partial x^2} + v_y\frac{\partial^2 h}{\partial y^2} + \frac{\lambda_x}{2}\left(\frac{\partial h}{\partial x}\right)^2 + \frac{\lambda_y}{2}\left(\frac{\partial h}{\partial y}\right)^2 - K\nabla^4 h + \eta(x, y, t), \tag{8.13}$$

which is an anisotropic version of the stochastic KS partial differential equation and describes the evolution of surfaces eroded by ion bombardment. The non-linear quadratic term with the coefficient λ was added to the original continuum KS equation. The KS equation was successfully used to describe complex spatiotemporal dynamics at the evolution of the propagation of perturbations to chemical waves and the dynamics of flame front modulations as well as the evolution of spatially uniform oscillating chemical reactions [36].

The lowest order non-linearities in (8.13) are quadratic and integrate the angular dependence of the sputter yield (referred to as the curvature-dependent sputtering yield). These provide the dependence of the local erosion velocity (sputter yield) in relation to the absolute surface slope (angle). The coefficients λ_x and λ_y are

Kardar-Parisi-Zhang non-linearities [151], which control the exponential growth of the linearly unstable modes. The coefficients are given by Cuerno and Barabási et al. [66] in Appendix O [see (O.7)-(O.13)].

As a result of the numeric integration of (8.14), Park et al. [150] have shown that surface roughness, which is proportional to ripple amplitude, increases exponentially with time before a crossover or saturation time τ_s, after which surface roughness grows at a significantly slower rate and then saturates (see also Appendix M, Fig. M.1), i.e. the anisotropic Kuramoto–Sivashinsky equation, (8.13) behaves like the Bradley–Harper equation for short sputtering times ($t \ll \tau_s$). The crossover time is given by

$$\tau_s \propto \frac{K}{v_{x,y}} \ln\left(\frac{v_{x,y}}{\lambda_{x,y}}\right). \tag{8.14}$$

With increased sputtering time, the non-linear coefficients λ_x and λ_y become increasingly dominant. The signs preceding λ_x, λ_y, v_x and ν_y indicate the topography at longer sputtering times.

For sputtering times at $t < \tau_s$ (linear regime), the non-linear terms are insignificant and ripples are formed, where the amplitude of these ripples grows exponentially over time. For $t > \tau_s$ (non-linear regime) the ripples formed during the linear growth regime can disappear through a transition to kinetic roughening when the product of $\lambda_x \cdot \lambda_y$ is positive. When the product of $\lambda_x \cdot \lambda_y < 0$, the ripples also disappear [150]. However, for very long sputtering times, after a second crossover time, a new ripple structure is formed in which the ripples are stable and rotated at an angle of $\theta_c = \tan^{-1}\sqrt{-\lambda_x/\lambda_y}$ and $\theta_c = \tan^{-1}\sqrt{-\lambda_y/\lambda_x}$ to the x-direction and the y-direction, respectively.

Kahng et al. [111] have explained the formation of spatially ordered uniform nanoscale dots and holes by sputtering on the basis of the Kuramoto–Sivashinsky equation. Under normal ion incidence, the coefficients in (8.14) can be simplified, so that the resulting Kuramoto–Sivashinsky equation is isotropic. Numerical simulations of the isotropic KS equation for normal incidence by Kahng et al. verified that dots are generated if $\lambda > 0$, otherwise holes are formed ($\lambda < 0$). Expanding this theory to the case of sputtering under oblique ion incidence with simultaneous sample rotation [107] offers an explanation for the self-organized formation of dots by ion sputtering for a large variety of ion-material combinations. The size of these nanostructures is independent of flux and temperature but can be controlled by tuning the ion energy.

Damped Kuramoto–Sivashinsky Equation

Experimental studies have shown that hexagonally ordered dot patterns can be produced under normal ion incidence as well as under oblique ion incidence with simultaneous sample rotation (see Sect. 8.1). The application of the anisotropic Kuramoto–Sivashinsky equation to reproduce these patterns was not successful in its isotropic limits [18, 36]. Chaté et al. [152] introduced a damping term of the form $-\alpha h$ in the Kuramoto–Sivashinsky equation [153]. Facsko et al. [113] have adopted this damped version of the anisotropic Kuramoto–Sivashinsky equation, given by

$$\frac{\partial h}{\partial t} = -v - \alpha h + v' \frac{\partial h}{\partial x} + v_x \frac{\partial^2 h}{\partial x^2} + v_y \frac{\partial^2 h}{\partial y^2} + \frac{\lambda_x}{2}\left(\frac{\partial h}{\partial x}\right)^2 + \frac{\lambda_y}{2}\left(\frac{\partial h}{\partial y}\right)^2 - K\nabla^4 h + \eta(x, y, t), \tag{8.15}$$

to reproduce the hexagonal dot pattern under normal ion incidence. Figure 8.43 shows the surface pattern calculated on the basis of the damped Kuramoto–Sivashinsky equation for two different times of ion bombardment, with a damping term of $\alpha = 0.24$ as well as the corresponding two-dimensional PSD. In the early time regime (Fig. 8.43a), no ordering is visible, i.e. the pattern is fully isotropic. In this linear regime, the pattern consists of chainlike structures, without clearly separated dots. In the crossover regime (Fig. 8.43b), however, where the nonlinear term begins to be relevant, the dot structure develops and isolates, forming hexagonally ordered domains. In this regime, the individual dots can be rearranged, while still keeping their shape. Therefore, the dots will be ordered in a close-packed arrangement in what appears to be a stationary solution. The damping term $-\alpha h$ effectively suppresses secondary instabilities for only a narrow range of values of $\alpha \in (0.2, 0.25)$. This damping term, interpreted as the effect of the re-deposition of sputtered particles, fails to fulfill the fundamental symmetry of translation invariance in erosion direction. Consequently, Facsko et al. [113] suggested replacing αh with $\alpha(h - \bar{h})$, with $\bar{h}$ being the spatial average over a particular sample area.

Vogel and Linz [148] have applied also the damped anisotropic Kuramoto–Sivashinsky to explain surface erosion under oblique incidence. They were able to reproduce many different types of patterning observed experimentally, i.e. hexagonally arranged dots or ripples, depending on the incidence angle, and have studied the transition between the different topographic states.

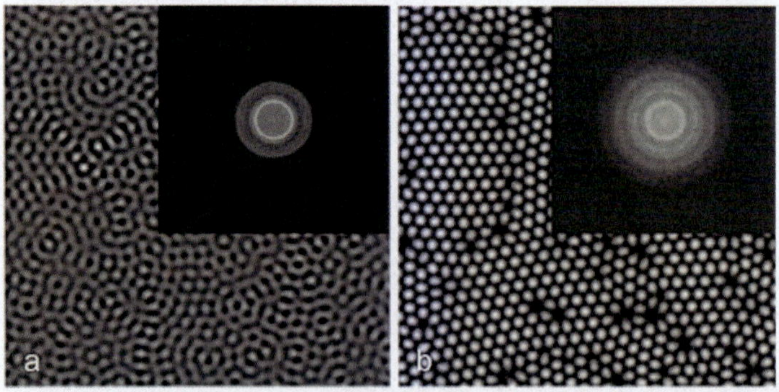

Fig. 8.43 Patterns calculated by the damped Kuramoto-Shivashinsky equation for $\alpha = 0.24$ and the corresponding 2-dimensional PSD (insets). Patterns for an early time regime (left) and stationary pattern with hexagonal ordering in the late time regime (right) are shown (figure adapted from [113])

Makeev-Barabási Non-Thermal Relaxation Model

A significant enhancement to the Bradley–Harper model was achieved by Makeev, Cuerno and Barabàsi [63, 115]. The higher-order derivatives in the expansion of the erosion rate normal to the surface produce terms that mimic surface diffusion. Makeev, Cuerno and Barabàsi introduced a new mechanism for the ion beam-induced surface diffusion, the so-called 'effective ion-induced surface diffusion', which acts as an additional smoothing mechanism. This newly created effective surface diffusion does not describe a true transport process and is not dependent on the temperature. It is merely produced by a higher order expansion of the erosion effect, and results in both roughening and smoothing of surfaces. Here, this extended Kuramoto–Sivashinsky equation is given by (for details of derivation, see [63])

$$\frac{\partial h}{\partial t} = -v + v'\frac{\partial h}{\partial x} + v_x\frac{\partial^2 h}{\partial x^2} + v_y\frac{\partial^2 h}{\partial y^2} + \frac{\lambda_x}{2}\left(\frac{\partial h}{\partial x}\right)^2 + \frac{\lambda_y}{2}\left(\frac{\partial h}{\partial y}\right)^2 - D_{xx}\frac{\partial^4 h}{\partial x^4} - D_{yy}\frac{\partial^4 h}{\partial y^4} + \eta(x, y, t). \tag{8.16}$$

The terms with coefficients D_{ii} in this equation (explicitly given in Appendix O, see (O.14) and (O.15)) are proportional to the fourth derivative of the height function. According to Makeev and Barabàsi [115], these terms complement the thermal diffusion on the surface and can provide an additional contribution to surface smoothing. Interestingly, these terms have the same mathematical form, $D\nabla^4 h$, as the thermal diffusion. The total diffusion constant is given by the sum of the thermal diffusion and the effective ion-induced diffusion constant. At a critical temperature, T_c, both diffusion constants are equal. At $T < T_c$, effective ion-induced diffusion is dominant, because this diffusion constant is independent of temperature. Facsko et al. [116] have compared both diffusion processes after low-energy Ar ion bombardment of III/V semiconductors at different temperatures. They found that effective ion-induced surface diffusion is the dominant surface relaxation process for energies between 75 eV and 1800 eV and temperatures ≤ 60 °C. Recently, Bradley [154] has shown that the effective ion-induced surface diffusion cannot be the only smoothing mechanism that operates at low temperatures and that derivatives in the expansion of the erosion rate higher than fourth order should have no effect on the surface relaxation processes.

General Continuum Equation

The most general non-linear equation that has been derived on the basis of Bradley and Harper's procedure has been published by Makeev, Cuerno and Barabàsi [63]. They have derived a stochastic nonlinear continuum equation to describe the topographic evolution of amorphous surfaces eroded by ion bombardment. Under the condition of the expansion of the surface curvature, higher order terms of the Taylor series expansion of the height function have been taken into account and have expanded the surface evolution equation. They obtained a very complex partial differential equation of the fourth order for the time evolution of the surface heights, containing numerous linear and non-linear terms:

$$\frac{\partial h}{\partial t} = -\upsilon + \upsilon' \frac{\partial h}{\partial x} + \xi_x \left(\frac{\partial h}{\partial x}\right)\left(\frac{\partial^2 h}{\partial x^2}\right) + \xi_y \left(\frac{\partial h}{\partial y}\right)\left(\frac{\partial^2 h}{\partial y^2}\right) + \upsilon_x \frac{\partial^2 h}{\partial x^2} + \upsilon_y \frac{\partial^2 h}{\partial y^2}$$
$$+ \Omega_1 \frac{\partial^3 h}{\partial x^3} + \Omega_2 \frac{\partial^3 h}{\partial x \partial y^2} - D_{xx} \frac{\partial^4 h}{\partial x^4} - D_{xy} \frac{\partial^4 h}{\partial x^2 \partial y^2} - D_{yy} \frac{\partial^4 h}{\partial y^4}$$
$$- K\nabla^4 h + \frac{\lambda_x}{2}\left(\frac{\partial h}{\partial x}\right)^2 + \frac{\lambda_y}{2}\left(\frac{\partial h}{\partial y}\right)^2 + \eta(x, y, t). \tag{8.17}$$

This stochastic partial differential equation for the surface height involves up to fourth order derivatives of the height. The explicit coefficients are provided in Appendix O (O.17)–(O.31) [63]. The physical relevance of the additional non-linear coefficient Ω is taken into consideration, while ξ_x and ξ_x are non-linearities that are responsible for the development of ripples with an asymmetric shape. For non-normal ion incidence, the highly non-linear character of the generalized continuum equation opens an extensive parameter space, which results in rather complex topographical states. For normal ion incidence, this equation can be reduced to the isotropic KS equation with: $\upsilon_x = \upsilon_y$, $\overline{\upsilon} = \xi_x = \xi_y = \Omega_1 = \Omega_2 = 0$, $\lambda_x = \lambda_y$, and $D_{xx} = D_{yy} = D_{xy}/2$.

Hydrodynamic Model

Low-energy ion bombardment of a silicon surface under oblique ion incidence exhibits the formation of a rippled surface with a periodicity (wavelength) lower than 100 nm, oriented normal to the direction of the incident ion beam. With increased time of ion bombardment (fluence), wavelength coarsening can be observed (see Fig. 8.44). It is apparent that the generalized KS equation (8.17) is not able to reproduce the wavelength coarsening, i.e. an additional physical mechanism must be involved that would describe this topographical evolution. Section 4.8 describes how ion bombardment with fluences above the threshold fluence of amorphization creates an amorphous thin film. Rudy and Smirnov [155] and Castro and Cuerno [46] have proposed describing the atomic processes within this near-surface film in terms of a viscous flow. Analogous to the description of the evolution of underwater sand ripples, Aste und Valbusa [156] have applied a hydrodynamic approach to describe the instability of surfaces under ion bombardment. This approach couples thickness of the mobile surface adatoms layer, R_f, with the viscous film and the height of the bombarded surface above a reference plane, h, using the following coupled rate equations:

$$\frac{\partial R}{\partial t} = (1 - \Phi)\Gamma_{ex} - \Gamma_{ad} + D_{sd} \frac{\partial^2 R_f}{\partial x^2} \tag{8.18}$$

and

$$\frac{\partial h}{\partial t} = -\Gamma_{ex} + \Gamma_{ad}, \tag{8.19}$$

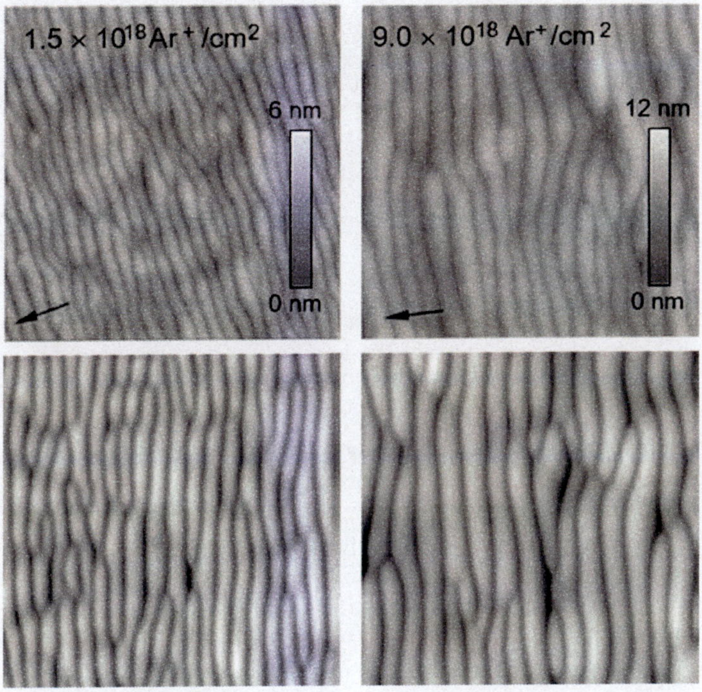

Fig. 8.44 AFM images (upper row) of fused silica after bombardment with 800 eV Ar ions at room temperature and an angle of ion incidence of 60° at two different ion fluences. The black arrows indicate the projection of the ion beam direction (figures adapted from [159]). Predicted topography (lower row) by numerical integration of (8.22) to an early and late instant in time (figures adapted from [160])

where D_{sd} is the surface diffusion coefficient (see Sect. 6.5.4). As is generally known, the sputtering process is characterized by the fact that atoms within the first layers of a solid do have enough energy to overcome the binding energy and leave the sample. Aste and Valbusa [156] and later, Muñoz-García et al. [64, 65], have assumed that a fraction of this excavated material is redeposited on the surface, where Γ_{ad} and Γ_{ex} are the rates of the production of mobile adatoms to the surface and of atom excavation from the surface, respectively. These magnitudes are dependent on R_f, the height, h, and its derivatives. $(1 - \Phi)$ is a measure of the fraction of sputtered atoms, which are redeposited and can be mobile on the surface (if $\Phi = 0$, then all atoms have been sputtered away). The local addition rate is given by [64]

$$\Gamma_{ad} = \gamma_o\left[R_f\left(1 + \gamma_{2x}\frac{\partial^2 h}{\partial x^2} + \gamma_{2y}\frac{\partial^2 h}{\partial y^2}\right) - R_{eq}\right], \tag{8.20}$$

where $1/\gamma_o$ is the average time in which adatoms are incorporated into a flat surface, $\gamma_{2x,y}$ are the anisotropic surface tensions, and R_{eq} represents the intrinsic fraction of the mobile atoms at the surface (i.e. without the fraction of the redeposited atoms). Taking into account only the lowest-order derivatives of the Taylor series expansion, the excavation rate for arbitrary angles of incidence and incidence of ions can be expressed by [64]

$$\Gamma_{ex} = \alpha_o\left[1 + \alpha_{1x}\frac{\partial h}{\partial x} + \alpha_{2x}\frac{\partial^2 h}{\partial x^2} + \alpha_{2y}\frac{\partial^2 h}{\partial y^2} + \alpha_{3x}\left(\frac{\partial h}{\partial x}\right)^2 + \alpha_{3y}\left(\frac{\partial h}{\partial y}\right)^2 + \cdots\right], \tag{8.21}$$

where α_o is the average excavation rate for a flat surface (= JY/N). The coefficients α_{1x} and $\alpha_{2x,y}$ serve the correction of the lowest linear-order approximation of the curvature-dependent sputtering yield, while the coefficients α_{3x} and α_{3y} serve the correction of the lowest non-linear-order approximation of the curvature-dependent sputtering yield [66]. In normal ion incidence, $\alpha_{1x} = 0$, $\alpha_{2x} = \alpha_{2y}$ and $\alpha_{3x} = \alpha_{3y}$.

Muñoz-García et al. [35, 56] have derived a generalized continuum equation, with the objective of providing a theoretical description of the main non-linear features of ripple dynamics. Based on the fact that typical rates for an erosion events ($\approx 1\ \mathrm{s}^{-1}$) are significantly lower than the rate of adatom hopping events ($\approx 10^{12}\ \mathrm{s}^{-1}$), it follows that the height h evolves on a much faster time scale than the R_f field. By eliminating R_f, these authors [64, 157] obtained a partial differential equation for the lowest non-linear order in a power expansion for the time-dependent height evolution. The time evolution of the topography, h(x, y, t), along each direction x and y, is then given by a highly non-linear equation. Where no variation of the field in the y-direction is allowed (one-dimensional model), the time evolution can be expressed by [65]

$$\frac{\partial h}{\partial t} = -v + \gamma_x\frac{\partial h}{\partial x} + v\frac{\partial^2 h}{\partial x^2} + \lambda_x^{(1)}\left(\frac{\partial h}{\partial x}\right)^2 + \Omega_x\frac{\partial^3 h}{\partial x^3} + \left(\frac{\partial h}{\partial x}\right)\left(\frac{\partial^2 h}{\partial x^2}\right) - K_{xx}\left(\frac{\partial^4 h}{\partial x^4}\right) - \lambda_{x,x}^{(2)}\frac{\partial^2}{\partial x^2}\left(\frac{\partial h}{\partial x}\right)^2 \tag{8.22}$$

where the coefficients in this equation are dependent on the parameters of the ion bombardment (ion energy, incidence angle, etc.). The coefficients $\lambda_x^{(1)}$ and $\lambda_{x,x}^{(2)}$ (provided in [65]) describe the local re-deposition of the sputtered atoms and the transport of redeposited atoms within the surface layer. Equation (8.22) is similar to the general continuum equation, (8.18), but the coefficients [provided in Appendix O, (O.32)–(O.38)] differ significantly from those of the general continuum equation. For normal ion incidence [158], (8.22) can be simplified, because the coefficients are isotropic.

The hydrodynamic model of surface height changing under ion bombardment not only confirms the features of the Bradley–Harper model, but also integrates the redistribution of the eroded material and illustrates some additional non-linear phenomena, such as the wavelength coarsening with increasing sputter time and non-uniform translation velocity of the evolving pattern [46]. For example, the coarsening of wavelength on fused silica surfaces after Ar ion bombardment as a function of the ion fluence (time of ion bombardment as shown in Fig. 8.44) can be reproduced. The ordered ripple coarsens in dependence on the ion fluence, i.e. the results of the numerical integration of (8.22) confirm the coarsening process (Fig. 8.44).

8.2.1.3 Carter and Vishnyakov Direct Mass Redistribution Model

The formation of ion beam-induced patterns on surfaces is frequently explained on the basis of the linear Bradley–Harper theory (see Sect. 8.2.1.1) or its non-linear extension, based on the Kuramoto–Sivashinsky equation (see Sect. 8.2.1.2). According to these continuum theories, pattern formation is based on the competition between roughening processes by ion beam-induced sputtering and smoothing processes by surface diffusion. However, the experimental results in particular are still debatable on the issue of the appearance of ripples as a function of the angle of ion incidence.

Carter and Vishnyakov [142] have proposed another linear continuum theory to explain the smoothing and formation of ripples after Xe ion bombardment of silicon at angles of ion incidence between 0° and 45° and temperatures between room temperature and 300 °C. They recall that ion impact is associated not only with the deposition of energy, but also with momentum transfers within the localized region (spike) [161]. Based on experimental studies of momentum dissipation by particles emitted from a bombarded surface [162], Littmark and Sigmund [163] determined the components of the momentum transfer on the target during the penetration of energetic ions into a solid. Figure 8.45 (left) shows the transferred momentum after perpendicular ion impact. Since the incident momentum P is normal to the surface, the momentum component is $P_y = 0$. In the near-surface region (close to the surface), the momentum transfer is perpendicular to the direction of ion impact, while for deeper penetrations, the transfer is parallel to the ion direction and increases to a maximum. Consequently, target atoms will be displaced both parallel and normal to the direction of the incoming ion (c.f. Fig. 5.12). It can be assumed that the displacement constitutes a ballistic diffusion process. For oblique ion bombardment (Fig. 8.45, right), where $\theta = 45°$, two momentum components appear (parallel and normal to the surface). The transverse momentum causes an effective ballistic movement parallel to the surface (for more details, see also [53, 63]). For low ion energies, it can be expected that the induced ballistic diffusion process occurs at the surface or in the near-surface region.

Carter and Vishnyakov [142] have recognized that (i) the gradient of atomic flux parallel to the surface can compensate for the curvature-dependent sputtering process and lead to smoothing, particularly for normal and near-normal incidences and (ii) the mass redistribution of surface atoms can significantly contribute to the formation of patterns at incidence angles $\geq 45°$. This effect could be confirmed using MD simulations by Moseler et al. [164] for low-energy (30 eV – 150 eV) ion beam induced smoothing on diamond-like carbon surfaces. These simulations revealed that the average net effect of each ion impact is a displacement along the surface, proportional to the angle of ion incidence. Carter and Vishnyakov [142] have defined a lateral flux of substrate atoms that is proportional to the ion flux $J\cos\varphi$ and sine of the local incident angle φ relative to the surface normal. According to Fig. 8.46, the projected ion flux in the x-direction is given by

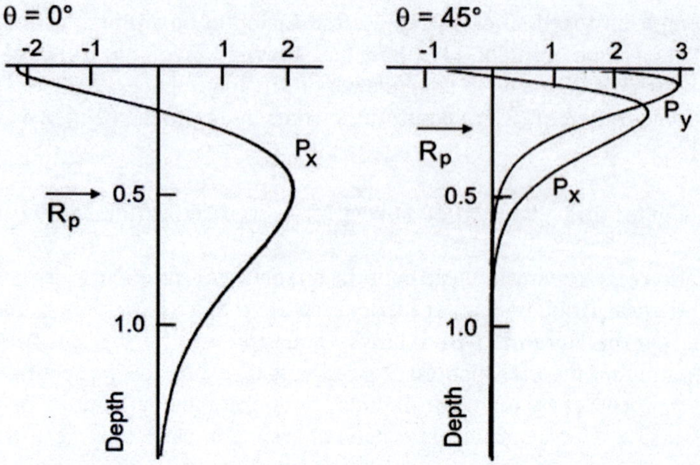

Fig. 8.45 Components P_x and P_y of the momentum P at an ion incidence angle of 0° (left) and 45° (right), for depths ≤ 0. The curves are normalized to unit incoming momentum. The mass ratio $M_2/M_1 = 1.5$ and the power cross sections are m = 1/2 (left) and m = 1/3 (right). R_p is the mean projected range. The abscissa unit is E^{2m}/NC (see Sect. 2.2.5). Modified figure is adapted from [163]

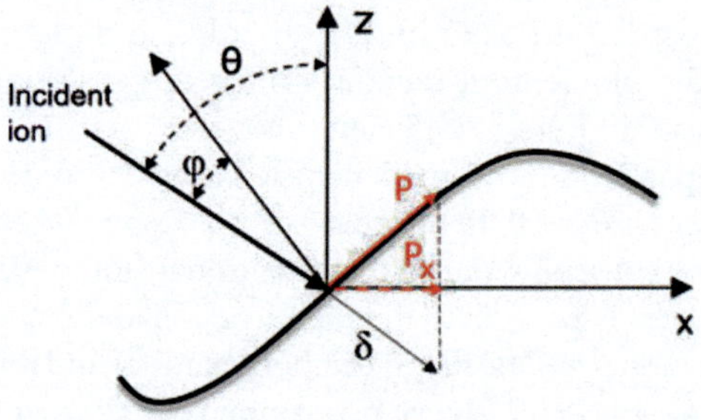

Fig. 8.46 Schematic illustration of the Carter–Vishnyakov mechanism in the laboratory coordination system, where δ is the mean displacement distance of recoils and P_x is the projection of the momentum P on the x-direction

$$J_x = J \sin\varphi / \cos(\theta - \varphi) = J \sin\varphi / \cos\alpha, \tag{8.23}$$

where $\alpha = \theta - \varphi$ and θ is the ion incidence angle with respect to the substrate surface. The incident ion flux J and the projection of the momentum, P, on the x-direction can be expressed by $P_x = \sin\varphi \cos(\theta - \varphi)$. The flux on the surface in x-direction is then given by

$$J_s = P_x J_x = J\delta \sin(\varphi)\cos(\varphi), \tag{8.24}$$

where δ is an empirical constant that gives the magnitude of net displacements distance of recoil atoms per incident ion (c.f. mean square static displacement, Sect. 4.9.2). According to MD simulations by Moseler et al. [164], the lateral atomic displacement can be expressed by

$$\delta = \left\langle \sum_{i=1}^{i=n} d^{(i)} \right\rangle, \tag{8.25}$$

where $d^{(i)}$ is the distance between the position of the recoiled atoms before and after the displacement projected on the surface plane and n is the number of recoiled atoms. The mass flux is then given by $\delta J = |J_s|$. For example, Madi et al. [30] have determined that $\delta = 28 \pm 5$ nm for 1 keV Ar ion impact on Si, and Norris et al. [47] have calculated that $\delta = 10$ nm for 250 eV Ar ion impact on Si. Moseler et al. [165] have proposed that δ can be roughly estimated by $\delta \approx \left(\mathrm{A}E^{1/2} + E^{4/3}\right)/\tan(90^\circ - \theta)$, where $\mathrm{A} = 1.4$ nm/keV$^{1/2}$ and $\mathrm{B} = 52$ nm/keV$^{4/3}$, while Babonneau et al. [51] have used $\delta \approx v_o a / J\Omega_o \cos\theta$, where a is again the ion range and Ω_o is the atomic volume of the mobile species participating in the surface mass transfer. Using the continuity equation of mass, the equation of motion in x-direction is defined as

$$\begin{aligned}\frac{\partial h}{\partial t} &= -\frac{1}{N}\frac{\partial J_s}{\partial x} = -\frac{J}{N}\delta\frac{\partial}{\partial x}[\sin(\varphi)\cos(\varphi)] = -\frac{J}{N}\delta\cos(2\varphi)\frac{\partial \varphi}{\partial x} \\ &= -\frac{J}{N}\delta\cos(2\varphi)\frac{\partial(\theta-\alpha)}{\partial x} \\ &= \frac{J}{N}\delta\cos(2\varphi)\frac{\partial \alpha}{\partial x},\end{aligned} \tag{8.26}$$

Using $\alpha = \tan^{-1}\frac{\partial h}{\partial x} \approx \frac{\partial h}{\partial x}$ (see Sect. 6.4.1), Carter and Vishnyakov have obtained a linear partial differential equation of the time-dependent changing of the topography along the x-direction

$$\frac{\partial h}{\partial t} = \frac{J}{N}\delta\cos(2\theta)\frac{\partial^2 h}{\partial x^2} = v_x^{redist}(\theta)\frac{\partial^2 h}{\partial x^2}, \tag{8.27}$$

where φ is averaged to θ and v_x^{redist} is the redistributive coefficient (in x-direction). The contribution of the ballistic drift on the mass redistribution is then given by multiplication of (8.27) with the total number of recoils produced by each ion (= 0.067 E/E_d, see Sect. 4.5.1), and can be given by

$$\frac{\partial h}{\partial t} = 0.067\frac{E}{E_d}\frac{J}{N}\delta\cos(2\theta)\frac{\partial^2 h}{\partial x^2} = N_x\frac{\partial^2 h}{\partial x^2}. \tag{8.28}$$

Davidovitch et al. [43, 166] have extended the Carter–Vishnyakov theory in two spatial dimensions. The assumption of an exclusively erosive response based on ellipsoidal cascade by Bradley and Harper to describe the height evolution has been replaced by the assumption of two Gaussian ellipsoids. In this way, the change of the height in y-direction can be determined by

$$\frac{\partial h}{\partial t} = \frac{J}{N}\delta \cos^2\theta \frac{\partial^2 h}{\partial y^2} = v_y^{redist}(\theta)\frac{\partial^2 h}{\partial y^2}. \tag{8.29}$$

According to Carter and Vishnyakov [142], the total change of the height in both directions is given by

$$\frac{\partial h(x, y.t)}{\partial t} = v_x^{redist}(\theta)\frac{\partial^2 h}{\partial x^2} + v_y^{redist}(\theta)\frac{\partial^2 h}{\partial y^2}. \tag{8.30}$$

The stability of the surface under ion bombardment is determined by the redistributive coefficient $v_{x,y}^{redist}$. The surface is stable when these coefficients are positive and instable when these coefficients are negative (see discussion regarding to the corresponding coefficients in the Bradley–Harper equation).

The Bradley–Harper equation of surface height under ion bombardment, (8.4), can now be completed by the terms of Carter and Vishnyakov, (8.27), and Davidovitch et al., (8.29), to be

$$\frac{\partial h(x, y, t)}{\partial t} = -v + v'\frac{\partial h}{\partial x} + v_x^t\frac{\partial^2 h}{\partial x^2} + v_y^t\frac{\partial^2 h}{\partial y^2} - K\nabla^4 h + \eta(x, y, t), \tag{8.31}$$

where the angle-dependent coefficients, $v_x^t(\theta)$ and $v_y^t(\theta)$, represent the sum of an erosion coefficient according to Bradley and Harper and a redistributive coefficient according to Carter–Vishnyakov and Davidovitch et al., i.e. $v_x^t = v_x + v_x^{redist}$ and $v_y^t = v_y + v_y^{redist}$, respectively.

In summary, based on the theory by Carter and Vishnyakov [37] and the extension by Davidovitch et al. [43, 166], it can be asserted that:

(i) Ion-bombardment-induced non-erosive collision cascade mass transport and redistribution can significantly influence the topography.
(ii) Rippled topography with increasing amplitude can be generated by direct momentum transfer from the incoming ion to the target atoms in the near-surface region. This momentum transfer results in net smoothing, particularly for normal and near-normal incidences, a net mass flux downhill on average for $\theta < 45°$, and a mass flux uphill at $\theta > 45°$ (note that φ is averaged to θ, Fig. 8.47).
(iii) The equations of motion by Bradley–Harper and Carter–Vishnyakov are similar. However, the coefficients $v_x = v_x^{redist}\cos(2\theta)$ and $v_y = v_y^{redist}\cos^2\theta$ differ significantly.
(iv) The positive and negative values of angle-dependent coefficients determine the linear stability of the bombarded surfaces. Patterns can be expected when these coefficients are negative for values of θ, where for values of θ where these quantities are both positive, smooth surfaces can be expected.

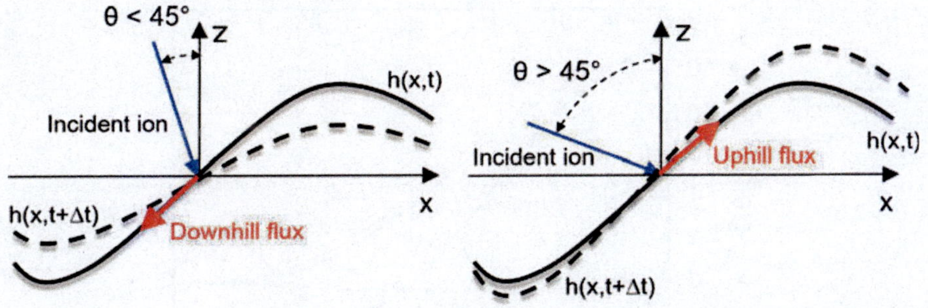

Fig. 8.47 Schematic drawing illustrating non-erosive mass redistribution under near-normal ion incidence (left) and under an ion incidence angle < 45° (right). The solid lines indicate the initial topography at the beginning of the ion bombardment and the dashed lines indicate the evolving surface after bombardment Δt

(v) The imprecise determination of the sum of the displacements, δ, of all recoiled atoms per incident ion complicates the application of this theory.

(vi) The Carter–Vishnyakov model cannot explain the formation of perpendicularly oriented ripples at large incidence angles.

Numerous simulation studies and experiments have demonstrated that mass redistribution can dominate the formation of patterns as compared to curvature-dependent sputtering (see the following sub-chapter).

8.2.1.4 Indirect Mass Redistribution Models

In the Bradley–Harper equation, (8.5), the stability of a surface undergoing ion beam erosion is described by the two curvature-dependent coefficients $v_x(\theta)$ and $v_y(\theta)$. The surface stability or instability is determined by the positive or negative values of these coefficients. Madi et al. [33] have decomposed the curvature coefficients $v_{x,y}$ into an ion-stimulated erosive coefficient, $\upsilon_{x,y}^{eros.}(\theta)$, and an ion-stimulated mass redistribution coefficient, $\upsilon_{x,y}^{redist.}(\theta)$. This procedure enables the distinction between erosive (sputter-induced erosion) and non-erosive (impact-induced mass redistribution) contributions to pattern formation. The sum of both contributions can be expressed by

$$\upsilon_{x,y}(\theta) = \upsilon_{x,y}^{eros.}(\theta) + \upsilon_{x,y}^{redist.}(\theta). \tag{8.32}$$

This issue can also be explicitly given with the Bradley–Harper expression of the angle-dependent erosive curvature coefficient (8.5) and the Carter–Vishnyakov expression of the redistributive coefficient, (8.28), by

$$\upsilon_x(\theta) = \upsilon_x^{eros.}(\theta) + \upsilon_x^{redist.}(\theta) = \frac{Ja}{N}Y(\theta)\Gamma_x(\theta) + \frac{J}{N}\gamma_{corr} \cdot \delta\cos(2\theta) \tag{8.33}$$

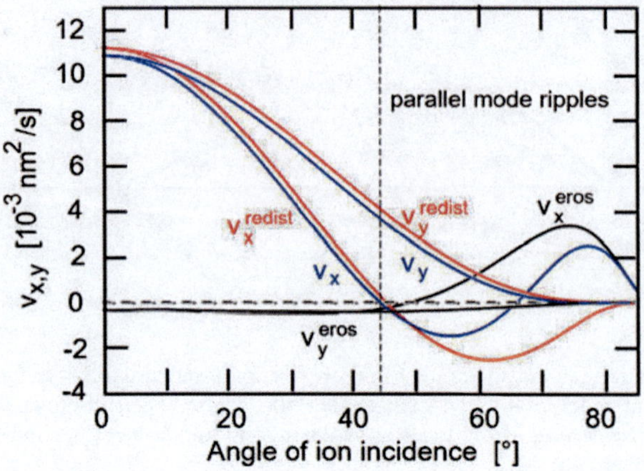

Fig. 8.48 Calculated coefficients $v_{x,y}(\theta)$, $v_{x,y}^{eros.}(\theta)$ and $v_{x,y}^{redist.}(\theta)$ for 1 keV Ar ion bombardment of silicon, where $J = 2 \times 10^{12}$ cm^{-2} s^{-1}, a = 3.4 nm, σ = 1.9 nm, μ = 1.6 nm, and δ = 28 nm. Figure is adapted from [35] and has been modified

and

$$v_y(\theta) = v_y^{eros.}(\theta) + v_y^{redist.}(\theta) = \frac{Ja}{N} Y(\theta)\Gamma_x(\theta) + \frac{J}{N}\gamma_{corr} \cdot \delta \cos^2\theta, \qquad (8.34)$$

where for the sputtering yield, Y(θ), the empirical expression for the large angle correction by Yamamura [see (5.55)], given by $Y(\theta) = Y(0)\cos^{-f}(\theta) \cdot \gamma_{corr}$, with $\gamma_{corr} = \exp\left[-\Sigma\left(\cos^{-1}\theta - 1\right)\right]$ is applied. The parameters f and Σ are determined by fitting of available experimental sputtering data [58].

As an example, Fig. 8.48 illustrates the variation of $v_{x,y}(\theta)$, $v_{x,y}^{eros.}(\theta)$ and $v_{x,y}^{redist.}(\theta)$ with the angle of ion incidence. The most important result is that the mass redistribution at surface or in the near-surface region significantly dominates the pattern formation. This result was able to be verified experimentally. Moreover, it could be demonstrated that parallel ripple mode occurs between approximately θ = 45° and θ = 65°. For large ion incidence angles, the contribution of $v_x^{eros.}(\theta)$ determines the curvature-dependent coefficient $v_x(\theta)$. But, the coefficient v_y is positive throughout the entire range of the incidence angle. Consequently, a change in the ripple orientation cannot be expected.

8.2.2 *Crater Function Formalism*

In the final chapters, the Bradley–Harper and Carter–Vishnyakov linear continuum theories were introduced, which based on curvature-dependent sputtering and mass redistribution, respectively. The predictions of both theories are dependent on phenomenological parameters that cannot be calculated [167]. One way to resolve this problem is the combination of the continuum theory with the result of MD simulations within the scope of crater function formalism. In Sect. 6.2.2.3 (c.f. Fig. 6.11), it was shown that a single ion impact can induce a crater through sputtering and local mass redistribution. On the basis of the average results of numerous ion impacts, a function, referred to as crater function, can be evolved to describe the effect of the ion bombardment on the surface topography. This means that MD simulations within crater function formalism provide the coefficients for the continuum equation of motion.

For this purpose, Norris et al. [47, 48] studied the Ar ion bombardment of silicon with 100 eV and 250 eV by MD simulation on the basis of crater function analysis. Crater function formalism is based on the equation of motion of surface. This equation is written in terms of the moments of the crater function F(x, y, θ), where the coordinates are so selected that x is parallel to the projection of ion beam direction and y is perpendicular to x. The point of ion impact is given by x = y = 0. The crater function describes the average height changes caused by ion impact. According to Norris et al. [47], the crater function at a surface point is the sum of the different mechanisms of the erosion by sputtering and local mass redistribution. The crater function is then given by

$$F(x, y, \theta) = F^{eros.}(x, y, \theta) + F^{redist.}(x, y, \theta). \tag{8.35}$$

The incorporation of ions during ion implantation is partially comprised in the crater function [168, 169].

The term $F^{eros.}(x, y, \theta)$ on the right-hand side of (8.35) represents the prompt regime. It is assumed that the single ion impact leads to an amorphization and sputtering within a picosecond time scale. On the left-hand side, the term $F^{redist.}(x, y, \theta)$ represents the gradual regime with a longer timescale. Within this regime, relaxation processes, such as Mullins surface diffusion (see Sect. 6.5.4) or ion enhanced viscous flow (see Sect. 6.5.6), occur. Thus, both mechanisms, sputtering and local mass redistribution, are integrated into crater function formalism.

Under the condition that local surface shape and local incidence angle are very slow in space compared to the crater function, the normal surface velocity of the bombarded surface can be described by the integro-differential equation [9, 47]

$$\frac{\partial h}{\partial t} = \int J(x')F(x - x')dx, \tag{8.36}$$

where $\boldsymbol{x}' - \boldsymbol{x}$ is the distance between point **x** and the point of ion impact **x**′. Norris et al. carried out a multi-scale analysis in which the integral in (8.36) is expanded in powers of ε into an infinite series of terms involving the moments of crater function F(x, y, θ), where ε is the ratio of the average ion penetration depth to a length scale associated with macroscopic relaxation. This formalism allows the separation of the spatial dependence of the crater function (prompt regime) from that of the surface topography (gradual regime). It is assumed by Norris et al. [47] that the ratio of micro- to macro-scale can be written as $\varepsilon = v\eta/\gamma$, where v is the normal erosion velocity, η is the viscosity, and γ is free surface energy (details are provided in [9,

47]). In the linear case, this process yields

$$\frac{\partial h}{\partial t} = \left[JM^{(0)}\right] - \varepsilon\nabla\left[JM^{(1)}\right] + \frac{1}{2}\varepsilon^2\nabla\cdot\nabla\left[JM^{(2)}\right] + \cdots, \tag{8.37}$$

where moments of crater function F(x, y, θ) in the one-dimensional case, $M^{(i)}$, are given by

$$M^{(0)} = \iint F(x, y, \theta)dxdy, \tag{8.38}$$

$$M_x^{(1)} = \iint F(x, y, \theta)x\,dxdy \quad \text{and} \quad M_y^{(1)} = \iint F(x, y, \theta)y\,dxdy, \tag{8.39}$$

$$M_{xy}^{(2)} = \iint F(x, y, \theta)xy\,dxdy, \quad \text{etc.} \tag{8.40}$$

Under the precondition that the variation of height is significantly larger than the dimension of the collision cascade, the time-dependent evolution of the surface under ion bombardment can be described by

$$\frac{1}{J}\frac{\partial h}{\partial t} = C_o(\theta) + C_1(\theta)\frac{\partial h}{\partial x} + C_{11}(\theta)\frac{\partial^2 h}{\partial x^2} + C_{22}(\theta)\frac{\partial^2 h}{\partial x^2}, \tag{8.41}$$

where the smoothing terms caused by thermally activated surface diffusion, $-K\nabla^4 h$ [see (6.52)], and the radiation-induced viscous flow, $-B\nabla^4 h$ [see (6.65)], can be added. The first two coefficients can be expressed by

$$C_o(\theta) = v(\theta) = -\mathrm{M}^{(0)}\cos\theta \tag{8.42}$$

and

$$C_1(\theta) = -\overline{v}(\theta) = \frac{\partial}{\partial\theta}\left(M^{(0)}\cos\theta\right). \tag{8.43}$$

By comparing the coefficients in (8.42) and (8.43), the similarity between the Bradley–Harper equation, (8.5), and the equation by Norris et al., (8.40), is obvious. Thus, the coefficients $C_o(\theta)$ and $C_1(\theta)$, respectively, describe the erosion rate and the contribution to the erosion rate that arises when the surface is sloped.

The angle-dependent coefficients C_{11} and C_{22} are dependent on the specific form of the crater function via the relations

$$C_{11}(\theta) = \frac{\partial}{\partial\theta}\left[\mathrm{M}^{(1)}(\theta)\cos\theta\right] \tag{8.44}$$

and

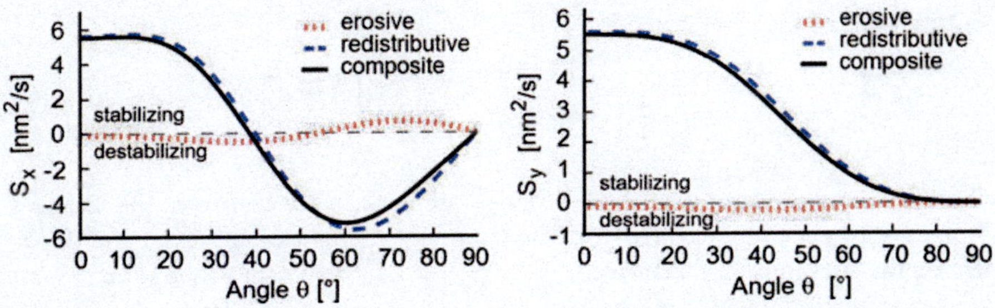

Fig. 8.49 Curvature-dependent coefficients $C_{11}(\theta) = S_y(\theta)$ and $C_{22}(\theta) = S_x(\theta)$, decomposed into their erosive and redistributive components as a function of the ion incidence angle for 250 eV Ar ion bombardment of silicon. Figure is taken from [48]

$$C_{22}(\theta) = \mathrm{M}^{(1)}(\theta)\cos\theta\cot\theta, \tag{8.45}$$

where these coefficients can be decomposed again into their erosive ($C_{11}^{eros.}$ and $C_{22}^{eros.}$) and redistributive components ($C_{11}^{redis.}$ and $C_{22}^{redis.}$), i.e.

$$C_{11}(\theta) = \frac{\partial}{\partial\theta}\left[M_{eros.}^{(1)}(\theta)\cos\theta\right] + M_{eros.}^{(1)}(\theta)\cos\theta\cot\theta\Big] \tag{8.46}$$

and

$$C_{22}(\theta) = \frac{\partial}{\partial\theta}\Big[M_{redis.}^{(1)}(\theta)\cos\theta\Big] + M_{redis.}^{(1)}(\theta)\cos\theta\cot\theta\Big]. \tag{8.47}$$

For the 250 eV Ar ion bombardment of silicon, Norris et al. [48] used MD simulations to study (8.36). Figure 8.49 shows the stability analysis based on the crater function framework according to Norris et al. [40]. The calculation demonstrates that the erosive moments are an order of magnitude smaller than the redistributive moments. This leads to a complete dominance of the coefficients by redistributive effects. The erosive contribution to the topographic changes seems to be irrelevant. For $\theta < 40°$, a smooth surface can be expected, while for $\theta > 40°$, the formation of ripples should be observable. According to (8.7), the growth rate can be determined by $R = -\left[C_{11}q_x^2 + C_{22}q_y^2\right] - K\left(q_x^2 + q_y^2\right)^2$.

Crater function formalism considers the influence of sputtering as well as mass redistribution, and also allows a determination of the critical angle for pattern formation. However, the shape-dependence of the crater was neglected by Norris et al. [40], as it is not possible to find the crater function for an arbitrarily shaped surface using MD simulations. For example, Perkinson et al. [170] have shown that measurement results for low-energy noble ion bombardment of Si and Ge are inconsistent with the predictions of crater function theory.

Consequently, Harrison and Bradley [167, 171] have extended crater function formalism by comprehension of dependence of the crater function on the curvature of the surface at the point of impact. They have derived that the curvature-dependent contributions of the crater function behave as

$$-\cos\theta\left[\frac{\partial M_{11}}{\partial K_{11}}\right]_{K_{11}=0} \quad \text{and} \quad -\cos\theta\left[\frac{\partial M_{22}}{\partial K_{22}}\right]_{K_{22}=0}, \tag{8.48}$$

where $K_{11} = \partial^2 h/\partial x^2$ and $K_{22} = \partial^2 h/\partial y^2$ are surface curvatures in the x- and y-directions, respectively [c.f. (6.44)]. The moments of crater function are also dependent on the curvature of the surface, i.e. $M_{11,22} = \iint F\left(x, y, \theta, K_{11,22}\right)dxdy$. Then, the curvature coefficients are given by

$$C_{11}(\theta) = -\frac{\partial}{\partial\theta}\left[\mathrm{M}^{(1)}(\theta)\cos\theta\right] - \cos\theta\left[\frac{\partial M_{11}}{\partial K_{11}}\right]_{K_{11}=0} \tag{8.49}$$

and

$$C_{22}(\theta) = -\mathrm{M}^{(1)}(\theta)\cos\theta\cot\theta - \cos\theta\left[\frac{\partial M_{22}}{\partial K_{22}}\right]_{K_{22}=0}. \tag{8.50}$$

The first terms on the right-hand side of (8.49) and (8.50) indicate a rise to a local angle of ion incidence that is dependent on the point of impact for a non-zero surface curvature. The second terms are a result of the curvature-dependence of the crater function in the x- and y-directions [167]. In Fig. 8.50, the coefficients C_{11} and C_{22} are compared as a function of the angle of ion incidence when the curvature dependence of the crater function is either taken into account or neglected. The angle θ is given in degrees and the coefficients are in units of $2\sqrt{2\pi}/\Lambda E a_\sigma^3 a_\mu$, where E is the total deposited energy, Λ is a constant (see (5.22) and [110]), $a_\sigma = a/\sigma$ and $a_\mu = a/\mu$, where a is the average energy deposition depth and σ and μ are the longitudinal and transverse straggling lengths, respectively.

It is obvious that the ratio of C_{11} given by Norris et al. [40], (8.45) to C_{11} given by Harrison and Bradley [59], (8.49), differs significantly. A similar statement applies to the coefficient C_{22}.

The stability/instability of the bombarded surface must once again be discussed on the basis of the signs of the coefficients C_{11} and C_{22}. Figure 8.50 also indicates that the angle for the switch from parallel to perpendicular-mode ripples is expected at about 51° if the curvature-dependence of the crater function is taken into account instead of about 67° if this dependence not considered.

The extended crater function formalism by Harrison and Bradley [167, 171] has explicitly demonstrated that crater function formalism yields the exact Bradley–Harper coefficients for the Sigmund model [see Appendix O, (O.39)–(O.44)]. If the curvature-dependence of the crater function is included, the smoothing effect can also be explained. When the curvature-dependence of the crater function is disregarded, the values of the coefficients can deviate substantially.

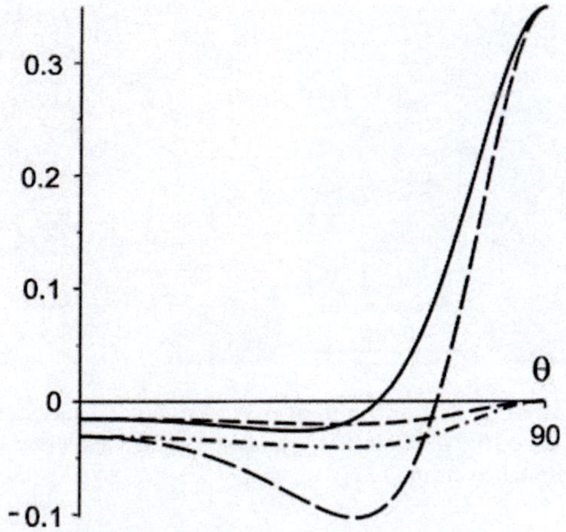

Fig. 8.50 The coefficients C_{11} and C_{22} as a function of the angle of ion incidence θ for 1 keV Ar ion bombardment of Si. The solid and dashed curves show the results for C_{11} and C_{22}, respectively, if the curvature dependence is taken into account. In contrast, the curve with long dashes and the dash-dotted curve show the results for C_{11} and C_{22}, respectively, if the curvature dependence of the crater function is neglected. The figure is adapted from [167]

8.3 Formation of Ripples on Polycrystalline Surfaces

Previously, it was assumed that the ion-irradiated material is either amorphous or monocrystalline. However, the formation of ripples has also been demonstrated on polycrystalline surfaces after low-energy ion irradiation. According to the results in Sects. 5.6 and 6.3.2, the sputtering yield is dependent on the orientation of the grains, where the most densely packed surface is characterized by the lowest sputtering yield. Consequently, it can be expected an orientation dependent ion etching rate. In early experiments on Cu (100) surfaces, ripple-like structures were observed before the formation of facets after oblique irradiation with Ne ions at different temperatures [172]. Later, the ion beam-induced patterning has been successfully observed on polycrystalline surfaces of metals such as Fe [173, 174], Ni [173, 175], Al [176], Cu [176], Co [176, 177], Ti [178], Ag [176, 179, 180] and Au [179, 180]. For example, Fig. 8.51 illustrates the ripple formation on polycrystalline Ti surfaces after Ga ion bombardment and an incident angle of 15°. The sputtered surface shows several grains with different topographies according to the different grain orientations (Fig. 8.51, left). The ripples formed on the surface of the grains are aligned along the different orientations (Fig. 8.51, right), where the alignment of the ripples vary in the grains. This shows that the crystallographic orientation appear to play an important role in the ripple evolution (orientation).

It is noteworthy that the formation of ripples could be observed not only after oblique ion irradiation, but also at normal incidence angle ion irradiation. For example, ripples could be detected on polycrystalline Ti [178], Cu [132], Ag [181], Au [182], and Sn [183] surfaces after normal ion bombardment, where the ripple

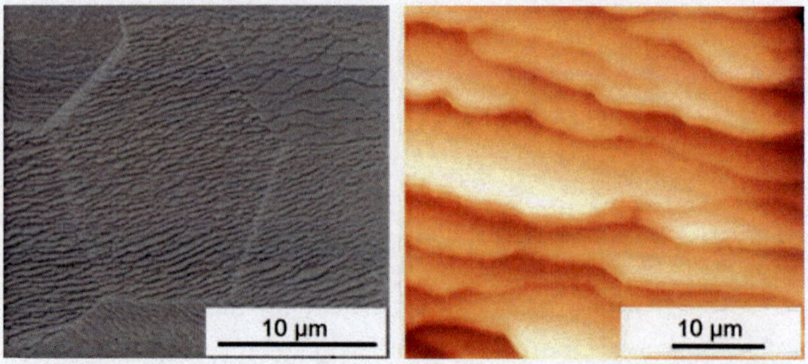

Fig. 8.51 SEM (left) and AFM images (right) of polycrystalline Ti surfaces after Ga ion bombardment with fluences of 2.40×10^{18} ions/cm^2 (left) and 4.63×10^{18} ions/cm^2 (right). The ion incident angle was 15° (figures are taken from [177])

direction was determined in some cases by the crystallographic orientation of the grains, but without considering the influence of possible contamination.

8.4 Application of Nanostructures Produced by Ion Beam Sputtering

Ion beam sputter machining of solid surfaces is a fundamental method for processing materials with pinpoint precision at the micro- and nanometer scale. In 1974, Taniguchi [184] coined the term 'nanotechnology' and introduced the basic concept of this new technology. In his keynote paper, he called the erosion of solid surfaces by ion beam sputtering as the most promising technique for structuring surfaces, performed at nanoscale ($\approx 10^{-9}$ m) and independent of type of material. Meanwhile, the method for using ion beam sputtering to design solid surfaces has undergone tremendous development, with the result that nanoscale size patterns can now be reproducibly manufactured on the surfaces of all solid material classes. The major advantages of this technique are its simplicity, cost-effectiveness, high output, and scalability. In comparison with lithography, ion beam sputter technology is an attractive option for producing nanostructures for applications in the fields of optics, semiconductor technology, microelectronics, and magnetic data storage. In the following, some selected examples of the application of low energy ion beam-induced nanostructures will be presented.

8.4.1 Quantum Dots

Quantum dots are a class of semiconductor materials in the order of nanometer size (< 50 nm) in which quantum confinement effects can be evidenced. As a result, they tightly confine electrons or electron–hole pairs, called excitons, in all three dimensions. The optoelectronic properties of quantum dots are determined by their size and shape. Consequently, these dots possess huge application potential in nanotechnology, particularly in optoelectronics.

The formation of ordered quantum dots on a surface, induced by low-energy ion beam sputtering, has been reported for Si and Ge as well as a variety of III–V compounds (see Sect. 8.4.1). For example, Facsko et al. [98] measured the low-temperature (15 K) photoluminescence (PL) spectrum of GaSb nanodots formed by low-energy Ar ion bombardment. The spectrum is characterized by a broad PL peak at about 1.6 eV (FWHM is 290 meV) and exhibits a broadened PL peak and some weak peaks due to the contribution of electronic states as the result of a high density of ion beam-induced defects. An option for reducing this contribution is the annealing of the quantum dots. For example, Fig. 8.52 shows PL spectra of a GaAs/AlGaAs heterostructure deposited on GaAs substrate by MBE (the inset in Fig. 8.52 shows the schematic of the heterostructure). The rippled topography was generated by low-energy Ar ion bombardment on the surface of a thin GaAs film on the AlGaAs layer. In a last step, an AlGaAs capping layer was deposited. The PL spectra before (as-grown) and after annealing at 600 °C for 90 s (annealed) are compared in Fig. 8.52. A significant improvement in the integrated PL signal was achieved, indicating the removal of defects by thermal annealing.

In spite of numerous attempts to improve the quality of quantum nanodots prepared by ion sputtering, it must be asserted that traditional fabrication techniques, such as lithographic techniques of spontaneous growth in the Stranski–Krastanov growth mode, have, up until now, produced better-quality quantum dots.

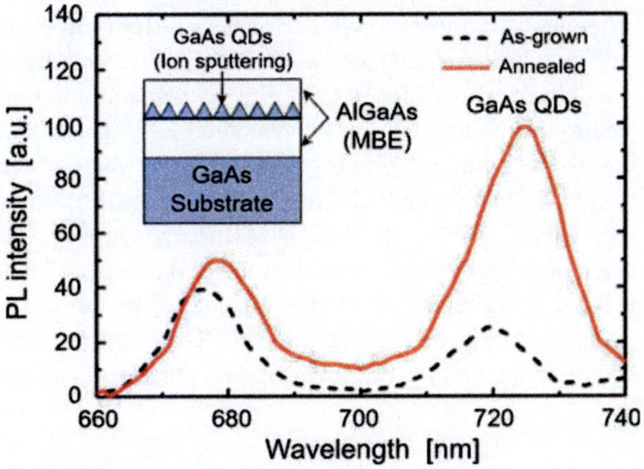

Fig. 8.52 Photoluminescence spectra of GaAs/AlGaAs quantum dots prepared using ion beam sputtering and MBE. The inset illustrates the schematic of the sample structure. The dotted curve (black) represents the PL spectrum of the sample before, and the solid curve (red) represents the PL spectrum after annealing at 600 °C for 90 s. PL spectra were measured at 5 K. Figure is taken from [117]

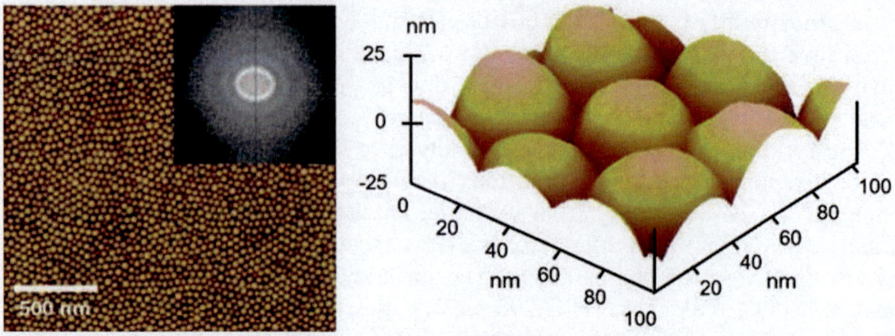

Fig. 8.53 Left: AFM image and corresponding 2D power spectral density of GaSb surface after Ar ion bombardment. Right: High-resolution AFM image of an area from the figure on the left side (figures taken from [187])

8.4.2 Templates for Deposition of Thin Films and Nanostructures

A frequent application for rippled and dotted surfaces produced by low-energy ion bombardment is the production of templates for developing patterns on deposited films or for the deposition of nanotubes, nanowires or nanoparticles, where those dimensions can be easily controlled by adjusting the dimension of the template geometry and the deposition conditions. For example, carbon nanotubes were deposited on rippled quartz substrates to study the nanotube positioning on the ripples [185] and para-hexaphenyl molecules were deposited on rippled $TiO_2(110)$ surfaces to investigate the influence of nanostructured surfaces on the growth of organic films [186]. Teichert and coworkers [187] introduced a very interesting application for nanostructured semiconductor surfaces as templates for nanomagnets. They used low-energy ion beam-induced hexagonally arranged GaSb nanodots (Fig. 8.53) for the fabrication of closed-packed nanomagnets. After 50 eV Ar ion bombardment of a GaSb(001) surface at 100 °C, hexagonally ordered nanodots formed, with an average diameter of 50 nm. 19 ML Pt/18 ML Co/19 ML Pt trilayers are deposited under oblique angles. Then, a calotte-like area on the tops of the dots will be exposed to the incident metal beam due to the partial shadowing caused by the surface topography (for the magnetic characterization, see Sect. 8.4.6). The resulting storage density of the nanomagnets located on the tops of the semiconductor dots is up to 0.2 Tbit/(in.)^2.

8.4.3 Nanometric Pattern Transfer

One option for creating nanostructures is the transfer of patterns (nanoripples or dots) onto a desired material. Realization requires the transfer of the pattern onto the final surface by either subtractive approaches, in which the undesired material

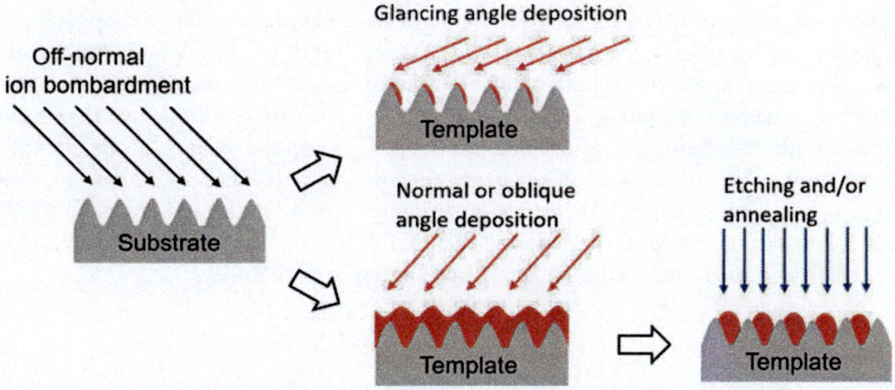

Fig. 8.54 Schematic of the preparation of optically active nanowires and nanoparticles

is removed by etching, or by additive means, if the desired material is added by selective deposition in the pattern resist. Sometimes, the two steps can be combined, as in the application of a focused ion beam to directly etch the surface.

As a low-cost approach for nanostructuring soft matter, Mele et al. [188] have proposed a combination of ion beam sputter induced nanostructuring and soft lithography. Glass surfaces were bombarded with 800 eV Ar ions at room temperature and an incidence angle of 35°. Under these conditions, rippled nanopatterns can be formed on the glass surface. The wavelength variation was nearly linear with a sputtering time between approximately 80 nm and 350 nm. The ripple pattern of this glass master was then transferred, using a standard replica molding procedure (see e.g., [189]) to an elastometric template. This template is used to transfer the ripple pattern to the thermoplastic film (here, a PMMA film on a silicon surface), with the objective of attaining a conformal map. When the polymer is heated to a temperature higher than the glass transition temperature, the polymer melt fills the space between the film and the replica. After subsequent cooling, the replica is peeled off. Dependent on the condition of the ion bombardment, patterns with nanometer resolution remain on the template.

Further, Bobek et al. [190, 191] have demonstrated that hexagonal GaSb nanodot patterns fabricated by ion beam erosion can be transferred onto buried Co nanodiscs (see Sect. 8.4.6), and Ou et al. [192] have shown that arrays of parallel-ordered crystalline Si nanowires on insulating SiO_2 substrate can be obtained by oblique Xe ion bombardment of silicon-on-isolator substrates.

8.4.4 Microelectronic Devices

Smirnov et al. [193] have reported on one of the first applications of rippled nanostructures in microelectronics. The authors discuss the opportunity for improving the performance of MOSFETs by reduction of the total channel length between the doped areas. They have proposed that the channel length be subdivided into shorter sectors, which can be viewed as a chain of numerous short-channel MOSFETS with a common gate. For this purpose, rippled nanostructures were produced by off-normal nitrogen ion bombardment of the amorphous Si layer on the surface of the channel region, where the formed silicon nitride acts as a nanomask of the subsequent reactive ion etching. The doping by As ion implantation is carried out through the nanomask, which results in a periodically doped channel field-effect transistor. This FET is characterized by a greater drain current and higher drift mobility.

8.4.5 Optically Active Nanostructures

In the following, the potential application of rippled nanostructures in the field of optics will be briefly presented. The direct application of rippled nanostructures and the use of ripples structures as templates will be separately discussed.

Optically Active Rippled Structures

The optical response of native-oxide-covered Si surfaces with a ripple wavelength between 20 nm and 50 nm but without any metallic coverage was studied by Perseschini et al. [194]. The linear response measured by reflection anisotropy spectroscopy exhibits a small peak at about 2.5 eV, where this signal decreases with an increase of the ripple wavelength. The non-linear optical response, using the second harmonic generation at a femtosecond excitation at 800 nm, decreased when the plane of light incidence was orthogonal to the ripple orientation.

Mussi et al. [195, 196] found that the ion beam-induced modification of the topography can be accompanied by the generation of electronic defects. They studied the formation of periodic nanostructures on LiF crystal surfaces under off-normal 800 eV Ar ion bombardment. The photoluminescence spectra are characterized by two broad Gaussian bands at about 680 nm (red) and 535 nm (green) in the visible spectra. These bands are ascribed to F_2 and $F_2{}^+$ active centers (two electrons bound to two and three anion vacancies, respectively). According to the authors, the results open up the possibility for fabricating optically active nanostructured substrates.

A noteworthy study on the enhancement of external quantum efficiency (EQE) over the total visible region absorption spectrum was published by Martella et al. [197]. They deposited Ag films on glass surfaces and sputtered under grazing angle conditions, with the objective of achieving rippled surfaces. Amorphous Si thin-film solar cells grown on these substrates show a strong improvement in spectral response with respect to the reference devices grown on unstructured glass substrates. The EQE was improved to an order of 15%, demonstrating the potential of this technology for efficient photon harvesting in thin-film solar cells.

Optically Active Metallic Nanostructures on Rippled Templates

The synthesis of metal nanoparticles and nanowires is of special interest because of their unique optical properties in the visible spectral region, caused by the excitation of collective oscillations of the electrons in the conduction band (known as 'localized surface plasmon resonance', LSPR) [198]. One possible method for fabricating ordered arrays of nanowires or aligned nanoparticles is the deposition of metals on rippled substrates.

The experiments to prepare optically active metal structures on rippled template surfaces are based on a procedure that is applied in different variations. A schematic of this procedure is shown in Fig. 8.54. After cleaning the substrate surface, the sputter experiments were performed in high or ultra-high vacuum chambers through bombardment of the surface with low-energy rare gas ions. In general, the temperature during the sputter process is lower than 150 °C. The ion incidence angle is fixed and for silicon should be > 50°. The rippled templates are fabricated on the amorphous or amorphized surface, exploiting the self-organization process induced by low-energy ion bombardment. The rippled templates are then deposited with a metal by thermal evaporation, RF or DC magnetron sputtering at a fixed angle to the global surface normal. On the one hand, evaporation is accomplished at a glancing angle ($\geq$ 80°) of the metal atom flux, whereby the rippled surface modulates the spatial distribution of the metal atom flux due the shadowing effect (see e.g., [199]). On the other hand, metal is deposited at normal or oblique incident angles (θ > 10°... 40°), where the direction of the metal flux with respect of the orientation of the ripples and the temperature during the deposition process determine the thickness distribution of the metal coverage on the rippled template (see e.g., [200]. A subsequent etching and/or annealing process leads to the spontaneous formation of nanowires or nanoparticles. For example, Fig. 8.55 shows aligned Ag nanoparticles and nanowires on rippled silicon. By controlling the metal particle flux, the nucleation density of these particles on the rippled surface, the agglomeration (clustering) of metal adatoms, and the film thickness can be determined.

Rippled silicon surfaces were preferentially used as templates to fabricate metal nanoparticles and nanowires aligned along the flanks of the ripples that face the particle flux during metal deposition. In detail, Ag [200–204], Au [201, 205], and Co nanoparticles [200] as well as Ag [199, 203, 206] and Au nanowires [204] were successfully synthesized. Typically, the particle diameter ranged between 20 nm and 100 nm and aligned particle chains with a lateral period down to 5...20 nm were formed.

Dichroic filters and similar optical components are based on the polarization-dependent shift in the plasmon polariton resonance, which can be realized by plasmonic coupling along the metal particle chains. A number of different research groups [200, 203, 206] have thus studied the plasmon resonance peaks of Ag nanoparticles and nanowires both perpendicular and parallel to the direction of the oriented ripples on Si surfaces.

As an example, Fig. 8.56 (left) shows a cross-sectional TEM micrograph of Ag particles deposited on rippled Si substrate. The particles show an oblate shape and are located in the ripple valleys, i.e. the Ag particles are aligned along the ripple valleys.

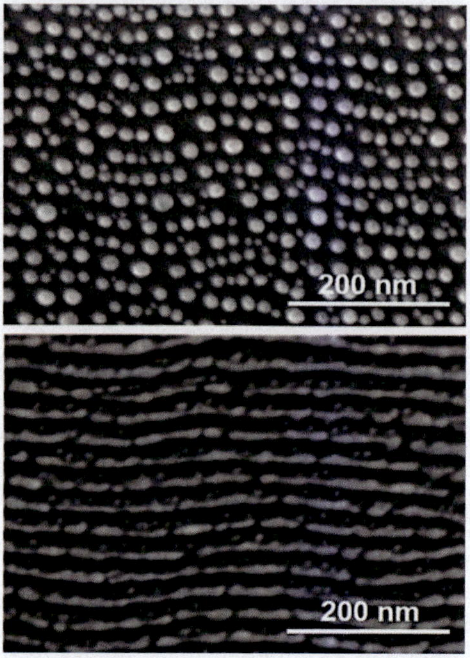

Fig. 8.55 SEM images of aligned silver nanoparticles (at the top) and nanowires (bottom) on rippled Si surfaces. Figures were taken from [205, 206]

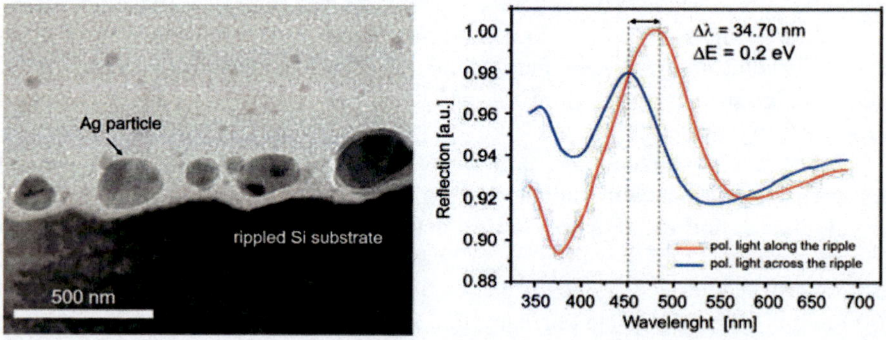

Fig. 8.56 Left: Cross-sectional TEM image of silver particles on the rippled silicon substrate. A number of smaller particles, having dissociated from the surface, are embedded in the carbon coating (figure taken from [203]). Right: Polarized reflection spectra for light polarized parallel (red line) and perpendicular (blue line) to the direction of the ripples (figure taken from [206])

The polarized reflection spectra (Fig. 8.56, right) were measured both parallel and perpendicular to the direction of the ripples and are characterized by a red shift of about 0.2 eV between light polarized parallel and perpendicular to the direction of the ripples. This effect is caused by the coupling of the plasmon resonance between adjacent particles within the valleys [186].

Ranjan et al. [202] have determined the dielectric functions of the aligned Ag nanoparticles. A variation in the periodicity of the ripple wavelength caused a modification of the plasmonic character of the Ag nanoparticles is visible in the imaginary part of dielectric function. It could be shown that biaxial anisotropy leads to an anisotropic LSRP. Consequently, it is possible to tune plasmonic behavior for potential application in photovoltaic systems or plasmonic back reflectors.

Camilio et al. [201] have performed a detailed investigation of the optical anisotropy of aligned Ag nanoparticles on rippled dielectric films as Al_2O_3, BN, and Si_3N_4. The Ag nanoparticles are ellipsoidal, with an effective thickness of 2 nm, where the major axis of these nanoparticles is parallel to the ripples. It could be demonstrated that the splitting of the surface plasmon resonance parallel and perpendicular to the ripples depends on the nature of the substrate, the ion energy during the nanopatterning, the distance between the nanoparticles, and also the size and shape of nanoparticles.

Surface-enhanced Raman spectroscopy (SERS) is a technology used to detect molecules, with a very high sensitivity on rough metal surfaces (down to the detection of single molecules) [207]. Self-aligned Ag nanoparticles and nanowires deposited on ripple-patterned Si surfaces can be used as surface-enhanced Raman scattering active substrates [205]. In particular, high SERS intensity can be observed in ordered elongated particles along the ripples due to the strong anisotropic coupling between Ag particles. Instead of silver, Schreiber et al. [208] used Au nanoparticles on rippled Si surfaces and also observed the bidirectional plasmonic response. The localized surface plasmon resonances in each direction are independent of each other and can be tuned using the fabrication parameters. A SERS enhancement factor of more than one thousand could be proven.

8.4.6 Magnetic Films and Nanostructures

The ideal type of magnetic ultra-high density storage device is positively characterized by such superior capabilities as a long life-cycle, small component size, large storage capability, and low power consumption. An interesting method for fabricating this type of nanomagnet-based magnetic recording device is ion beam-induced nanopatterning.

For this purpose, several procedures were applied for the preparation of periodic nanostructures (Fig. 8.56). Frequently, thin magnetic films are deposited on nanostructured non-magnetic substrates (procedure A in Fig. 8.57). Then, the periodic surface morphology is transferred to the deposited films, where the thickness of the film in dependence on the temperature during film deposition varies locally. As early as in 2002, there were reports on the fabrication of magnetic nanostructures on surfaces using ion bombardment. Chen et al. [209] prepared periodic nanostructures using sputtering of Co/Pd multilayers on nanostructured non-magnetic GaSb(100) surfaces (dots) at 420 eV Ar ion bombardment and a fluence of about 6×10^{18} ions/cm^2. The thin magnetic films on the nanodots then show the same periodic morphology as the non-deposited nanodots. Another possibility [190, 191] is the transfer of dot patterns on GaSb, prepared by Ar ion irradiation, onto a buried metal film (method B in Fig. 8.57). A heterostructure GaSb/5 nm Co/GaSb was bombarded, with the objective of the formation of a self-organized arrangement of nanodots. The erosion process is continued until the buried metallic layer is intersected. An array of disc- or pillar-shaped metallic dots is thus formed. Magnetic nanoparticles with a lateral periodicity of 44 nm were successfully produced. A procedure for preparing rippled magnetic films was presented by Bisio et al. [210]. They produced nanoripples using 1 keV Ar ion bombardment at a glancing ion incidence angle of 70° on Fe films deposited on flat Ag(001) surfaces (procedure C in Fig. 8.57).

According to the procedure (A) and (B) in Fig. 8.57, nanomagnets were successfully prepared on the basis of highly regular dot patterns, where nanodot densities up to 1 Tb/(in.)^2 could be realized [190, 191, 209].

It is common knowledge that magnetic anisotropies in thin magnetic films are caused when the translational invariance is broken [211]. Consequently, periodic surface patterning by ion beam irradiation of magnetic materials should create magnetic anisotropy in dependence on the orientation of the ripples. In general, the magnetization was investigated by means of magneto-optic Kerr effect (MOKE) magnetometry. For instance, Fig. 8.58, left, shows the hysteresis loop of an 18 nm thick Co film deposited on a rippled Si substrate [212]. Two MOKE measurements are compared as a function of the applied magnetizing field strength along and normal to the wave-vector of the ripples. It is obvious that the Co film possesses a uniaxial magnetic anisotropy with easy axis of magnetization in a direction normal to the wave-vector of the ripples, resulting in a preferential coalescence normal to the ripples of the growing Co islands during deposition. Uniaxial magnetic anisotropy has been also observed, e.g., for Fe films on Ag(001) substrates [210] and for ferromagnetic Permalloy ($Ni_{81}Fe_{19}$), Fe and Co films on rippled Si(001) [213, 214].

The magnetic anisotropy is strongly dependent on film thickness and the wavelength of the ion bombardment-induced ripples. Bisio et al. [11] found that in Fe films on silver substrates, the uniaxial magnetic anisotropy decreases with the film thickness and later, Liedke et al. [213] observed the same for Fe and other ferromagnetic films on silicon. Fassbender et al. [214] found that the uniaxial magnetic anisotropy in

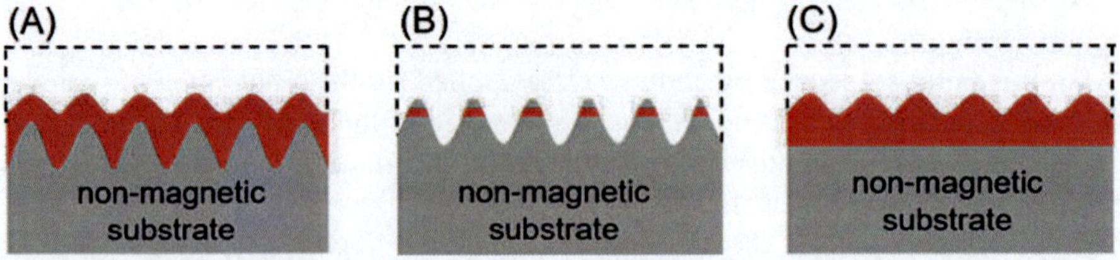

Fig. 8.57 Schematic presentation of the procedures for preparing periodic patterning of magnetic surface structures (red represents the magnetic material)

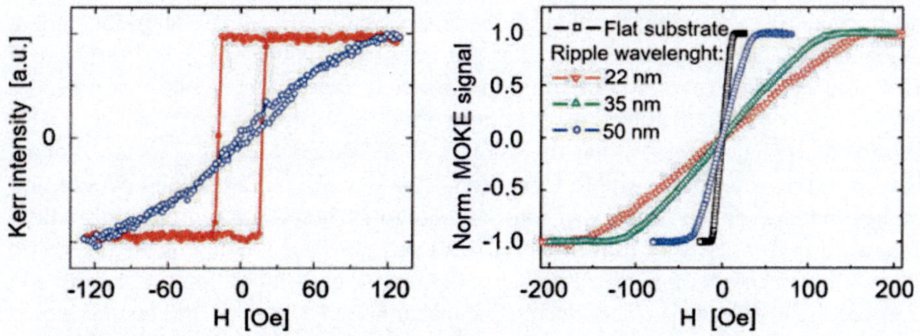

Fig. 8.58 Left: MOKE hysteresis loops of 18 nm Co film on rippled Si substrate taken with external field H along (open blue circles) and normal (closed red circles) to the wave-vector of the ripples. Figure is adapted from [212]. Right: MOKE measurements for Permalloy films on rippled silicon oxide substrates with various wavelengths. Figure adapted from [214]

Permalloy films on rippled SiO_2 substrates is strongly dependent on the ripple wavelength of the substrate material, i.e. the magnetic anisotropy is dependent on the ion energy that is used to create nanoripples. An example is given in Fig. 8.58, right. The MOKE measurements show that the saturation field, i.e. the applied magnetic field at which the magnetic moment reaches saturation, is strongly dependent upon the ripple wavelength. With decreasing wavelength, the saturation field strength increases, i.e. the induced magnetic anisotropy also increases. For ripple wavelengths > 60 nm, no significant magnetic anisotropy was proved. These results could be confirmed for Permalloy, Fe, and Co films on silicon [213].

The influence of rippled surfaces on antiferromagnetic interlayer exchange coupling has also been studied [215]. For this purpose, polycrystalline Fe/Cr/Fe layer stacks were deposited on rippled Si surfaces and the magnetic properties were determined by MOKE magnetometry. For 22 nm rippled substrates, a strong uniaxial anisotropy could be proven, while for longer wavelengths this effect becomes weaker, i.e. it is possible to tailor interlayer exchange coupling using wavelength selection.

8.4.7 Wettability of Rippled Surfaces

Wetting is the ability of liquids to remain in contact with solid surfaces. It is the result of intermolecular interactions when two media (solid and liquid) are brought together. The degree of wettability can be determined by measurement of the contact angle [216]. A significant influencing factor is the topography of the solid surface. Thus, it is important to study the wettability of ion beam-induced rippled surfaces for potential applications.

There are only a few known experimental investigations that have measured the wetting of liquids on rippled surfaces. Metya et al. [217] observed that smooth hydrophilic mica surfaces are transformed into hydrophobic rippled surfaces after off-normal 500 eV Ar ion bombardment. Interestingly, the contact angle is almost independent of the roughness as well as the ion fluence between 1×10^{17} and 2.1×10^{20} ions/cm^2. Wetting experiments on silicon surfaces after significantly higher ion energy bombardment also reveal the hydrophilic-hydrophobic transition [218]. Garg et al. [219] observed that the contact angle varied according to the Wenzel law up to the onset of the ripple formation. This deviation is explained by a reduction of the surface-free energy of ion-bombarded surfaces due to ripple formation. The wetting/dewetting of thin films on solid surfaces was studied by Petersen and Mayr [220]. They compared the dewetting behavior of thin Ni films on smooth and rippled silicon dioxide surfaces during thermal annealing. The Ni films on the rippled surfaces decrease faster with increasing temperature than do Ni films on smooth Si surfaces. It is assumed that the additional driving force of curvature-induced diffusion on the ripples surface leads to the formation of large voids even at low annealing temperatures. Subsequently, the film material is organized in nanorods or nanowires along the valleys.

8.5 List of Symbols

Symbol	Notation
C_i	Concentration of the component i
E	Ion energy
E_d	Displacement energy
E_{sd}	The activation energy of the surface diffusion
F(y, x, θ)	Crater function
F_D	Deposited energy depth distribution function
J	Ion flux
N	Atomic number density
N_r	Total number of recoils produced by each ion
P	Momentum
R	Growth rate
R_x, R_y	Radii of curvature
T	Temperature
$\nu_x^{eros} \cdot (\theta)$	ion-stimulated erosive coefficient
$\nu_y^{redist}(\theta)$	ion-stimulated mass redistribution coefficient
Y	Sputtering yield
a	Ion range
f	Spatial frequency
t	Time

(continued)

(continued)

Symbol	Notation
v	Erosion velocity (rate)
q	Wave number
w	Root mean square (rms) roughness
Λ	Material constant
Φ	Ion fluence
Φ_o	Threshold fluence for appearance of nanoripples
Ω_o	Atomic
α	Roughness exponent or damping term
β	Growth exponent
γ	Free surface energy
η	Viscosity or white noise
θ	Angle of ion incidence
θ_c	Ion incident angle of the stability-instability transition
λ	Wavelength of the ripples
λ_c	Characteristic wavelength
μ	Transversal straggling
ν	Largest absolute value of the negative surface tension coefficients
ν_x	Effective surface tension coefficient parallel to the projection of the ion beam onto the surface
ν_y	Effective surface tension coefficient perpendicular to the projection of the ion beam onto the surface
ξ	Correlation length
σ	Longitudinal straggling
φ	Angle between of the beam direction and the local normal to the surface
δ	Lateral atomic displacement of recoil atoms per incident ion

References

1. W.L. Chan, E. Chason, Making waves: kinetic processes controlling surface evolution during low energy ion sputtering. J. Appl. Phys. **101**, 121301 (2007)
2. B. Ziberi, M. Cornejo, F. Frost, B. Rauschenbach, Highly ordered nanopatterns on Ge and Si surfaces by ion beam sputtering. J. Phys. Condens. Matter. **21**, 224003 (2009)
3. T.K. Chini, D.P. Datta, S.R. Bhattacharyya, Ripple formation on silicon by medium energy ion bombardment. J. Phys. Condens. Matter. **21**, 224004 (2009)
4. J. Muñoz-García, L. Vázquez, R. Cuerno, J.A. Sánchez-García, M. Castro, R. Gago, Self-organized surface nanopatterning by ion beam sputtering, in *Toward Functional Nanomaterials*, ed. by Z. Wang, Lecture Notes in Nanoscale Science and Technology, vol. 5 (Springer, New York, 2009), pp. 323–398

5. A. Keller, S. Facsko, Ion-induced nanoscale ripple patterns on Si surfaces: Theory and experiment. Materials **3**, 4811–4841 (2010)
6. M. Cornejo, J. Völlner, B. Ziberi, F. Frost, B. Rauschenbach, Ion beam sputtering: a route for fabrication of highly ordered nanopatterns, in *Fabrication and Characterization in the Micro-Nano Range*. ed. by F. Lasagni, A. Lasagni (Springer, Berlin, 2011), pp. 69–94
7. T. Som, D. Kanjilal (eds.), *Nanofabrication by Ion-Beam Sputtering: Fundamentals and Applications* (Pan Stanford Publishing, Boca Raton, 2013)
8. J. Muñoz-García, L. Vázquez, M. Castro, R. Gago, A. Redondo-Cubero, A. Moreno-Barrado, R. Cuerno, Self-organized nanopatterning of silicon surfaces by ion beam sputtering. Mater. Sci. Engng. R **86**, 1–44 (2014)
9. S.A. Norris, M.J. Aziz, Ion-induced nanopatterning of silicon: toward a predictive model. Appl. Phys. Rev. **6**, 011311 (2019)
10. B. Rauschenbach, F. Frost, Destructive and constructive routes to prepare nanostructures on surfaces by low-energy ion beam sputtering. Proc. SPIE **9929**. *Nanostructured Thin Films IX*, 99290A (2016)
11. P. Haymann, J.J. Trillat, Action des ions argon de faibles energies sur des surfaces duranium. Compt. Rend. **248**, 2472–2475 (1959)
12. R.L. Cunningham, P. Haymann, C. Lecomte, W.J. Moore J.J. Trillat, Etching of surfaces with 8-Kev argon ions. J. Appl. Phys. **31**, 839–842 (1960)
13. M. Navez, D. Chaperot, C. Stella, Microscopie electronique - etude de l'attaque du verre par bombardement ionique, Comptes Rendus Hebdomadaires Des Seances De L Academie Des Sciences, Paris **254**, 240–248 (1962)
14. R.S. Nelson, D.J. Mazey, Surface damage and topography changes produced during sputtering. Rad. Eff. **18**, 127–134 (1973)
15. J.L. Whitton, G. Carter, M.J. Nobes, J.S. Williams, The development of cones and associated features on ion bombarded copper. Rad. Eff. **32**, 129–133 (1977)
16. G. Carter, N.J. Nobes, F. Paton, J.S. Williams, J.L. Whitton, ion bombardment induced ripple topography on amorphous solids. Rad. Eff. **33**, 65–77 (1977)
17. A. Zalar, Improved depth resolution by sample rotation during Auger electron spectroscopy depth profiling. Thin Solid Films **124**, 223–230 (1985). Sample rotating in Auger electron spectroscopy depth profiling. J. Vac. Sci. Technol. A **5**, 2979–2980 (1987)
18. Y. Zhao, G.-C. Wang, T.-M. Lu, *Characterization of Amorphous and Crystalline Rough Surface: Principles and Applications* (Academic Press, London, 2001)
19. G. Ozaydin, A.S. Özcan, Y. Wang, K.F. Ludwig, H. Zhou, R.L. Headrick, D.P. Siddons, Real-time X-ray studies of Mo-seeded Si nanodot formation during ion bombardment. Appl. Phys. Lett. **87**, 163104 (2005)
20. B. Ziberi, F. Frost, T. Höche, B. Rauschenbach, Ripple pattern formation on silicon surfaces by low-energy ion-beam erosion: experiment and theory. Phys. Rev. B **72**, 235310 (2005)
21. F. Frost, B. Ziberi, A. Schindler, B. Rauschenbach, Surface engineering with ion beams: from self-organized nanostructures to ultra-smooth surfaces. Appl. Phys. A **91**, 551–559 (2008)
22. R.S. Robinson, S.M. Rossnagel, Ion beam-induced topography and surface diffusion. J. Vac. Sci. Technol. **21**, 790–797 (1982)
23. T. Allmers, M. Donath, G. Rangelov, Pattern formation by erosion sputtering on GaSb: transition from dot to ripple formation and influence of impurities. J. Vac. Sci. Tech. B **24**, 582–586 (2006)
24. B. Ziberi, F. Frost, B. Rauschenbach, Pattern transitions on Ge surfaces during low-energy ion beam erosion. Appl. Phys. Lett. **88**, 1731115 (2006)
25. D. Carbone, A. Alija, O. Plantevin, R. Gago, S. Facsko T.H. Metzger, Early stage of ripple formation on Ge(001) surfaces under near-normal ion beam sputtering. Nanotechnology **19**, 035304 (2008)
26. B. Ziberi, F. Frost, M. Tartz, H. Neumann, B. Rauschenbach, Ripple rotation, pattern transitions, and long range ordered dots on silicon by ion beam erosion. Appl. Phys. Lett. **92**, 063102 (2008)

27. B. Rauschenbach, B. Ziberi, F. Frost, Self-organized pattern formation by ion beam erosion, in: *Proceedings of the 2nd IEEE-International Nanoelectronics Conference* (INEC, Shanghai, 2008), pp. 93–97
28. G. Ozaydin-Ince, K.F. Ludwig, In situ X-ray studies of native and Mo-seeded surface nanostructuring during ion bombardment of Si(100). J. Phys. Condens. Matter. **21**, 224008 (2009)
29. C.S. Madi, B. Davidovitch, H. Bola George, S.A. Norris, M.P. Brenner, M.J. Aziz, Multiple bifurcation types and the linear dynamics of ion sputtered surfaces. Phys. Rev. Lett. **101**, 246102 (2008)
30. C.S. Madi, H.B. George, M.J. Aziz, Linear stability and instability patterns in ion-sputtered silicon. J. Phys. Condens. Matter **21**, 224010 (2009)
31. C.S. Madi, M.J. Aziz, Multiple scattering causes the low energy–low angle constant wavelength topographical instability of argon ion bombarded silicon surfaces. Appl. Surf. Sci. **258**, 4112–4115 (2012)
32. J.C. Perkinson, C.S. Madi, M.J. Aziz, Nanoscale topographic pattern formation on Kr^+-bombarded germanium surfaces. J. Vac. Sci. Technol. A **31**, 021405 (2013)
33. C.S. Madi, E. Anzenberg, K.F. Ludwig, Jr., Michael J. Aziz, Mass redistribution causes the structural richness of ion-irradiated surfaces. Phys. Rev. Lett. **106**, 066101 (2011). Erratum, Phys. Rev. Lett. **110**, 069903 (2013)
34. E. Anzenberg, C.S. Madi, M.J. Aziz, K.F. Ludwig, Jr., Time-resolved measurements of nanoscale surface pattern formation kinetics in two dimensions on ion-irradiated Si. Phys. Rev. B **84**, 214108 (2011)
35. M. Teichmann, Musterbildung auf Si- und Ge-Oberflächen durch niederenergetische ionenstrahlerosion. Dissertation, University Leipzig (2015)
36. S. Macko, F. Frost, B. Ziberi, D.F. Förster, T. Michely, Is kev ion-induced pattern formation on Si(001) caused by metal impurities? Nanotechnology **21**, 85301 (2010)
37. M. Engler, Self-organized ion beam pattern formation on Si(001). Dissertation, Universität zu Köln (2015)
38. T. Basu, J.R. Mohanty, T. Som, Unusual pattern formation on Si(100) due to low energy ion bombardment. Appl. Surf. Sci. **258**, 9944–9994 (2012)
39. D. Chowdhury, D. Ghose, Tuning of nanopatterns on Si (100) substrate: role of ion beam processing parameters during sputtering. Appl. Sci. Lett. **22**, 105–110 (2016)
40. M. Castro, R. Gago, L. Vázquez, J. Muñoz-García, R. Cuerno, Stress-induced solid flow drives surface nanopatterning of silicon by ion-beam irradiation. Phys. Rev. B **86**, 214107 (2012)
41. E. Anzenberg, J.C. Perkinson, C.S. Madi, M.J. Aziz, K.F. Ludwig, Jr., Nanoscale surface pattern formation kinetics on germanium irradiated by Kr+ ions. Phys. Rev. B **86**, 245412 (2012)
42. H. Hofsäss, K. Zhang, Surfactant sputtering. Appl. Phys. A **92**, 517–524 (2008)
43. B. Davidovitch, M.J. Aziz, M.P. Brenner, On the stabilization of ion sputtered surfaces. Phys. Rev. B **76**, 205420 (2007)
44. Y. Kuramoto, T. Tsuzuki, Persistent propagation of concentration waves in dissipative media far from thermal equilibrium. Prog. Theor. Phys. **55**, 356–369 (1976). G.I. Sivashinsky, Nonlinear analysis of hydrodynamic instability in laminar flames—I. Derivation of basic equations. Acta Astronaut. **4**, 1177–1206 (1977)
45. J. Erlebacher, M.J. Aziz, E. Chason, M.B. Sinclair, J.A. Floro, Nonlinear amplitude evolution during spontaneous patterning of ion-bombarded Si(001). J. Vac. Sci. Technol. A **18**, 115–120 (2000)
46. M. Castro, R. Cuerno, Hydrodynamic approach to surface pattern formation by ion beams. Appl. Surf. Sci. **258**, 4171–4178 (2012)
47. S.A. Norris, M.P. Brenner, M.J. Aziz, From crater functions to partial differential equations: a new approach to ion bombardment induced nonequilibrium pattern formation. J. Phys. Condens. Matter. **21**, 224017 (2009)
48. S.A. Norris, J. Samela, L. Bukonte, M. Backman, F. Djurabekova, K. Nordlund, C.S. Madi, M.P. Brenner, M.J. Aziz, Molecular dynamics of single-particle impacts predicts phase diagrams for large scale pattern formation. Nat. Comm. **2**, 276–281 (2011)

49. M. Teichmann, J. Lorbeer, B. Ziberi, F. Frost, B. Rauschenbach, Pattern formation on Ge by low energy ion beam erosion. New J. Phys. **13**, 103029 (2013)
50. J. Lorbeer, Selbstorganisierte Nanostrukturierung oxidischer Oberflächen mittels niederenergetischer ionenstrahlerosion. Dissertation, University Leipzig (2015)
51. D. Babonneau, E. Vandenhecke, S. Camelio, Formation of nanoripples on amorphous alumina thin films during low-energy ion-beam sputtering: experiments and simulations. Phys. Rev. B **95**, 085412 (2017)
52. M. Teichmann, J. Lorbeer, F. Frost, B. Rauschenbach, Ripple coarsening on ion beam-eroded surfaces. Nanoscale Res. Lett. **9**, 439–446 (2014)
53. K. Zhang, M. Brötzmann, H. Hofsäss, Surfactant-driven self-organized surface patterns by ion beam erosion. New J. Phys. **13**, 013033 (2011)
54. Y. Liu, D. Hirsch, R. Fechner, Y. Hong, S. Fu, F. Frost, B. Rauschenbach, Nanostructures on fused silica surfaces produced by ion beam sputtering with Al co-deposition. Appl. Phys. A **124**, 73 (2018)
55. T. Basu, D. Datta, T. Som, Transition from ripples to faceted structures under low-energy argon ion bombardment of silicon: understanding the role of shadowing and sputtering. Nanoscale Res. Lett. **8**, 289–296 (2013)
56. K. Zhang, H. Hofsäss, F. Rotter, M. Uhrmacher, C. Ronning, J. Krauser, Morphology of Si surfaces sputter-eroded by low-energy Xe-ions at glancing incident angle. Surf. Coat. Technol. **203**, 17–18 (2009)
57. C.S. Madi, H.B. George, M.J. Aziz, Linear stability and instability patterns in ion-sputtered silicon. J. Phys. Consens. Matter **21**, 224010 (2009)
58. A. Keller, S. Roßbach, S. Facsko, W. Möller, Simultaneous formation of two ripple modes on ion sputtered silicon. Nanotechnology **19**, 135303 (2008)
59. P.F.A. Alkemade, Propulsion of ripples on glass by ion bombardment. Phys. Rev. Lett. **96**, 107602 (2006)
60. D. Chowdhury, D. Ghose, S.A. Mollick, B. Satpati, S.R. Bhattacharyya, Nanorippling of ion irradiated GaAs (001) surface near the sputter-threshold energy. Phys. Stat. Sol. (b) **252**, 811–815 (2015)
61. P. Karmakar, D. Ghose, Nanoscale periodic and faceted structures formation on Si(100) by oblique angle oxygen ion sputtering. Nucl. Instr. Meth Phys. Res. B **230**, 539–544 (2005)
62. C.C. Umbach, R.L. Headrick, K.-C. Chang, Spontaneous nanoscale corrugation of ion-eroded SiO_2: the role of ion-irradiation-enhanced viscous flow. Phys. Rev. Lett. **87**, 246104 (2001)
63. M.A. Makeev, R. Cuerno, A.-L. Barabási, Morphology of ion-sputtered surfaces. Nucl. Instr. Meth. Phys. Res. B **97**, 185–227 (2002)
64. J. Muñoz-García, M. Castro, R. Cuerno, Nonlinear ripple dynamics on amorphous surfaces patterned by ion beam sputtering. Phys. Rev. Lett. **96**, 08610 (2006)
65. J. Muñoz-García, R. Cuerno, M. Castro, Coupling of morphology to surface transport in ion-beam irradiated surfaces: oblique incidence. Phys. Rev. B **78**, 205408 (2008)
66. R. Cuerno, A.-L. Barabási, Dynamic scaling of ion-sputtered surfaces. Phys. Rev. Lett. **74**, 4746–4749 (1995)
67. G.K. Wehner, D.J. Hajicek, Cone formation on metal targets during sputtering. J. Appl. Phys. Phys. **42**, 1145–1149 (1971)
68. S. Rossnagel, R. Robinson, Surface diffusion activation energy determination using ion beam microtexturing. J. Vac. Sci. Technol. **20**, 195–198 (1982). H.R. Kaufman, R.S. Robinson, Ion beam texturing of surfaces. J. Vac. Sci. Technol. **16**, 175–178 (1979)
69. W.R. Hudson, Ion beam texturing. J. Vac. Sci. Technol. **14**, 286–289 (1977)
70. B. Ziberi, F.Frost, M. Tartz, H. Neumann B. Rauschenbach, Importance of ion beam parameters on self-organized pattern formation on semiconductor surfaces by ion beam erosion. Thin Solid Films **459**, 106–110 (2004)
71. J.A. Sánchez-García, L. Vázquez, R. Gago, A. Redondo-Cubero, J.M. Albella, Z. Czigány, Tuning the surface morphology in self-organized ion beam nanopatterning of Si(001) via metal incorporation: from holes to dots. Nanotechnology **19**, 355306 (2008)

72. J.A. Sánchez-García, R. Gago. R. Caillard, A. Redondo-Cubero, J.A. Martin-Gago, F.J. Palomares, M. Fernández, L. Vázquez, Production of nanohole/nanodot patterns on Si(001) by ion beam sputtering with simultaneous metal incorporation. J. Phys. Condens. Matter **21**, 224009 (2009)
73. M. Cornejo, B. Ziberi. C. Meinecke, D. Hirsch, J.W. Gerlach, T. Höche, F. Frost, B. Rauschenbach, Self-organized patterning on Si(001) by ion sputtering with simultaneous metal incorporation. Appl. Phys. A **102**, 593–599 (2011)
74. S. Macko, F. Frost, M. Engler, D. Hirsch, T. Höche, J. Grenzer, T. Michely, Phenomenology of iron-assisted ion beam pattern formation on Si(001). New J. Phys. **13**, 073017 (2011)
75. A. Redondo-Cubero, R. Gago, F. Palomares, A. Mücklich, M. Vinnichenko, L. Vázquez, Nanopatterning dynamics on Si(100) during oblique 40-keV Ar+ erosion with metal codeposition: morphological and compositional correlation. Phys. Rev. B **86**, 085436 (2012)
76. J. Zhou, M. Lu, Mechanism of Fe impurity motivated ion-nanopatterning of Si (100) surfaces. Phys. Rev. B **82**, 125404 (2010)
77. G. Carter, The physics and applications of ion beam erosion. J. Phys. D Appl. Phys. **34**, R1–R22 (2001)
78. S. Bhattacharjee, P. Karmakar, V. Naik, A. Sinha, A. Chakrabarti, The role of carbon in ion-beam nano-pattering of silicon. Appl. Phys. Lett. **103**, 181601 (2013)
79. R. Gago, L. Vázquez, O. Plantevin, J.A. Sánchez-García, M. Varela, M.C. Ballesteros, J.M. Albella, T.H. Metzger, Temperature influence on the production of nanodot patterns by ion beam sputtering of Si(001). Phys. Rev. B **73**, 155414 (2006)
80. Z.M. Wang, L.P.H. Juergens, J.Y. Wang, E.J. Mittemeijer, Fundamentals of metal-induced crystallization of amorphous semiconductors. Adv. Eng. Mater. **11**, 131–135 (2009). D.N. Lee, S.B. Lee, Solid-phase crystallization of amorphous silicon films, in *Advanced Topics in Crystallization*, ed. by Y. Mastai, InTech, Rijeka, 2015, pp. 205–234
81. K. Zhang, M. Brötzmann, H. Hofsäss, Sharp transition from ripple patterns to a flat surface for ion beam erosion of Si with simultaneous co-deposition of iron. AIP Adv. **2**, 032123 (2012)
82. O. El-Atwani, S. Gonderman, A. DeMasi, A, Suslova, J. Fowler, M. El-Etwani, K. Ludwig, J.P. Allain, Nanopatterning of metal-coated silicon surfaces via ion beam irradiation: real time X-ray studies reveal the effect of silicide bonding. J. Appl. Phys. **113**, 124305 (2013)
83. S.K. Vayalil, A. Gupta, S.V. Roth, V. Ganesan, Investigation of the mechanism of impurity assisted nanoripple formation on Si induced by low energy ion beam erosion. J. Appl. Phys. **117**, 024309 (2015)
84. M. Engler, F. Frost, S. Müller, S. Macko, M. Will, R. Feder, D. Spemann, R. Hübner, S. Facsko, T. Michely, Silicide induced ion beam patterning of Si(001). Nanotechnology **25**, 115303 (2014)
85. A.R. Miedema, A simple model for alloys. I. Rules for the alloying behaviour of transition metal. Philips Techn. Rev. **33**, 149–160 (1973). A simple model for alloys. Il, The influence of ionicity on the stability and other physical properties of alloys. Philips Techn. Rev. **33**, 196–202 (1973)
86. B. Moon, S. Yoo, J.-S. Kim, S.J. Kang, J. Muñoz-García, R. Cuerno, Ion-beam nanopatterning of silicon surfaces under codeposition of non-silicide-forming impurities. Phys. Rev. B **93**, 115430 (2016)
87. C. Teichert, C. Hofer, G. Hlawacek, Self-organization of nanostructures in inorganic and organic semiconductor systems. Adv. Eng. Mat. **8**, 1065–1073 (2006)
88. H. Hofsäss, K. Zhang, A. Pape, O. Bobes, M. Brötzmann, The role of phase separation for self-organized surface pattern formation by ion beam erosion and metal atom co-deposition. Appl. Phys. **111**, 653–664 (2013)
89. M. Yoshitake, Y. Yamauchi, C. Bose, Sputtering rate measurements of some transition metal silicides and comparison with those of the elements. Surf. Interface. Anal. **36**, 801–804 (2004)
90. S. Berg, A.M. Barklund, B. Gelin, C. Nender, I.V. Katardjiev, Atom assisted sputtering yield amplification. J. Vac. Sci. Technol. A **10**, 1592–1596 (1992). C. Nender, I.V. Katardjiev, J.P. Biersack, S. Berg, A.M. Barklund, Numerical and experimental studies of the sputter yield amplification effect. Radiat. Eff. Def. Solids **25**, 281–291 (1993)

91. J.M.E. Harper, S. Motakef, D. May, Highly selective sputtering of silicon from $TiSi_2$ at elevated temperatures. Appl. Phys. Lett. **60**, 1196–1198 (1992)
92. V.B. Shenoy, W.L. Chan, E. Chason, Compositionally modulated ripples induced by sputtering of alloy surfaces. Phys. Rev. Lett. **98**, 25610 (2007)
93. R.M. Bradley, P.D. Shipman, Spontaneous pattern formation induced by ion bombardment of binary compounds. Phys. Rev. Lett. **105**, 145501 (2010). A surface layer of altered composition can play a key role in nanoscale pattern formation induced by ion bombardment. Appl. Surf. Sci. **258**, 4161–4170 (2012)
94. P.D. Shipman, R.M. Bradley, Theory of nanoscale pattern formation induced by normal incidence ion bombardment of binary compounds. Phys. Rev. B **84**, 085420 (2011)
95. R.M. Bradley, Nanoscale patterns produced by ion erosion of a solid with codeposition of impurities: the crucial effect of compound formation. Phys. Rev. B **87**, 205408 (2013)
96. R.M. Bradley, Morphological transitions in nanoscale patterns produced by concurrent ion sputtering and impurity co-deposition. J. Appl. Phys. **119**, 134305 (2016)
97. S.A. Norris, Ion-assisted phase separation in compound films: an alternate route to ordered nanostructures. J. Appl. Phys. **114**, 204303 (2013)
98. S. Facsko, T. Dekorsy, C. Koerdt, C. Trappe, H. Kurz, A. Vogt, H.L. Hartnagel, Formation of ordered nanoscale semiconductor dots by ion sputtering. Science **285**, 1551–1553 (1999)
99. F. Frost, A. Schindler, F. Bigl, Roughness evolution of ion sputtered rotating InP surfaces: pattern formation and scaling laws. Phys. Rev. Lett. **85**, 4116–4119 (2000)
100. R. Gago, L. Vázquez, R. Cuerno, M. Varela, C. Ballesteros, J.M. Albella, Production of ordered silicon nanocrystals by low-energy ion sputtering. Appl. Phys. Lett. **78**, 3316–3318 (2001)
101. A. Zalar, Improved depth resolution by sample rotation during Auger electron spectroscopy depth profiling. Thin Solid Films **124**, 223–230 (1985)
102. J. Muñoz-García, R. Cuerno, M. Castro, Coupling of morphology to surface transport in ion-beam-irradiated surfaces: normal incidence and rotating targets. J. Phys. Condens. Matter **21**, 224020 (2009)
103. B. Ziberi, F. Frost, B. Rauschenbach, T. Höche, Highly ordered self-organized dot patterns on Si surfaces by low-energy erosion. Appl. Phys. Lett. **87**, 033113 (2005)
104. B. Ziberi, Ion beam induces pattern formation on Si and Ge surfaces. Dissertation. University Leipzig (2007)
105. F. Frost, B. Ziberi, T. Höche, B. Rauschenbach, The shape and ordering of self-organized nanostructures by ion sputtering. Nucl. Instr. Meth. Phys. Res. B **216**, 9–19 (2004)
106. B. Ziberi, F. Frost, T. Höche, B. Rauschenbach, Ion-induced self-organized dot and ripple patterns on Si surfaces. Vacuum **81**, 155–159 (2005)
107. F. Frost, B. Rauschenbach, Nanostructuring of solid surfaces by ion-beam erosion. Appl. Phys. A **77**, 1–9 (2003)
108. B. Ziberi, F. Frost, B. Rauschenbach, Self-organized dot patterns on Si Surfaces during noble gas ion beam erosion. Surf. Sci. **600**, 3757–3761 (2006)
109. B. Ziberi, F. Frost, T. Höche, B. Rauschenbach, Dot Pattern Formation on Silicon Surfaces by Low-Energy Ion Beam Erosion, ed. by E. Chason, G.H. Gilmer, H. Huang, E. Wang, vol. 849 (MRS Proceed. 2004), KK6.2
110. R.M. Bradley, J.M.E. Harper, Theory of ripple topography induced by ion bombardment. J. Vac. Sci. Technol. A **6**, 2390–2396 (1988)
111. B. Kahng, H. Jeong, A.-L- Barabási, Quantum dot and hole formation in sputter erosion. Appl. Phys. Lett. **78**, 805–807 (2001)
112. J.F. Ziegler, J.P. Biersack, U. Littmark, *The Stopping and Range of Ions in Solids* (Pergamon Press, New York, 1985)
113. S. Facsko, T. Bobek, A. Stahl, H. Kurz, T. Dekorsy, Dissipative continuum model for self-organized pattern formation during ion-beam erosion. Phys. Rev. B **69**, 153412 (2004)
114. D. Chowdhury, D. Ghose, B. Satpati, Production of ordered and pure Si nanodots at grazing ion beam sputtering under concurrent substrate rotation. Mater. Sci. Eng. B **179**, 1–5 (2014)

115. M.A. Makeev, A.-L. Barabási, Ion-induced effective surface diffusion in ion sputtering. Appl. Phys. Lett. **71**, 2800–2802 (1997)
116. S. Facsko, H. Kurz, T. Dekorsy, Energy dependence of quantum dot formation by ion sputtering. Phys. Rev. B **63**, 165329 (2001)
117. Y. Wang, S.F. Yoon, C.Y. Ngo, C.Z. Tong, C.Y. Liu, Low energy Ar+ sputtering-induced GaAs quantum dot formation and evolution. J. Appl. Phys. **106**, 024301 (2009)
118. L. Ling, W. Li, L. Qi, M. Lu, X. Yang, C. Gu, Nanopatterning of Si(110) surface by ion sputtering: an experimental and simulation study. Phys. Rev. B **71**, 155329 (2005)
119. M. Lu, X.J. Yang, S.S. Perry, J.W. Rabalais, Self-organized nanodot formation on MgO(100) by ion bombardment at high temperatures. Appl. Phys. Lett. **80**, 2096–2098 (2002)
120. W.-B. Fan, W.-Q. Li, L.-J. Qi, H.-T Sun, J. Luo, Y.-Y Zhao M. Lu, On the role of ion flux in nanostructuring by ion sputter erosion. Nanotechnol. **16**, 1526–1529 (2005)
121. E. Trynkiewicz, B.R. Jany, D. Wrana, F. Krok, Thermally controlled growth of surface nanostructures on ion-modified AIII–BV semiconductor crystals. Appl. Surf. Sci. **427**, 349–356 (2018)
122. S.K. Tan, R. Liu, C.H. Sow, A.T.S. Wee, Self-organized nanodot formation on InP(100) by oxygen ion sputtering. Nucl. Instr. Meth. Phys. Res. B **248**, 83–89 (2006)
123. I. Sulania, A. Tripathi, D. Kabiraj, M. Lequeux, D. Avasthi, Surface patterning on indium phosphide with low energy Ar atoms bombardment: an evolution from nanodots to nanoripples. Adv. Mat. Lett. **1**, 118–122 (2010)
124. S.K. Tan, A.T.S. Wee, Self-organized nanodot formation on InP(100) by argon ion sputtering at normal incidence. J. Vac. Sci. Technol. B **24**, 1444–1448 (2006)
125. D. Paramanik, S. Varma, Studies of InP nano dots formation after keV Ar+ irradiation. Nucl. Instr. Meth. Phys. Res. B **266**, 1257–1260 (2008)
126. T. Bobek, S. Facsko, H. Kurz, T. Dekorsy, M. Xu, C. Teichert, Temporal evolution of dot patterns during ion sputtering. Phys. Rev. B **68**, 085324 (2003)
127. M. Xu, C. Teichert, How do nano islands induced by ion sputtering evolve during the early stage of growth? J. Appl. Phys. **96**, 244–2248 (2004)
128. S. Facsko, T. Bobek. H. Kurz, T. Dekorsy, S. Kyrsta, R. Cremer, Ion-induced formation of regular nanostructures on amorphous GaSb surface. Appl. Phys. Lett. **80**, 130–132 (2001)
129. R. Gago, L. Vázquez, O. Plantevin, T. Metzger, J. Mu˜noz-García, R. Cuerno, M. Castro, Order enhancement and coarsening of self-organized silicon nanodot patterns induced by ion-beam sputtering. Appl. Phys. Lett. **89**, 233101 (2006)
130. R. Kree, T. Yasseri, A.K. Hartmanns, The influence of beam divergence on ion beam-induced surface patterns. Nucl. Instr. Meth. Phys. Res. B **267**, 1407–1411 (2009)
131. U. Valbusa, C. Boragno, F. Buatier de Mongeot, Nanostructuring surfaces by ion sputtering. J. Phys.: Condens. Matter **14**, 8153–8175 (2002)
132. M. Ritter, M. Stindtmann, M. Farle, K. Baberschke, Nanostructuring of the Cu(001) surface by ion bombardment: a STM study. Surf. Sci. **348**, 243–252 (1996)
133. G. Constantini, F. Buatier de Mongeot, C. Boragno, U. Valbusa, Is ion sputtering always a "negative homoepitaxial deposition"? Phys. Rev. Lett. **86**, 838–841 (2001)
134. T. Michely, G. Comsa, The scanning tunneling microscope as a means for the investigation of ion bombardment effects on metal surfaces. Nucl. Instr. Meth. Phys. Res. B **82**, 207–219 (1993)
135. J. Naumann, J. Osing, A.J. Quinn, I.V. Shvets, Morphology of sputtering damage on Cu(111) studied by scanning tunneling microscopy. Surf. Sci. **388**, 212–219 (1997)
136. Q. Wei, X. Zhou, B. Joshi, Y. Chen, K.-D. Li, Q.H. Wei, K. Sun, L. Wang, Self-assembly of ordered semiconductor nanoholes by ion beam sputtering. Adv. Mater. **21**, 2865–2869 (2009)
137. X. Ou, A. Keller, M. Helm, J. Fassbender, S. Facsko, Reverse epitaxy of Ge: ordered and faceted surface patterns. Phys. Rev. Lett. **111**, 016101 (2013)
138. D. Chowdhury, D. Ghose, S.A. Mollick, Homoepitaxy of germanium by hyperthermal ion irradiation. Vacuum **107**, 23–27 (2014)
139. I. Koponen, M. Hautala, O.-P. Sievänen, Simulations of ripple formation on ion-bombarded solid surfaces. Phys. Rev. Lett. **78**, 2612–2615 (1997)

140. R. Cuerno, H.A. Makse, S. Tomassone, S.T. Harrington, H.E. Stanley, Stochastic model for surface erosion via ion sputtering: dynamical evolution from ripple morphology to rough morphology. Phys. Rev. Lett. **75**, 4464–4469 (1995)
141. P. Sigmund, A mechanism of surface micro-roughening by ion bombardment. J. Mater. Sci. **8**, 1545–1553 (1973)
142. G. Carter, V. Vishnyakov, Roughening and ripple instabilities on ion-bombarded Si. Phys. Rev. B **54**, 17647–17653 (1996)
143. S.A. Norris, Stress-induced patterns in ion-irradiated silicon: model based on anisotropic plastic flow. Phys. Rev. B **86**, 235405 (2012)
144. W.W. Mullins, Theory of thermal grooving. J. Appl. Phys. **28**, 333–339 (1957)
145. W.W. Mullins, Flattening of a nearly plane solid surface due to capillarity. J. Appl. Phys. **30**, 77–83 (1959)
146. M. Feix, A.K. Harmann, R. Kree, J. Moñoz-Garcia, R. Cuerno, Influence of collision cascade statistics on pattern formation of ion-sputtered surfaces. Phys. Rev. B **71**, 125407 (2005)
147. B. Ziberi, F. Frost, B. Rauschenbach, Formation of large-area nanostructures on Si and Ge surfaces during low energy ion beam erosion. J. Vac. Sci. Technol. A **24**, 1344–1348 (2006)
148. S. Vogel, S. Linz, How ripples turn into dots: modeling ion-beam erosion under oblique incidence. Europhys. Lett. **76**, 884–890 (2006)
149. D.A. Pearson, R.M. Bradley, Theory of terraced topographies produced by oblique-incidence ion bombardment of solid surfaces. J. Phys. Condens. Matter **27**, 015010 (2015)
150. S. Park, B. Kahng, H. Jeong, A.-L. Barabási, Dynamics of ripple formation in sputter erosion: nonlinear phenomena. Phys. Rev. Lett. **83**, 3486–3489 (1999)
151. M. Kardar, G. Parisi, Y.-C. Zhang, Dynamic scaling of growing interfaces. Phys. Rev. Lett. **56**, 889–892 (1986)
152. H. Chatè, P. Manneville, Transition to turbulence via spatio-temporal intermittency. Phys. Rev. Lett. **58**, 112–116 (1987)
153. M. Paniconi, K. Elder, Stationary, dynamical, and chaotic states of the two dimensional damped Kuramoto-Sivashinsky equation. Phys. Rev. E **56**, 2713–2721 (1997)
154. R.M. Bradley, Exact linear dispersion relation for the Sigmund model of ion sputtering. Phys. Rev. B **84**, 075413 (2011)
155. A.S. Rudy, V.K. Smirnov, Hydrodynamic model of wave-ordered structures formed by ion bombardment of solids. Nucl. Instr. Meth. Phys. Res. B **159**, 52–59 (1999)
156. T. Aste, U. Valbusa, Surface instabilities in granular matter and ion-sputtered surfaces. Phys. A 332, 548–558 (2004). Ripples and ripples: from sandy deserts to ion-sputtered surfaces. New J. Phys. **7**, 122 (2005)
157. R. Cuerno, M. Castro, J. Muñoz-García, R. Gago, L. Vázquez, Universal non-equilibrium phenomena at submicrometric surfaces and interfaces. Eur. Phys. J. Special Topics **146**, 427–441 (2007)
158. M. Castro, R. Cuerno, L. Vázquez, R. Gago, Self-organized ordering of nanostructures produced by ion-beam sputtering. Phys. Rev. Lett. **94**, 016102 (2005)
159. D. Flamm, F. Frost, D. Hirsch, Evolution of surface topography of fused silica by ion beam sputtering. Appl. Surf. Sci. **179**, 95–101 (2001)
160. R. Cuerno, M. Castro, J. Muñoz-García, R. Gago, L. Vázquez, Universal non-equilibrium phenomena at submicrometric surfaces and interfaces. Eur. Phys. J. Spec. Top. **146**, 427–441 (2007)
161. M.W. Sckerl, P. Sigmund, M. Vicanek, Particle fluxes in atomic collision cascade. Det. Kgl. Danske Vid. Selskab. Mat.-Fys. Medd. **44** (1996). M.W. Sckerl, M. Vicanek, P. Sigmund, Momentum in atomic collision cascades. Nucl. Instr. Meth. Phys. Res. B **102**, 86–92 (1995)
162. G.K. Wehner, Velocities of sputtered atoms. Phys. Rev. **114**, 1270–1272 (1959). K. Kopitzki, H.-E. Stier, Mittlere Geschwindigkeit der bei der Kathodenzerstäubung von Metallen ausgesandten Partikel. Z. Naturf. **17a**, 346–352 (1962)
163. U. Littmark, P. Sigmund, Momentum deposition by heavy-ion bombardment and an application to sputtering. J. Phys. D Appl. Phys. **8**, 241–245 (1975)

164. M. Moseler, P. Gumbsch, C. Casiraghi, A.C. Ferrari, J. Robertson, The ultrasmoothness of diamond-like surfaces. Science **309**, 1545–1548 (2005)
165. M. Moseler, O. Rattunde, J. Nordiek, H. Haberland, On the origin of surface smoothing by energetic cluster impact: molecular dynamics simulation and mesoscopic modeling. Nucl. Instr. Meth. Phys. Res. B **164–165**, 522–536 (2000)
166. B. Davidovitch, M.J. Aziz, M.P. Brenner, Linear dynamics of ion sputtered surfaces: instability, stability and bifurcations. J. Phys. Condens. Matter **21**, 224019 (2009)
167. M.P. Harrison, R.M. Bradley, Crater function approach to ion-induced nanoscale pattern formation: craters for flat surfaces are insufficient. Phys. Rev. B **89**, 245401 (2014)
168. H. Hofsäss, K. Zhang, O. Bobes, Self-organized surface ripple pattern formation by ion implantation. Appl. Phys. **120**, 135308 (2016)
169. G. Hobler, D. Maciążek, Z. Postawa, Crater function moments: role of implanted noble gas atoms. Phys. Rev. B **97**, 155307 (2018)
170. J.C. Perkinson, E. Anzenberg, M. J. Aziz, K.F Ludwig Jr, Model independent test of the truncated crater function theory of surface morphology evolution during ion bombardment. Phys. Rev. B **89**, 115433 (2014)
171. M.P. Harrison, Modeling and controlling nanoscale patterns formed by bombardment with broad beam. Dissertation, Colorado State University (2017)
172. J.J. Elich, H.E. Roosendaal, H.H. Kersten, D. Onderdelinden, J. Kistenmaker, J.D. Elen, Relation between surface structures and sputtering ratios of copper single crystals. Rad. Eff. **8**, 1–11 (1971). See also H.E. Roosendaal, Sputtering yields of single crystalline targets, in *Sputtering by Particle Bombardment I*, ed. by R. Behrisch, Topics in Applied Physics, vol. 47 (Springer, Berlin, Heidelberg, 1981), pp. 219–256
173. D. Ghose, Ion beam sputtering induced nanostructuring of polycrystalline metal films. J. Phys. Condens. Matter. **21**, 224001 (2009)
174. K. Zhang, M. Uhrmacher, H. Hofsäss, J. Krauser, Magnetic texturing of ferromagnetic thin films by sputtering induced ripple formation Fe. J. Appl. Phys. **103**, 83507 (2008)
175. T. Skerén, K. Temst, W. Vandervorst, A. Vantomme, Ion-induced roughening and ripple formation on polycrystalline metallic films. New J. Phys. **15**, 093047 (2013)
176. P. Mishra, D. Ghose, The rotation of ripple pattern and the shape of the collision cascade in ion sputtered thin metal films. J. Appl. Phys. D **104**, 94305 (2008)
177. J.M. Colino, M.A. Arranz, Control of surface ripple amplitude in ion beam sputtered polycrystalline cobalt films. Appl. Surf. Sci. **257**, 4432–4438 (2011)
178. H.X. Qian, W. Zhou, Self-organization of ripples on Ti irradiated with focused ion beam. Appl. Surf. Sci. **258**, 1924–1928 (2012)
179. A. Toma , B.S. Batic, D. Chiappe, C. Boragno, U. Valbusa, N. Godec, M. Jenko, F.B. de Mongeot, et al., Patterning polycrystalline thin films by defocused ion beam: the influence of initial morphology on the evolution of self-organized nanostructures. J. Appl. Phys. **104**, 104313 (2008)
180. P. Gailly, C. Petermann, P. Tihon, K. Fleury-Frenette, Ripple topography and roughness evolution on surface of polycrystalline gold and silver thin films under low energy Ar-ion beam sputtering. Appl. Surf. Lett. **258**, 7717–7725 (2012)
181. S. Rusponi, C. Boragno, U. Valbusa, Ripple structure on Ag (110) surface induced by ion sputtering. Phys. Rev. Lett. **78**, 2795–2797 (1997)
182. M.V.R. Murty, T. Curcic, A. Judy, B.H. Cooper, A.R. Woll, J.D. Brock, S. Kycia, R.L. Headrick, Real-time X-ray scattering study of surface morphology evolution during ion erosion and epitaxial growth of Au (111). Phys. Rev. B **60**, 16956–16964 (1999)
183. H.X. Qian, W. Zhou, Y.Q. Fu, B.K.A. Ngoi, G.C. Lim, Crystallographically-dependent ripple formation on Sn surface irradiated with focused ion beam, Appl. Surf. Sci. **240**, 140–145 (2005). H.X. Qian, W. Zhou, X.R. Zeng, Dwell time dependent morphological transition and sputtering yield of ion sputtered Sn. J. Phys. D Appl. Phys. **43**, 345302 (2010)
184. N. Taniguchi, On the basic concept of nanotechnology, in *Proceedings of the International Conference on Production Engineering (ICPE)* (Tokyo, 1974), pp.18–23

185. F. Granone, V. Mussi, A. Toma, S. Orlanducci, M.L. Terranova, C. Boragno, F. Buatier de Mongeot, U. Valbusa, Ion sputtered surfaces as templates for carbon nanotubes alignment and deformation. Nucl. Instr. Meth. Phys. Res. B **230**, 545–550 (2005)
186. M. Kratzer, K. Szajna, D. Wrana, W. Belza, F. Krok, C. Teichert, Fabrication of ion bombardment induced rippled TiO_2 surfaces to influence subsequent organic thin film growth, J. Phys.: Condens. Matter **30**, 283001 (2018). M. Kratzer, D. Wrana, K. Szajna, F. Krok, C. Teichert, Island shape anisotropy in organic thin film growth induced by ion-beam irradiated rippled surfaces. Phys. Chem. Chem. Phys. **16**, 26112–26118 (2014)
187. C. Teichert, J.J. Miguel, T. Bobek, Ion beam sputtered nanostructured semiconductor surfaces as templates for nanomagnet arrays. J. Phys. Condens. Matter **21**, 224025 (2009)
188. E. Mele, F. Di Benedetto, R. Cingolani, D. Pisignano, A. Toma, F.B. de Mongeot, R. Buzio, C. Boragno, G. Firpo, V. Mussi, U. Valbusa, Nanostructuring polymers by soft lithography templates realized via ion sputtering. Nanotechnology **16**, 2714–2717 (2005)
189. Y. Xia, G.M. Whitesides, Soft lithography. Angew. Chem. Int. Ed. **37**, 550–575 (1998)
190. T. Bobek, H. Kurz, Fabrication of metallic nanomasks by transfer of self-organized nanodot patterns from semiconductor material into thin metallic layers. Nucl. Instr. Meth. Phys. Res. B **257**, 771–776 (2007)
191. T. Bobek, N. Mikuszeit, J. Camarero, S. Kyrsta, L. Yang, M.A. Nino, C. Hofer, L. Gridneva, D. Arvanitis, R. Miranda, J.J. de Miguel, C. Teichert, H. Kurz, Self-organized hexagonal patterns of independent magnetic nanodots. Adv. Mat. **19**, 4375–4380 (2007)
192. X. Ou, R. Kögler, X. Wei, A. Mücklich, X. Wang, W. Skorupa, S. Facsko, Fabrication of horizontal silicon nanowire arrays on insulator by ion irradiation. AIP Adv. **1**, 042174 (2011)
193. V.K. Smirnov, D.S. Kibalov, O.M. Orlov, V.V. Graboshnikov, Technology for nanoperiodic doping of a metal-oxide-semiconductor field-effect transistor channel using a self-forming wave-ordered structure. Nanotechnology **14**, 709–715 (2003). V.K. Smirnov, D.S. Kibalov, Methods of formation of a silicon nanostructure, a silicon quantum wire array and devices based thereon. US Patent 6,274,007 (2001)
194. L. Persechini, M. Ranjan, F. Grossmann, S. Facsko, J.F. McGilp, The linear and nonlinear optical response of native-oxide covered rippled Si templates with nanoscale periodicity. Phys. Stat. Sol. (b) **249**, 1173–1177 (2012)
195. V. Mussi, F. Granone, C. Boragno, F. Buatier de Mongeot, U. Valbusa, V. Dodecaneso, T. Marolo, R.M. Montereali, Surface nanostructuring and optical activation of lithium fluoride crystals by ion beam irradiation. Appl. Phys. Lett. **88**, 103116 (2006)
196. V. Mussi, R.M. Montereali, E. Nichelatti, C. Boragno, F. Buatier de Mongeot, U. Valbusa, Broad band light-emitting nanostructured substrates by ion beam irradiation of lithium fluoride crystals. Surf. Sci. **601**, 2746–2749 (2007)
197. C. Martella, D. Chiappe, P. Delli Veneri, L.V. Mercaldo, I. Usatii, F. Buatier de Mongeot, Self-organized broadband light trapping in thin film amorphous silicon solar cells. Nanotechnology **24**, 225201 (2013)
198. U. Kreibig, M. Vollmer, *Optical Properties of Metal Clusters* (Springer, Berlin, 1995)
199. A. Toma, D. Chiappe, D. Massabò, C. Boragno, F.B. de Mongeot, Self-organized metal nanowire arrays with tunable optical anisotropy. Appl. Phys. Lett. **93**, 163104 (2008)
200. T.W.H. Oates, A. Keller, S. Noda, S. Facsko, Self-organized metallic nanoparticle and nanowire arrays from ion-sputtered silicon templates. Appl. Phys. Lett. **93**, 063106 (2008)
201. S. Camelio, D. Babonneau, D. Lantiat, L. Simonot, F. Pailloux, Anisotropic optical properties of silver nanoparticle arrays on rippled dielectric surfaces produced by low-energy ion erosion. Phys. Rev. B. **80**, 155434 (2009)
202. M. Ranjan, M. Bhatnagar, S. Mukherjee, Localized surface plasmon resonance anisotropy in template aligned silver nanoparticles: a case of biaxial metal optics, J. Appl. Phys. **117**, 103106 (2015)
203. T.W.H. Oates, A. Keller, S. Facsko, A. Mücklich, Aligned silver nanoparticles on rippled silicon templates exhibiting anisotropic plasmon absorption. Plasmonics **2**, 47–50 (2007)
204. M. Bhatnagar, M. Ranjan, K. Jolley, A. Lloyd, R. Smith, S. Mukherjee, Thermal dynamics of silver clusters grown on rippled silica surface. Nucl. Instr. Meth. Phys. Res. B **393**, 5–12 (2017)

205. M. Ranjan, S. Facsko, Anisotropic surface enhanced Raman scattering in nanoparticle and nanowire arrays. Nanotechnology **23**, 485307 (2012)
206. M. Ranjan, S. Facsko, M. Fritzsche, S. Mukherjee, Plasmon resonance tuning in Ag nanoparticles arrays grown on ripple patterned templates. Microelectr. Eng. **102**, 44–47 (2013)
207. S.K. Srivastava, A. Shalabney, I. Khalaila, C. Grüner, B. Rauschenbach, I. Abdulhalim, SERS biosensor using metallic nano-sculptured thin films for the detection of endocrine disrupting compound biomarker vitellogenin. Small **10**, 3579–3587 (2014)
208. B. Schreiber, D. Gkogkou, L. Dedelaite, J. Kerbusch, R. Hübner, E. Sheremet, D.R.T. Zahn, A. Ramanavicius, S. Facsko, R.D. Rodriguez, Large-scale self-organized gold nanostructures with bidirectional plasmon resonances for SERS. RSC Adv. **8**, 22569–22576 (2018)
209. Y.J. Chen, J.P. Wang, E.W. Soo, L. Wu, T.C. Chong, Periodic magnetic nanostructures on self- assembled surfaces by ion beam bombardment. Appl. Phys. Lett. **91**, 7323–7325 (2002)
210. F. Bisio, R. Moroni, F. Buatier de Mongeot, M. Canepa, L. Mattera, Isolating the step contribution to the uniaxial magnetic anisotropy in nanostructured Fe/Ag(001) films. Phys. Rev. Lett. **96**, 057204 (2006)
211. U. Gradmann, Ferromagnetism near surfaces and in thin films. Appl. Phys. A **3**, 161–178 (1974). Magnetism of surfaces and interfaces. J. Magn. Mag. Mater. **6**, 173–182 (1977)
212. K.V. Sarathlal, D. Kumar, A. Gupta, Growth study of Co thin film on nanorippled Si(100) substrate. Appl. Phys. Lett. **98**, 123111 (2011)
213. M.O. Liedke, M. Korner, K. Lenz, M. Fritzsche, M. Ranjan, A. Keller, E. Čižmár, S. A. Zvyagin, S. Facsko, K. Potzger, J. Lindner, J. Fassbender, Crossover in the surface anisotropy contributions of ferromagnetic films on rippled Si surfaces. Phys. Rev. B **87**, 024424 (2013)
214. J. Fassbender, T. Strache, M.O. Liedke, D. Markó, S. Wintz, K Lenz, A. Keller, S. Facsko, I. Mönch, J. McCord, Introducing artificial length scales to tailor magnetic properties. New J. Phys. **11**, 125002 (2009)
215. M. Körner, K. Lenz, M.O. Liedke, T. Strache, A. Mücklich, A. Keller, S. Facsko, J. Fassbender, Interlayer exchange coupling of Fe/Cr/Fe thin films on rippled substrates. Phys. Rev. B **80**, 214401 (2009)
216. P.-G. De Gennes, Wetting: statics and dynamics. Rev. Mod. Phys. **57**, 827–863 (1985)
217. A. Metya, D. Ghose, N.R. Ray, Development of hydrophobicity of mica surfaces by ion beam sputtering. Appl. Surf. Sci. **293**, 18–23 (2014)
218. T. Kumar, U.B. Singh, M. Kumar, S. Ojha, D. Kanjilal, Tuning of ripple patterns and wetting dynamics of Si (100) surface using ion beam irradiation. Curr. Appl. Phys. **14**, 312–317 (2014)
219. S.K. Garg, D.P. Datta, J. Ghatak, I. Thakur, K. Khare, D. Kanjilal, T. Som, Tunable wettability of Si through surface energy engineering by nanopatterning. RSC Adv. **6**, 48550–48557 (2016)
220. J. Petersen, S.G. Mayr, Dewetting of Ni and NiAg solid thin films and formation of nanowires on ripple patterned substrates. J. Appl. Phys. **103**, 023520 (2008)

Chapter 9
Ion Beam Deposition and Cleaning

Abstract Direct deposition of low-energy ionized atoms or molecules (IBD) onto a substrate has key advantages in terms of controlling layer properties. In general, this technique is based either on the deceleration of high-energy, mass-separated ion beams or the generation of mass-separated, low-energy ion beams. The growth of the layers is primarily determined by the balance between two energy-dependent effects, deposition (sticking of deposited atoms or molecules) and sputtering. Processes such as sputtering, ion reflection, and athermal generation of defects define an energy window for successful deposition of the layers. The application of molecular ions for layer deposition is characterized by the fact that these ions are fragmented above a threshold energy. The potential of IBD with hyperthermal particles can be advantageously used in the synthesis of both stable and metastable phases in the deposited layers. As examples, the deposition of carbon and diamond-like carbon films, epitaxial silicon and germanium films, metal films and also compound films can be mentioned. High-quality thin organic films can be deposited by a specific variant of the IBD technique, the electrospray deposition. This technique, based on soft landing, allows the non-destructive deposition of large organic molecules on the surface of a substrate, when the energy of these molecules is lower than the activation energy for collision induced dissociation. This technique is presented and the deposition of selected organic layers is demonstrated. Finally, a method for cleaning surfaces by low energy ion bombardment is reviewed. It is characterized by the removal of surface contamination, adsorbates or compound layers on the surface without significantly damaging the underlying structure.

In certain applications, ion beam techniques have been recognized as a method to deposit high-quality thin films. Ion beam deposition (IBD) is a physical method which uses hyperthermal (1–100 eV) and low-energy (100–1000 eV) ionized atoms or molecules to deposit thin films directly onto substrates (for comprehensive overviews see [1–6]). The direct deposition of low-energy ions onto a substrate has some distinctive advantages like the precise control of particle energy, the high purity of the ion beam deposited material due an ion mass selection process, the low temperatures

B. Rauschenbach, *Low-Energy Ion Irradiation of Materials*, Springer Series in Materials Science 324, https://doi.org/10.1007/978-3-030-97277-6_9

used by the IBD process, the films anchoring in the substrate, the high lateral film uniformity and the opportunity to control the film properties. The most important disadvantage of IBD is the slow growth rate due to low ion current densities at low ion energies.

Initially, Almén and Bruce [7] have experimented with the deposition of low-energy ions on solid surfaces. Later, Fontell and Arminen [8] have studied the enrichment (collection) of implanted atoms in the near-surface region and the formation of layers after mass-separated metal ion implantation with energies between 10 eV and 110 keV. The highest energy at which the build-up of layers was determined lies between 300 eV and 1500 eV, depending on the ion species. It could be concluded that self-sputtering during deposition limits layer formation. In 1971, Fair [9] has designed and constructed the first ion beam deposition system for thin films.

In principle, a distinction must be made between two different vacuum process variants. On the one side, direct ion beam deposition is based on the deposition of ionized atoms on substrate surfaces with the objective to form high-quality thin inorganic films without significant defects. (Sect. 9.1). The fundamentals of film formation by deposition of energetic ions (Sect. 9.2) and the synthesis of coatings (Sect. 9.3) are presented. On the other side, the preparation of thin organic films is hardly possible, because of the thermally instability of organic molecules. In 1977, Cooks and coworkers [10] have proposed a new technique, the soft landing of ionized polyatomic molecules (Sect. 9.4) for the preparation of organic films. Finally, it is shown that the inverse process (removal of near-surface regions by ion irradiation instead of layer deposition) can be successfully applied to clean surfaces (Sect. 9.5).

9.1 Deposition by Direct Low-Energy Ion Bombardment

From an experimental point of view, direct low-energy ion thin film deposition can be realized by deceleration of high-energy ion beams or by generation of low-energy ion beams.

In the first case the energy of the ions must be decelerated before the ions reach the substrate. Consequently, special ion optics are necessary to reduce the ion energy significantly and to avoid a focusing or defocusing of the ion beam (see e.g., [5] and references herein).

An alternative approach is the thin film formation by direct ion beam deposition by means of special low-energy ion sources. These broad beam ion sources are capable to produce low-energy ions with high ion current densities. This represents a major challenge, due to the space-charge effect as result of the Coulomb repulsion of the ions in the beam [11], which limits the ion current density. Another challenge is the fact that ion generation efficiency decreases with decreasing ion energy [12].

For IBD different types of low-energy ion sources are used [1, 6]. The active principle of these ion sources are primarily electron impact ionization, radio-frequency discharge or the electron cyclotron resonance. The application of broad beam ion sources is associated with some disadvantages. The UHV conditions cannot be realized because higher pressures are required to operate the ion sources implemented in the vacuum system. The applied ion sources generate multiple ionized atomic and molecular species simultaneously and are characterized by a broad energy distribution of the ion beam (see Sect. 9.1.1). Therefore, ion mass filters (quadrupole mass spectrometer, Wien filter, etc.) are used, equipped with entry and exit ion optics, ion beam deflection, as well as ion beam current monitoring in order to control the deposition process. A mass-separated ion beam can also be generated by ion implantation equipment with energies up to some hundred electron-volts and an analyzing magnet. Then, the extracted, mass-separated ion beam is decelerated in front of the substrate in the UHV chamber to reduce the ion energy down to $\leq$ 100 eV [13].

General designs of the IBD systems are discussed by Miyake [6]. In Fig. 9.1 the main ion beam deposition configurations are schematically presented. The low energy ionized atoms are produced by broad beam ion sources and deposited onto the substrate directly without mass-separation (IBD) or after mass separation by analyzing magnets (MS-IBD) with the objective to fabricate an isotopically pure ion beam. Ions are generated in an ion source or cathodic vacuum arc and passed through an electromagnetic mass filter. In this filter, ions with the desired m/e ratio are selected and focused by electromagnetic lenses onto the substrate which is placed in the UHV camber. The advantages of MS-IBD is the production of films under well controlled conditions. However, the disadvantages are low deposition rates and high costs. On this way, both reactive and inert ion species can be provided for the film growth. Another IBD alternative is the reactive ion beam deposition (R-IBD) arrangement. Here, the deposition chamber is maintained in a reactive gas environment (oxygen, nitrogen, etc.) with a pressure in the range between 10^{-4} Pa and 10^{-2} Pa. In contrast to the electromagnetic mass separation technique, the generation of ionized particles with energies up to some kilo-electron-volts in an ion

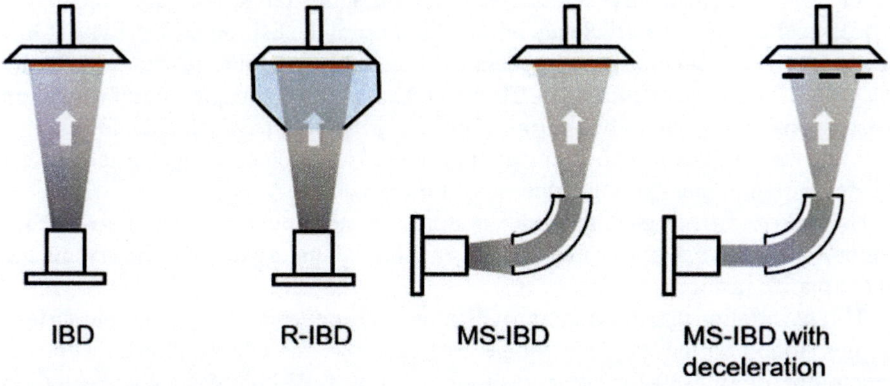

Fig. 9.1 Configurations of the direct ion beam deposition equipment. IBD—ion beam deposition without mass separation, R-IBD—reactive ion beam deposition, MS-IBD—ion beam deposition with mass separation, MS-IBD with deceleration—ion beam deposition with mass separation, where the high energy ions are decelerated close to the substrate

implantation system and subsequently deceleration immediately close to substrate can be used (MS-IBA with deceleration).

As early as 1982, Thomas and coworkers [14] have compiled the following requirements on a IBD apparatus:

(i) capability of producing stable, relatively intense ion beams of a variety of materials,
(ii) presence of mass separation facilities in the beam line for beam purification,
(iii) capability of delivering mass analyzed ion fluxes with extremely low energies and low energy spread,
(iv) deposition under UHV conditions,
(v) presence of facilities for in situ analysis of both the surface purity and the crystallinity of the substrate and growing film and
(vi) temperature control of the substrate during deposition.

To produce compound films with a particular composition ratio, a sequential direct ion beam deposition technique is used [15]. The number of ions of the first component delivered to substrate is counted. When the desired ion fluence (concentration) is achieved, the deposition process can be instantly turned off. Then the ions of the second component are deposited in a similar manner, where high ion energies (> 100 eV) are commonly used. The higher ion energy is a favorable feature in the interface mixing process and the uniformity of the sequentially deposited layers. Therefore, the final composition is directly related to the fluence and the duty cycle of each deposited species.

9.1.1 Process of Direct Ion Beam Deposition

In Fig. 9.2 schematically summarized the fundamental ion-surface interaction processes that affect the efficiency of direct ion beam deposition of thin films at low energies. The direct deposition (process I in Fig. 9.2), the sputtering (process II) and the process of interface mixing (VIII) which happens at the beginning of direct ion beam deposition, essentially determine the film growth. In particular, the ion energy, the temperature during deposition and the angle of ion incidence significantly affect the deposition process and the kinetics of film growth.

Using low ion energies, the ionized atoms are accumulated in the near-surface region. The accelerated ions must possess a minimal energy to enter the crystal and to thermalize inside.

This penetration threshold energy will be strongly dependent on the atomic surface configuration and the incidence angle. For example, the energy threshold for the penetration of W(100) for noble gas ions is about 8 eV [16]. With increasing ion energy the penetration probability of the incident species and also the sputtering yield are raised.

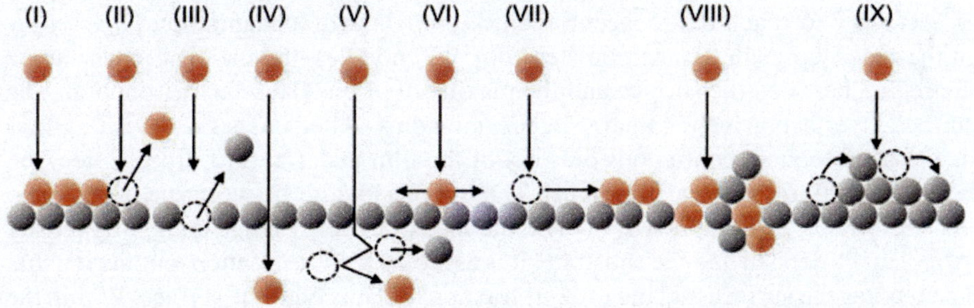

Fig. 9.2 Schematic presentation of the direct ion beam deposition processes: (I) direct ion beam deposition and insertion of the incident particle into a step edge, (II) self-ion or re-sputtering, (III) sputtering of the substrate atoms, (IV) implantation, (V) formation of interstitials and vacancies at the surface and in the subsurface region, (VI) athermal surface diffusion, (VII) formation of dimers, trimers, etc., (VIII) ion beam-induced interface mixing, (IX) ion impact up-step and down-step diffusion

The energy of penetrating ions dissipates in a small near-surface volume and generates a high-excited region for very short times (see Sect. 4.5). Their interaction with surface atoms leads to the sputtering of substrate surface atoms (process III, Fig. 9.2) or to self-atom sputtering (process II), the formation of defects (interstitials and vacancies, process V) and the collisional interface mixing (process VIII).

Herbots et al. [17] have described the growth of thin films by ion beam deposition in three steps under the assumption of a Gaussian-like distribution of the implanted species characterized by the mean projected range R_p and the mean projected range straggling ΔR_p. In the first process step, called implantation step (Fig. 9.3, left), ions are implanted into the near-surface region of the substrate material until the maximum atomic concentration, N_{max}, is reached. According to (3.40), a fluence of the deposited ions

$$\Phi = N_I \sqrt{2\pi} \Delta R_p \tag{9.1}$$

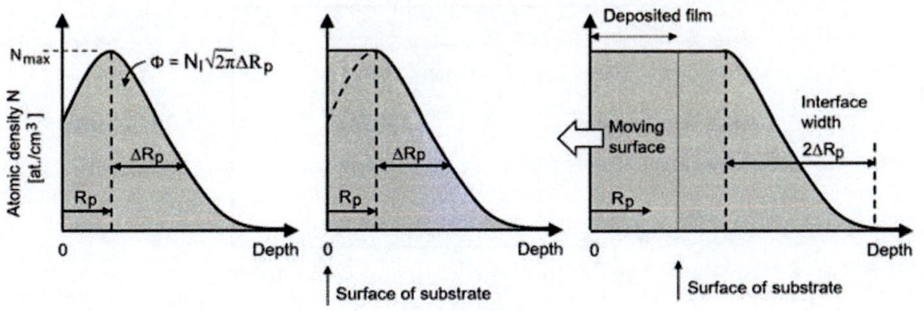

Fig. 9.3 Schematic process steps of ion beam deposition according to Herbots et al. [17]

is necessary to reach this concentration, i.e. implanted concentration $N_I = N_{max}$. In the next step, called transition step (Fig. 9.3, middle), the incident ions saturate the region between the surface and the maximum of the Gaussian distribution. The surface layer exclusively contains deposited atoms. After this step, the film grows due to a subsequent continuous delivery of the film material (last process step, ion beam deposition step Fig. 9.3, right). The interface width between the deposit and the substrate is given by $2\Delta R_p$, where the mean projected range straggling mostly differs in the deposit and the substrate. It is assumed that the fraction α of the ion flux J penetrates the surface and the $(1 - \alpha)$ fraction remains onto the surface. Within the time interval Δt of the bombardment time, the film grows outwards by

$$\Delta x = \frac{(1-\alpha)J}{N}\Delta t \tag{9.2}$$

where N is the atomic number density. The number of deposited adatom can be significantly reduced, because the incident ions (i) can be reflected and (ii) the pre-deposited adatoms can be re-sputtered [18]. The reflection is related to the sticking of the incoming ions by $s = 1 - R$, where R is the reflection probability (coefficient). The sticking probability is defined as the number of the adsorbed atoms to the total number of impinging atoms. According to Rabalais and Marton [19], the sticking probability decreases with the kinetic energy up to an energy of about 10 eV…100 eV and then increases again (see Fig. 9.4). Unfortunately, the particle sticking or probability is small for the energy range of the direct ion beam deposition. For comparison, the sticking coefficient is generally very high for surfaces in a normal pressure and room temperature environment (0.8...0.9, see dashed line in Fig. 9.4) and high (0.7...0.8) for thermal deposition. A few studies have investigated the reflection of low-energy ions as function of the ion energy and the incident angle. For example, MD simulations are carried out, which have studied the reflection of Si ions from Si surfaces [20] and

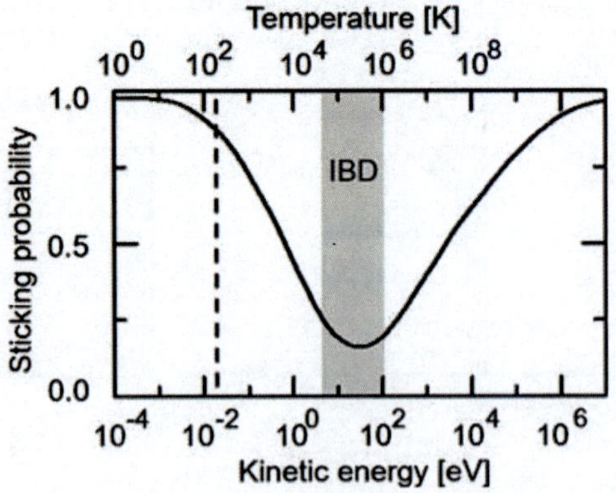

Fig. 9.4 Schematic plot of the particle sticking probability in dependence of the kinetic energy for light ions. The dashed line represents the kinetic energy equivalent to the thermal energy of the particles at room temperature (modified version of a figure adapted from [19])

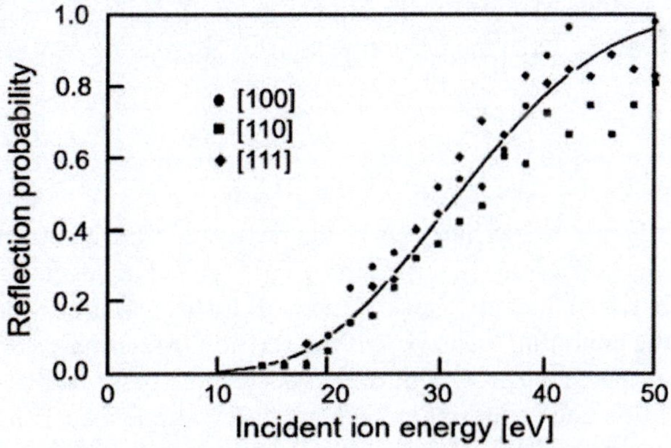

Fig. 9.5 Reflection probability as a function of the incident energy at a fixed incident angle of 80° (figure adapted from [21])

Cu ions from Cu surfaces [21]. Depending on the ion energy and the incidence angle, the reflection probability varies between 0 and 1. For example, Fig. 9.5 shows the reflection probability of copper ion on Cu surfaces for incident energies up to 50 eV. The MD simulations demonstrate the ion impact on several Cu crystal surfaces under an angle of 80°. At very low particles energies ($\leq$10 eV), the reflection probability is negligible. Then, the reflection probability increases strongly with increasing ion energy. The influence of the crystallographic orientation of the Cu surface is only small. The reflection probability as function of the incident angle at fixed ion energy shows a similar behavior (see Chap. 5). For small incident angles up to about 30° the reflection probability insignificant. Above 30° the reflection probability increases and reaches a reflectivity of about 1 after angles of incidence between 65° and 80° [21].

Re-sputtering of the growing film is a second process, which can lead to significant reduction of the number of deposited adatoms. In Sect. 5.10 the mechanisms and the dependence of the sputtering on the ion energy and the incidence angle is described. Above a threshold energy (between 10 eV and 50 eV) re-sputtering occurs and increases with increasing incident energy. Approximately, the sputtering yield increases linearly with root of the ion energy for low-energy ion bombardment, (see Sect. 5.3). Todorov and Fossum [22] have studied the influence of the ion sputtering on the growth of thin silicon oxide films prepared by low-energy oxygen bombardment. For low ion energies (< 60 eV), a logarithmic increase of the oxide thickness with the ion fluence could be observed. At higher energies ($\geq$ 100 eV) the sputtering of the growing film leads to a self-limiting growth. Also, Amano [23] has shown that the film thickness of the deposited magnesium and lead films significantly decreases as the ion energy increases. On the one side, the incident ions deplete the substrate atoms in the near-surface region (thickness is about $2R_p$ of the incident ions, see Fig. 9.3) and on the other side the self-sputtering of the deposited atoms reduces the deposition rate. Therefore, it is evident that the energy-dependent sputtering plays an important role in the film growth by direct ion beam deposition.

When the reflection and the re-sputtering are considered the growth rate of thin films by IBD, the growth velocity, v_{IBD}, can be expressed by

$$v_{IBD} = \frac{dx}{dt} = \frac{J}{N}(s - Y_{sis}), \tag{9.3}$$

where x is the thickness of the deposited film, J is the ion current density (flux) and Y_{sis} is the self-ion sputtering yield. Typical deposition rates of IBD techniques range between 0.02 nm/s and 0.2 nm/s. Film growth is only possible when $Y_{sis} < s$ or under the condition that $s = 1$, if the self-ion sputtering yield is smaller than 1.

Thus, the growth of thin layers can be described as a competition between deposition and sputtering, which are energy-dependent effects. The balance between the deposition and sputter removal of the film limits the growth rate. But, experimental results based only on this simple balance cannot quantitatively explain the film thickness dependence upon ion energy or the phase formation and film stoichiometry. A deeper insight into the mechanism of the film formation by ion beam deposition is necessary. Herbots et al. [24, 25] have analyzed in detail the process of interaction in low-energy deposition of ions with the substrate. In a first process step replacement and relocation events are taken into account. Replacement is when a neutralized ion recombines into a lattice position, and relocation is when a substrate atom is removed from a lattice position. These events occur in a time scale of 10^{-12} s. Both processes correspond to a low-energy ion beam mixing effect and lead to a mass transport over some interatomic distances. Finally, these processes result into a surface growth or desorption from the surface of the growing film. In a second step, the replacements and relocations are coupled with the defect diffusion and recombination [26]. An example of the simulated surface motion and growth rate is shown in Fig. 9.6. The location of the surface is given after 1 s, 10 s, and 20 s during Si-deposition with

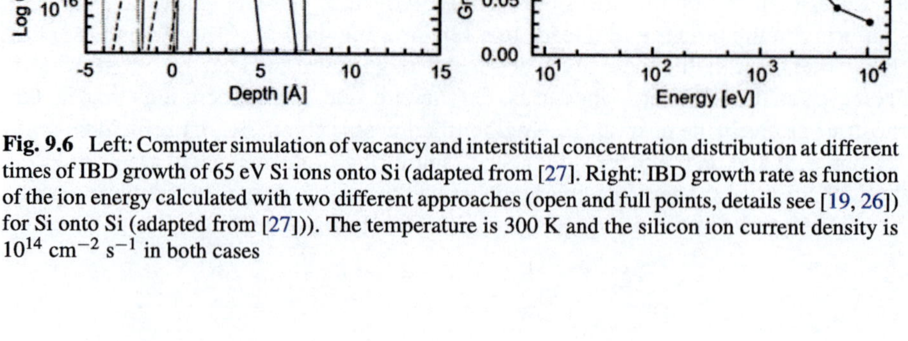

Fig. 9.6 Left: Computer simulation of vacancy and interstitial concentration distribution at different times of IBD growth of 65 eV Si ions onto Si (adapted from [27]. Right: IBD growth rate as function of the ion energy calculated with two different approaches (open and full points, details see [19, 26]) for Si onto Si (adapted from [27])). The temperature is 300 K and the silicon ion current density is 10^{14} cm^{-2} s^{-1} in both cases

$J = 10^{14}$ Si-ions $cm^{-2}s^{-1}$. Herbots et al. [24] have developed a dynamic continuum model for the diffusion and recombination of interstitials and vacancies in conjunction with TRIM simulation (see Sect. 3.5.2) to perform calculations of the ion beam deposition. The IBD growth process can be obtained by calculating the flux of interstitials reaching the surface. The interstitial diffusion is the determining factor of the growth mechanism. Three different diffusion paths could be identified [26]. First, the interstitials can recombine with the surface. This leads to net growth of the film (see dotted lines in Fig. 9.6, left). Second, the interstitials can combine with vacancies, which leads to a vacancy annihilation. Thus, a dynamic equilibrium is reached if the vacancy annihilation by diffusing interstitials is equal to the generation of vacancies by ion bombardment. As shown in Fig. 9.6 (left), no vacancies exist behind the surface. This is due to low diffusivity of the vacancies (e.g., for Si at 300 K is $D_{interst.} \approx 2 \times 10^{-12}$ $cm^2 s^{-1}$ and $D_{vacancy} \approx 10^{-34}$ $cm^2 s^{-1}$ [28]) and their complete annihilation by interstitials. Thirdly, a small fraction of the interstitials diffuses into the substrate and form one-dimensional defects (dislocations). Figure 9.6 (right) shows the growth rate of Si films after Si ion irradiation with 10^{14} ions cm^{-2} s^{-1} in dependence of the ion energy. The growth rate is calculated by a dynamic continuum model for the diffusion and recombination of interstitials and vacancies and the net material deposition given by TRIM is compared (details see [24, 27]). Below 100 eV, the growth rate is independent of the ion energy. Above 100 eV, the growth rate begins to decrease. This cannot be exclusively explained by an increased sputtering rate, because the growth rate drops about twice as fast as sputtering loss [27]. An enlarging fraction of interstitials remains in the substrate with increase of the ion energy. Thus, the fraction of interstitials reaching the surface decreases at higher energies, reducing the growth rate. Consequently, ion energies maintaining the sputtering yield below one is not sufficient to obtain a significant film growth. For the growth of thin films by IBD, it is also important to manage defect distribution and recombination.

9.1.2 Deposition of Polyatomic Ions

In the last subchapter, the process of the direct ion beam deposition was introduced under the precondition that monoatomic ions or molecules are used (e.g., N^+, $N_2{}^+$, O^+, $O_2{}^+$, etc.) . Frequently, polyatomic ions or ionized organic molecules will be applied to realize ambitious compound layers. When molecules are included in the process of direct ion beam deposition, numerous additionally processes must be considered, which competitively occur in energy regimes up to some ten electron-volts (see [29, 30]). In Fig. 9.7 are summarized the most important polyatomic ion-surface interactions (in addition to the monoatomic ion-surface interactions, see Fig. 9.2).

(I) The *elastic scattering* (process I in Fig. 9.7) is a collision process characterized by conservation of the kinetic energy of the incident molecule, but changing of the direction of movement of the particle after the collision (details see Sect. 2.2).

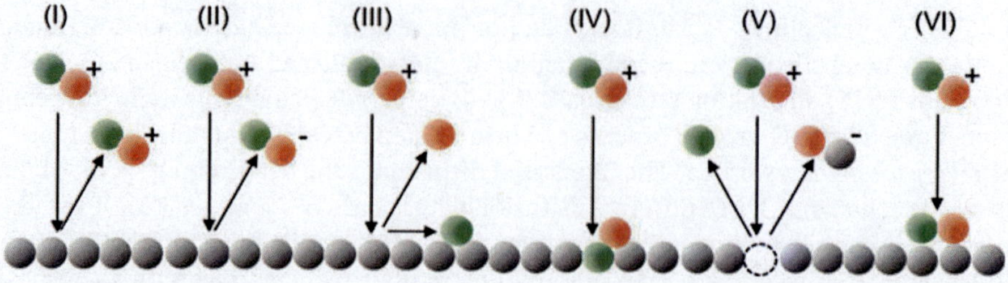

Fig. 9.7 Schematic of the polyatomic ion surface processes. (I) elastic scattering of polyatomic molecules (reflection), (II) charge exchange processes, (III) surface-induced dissociation, (IV) reactive transformation of the surface, (V) chemical sputtering of polyatomic molecules, (VI) soft landing of polyatomic molecules

(II) *Charge exchange process* between incident ionized molecule and surface (process II)

If polyatomic ions collide with a substrate surface the resulting product ion can have an opposite polarity or is neutralized. Charge inversion reactions, like electron transfer and charge stripping processes, mainly occur in the keV energy range, because at these energies electron stripping and electron transfer reactions have a significant cross-section [31]. Experimental studies have demonstrated that charge inversion processes (resonant or Auger electron transition) at low incident energies, i.e. inversion of positive to negative ions and inversion of negative to positive ions as well as formation of neutralized atoms, are strongly dependent on the nature of the outermost atomic layers of the surface [30]. The charge inversion process can be associated with a fragmentation of the polyatomic ion species. Charge inversion of multiply charged ions can be observed when dissociation activated low-energy collision takes place under multiple collision conditions.

(III) *Surface induced dissociation* (process III)

Inelastic collisions can also result in a fragmentation of polyatomic ions, known as surface induced dissociation. The translational energy of the incident molecule is transferred into rotational-vibrational energy by the impulsive collision with the surface atoms, which leads to a subsequent fracture of the bonds within the molecules. With regard to small molecules (diatomic molecules) this momentum induced process is very fast ($\approx 10^{-14}$ s, i.e. in the timescale of a part of a vibrational period) and can undergo further electron transfer processes before the molecule fragments are scattered to back in the vacuum or is trapped on the surface [32]. In contrast, larger polyatomic molecules are more stable. It could be demonstrated by MD simulations and STM studies that the stability of large molecules decreases with an increase of the hyperthermal incident energy [33]. The occurrence of the molecule fragments after surface collision, characterized by a threshold energy, is frequently higher than the dissociation energy of the molecules.

Alternatively, an incident molecule is dissociated in fragments if this molecule is neutralized as result of the collision process. Then, this molecule becomes instable, due to repulsive electronic state within the molecule. This electronic dissociation could be mainly proven for small molecules like H_2^+, N_2^+, O_2^+, etc. [32].

(IV) *Chemical transformation of the surface* (process IV)

A molecule can be considered to be an energetic particle, which can be implanted below the surface in dependence on energy, incident angle, and the mass ratio of the target material and the molecule mass (see Sect. 3.5). As a result of the implantation process, a fragmentation of the molecules can be expected, where those remain as implanted species or react with substrate atoms to form a compound.

(V) *Chemical sputtering* (process V)

The ion bombardment of chemically reactive ion species can cause chemical reactions at the surface or in the near-surface region and results in particles which are weakly bounded to the surface and can therefore be desorbed. This process, called chemical sputtering [34], can be expected after hyperthermal low-energy polyatomic ion bombardment. In particular, the kinetic energy of the incident particle is transformed through the collision process into a translational and later a vibrational excitation. In contrast to physical sputtering (Chap. 5), the process of chemical sputtering is strongly facilitated by temperature.

(VI) *Soft landing* (process VI)

Soft landing allows a destruction-free deposition of organic or biological molecular films, which cannot be fabricated by conventional physical or chemical deposition vapor techniques, because their thermal instability and/or low vapor pressure (details see Sect. 9.4) [10, 29, 30].

9.1.3 Role of Ion Energy

(1) *Plasma potential*

IBD is characterized by ion energies less than some hundred electron-volts. Frequently, the published ion energy values are based only on the source-bias voltage. Excess energies due to the source-plasma potentials are neglected, although is known that the electron emission from the internal source walls will also affect the potential value [35]. The applied power supply voltage for ion acceleration is identical to the chamber potential with respect to ground, called the ion-source voltage eV_s (Fig. 9.8). Then, the energy of the extracted ions is given by eV_s, where V_s is the ion source-bias potential and equal to the power-supply voltage. Because the ion source wall under ion and electron bombardment emits secondary electrons back into the plasma, an increased positive current or a decreased negative current can be expected.

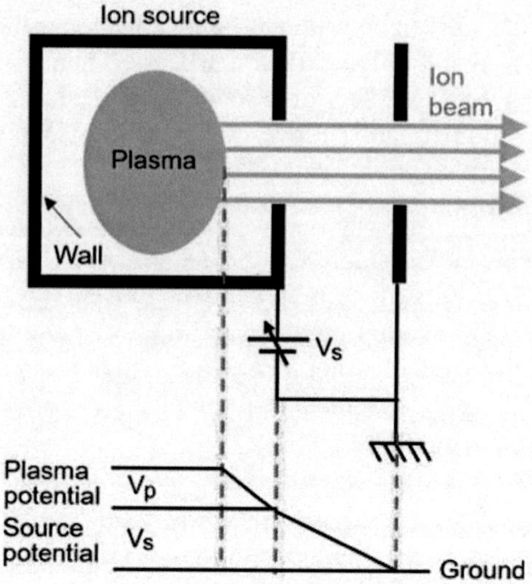

Fig. 9.8 Schematic of the potential relationship in a plasma-based ion source (figure is taken from [36] and modified)

Consequently, an excess energy due to the source-plasma potential, V_p, should be expected. Thus, the energy of the extracted ion beam is determined not by eV_s, but by $e(V_s + V_p)$. Sakudo et al. [36] have determined the plasma floating potential for this boundary condition to

$$V_p = \frac{k_B T_e}{2e}\left[\ln\left(\frac{M}{1.26 \cdot 10^{-3}}\right) - 2\ln\left(\frac{1+\gamma_I}{1-\gamma_e}\right)\right], \tag{9.4}$$

where M is the mass of the ion, T_e is the electron temperature, e is the elementary charge. The γ_I and γ_e are the secondary electron coefficients of the wall material for ion and electron bombardment, respectively. When either one or both coefficients have any value, the plasma potential decreases, where both coefficients are dependent on the primary particle energy and the wall material.

When the plasma potential is comparable to the ion source potential (applied supply voltage), the plasma potential cannot be neglected. This fact can be significant at the low energies used in the IBD technology. Then, the plasma potential should be added to the source plasma potential to evaluate the precise ion energy.

(2) *Lower and upper energy threshold for IBD film growth*

For low energies, it can be assumed that the ions are neutralized upon contact with the substrate surface. Kasi et al. [37] have shown that the neutralization efficiency is >99% after C^+ and O^+ ion bombardment of nickel for ion energies between 50 eV and 200 eV. The neutralization takes place at the surface or within the first monolayers by Auger or charge exchange resonant processes (see Sect. 9.1.1).

The energy of these neutralized particles is an important parameter to control the IBD process, because this parameter influences many processes relating to the film growth. In detail, (i) the re-sputtering of the deposited film atoms, (ii) the reflectivity of incoming particles, and (iii) the athermal generation of the point defects at the surface and the near-surface region must be considered:

(i) The *sputtering yield* decreases when the ion energy decreases (details see Sect. 5.3). Below a threshold ion energy, the sputtering yield will be less 1 (see Sect. 9.1.1), i.e. film growth by IBD can be expected.
(ii) In contrast to the re-sputtering of deposited films, the *reflection* of low-energy incident ions, i.e. ion energies between some ten and some few thousand electron-volts, are only superficially understood. In principle, the reflection of ions/molecules reduces the number of particles which contribute to film growth. No reflection was observed for incident energies below a threshold incident energy (e.g., see Fig. 9.5). However, when the incident energy is increased, the probability of reflection of pre-deposited atoms rapidly grows (Fig. 9.5). The probability of reflection is increased also in connection with the angle of incidence and is greatest for glancing angle ion impacts.
(iii) An *athermal generation of point defects* can be observed when the transferred kinetic ion energy is higher than the displacement energy. Then, neutralized ions are capable to penetrate below the surface of the substrate or the growing film. Before the penetrated particle comes to rest, substrate atoms are displaced. This process is linked with the formation of point defects (vacancies and interstitials). It has been observed that for low ion energies ($\leq$ 100 eV) the adatom yields are significantly higher than sputtering yields [38].

In general, the ion energy used for IBD processes so low, that a full collision cascade cannot be formed. This process can be described by a binary collision approximation (see Sect. 3.5.2). But, by choosing choice an appropriate interaction potential (universal potential, Born–Mayer potential, etc.), the surface binding energy U_s and the displacement energy E_d must be considered critically. Brice et al. [39] and Ma [40] have proposed that the surface displacement energy $E_d^{(s)}$ is one-half of the bulk displacement energy $E_d^{(b)}$, i.e. $E_d^{(s)} = \frac{1}{2} E_d^{(b)}$, and Herbots et al. [24] have assumed that the surface binding energy U_s is given by $U_s = \Delta H - BE$, where ΔH is the sublimation energy and BE is the binding energy of the deposited material. With a decrease of ion energy, fewer point defects are generated. Because the average ion range and the straggling also decrease in line with decreasing the ion energy, the defect distribution is positioned closer to the surface. The point defects formed by these collisions are mobile even at room temperature and capable to form one-dimensional defects (e.g., dislocations) [41]. Therefore, a lower threshold energy exists for the growth of films by IBD. For example, Marton et al. [42] have shown that Kr ions with an energy smaller than 45–50 eV do not penetrate below a graphite surface and Kasi et al. [43] have found that the formation of diamond-like carbon films could not be produced by carbon ion bombardment with kinetic energies smaller than 10 eV (see Sect. 9.3.1).

But, there is also an upper energy threshold, which defines the limits for the sputtering yield and the defect production rate which can still be accepted so that a thin film just still grows.

(3) *Threshold energy for fragmentation of molecules*

In the context of direct ion beam deposition, thin films can be coated onto substrates with or without mass-separation (see Fig. 9.1). In the first case, molecular ions exist in the ion beam in addition to monatomic ions. In the second case, an ion beam of only molecular ions can be used. The interaction of small diatomic molecules with the surface of the growing film results in a co-occurrence of atomic and molecular ion species which influence the growth of film in different ways [44]. Hence, it is useful to know the threshold energy which causes the atomic species to appear after a dissociation (or fragmentation) of molecular species on the surface. It can be assumed that dissociation induced by collision (see Sect. 9.1.2) is the dominant process for low incident energies. Therefore, in diatomic ion to solid surface scattering experiments dissociatively scattered ion products can be observed in addition to non-dissociatively scattered origin ions. For example, Akazawa and Murata [45] have studied the dissociative scattering of N_2^+, CO_2^+ and CO^+ from Pt(100) surfaces at low energies and found a correlation between the dissociation energy and the threshold energy for the emergence of fragments after collision. It could be observed that threshold energies are about four to five times higher than dissociation energies for the ground states (e.g., threshold energy of N^+ is 40 eV [45], while the dissociation energy of N_2 is 9.76 eV).

If a diatomic molecule dissociates, the kinetic energy is distributed to the fragments according to its mass ratio. Kasi et al. [47] for example, have shown that the energy of the fragments C and O (mass of carbon/mass of oxygen $\approx$ 0.75) after dissociation by collision of 100 eV CO^+ on Ni(111) surface was 42 eV and 58 eV, respectively (i.e. ratio of the energies was 0.725).

9.2 Film Growth by Direct Ion Beam Deposition

The kinetics of the film growth under deposition with hyperthermal ions ($\leq$ 100 eV) are fundamentally different from the thermal film deposition ($\approx$ 0.2–0.4 eV). The highly increased ion energy causes the formation of vacancy-adatom pairs and atomic insertions. It also influences the diffusion behavior of the deposited adatoms, island size distribution, island density and the growth mode (see Fig. 9.2). The description of the atomic processes related to the growth of films under direct hyperthermal ion bombardment is very similar to that of film growth by ion beam-assisted deposition (see Sect. 10.3).

9.2.1 *Experimental Studies for Film Growth by Ion Beam Deposition*

In early sputter experiments, an increase in the area density of islands on different substrate materials during sputtering of metal targets by high-energy noble gas ion bombardment could be observed. Chapman and Campbell [47], for example, have observed an enhanced nucleation rate for Au islands on NaCl(001) substrates when the Au atoms have energies of about 20 eV instead of 200 eV. Also, Lane and Anderson [48] found an increase in area density of Au islands after 2 keV Ar ion sputtering of an Au target. It was assumed that this effect is a result of strongly binding sites due to ion beam-induced defects. They observe also major reductions in nucleation rate when temperature is increased, due to the thermal desorption of monomers.

In contrast, other studies have reported a decrease of area density and a larger average grain size under very low-energy ion bombardment. Hazan et al. [49] have deposited *In* atoms by thermal deposition and by partially ionized *In* atoms on amorphous Si_3N_4 substrates and have determined the island density, size distribution and film thickness as function of the In ion energy between 0 eV (thermal evaporation) and 300 eV. The increasing of the average grain sizes (increase of the island diameter from 6.5 nm…8 nm to 13 nm…50 nm) and a decrease of island area density is explained by the depletion of clusters due to sputtering and ion beam-induced dissociation (see Fig. 9.2, processes IV and V). Esch et al. [50] and Kalff et al. [51] have studied the homo-epitaxial deposition of Pt under assisted Ar-ion bombardment (400 eV and 4 keV). They have observed that the Pt island number density exactly corresponded to the number density of nucleation centers generated by ion bombardment. In the presence of ion irradiation, the size of these islands is much smaller but their density much larger. It was also found that vacancy clusters do not act as nucleation centers.

Pomeroy et al. [52] and Degroote et al. [53] have studied the influence of deposition energy at ambient temperature on the homoepitaxial and heteroepitaxial film growth in detail. For the hyperthermal deposition of Cu thin films on Cu(111) crystal surfaces a strong morphological dependence on the incident energy could be observed. Figure 9.9 shows that films grown at 20 eV are smooth due to high adatom diffusion and atomistic insertion mechanisms. Films grown at 40 eV exhibit a high density of vacancy islands of nearly uniform size, whose presence is attributed to the onset of adatom/vacancy production near 20 eV. With increasing energy, the vacancy islands grow coarser at lower levels. The transition from a predominantly two-dimensional morphology decorated with vacancy islands at 60 eV to a plateau-like three-dimensional morphology at 100 eV is attributed to the onset of sputter roughening during growth. MD simulations [54, 55] suggest an optimal energy window for film growth at 25 eV in a diffusion-limited regime.

Degroote et al. [53] have reported the effect of deposition energies of Co ions between 5 eV and 30 eV on Ag(001) substrates. For low energies (5 eV) the topography of the Co sub-monolayers is similar to thermal deposition. The island area

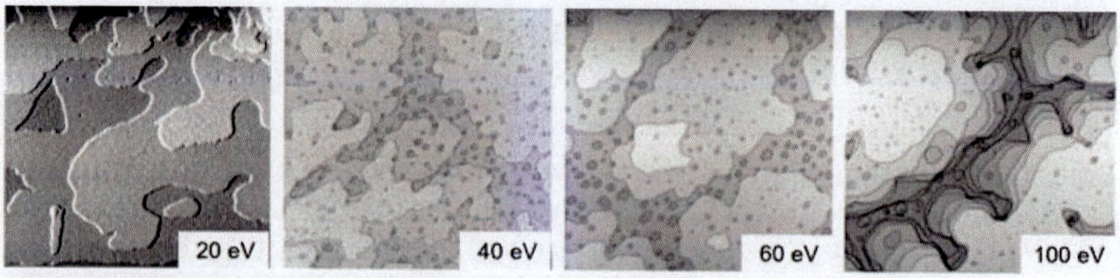

Fig. 9.9 STM images of Cu films deposited on Cu(111) by direct ion deposition at room temperature. The step edges are single atom height steps and the darker colors represents lower levels. The images are 200 × 200 nm^2 (figures adapted from [52])

density strongly increases at 10 eV Co particle energy. Simultaneously, a decrease of average grain size and island heights could be determined. These results are explained by an ion impact-induced island fragmentation and an increase in the number of surface-confined Co clusters (see Fig. 9.2, processes IV and VII).

9.2.2 Growth Processes

Similar to ion beam-assisted deposition (Chap. 10), the growth and morphology of thin films generated by direct ion beam deposition is discussed with regard to additional effects caused by the interaction of energetic particles with previously deposited single adatoms and adatom islands. Atomistic IBD processes during thin film growth have been studied in experiments using AFM/STM, electron microscopic methods and computational approaches. These studies are focused on three topics:

(i) The impingement of energetic particles induces numerous atomic rearrangement processes during film growth at the surface and in the near-surface region (see Fig. 9.2), which leads to the formation of vacancies, interstitials and adatoms (9.2.2.1).
(ii) The ion impact induces additional athermal diffusion processes of generated point defects or previously deposited adatoms besides the thermally activated processes caused by the heating of the substrate. These athermal processes are separated in ballistic induced diffusion, long range biased diffusion in impact direction and localized heating (transient diffusion) (Sect. 9.2.2.2).
(iii) Growth and fragmentation of islands on the surface of the growing film can be markedly influenced by energetic particles (Sect. 9.2.2.3).

9.2.2.1 Point Defect Generation

The application of energetic ions to deposit thin films is attractive because deposition parameters like ion energy, fluence, temperature etc. can be used to control morphology and crystalline quality during film growth. The minimization of defect concentration at the surface and in the near-surface region as well as enhancing growth stimulating processes, e.g., enhanced surface diffusion, are of crucial significance.

Generally, damages caused by ion bombardment can be described by defining the number of displaced atoms per incoming ion (Sect. 4). Computer simulation codes and analytical calculation based on the linear Boltzmann transport theory have been applied for this purpose. In this context, the point defect generation rate (vacancies and/or adatoms per incident ion) is of vital importance for direct ion beam-induced film growth. It is known that a certain threshold energy is necessary to displace an atom from its lattice site by ion bombardment. Brice et al. [39] and Ma et al. [56] assume that this displacement energy is one-half of the bulk displacement energy (details see Sect. 10.3.5). They were able to estimate the displacement energy at the surface and the bulk as well as the limit for the ion energy above which more defects are generated in the bulk than in the surface. Also, the number of vacancies varies with the square-root of ion energy. Mohajerzadeh and Selvakumar [57] have used an approximation for the nuclear stopping power to obtain an analytic solution for the calculation of the deposited energy. On this basis, they were also able to determine the deposited energy, the number of vacancies generated per incident ion, and a range of ion energies which enhances the crystalline quality of the epitaxial films without damaging the grown film. For Ar ion bombardment of silicon, for example, it could be ascertained that for ion energies between 20 eV and 40 eV the displacements primarily occur at the surface, i.e. energies below 40 eV cause no vacancies in the Si bulk. For energies from 40 eV to about 200 eV, the number of vacancies per incident ion in the near-surface region is higher than the number of vacancies in the bulk. For energies ≥ 200 eV, one vacancy in the surface and one in the bulk per incident ion is generated [51]. Experimental studies and MC simulations have shown that the displacement energy at the surface can actually be smaller. Kuboto et al. [58] and Chang et al. [59], for example, have found a surface displacement energy of 37.5 eV and 35 eV for silicon, respectively (bulk displacement energy ranged between 10 eV and 70 eV, see Appendix H). Laurens et al. [60] have observed that *In* atoms with energies > 6 eV are capable to displace Cu atoms from their lattice positions. RBS and channeling studies have also demonstrated that the number of defects in Si films prepared by direct ion beam deposition of ^{28}Si ions are very sensitive to changes in both ion energy and substrate temperature [61]. Therefore, an ideally suited energy window for the growth of Si films could be proposed. As is well know, that the displacement energies for different defect species (vacancies, interstitials, etc.) are different [62]. A detailed analysis of the adatom generation, the distribution of adatoms per impact, the radial distribution of the adatoms and the mean size of adatom and vacancy clusters were carried out by Morgenstern et al. [63] for 4.5 keV noble gas ion impacts on Pt(111) at very low temperatures. The Ne single ion impacts produce only isolated adatoms with a number distribution of adatoms per impact close to the Poisson distribution, where the adatom generation rate is compatible with a linear cascade model (Fig. 9.10).

According to STM observation, 4.0 ± 0.2 adatoms per Ne ion impact are created on average. By application of noble gas ions with a higher ion mass, adatom clusters with number distributions much broader than the corresponding Poisson distribution

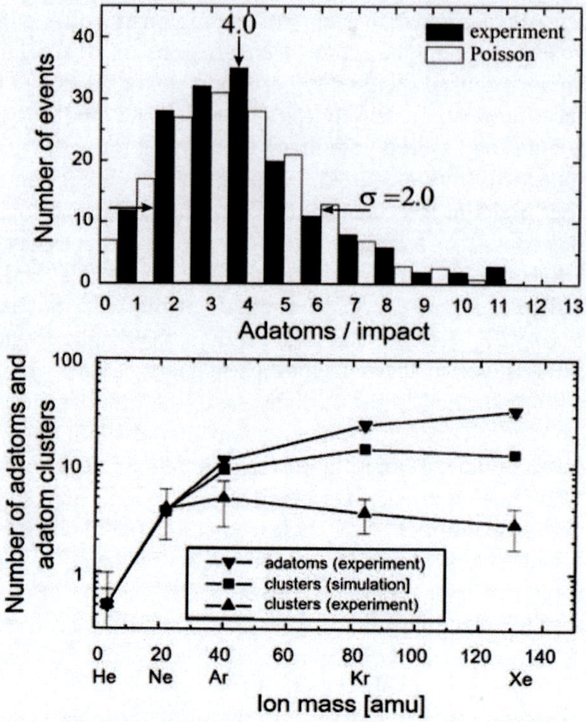

Fig. 9.10 Above: Number distribution (black bars) of adatoms per 4.5 keV Ne ion impact in Pt(111). The vertical and horizontal arrows indicate the average number of generated adatoms per impact and standard deviation, respectively. For comparison a Poisson distribution is included (white bars). Below: Number of adatoms and adatom clusters as function of the ion mass. The ion energy is 4.5 keV. The experimental results are compared with results by simulations (figures adapted from [63])

are generated. For impacts of higher energy densities (Fig. 9.10), more adatoms (for example, the average number of adatoms per Xe ion is 53 $\pm$ 2.6) [63] and surface vacancy clusters are produced.

The number distribution is significantly broader compared to a Poisson distribution, indicating that the adatom formation is mainly caused by collective effects. Based on MD simulations, Ghaly et al. [64] have assumed that this collective effect is a viscous flow of material from a molten core in the center of the impact region towards the surface. A successively increased fluence induces an increased ion-induced adatom mobility at low energy densities, resulting in adatom island formation at low temperatures. Stable subsurface vacancy clusters are only formed at higher energies (6.5 keV Xe ion irradiation).

It was found that several processes triggered by ion impact happen simultaneously. Pomoroy et al. [52, 54] have described a hierarchy of different atomic processes induced by ion impacts, here, for the direct hyperthermal Cu ion beam deposition (Fig. 9.11). Dependent on ion energy, a general image of the energetic sequence

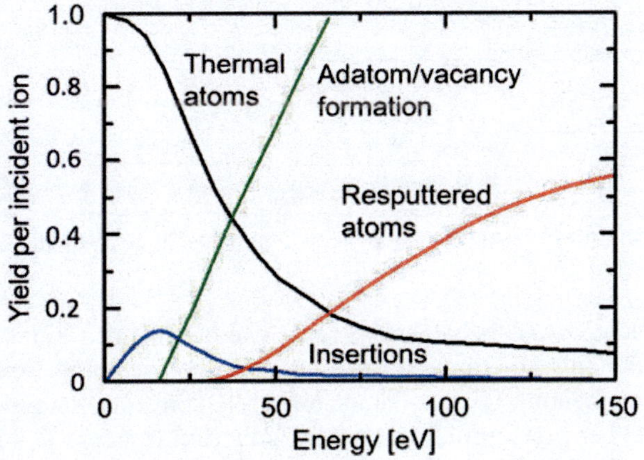

Fig. 9.11 Schematic presentation of the hierarchy of impact induced processes as function of the ion energy (figure is taken from [54] and modified)

of microscopic processes in copper [54], but also in silver and platinum [45] has been developed. Firstly, insertion mechanisms (Fig. 9.2, process I) are active above a step edge as low as 3 eV. Incident particles are deposited deeper into islands with creasing ion energy. Consequently, the yield per incident incident particle increases and reaches a maximum at 15 eV. For energies at about 20 eV, the adatom-vacancy pair formation increases significantly, reaching a yield of one pair per incident particles at 60 eV and two pairs at 150 eV (Fig. 9.2, process V). Re-sputtering (Fig. 9.2, process II) begins at a threshold energy of more than 40 eV. For thermal deposition ($\approx$0.2…0.4 eV), each deposited adatom therefore has a yield of one. Increasing energies of incident particles reduces the thermal deposited adatom yield by energetically activated non-equilibrium processes.

9.2.2.2 Athermal Surface Diffusion

The deposited particles after IBD diffuse over the surface of the growing film. Because diffusing adatoms are the same chemical species as the film this process is referred to as self-diffusion. Without additional ion bombardment or deposition of energetic particles, the diffusion of adatoms on the surface is determined by temperature and obeys a conventional Arrhenius behavior, i.e., conventional thermal diffusion. Additional ion bombardment or deposition of particles with energies higher than the thermal energies, generates point defects at the surface or in the near-surface region (Fig. 9.2), and can influence the surface diffusion, comprehensively. Consequently, the influence of the adatom diffusion on the growth of thin films produced by IBD must be discussed in the light of the ballistic, biased and the transient diffusion (Fig. 9.12).

(1) *Ballistic diffusion*

In general, surface diffusion is caused by simply ion beam heating as well as by increased substrate temperatures or by the impact of hyperthermal particles on surface

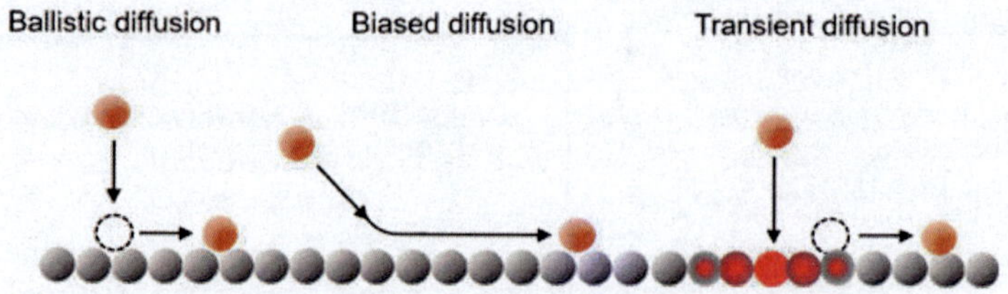

Fig. 9.12 Schematic presentation of diffusion processes influenced by additional ion impact

adatoms. This ballistic process can be assumed to be independent of temperature. For this athermal ballistic process, the incoming ions can provoke an additional momentum to the adatom, which is converted into enhanced mobility, leading to either preferential desorption/sputtering (process II, Fig. 9.2) or enhanced surface diffusion (process VI, Fig. 9.2).

If the ion energy is high enough to displace atoms on the surface of the growing film, the ballistic induced diffusivity is approximately given by [65]

$$D_b \approx 0.07J\frac{\lambda^2 S_n(E, x = 0)}{E_d} = \upsilon_b \lambda^2, \tag{9.5}$$

where J is the ion current density, $S_n(E, x = 0)$ is the nuclear stopping cross-section at the surface (depth $x = 0$), E_d is the adatom displacement energy, λ is the mean square jump distance of the recoiling atom (in order of several atomic spacings) and ν_b is the jump frequency of the ballistic generated adatom. This expression is equivalent to the expression for thermal diffusion ($D_{th} = \nu_{th}\lambda^2$ with thermal frequency factor, ν_{th}, of the order of about 10^{13} s^{-1}). Therefore, the ballistic frequency factor must be much larger than the thermal frequency factor ($\nu_b \gg \nu_{th}$), for the case that ballistic diffusion dominates the thermal diffusion. This situation can only be realized if the temperature is low or/and the ion current density is large.

Also, the kinetic energy of impinging ions influences the diffusivity. It can be expected that the kinetic energy of incident ions is not completely lost after the first collision with film atoms in the upper row of the growing film or bombarded substrate. The remaining energy after the first collision can be also dissipated by further collisions. Müller [66] has studied the growth as function of the incident kinetic energy of arriving adatoms by employing a two-dimensional molecular dynamic simulations. At very low energies (thermal energy) most of the arriving atoms move a distance of one lattice spacing after the first collision, only. At higher kinetic energies (hyperthermal energy) the mobility of the atoms is enhanced. The kinetic energy of the adatoms is significantly increased, the films become denser and the number of defects is reduced.

Ditchfield and Seebauer [67, 68] have carried out measurements of surface diffusion under low energy Ar ion bombardment during Ge deposition on Si(111) and have discussed the results in dependence on the substrate temperature. For low temperatures (< 730 °C) the pre-exponential factor of the surface diffusion, D_o, increases for ion energies < 15 eV. They explain this circumstance by a steady-state balance between ion beam-induced generation of surface defects and a thermally activated recovery. In the high-temperature regime (> 730 °C), both the pre-exponential factor and the activation energy of the surface diffusion decreases in line with increasing ion energy. It is assumed that the defects generated by the ion beam are preserved and act as preferential sinks for adatoms.

The influence of the ion masses up to 131 amu (Xe) on surface diffusion of germanium on silicon in the low-temperature regime have also been studied [67]. Identical to the case of thermal diffusion, the activation energy remains constant (Fig. 9.13). In contrast, the value of the pre-exponential factor D_o grows with the energy by up to a factor of five for ion energies above 15 eV. The increase follows the square-root dependence of ion energy [66]. The pre-exponential factor D_o also increases proportionally to the square-root of the ion mass: $E^{1/2} - (15\text{ eV})^{1/2}$ (Fig. 9.13), in the low-temperature regime (a factor of ten over the case of the thermal diffusion, approximately) [67]. The studies by Ditchfield and Seebauer indicate that (i) the diffusion under low-energy bombardment is determined by the interaction of generated point defects and (ii) the diffusion of the adatoms under ion bombardment cannot be described by a simple superposition of the thermal and the ion beam-influenced contributions.

(2) *Biased diffusion*

Experimental studies have shown that atoms arriving after hyperthermal deposition often diffuse across the surface before coming to rest at island edges, which cannot explained with a thermal activated diffusion process [69]. This phenomenon is called biased diffusion and occurs during oblique deposition where the arriving ion is capable to induce high momentum of the adatoms. Then, this adatom diffuses over long distances across the surface. Laurens et al. [60] have demonstrated this phenomenon by performing MD simulations as well as perturbed angular correlation experiments on the system *In* on Cu. *In* ions with energies less than 6 eV (*In* atoms

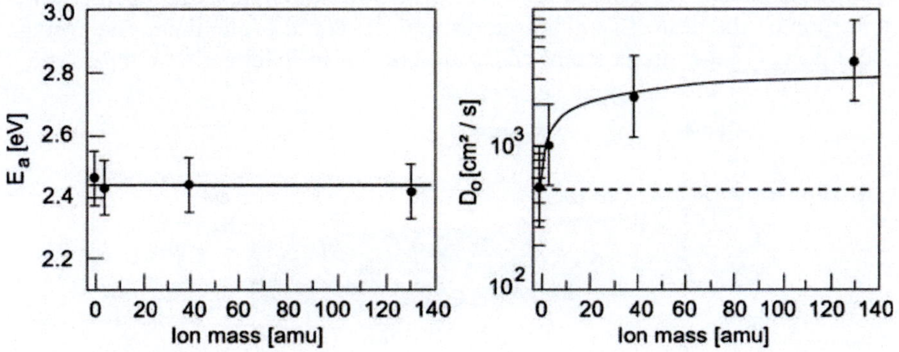

Fig. 9.13 The activation energy of diffusion E_A (left) and the pre-exponential factor D_o (right) in dependence of the ion mass for the low-temperature regime. The dotted line represents square-root fits (figure adapted from [67])

are unable to displace Cu atoms) land at regular adatom sites after perpendicular bombardment. At incoming angles up to 40° with the surface normal, these adatomic sites are within one or two lattice spacings from the point of impact. However, at larger incident angles, the indium atoms can travel up to 30 lattice spacings before coming to rest. Zhou and Wadley [70] have also simulated the biased diffusion for incident energies ≤ 20 eV and incident angles $\leq 85°$, based upon a three-dimensional molecular dynamic model analysis. This model used an embedded atom potential to determine the interacting forces between atoms during the impact. Figure 9.14 illustrates the biased diffusion and shows the position of the particle reaching the surface under an angle of 80°, and with an impact energy of 20 eV, the positions of the adatom after time intervals of 0.1 ps, and the position where the adatom come to rest. The authors have calculated the biased diffusion distance in dependence of the incident angle and energy for three low-index Cu crystal directions. For energies $\geq$10 eV, the bias diffusion distance, d_s, increases in line with the incident angle and reaches a maximum at about 80°, where the longest bias distance is observed for the most densely packed (111) surface ($d_s \approx 3$ nm for $E = 10$ eV and $d_s \approx 8$ nm for $E = 20$ eV). Otherwise, the adatom bias diffusion distance monotonically increases with increasing ion energy. For ion energies > 20 eV and incident angles $\geq$ 70° distances sometimes exceed 0.01 μm.

(3) *Transient diffusion*

Transient mobility is defined as the ability of adatoms to transform kinetic energy gained by adsorption into movement parallel to the surface and coming to rest, for example, at growing island edges [71]. The energy of the incoming adatom is transferred into vibration of the crystal at the impact region. Then, this vibration energy is dissipated by thermal conduction via the lattice and can enhanced the mobility of the adatoms. This is a possible explanation for layer-by-layer growth of thin metal films at low temperatures [68]. Zhou and Wadley [70] use three-dimensional molecular dynamics simulations to analyze the local heating after hyperthermal ion impacts. According to these authors, this heating effect is responsible for the activation of an athermal adatom diffusion. Fig. 9.15 (left) shows the transferred energy of the impact ion on the nearest-neighbor atoms in dependence of the time. The average kinetic energy of the surface atoms $\langle E \rangle$, can be related to an 'effective' temperature T

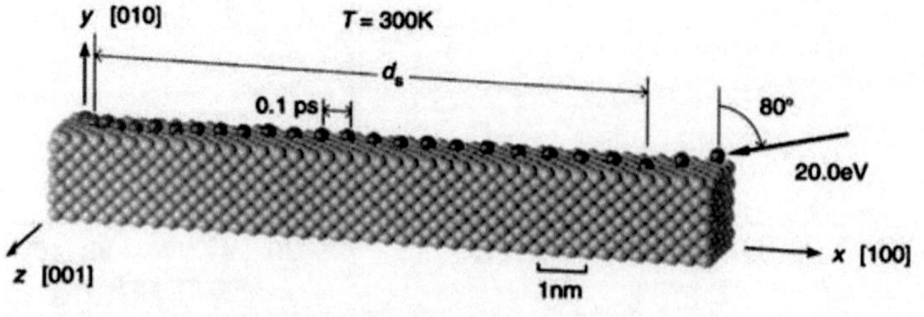

Fig. 9.14 Schematic picture of the adatom biased diffusion on a [100] surface at room temperature. The incident angle is 80° and the incident energy is 20 eV. The bias diffusion distance or skipping distance d_S is shown (figure adapted from [70])

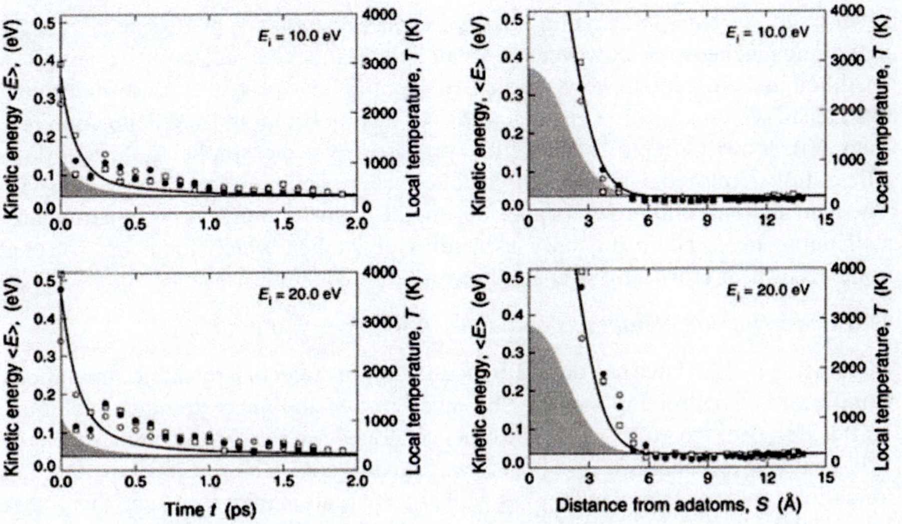

Fig. 9.15 Temperature of the impact atoms nearest-neighbor atoms as a function of time after the impact (left) and the distance, S, from the incoming adatom impact site (right) for two ion energies E_I. The substrate temperature was 300 K and the incident angle 0°. The three different symbols correspond to MD simulations of impact with three different low-index crystal surfaces. The lines are best-fit curves, and the shaded regions indicate the contribution of the latent heat release, i.e. the heating deduced from a zero incident-energy impact calculation (figures adapted from [70])

described by the relationship $T = 2\langle E\rangle/3k_B$. The deposited heat is obviously highly localized in both space and time. Extremely high temperatures, partially above the melting point, are reached. Then these temperatures drop in tenths picoseconds. The spatial distribution is shown in Fig. 9.15 (right). This high temperature affects only a small number of atoms within one or two atomic spacings. Nevertheless, significant vibrational energy (i.e. heating) should be sufficient to result in enhanced surface mobility.

9.2.2.3 Formation and Fragmentation of Islands

In general, thermal deposition is characterized by randomly deposited atoms, which have insufficient mobility to diffuse to the lowest energy sites. Consequently, the surface energy is increased by expansion of the surface area (increased roughness) as well as expansion of the number of surface defects. An additional input of energy (e.g., by higher kinetic energy of the incoming atoms) allows the surface to evolve into a lower energy state (atoms occupied regular lattice sites). This occurs by numerous ion beam-induced processes (Fig. 9.2). An atomic reassembly of a surface under these constrained conditions can result in the formation/breakup of flat islands (terraces), or the athermal diffusion or the vacancy annihilation in the subsurface region. The driving force is the formation of a flat surface or layer-by-layer thin film arrangement, because such surfaces tend to minimize the surface energy. The precise details for the underlying mechanisms, however, are yet unknown and is subject to further analyses.

Although being intensively studied to identify key processes and underlying mechanism which are very similar to those of the ion beam-assisted deposition (see Chap. 10), growth kinetics of thin films prepared by direct ion beam deposition is still not fully explored. Gaining insights into the interaction of low-energy ions with generated adatoms and grown islands during the deposition process is a particularly challenging task. There are only a handful of studies which focus on selected problems of film growth by IBD are known:

(1) *Redistribution of adatoms and dimer formation on the surface*

Adamovic et al. [72] have performed embedded-atom molecular dynamic simulations with the aim of optimizing the layer-by-layer growth and microstructure evolution. For this purpose, the effect of Pt adatoms, generated after self-ion bombardment of Pt(111) at low ion energies (5–50 eV), was investigated. The results show that ion irradiation enhances lateral adatom as well as terrace atom migration, and that dimer formation occurs for all energies. For ion energies $\geq$15 eV, atomic intermixing events involving incident and terrace atoms are observed, while residual lattice defects such as vacancies or interstitials are only observed at energies $\geq$ 40 eV. If Ar ions instead Pt ions are used as the incident energetic species, the threshold energy increases due to reduced momentum transfer. However, the same trends and general results can be expected.

(2) *Island growth*

Jacobsen et al. [55] have used kinetic Monte Carlo simulations to study the early stages of island growth under ion bombardment. It is assumed that ion beam-induced collision effects like adatom/vacancy formation, atom insertion, atom pile up, island breakups and horizontal movements of adatoms (see Fig. 9.2) are influenced by ion impact. The initial stage of the growth already allows to distinguish between layers fabricated by thermal deposition and layers produced by IBD. For thermal deposition, coalescence of the islands starts at 0.4–0.5 ML coverage, and island density subsequently drops down. The islands tend to be equally spaced, with spacing determined by the diffusion of the adatoms. Another island evolution can be observed for low-energy ion deposition. Figure 9.16 illustrates coalescence of parts of the island in the early stages of the monolayer growth, with a coverage between $\overline{\theta} =$ 0.056 ML and 0.15 ML. In contrast to thermal deposition, Fig. 9.15 also shows the coalescence of islands as early as 1 ML. The little island (coverage $\overline{\theta} = 0.051$ ML) first divides into two and then into three pieces, which then coalesce again before a coverage of 0.15 ML. Such early coalescence can be expected for islands formed by an ion beam breaking off pieces of an existing island. This is generally not observed for thermal deposition. Thus, island density and step density can be drastically increased by breaking existing islands.

Also, the growth of several atomic layers on the substrate differs significantly in the deposition of thermal and energetic particles. After thermal deposition of atoms on the surface of islands, these adatoms will stay there, because thermally

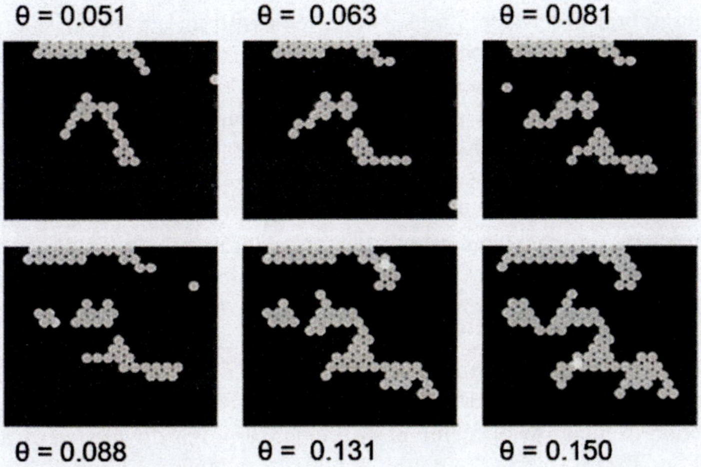

Fig. 9.16 Results of the kinetic Monte-Carlo simulation of an island at low Ag coverage, θ, for 11 eV Ag ion deposition on Ag(111) substrate by 60 K (figure adapted from [55])

activated mobility of these particles at low temperatures is dramatically limited. Hence, new layers are formed on the top of the islands. In contrast, the growth by energetic particle deposition is characterized by collision-induced effects which are mentioned above. For low-energies ($\leq$ some ten electron-volts) the ballistic collision sequences reinforce the tendency towards a distinct layer-by-layer growth depending on impact parameters. Jacobsen et al. [55] found an ideal energy for layer-by-layer growth (25 eV for Ag on Ag(111) surfaces), which correlates to the fact that the impact induced downward interlayer transport has a maximum.

Experimental studies by Degroote et al. [53] have shown that Co island density is strongly increased on Ag(001) surfaces for very low ion energies (10 eV) in combination with a decrease of island size and height. For higher ion energies the island density is reduced, but still higher compared to the density after thermal deposition. It is assumed that the mechanism causing a decrease of the island density is the ion impact-induced island dissolution.

(3) *Island fragmentation*

Carter [59] and Sillanpaä and Koponen [73] have shown that, compared to the results obtained in thermal deposition, in the case of deposition with energetic ions the total density of islands increases, while the average island size decreases. This behavior is in qualitative agreement with experimental findings for the direct deposition of energetic Cu ions on Mo(110) [74].

The behavior of Co islands on copper surfaces under rare ion bombardment were studied using molecular dynamic simulations in dependence of the incident angle and the ion fluence [75]. This study is focused on the assembly mechanisms activated by impact of ions. It was found that the incidence angle has a strong influence on the in-plane migration of surface atoms. Using ions with an atomic mass comparable to that of Co atoms, it is possible to select the desirable assembly mechanisms by variation of ion fluence and incident angle.

Huhtamäki et al. [76] have proposed a phenomenological model for the growth of thin films in low-energy ion deposition which is taking into account up-step and down-step diffusion of adatoms due to ion impacts. This model enables rapid identification of different film growth modes and helps to detect of the transition from 3-dimensional to 2-dimensional growth mode. According to Fu and Wagner [77] the growth rate (change of coverage with time) of any particular nth-layer is assumed to be proportional to the flux of sticking adatoms and to the difference of coverage between the (n−1)th- and the nth-layer. The terms for up-step and down-step diffusion can be combined by introducing a diffusion-corrected coefficient, f, which accounts for the differences in diffusion behavior. This diffusion-corrected coefficient is a measure of the fraction of adatoms diffusing up- or down-step. For down-step diffusion this is $0 < f < 1$ and for up-step diffusion this is $f < 0$, i.e. a layer-by-layer like growth and an island like growth can be expected, respectively. Figure 9.17 shows the coverage of the first 20 layers for both cases $f = 1$ (left) and $f = -1$ (right). It is clear, that for $f = 1$ a layer is almost complete before the next layer starts the growth. In contrast, for $f = -1$ no completely closed layer can be observed. In a more generalized form, Huhtamäki et al. [76] have used this model to quantify the degree of the layer-by-layer growth under low-energy ion bombardment. They predict bombarding ion energy between 5 eV and 25 eV as optimum for the layer-by-layer growth.

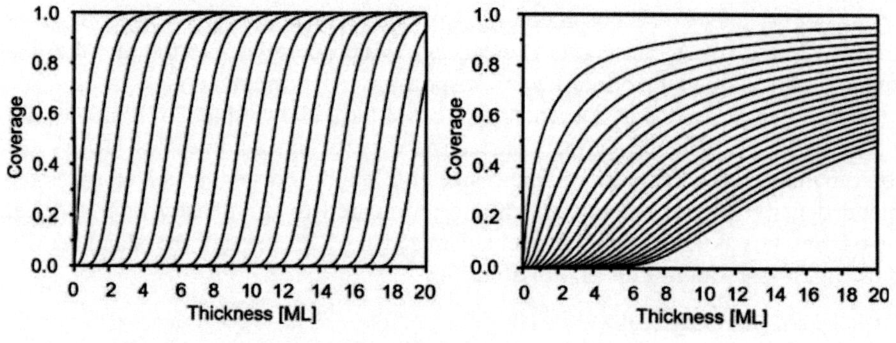

Fig. 9.17 Coverage of the first 20 layers as function of the film thickness in numbers of multilayers (ML) calculated for the diffusion-correlated coefficients $f = 1$ (layer-by-layer like growth, left) and $f = -1$ (island like growth, right) in the frame of the diffusion-corrected simultaneous multilayer model (figure adapted from [77])

9.3 Synthesis of Films by Direct Ion Beam Deposition

The potential of direct ion beam deposition with hyperthermal particles to modify the growth kinetic and to enable the evolution of both stable and metastable materials can be readily exploited in the synthesis of different phases. The properties of these phases such as composition, stoichiometry, thickness, etc. can be precisely controlled by variation of the ion energy, ion flux, temperature and directionality.

9.3.1 Carbon and Diamond-Like Carbon Films

Graphite, amorphous carbon and diamond have properties that are outstanding importance for various applications. Solid carbon manifests in different allotropic modifications, of which diamond and graphite are the well-known crystalline modifications. The structure of graphite is characterized by sp^2-bonds in a plane with each of the three nearest carbon neighbors. Diamond has a zinc blende structure with a four-fold sp^3 covalent bonds, where the carbon atoms form tetrahedral structure units with each four nearest carbon atom neighbors. Amorphous carbon is often referred to as diamond-like carbon (DLC), which is subdivided into different classes (details see [78]), amongst others hydrogen-free amorphous carbon (a-C films for dominating sp^2-bonds), tetrahedrally bonded amorphous carbon (ta-C for dominating sp^3 bonds) and hydrogenated amorphous carbon films (a-C:H with >3 atom-% hydrogen). These materials have been successfully prepared by IBD and are of special interest because due to properties like hardness, optical transparency and electrical resistivity are near those of diamond.

The general concept of the formation of sp^3 bonds is based on the ion bombardment induced density increase. During ion bombardment, the bond hybridization is controlled by the local density, where sp^2 and sp^3 bonds are preferentially formed at low density and high density, respectively. The high density is generated by the penetration of ions into the first atomic layers of the growing film and also leads to the formation of interstitial sites. Increasing the concentration of implanted ions increases the local density without density relaxation. The increased density causes an appropriated hybridization, which results in regenerated local bonding states (sp^3 bonds). A fraction of the kinetic ion energy dissipates into a thermal spike (see Sect. 4.6.2) extremely fast (about 10^{-11} s). This energy is able to trigger a density relaxation process, means that both the increase of the density and the relaxation process are dependent on ion energy. Therefore, it can be expected that the optimal ion energy to generate a maximum density changing is above the penetration threshold energy.

Aisenberg and Chabots [79] were the first to demonstrate the formation of DLC films using energetic carbon ions without hydrogen. Carbon ions were generated by Ar ion sputtering of graphite, where the kinetic energy of ions ranged between 40 eV and 100 eV. Since this first demonstration the production of DLC films by direct ion beam deposition were intensively studied (see Table 9.1).

Table 9.1 Carbon and diamond-like films prepared by direct ion beam deposition (DLC—diamond-like carbon, RT—room temperature, LNT—temperature of liquid nitrogen, st.st.—stainless steel)

Ion species	Substrate	Ion energy (eV)	Substrate temp. (°C)	Results	References
C^+	Si, st.st., st.st.	40	RT	First proof of DLC by IBD	1
	Si, NaCl, KCl, Ni	50–100	RT	Polycrystalline films of cubic diamond	2
		20–100		Very hard, inert and transparent DLC films	3
	Si, Al, glass	300–600	RT	Tetrahedrally bonded carbon atoms	4
	diamond	< 1000	500–800	Epitaxial DLC films with carbonaceous inclusions	5
	Al, Ta, Ti, WC, TiN	500		Thick, wear-resistance and hard DLC layers	6
		500		Atomic density of DLC films is agree with the density of diamond	7
	Si, Ni, Ta, W, Au	10–300	RT	Diamond-like structure of films, optimal C^+ ion energy for DLC formation is 30–175 eV, carbide interface layer,	8
	Si, Ni, Ta, W, Au	1–300	RT	Carbide structure at the interface between film and substrate diamond-like structure for ion energies > 10 eV and < 180 eV	9

(continued)

Table 9.1 (continued)

Ion species	Substrate	Ion energy (eV)	Substrate temp. (°C)	Results	References
	Si	100	25–800	Amorphous films at RT, microcrystalline graphite at higher temperatures	10
	graphite	30–300		Two stage process: (i) damaging of the graphite lattice and then (ii) formation of sp^3-hybride bondings	11
	Si(100)	≤ 1000		Formation of ta-C films	12
	Si	≤ 450	≤ 45	Tetrahedrally bonded amorphous carbon, max. sp^3 fraction at 140 eV	13
	Si	100	RT	Formation of DLC films, Doping of these films by deposition of Cu^+, Al^+, B^+, and N^+ ions	14
	Si	50, 120, 500	RT	Formation of DLC	15
	Si	10–300	RT	Diamond-like properties for 30–300 eV	16
	Si(100)	10–2000	RT-250	Highest sp^3 content (about 80%) occurs between 50 and 600 eV	17
	Si	5–20,000	RT	Films with significant sp^3 content (>40%) are formed between 30 eV and 10,000 eV	18
	Si	80–120	RT-200	Diamond and DLC films, (110) plane of diamond crystals parallel to Si surface	19

(continued)

Table 9.1 (continued)

Ion species	Substrate	Ion energy (eV)	Substrate temp. (°C)	Results	References
	300 nm SiO_2 on Si	800	≈ 127 ,≈ 477	Smooth DLC films at 400 K polymer-like carbon films at 750 K	20
	Si(100)	5–500		Carbon films, lowest roughness for 100 eV	21
C^+, $C_2{}^+$, $C_3{}^+$	KCl, Al_2O_3, Si, metals	10–1000	LNT-325	Using mass-separated C ions, single-crystal DLC films are formed	22
C^-, C^{2-}	Si	10–1000	RT, 400, 800	Transparency and electrical resistivity strongly depended on ion energy	23
C^+, $CH_3{}^+$	Si	15–500	25–800	Amorphous carbon films after deposition at RT and microcrystalline graphite for T > RT no formation of diamond	24
$CH_3{}^+$	Si, glass	100–1600		Amorphous, hydrogenated carbon (a-C:H) films	25
$CH_4{}^+$	Si, Ni, KCl	30–100	RT	Hard polycrystalline films	26
	sapphire	100		DLC films	27
	Si	750		DLC films are deposited	28
$CH_4{}^+$, Ar^+	Si, DLC, SiO_x	100		DLC films, threshold in C/At ratio is given for growth on SiO_x	29

(continued)

Table 9.1 (continued)

Ion species	Substrate	Ion energy (eV)	Substrate temp. (°C)	Results	References
$C_6H_6^+$, $C_6H_5^+$,		100–300	RT	Amorphous films with DLC precipitates	30
C^+, N^+ or N_2^+	Si(100)	5–100	RT	Films formed with a composition above C_3N_4	31
C^+, N^+	graphite	30–350	RT, 255	Graphite-like and C_3N_4-like phases	32

[1] S. Aisenberg, R. Chabot, J. Vac. Sci. Technol. 8 (1971) 112 and J. Appl. Phys. 42 (1972) 2953
[2] E.G., Spencer et al., Appl. Phys. Lett. 29 (1976) 118.
[3] T.J. Moavec, T.W. Orent, J. Vac. Sci. Technol. 18 (1981) 226.
[4] T. Miyazawa et al., J. Appl. Phys. 55 (1984) 188.
[5] J.H. Freeman et al., Vacuum 34 (1984) 305.
[6] J. Koskinen et al., Appl. Phys. Lett. 47 (1985) 941.
[7] A. Anttila et al., Thin Solid Films 136 (1986) 129.
[8] S.R. Kasi et al., J. Chem. Phys. 88 (1988) 5914 and Phys. Rev. Lett. 59 (1987) 75.
[9] S. Kasi, et al., J. Chem. Phys. 88 (1988) 5914.
[10] F. Qin, et al., Rev Sci. Instr. 62 (1991) 2322.
[11] H.J. Steffen, et al., Phys. Rev. B 44 (1991) 3981.
[12] D.R. McKenzie, et al., Phys. Rev. Lett. 67 (1991) 773.
[13] P.J. Fallon, et al. Phys. Rev. B 48 (1993) 4777.
[14] H. Hofsäss et al., Diamond Rel. Mater. 3 (1994) 137.
[15] J. Kulik, et al., J. Appl. Phys. 76 (1994) 5063.
[16] Y. Lifshitz, et al., Diamond Rel. Mater. 3 (1994) 542.
[17] J. Kulik et al., Phys. Rev. B 52 (1995) 15,812 and Y. Lifshitz, et al., Diamond Rel. Mater. 4 (1995) 318.
[18] E. Grossman, et al., Appl. Phys. Lett. 68 (1996) 1214.
[19] H. Ohno, et al., Nucl. Instr. Meth. in Phys. Res. 148 (1999) 673.
[20] V. Kopustinskas et al., Vacuum 72 (2004) 193.
[21] K. Ito, et al., Nucl. Instr. Meth in Phys. Res. B 59/60 (1991) 321.
[22] V.M. Puziko, A.V. Semenov, Surf. Coat. Technol. 47 (1991) 445.
[23] J. Ishikawa et al., J. Appl. Phys. 61 (1987) 2509 and Nucl. Instr. Meth. in Phys. Res. B 21 (1987) 205.
[24] W.M. Lau et al., J. Appl. Phys. 70 (1991) 5623.
[25] R. Locher, et al., Surf. Coat. Technol. 47 (1991) 426.
[26] E.F. Thaikovsky et al., Sov. Phys., Crystallogr. 26 (1981) 122.
[27] T. Mori, Y. Namba, J. Vac. Sci Technol. A 1 (1983) 23.
[28] V. Palshin, et al., Thin Solid Films 270 (1995) 165.
[29] D. Nir, M. Mirtich, J. Vac Sci Technol. A 4 (1986) 2954.
[30] C. Weissmantel, et al., Thin Solid Films 72 (1980) 19.
[31] D. Marton, et al., Nucl. Instr. Meth in Phys. Res. B 90 (1994) 277.
[32] K.J. Boyd et al., J. Vac. Sci. Technol. A 13 (1995) 2110.

The properties of carbon films deposited by direct ion beam deposition are strongly influenced by ion irradiation conditions like ion energy, temperature during ion bombardment as well as the built-up stress state of the generated films.

(1) *Kinetic energy of carbon ions*

Kinetic energy of carbon ions and the temperature play a key role in determining the ratio between the fractions of sp^3 and sp^2 bonds in films prepared by IBD. The influence of the angle of incidence and deposition rate on this ratio could be proved in parts. The sp^3 bond fraction can be determined by electron spectroscopy methods (e.g., by Auger electron spectroscopy, electron energy loss spectroscopy, nuclear magnetic resonance or by diffraction techniques).

A generally accepted requirement for the formation of a higher fraction (> 10%) of sp^3 C–C bonds or the transition of sp^2 bonds into sp^3 bonds is that a certain threshold energy of the incident carbon ions is necessary. Analysis of the experimental results (e.g., see Table 9.2) shows that this threshold energy ranged between 20 eV and 40 eV. In a rough approximation this energy corresponds to the threshold displacement energy E_d of graphite (see Appendix H), i.e. that carbon ions with energy E_d are capable to displace atoms of the deposited carbon film and penetrate beneath the surface.

In an early study, Ishikawa et al. [80] have demonstrated that the atomic number density of carbon film prepared by mass-selected negative carbon ion beam deposition was higher than that of graphite but lower than that of diamond. It has relatively high values in the range of ion energies between 100 eV and 200 eV. Fallon et al. [81] have investigated amorhous carbon films prepared by C ions produced by a filtered cathodic carbon arc. They found a broad maximum of sp^3 fraction at an ion energy of about 140 eV. By increasing the ion energy, the sp^3 bond fraction decreases significantly. McKenzie et al. [82] have monitored significantly smaller carbon ion energies (about 20–40 eV) for the deposition of diamond-like carbon films (sp^3 rich films). In contrast, Hirvonen et al. [83] have prepared films with diamond-like properties for higher carbon ion energies (100–700 eV) and Lifshitz et al. [84] have observed such films in the energy region from 30 eV to 300 eV. For lower energies than 30 eV a transition to graphitic properties could be determined. Over a wide energy range (5–20 keV), Grossman et al. [85] and Marton et al. [86] have studied sp^3 bond fraction in dependence of the carbon ion energy. They found a threshold energy of 30 eV for the formation of significant amount of sp^3 bonds and a sharp increase of the sp^3 fraction with increasing ion energy, for energies > 30 eV (Fig. 9.18). The optimal ion energy region for maximal sp^3 fraction is between 50 eV and 600 eV. Above 2 keV, the sp^3 fraction decreases with increasing ion energy. If the substrate temperature is increased from room temperature up to T > 150 °C, the sp^3 content and density of the films significantly decreases for a carbon ion energy of 120 eV. This observation shows the importance of the radiation-induced diffusion [76]. Kulik et al. [87] determined the sp^3 fraction in amorphous carbon films grown by mass-selected carbon ion beam deposition by elastic and inelastic electron scattering. The highest sp^3 fractions (about 80%) occur for films grown with carbon ion energies between

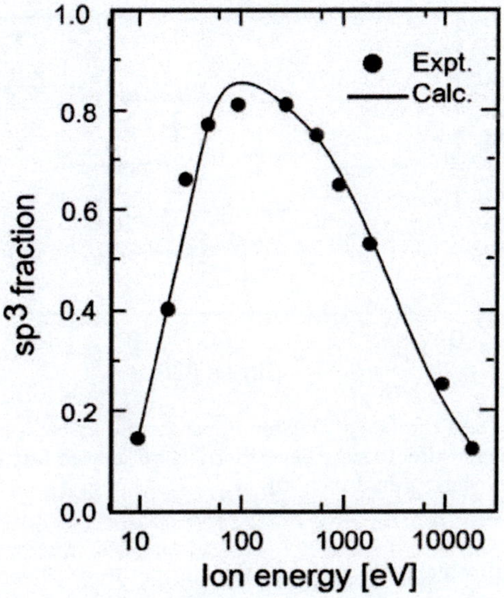

Fig. 9.18 Energy dependence of the tetrahedral sp^3 fraction of DLC films produced by IBD at room temperature (figure adapted from [86]). The experimental data (points) are from [85]

60 eV and 600 eV. For energies above 600 eV, the sp^3 fraction decreases but then remains at about 50%.

In spite of many remaining uncertainties, it can be assessed that the sp^3 bond fraction in dependence of the carbon ion energy shows a typical trend. Nearly all experiments have shown a threshold energy for the detection of significant sp^3 bonds and a strong increase of the sp^3 fraction with increasing carbon ion energy. According to the different studies at room temperature, the optimal ion energy for a maximal sp^3 content (> 80%) ranged between 30 eV and 600 eV. For further increase of the ion energy, the sp^3 fraction decreases (c.f. Fig. 9.18).

(2) *Compressive film stress*

According to Thornton and Hoffmann [88], who state that ion implantation leads to an increase of compressive stress (see Sect. 10.6.7), the relation between stress in carbon films and the kinetic energy of carbon ions have been thoroughly investigated. Numerous studies have shown, that high sp^3 bond fractions in tetrahedrally amorphous bonded carbon films are coupled with high compressive stress state, immediately [81, 89, 90]. Some experimental results are shown in Fig. 9.19. The data represent a non-linear dependence of the sp^3 fraction on the compressive stress, where the relation by Davis [91] (see 10.88 and 10.93) fits the data reasonably well. Compressive stress increases with the sp^3 fraction in films up to 11 GPa. McKenzie and coworkers [90] have converted the measured biaxial stress σ into the hydrostatic pressure, $p = 2/3\sigma$. Then, the obtained hydrostatic pressures for different carbon

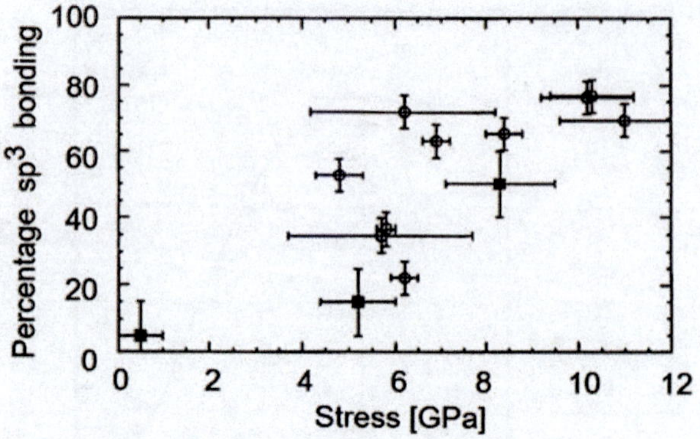

Fig. 9.19 Experimental data for the sp^3 fraction in percent for carbon films as a function of the compressive stress measured after the film deposition (figure adapted from [89]). The open circles are from [81] and the full squares are from [90]

ion energies can be plotted into the pressure–temperature phase diagram of carbon in which the Berman-Simon curve separates the field of existence of diamond and graphite. For room temperature carbon ion deposition, the diamond region is entered for ion energies > 15 eV.

(3) *Deposition temperature*

It was demonstrated that roughness and density of carbon films deposited by mass-separated carbon ion beam deposition are independent of temperatures smaller than a transition temperature. Above this temperature an abrupt transition for these properties was observed [92, 93]. For example, Chhowalla et al. [94] have shown that the structure of ta-C films is strongly dependent on the deposition temperature above a transition temperature of about 200 °C. Furthermore, it was found that this temperature falls with increasing carbon ion energy. In accordance with these results, a sharp changing of the carbon film properties reflects the transition of the amorphous structure with a high sp^3 bond fraction to the graphitic sp^2 films. Raman spectroscopy studies [95] of carbon films prepared by mass-separated carbon ion deposition on Si substrates at room temperature and at elevated temperatures have also proven this transition process.

Phenomenological Interpretation of Carbon Phase Formation by Direct Ion Deposition

In principle, there are two different types of models for the formation of carbon phases. On the one hand there are models which explain the formation of sp^3 bonds on the base of a high compressive stress states. On the other hand, there are models which interpret the sp^2-to-sp^3 transition as result of the ion bombardment or/and annealing processes. The interpretation of the experimentally observed results is ideally based on the known phenomenological models like the high compressive stress model [89, 90, 96], sputtering model [97], cylindrical thermal spike model

[98], subplantation model [91, 99, 100], semi-quantitative subplantation model [86, 101] and the shock wave model [102, 103].

- Spencer et al. [104] have proposed a first *sputter model* to explain the formation of DLC films. According to these authors the direct carbon ion beam deposition process results in the formation of all kinds of C–C bonds, i.e. sp-, sp^2- and sp^3-type bonds. C–C bonds with smaller bond energy (e.g., sp- and sp^2-type bonds) are preferentially sputtered during IBD processes, while C–C bonds with higher bond energy (sp^3-type bonds) remains in the growing carbon films.

- Lifshitz et al. [105] explained the formation of diamond-like carbon in line with the *subplantation model* (see Sect. 10.6.7) in which the rigid, dense DLC phase is generated by shallow implantation of hyperthermal carbon ions beneath the surface. They compare the displacement energy of different carbon phases with the transferred energy from the incident ion to the substrate atom. The transferred energy T [see (2.35) and (2.36)] is dependent on the mass of the collision partners, the incident angle and the incident ion energy. Among other factors, the lower and upper threshold energy can be explained based on this model. It is well known that the displacement energy of the two carbon phases, graphite and diamond, differ significantly. The denser metastable carbon phase diamond has a displacement energy (according to [98] here referred to as $E_d(H)$, values see Appendix H) which is higher than the displacement energy of the stable phase graphite (here referred to as $E_d(L)$).
For low ion energies, i.e. $T < E_d(L)$, Lifshits et al. [105] have assumed that the so-called mold effect is active. Here, the replacements are carried out in the newly generated phase. The ion bombardment induced electronic and phonon excitations into the thermal spike (see Sect. 4.5.2) exist for times in order of 10^{-11} s. This is too short for a complete thermalization of energy deposited by the carbon ion. Consequently, the thermal spike causes rearrangements in substrate and leads to the formation of a stable phase (graphite).
For medium ion energies, $E_d(L) < T < E_d(H)$, atomic replacements within the low-density phase will take place, because $E_d(L) < T$. Replacements in the high-density phase are not permanent, because $E_d(H) > T$. The evolution of the high-density phase is preferred at the cost of the low-density phase.
For high energies, $E_d(H) < T$, the formation of both phases can be expected. With this model, the existence of low and high energy limits for the phase formation by IBD can be established. Molecular dynamics calculations, for example, have confirmed that the density of diamond-like carbons films after 40 eV carbon ion deposition have a maximum [106], i.e. the optimal diamond-like carbon deposition by IBD should be expected within a certain energy window. On the other side, the experimentally detected saturation of the sp^3 bond fraction for high carbon ion energies demonstrates the influence of the thermal spike. According to Davis [91] and Robertson [99], the thermally induced displacement as a result of the thermal spike formation causes a relaxation of the sp^3 bonds into sp^2 bonds.

The orientation dependence of the displacement energy, the small difference between the displacement energy in the low- and high-density phases (e.g., in the case of the graphite and the diamond phases) and the influence of surface effects on the replacement are not considered in the subplantation model. Experimental studies have shown that the discrepancy between the displacement threshold energies for graphite and diamond is very small [107].

- Marton and coworkers [86] have used the *semi-quantitative subplantation model* by Boyd et al. [101] (see Sect. 10.6.7) to calculate the energy dependence of sp^3 fraction in carbon films deposited by IBD over a broad energy range between 10 eV and 20 keV. Fig. 9.18 (solid line) shows the energy dependence tetrahedral sp^3 fraction of DLC films, where a penetration threshold energy of 8 eV was assumed. This model considers effects of penetration threshold, radiation-enhanced diffusion, and densification, but also displacement threshold and damage production with the objective to provide the content of sp^3 bonds as function of the ion energy. In the frame of the static version of the semi-quantitative subplantation model [101], the sp^3 bond fraction, f, can be expressed as the difference between the penetration probability [first term in (9.6)] and the defect generation rate [second term in (9.6)] by

$$f = \alpha \ln\left(\frac{E}{E_p}\right) - \beta f\left(\frac{E - E_d}{E_d}\right) + f_o, \qquad (9.6)$$

where α and β are fit parameter, E is the ion energy, E_p is the penetration threshold energy (here 8 eV), E_d is the displacement energy (here 35 eV) and f_o is an energy-independent constant (here 0.05). Marton et al. [86] have used the parameters $\alpha = 0.393$ and $\beta = 0.017$ to reproduce the experimental data points (see solid line in Fig. 9.18). It is evident that this model is capable to describe the energy dependence of the sp3 bond content over a broad range of ion energies, approximately.

- In the *cylindrical thermal spike* model [98], the energy loss in collision effects and also in phonons and electronic excitations is considered (see Sect. 4.6.2). The ratio of the atomic rearrangements to the number of atoms within the thermal spike is the crucial parameter characterizing the ability of the spike volume for rearrangement. When this ratio is > 1, the optimum condition for a diamond-like film growth is given. In dependence on the ion energy this ratio reaches a maximum around 100 eV and then decreases for higher energies. The rearrangement ratio is significantly larger than one only for ion energies between about 50 eV and 500 eV, i.e. for this energy range the highest sp^3 bond fraction can be expected.

- According to the *stress model* by McKenzie et al. [90], a precondition for the formation of significant sp^3 contents by shallow deposition of energetic carbon ions is the existence of high compressive stresses (more than 10 GPa). Marks et al. [108] and Zhang et al. [109] have studied the relationship between ion energy and stress in growing carbon films under low-energy carbon ion bombardment. These

MD simulations led to the conclusion that the energetic ions were deposited on the surface of the film but did not penetrate below the surface layer. In a good agreement with experimental measurements, Marks et al. [108] observe the transition from tensile stress to compressive stress with increasing ion energy and calculate a maximum of compressive stress at around 30 eV. Zhang et al. [109] have observed interface mixing of the substrate atoms and the newly deposited atoms below the film surface, if the kinetic energy of incident ions is sufficiently high and have found a broad maximum of compressive stress between 20 eV and 100 eV as well as a maximum of sp^3 content for 60 eV carbon ions. Consequently, this model explains the compressive stress in the carbon films as a balance between incident ion-induced compression and the thermal spike annealing contradicting the subplantation model.

- According to the *shock wave model* [102], ions injected into the growing film initiate quasi-shock waves by a collective collision process. This shock wave was propagated with high velocity (hypervelocity) and its energy dissipates rapidly into the surrounding solid. In contrast to the thermal spike model, in shock wave model involves both mass and energy (thermal) transport processes. During the shock wave propagation through a medium, pressure and temperature are dramatically increased so a superheating liquid state can be achieved. As a result of these high-temperature/high-pressure conditions an enhanced formation of sp^3 bonds can be expected [103]. During the cooling down process of the shock wave induced spike, some of the generated sp^3 bonds are retained in dependence on the ion energy. Calculations based on the Rankine-Hugoniot shock wave equation indicate extreme local pressures ($\geq$ 100 GPa) and temperatures ($\geq 10^3$ K) for keV carbon ion bombardment of carbon films [103].

9.3.2 *Epitaxial Silicon and Germanium Films*

The direct Si and Ge ion beam deposition has been investigated intensively with the aim to achieve a formation of epitaxial thin films with sharp interfaces at low temperatures, i.e. the interface mixing between growing film and substrate should be avoided as far as possible. Table 9.2 summarizes several results of the direct Si and Ge ion deposition together with main experimental conditions.

Zuhr et al. [110] and Herbots et al. [15] have demonstrated a homoepitaxial growth of Si films down to 375 °C after 30 eV Si ion deposition, while Orrman-Rossiter et al. [111] have shown that the quality of the epitaxial Si films can be improved by reducing of the Si ion energy down to 10 eV. Silicon films prepared by Si IBD with a crystalline quality comparable to Si bulk single crystal were produced with ion energies less than 25 eV at 320 °C [112].

A concurrent low energy rare gas ion bombardment during the conventional thermal Si deposition (e.g., by molecular beam epitaxy, MBE) leads to similar results. Murty et al. [113] have demonstrated that simultaneous Ar ion bombardment (energy

Table 9.2 Silicon and germanium films prepared by direct ion beam deposition (RT—room temperature)

Ion species	Substrate	Ion energy [eV]	Substrate temp. [°C]	Results	References
Si^-	Glass	200	500	50 eV is effective to deposit Si thin film with small grain	1
$^{29}Si^-$	Si(111)	30		Epitaxial Si films	2
Si^+	Si(100), Si(111), Ge(100)	50–100	RT-627	Epitaxial Si films at 127 °C	3
Si^{2+}	Au onto Si	200	RT	Amorphous silicon oxide films	4
$^{28}Si^+$	Si(100)	8–80	40–500	Epitaxial Si films at 20 ± 10 eV and about 160 °C	5
	Si(100)	200	≤740	Homoepitaxial Si film growth at 740 °C and an ion energy of 200 eV	6
$^{28}Si^+$, $^{30}Si^+$	Si(100)	10–200	350–550	Homoepitaxial Si films are grown using energies between 10 and 40 eV	7
		10–100	350–550	Homoepitaxial Si films are fomed	8
$^{30}Si^+$	Si(100)	≤ 850	RT-600	Epitaxial films at both 600 °C and 400 °C, at 400 °C epitaxial films on n-type Si, only	9
				Low temperature limits for quality epitaxial growth is about 375 °C	10
	Si, SiO_2	40–200	< 627	Epitaxial Si films at 375 °C	11
	Si(100)	40–200	RT-625	Amorphous films for RT deposition, polycrystalline films for T between 425 °C and 625 °C	16

(continued)

Table 9.2 (continued)

Ion species	Substrate	Ion energy [eV]	Substrate temp. [°C]	Results	References
$^{74}Ge^+$	Ge(100)	10–200	< 627	Epitaxial Ge films at 375 °C	11
		≤ 850		Epitaxial films for 40 eV	9
				temperature limits for quality epitaxial growth is about 200 °C	8
	GaAs	40	270–410	Epitaxial oriented Ge layers	12, 10
	Si(100)	40–200	RT-625	Amorphous films for RT deposition, polycrystalline films for T between 425 °C and 625 °C	16
$^{76}Ge^+$	Ge(100)	30	200–400	200 °C is the low-temperature for epitaxial growth of Ge onto Ge	10
Ge^+	Si (100)	100–200	300	Heteroepitaxial Ge films ≥ 230 °C and an ion energy of 100 eV	6
	Si(100)	100	≤ 350	Epitaxial films onto Ge at 300 °C	13
	Si(111), Ge(111)	100	?	Amorphous films for T ≤ 200 °C	14
	Si(111)	25–300	RT	Crystalline films	15

[1] D. Kim, Nucl. Instr. Meth in Phys. Res. B 269 (2011) 2017
[2] F. Gorris et al., phys. stat. sol. (a) 173 (1999) 167.
[3] P.C. Zalm, L.J. Beckers, Appl. Phys. Lett. 41 (1982) 167.
[4] J. Yanagisawa, et al., Nucl. Instr. Meth in Phys. Res. B 127/128 (1997) 893 and B 148 (1999) 42.
[5] J.W. Rabalais, et al., Phys. Rev. B53 (1996) 10,781.
[6] K. Miyake, T. Tokuyama, Thin Solid Films 92 (1982) 123.
[7] K.G. Orrman-Rossiter, et al., Phil. Mag. Lett. 61 (1990) 311 and Nucl. Instr. Meth in Phys. Res. B 59/60 (1991) 197.
[8] A.H. Al-Bayati et al., Nucl. Instr. Meth in Phys. Res. B 63 (1992) 109.
[9] B.R. Appleton et al., Nucl. Instr. Meth in Phys. Res. B 19/29 (1987) 975.
[10] R.A. Zuhr et al., Nucl. Instr. Meth in Phys. Res. B 37/38 (1989) 16.
[11] H. Herbots, et al., Nucl. Instr. Meth in Phys. Res. B 13 (1986) 250.
[12] T.E. Hayes, et al., J. Vac. Sci. Technol. A 7 (1989) 1372 and Nucl. Instr. Meth. Phys. Res. B37/38 (1989) 16.
[13] T. Tokuyama, et al., Nucl. Instr. Meth. 182–183 (1981) 241.
[14] K. Yagi et al., Jap. J. Appl. Phys. 16 (1977) 245.
[15] G.E. Thomas, et al., J. Cryst. Growth 56 (1982) 557.
[16] R.A. Zuhr, et al., J. Vac. Sci. Technol. A5 (1987) 2135.

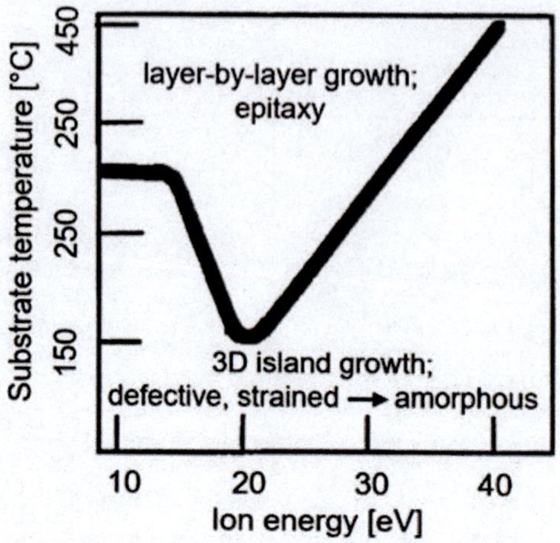

Fig. 9.20 Substrate temperature versus ion energy phase diagram for silicon homoepitaxy by IBD (figure adapted from [115])

between 50 eV and 70 eV) during Si MBE results in an increased epitaxial Si film thickness for low temperatures (< 400 °C) and an improved surface smoothing. Epitaxial Si films were also produced at temperatures < 100 °C, if the substrate surface can be reconstructed by Ar ion bombardment. In contrast to epitaxial Si films deposited by thermal deposition (energy about 0.2 eV), Lee et al. [114] have shown that significantly increased Si(001) epitaxial thickness (10 nm–1.2 μm) can be built up when high-energy Si ions (average energy about 18 eV) are used at temperatures from 80 °C to 300 °C. Rabalais et al. [115] have systematically studied the homoepitaxial film growth on Si(100) by direct $^{28}Si^{+}$ ion beam deposition with energies between 8 eV and 80 eV and temperatures between 40 °C and 500 °C. They have confirmed that the growth mode (see Fig. 9.20), the number of defects and the crystalline quality is strongly dependent on substrate temperature as well as ion energy. Layer-by-layer growth (two-dimensional growth) could be observed for temperatures down to 160 °C. Below this temperature island growth with a transition to an amorphous phase occurs. The growth rate at 290 °C and 20 eV is about 200 times faster than this rate for solid phase epitaxy. An optimum ion energy for achieving two-dimensional epitaxial growth is given by about 20 ± 10 eV at 160 °C. In good accordance, Ma [40, 56] has obtained lower and upper limitations for by energy of Si homoepitaxy by IBD between 18 eV and 38 eV. Figure 9.20 shows a sharp minimum which corresponds to a minimum of the number of defects. This is evidence that ion irradiation enhances epitaxial growth within an optimum energy window. According to Rabalais et al. [115], an increase of ion energy is detrimental to epitaxy, because it increases the number of permanent defects, the sputtering and atomic interface mixing. As temperature rises, thermal vibrations, atomic mobility, and the ability to anneal the ion beam-induced defects are enhanced, so that the narrow energy window observed at low temperatures becomes wider as the temperature increases.

Boyd et al. [116], and Marton et al. [117] have applied the semi-quantitative subplantation model [101] to interpret the experimental results for Si ion beam homoepitaxy. They calculate a generalized temperature-ion energy phase diagram for the silicon homoepitaxy under both thermal (MBE) and hyperthermal conditions (IBD), which shows more details (four different growth regions) compared to the phase diagram constructed by Rabalais et al. [115] on the base of experimental measurements.

9.3.3 Metal Films

The investigation to direct deposition of metal ions on surfaces has two aims. On the one hand, to clarify the fundamental mechanisms of the layer evolution and their growth and, on the other hand, to determine the conditions needed for the production of high-quality metallic contacts and stoichiometic metal silicide films on metal and semiconductor surfaces. Table 9.3 lists some experimental studies dealing with the energetic deposition of metal films on different substrate materials.

Examples of fundamental investigations of metal film growth during deposition of energetic metal particles are amongst others: the formation the Au films onto NaCl [48], In films onto Si and Ge [118] as well as Cu [60] on Ni, Co films on Ag [53] and Cu films on Cu [54], Ag on Ag [55] and Ni [119], and Pt on Pt [50, 51, 63]. Some studies within the scope of direct metal ion deposition have dealt with optimal substrate temperature and energy range to produce pure metal films. Amano [23], for example, have observed an optimal deposition energy between 50 eV and 70 eV for Pb and Mg deposition on carbon and NaCl. The best energies for room temperature epitaxy of Ag films on Si have been determined to be between 25 eV and 100 eV [14], of oriented Al films on single crystal Si(111) to be between 30 eV and 120 eV [120] and for Ca films on Si(100) deposited at 400 °C to be about 20 eV [121]. Thin films of transition metal silicide have been synthesized on Si substrates by IBD of metal ions [122]. Co, Fe, Ni, Ti, and W silicide have been formed at substrate temperatures of some hundred degrees. The stoichiometric silicide films have a thickness of up to 300 nm. For example, $TiSi_2$ films have been formed at 100 eV and 550 °C and TiSi films for temperatures at 400 °C. Cobalt silicide films were found after 40 eV Co ion deposition at 500 °C. Bousetta et al. [123] have synthesized $CoSi_2$ at room temperature with an ion energy of 40 eV.

The advantages of this technique over other methods of metal and silicide film formation are the perfect control of thickness by ion current integration, high purity due to the mass analysis within the deposition process, and control of incident ion energy which permits formation of the silicide phases at low temperatures, thereby minimizing the thermal budget and the associated dopant diffusion in the underlying substrate [122].

Table 9.3 Metallic films prepared by direct ion beam deposition (RT—room temperature)

Ion species	Substrate	Ion energy [eV]	Substrate temp. [°C]	Results	References
Ag^+	Al(110)	30–300		Coalescence and island growth can be observed	1
	Ni(100)	20	RT	Monolayer Ag films with fcc(111) structure at RT	2
	Si(111)	50	RT	Crystalline Ag films	3
	Si(111)	25–125	RT	Crystalline films grown with the orientation Ag[111] II Si[111] and Ag[110] II Si[110]	4
	Si(111)	20–100	RT	Ag film growth rate is 50 nm/min	5
	Si(111)	≤ 250	RT	Film quality increases with ion energy	6
	Si(111), quartz	≤ 1000		Enhanced microhardness and adhesion	7
Pb^+	C, NaCl	24–500		50 eV is the optimal ion energy to form crystalline Pb films, for energies > 50 eV, a preferred [100] orientation can be observed	8, 9
Mg^+	C	24–500		100 eV is the optimal ion energy to form Mg films	9, 10
Cu^+	Cu(111)	20–100	RT	Roughness depends E_I	11
	Ge(100), Si(100)	20–100		Maximum of defect concentration at 35 eV	12
	Si(111)	≤ 250	RT	Film quality increases with ion energy	6
Fe^+	Si(111)	< 100	RT	Single crystalline Fe films	13
$^{56}Fe^+$, $^{57}Fe^+$	Si	50–500	RT	Smooth polycrystalline bcc Fe films, corrosion resistance is depending on ion energy	14

(continued)

Table 9.3 (continued)

Ion species	Substrate	Ion energy [eV]	Substrate temp. [°C]	Results	References
Ca^+	Ge(111)	< 100	RT		13
Pd^+	Si(111)	10–400	RT	50–100 eV is the optimal energy for film formation	15
$^{50}Co^+$	Si(111)	50	RT, 400	Co-silicide formation after post-deposition annealing	16
Co	Si(111)	0.1–155	RT	Optimal energy window (≈ 25 eV) for smooth surfaces	x
In^+	Cu(17,1,1)	5–100	80	For < 6 eV no displacements, In atoms occupies vacancies between 6 and 20 eV, Cu interstitials are produced for energies > 20 eV	18
Al^+	Si(111)	10–120	RT-300	Oriented metal films	19
Cr^+	Glass	< 500		Polycrystalline films	20

[1] T. Tsukizoe, et al., J. Appl. Phys. 48 (1977) 4770
[2] S.S. Todorov, et al., Surf. Sci. 429 (1999) 63.
[3] J.J. Vrakking, et al., Thin Solid Films 92 (1982) 131.
[4] G.E. Thomas, et al., J. Cryst. Growth 56 (1982) 557.
[5] A.E.T. Kuiper, et al., J. Cryst. Growth 45 (1978) 332.
[6] S. Iida et al., Nucl. Instr. Meth. Phys. Res. B 121 (1997) 162.
[7] X. Pan et al., Nucl. Instr. Meth. in Phys. Res. B 37/38 (1989) 858.
[8] J. Amano, R.P.W. Lawson, J. Vac. Sci. Technol. 14 (1977) 690 and J. Vac. Sci. Technol. 14 (1977) 831.
[9] J. Amano, Thin Solid Films 92 (1982) 115.
[10] J. Amano, R.P.W. Lawson, J. Vac. Sci. Technol. 14 (1977) 695 and J. Vac. Sci. Technol. 14 (1977) 836.
[11] J.P. Pomeroy, et al., MRS Proceed. Vol. 648 (2000) P7.3.
[12] B.W. Karr, et al., J. Appl. Phys. 80 (1996) 6699.
[13] N. Sasaki, et al., Thin Solid Films 281–282 (1996) 175–178.
[14] K. Miyake, et al., Proceed. MRS, Volume 279 (1992) 787 and Surf. Coat. Technol. 65 (1994) 208.
[15] I. Yamada, et al., Nucl. Instr. Meth in Phys. Res. B 6 (1985) 439.
[16] A. Bousetta, et al., Nucl. Instr. Meth in Phys. Res. B 56–58 (1992) 480.
[17] K. Vanormelingen et al., Surf. Sci. 561 (2004) 147.
[18] C.R. Laurens, et al., Phys. Rev. Lett. 78 (1997) 4057.
[19] R.A. Zuhr, et al., Nucl. Instr. Meth. Phys. Res. B 59/60 (1991) 308 and Report CONF-00936–12 (1991).
[20] B.A. Probyn, J. Phys. D: Appl. Phys. 1 (1968) 457.

9.3.4 Compounds Films

The direct deposition of high-quality multicomponent films on different substrates was realized by different version of the mass separation IBD technology (Fig. 9.21):

(a) Dual ion beam system including deceleration and beam switching [124],
(b) Combination of two different ion beam lines and deceleration systems [125, 126],
(c) Combination of mass-separated IBD system with MBE [127, 128],
(d) Simultaneous [129, 130] or sequential [131] deposition of two different species from the same ion source by switching the mass separator to pass the different elements.

In Table 9.4 any experimental studies which have dealt with the energetic deposition of compound films on different substrate materials are summarized. Oxides, nitrides and carbides have been deposited, preferentially. Simultaneous irradiation with a mass-separated refractory metal ion beam (e.g., Ta, W, or Mo) and a reactive gas ion beam (oxygen or nitrogen ions) enabled the preparation of refractory compounds such as Ta_2O_5, WO, MoO_2 or TaN films on a Si substrate at room temperature [125, 132]. Moreover, crystallinity and stoichiometry of the deposited film are dependent on the deposition energy.

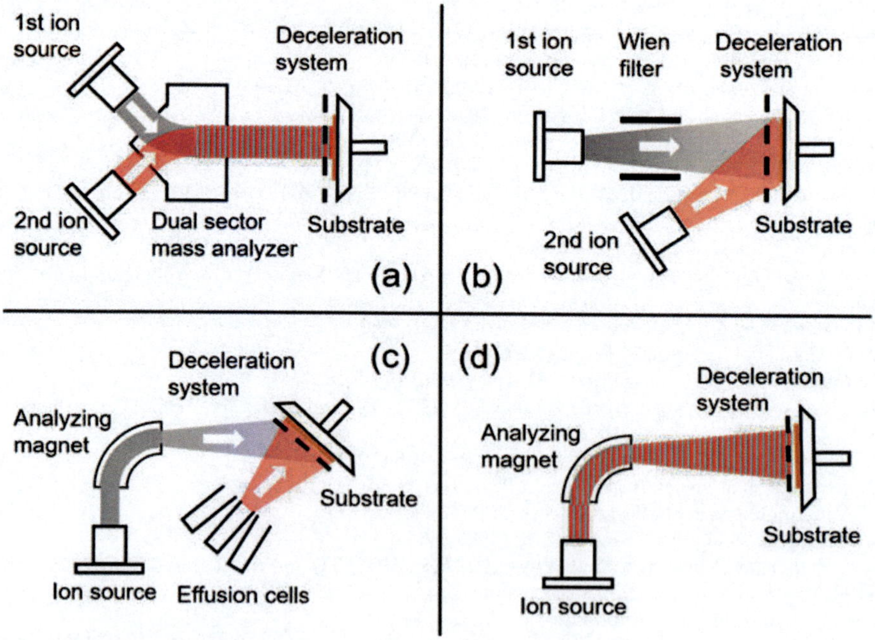

Fig. 9.21 Schematic diagrams of the different variants of the mass selection IBD for the formation of compound films

Table 9.4 Compound films prepared by direct ion beam deposition (Sap—sapphire, R—arrival ratio, x—composition ratio, ITO—indium tin oxide)

Film	Ion species	Substrate	Ion energy [eV]	Results	References
SiC	$SiC_3H_9^+$	Si(100)	10, 100, 200	Silicon carbide film for 20 eV, 3C-SiC films and diamond like carbon for 100 eV and 200 eV	1
	^{13}C, ^{30}Si	Si	40	Formation of 3C-SiC films	2
	C^+	Si(100)	50–300	Epitaxial 3C-SiC between 50 and 100 eV and 820 and 1000 °C	3
	$SiCH_3^+$	Si(111)	20–100	2H-SiC films at 20–50 eV 3C-SiC films at 60–100 eV	4
GaN	$^{69}Ga^+$,$^{14}N^+$	Si(111), Sap($1\bar{1}02$)	100	Stoichiometric crystalline GaN film at 600 °C and R between 1.5 and 2	5
InN	In^+, N_2^+	Si(100)	50	Non-stoichiometric indium nitride	6
TaN_x	N^+, N^+_2	Si	60–200	Film formation at RT	7
CN	N^{2+}, N^+ & C^-_1, C^-_2	Si(100),a-SiN_x	80, 125	Formation of amorphous carbon nitride films at RT	8
Ta_2O_5	Ta^+, O^+, O^+_2	Si	100–200	Pure and stoichiometric Ta_2O_5 films prepared at RT	9
ITO		glass	25–200	Polycrystalline ITO films at RT	10
$TiSi_2$	Ti^+	Si(111)	10–500	Stoichiometric films grown by 50 eV and 600 °C	11
InP	P^+	InP(100)	50–300	Homoepitaxial growth of InP films between 200 and 500 °C and between 50 and 300 eV	12, 13

(continued)

Table 9.4 (continued)

Film	Ion species	Substrate	Ion energy [eV]	Results	References
			100, 200	Homoepitaxial growth of crystalline films between 210 and 480 °C	14
		Si(100)	100	Mirror-like single crystal films at 375 °C	15
GaAsP	P^+	GaAs(100)	50–300	Vary between 0.5 and 0.67 with increasing ion energy for $GaAs_xP_{1-x}$	13
GaAs	$^{69}Ga^+$,$^{75}As^+$	Si(100), Ge(100)	30, 40	Stoichiometric, single crystalline GaAs films by 30 eV and 400 °C	16
	Ga^+, As^+	Si(100)	40–240	Single crystalline GaAs films by 40 eV	17
	As^+	GaAs	100, 200	Homoepitaxial growth of crystalline films between 220 and 450 °C	14
PbMg alloy	Pb^+, Mg^+	C	50	Polycrystalline alloy films	18

[1] S. Yoshimura et al., Nucl. Instr. Meth in Phys. Res. B 420 (2018) 6
[2] S.P. Withrow, et al., Vacuum 11/12 (1989) 1065.
[3] T. Miyazawa, et al., Appl. Phys. Lett. 45 (1984) 380.
[4] T. Matsumoto et al., Thin Solid Films 464–465 (2004) 103.
[5] F. Qin, et al., Rev Sci. Instr. 62 (1991) 2322.
[6] W.M. Lau, et al., Nucl. Instr. Meth. in Phys. Res. B 59/60 (1991) 316.
[7] Y. Yoshida, et al., Nucl. Instr. Meth. in Phys. Res. B 37/38 (1989) 866.
[8] D.Y. Lee, et al., Thin Solid Films 355-356 (1999) 239.
[9] Y. Yoshida et al, Jpn. J. Appl. Phys. 27 (1988) 140 and T. Ohnishi et al., Nucl. Instr. Meth. in Phys. Res. B 37/38 (1988) 850.
[10] D. Kim, S. Kim, Thin Solid Films 408 (2002) 218.
[11] S.M. Lee, et al., Nucl. Instr. Meth. in Phys. Res. B 157 (1999) 220.
[12] S. Maruno, et al, J. Cryst. Growth 81 (1987) 338 and Y. Morishita, et al., J. Cryst. Growth 88 (1988) 215.
[13] S. Murano, et al, Surf. Sci. 201 (1988) 335.
[14] S. Shimizu, et al. Jap. J. Appl. Phys. 24 (1985) 1130 and Jap. J. Appl. Phys. 24 (1985) L 115.
[15] S. Himizu, S. Komiya, J. Cryst. Growth 81 (1987) 243.
[16] T.E. Haynes, et al., Appl. Phys. Left. 54 (1989) 1439.
[17] S. Tamura, et al., Nucl. Instr. and Meth. B37/38 (1989) 862.
[18] J. Ahn et al., Nucl. Instr. Meth. in Phys. Res. B 17 (1986) 37.

Hayes and coworkers [129] have studied the feasibility of growing isotopically pure, single-crystal compound semiconductor layers at relatively low temperatures (400 °C) by deposition of alternating, fully ionized, low-energy beams. For example, GaAs films were grown by IBD using repetitive switching between $^{69}Ga^+$ and $^{75}As^+$ ion beams extracted from a single ion source [129] and InN films are deposited by a combination of mass-separated IBD system with MBE (see Fig. 9.21d) [133]. Single-crystalline GaAs films with lattice parameters near the bulk crystal were deposited on Si substrates by using Ga^+ ion beams simultaneously with As ion beams [114].

In particular, the formation of the hexagonal compound semiconductor 3C-SiC (β-SiC) at low temperature has been intensively studied using different carbon containing ion beams [130, 134]. Matsumoto et al. [134] have deposited SiC films on Si(111) substrates for the purpose of investigating momentum effects on the film growth. With an ion energy ranged between 60 eV and 100 eV 3C-SiC(111) was grown at 750 °C. At the same temperature, 2H-SiC was prepared by irradiation of molecular ions between 20 eV and 50 eV. Probably, the momentum of the energetic ions influences the formation of the crystal structure.

The formation of oxides and nitrides of Si, Ge and $Si_{1-x}Ge_x$ has been studied in depth using the implantation of O^+ and N^+ ions beams and the subsequent chemical reaction of the ion species with target atoms [24–27]. The experimental results demonstrate a strong dependence of the phase formation, the stoichiometry and the film thickness on the ion energy. On the base of these results different models with special insistence on the role of the ion energy were developed, which discuss the influence of implantation and sputtering of the implanted ion species, the displacement processes and the oxygen diffusion. The oxide formation by oxygen ion beam deposition shows increasing oxide film thickness with higher ion energy and can be interpreted in three stages. Frist, the formation of substoichiometric oxides, then the conversion into stable oxides (e.g., SiO_2) und finally a saturation of the growth rate, i.e. formation of oxide films for all energies and temperatures. SiO_2 films, for example, show a high-quality interface and no damages in the Si substrate. The oxide formation after oxygen ion deposition on germanium is similar to the one observed for Si oxides [24]. Initially, the growth of stoichiometric GeO_2 can be observed until the growth rate is saturated, i.e., the formation of germanium oxide films is completed. This process is strictly dependent on the oxygen ion energy. Below 200 eV, GeO_2 is formed while at energies $\geq$ 500 eV GeO is only formed. For high energies (> 200 eV) the sputtering rate of GeO_2 is higher than the formation rate.

Direct nitrogen ion deposition on crystalline Si surfaces leads to the formation of Si_3N_4, which grows on an amorphous Si interlayer between the silicon nitride film and the Si substrate. The thickness of the amorphous layer grows with decreasing ion energy and increases with increasing ion fluence. In contrast to film oxidation, it was established that diffusion is the determining factor of silicon nitride formation. The stoichiometry of Si nitrides is depended on both ion energy and temperature. Germanium nitrides could be observed for nitrogen ion energies smaller than 100 eV only. It is assumed that the small enthalpy of formation for germanium nitrides and negative dissociation energy for germanium nitrides prevent the phase formation and facilitate the dissociation.

9.4 Molecular Thin Film Deposition by Soft Landing

In the last subchapters were demonstrated that the IBD technique is becoming one of the common thin film deposition techniques for the preparation of high-quality thin inorganic films. The different variants of this technique rely on the generation of atomic and/or molecular ions in an ion source, the acceleration to hyperthermal or low energies and the deposition of these particles on a substrate surface.

However, this technique is not applicable to a large fraction of functional organic molecules that cannot enter the gas phase due to their thermal instability or lack of vapor pressure. Presently, only small molecules can be deposited by means of the molecular beam epitaxy. Larger molecules typically exhibit lower vapor pressures and are unstable at elevated temperatures, with the consequence that they decompose before sublimation. Thus, the volatility of the atoms and molecules presents an intrinsic limitation, which particularly prevents large molecules such as polymers, dendrimers, biomolecules, and many other complex synthetic functional molecules with large mass. Consequently, the deposition of larger organic molecules including formation of molecular layers on a substrate is unrealistic.

In a seminal study by Cooks and coworkers [10], a new technique to modify surfaces and to deposit organic films of polyatomic ions was introduced. This technique, called soft landing (see Sect. 9.1.2), is characterized (i) by generation of mass separated polyatomic ions and (ii) deposition of these molecular ions. These ionized molecules are produced by methods as electrospray (ESI) or matrix-assisted laser desorption/ionization (MALDI), where the ESI provides a continuous and intensive ion current in the nA range. Polyatomic ions produced by these techniques can be non-destructively deposited on the surface of a substrate, when the energy is lower than the activation energy for collision induced dissociation. Typically, energies of the ionized molecules formed into an ion beam are in the hyperthermal energy range up to any ten electron-volts, i.e. comparable to the chemical binding energies. For example, in the last decade numerous large complex ions including metalorganic complexes [135, 136], proteins [137–141], peptides [142–148], DNA [149], viruses [150, 151], saccachrides [152–154] as well as metallic and semiconducting nanoparticles [142, 151] could be deposited without fragmentation. Consequently, a thin film deposition technique with huge application potential is available.

9.4.1 Deposition of Electrospray Ion Beams

Electrospray ion beam deposition (ES-IBD) in the vacuum is a method capable to create intact gas phase ions of large molecules and deposit them on solid substrates in vacuum. With the objective to create a beam of ionized molecules, a differentially pumped vacuum system must be available, which is used (i) to generate ionized molecules at ambient pressure conditions by electrospray ionization, (ii) to control the selected ion mass by a mass separator and (iii) to guide the intact molecules to a sample surface in (ultrahigh) vacuum, where the molecular ions are deposited on a substrate [29, 155].

9.4.1.1 Electrospray Ionization

Fenn et al. [156] introduced electrospray ionization to ionize chemical species intact from solution by multiple charging for the mass spectrometry. The ionization is soft because no fragmentation of molecules is caused by ionization. In general, a dilute analyte solution is injected by a mechanical syringe pump through a hypodermic needle or steel capillary (Fig. 9.22). A high voltage of some kilovolts is applied to tip of the capillary relative to the vacuum inlet conduit. Depending on the polarity an electron flow occurs towards or from the emitter needle, which results in positive or negative charging of the solution in positive or negative ion mode, respectively. The accumulated charge drifts into the high-voltages capillary in direction of it's apex where a so-called Taylor cone is formed [157]. Here, under the influence of high electrical field (~10^6 V/m), an aerosol jet of highly charged electrospray droplets is emitted. These droplets become instable as they shrink due to rapid evaporation of solvent and are resolved into smaller droplets by Coulomb explosion on the way to the counter electrode. Two counteractive forces are responsible for the dimension of the final droplets [158, 159]. On the one hand, the surface tension of the charged droplets intends to conserve the spherical shape of the droplets. On the other hand, the Coulomb force of repulsion between the identical charges at the surface of the droplets destabilizes these droplets. Consequently, the resulting size of the droplets is given by the so-called Rayleigh limit, defined as the condition where surface tension is overcome by the Coulomb force of repulsion. This process of Coulomb fission is repeated multiple times, until finally nanometer sizes droplets or desolvated, charged molecules/clusters are obtained [160]. The time of few milliseconds between Coulomb fission events is dependent on the applied voltage, nature of the solvents, flow rate, and the distance between the nozzle and the counter-electrode.

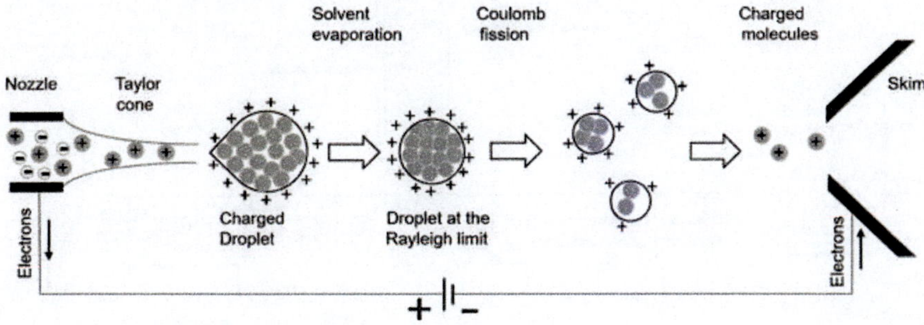

Fig. 9.22 Scheme of the electrospray ionization

9.4.1.2 Experimental Setup of Electrospray Deposition

In order to perform deposition experiments with large organic and biological molecules the experimental setup is characterized by a combination of different techniques: electrospray ionization as the ion source, differential pumping to overcome the gap between atmospheric pressure and ultra-high vacuum, ion optics to guide the beam, a mass spectrometer to monitor the beam composition, and finally a gentle vacuum deposition.

Figure 9.23 shows a schematic picture of an electrospray deposition system. Ions generated by pneumatically assisted electrospray ionization are transferred into vacuum through a metallized glass capillary. With the object of a better nebulization and to align the spray, dry gas (frequently N_2 gas) flows around the spray needle and the heated capillary, respectively (see green arrows in Fig. 9.23).

In the first vacuum stage, the ion beam is skimmed after supersonic expansion into vacuum at a pressure of some hundred Pa. In this chamber the voltage difference between nozzle and skimmer defines the breakup potential (see Fig. 9.22). This parameter can be used to adjust the collision intensity of the ions with the background gas, which may lead to nozzle-skimmer fragmentation. Alternatively, in more recent instruments, here ion funnels are employed to collimate the ion beam, yielding larger overall ion currents [162–164].

Several quadrupole arrangements in the next vacuum stages are applied to form an ion beam and guide the ionized molecules, where the last quadrupole is used to select the mass-to-charge ratio of particles that will be transmitted. In the last vacuum stage a mass spectrometer (e.g., time-of-flight mass spectrometer) and the sample holder are installed. Typically, the mass ranges of a linear TOF-mass spectrometers varies between 1 Th and 20 000 Th (1 Th = 1amu/e, where Th is the unit Thomson

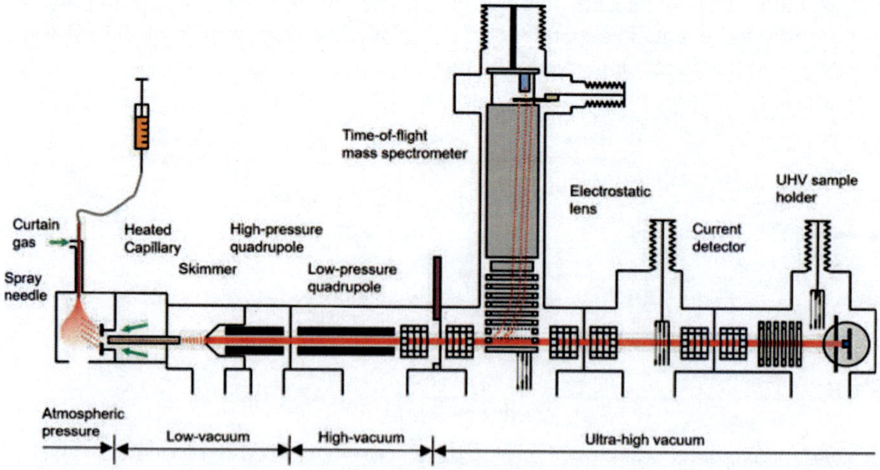

Fig. 9.23 Schematic of the electrospray deposition setup (figure adapted from [161])

and amu (or u) is the unified atomic mass) at mass resolution $M/\Delta M \approx 1000$. Sample holders are individually designed and introduced through a load-lock. Frequently, the deposition chamber is equipped with surface analysis techniques (e.g., scanning tunneling microscopy) for an in-situ characterization of the deposited material. The total fluence of the deposited molecular charges is accurately measured on the sample with an electrometer with an accuracy in the pA range. The kinetic energy of the ionized molecules can be controlled by applying a bias voltage to the sample.

9.4.2 Examples for the Deposition of Molecular Films

Besides the successful application of the soft landing of molecules of individual non-volatile molecules on solid surfaces into the UHV for numerous analytical applications (see [165]), this deposition technique can be also applied to fabricate very thin molecular films, effectively [166].

In early studies, Usui et al. [167] have prepared anthracene films on glass substrates with a preferential orientation (001) plane parallel to the substrate and polyethylene films with a high density and lattice dimension significantly close to that of single crystal polyethylene. Based on MD simulation, Qi et al. [168] could demonstrated that hyperthermal collisions of ethylene and acetylene clusters with diamond surfaces indicate stable networked polymer structures.

Another example is the formation of a molecular double layer of soft landed sodium dodecyl sulfate (SDS) on graphite and silicon oxide surfaces [169]. SDS films of submonolayer up to multilayer coverage grown as stacked layers with increasing deposition duration as flat compact islands of some nanometer heights (Fig. 9.24). The island distribution is homogeneous in size and density. The closed molecular film is the result of diffusion and agglomeration of islands with a lateral dimension between 50 nm and 500 nm. It is known that SDS layers grown in an oriented arrangement, where the molecules are ordered in a head-to-head manner (Fig. 9.24d) due to the missing hydrophilic-hydrophobic interaction in vacuum The resulting bilayer order leads to a thickness of the bilayers of about 3.8 nm. Further analysis with X-ray diffraction and scanning tunneling microscopy show that these layers are crystalline.

Kappes and coworkers [170] have deposited thin films of fullerenes by reactive soft landing of mass-selected stable C_n cations at room temperature. The molecular building blocks were generated by electron impact induced ionization of sublimed C_{70} or C_{60}. Figure 9.25 shows AFM images C_{58} and C_{60} layers deposited onto HOPG under otherwise identical conditions. Whereas C_{60} forms compact islands terminated by smooth rims always associated with HOPG step edges, C_{58} arranges in fractal-like islands which decorate not only step edges but also flat. Interestingly, annealing experiments indicate that C_{58} films are thermally more stable than films of C_{60}. The morphology of the resulting films was shown to be strongly dependent on the kinetic energy of the ions and the temperature of the surface. Larger islands were formed at low kinetic energies and low surface temperatures indicating that the kinetic energy

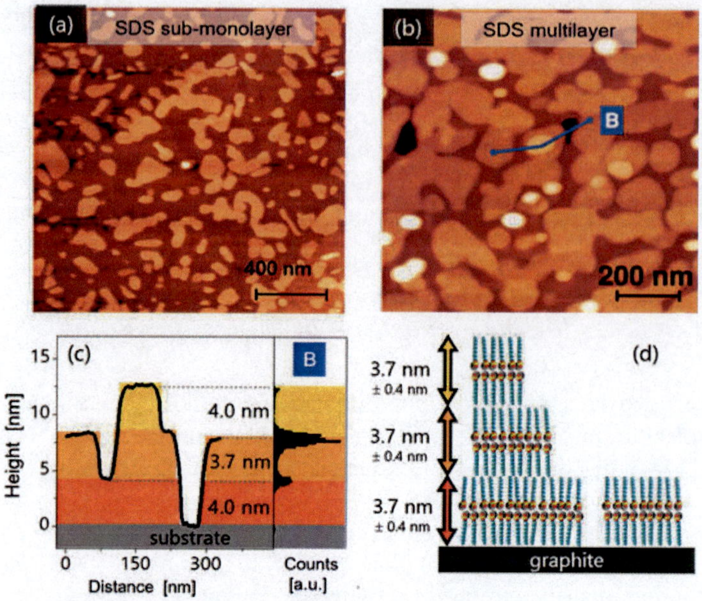

Fig. 9.24 **a** SDS layers on graphite substrate (adapted from [169]), **b** AFM image of a sub-monolayer, **c** AFM image of multilayer SDS islands (cross section indicated; inset: edge filtered image; red arrows pointed out straight edges), **d** cross section (indicated B) of the SDS layer arrangement and the height histogram derived from the whole AFM image, **e** schematic model of layer formation for SDS on SiO_x

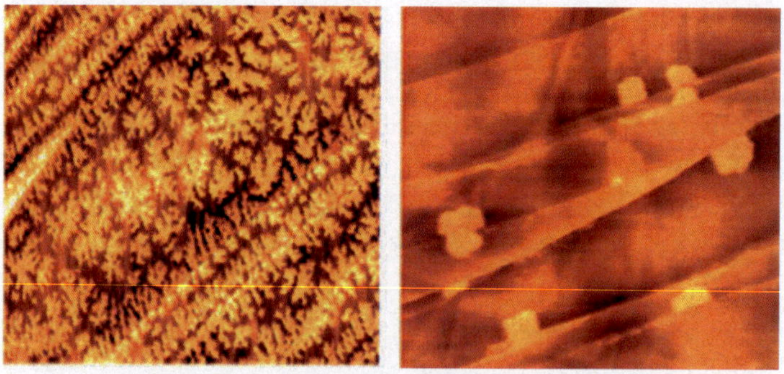

Fig. 9.25 AFM images of the C_{58} (left) and C_{60} films onto HPOG (image size is $3 \times 3\ \mu m^2$). Both films were deposited at room temperature, an ion energy of 1 eV and an ion flux of 10^{11} ions/cm^{2}s. The thickness of the films was about 1 nm (figure is adopted from [171])

of the ion must be efficiently dissipated prior to the formation of covalent bonds with other molecules present on the surface.

It is obvious, that versatility of soft landing technique has triggered the preparation of thin molecular organic and inorganic films in the last years.

9.5 Ion Beam-Induced Cleaning of Surfaces

In the previous chapters, it was demonstrated that film growth can be realized if the ion energy is chosen such that the sputtering yield during IBD process is less than one and the reflection of the layer-forming ions is small, i.e. the growth of thin layers can be understood as a competition between deposition and sputtering, both of which are energy-dependent effects. Consequently, an increase in ion energy (i.e. an increase of sputtering yield) will result in a suppression of film growth and ultimately to erode the near-surface regions. Based on this concept, the method of surface cleaning by ion irradiation was developed. It is characterized by the removal of surface contamination, adsorbates or compound layers (e.g., oxides) on the surface by low-energy ion bombardment without damaging the underlying structure.

In general, surfaces are characterized by adsorbed water vapor, hydrocarbons and other gas atoms and the structure of the surface is disturbed by defects. These layers can be caused by adsorption of the residual gas from the environment and impurities deposited on the surface at elevated temperatures, or by other pretreatment of the material. Adsorbates can be physically or chemically adsorbed. The physisorption is due to Van der Waals forces with typical binding energy < 0.5 eV, while chemisorption is characterized by bonds via exchange or sharing of electrons with typical binding energies of a few electron-volts.

For numerous applications, however, clean and undisturbed surfaces are an absolute necessity. In the past, low-energy sputtering, often in combination with thermal processes, has proven to be an ideal method to achieve near-perfect surfaces. This procedure was first applied in 1950 (see e.g., [172]) and is based on the idea that adsorbates or contaminated films on the surface are removed layer-by-layer. The momentum transfer as a result of an ion impact can either knock-off the adsorbed particles directly, or form a collision cascade, which then can knock-off the adsorbed atoms or molecules on the surface. Thermal treatment is frequently necessary to anneal the defects caused by ion bombardment. Unlike other cleaning processes, the ion sputtering process is easy to incorporate into high vacuum systems. Important system parameters such as ion species (reactive or inert), flux and energy can be independently controlled.

9.5.1 Ion Beam-Induced Cleaning Process

The energy necessary to desorb a particle from the adsorbate can either come from the solid (thermal desorption) or from external sources. As an example for the latter, ion beams are employed to cause ion beam-stimulated desorption. While the thermal desorption of adsorbates is determined by the heat of adsorption (which corresponds approximately to the depth of the surface potential), the ion beam-induced desorption is governed by the parameters of the impact process (ion energy, incident angle, masses of the collision partners, etc.). For the quantification of desorption by ion bombardment, the desorption yield is used, i.e. the average number of desorbed molecules or atoms per incident ion.

Noble gas ions with energies up to a few kilo electron-volts are probably most often used for sputter cleaning. The removal of adsorbed atoms or contaminants is based on the sputtering effect (see Chap. 5). The ability to precisely control this process is extremely advantageous. Disadvantages can be the implantation of the incident ions, the generation of irradiation-induced defects and recoil implantation of adsorbed atoms into the substrate as well as possible interface mixing at the interface between the adsorbed layer and the substrate (Fig. 9.26). In order to avoid these unwanted effects, various thermal procedures which are often carried out in sequence or sometimes simultaneously, have been used to achieve clean surfaces. Frequently, undesirable processes cannot be completely avoided. Therefore, only optimal conditions for cleaning surfaces by ion irradiation will be published.

The cleaning procedure starts with ion bombardment of the adsorbed surface. Typical ion energies between 100 eV and a few kilo-electron-volts and ion current densities smaller than 100 $\mu A/cm^2$ are used. After a certain time of the ion bombardment, the procedure is stopped and any remaining contaminations as well as the atomic structure of the surface is controlled. It should be noted that the resolution of all surface analysis methods is limited (e.g., the sensitivity of AES for contamination is usually not better than 10^{-3} monolayers and LEED pattern does not give much information about long-range order extending over distances longer than the coherence length of the electrons (typically 10 nm [173]). A subsequent thermal treatment is necessary to remove embedded and adsorbed noble gas atoms, to anneal the irradiation-induced defects and to restore the crystallographic structure of the surface. In general, the recovery process is carried out at homologous temperatures $> 0.6 \ldots 0.85$. The whole procedure may be performed several times. The process has proven successful for single-component materials. For multi-component materials, such as alloys, oxides or compound semiconductors, ion irradiation often generates enrichment of a component near the surface (preferential sputtering, Sect. 5.8) and

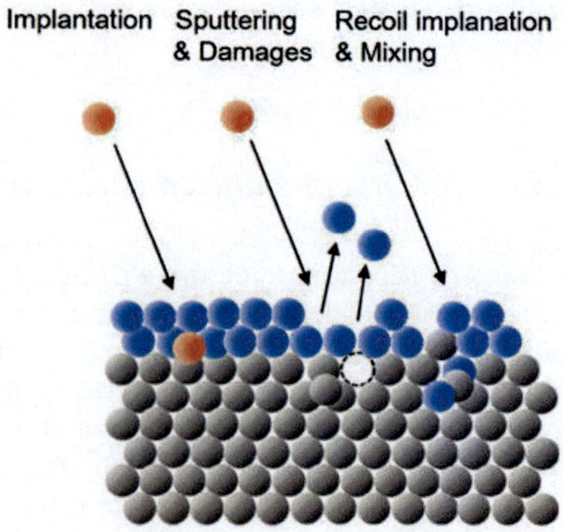

Fig. 9.26 Schematic representation of the removal of the adsorbed layer and the most common unwanted processes occurring in this context (see text for details)

thermal treatment causes diffusion processes. This means that ion bombardment cleaning can only be used to a certain extent for these materials.

A distinction is made between two basic ion beam-assisted cleaning approaches [174]. On the one hand, initial cleaning of the surface before the next process step (e.g., before deposition of epitaxial layers) and, on the other hand, continuous cleaning during the various process steps (e.g., during the deposition process).

The experimental studies of ion beam-stimulated cleaning of surfaces are mainly focused on (i) desorption of adsorbates and (ii) removal of oxides. In the following, some selected examples of surface cleaning with low-energy ion irradiation for both cases will be presented.

With the aim of cleaning Ge and GaAs wafers from adsorbed contamination (carbon, water, etc.) and the native oxide film, the surface of these materials were irradiated with hydrogen ions (300 eV, 4.5 $\mu A/cm^2$) under high-vacuum conditions (details see [175]). During the ion bombardment, the wafers were heated on 300 ± 20 °C. The surface composition of Ge was investigated before and after bombardment using X-ray photoelectron spectroscopy (XPS) with a monochromatic MgKα radiation. Figure 9.27 shows the XPS peaks of germanium, oxygen and carbon before H ion bombardment, after 10 min and after 25 min H ion bombardment. Strong peaks for carbon and oxygen are observed on the sample surface before H ion bombardment. The peaks of carbon and oxygen after hydrogen ion bombardment at elevated temperature disappear, completely. Consequently, due to hydrogen exposure at elevated temperature, the Ge samples were free of absorbent contamination, indicating clean and active surfaces.

This ion beam technique is capable of removing oxides from GaAs surfaces without inducing significant near surface damages and/or changes of the surface composition. XPS studies have shown that as-received surfaces are characterized

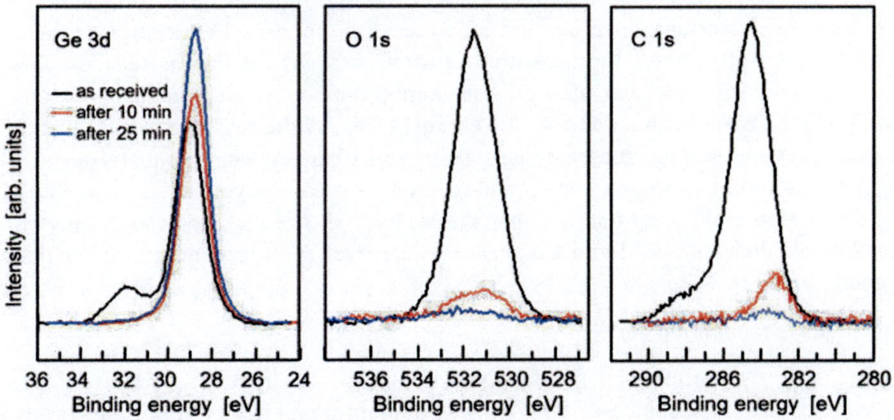

Fig. 9.27 Ge 3d (left), O 1 s (middle) and C 1 s (right) core level spectra of Ge surface before hydrogen ion bombardment, after H ion bombardment at 300 °C for 10 min and after H ion bombardment for 25 min. The ion energy was 300 eV, the ion incident angle was 22° and the ion current density 4.5 $\mu A/cm^2$

by Ga and As oxides [175]. A large change in surface composition as well as the existence of oxygen can be detected after hydrogen ion bombardment. In detail, after H ion bombardment at 150 °C for 410 s, both As and Ga oxides are completely disappeared. The mechanisms for the removal of oxides from the as-received GaAs (001) surface by bombardment with hydrogen ions are the chemical reaction between ionic or atomic hydrogen as well as As and Ga oxides. On the one hand, these reactions lead to the formation of molecular arsenic, which is volatile under vacuum conditions, and on the other hand, Ga reacts with As to form GaAs.

Especially in the production of Si-based devices, the cleaning of Si surfaces by removing the oxide layer is absolutely necessary before further technological steps, such as the deposition of layers or doping. There are several methods to obtain atomically clean surfaces. Among them, cleaning by low-energy ion irradiation is expected to be very promising. Recently, Sharopov et al. [176] proposed a technology for low-temperature ion beam-assisted cleaning of Si wafers leading to atomically clean surfaces. These atomically clean surfaces (7×7 Si structure) were obtained by 2.5 keV bombardment with positive cesium ions and ion fluences between 10^{13} ions/cm^2 and 10^{15} ions/cm^2 followed by thermal treatment at a temperature of 650 °C in ultra-high vacuum.

Similarly, thin oxide layers on the surface of other materials such as Si [177–180], SiGe [177] and metals [181] have been successfully performed at elevated temperatures without introducing permanent damage using ion beam-induced cleaning.

The ion beam-stimulated desorption was studied in dependence of the parameter of ion bombardment, where, in general, the desorption yield was studied as a function of the following parameters.

(1) *Energy of the incident ion*

In many experiments it was found that with increasing energy of the incidence particles the desorption cross-section increases and therefore the desorption yield. Figure 9.28 (left) shows the measured desorption yield for commonly adsorbed molecules on steel surfaces after Ar ion bombardment. As expected, the desorption yield increases with increasing ion energy, whereby the increase is significantly smaller for heavier molecules. Similar results were obtained for other metal surfaces such as aluminum, titanium, copper and Inconel. For ion energies larger than 3 keV, the desorption yield is saturated. Other studies have shown that the desorption yield for these molecules on Al and Cu surfaces decreases with ion energy higher than some keV [182].

(2) *Incidence ion fluence*

Figure 9.28, right, illustrates the desorption yield for different adsorbed molecules on Al and Cu samples due to 5 keV Ar ion bombardment [183]. For both materials, the desorption yield decreases as a function of the gas species as follows: $H_2 > CO > CO_2 > C_2H_6 > CH_4$. This behavior has also been observed at 3 keV and 7 keV Ar ion bombardment. Note that the desorption yield tends to remain constant for high ion fluences. This is the result of a balance between re-adsorption gas species from the

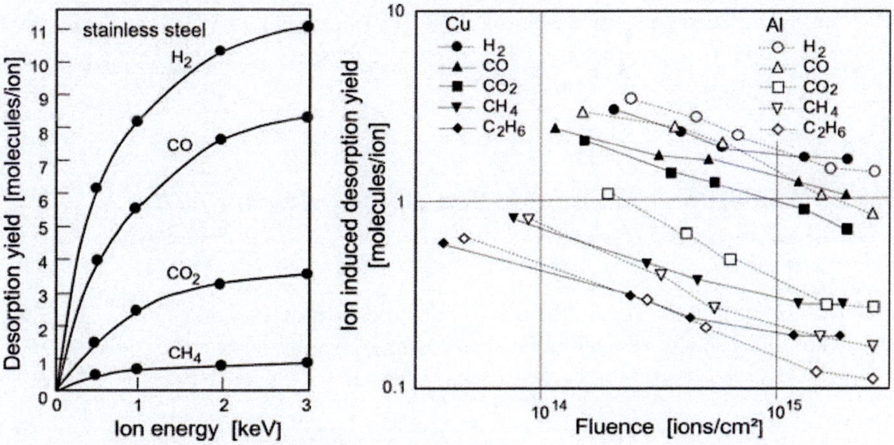

Fig. 9.28 Left: Desorption yield of several molecular adsorbates in dependence of Ar ion energy for bombardment of stainless steel (figure is adapted from [182]). Right: Ion-induced desorption yield as function of ion fluence. Aluminum and copper samples as-received bombarded with argon ions at 5 keV and incident current between 0.05 μA and 0.5 μA (figure is adapted from [183])

residual vacuum and the ion beam-induced desorption. It should be also noted that the total desorption yield for as-received Cu and Al in the order of 8 gas molecules per incident Ar ion is approximately halved after annealing at 200 °C for 24 h.

(3) *Mass of the incident ions*

Hilleret [184] has measured the influence of the incident ion mass on the desorption yield for stainless steel at different sample temperatures and after annealing at 300 °C. With increasing ion mass, the desorption yield also increases, regardless whether the samples were thermally treated or not. This trend was also evident in the studies by Achard [185] on the desorption behavior of aluminum and titanium after 2 keV N_2 and Kr ion bombardment.

(4) *Temperature dependence of ion-stimulated desorption*

The desorption of gas molecules per incident Kr ion for stainless steel, Inconel, titanium alloy, copper and aluminum were studied in detail as function of the annealing temperature ranging between 250 °C and 600 °C [186]. Within this temperature range, for all metals except copper, the ion-induced desorption yield for the C and O based molecules change between one and three orders of magnitude. For Cu, the decrease of the ion beam-induced desorption is significantly stronger.

It should be noted, that in addition to the initial coverage of the surface with an adsorbate layer, the surfaces are exposed to a continuous flow of atoms from the residual gas. This residual gas flux causes a continuous re-formation of the adsorbate layer, which can only be prevented by a sufficiently high ion current and low residual gas pressure. Assuming that typical ion current densities vary between 10 and 100 μA/cm^2 and that the surface density of an adsorbed monolayer is about

10^{15} atoms/cm^2 and has a thickness of about 0.5 nm, then a residual gas pressure of 10^{-6} to 10^{-7} Pa (i.e. pressure in UHV range) is necessary to avoid a deposition rate of the adsorbate layer > 0.05 nm/s [187].

9.5.2 *Models for Cleaning by Ion Bombardment of Thin Adsorbate Layers*

In a first approximation (i.e. in the case that the desorption processes of the adsorbed atoms are independent of each other) the coverage is defined by

$$\overline{\theta} = \frac{N_{ad}}{N^o_{ad}}, \tag{9.7}$$

where N_{ad} is the number of adsorbed atoms (or molecules) per unit area on the surface of the substrate and N^o_{ad} is the number of adsorbed atoms (or molecules) per unit area at maximum monolayer coverage (number of lattice sites per unit area at the surface).

9.5.2.1 Initial Cleaning of Adsorbed Monolayers

The description of the ion beam-assisted initial cleanup process is well known for extremely thin adsorbate layers (see e.g., [188–190]). The time of the erosion of a monoatomic adsobate layer can be estimated using the relation $t_s = N^{2/3}/JY$, where N is the atomic number density, J is the ion current density and Y is the sputtering yield. Assuming that the area density $N^{2/3} = 10^{15}$ atoms/cm^2 and an ion current density of 10 μA/cm^2, an irradiation time $t_s \approx 2.7$ min is required for a sputtering yield of 1 and a time $t_s \approx 1.3$ min is necessary for a sputtering yield of 2 to completely remove an adsorbed monolayer.

It can be assumed that the erosion of adsorbed atoms on a substrate by ion bombardment can be described by the following equation

$$-\frac{dN_{ad}}{dt} = J\sigma_{sp} N_{ad}(t), \tag{9.8}$$

where N_{ad} is given by (9.7), J is the current density of the incident ions and σ_{sp} is the sputtering or desorption cross-section. Thus, the ion beam-induced erosion of the adsorbed layer is given by

$$N_{ad} = N^o_{ad} \exp(-J\sigma_{sp} t) = N^o_{ad} \exp(-\sigma_{sp}\phi), \tag{9.9}$$

where N^o_{ad} is the initial surface concentration of adsorbed atoms (typically in the range of 10^{15} atoms/cm^2) and Φ (= J·t) is the ion fluence.

The value of sputtering cross-sections can be obtained from measurements such as AES, ISS, XPS etc. that monitor the surface composition. Examples are given in Fig. 9.29. Chemisorbed nitrogen on W crystals was sputtered by noble gas ion bombardment in the energy range between 25 eV and 500 eV [188]. An exponential decay can be detected over almost two orders of magnitude. This behavior is described by (9.9). Then, the sputtering cross-section can be determined from the slope of the N_{ad}/N^o_{ad} ratio versus the ion fluence (see Fig. 9.29).

Assuming that the sputtering yield does not vary with the coverage of the surface by an adsorbate, the cross-section is related to the sputtering yield through the surface density, N_{ad}, is given by [190].

$$Y = \sigma_{sp} N_{ad}. \tag{9.10}$$

Winters and Sigmund [188] have assumed that three mechanisms contribute to the sputtering of an adsorbed thin layer on a substrate (Fig. 9.30). Therefore, three different contributions to the total sputtering cross-section, σ_{sp}, could be expected, where the cross-sections were determined by using a power potential for the interaction between two collision partners (see 2.76 in Sect. 2.2.4):

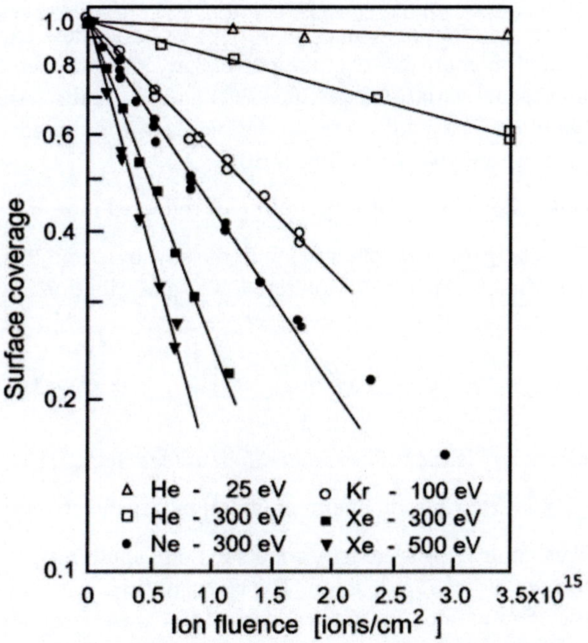

Fig. 9.29 Surface nitrogen coverage N_{ad}/N^o_{ad} on tungsten versus ion fluence (figure is taken from [188] and modified)

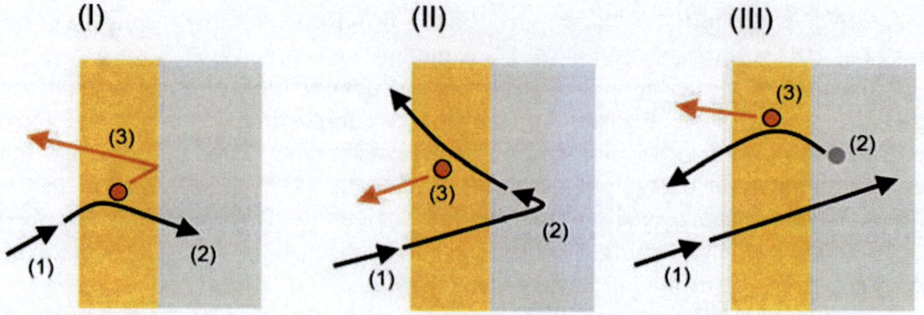

Fig. 9.30 Schematic of the sputtering mechanisms of substrates (grey) with an adsorbed layer, where (1) is the incident ion, (2) is the substrate and (3) is the adsorbed layer

(I) Contribution of direct knock-off by the incoming ion, σ_I:

The incoming ion hits an atom of the adsorbed layer and is reflected from the substrate lattice. At oblique ion incidence, the adsorbed atom can be knocked off the surface directly. The sputtering cross-section for the direct knock-off contribution is given by

$$\sigma_I = \frac{1}{\cos\theta}\frac{C_{13}}{m_{13}}\left(EU_{s,3}\right)^{-m_{13}}\left[1 - x^{-m_{13}}\right], \tag{9.11}$$

where θ is the incidence angle, $U_{s,3}$ is the surface energy of the adsorbed layer, E is the ion energy, C_{13} is the power law constant (see 2.77), $x = \gamma_{13}\, E/U_{s,3}$, m_{13} is power law variable (see Table 2.2) and γ_{13} is the transfer energy efficiency factor. Numbers 1, 2 and 3 refer to the incident ion, the adsorbed atom and the substrate atom, respectively (c.f. Fig. 9.30).

(II) Contribution by sputtering of reflected ions, σ_{II}:

The energetic ion penetrates into the substrate and can be reflected. On its way out, adsorbed atoms can be sputtered. This contribution is given by

$$\sigma_{II} = \sigma_I R_{12}\left[4 - \frac{4\ln x}{3x^{1/3} - 3}\right]\cos\theta, \tag{9.12}$$

where R_{12} is the reflection coefficient (see Sect. 5.10).

(III) Contribution due to momentum transfer from sputtered substrate atoms, σ_{III}:

This contribution describes the substrate sputtering as result of a collision cascade formed by the incident ion. The outward flux of the atoms involved into the cascade transfer momentums to the adsorbed atoms which can be sputtered. This contribution can be determined using Sigmund's sputtering theory and is given by

$$\sigma_{III} = 4\gamma_{23} C_{23} \frac{U_{s,2}}{U_{s,3}} \frac{1 - (1 + \ln y)/y}{(1 - \gamma_{23}/y)^2}, \tag{9.13}$$

where Y is the substrate sputtering yield and $y = \gamma_{12}\gamma_{23}(E/U_{s,3})$. The constant C_{23} can be expressed by

$$C_{23} = \frac{1}{2}\pi\lambda_m a \left(\frac{M_2}{M_3}\right)^m \left(\frac{2Z_2Z_3e^2}{a_B}\right), \tag{9.14}$$

where m is the parameter of the inverse power potential, λ_m is a constant given in Table 2.2, and a_B is the Bohr screening length (Sect. 2.1.1.1).

Consequently, the total sputtering cross-section or desorption cross-section is given by

$$\sigma_{Sp} = \sigma_I + \sigma_{II} + \sigma_{III}. \tag{9.15}$$

Typically, total sputtering cross-section between 10^{-14} cm^2 and 10^{-17} cm^2 were determined [191]. The first two contributions to the cross-sections decrease with increasing ion energy while the last contribution increases with increasing ion energy. Consequently, only a slight energy dependence of the total sputtering cross-section can be observed. Winters and Taglauer [192], for example, have compared the experimental values of sputtering cross-section for adsorbed nitrogen on Mo(100) and W (100) with the values calculated by (9.15). They found that a monolayer adsorbate, particularly when this is a monatomic species, can be described by the existing model. The desorption yield can be obtained by multiplying the cross-section by the number of adsorbed atoms per unit area according to (9.10).

9.5.2.2 Effects Connected with Sputter-Induced Cleaning

It can be expected that the cleaning of surfaces by ion irradiation leads to some side-effects (c.f. Fig. 9.26). An effective and practicable ion beam-induced cleaning technology is characterized by a significant reduction or elimination of these side effects. The most significant side effects are [192]:

(i) Ion implantation and associated radiation damages

Ion bombardment of surfaces with the aim of cleaning is also associated with the implantation of these ions and the formation of defects on and beneath the surface (see Sect. 3.5 and Chap. 4). Concentrations of implanted particles up to a few atomic percent can be achieved.

Therefore, ion bombardment at higher temperatures or a subsequent thermal treatment is often carried out after ion irradiation. This procedure aims at desorbing implanted ions and annealing radiation damages.

(ii) Recoil implantation

It should also be noted that ion irradiation not only leads to sputtering of adsorbed atoms and molecules into the vacuum, but also to displacement of adsorbed atoms into the substrate. This recoil implantation process is characterized by an additional recoil cross-section, σ_R. According to Morita et al. [193] and Inoue [194], this process is particularly important for heavy adsorbates on light substrates. Then, the simple equation of the surface coverage after ion bombardment, (9.9), must be expanded to an expression containing the recoil implantation,

$$N_{ad} = N_{ad}^{o} \frac{\sigma_R}{\sigma_{Sp} + \sigma_R} \{1 + \exp[-(\sigma_{sp} + \sigma_R)]\}. \tag{9.16}$$

For example, an experimental study of sputtering of CO from Ni surfaces with 1 keV Ne ions [195] has shown that after removal of about 90% of the initial coverage a deviation from the exponential decrease (smaller slope) due to recoil implantation can be observed (c.f. Fig. 9.29). This indicates that about 5–10% of the adsorbed particles removed from the surface are displaced inward into the crystal.

(iii) Preferential sputtering

For ion beam-assisted cleaning of multicomponent material surfaces an additional effect must be considered, the preferential sputtering (details see Sect. 5.8). It has been shown that the cleaning of surfaces of multicomponent substrates by ion bombardment inevitably leads to a change of composition of the surface and the region near the surface. Therefore, surfaces of these materials must be cleaned by other methods.

9.5.2.3 Cleaning of Adsorbed Multilayers

In the previous subsections, it was assumed that the adsorbed layer consists of a monolayer or if an adsorption layer consists of several layers, the adsorbed multilayers can be removed layer-by-layer. This is an idealized view, because the atoms of the underlying layer (second monolayer) are already partially removed even before the atoms of the overlying layer (first monolayer) are completely removed. A simple model for successive layer removal was proposed by Benninghoven [196]. Assuming that (i) the solid is separated into monolayers (sputtering is possible only from the upper atomic layer), (ii) the sputtering is independent of the sputtering time and (iii) a uniform ion bombardment takes place, the remaining portion of the adsorbed layer n at time t is given by

$$\left(\frac{N_{ad}}{N_{ad}^{o}}\right)_n = \bar{\theta}_n = \frac{1}{(n-1)!}\left(\frac{t}{\tau}\right)^{n-1} \exp\left(-\frac{t}{\tau}\right), \tag{9.17}$$

where coverage $\left(N_{ad}/N_{ad}^{o}\right)_n$ being the fraction of the atoms belonging to the atomic layer n (for the top or first monolayer is n = 1, second layer is n = 2, etc.). This means that the time, τ, necessary to remove a monolayer can be expressed by

$$\tau = \frac{N^o_{ad}}{YJ} = \frac{1}{\sigma_{sp} J}. \tag{9.18}$$

Equation (9.17) can be solved for different coverages $\left(N_{ad}/N^o_{ad}\right)$ in order to establish a real description of the desorption of multilayers.

Monte Carlo computer simulations (see Sect. 3.6.2) are often applied to study the processes of ion beam-stimulated cleaning of surfaces [197]. Important parameters in these simulations are the energies required to displace an adsorbed atom from its surface position and the cutoff energy, which defines a particle having come to rest. These energies are usually on the order of a few electron-volts. For example, the process of removing thin carbon films from a copper surface were analyzed by of two different Monte Carlo simulation methods [198]. These simulations have confirmed that the cleaning process is very effective when the incidence angle of the ions varies between 60° and 70° and the ion energy between 500 eV and 2000 eV.

9.6 List of Symbols

Symbol	Notation
C	Power law constant
D_b	Ballistic induced diffusion coefficient
D_{th}	Thermal diffusion coefficient
E	Ion energy
E_d	(Adatom) displacement energy
J	Ion flux (ion current density)
N	Atomic number density
N_{ad}	Number of adsorbed atoms or molecules per unit area
$N^o{}_{ad}$	Number of adsorbed atoms (or molecules) per unit area at maximum monolayer coverage
N_I	Implanted concentration
N_{max}	Maximum atomic concentration
M_I	Ion mass
R_p	Mean projected range
ΔR_p	Projected range straggling
$S_n(E)$	Nuclear stopping cross section at the surface
T_e	Electron temperature
U_s	Surface binding energy
V_p	Plasma potential
V_s	Ion source—bias potential
Y	Sputtering

(continued)

(continued)

Symbol	Notation
Y_{sis}	Self-ion sputtering
a_B	Bohr screening length
m	Power law variable
s	Sticking coefficient (sticking probability)
v_{IBD}	Growth rate of direct ion deposited film
v_b	Jump frequency of ballistic generated adatoms
v_{th}	Jump frequency of thermal activated atoms
Φ	Fluence of the deposited species (ions)
γ	Transfer energy efficiency factor
γ_e	Secondary electron coefficients at electron bombardment
γ_I	Secondary electron coefficients at ion bombardment
θ	Angle of ion incidence
$\overline{\theta}$	Coverage
λ	Jump distance
λ_m	Power law fitting variable
σ_{sp}	Sputtering or desorption cross-section

References

1. J.M.E. Harper, Ion beam deposition, in *Thin Film Processes.* ed. by J.L. Vossen, W. Kern (Academic Press, New York, 1978), pp. 175–206
2. K. Miyaka, T. Tokuyama, Direct ion beam deposition, in *Ion Beam Assisted Film Deposition*, ed. by T. Itoh, (Elsevier Science, Amsterdam 1989), pp. 289–317
3. J. J. Cuomo, S. M. Rossnagel, H. R. Kaufman, (Eds.) Handbook of ion beam processing, Noyes Publ. Westwood 1989
4. D.G. Amour, Ion beam deposition. Nucl. Instr. Meth. Phys. Res. B **89**, 325–331 (1994)
5. D. Marton, Film deposition from low-energy ion beams, in *Low-Energy Ion-Surface Interaction*, ed. by J.W. Rabalais, (Wiley, Chichester 1994), pp. 481–534
6. K. Miyaka, Ion-beam deposition, in *Fundamentals and the Present Status of Purification of Metals.* ed. by Y. Wasada, M. Isshiki (Springer, Berlin, Heidelberg, 2002), pp. 203–222
7. O. Almén, G. Bruce, Collection and sputtering experiments with noble gas ions. Nucl. Instr. Meth. **11**, 257–278 (1961)
8. A. Fontell, E. Arminen, Direct collection of some metal ions in electromagnetic isotope separator and related surface effects. Canad. J. Phys. **47**, 2405–2414 (1969)
9. R.B. Fair, Analysis and design of ion-beam deposition apparatus. J. Appl. Phys. **42**, 3176–3181 (1971)
10. V. Franchetti, B.H. Solka, W.E. Baitinger, J.W. Amy, R.G. Cooks, Soft landing of ions as a means of surface modification. Int. J. Mass Spectrom. Ion Process **23**, 29–35 (1977)
11. J.M.E. Harper, Effects of beam, target, and substrate potentials in ion beam processing. Thin Solid Films **92**, 107–114 (1982)
12. B. Wolf (ed.), *Handbook of Ion Sources* (CRC Press, Boca Raton, 1995)

13. N. Sasaki, S. Shimizu, S. Ogata, Thin film growth using a high-current, mass-separated low energy ion beam deposition system. Thin Solid Films **281–282**, 175–178 (1996)
14. G.E. Thomas, L.J. Beckers, J.J. Vrakking, B.R. de Koning, Ion beam epiplantation. J. Cryst. Growth **56**, 557–575 (1982)
15. J. Ahn, R.P.W. Lawson, K.M. Yoo, K.A. Stromsmoe, M.J. Brett, Deposition of metastable binary alloy thin films using sequential ion beams from a single ion source. Nucl. Instr. Meth Phys. Res. B **17**, 37–45 (1986)
16. A. van Veen, Ion trapping and cluster growth, in *Erosion and Growth of Solids Stimulated by Atom and Ion Beams*, ed. by G. Kiriakidis, G. Carter, J.L. Whitton, (Martinus Nijhoff Publ., Dordrecht 1986), pp. 200–221
17. N. Herbots, B.R. Appleton, T.S. Noggle, R.A. Zuhr, S.J. Pennycook, Ion-solid interaction during ion beam deposition of ^{47}Ge and ^{30}Si on Si at very low energies (0–200 eV) range. Nucl. Instr. Meth Phys. Res. B **13**, 250–258 (1986)
18. S.M. Rossnagel, Directional and preferential sputtering -basedphysical vapor deposition. Thin Solid Films **263**, 1–12 (1995)
19. J.W. Rabalais, D. Marton, Atomic collisions in surface chemistry and film deposition. Nucl. Instr. Meth Phys. Res. B **67**, 287–295 (1992)
20. B.W. Dobson, Atomistic simulation of silicon beam deposition Phys. Rev. B **36**, 1068–1074 (1987) and Molecular-dynamics simulation of low-energy beam deposition of silicon, J. Vac. Sci. Technol. B **5**, 1393–1398 (1987)
21. X.W. Zhou, H.N.G. Wadley, Hyperthermal vapor deposition of copper: reflection and resputtering effects. Surf. Sci. **431**, 58–73 (1999)
22. S.S. Todorov, E.R. Fassum, Oxidation of silicon by a low-energy ion beam: experiment and model. Appl. Phys. Lett. **52**, 48–50 (1988)
23. J. Amano, Direct ion beam deposition for thin film formation. Thin. Solid Films **92**, 115–122 (1982)
24. N. Herbots, O.C. Hellman, P. Ye, X. Wang, O. Vancauwenberghe, Chemical reactions during thin-film synthesis: Ion-beam oxidation (IBO) and ion-beam nitration (IBN) of semiconductor surfaces, in *Low-Energy Ion-Surface Interaction*, ed. by J.W. Rabalais, (Wiley, Chichester 1994), pp. 387–480
25. O. Vancauwenberghe, N. Herbots, O.C. Hellman, Role of ion energy in ion beam oxidation of semiconductors: experimental study and mode, J. Vac, Sci. Technol. A **10**, 713–718 (1992) and O. Vancauwenberghe, On the growth of semiconductor-based epitaxial and oxide films from low energy ion beams, Dissertation, MIT, Cambridge, Mass. 1991
26. O. Vancauwenberghe, N. Herbots, O.C. Hellman, A quantitative model of point defect diffusivity and recombination in ion beam deposition and combined ion and molecular vapor deposition. J. Vac. Sci. Technol. B **9**, 2027–2033 (1991)
27. N. Herbots, O. Vancauwenberghe, O.C. Hellman, Y.C. Joo, Atomic collisions, elastic recombination, and thermal diffusion during thin-film growth from low-energy ion beams. Nucl. Instr. Meth Phys. Res. B **59**(60), 326–331 (1991)
28. T.Y. Tan, U. Gösele, Point defects, diffusion processes, and swirl defect formation in silicon. Appl. Phys. A **37**, 1–17 (1985)
29. V. Grill, J. Shen, C. Evans, R.G. Cooks, Collisions of ions with surface at chemically relevant energies: Instrumentation and phenomena. Rev. Sci. Instr. **72**, 3149–3179 (2001)
30. B. Gologan, J.R. Green, J. Alvarez, J. Laskin, G. Cooks, Ion/surface reactions and soft-landing. Phys. Chem. Chem. Phys. **7**, 1490–1500 (2005)
31. S. Hayakawa, Charge inversion mass spectrometry: dissociation of resonantly neutralized molecules. J. Mass Spectrom. **39**, 111–135 (2004)
32. D.C. Jacobs, Reactive collisions of hyperthermal energy molecular ions with solid surfaces. Annu. Rev. Phys. Chem. **53**, 379–407 (2002)
33. L. Krumbein, K. Anggara, M. Stella, T. Michnowicz, H. Ochner, S. Abb, G. Rinke, A. Portz, M. Dürr, U. Schlickum, A. Baldwin, A. Floris, K. Kern, S. Rauschenbach, Fast molecular compression by a hyperthermal collision gives bond-selective mechanochemistry. Phys. Rev. Lett. **126**, 056001 (2001)

34. W. Jacob, J. Roth, Chemical sputtering in *Sputtering by Particle Bombardment, Experiments and Computer Calculations from Threshold to MeV Energies*, ed. by R. Behrisch, W. Eckstein, (Springer, Berlin, 2007), pp. 329–400
35. N. Sakudo, K. Hayashi, Exact energy values of low-energy ion beams. Rev. Sci. Instr. **67**, 1218–1220 (1996)
36. N. Sakudo, K. Hayashi, N. Ikenaga, N. Sakaguchi, K. Moriike, K. Fujimura, M. Okada, T. Maesaka, Factors determining energy values of ion beams for ion-beam deposition. Nucl. Instr. Meth. Phys. Res. B **148**, 53–57 (1999)
37. S.R. Kasi, M.A. Kilburn, H. Kang, J.W. Rabalais, L. Tavernini, P. Hochmann, Interaction of low energy reactive ions with surfaces. III. Scattering of 30–200 eV Ne^+, O^+, C^+, and CO^+ from Ni(111), J. Chem. Phys. **88**, 5902–5913 (1988)
38. T. Michely, C. Teichert, adatom yields, sputterings, and damage patterns of single-ion impacts on Pt(111). Phys. Rev. B **50**, 11156–11566 (1994)
39. D.K. Brice, J.Y. Tsao, S.T. Picraux, Partitioning of ion-induced surface and bulk displacements. Nucl. Instr. Meth. Phys. Res. B **44**, 68–78 (1989)
40. Z.Q. Ma, Optimal energies for ion-assisted growth of IVA thin films. Intern. J. Mod. Phys. B **21**, 4299–4322 (2007)
41. H.D. Hagstrum, Studies of adsorbate electronic structure using ion neutralization and photoemission spectroscopies, in *Electron and Ion Spectroscopy of Solids*, ed. by L. Fiermanns, J. Vennik, W. Dekeyser, (Plenum Press, New Your 1978), pp. 273–323
42. D. Marton, K.J. Boyd, T. Lytle, J.W. Rabalais, Auger-electron spectroscopy of krypton subplanted in graphite. Surf. Sci. **282**, 113–121 (1993)
43. S.R. Kasi, H. Kang, J.W. Rabalais, Interaction of low energy reactive ions with surfaces. IV. Chemically bonded diamond-like films from ion-beam deposition. J. Chem. Phys. **88**, 5914–5924 (1988)
44. M. Mensing, P. Schumacher, J.W. Gerlach, S. Herath, A. Lotnyk, B. Rauschenbach, Influence of nitrogen ion species on mass-selected low energy ion-assisted growth of epitaxial GaN thin films. Appl. Surf. Sci. **498**, 143830 (2019)
45. H. Akazawa, Y. Murata, Interaction of reactive ions with Pt(100). II. Dissociative scattering of molecular ions near the threshold energy region. J. Chem Phys. **92**, 5561–5568 (1990)
46. S.R. Kasi, H. Kang, C.S. Sass, J.W. Rabalais, Inelastic processes in low-energy ion surface collisions, Surf. Sci. Rep. **10**, 1–104 (1989)
47. B.N. Chapman, D.S. Campbell, Condensation of high-energy atomic beams. J. Phys. C **2**, 200–209 (1969)
48. G.E. Lane, J.C. Anderson, The nucleation and initial growth of gold films deposited onto sodium chloride by ion-beam sputtering. Thin Solid Films **26**, 5–23 (1975). Features of initial growth of gold films deposited on rock-salt substrates by ion beam sputtering. Thin Solid Films **57**, 277–283 (1979)
49. M.-A. Hasan, S.A. Barnett, J.-E. Sundgren, J.E. Greene, Nucleation and initial growth of In deposited on Si_3N_4 using low-energy (<300 eV) accelerated beams in ultrahigh vacuum. J. Vac. Sci. Technol. A **5**, 1883–1887 (1987)
50. S. Esch, M. Bott T. Michely, G. Comsa, Nucleation of homoepitaxial films grown with ion assistance on Pt(111). Appl. Phys. Lett. **67**, 3209–3211 (1995). S. Esch, M. Breeman, M. Morgenstern, T. Michely, G. Comsa, Nucleation and morphology of homoepitaxial Pt(111)-films grown with ion beam assisted deposition. Surf. Sci. **365**, 187–204 (1996)
51. M. Kaff, M. Breeman, M. Morgenstern, T. Michely, G. Comsa, Effect of energetic particles on island formation in sputter deposition of Pt on Pt(111). Appl Phys Lett. **70**, 182–184 (1997)
52. J. M. Pomeroy, A. Couture, J. Jacobsen, C.C. Hill, J.P. Sethna, B.H. Cooper, J.D. Brock, STM characterization of Cu thin films grown by direct ion deposition. Vol. 648, Proceed. MRS (2000), P7.3
53. B. Degroote, A. Vantomme, H. Pattyn, K. Vanormelingen, Hyperthermal effects on nucleation and growth during low-energy ion deposition. Phys. Rev. B **65**, 195401 (2002)
54. J. M. Pomeroy, J. Jacobsen, C.C. Hill, B.H. Cooper, J.P. Sethna, Kinetic Monte Carlo–molecular dynamics investigations of hyperthermal copper deposition on Cu(111), Phys. Rev. B **66**, 235412 (2002)

55. J. Jacobsen, B.H. Cooper, J.P. Sethna, Simulations of energetic beam deposition: from picoseconds to seconds. Phys. Rev. B **58**, 15847–15865 (1998)
56. Z.Q. Ma, Y.F. Zheng, B.X. Liu, Ion-induced surface and bulk displacement threshold for epitaxial growth. Phys. Stat. Sol. (a) **169**, 239–248 (1998). Sketching of preferred energy regime for ion beam assisted epitaxy. App. Surf. Sci. **137**, 184–190 (1999)
57. S. Mohajerzadeh, C.R. Selvakumar, A simple model for low energy ion-solid interactions. J. Appl. Phys. **81**, 3003–3006 (1997)
58. N.A. Kuboto, D.J. Economou, S.J. Plimpton, Molecular dynamics simulations of low-energy (25–200 eV) argon ion interactions with silicon surfaces: sputter yields and product formation pathways. J. Appl. Phys. **83**, 4056–4063 (1998)
59. J.P. Chang, J.C. Arnold, G.C.H. Zau, H.-S. Shin, H.H. Sawin, Kinetic study of low energy argon ion-enhanced plasma etching pf polysilicon with atomic/molecular chlorine. J. Vac. Sci. Technol. A **15**, 1853–1863 (1967)
60. C.R. Laurens, M.F. Roşu, F. Pleiter, L. Nielsen, Soft-landing ion deposition of isolated radioactive probe atoms on surfaces: a novel method. Phys. Rev. Lett. **78**, 4075–4078 (1997)
61. A.H. Al-Bayati, K.J. Boyd, D. Marton, S.S. Todorov, J.W. Rabalais, Z.H. Zhang, W.K. Chu, Substrate temperature dependence of homoepitaxial growth of Si using mass selected ion beam deposition. J. Appl. Phys. **76**, 4383–4389 (1994)
62. M.V. Murty, H.A. Atwater, Defect generation and morphology of (001) Si surface during Ar-ion bombardment. Phys. Rev. B **45**, 1507–1510 (1992)
63. M. Morgenstern, T. Michely, G. Comsa, Collective effects in the adatom production by 4.5 keV rare-gas impacts on Pt(111): a low temperature scanning tunnelling microscopy analysis. Phil. Mag. A **79**, 775–794 (1999)
64. M. Ghaly, K. Nordlund, R.S. Averback, Molecular dynamics investigations of surface damage produced by kiloelectronvolt self-bombardment of solids. Phil. Mag. A **79**, 795–820 (1999)
65. G. Carter, Island-density and size dynamics in ion-assisted atomic deposition. Vacuum **55**, 235–247 (1999). See also H.H. Anderson, The depth resolution of sputter profiling, Appl. Phys. **18**, 131-140 (1979)
66. K.-H. Müller, Role of incident kinetic energy of adatoms in thin film growth. Surf. Sci. **184**, L375–L382 (1987)
67. R. Ditchfeld, E.G., Seebauer, Semiconductor surface diffusion: effects of low-energy ion bombardment, Phys. Rev. B **63**, 125317 (2001)
68. R. Ditchfeld, E.G., Seebauer, Direct measurement of ion-influenced surface diffusion. Phys. Rev. Lett. **82**, 1185–1188 (1999)
69. W.F. Engelhoff, I. Jacob, Reflection high-energy electron diffraction (RHEED) oscillations at 77 K. Phys. Rev. Lett. **63**, 921–924 (89). R. Kunkel, B. Poelsema, L. K. Verheij, G. Comsa, Reentrant layer-by-layer growth during molecular-beam epitaxy of metal-on-metal substrates. Phys. Rev. Lett. **65**, 733–736 (1990)
70. X.W. Zhou, H.N.G. Wadley, Hyperthermal vapor deposition of copper: athermal and biased diffusion effects. Surf. Sci. **431**, 42–57 (1999)
71. D.E. Sanders, A. E. DePristo, Metal/metal homo-epitaxy on fcc (001) surfaces: Is there transient mobility of adsorbed atoms? Surf. Sci. **254** (1991) 341–353 (1991). P. Stoltze and J.K. Nørskov, Accommodation and diffusion of Cu deposited on flat and stepped Cu(111) surfaces. Phys. Rev. B **48** (1993) 5607–5611
72. D. Adamovic, E.P. Münger, V. Chirita, L. Hultman, J.E. Greene, Low-energy ion irradiation during film growth: Kinetic pathways leading to enhanced adatom migration, Appl. Phys. Lett. **86**, 211915 (2005). D. Adamovic, V. Chirita, E.P. Münger, L. Hultman, J.E. Greene, Enhanced intra- and interlayer mass transport on Pt(111) via 5–50 eV Pt atom impacts on two-dimensional Pt clusters. Thin Solid Films **515**, 2235–2243 (2006)
73. J. Sillanpaä, I. Koponen, Island growth in ion beam assisted metal-on-metal deposition modelled by rate equations. Nucl. Instr. Meth. Phys. Res. B **142**, 67–76 (1998)
74. E.F.C. Haddeman, B.S. Bunnik, B.J. Thijsse, Thin film growth and ion-beam modification: MD simulations going beyond simple systems, in *Fundamental Mechanisms of Low-Energy-Beam-Modified Surface Growth and Processing*, ed. by S.C. Moss, E.H. Chason, B.H. Cooper,

J.M.E. Harper, T.D. de la Rubia, M.V.R. Murty, Vol. 585, Mater. Rec. Soc., Symp. Proceed. (2000), pp. 103–106
75. J.J. Quan, X.W. Zhou, H.N.G. Wadley, Atomic assembly of metal surfaces and interfaces. Surf. Sci. **600**, 4537–4547 (2006)
76. T. Huhtamäki, M.O. Jahma, I.T. Koponen, A simple model for quantifying the degree of layer-to-layer growth in low energy ion deposition of thin films. Nucl. Instr. Meth. in Phys. Res. B **264**, 55–60 (2007)
77. Q. Fu, T. Wagner, Diffusion-corrected simultaneous multilayer growth model, Phys. Rev. Lett. **90**, 106105 (2003)
78. B. Schultrich, *Tetrahedrally Bonded Amorphous Carbon Films*, Springer-Series in Materials Sciences, Vol. 263 (Springer, Berlin, 2018)
79. S. Aisenberg, R. Chabot, Ion-beam deposition of thin films of diamondlike carbon. J. Appl. Phys. **42**, 2953–2958 (1972). S. Aisenberg, R. Chabot, Physics of ion plating and ion beam deposition. J. Vac. Sci. Technol. **10**, 104–107 (1973)
80. J. Ishikawa, Y. Takeiri, K. Ogawa, T. Takagi, Transparent carbon films prepared by mass-separated negative-carbon-ion-beam deposition. J. Appl. Phys. **61**, 2509–2515 (1987). J. Ishikawa, K. Ogawa, K. Miyata, H. Tsuji, T. Takagi, Negative ion source (NIABNIS) and preparation of transparent carbon films by negative carbon ion beam deposition. Nucl. Instr. Meth. Phys. Res. B **21**, 205–208 (1987)
81. P.J. Fallon, V.S. Veerasamy, C.A. Davis, J. Robertson, G.A.J. Amaratunga, W.I. Milne, J. Koskinen, Properties of filtered-ion-beam-deposited diamondlike carbon as a function of ion energy, Phys. Rev. B. **48**, 4777–4782 (1993). Erratum: Phys. Rev. B **49**, 2287 (1993)
82. D.R. McKenzie, D. Muller, B.A. Pailthorpe, Z.H. Wang, E. Kravtchinskaja, D. Segal, P.B. Lukins, P.D. Swift, P.J. Martin, G. Amaraunga, P.H. Gaskell, A. Saeed, Properties of tetrahedral amorphous carbon prepared by vacuum arc deposition. Diamond Rel. Mater. **1**, 51–59 (1991)
83. J.P. Hirvonen, J. Koskinen, R. Lappalainen, A. Anttila, Preparation and properties of high density, hydrogen free hard carbon films with direct ion beam or arc discharge deposition, in *Properties and Characterization of Amorphous Carbon Films*, ed. by J.J. Pouch, S.A. Alterovitz, (Trans Tech Publications, Zürich 1990) pp. 197–216
84. Y. Lifshitz, G.D. Lempert, S. Rotter, I. Avigal, C. Uzan-Saguy, R. Kalish, D. Marton, J.W. Rabalais, The effect of ion energy on the diamond-like/graphitic (sp^3/sp^2) nature of carbon films deposited by ion beams. Diamond Rel. Mater. **3**, 542–546 (1994)
85. E. Grossman, G.D. Lampert, J. Kulik, D. Marton, J.W. Rabalais, Y. Lifshitz, Role of ion energy in determination of the sp^3 fraction of ion beam deposited carbon films, Appl. Phys. Lett. **68**, 1214–1216 (1996)
86. D. Marton, K.J. Boyd, I.W. Rabalais, Y. Lifshitz, Semiquantitative subplantation model for low energy ion interactions with surfaces. II. Ion beam deposition of carbon and carbon nitride. J. Vac. Sci. Technol. A. **16**, 455–462 (1998)
87. J. Kulik, G.D. Lempert, E. Grossman, D. Marton, J.W. Rabalais, Y. Lifshitz, sp3 content of mass-selected-ion-beam-deposited carbon films determined by inelastic and elastic electrons scattering. Phys. Rev. B **52**, 151812–215822 (1995)
88. J.A. Thornton, D.W. Hoffman, Stress-related effects in thin films. Thin Solid Films **171**, 5–31 (1989)
89. D.R. McKenzie, Tetrahedral bonding in amorphous carbon. Rep. Prog. Phys. **59**, 1611–1664 (1996)
90. D.R. McKenzie, D. Muller, B.A. Pailthorpe, Compressive-stress-induced formation of thin-film tetrahedral amorphous carbon. Phys. Rev. Lett. **67**, 773–776 (1991)
91. C.A. Davis, A simple model for the formation of compressive stress in thin films by ion bombardment. Thin Solid Films **226**, 30–34 (1993)
92. Y. Lifshitz, G.D. Lempert, E. Grossman, Substantiation of subplantation model for diamond-like film growth by atomic force microscopy. Phys. Rev. Lett. **72**, 2753–2757 (1994)
93. Y. Lifshitz, G.D. Lempert, E. Grossman, I. Avigal, C. Uzan-Saguy, R. Kalish, J. Kulik, D. Marton, J.W. Rabalais, Growth mechanisms of DLC films from C^+ ions: experimental studies. Diamond Rel. Mater. **4**, 318–323 (1995)

94. M. Chhowalla, J. Robertson, C.W. Chen, S.R.P. Silva, C.A. Davis G.A.J. Amaratunga, W.I. Milne, Influence of ion energy and substrate temperature on the optical and electronic properties of tetrahedral amorphous carbon (ta-C) films. J. Appl. Phys. **81**, 139–145 (1997)
95. W.M. Lau, I. Bello, X. Feng, L.J. Huang, Q. Fuguang, Y. Zhenyu, R. Zhizhang, S.-T. Lee, Direct ion beam deposition of carbon films on silicon in the ion energy range of 15–500 eV. J. App. Phys. **70**, 5623–5628 (1991)
96. D.R. McKenzie, Generation and applications of compressive stress induced by low energy ion beam bombardment. J. Vac. Sci. Technol. B **11**, 1928–1935 (1993)
97. I. Koponen, M. Harkovirta, R. Lappalainen, Modeling the ion energy dependence of the sp2/sp3 bonding ratio in amorphous diamondlike films produced with a mass-separated ion beam. J. Appl. Phys. **78**, 5837–5839 (1995)
98. H.C. Hofsäss, H. Feldermann, R. Merk, M. Sebastian, C. Ronning, Cylindrical spike model for the formation of diamondlike thin films by ion deposition. Appl. Phys. A **66**, 153–181 (1998)
99. J. Robertson, Deposition mechanisms for promoting sp3 bonding in diamond-like carbon, Diam. Relat. Mater. **2**, 984–989 (1993). The deposition mechanism of diamond-like a-C and a-C:H, Diam. Relat. Mater. **3**, 361–368 (1994)
100. J. Robertson, Diamond-like amorphous carbon. Mater. Sci. Eng. R **37**, 129–281 (2002)
101. K.J. Boyd, D. Marton, J.W. Rabalais, S. Uhlmann, T. Frauenheim, Semiquantitative subplantation model for low energy ion interactions with surfaces. I. Noble gas ion-surface interactions. J. Vac. Sci. Technol. A **16**, 444–462 (1998)
102. G. Carter, Spike and shock processes in high energy deposition density atomic collision events in solids. Rad. Eff. Lett. **43**, 193–199 (1979). A semi quantitative approach to impact induced shock processes in solids. Rad. Eff. Lett. **50**, 105–109 (1980)
103. B. Rauschenbach, K. Hohmuth, A simple approach to the analysis of ion collision cascade in solids based on the shock wave model. Phys. Status Solidi (a) **75**, 159–168 (1983). Analysis of heavy-ion sputtering based on the shock wave model. Phys. Status Solidi (a) **79**, 395–401 (1983)
104. E.G. Spencer, P.H. Schmidt, D.C. Joy, F.J. Sansalone, Ion-beam-deposited polycrystalline diamond-like films. Appl. Phys. Lett. **29**, 118–120 (1976)
105. Y. Lifshitz, S.R. Kasi, J.W. Rabalais, Subplantation model for film growth from hyperthermal species: application to diamond. Phys. Rev. Lett **62**, 1290–1293 (1987). Subplantation model for film growth from hyperthermal species. Phy. Rev. B **41**, 10468–10480 (1990)
106. H.-P. Kaukonen, R.M. Nieminen, Molecular-dynamics simulation of the growth of diamond-like films by energetic carbon-atom beams. Phys. Rev. Lett. **68**, 620–623 (1992)
107. H.J. Steffen, D. Marton J.W. Rabalais, Displacement energy threshold for Ne^+ irradiation of graphite, Phys. Rev. Lett. **68**, 1726–1729 (1992). J. Koike, D.M. Parkin, T.E. Mitchell, Displacement threshold energy for type IIa diamond. Appl. Phys. Lett. **60**, 1450–1452 (1992)
108. N.A. Marks, D.R. McKienzie, B.A. Pailthorpe, Molecular-dynamics study of compressive stress generation. Phys. Rev. B **53**, 4117–4127 (1996)
109. S. Zhang, H.T. Johnson, G.J. Wagner, W.K. Liu, K.J. Hsia, Stress generation mechanisms in carbon thin films grown by ion-beam deposition. Acta Mater. **51**, 5211–5222 (2003)
110. R.A. Zuhr, B.R. Appleton, N. Herbots, B.C. Larson, T.S. Noggle, S.J. Pennycook, Low-temperture epitaxy of Si and Ge by direct ion beam deposition. J. Vac. Sci. Technol A **5**, 2135–2139 (1987). R.A. Zuhr, S.J. Pennycook, T.S. Noggle, N. Herbots, T.E. Haynes, B.R. Appleton, Ion beam deposition in materials research. Nucl. Instrum. Methods Phys. Res. B **37/38**, 16–21 (1989)
111. K.G. Orrman-Rossiter, D. R. G. Mitchell, S.E. Donnelly, C.J. Rossouw, S.R. Glanvill, P.R. Miller, A.H. Al-Bayati, J.A. van den Berg, D.G. Armour, Evidence for competing growth phases in ion-beam-deposited epitaxial silicon films, Philos. Mag. Lett. **61**, 311–318 (1990). K.J. Orrman-Rossiter, A.H. Al-Bayati, D.G. Armour, S.E. Donnely, J.A. van den Berg, Ion beam deposited epitaxial thin silicon films, Nucl. Instrum. Methods Phys. Res. B **59/60**, 197–202 (1991)

112. M. Matsuoka, S. Tohno, Ion beam epitaxy of silicon films in an ultrahigh vacuum using a sputtering-type metal ion source. J. Vac. Sci. Technol. A **13**, 305–331 (1995)
113. M.V.R. Murty, H.A. Atwater, A.J. Kellock, J.E.E. Baglin, Very low temperature (<400 °C) silicon molecular beam epitaxy: The role of low energy ion irradiation, Appl. Phys. Lett. **62**, 2566–2568 (1993). M.V. R. Murty, H.A. Atwater, Crystal-state - amorphous-state transition in low-temperature silicon epitaxy. Phys. Rev. B **49**, 8483–8486 (1994)
114. N.-E. Lee, G.A. Tomasch, J.E. Greene, Low-temperature Si(001) epitaxy using low-energy ($<E> \approx 18$ eV) Si atoms. Appl. Phys. Lett. **63**, 3236–3238 (1994)
115. J.W. Rabalais, A.H. Al-Bayati, K.J. Boyd, D. Marton, J. Kulik, Z. Zhang, W.K. Chu, Ion-energy effects in silicon ion-beam epitaxy. Phys. Rev. B **53**, 10781–10792 (1996)
116. K.J. Boyd, D. Marton, J.W. Rabalais, S. Uhlmann, T. Frauenheim, Semiquantitative subplantation model for low energy ion interactions with surfaces. III. Ion beam homoepitaxy of Si, J. Vac. Sci. Technol. A **16**, 463–471 (1998)
117. D. Marton, K.J. Boyd, J.W. Rabalais, Hyperthermal particle enhanced silicon epitaxy. Chem. Phys. Lett. **283**, 215–220 (1998)
118. Z. Wang, E. Seebauer, Temperature-dependent energy thresholds for ion-stimulated defect formation in solids: effects of ion mass and adsorbate–substrate pairing, Surf. Sci. **601**, 2453–2458 (2007). Z. Wang, E. Seebauer, Temperature-dependent energy thresholds for ion-stimulated defect formation in solids. Phys, Rev. Lett. **95**, 015501 (2005)
119. S.S. Todorov, H. Bu, K.J. Boyd, J.W. Rabalais, C.M. Gilmore, J.A. Sprague, Ion beam deposition of 107Ag(111) films on Ni(100). Surf. Sci. **429**, 63–70 (1999)
120. R.A. Zuhr, T.E. Hayes, M.D. Galloway, S. Tanaka, A. Yamada, I. Yamada, Formation of aluminium films on silicon by ion beam deposition. A comparison with ionized cluster beam deposition, Oak Ridge Nat. Laboratory Report CONF-900936 (1991)
121. S. Shimizu, N. Sasaki, S. Ogata, O. Tsukakoshi, S. Seki, H. Yamakawa, Thin film growth using low-energy multi-ion beam deposition system, in *Low Energy Ion Beam and Plasma Modification of Materials*, ed. by M. Gorbatkin, J.M.E. Harper, J.R. McNeil, K. Miyake, Mater. Res. Soc., Symp. Proceed. Vol. 223 (1991) pp. 347–352
122. R.A. Zuhr, S.J. Pennycook, T.E. Haynes, O.W. Holland, Metal silicides formed by direct ion beam deposition. Mater. Res. Soc. Symp. Proceed. **128**, 47–54 (1988)
123. A. Bousetta, A.H. Al Bayati, J.A. van den Berg, D.G. Armour, Structural and electrical properties of Co grown on Si (111) by low energy ion beam deposition. Appl. Surf. Sci. **56–58**, 480–485 (1992)
124. O. Tsukakoshi, S. Shimizu, S. Ogata, N. Sasaki, H. Yamakawa, A high-current low-energy multi-ion beam deposition system. Nucl. Instr. Meth Phys. Res. B **55**, 355–358 (1991)
125. T. Ohnishi, Y. Yoshida, Y. Hirofuji, H. lwasaki, Ta_2O_5 film formation by double ion beam method. Nucl. Instr. Meth. B **37/38**, 850–853 (1989)
126. S. Tamura, M. Hyouzho, K. Yokota, S. Katayama, Growth of crystalline GaAs films on Si substrates by Ga and As ion beams. Nucl. Instr. Meth. B **37**(38), 862–865 (1989)
127. Y. Morishita, S. Maruno, T. Isu, Y. Nomura, H. Ogata, Electrical and optical properties of InP grown by MBE using P^+ ion beam. J. Cryst. Growth **88**, 215–220 (1988)
128. A. Gottdang, K. Eich, A. Hassenbürger, W.H. Schulte, B. Cleff, D.J.W. Mous, R. Koudijs, G.F.A. van der Walle, J. Politiek, A dual source low-energy ion implantation system for use in silicon molecular beam epitaxy. Nucl. Instr. Meth in Phys. Res. B **55**, 310–313 (1991)
129. T.E. Haynes, R.A. Zuhr, S.J. Pennycook, B.R. Appleton, Heteroepitxy of GaAs on Si and Ge using alternating, low-energy ion beams. Appl. Phys. Left. **54**, 1439–1441 (1989)
130. S.P. Withrow, K.L. Moore, R.A. Zuhr, T.E. Haynes, Ion beam deposition of β-SiC layers onto α-SiC substrates. Vacuum **39**, 1065–1068 (1989)
131. J. Ahn, R.P.W. Lawson, K.M. Yoo, K.A. Stromsmoe, M.J. Brett, Deposition of metastable binary alloy thin films using sequential ion beams from a single source. Nucl. Instr. Meth in Phys. Res. B **17**, 37–45 (1986)
132. Y. Yoshida, T. Ohnishi, T. Sekihara, Y. Hirofuji, Low-energy double-ion-beam deposition system, Jpn. J. App. Phys. **27**, 140–143 (1988). Y. Yoshida, T. Ohnishi, Y. Hirofuji, H. Iwasaki, T. Ikeda, low-energy double ion-beam deposition of compound films. Nucl. Instr. Meth. in Phys. Res. B **37/38**, 866–869 (1989)

133. S. Maruno, Y. Morishita, T. Isu, Y. Nomura, H. Ogata, Growth mechanisms of molecular beam epitaxy of InP and $GaAs_xP_{1-x}$. Sci. **201**, 335–344 (1988)
134. T. Matsumoto, M. Kiuchi, S. Sugimoto, S. Goto, Momentum effects of molecular ion beam in Sic crystal growth. Thin Solid Films **464–465**, 103–106 (2004)
135. M. Mitsui, S. Nagaoka, T. Matsumoto, A. Nakajima, Soft-landing isolation of vanadium-benzene sandwich clusters on a room-temperature substrate using n-alkanethiolate self-assembled monolayer matrixes. J. Phys. Chem. B **110**, 2968–2971 (2006)
136. G.E. Johnson, J. Laskin, Preparation of surface organometallic catalysts by gas-phase ligand stripping and reactive landing of mass-selected ions. Chem. Eur. J. **16**, 14433–14438 (2010)
137. B. Gologan, Z. Takats, J. Alvarez, J.M. Wiseman, N. Talaty, Z. Ouyang, R.G. Cooks, Ion soft-landing into liquids: protein identification, separation, and purification with retention of biological activity. J. Am. Soc. Mass Spectrom. **15**, 1874–1884 (2004)
138. T.A. Blake, O.Y. Zheng, J.M. Wiseman, Z. Takats, A.J. Guymon, S. Kothari, R.G. Cooks, Preparative linear ion trap mass spectrometer for separation and collection of purified proteins and peptides in arrays using ion soft landing. Anal. Chem. **76**, 6293–6305 (2004)
139. M. Volny, W.T. Elam, A. Branca, B.D. Ratner, F. Turecek, Preparative soft and reactive landing of multiply charged protein ions on a plasma-treated metal surface. Anal. Chem. **77**, 4890–4896 (2005)
140. V.A. Mikhailov, T.H. Mize, J.L.P. Benesch, C.V. Robinson, Mass-selective soft-landing of protein assemblies with controlled landing energies. Anal. Chem. **86**, 8321–8328 (2014)
141. Z. Deng, N. Thontasen, N. Malinowski, G. Rinke, L. Harnau, S. Rauschenbach, K. Kern, A close look at proteins: submolecular resolution of two- and three-dimensionally folded cytochrome at surfaces. Nano Lett. **12**, 2452–2458 (2012)
142. S. Rauschenbach, F.L. Stadler, E. Lunedei, N. Malinowski, S. Koltsov, G. Costantini, K. Kern, Electrospray ion beam deposition of clusters and biomolecules. Small **2**, 540–547 (2006)
143. X.L. Yang, P.S. Mayer, F. Turecek, Preparative separation of a multicomponent peptide mixture by mass spectrometry. J. Mass Spectrom. **41**, 256–262 (2006)
144. O. Hadjar, P. Wang, J.H. Futrell, J. Laskin, Effect of the surface on charge reduction and desorption kinetics of soft landed peptide ions. J. Am. Soc. Mass Spectrom. **20**, 901–906 (2009)
145. J. Laskin, P. Wang, O. Hadjar, Soft-landing of peptide ions onto self-assembled monolayer surfaces: an overview. Phys. Chem. Chem. Phys. **10**, 1079–1090 (2008)
146. S. Rauschenbach, G. Rinke, R. Gutzler, S. Abb, A. Albarghash, D. Le, T.S. Rahman, M. pDürr, L. Harnau, K. Kern, Two-dimensional folding of polypeptides into molecular nanostructures at surfaces. ACS Nano **11**, 2420–2427 (2017)
147. S. Abb, L. Harnau, R. Gutzler, S. Rauschenbach, K. Kern, Two-dimensional honeycomb network through sequence-controlled self-assembly of oligopeptides. Nature Comm. **7**, 10335 (2016)
148. B.B. Feng, D.S. Wunschel, C.D. Masselon, L. Pasa-Tolic, R.D. Smith, Retrieval of DNA using softlanding after mass analysis by ESI-FTICR for enzymatic manipulation. J. Am. Chem. Soc. **121**, 8961–8962 (1999)
149. G. Siuzdak, B. Bothner, M. Yeager, C. Brugidou, C.M. Fauquet, K. Hoey, C.-M. Change, Mass spectrometry and viral analysis. Chem. Biol. **3**, 45–48 (1996)
150. S.D. Fuerstenau, W.H. Benner, J.J. Thomas, C. Brugidou, B. Bothner, G. Siuzdak, Mass spectrometry of an intact virus. Angew. Chem. **113**, 559–562 (2001)
151. Y. Tani, S. Kobayashi, H. Kawazoe, Characterization of electrospray ion-beam-deposited CdSe/ZnS quantum dot thin films from a colloidal solution. J. Vac. Sci. Technol. A **26**, 1058–1061 (2008)
152. K. Anggara, Y. Zhu, G Fittolani, Y. Yu, T. Tyrikos-Ergas, M Delbianco, S. Rauschenbach, S. Abb, K. Kern, Identifying the origin of local flexibility in a carbohydrate polymer. Proceed. Nat. Acad. Sci. (PNAS) **118**, 2102168118 (2021)
153. X. Wu, M. Delbianco, K. Anggara, T. Michnowicz, A. Pardo-Vargas, P. Bharate, S. Sen, M. Pristl, S. Rauschenbach, U. Schickum, S. Abb, P.H. Seeberger, K. Kern, Imaging single glycans. Nature **582**, 375–378 (2020)

154. S. Abb, N. Tarrat, J. Cortés, B. Andriyevsky, L. Harnau, J.C. Schön, S. Rauschenbach, K. Kern, Carbohydrate self-assembly at surfaces: STM imaging of sucrose conformation and ordering on Cu (100). Angew. Chem. **131**, 8424–8428 (2019)
155. S. Banerjee, S. Mazumdar, Electrospray ionization mass spectrometry: A technique to access the information beyond the molecular weight of the analyte. Inter. J. Analyt. Chem. 2012, **282574** (2012)
156. J.B. Fenn, M. Mann, C.K. Meng, S.F. Wong, C.M. Whitehouse, Electrospray ionization for mass spectrometry of large biomolecules. Science **246**, 64–71 (1989)
157. G. Taylor, Disintegration of water drops in an electric field. Proceed. Royal Soc. London, A **280**, 383–397 (1964)
158. P. Kebarle, L. Tang, From ions in solution to ions in the gas phase: the mechanism of electrospray mass spectrometry. Analyt. Chem. **65**, 972A-986A (1993)
159. P. Kebarle, M. Peschke, On the mechanisms by which the charged droplets produced by electrospray lead to gas phase ions. Analyt. Chim. Acta **406**, 11–35 (2000)
160. L. Konermann, E. Ahadi, A.D. Rodriguez, S. Vahidi, Unraveling the mechanism of electrospray ionization. Anal. Chem. **85**, 2–9 (2013)
161. S. Rauschenbach, Electrospray ion beam deposition and mass spectroscopy of nonvolatile molecules and nanomaterials, Dissertation, École Polytechnique Fédérale de Lausanne (2008)
162. O. Hadjar, P. Wang, J.H. Futrell, Y. Dessiaterik, Z. Zhu, J.P. Cowin, M.J. Iedema, J. Laskin, Design and performance of an instrument for soft landing of biomolecular ions on surfaces. Analy. Chem. **79**, 6566–6574 (2007)
163. K.D.D. Gunaratne, V. Prabhakaran, Y.M. Ibrahim, R.V. Norheim, G.E. Johnson, J. Laskin, Design and performance of high-flux electrospray ionization source for ion soft landing. Analyst **140**, 2957–2963 (2015)
164. Y.M. Warneke, M.E. Belov, G.A. Anderson, Design and performance of a dual-polarity instrument for ion soft landing. J. Laskin, Anal. Chem. **91**, 5904–5912 (2019)
165. G.E. Johnson, D. Gunaratne, J. Laskin, Soft- and reactive landings of ions onto surfaces: concepts and applications. Mass Spectrom. Rev. **35**, 439–479 (2016)
166. J. Warneke, M.E. McBriarty, S.L. Riechers, S. China, M.H. Engelhard, E. Aprá, R.P. Young, N.M. Washton, C. Jenne, G.E. Johnson, J. Laskin, Self-organizing layers from complex molecular anions. Nature Comm. **9**, 1889 (2018)
167. H. Usui, I. Yamada, T. Takagi, Anthracene and polyethylene thin-film depositions by ionized cluster beam. J. Vac. Sci. Technol. A **4**, 52–60 (1986)
168. Qi, L., Sinnott, S.B.: Generation of 3D hydrocarbon thin films via organic molecular cluster collisions, Sur. Sci. **398**, 195–202 (1998). L. Qi, W.L. Young, S.B. Sinnott, Effect of surface reactivity on the nucleation of hydrocarbon thin films through molecular-cluster beam deposition. Sur Sci. **426**, 83–91 (1999)
169. S. Rauschenbach, G. Rinke, N. Malinowski, R.T. Weitz, R. Dinnebier, N. Thontasen, Z. Deng,, T. Lutz,, P.M. de Ameida Rollo, G. Costantini, L. Harnau, K. Kern, Crystalline inverted membranes grown on surfaces by electrospray ion beam deposition in vacuum. Adv. Mater. **24**, 2761–2767 (2014)
170. S.-S. Jester, D. Löffler, P. Weis, A. Böttcher, M.M. Kappes, Morphology of C_n thin films ($50 \leq n \leq 60$) on graphite: Inference of energy dissipation during hyperthermal deposition. Surf. Sci. **603**, 1863–1872 (2009)
171. A. Böttcher, P. Weis, S.-S. Jester, D. Löffler, A. Bihlmeier, W. Klopper, M.M. Kappes, Solid C_{58} films. Phys. Chem. Chem. Phys. **7**, 2816–2820 (2005)
172. H.E. Fransworth, R.E. Schlier, T.H. George, R.M. Burger, Ion bombardment-cleaning of germanium and titanium as determined by low-energy electron diffraction. J. Appl. Phys. **26**, 252–253 (1956)
173. H. Lüth, *Solid Surfaces* (Springer, Heidelberg, Interfaces and Thin Films, 2010)
174. H.R. Kaufman, J.M.E. Harper, Ion doses for low-energy ion-assist applications. J. Vac. Sci. Technol. A **22**, 221–224 (2004)
175. N. Razek, A. Schindler, B. Rauschenbach, UHV-Direct bonding of semiconductor wafers at room temperature using hydrogen ion beam surfaces cleaning, Proceed. ECS Meeting 2005,

paper MA2005-01, 508.N. Razek, K. Otte, T. Chassé, D. Hirsch, A. Schindler, F. Frost, B. Rauschenbach, GaAs surface cleaning by low energy hydrogen ion beam treatment. J. Vac. Sci. Technol. A **20**, 1492–1497 (2002)
176. U.B. Sharopov, K. Kaur, M.K. Kurbanov, D.S. Saidov, E.T. Juraev, M.M. Sharipov, Controlling the low-temperature ionic purification of a silicon surface by electron spectroscopy. Silicon **14**, 4661–4667 (2022).
177. S. Mohajerzadeh, C.R. Selvakumar, D.E. Brodiet, M.D. Robertson, J.M. Corbett, Study of in-situ surface cleaning for Si and SiGe epitaxy on Si with a novel ion beam vapor/assisted deposition technique. Progr. Surf. Sci. **50**, 347–356 (1995)
178. P. Rabinzohn, G. Gautherin, B. Agius, C. Cohen, Cleaning of Si and GaAs crystal surfaces by ion bombardment in the 50–1500 eV range. J. Electrochem. Soc. **131**, 905–914 (1984)
179. l. Bello, W.H. Chang, W.M. Lau, Mechanism of cleaning and etching Si surfaces with low energy chlorine ion bombardment. J. Appl. Phys. **75**, 3092–3097 (1994)
180. J.H. Comfort, L.M. Garverick, R. Reif, Silicon surface cleaning by low dose argon-ion bombardment for low temperature (750°C) epitaxial silicon deposition. I. Process considerations. J. Appl. Phys. **62**, 3388–3397 (1987)
181. T. Choudhury, S. Saied, J.L. Sullivan, A.M. Abbot, Reduction of oxides of iron, cobalt, titanium and niobium by low-energy ion bombardment. J. Phys. D: Appl. Phys. **22**, 1185–1195 (1989)
182. A.G. Mathewson, Ion induced desorption coefficients for titanium alloy, pure aluminum and stainless steel, Report CERN-ISR-VA/76-5 (1976)
183. M.P. Lozano, Ion-induced desorption yield measurements from copper and aluminum. Vacuum **67**, 339–345 (2002)
184. N. Hilleret, Influence de la nature des ions incidents sur les taux de desorption par bombardement ionique de molécules adsorbées sur une surface d'acier inoxydable, Report CERN-ISR-VA/78-10 (1978)
185. M.-H. Achard, Desorption des gaz induite par des électrons et des ions de l'acier inoxydable, du cuivre OFHC, du titane et de l'aluminium purs, Report CERN-ISR-VA/76-34 (1976)
186. M.-H. Achard, R. Calder, A. Mathewson, The effect of bakeout temperature on the electron and ion induced gas desorption coefficients of some technological materials. Vacuum **29**, 53–65 (1979)
187. D.G. Armour, P. Bailey, G. Sharples, The use of ion beams in thin film deposition. Vacuum **36**, 769–775 (1986)
188. H.F. Winters, P. Sigmund, Sputtering of chemisorbed gas (nitrogen on tungsten) by low-energy ions. J. Appl. Phys. **45**, 4760–4766 (1974)
189. E. Taglauer, Surface cleaning using sputtering. Appl. Phys. A **51**, 238–251 (1990)
190. G. Carter, D.G. Armour, The interaction of low energy ion beams with surfaces. Thin Solid Films **80**, 13–29 (1981)
191. H. Niehus, W. Heiland, E. Taglauer, Low-energy scattering at surfaces. Suf. Sci. Rep. **17**, 213–303 (1993)
192. H.F. Winters, E. Taglauer, Sputtering of chemisorbed nitrogen from single-crystal planes of tungsten and molybdenum. Phys. Rev. B **35**, 2174–2187 (1987)
193. Morita, K., Mori, H., Horino, Y.: Partial cross-section for sputtering of monolayer impurity atoms on metal surfaces. Nucl. Instr. Meth. B **1**, 407–410 **8** (1986)
194. M. Inoue, S. Tanaka, J. Yukava, K. Morita, Ion impact desorption of metal atoms from Si(111)-$\sqrt{3}$ x$\sqrt{3}$ -metal surfaces. Nucl. Instr. Meth. B **58**, 411–416 (1991)
195. E. Taglauer, W. Heiland, R.J. MacDonald, The study of sputtered effects in oxides and metal-adsorbed-gas systems using combined analytical techniques. Surf. Sci. **90**, 661–675 (1979)

196. A. Benninghoven, Die Analyse monomolekularer Festkörperoberflächenschichten mit Hilfe der Sekundärionenemission. Z. Physik **230**, 403–417 (1970)
197. Taglauer, E.: Desorption, in data compendium for plasma-surface interactions. In: Langley, R.A., Bohdansky, J., Eckstein, W., Mioduszewski, P., Roth, J., Taglauer, E., Thomas, E.W., Verbeek, H., Wilson, K.L., (eds.) Nucl. Fusion **24**, 43–54 (1984)
198. T. Sikola, L. Dittrichova, J. Spousta, J. Zlamal, J. Stefka, Cleaning of metal surfaces by a broad beam ion source. Nucl. Instr. Meth. Phys. Res. B **127**(128), 865–868 (1997)

Chapter 10
Ion Beam-Assisted Deposition

Abstract In this chapter, the ion beam-assisted deposition (IBAD) method is presented, which is characterized by the deposition of material in high vacuum and the simultaneous bombardment of the surface with hyperthermal or low-energy ions. The individual sub-processes of the layer deposition, such as the generation of the atoms and ions, the transport of the particles through the vacuum and the processes on the surface during the deposition of the atoms under low-energy ion beam irradiation are discussed. The layer growth under ion irradiation is described. In detail, the significance of the ion energy, the temperature and the ion-to-atom arrival ratio for the layer growth is presented. Individual aspects such as epitaxial growth, evolution of topography, grain size, texture, biaxial orientation, mechanical layer stresses, and densification as a result of ion beam-assisted deposition are considered. The explanations and models for these aspects known in the literature are presented and critically discussed. A decisive advantage of IBAD over conventional deposition methods is the controlled fabrication of thin compound films, in particular nitride and oxide films. Examples of the formation of various nitride and oxide layers are presented.

Ion beam-assisted deposition (IBAD) is a high-vacuum-based physical vapor deposition method that supports the deposition of thin films by bombardment of energetic ions, with the aim of changing the thin film growth process and the film properties (for reviews, see [1–7]). This method is characterized by an independently controllable and simultaneous flux of ions and atoms to the substrate surface. Mattox (see e.g., [8]) was the first to recognize the positive role of energetic ions within the plasma on the modification of the films grown. Subsequently, Pranevicius [9] and Weissmantel et al. [10] carried out the first ion beam-assisted deposition experiments. The driving force behind the extensive development of this technique is the potential for controlling film properties, specifically, morphology, density, texture, epitaxy, composition, and internal stress.

Generally, the IBAD process is divided into two categories: *non-reactive* and *reactive ion beam-assisted deposition*. For the non-reactive ion beam-assisted deposition process, energetic noble gas ions are used, which introduce their momentum into the growing layer. The formation of compounds is the preferential aim of reactive ion beam-assisted deposition, where most studies have been focused on the formation

B. Rauschenbach, *Low-Energy Ion Irradiation of Materials*, Springer Series in Materials Science 324, https://doi.org/10.1007/978-3-030-97277-6_10

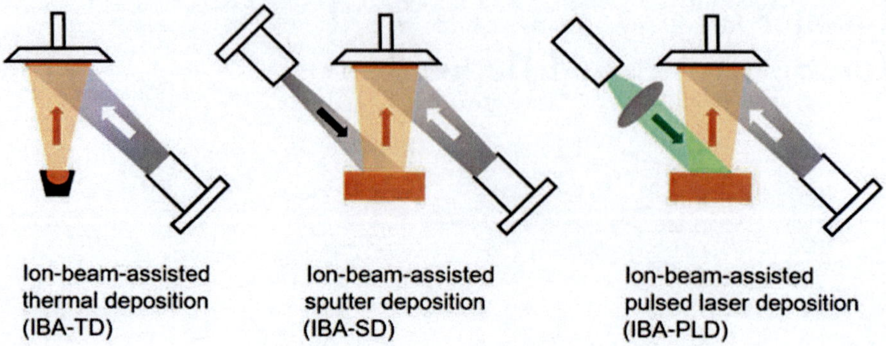

Fig. 10.1 Schematic configurations of ion beam-assisted deposition equipment

of oxides and nitrides by oxygen and nitrogen ion beam bombardment, respectively (see Sect. 10.5).

The typical energy of the assisted ion beam varies between 100 eV and 3000 eV. However, hyperthermal ion energies (10 eV…100 eV) as well as higher ion energies (up to some tens of keV) are also utilized. Usually, deposition rates or deposition velocity are experimentally aspired to be between 0.1 nm/s and 10 nm/s. Assuming that a stoichiometric compound $A_{0.5}B_{0.5}$ is to be synthesized and the stitching probability for both components is one, it will take about 8 seconds to deposit a monolayer at an ion current density of 10 $\mu A/cm^2$ (1 $\mu A/cm^2 \approx 6.25 \cdot 10^{12}$ ions/cm^{2}s). Typical configurations of ion beam-assisted deposition are illustrated in Fig. 10.1. The film materials are typically generated by electron beam or thermal evaporation (left), sputtering (middle), or pulsed laser beam irradiation (right). The ion bombardment is frequently realized by Kaufman sources, end Hall sources, or microwave ion sources (for details, see e.g., [2]). Over the past several decades, a variety of IBAD configurations have been presented and successfully applied (see e.g., [11]).

The atomic species can be supplied from a number of different sources. The production modality of the atomic species significantly determines the properties of the thin films deposited. In particular, the energy and angular distribution of the generated atomic species are dependent upon the production technique. Figure 10.2 provides a helpful comparison of the energy and angular distribution of atomic species produced by different methods. Using thermal and electron beam evaporation, the energy of the atomic particles ranged between 0.1 eV and 0.4 eV. In contrast, the energy of the atomic species produced by sputtering or laser ablation is wide-stretched (up to 100 eV), with a maximum at some electron-volts (Fig. 10.2, left). Consequently, a distinction can be made between the production of low-energy atomic species with a narrow energy distribution by thermal and electron evaporation and of high-energy atomic species with a broad energy distribution by (magnetron) sputter or laser ablation techniques. The emission characteristics of the atomic particle sources also differ significantly (Fig. 10.2, right). Under the condition of a small-surface source and a relatively long distance between the source and the substrate

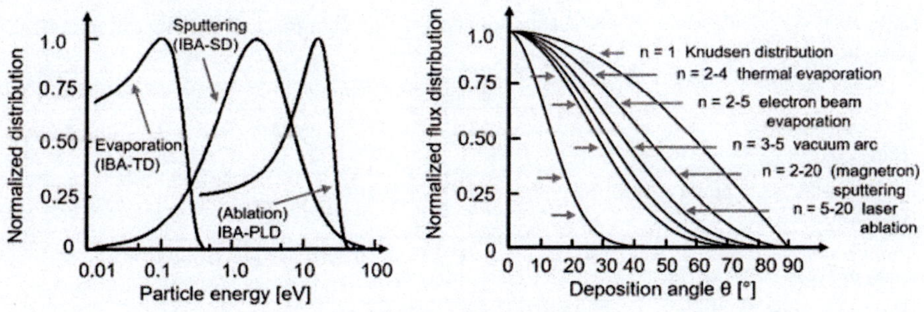

Fig. 10.2 A schematic comparison of typical energy (left) and typical angular distributions (right) of the atomic species for the most frequently used IBAD methods. For comparison, the thermal energy of an ideal gas at room temperature is also given

in comparison to the surface area of the particle source, the angular distribution of the emitted atomic species can be described by the Knudsen cosine distribution [12]. This characteristic is consistent with $\cos^n\theta$ distribution, where θ is defined as the angle between the surface normal and the direction of the incidence beam of atoms and n is a real number (see Appendix K for details). While the IBAD technique based on thermal and electron beam evaporation (see Fig. 10.2, right) is characterized by a small particle beam divergence, ion beam-assisted sputter (IBA-SD) and ion beam-assisted pulsed laser deposition (IBA-PLD) are marked by comparably larger particle divergence. Increasing beam divergence increases the probability of (i) non-perpendicular incidence of atomic species, (ii) significantly lower density of the deposited layer compared to the bulk material (iii) noticeably altered layer morphology (e.g., formation of voids, columns, etc.), and (iv) significantly increased layer roughness.

The different IBAD methods operate under unequal vacuum ranges. As the working pressure increases, the probability of contamination and the incorporation of residual gas atoms also significantly increases. The residual gas flux at room tempera-ture can be calculated by the Hertz-Knudsen equation (see Appendix K) to $J_R \approx 7.1 \times 10^{18}$ p, where the average mass of the residual gas atoms/molecules is known [see also (10.2)], J_R is given in units of [cm^{-2} s^{-1}] and the residual gas pressure p is given in units of [Pa]. On this basis, the residual gas flux can be estimated for each IBAD method (see Table 10.1), where typical working pressures were assumed. If the particle flux during deposition is significantly larger than the residual gas flux (in both cases, the sticking probabilities are assumed to be one), the contamination of the growing film can be minimized when the film-forming particle flux is comparable or larger than the residual gas flux. With the exception of ion beam-assisted sputter deposition (working gas pressure determined by the integrated sputter ion source), this situation is achieved during IBAD under ultra-high vacuum conditions. This means that IBA-TD and IBAD-PLD techniques are characterized by greatly reduced contamination and incorporation of impurities. This can also be achieved for the other preparation methods if the production of the ions (ion source) and the

Table 10.1 Residual gas flux on the substrate surface at room temperature (RT) for different IBAD methods and their typical working pressures

Method	Typ. working pressure	Typ. residual gas flux
Ion beam-assisted deposition (IBA-TD)	$>10^{-4}$ Pa ($>10^{-6}$ mbar)	$>7.1 \times 10^{14}$ cm^{-2} s^{-1}
Ion beam-assisted sputter deposition (IBA-SD)	$>10^{-1}$ Pa ($>10^{-3}$ mbar)	$>7.1 \times 10^{17}$ cm^{-2} s^{-1}
Ion beam-assisted pulsed laser deposition (IBA-PLD)	$>10^{-6}$ Pa ($>10^{-8}$ mbar)	$>7.1 \times 10^{12}$ cm^{-2} s^{-1}

deposition of the ions and the atomic particles is realized in different interconnected vacuum systems.

Numerous variations of the IBAD technology are known to exist. One of these technologies is direct ion beam deposition (IBD), which uses hyperthermal ions of different materials to deposit thin films (for details, see Chap. 9). The advantages of IBD are precise mass selection of the atomic species, the fluence, and the uniformity of the deposited layers. Several sputter technologies have also been used successfully. One of the most popular sputtering technologies is the magnetron sputtering technique [13]. This technique involves a gaseous plasma generated in a space containing the target material to be deposited. The surface of this material is sputtered by energetic ions within the plasma and the sputtered atoms travel through the gas environment and deposit as thin film onto a substrate. In contrast to the IBA-TD technique, the advantages of the (unbalanced and pulsed) magnetron sputter technique are the rapid, cost-efficient, and large-area production of thin films. Disadvantages include the high kinetic energy of the sputtered particles, the high working pressure, and the different charge states of the ions used.

10.1 Ion Beam-Assisted Deposition Process

The process of ion beam-assisted deposition can be subdivided into three steps:

(1) generation of ions and atoms in separate sources,
(2) transport of the particles through a vacuum or a gaseous atmosphere,
(3) simultaneous deposition of both particle species on the substrate surface.

10.1.1 Generation of Ions and Atoms

Ions Extracted from Ion Source

Regardless of the concrete arrangement in the IBAD experiments, the preparation of monoatomic or/and polyatomic ion species is a necessary prerequisite for IBAD experiments. The choice of the ion source in an IBAD equipment is determined by a variety of factors including ion energy, ion species, ion current, cleanliness, ion beam diameter, beam homogeneity, ability to adapt to the vacuum system, working pressure, etc. The most widely used low-energy ion sources in IBAD systems are the so-called broad beam ion sources (Kaufman sources, end-Hall and radio frequency ion sources; for more details see [2, 13, 14]). The ion source generates an incident ion current density, J, that typically consists of several ion species in different charge states. Consequently, the incident ion current density J_I (ions per unit area and time) is given by

$$J_I = \frac{J}{e}\sum_i n_i \epsilon_i, \tag{10.1}$$

where n_i is the fractional component of each ion species and ε_i is the number of atoms/ion species (e.g., one for N^+ or two for $N_2{}^+$, etc.) [15]. Thes total ion current or ion current density can be measured, e.g., using Faraday cups.

In almost all IBAD experiments, ions generated in the ion source are used to directly bombard the growing layer. However, the incident ion beam is not perfectly collimated. The energy of the emitted particles is characterized by a more-less broad energy distribution. Each small surface element of the bombarded target thus sees an incident beam with an angular spread of about the nominal incidence angle. This angular spread will be of particular importance when the erosion rate changes rapidly with the incident angle (see Chap. 5) or thin films and nanostructures are deposited under non-normal incident angles (see Chaps. 8 and 11). Figure 10.3 shows the calculated angular distribution of Xe ions produced within a Kaufman-type broad beam ion source equipped with a multi-aperture two-grid ion optical system [16]. The ion energy and the ion current density are constant. The angular distribution (beam divergence) can be partially controlled by the voltage applied on the ion optical system of the ion source. While the applied voltage on the first grid determines the energy of ions (here, 2 keV), the voltage on the second grid, the accelerator grid U_{acc}, is used to manipulate the angular distribution under which the ions leave the aperture. For example, the width of the distribution (FWHM) increases with $-U_{acc}$ from 3.3° at 200 V up to 7.3° for 1000 V.

The most frequently used broad ion beam sources are characterized by particle emission without mass and energy separation, although it is known that these ion sources emit atomic and molecular ions as well as neutral atoms and molecules of different masses. For example, a detailed diagnostic of the particle mass flux can be carried out through application of a quadrupole mass spectrometer. Figure 10.4 demonstrates the mass distributions of nitrogen ions generated by a Kaufman source. Essentially, molecular singly and doubly charged nitrogen ions and atomic nitrogen ions can be expected, but molecular nitrogen ions with other charge states and nitrogen isotopes are also produced. The ratio of the main nitrogen components N^+/N^{2+} in the ion beam ranged between 2/3 and 1/4 and is comparable to that for the Kaufman ion sources measured by Van Vechten et al. (usually varies between about 1/12 to 1/4, and is 2/3 at maximum) [17]. Retarding field or band pass filters

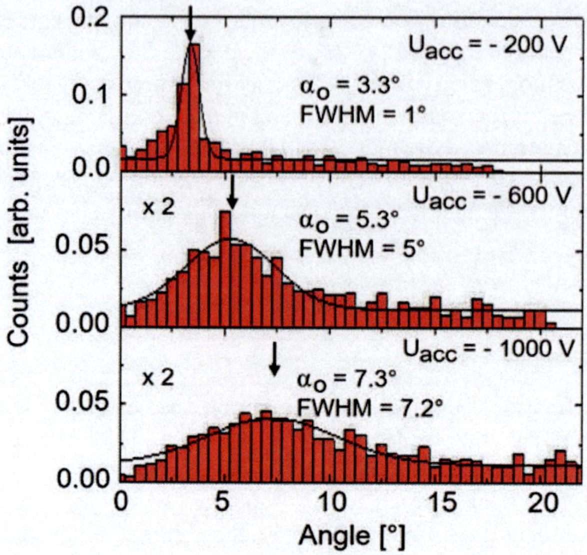

Fig. 10.3 Angular distributions of ions for different U_{acc} with $E = 2000$ eV and a plasma density of 2×10^{10} cm^{-3}. The plots are calculated using the simulation code IGUN [16]. The distributions are normalized to the number of ions. The solid curves represent a Gaussian fit, α_o the maximum and FWHM the width of the distribution. The arrows indicate the position of the distribution maximum

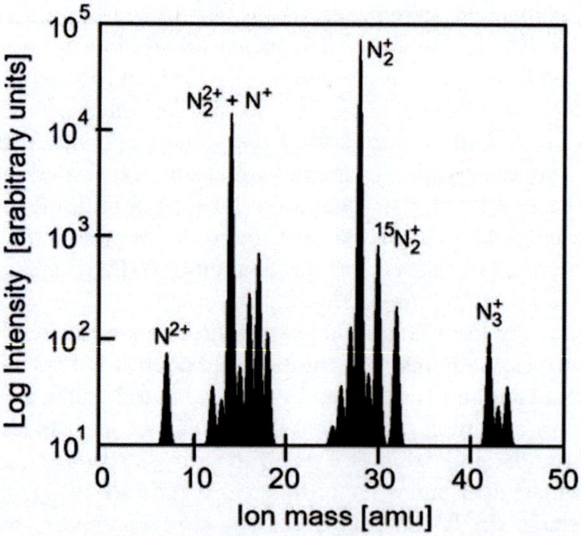

Fig. 10.4 Mass spectrum of a nitrogen ion beam generated by a Kaufmann ion source. The spectra were measured by energy-resolved mass spectrometry at the commercial broad beam ion source KF/F 40 of the company IOT GmbH Leipzig

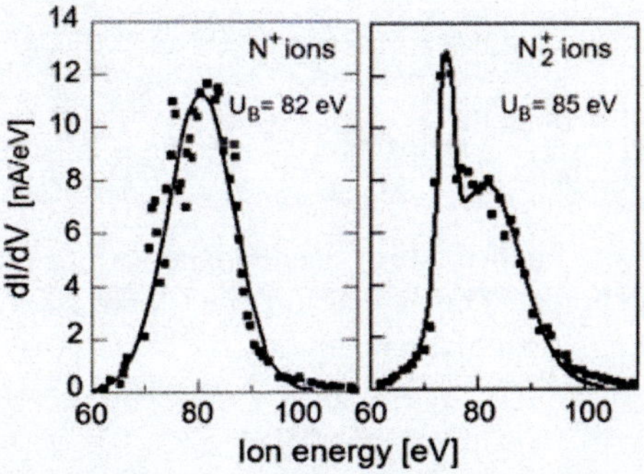

Fig. 10.5 Kinetic energy distribution for N^+ and $N_2{}^+$ generated by a Kaufmann ion source and measured by using RFA (retarding field analyzer) at the position of the sample surface. The beam voltages UB are very similar (82 and 85 eV). The FWHM is in both distributions is smaller than 20 eV

are applied in combination with mass-spectrometric methods to analyze the energy distribution of the generated ions. For example, the retarding field analyzer (RFA) determines the velocity distribution of ions by calculation of the deviation of the particle current in dependence on the applied retarding potential. Figure 10.5 shows an example of the kinetic energy distributions of atomic and molecular nitrogen ions generated by a Kaufman source. Consequently, in the majority of IBAD experiments, the used ion beam is characterized by a number of different atomic and molecular ions of an ion species, with different charge states. The properties of thin films prepared by IBAD should therefore also take this aspect into account. The application of mass and energy-selected ion beam-assisted methods can overcome this problem (see e.g., [18]).

The energy of hyperthermal (< 100 eV) and low-energy ions (< some keV) is frequently given by multiplying the applied power supply voltages by the electron charge. But for these small units of energies, ion energies must be given by the potential difference between the target and the source plasma rather than the ion chamber, since the plasma usually has a slightly higher potential than the ion chamber (plasma floating potential; for details, see Sect. 9.1.3).

It should be also noted that the generated ions flying through a gaseous atmosphere outside the ion source are subject to charge transfer processes (see Sect. 10.1.2.3). A fraction of the positively charged ions then become fast neutral atoms, which cannot be detected by Faraday cups or RFA's. They can, however, substantially affect the results of the ion beam-induced material modification processes.

Atoms Generated by Evaporation or Sputtering

The number of evaporated atoms hitting the surface of the substrate per unit area and time is given by Hertz-Knudsen equation (see Appendix K) as

$$J_A = \frac{p}{\sqrt{2\pi M_A k_B T}} \quad \text{resp.}$$
$$J_A\left[\frac{atoms}{\text{cm}^2\ \text{s}}\right] = 2.6 \times 10^{24} \frac{p[\text{Pa}]}{\sqrt{2\pi M_A[\text{amu}]T[\text{K}]}} \tag{10.2}$$

and referred to as the atom fluence rate of the deposited atoms (atoms per unit area and time) or the impingement rate. The deposition rate (deposition velocity) of deposited atoms is v_A [cm/s], and is then given by

$$v_A = J_A/N, \tag{10.3}$$

where N is the atomic number density (atoms per unit volume) of the deposited film material. Usually, the film thickness is measured in-situ using a vibrating quartz crystal microbalance method or with other analysis methods. For a precise measurement, dimensionless tooling factors must be introduced in (10.2) to account for differences between the position of deposited thin film and the quartz crystal with respect to the deposition source, and of the Faraday cup with respect to the ion source in (10.1), respectively.

IBAD experiments are preferably performed with evaporated or sputtered particles. The fraction dN_{ev}/N_{ev} of the evaporated atoms leaving the evaporation source (e.g., an effusion cell) with a kinetic energy between E_{ev} and $E_{ev} + dE_{ev}$ is given by

$$\frac{dN_{ev}}{N_{ev}} = F(E_{ev})dE_{ev}, \tag{10.4}$$

where $F(E_{ev})$ is the distribution function and can be expressed by the Maxwell–Boltzmann distribution as

$$F(E_{ev}) = 2\sqrt{\frac{E_{ev}}{\pi(k_B T)^3}} \exp\left(-\frac{E_{ev}}{k_B T}\right). \tag{10.5}$$

The maximum kinetic energy of evaporated atoms from an evaporation source is usually smaller than 0.2 eV and characterized by very narrow energy distribution. For example, Fig. 10.6 shows the normalized energy distribution of Fe particles produced by evaporation. This figure indicates that the kinetic energy of the evaporated atoms is about 0.1 eV and the energy distribution is very narrow. According to Thompson [19], the kinetic energy distribution of the sputtered atoms, E_{sp}, immediately after leaving the target is given by probability density function (see Sect. 5.7.1).

$$F(E_{sp}) = \begin{cases} K\frac{E_{sp}\left[1-\sqrt{(U_s+E_{sp})/\gamma_{sp}E}\right]}{(U_s+E_{sp})^3} & for \quad 0 \le E_{sp} \le \gamma_{sp}E \\ 0 & for \quad E_{sp} > \gamma_{sp}E, \end{cases} \tag{10.6}$$

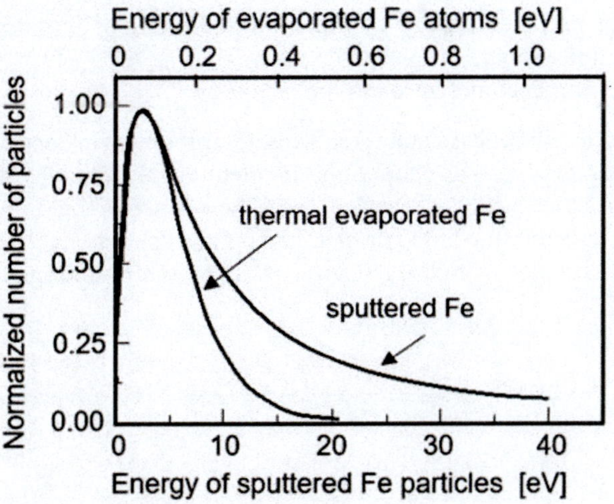

Fig. 10.6 Normalized energy distribution of Fe atoms after evaporation and sputtering

where the constant K is a normalization coefficient, E is the primary ion energy that creates the sputtered atoms by bombardment of the target, U_s is the surface binding energy of the sputtered material (see Sect. 5.4.1), and $\gamma_{sp} = 4M_i M_{sp}/(M_i + M_{sp})^2$, where M_i is the mass of the incident ion and M_{sp} is the mass of the sputtered target atom. Sputtered particles are generated when the maximum recoil energy is less than the energy of sputtered atoms (i.e. $E_{sp} \leq \gamma_{sp} E$).

Serikov and Nanbu [20] introduced a majorant function $F_{maj}(E_{sp})$, such that $F_{maj}(E_{sp}) \geq F(E_{sp})$ for any $E_{sp} \geq 0$ and determined normalization coefficient K (see also Sect. 5.7.1). With this, the energy distribution of the sputtered particles leaving the sputter target can be expressed as (c.f. 5.74)

$$F(E_{sp}) = 2\left(1 + \frac{U_s}{\gamma_{sp} E}\right)^2 \frac{U_s E_{sp}}{(U_s + E_{sp})^3} \quad for\ 0 \leq E_{sp} \leq \gamma_{sp} E. \tag{10.7}$$

In contrast to the energy distribution of evaporated atoms, the energy distribution of sputtered particles in Fig. 10.6 is characterized by a maximum at higher energies (some eV) and a large distribution tail in the higher energy range (up to 100 eV). In a rough approximation, the energy of sputtered particles is about ten times higher than that of evaporated particles. A more detailed consideration of Thompson formula leads to the following results (see also Sect. 5.7):

(i) the position of the maximum is half of the surface binding energy,
(ii) the distribution rises linearly to the peak at ½U_s and falls proportional to $1/E_{sp}^2$,

(iii) the most probable energy of the sputtered particles is independent of the incident ion energy,
(iv) the spectrum becomes zero at $\gamma_{sp} E$.

The angular distributions of sputtered particles, presented in Sect. 5.7, will depend upon various factors, such as energy and incident angle of bombarding ions, crystallographic structure of target materials, and the surface topography of the sputter target. If θ_{sp} is the emission angle measured from the sputter target normal, then the sputtered atoms are leaving the target surface with an energy and angular distribution [19] given by

$$F(E_{sp}, \theta_{sp}) = 2\left(1 + \frac{U_s}{\gamma_{sp} E}\right)^2 \frac{U_s E_{sp}}{(U_s + E_{sp})^3} \frac{\cos\theta_{sp}}{\pi} \approx \frac{2U_s E_{sp}}{(U_s + E_{sp})^3} \frac{\cos\theta_{sp}}{\pi}, \tag{10.8}$$

where the emission angle of sputtered particles, θ_{sp}, is given by (5.67) for not-too-high energies of the sputtered particles.

10.1.2 Traveling Through the Gas Environment

Ion beam-assisted deposition experiments are usually carried out under high or ultra-high vacuum conditions. However, the utilization of ion sources or a sputter gas atmosphere can lead to a significant increase in the residual gas pressure during the IBAD process. The transport of atoms or molecules through a background gas leads to changes in the energetic and angular distribution of the extracted material particles. This distribution may be very different from the Maxwell–Boltzmann distribution for particles generated under high- or ultra-high vacuum conditions (e.g., by thermal evaporation using a Knudsen cell) and is determined by the dissipation of the particle energy by collisions, referred to as 'thermalization'.

10.1.2.1 Mean Free Path

It is meaningful to know the average distance of the particles between these collisions, referred to as the mean free path. Taking into account the relative motion of all gas particles, the mean free path is given by (see Appendix I)

$$\lambda = \frac{1}{\pi\sqrt{1 + \frac{M_{sp}}{M_g}} N_g \sigma_{sp,g}^2} \quad (\text{approximation}: \lambda = 1/N_g \sigma_{sp,g}), \tag{10.9}$$

where N_g is the atomic density of gas particles and $\sigma_{sp,g}$ is the collision diameter for binary collision between sputtered particles and particles of the background gas, and can be expressed by the sum of the Lennard–Jones radii of the interacting particles. The quantity λ is proportional to the temperature and inversely proportional to the gas pressure (see Fig. I.2 in Appendix I). It should be noted that the mean free path estimated by (10.9) is a simple valuation, because the gas particles are assumed to be hard spheres and the (attractive or repulsive) potential between the particles is not taken into consideration.

Robinson [21] has derestricted the assumption that the cross-section is independent of the energy of the sputtered particles. Figure 10.7 shows the calculated cross-sections for energy and momentum transfer in binary collisions between noble gas atoms for low energies ($\leq$ 1 keV). Consequently, the mean free path is also dependent on the energy of the sputtered particles and increases rapidly with increasing energy. For example, the mean free path of 1000 eV Ar atoms in Ar atmosphere at 4 Pa and room temperature is about 23 times larger than the mean free path of thermal Ar atoms ($\approx$ 0.2 eV) in an identical environment.

Using (10.9), it is possible to determine whether collision processes account for the generated particle on the way through the residual gas, where the number of collisions is given by the source-substrate distance divided by the mean free path (see 10.14). Under low pressure conditions (< 0.1 Pa), the mean free path of particles is commensurate to or larger than the dimension of a common process chamber (see Fig. I.2 in Appendix I). The particles then have no or only a few collisions with the residual gas atoms. The particles reach the substrate surface without thermalization, i.e. without energy loss.

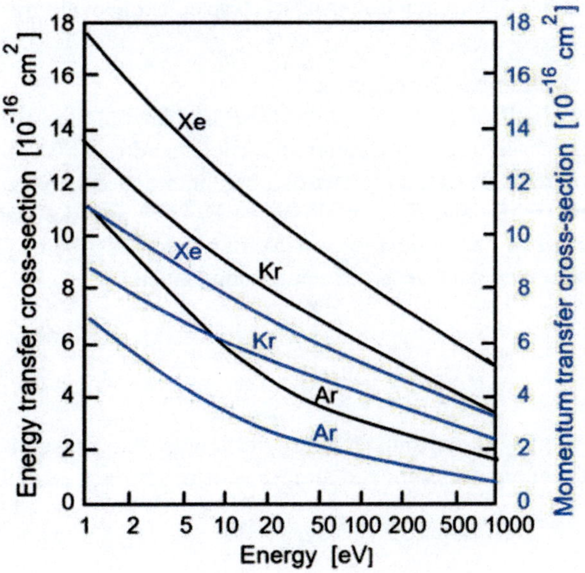

Fig. 10.7 Energy and momentum transfer cross-section for noble gas atoms as a function of the energy

In contrast, most sputtering experiments are conducted with pressures > 0.1 Pa. Consequently, multiple collisions are likely because λ is significantly smaller than the distance between the source and the substrate. Therefore, owing to collision events with gas particles in the sputter experiment, a clearly enlarged beam divergence can be also expected.

10.1.2.2 Thermalization

When the pressure is high enough, the evaporated or sputtered atoms transfer their energy to the background gas while traveling through the gas atmosphere by way of collisions that occur before they reach the substrate surface. After this thermalization process, the energy of these particles can no longer be described using the Thompson distribution. On the other hand, the number of collisions is not sufficient to achieve total thermalization (Maxwell–Boltzmann distribution). Especially for sputtered particles, a detailed analysis of the energy loss during transport is necessary.

Several approaches exist to describe the thermalization process, in particular, that of sputtered particles. These concepts are based on the following assumptions:

(i) The concentration of the generated particles by evaporation or sputtering is significantly smaller in comparison to the concentration of the background gas particles.
(ii) Only collisions between the generated particles and the background gas atoms are taken into consideration. Collisions between the generated particles themselves is ignored.
(iii) The interaction is characterized exclusively by elastic collisions.

Mechanical Description

A significant increase in the working gas pressure in deposition chamber collision processes between sputtered particles and residual gas atoms become increasingly more likely. Assuming hard-sphere interaction between a sputtered particle with the mass M_{sp} and the initial energy $E_{sp}{}^{before}$ and the stationary gas particle with the mass M_g, according to Gras-Marti and Valles-Abarca [22], the average energy of the sputtered particle after interaction is given by

$$\left\langle E_{sp}^{after} / E_{sp}^{before} \right\rangle = 1 - 2\frac{M_{sp} M_g}{\left(M_{sp} + M_g\right)^2} = 1 - \frac{\gamma}{2}, \tag{10.10}$$

where γ is the transfer energy efficiency factor (see 2.28) and E^{before} and E^{after} are the energies before and after the collision, respectively.

Lethargy Concept

Based on the neutron cooling theory developed by Fermi [23] and Keywell [24], lethargy has been introduced, i.e. an average energy quantity, expressed by

$$\ln\left(E_{sp}^{after}/E_{sp}^{before}\right) = 1 - \frac{\left(M_{sp} - M_g\right)^2}{2M_{sp}M_g}\ln\left|\frac{M_{sp} + M_g}{M_{sp} - M_g}\right| \equiv \xi, \tag{10.11}$$

or approximated by

$$\left(E_{sp}^{after}/E_{sp}^{before}\right) = 1 - \frac{2\gamma}{(1+\gamma)^2}, \tag{10.12}$$

which approximates the ratio of the kinetic energy of hard spheres before and after a collision. With this, the energy of a sputtered atom as it leaves the sputter target, $E_{sp,0}$, will be reduced to E_n after n collisions:

$$E_n = E_{sp,0}\exp(-n\xi). \tag{10.13}$$

where n is given by (see I.2)

$$n = \Lambda/\lambda = \Lambda\frac{p\sigma}{k_B T}, \tag{10.14}$$

where Λ is the distance between the source of the sputtered particles and the substrate surface and λ is the mean free path (see Appendix I). The average kinetic energy of the condensed atoms on the substrate surface after n collisions becomes [25]

$$\langle E_n\rangle = \left(E_{sp} - k_B T_g\right)\exp(n\xi) + k_B T_g, \tag{10.15}$$

where T_g is the sputter gas temperature (c.f. 5.73) and identical with the ambient temperature in a first approximation).

Consequently, the kinetic energy of the condensed particles without collisions on the way through the gas ambit is unchanged, while with increased working pressure, the kinetic energy is significantly reduced. Westwood [26] has pointed out that not only the particle energy loss varies with the mass ratio ξ, but also the scattering angle is dependent on this ratio, because high-energy ions have $\sqrt{2}$ fewer collisions than fully thermalized atoms. Then, the scattering angle is given by

$$\theta_{lab} = \arccos\left[\frac{1}{2}\left(\frac{M_g}{M_{sp}} + 1\right)\exp\left(-\frac{\xi}{2}\right) - \frac{1}{2}\left(\frac{M_g}{M_{sp}} - 1\right)\exp\left(\frac{\xi}{2}\right)\right]. \tag{10.16}$$

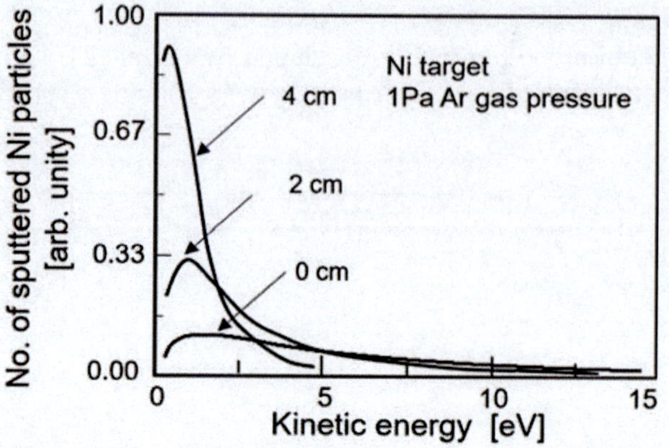

Fig. 10.8 Energy distribution of sputtered Ni particles as a function of the distance from the Ni sputter target in a 1 Pa Ar sputter gas atmosphere

It should be noted that the simple mechanical descriptions of the hard-sphere interaction or the lethargy concept are only approximations [22].

Monte Carlo Simulation

Another approach to studying the system consisting of atoms sputtered from a target that then travel through a background gas toward the substrate surface is based on the Monte Carlo simulations (see Sect. 5.7). Meyer et al. [25] and later, Gras-Marti and Valles-Abarca [22], have developed models to describe the energy dissipation process of sputtered particles in a sputter gas atmosphere on the basis of a hard-sphere potential and an approximate power-law potential with constant power, respectively. An example is shown in Fig. 10.8. From the comparison with Fig. 5.15, it can be seen that the energy distribution of the sputtering particles is identical to the sputtering gas atmosphere in the completely thermalized state. Other authors (see e.g., [27]) have conducted a Monte Carlo simulation using appropriate collision cross-sections and taking the total energy tail of the sputtered particles fully into account for the calculation of the thermalization process and the average collision number.

10.1.2.3 Charge Exchange

The charge exchange between ions in the beam and atoms and molecules in the residual gas is significantly influenced by interactions between ions, atoms and molecules. Important charge exchange processes include the neutralization dominant at low energies and the electron stripping at higher ion energies, where these processes are dependent on ion species, the initial charge state and energy, and the residual gas composition. Stripping processes are probable when the ion velocity exceeds the orbital electron velocity [28]. The exchange processes (neutralization) are based on the electron exchanges between the accelerated ions as well as between the atoms or molecules of the residual gas. The neutralization of low-energetic ions may occur via the following processes (here, X^+ and X^o are charged ions and neutralized atoms, respectively):

(a) resonance charge transfer between energetic ions from ion source, X^+ (fast) and neutral gas outside the ion source, X^o (thermal) by
X^+ (fast) + X^o (thermal) → X^o (fast) + X^+ (thermal), (cross-section ≈ 10^{-12} cm^2)

(b) radiative recombination of positive ions and electrons by
X^+ (fast) + e → X^o (fast) + hν, (cross-section ≈ 10^{-20} cm^2)

(c) dissociative recombination of molecular gas ions by
X_2^+ (fast) + e → X^o + X^o (cross-section ≈ 10^{-14} cm^2).

The resonance charge transfer processes thereby primarily determine the fraction of neutralized atoms in the ion beam. With this exchange process, the momentum of the accelerated ions is not changed (conservation of momentum). Thus, the affected particles will produce the same collision effects as ions that have not experienced such a charge exchange process. However, the ion current density measured by the Faraday cup is changed. The change in ion current density upon neutralization can be expressed by the relation

$$dJ_I/dJ_{I,o} = -\sigma_{ce} N_g \Lambda, \tag{10.17}$$

where J_I is the measured ion current density, $J_{I,o}$ the generated initial ion current density directly in front of the ion source, σ_{ce} is the cross-section of charge exchange, N_g is volume concentration (atomic number density) of residual gas atoms, and Λ is again the distance between ion source and the substrate surface. The ion current is then given by

$$J_I = J_{I,o} \exp\left(-\sigma_{ce} N_g x\right) = J_{I,o} \exp(-\sigma_{ce} x p / k_B T), \tag{10.18}$$

where p is the ambient gas pressure in the vacuum chamber and T is the temperature. Consequently, in order to minimize the neutral fraction of ion beams in a vacuum chamber, the gas pressure should be decreased or the distance x (ranging between 0 and Λ) between the ion source and the bombarded sample should be shortened. The current of the fast neutrals J_N after neutralization on the sample (detector), along with the beam of the charged ions, can be approximately determined to be

$$J_N = J_{I,o} - J_I = J_{I,o}\left[1 - \exp(-\sigma_{ce} x p / k_B T)\right]. \tag{10.19}$$

The cross-section σ_{ce} for most sputter and etching gases at collision energies smaller than some keV is unknown. For the neutralization of Ar ions in Ar gas, a cross-section of $\sigma_{ce} = 3 \times 10^{-15}$ cm^2 is assumed [29]. In Fig. 10.9 the fraction of neutral Ar gas atoms in an Ar ion beam as function of the Ar ion gas pressure and the distance between ion source and the substrate is shown. It is obvious that for all methods with a working pressure > 10^{-3} Pa, a significant fraction of neutral atoms

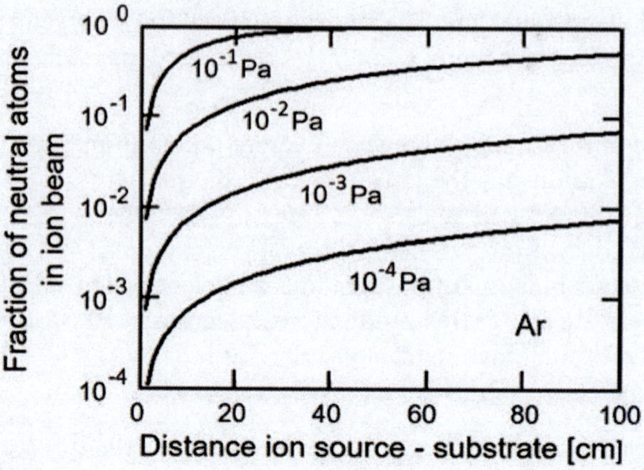

Fig. 10.9 Fraction of neutral Ar atoms in an Ar ion beam in dependence on working Ar gas pressure and the distance between the ion source and the substrate. A cross-section of 3×10^{-15} cm^2 is assumed

must be taken into account. For example, the neutral sputtering gas atoms sticking the substrate may be modified the topography, phase formation, defect concentration etc. in a completely uncontrolled manner.

10.1.2.4 Impingement or Arrival Ratio

The process of ion beam-assisted deposition is characterized by the simultaneous arrival of two particle fluxes, the ion flux and the atom flux, on the substrate surface (Fig. 10.10). When the diameter of the orifices of the particle sources (crucible) are small compared to the distances of this atom source to a planar substrate, the Knudsen cosine angular dependence of the flux can be expected (see Appendix K). Consequently, a flux distribution of atomic species (i.e. the variation of deposited film thickness) on the substrate surface is given by [30]

$$J_A = J_{A,0}\left[1 + \left(\frac{r}{\Lambda}\right)^2\right]^{-\left(\frac{n+3}{2}\right)}, \tag{10.20}$$

where r is the radial distance from that coaxial point, Λ is the distance between the source and substrate, and $J_{A,0}$ is the maximum of atom flux coaxial with the source (at point 0 in Fig. 10.10).

The angular distribution of ions emitted from the ion source (divergence) is assumed to be Gaussian, with a standard deviation angle φ_{div} (c.f. 6.56 and 6.57). The ion flux distribution on the substrate surface is then given by

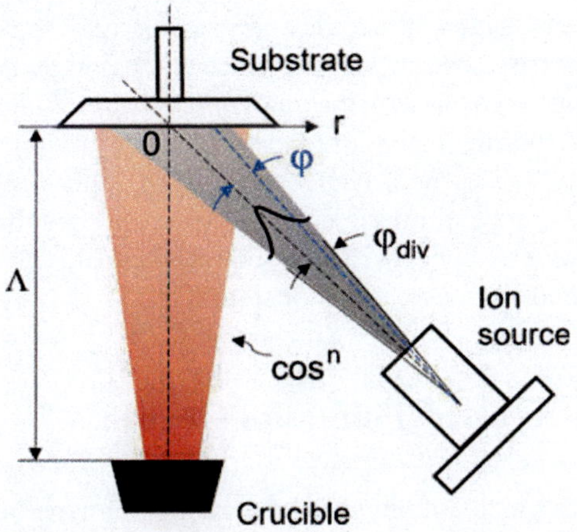

Fig. 10.10 Schematic representation of the process of ion beam-assisted deposition. The color scales correspond to the variation of the particle flux density

$$J_I = J_{I,0} \exp\left[-\frac{1}{2}\left(\frac{\varphi}{\varphi_{div}}\right)^2\right], \tag{10.21}$$

where $J_{I,0}$ is the maximum of ion flux coaxial with the source (at point 0 in Fig. 10.10) and φ is the angle between the center of the ion source and a point of interest on the substrate surface (blue dashed lines in Fig. 10.10). Consequently, the angular flux distribution of both particle fluxes along the surface of the substrate causes local variations of the thickness and also to the physical and chemical properties of the deposited films.

The impingement ratio, also referred to as the arrival ratio or ion-to-atom ratio, is defined as the ratio between the number of energetic particles (ions, charged molecules) per unit area and time and the number of deposited atoms effectively remaining in the growing film. The impingement ratio can then be expressed by

$$R = J_I / J_A. \tag{10.22}$$

The ion-to-atom ratio is a significant measure, because on the basis of this parameter, the changes in quality in films prepared by ion beam-assisted deposition can be correlated as a function of deposition and bombardment magnitudes.

Sputter and reflection processes on the growing film as the result of the incoming particles sustainably affect this impingement ratio. To take into account these surface interaction processes, Van Vechten et al. [17] proposed an approach to correcting this ratio. This corrected impingement ratio is given by

$$R_c = \frac{\langle\delta\rangle(1+\beta)}{eN}\frac{\zeta_I}{\zeta_A}\frac{J_I}{J_A} = K\frac{J_I}{J_A} = KR, \tag{10.23}$$

where $\langle\delta\rangle$ is the average value of the ionic charge per atom, $1+\beta$ is a term containing the charge exchange cross-section, e is the electron charge, N is the atomic number density of the film (e.g., measured by the quartz mass balance monitor), and ζ_I and ζ_A are the dimensionless tooling factors used to account for differences in the placement of substrate and Faraday cup with respect to the ion source and for differences in the placement of substrate and quartz crystal with respect to the deposition source, respectively. All parameters in (10.23) are independent of both the J_I and J_A fluxes and can be summarized in a correction constant K.

10.2 Ion Beam-Assisted Thin Film Growth

Thin film formation as a result of physical vapor deposition methods has been studied extensively and in detail. Some of the important reviews and books are cited here [31–34], which may be helpful to those with an interest in the general issue of thin film growth. The influence of low-energy ion bombardment during ion beam sputtering or direct ion beam deposition has also been extensively discussed (see e.g., [35]). The following subchapters are focused on the role of low-energy ion bombardment of the growing film within the framework of ion beam-assisted deposition.

As shown in Fig. 10.11, numerous effects can significantly contribute to affecting the IBAD process and must be taken into account in interpreting the film properties after IBAD. As is generally known, the ion beam-assisted film formation process

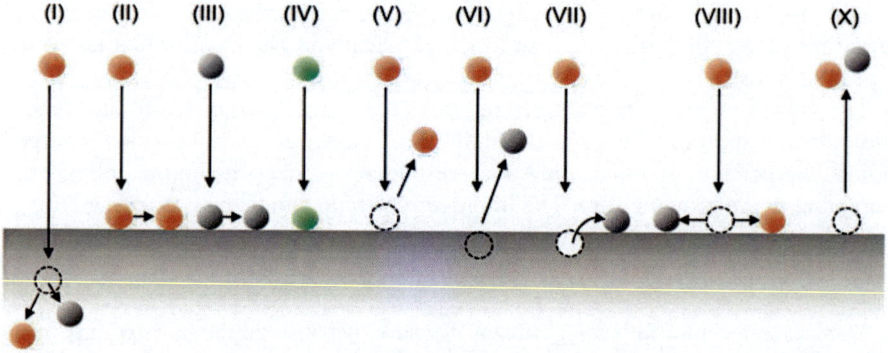

Fig. 10.11 Schematic representation of the elementary processes involved in ion beam-assisted deposition with low-energy ionized atoms (ions). The energetic ions, the deposited atoms and residual gas atoms/molecules are shown in orange, gray and green colors. respectively. (I) implantation of ions and formation of defects (interstitials and vacancies), (II) direct ion deposition (soft-landing) on surface and thermal diffusion of the energetic ion, (III) physical vapor deposition of atoms and thermal diffusion of the deposited atom, (IV) deposition (incorporation) of residual gas atoms/molecules, (V) reflection of energetic ions, (VI) sputtering of substrate and/or film atoms, (VII) formation of adatoms and vacancies at the surface, (VIII) ballistically induced surface diffusion, (IX) desorption of atoms or neutralized ions

is based on physical deposition of the film material with simultaneous ion irradiation. The motivation for performing additional ion bombardment is to expand the application fields for thin films and to sustainably control the film properties.

The first-order parameter of the IBAD process is the energy of the applied ions [1]. According to Fig. 10.11, the energy of the accelerated ions (or ionized molecules) determines the implantation process (I) or the ion direct deposition process (II, see also Chap. 9), the dissociation of molecules (V), the reflection of the incident ions (V), and the sputtering of the substrate or the growing film (VI) as well as the ballistic diffusion on the surface (VIII). Consequently, the choice of ion energy is highly significant. The interaction of the accelerated ions with the substrate at the beginning of the IBAD process and later, with the growing film, can be approximately described using screened Coulomb potential (see Chap. 2). However, for lower energies (e.g., hyperthermal energies), this approximation is not appropriate. In this energy range, both attractive as well as repulsive interactions account for interaction potential. Consequently, the prediction of both the ion range, including the spatial distribution, as well as the depth-dependent distribution of the ion beam-induced interstitials and vacancies (damage distribution) become increasingly inaccurate with decreasing ion energy.

The position of the implanted ion in rest can be a stable lattice position of a formerly replaced lattice atom, an interstitial position, or an instable (mobile) interstitial position. It has been frequently observed that the experimental ion range data tend to be somewhat higher than theoretical predictions (more than 30%) in the lower-energy region, $0.03 \leq \varepsilon \leq 0.3$ [36]. However, it is also generally accepted that the Born–Mayer potential in the hyperthermal energy range and the universal potential for low energies >1 keV provide more reasonable results.

10.2.1 Deposition Without Assisted Ion Beam Bombardment

In conventional thin film deposition, atoms or molecules impinge on the substrate surface from the vapor phase. These particles enter the adsorbed states and are able to desorbe and to diffuse over the surface. These so-called ‘adatoms’ combine with one another and form clusters. In dependence on the temperature and the pressure as well as the attachment and the detachment of adatoms, the clusters may grow or decay in size. Under steady-state conditions, for a critical cluster size, the probability of growth is greater than the probability of decay. All clusters larger than a critical cluster are indicated as stable clusters. These stable clusters grow continuously by capture of other diffusing adatoms or evaporated atoms directly from the vapor until these clusters form a continuous film.

According to Bauer [37], three growth modes can be distinguished as a function of the coverage $\overline{\theta}$ in monolayers (ML) (Fig. 10.12). The island growth mode, or Volmer-Weber growth mode, or 3D growth mode is characterized by discrete three-dimensional islands on the surface of the substrate, because the attractive interaction of the deposited atoms with each other is stronger than that with the atoms of the

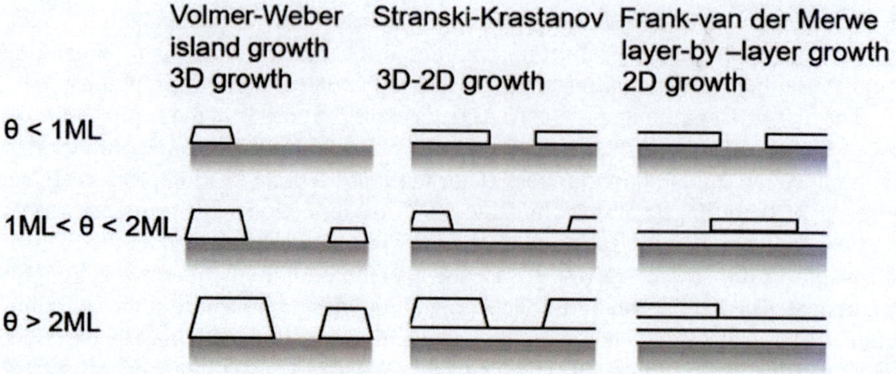

Fig. 10.12 Schematic representation of the growth modes as function of the coverage $\overline{\theta}$ in ML (monolayers)

substrate. Ultimately, the islands grow in size until they intersperse with each other to form a continuous film. If the interaction forces between the deposited atoms and the atoms of the substrate are stronger than those between neighboring film atoms, the growth mode is denoted as Frank-van der Merwe, or layer-by-layer, or 2D growth mode. The Stransky–Krastanov, or 2D-3D growth mode combines the features of layer-by-layer growth and discrete three-dimensional growth. Bauer [37] has also used the Young relation to classify the growth modes of thin films into three categories. The relation between the surface tension of a solid, γ_s, the deposited film, γ_f, the solid-film interface γ_{sf} and the contact angle Θ can be expressed as

$$\gamma_s - \gamma_{sf} = \gamma_f \cos \Theta . \tag{10.24}$$

In the presence of total wetting (given for the layer-by-layer growth), $\Theta = 0$ and $\gamma_f + \gamma_{sf} \leq \gamma_s$, and for island growth, $\Theta = 1$ (no wetting) and $\gamma_f + \gamma_{sf} > \gamma_s$. Consequently, the desired layer-by-layer growth can only be obtained for so-called "rigid" substrates, where the substrate imposes the lattice constant onto the deposited film and the lattice misfit is small. For films grown in the three-dimensional island mode, the reverse condition must be satisfied. There are two possible reasons for the transition from two-dimensional growth to the three-dimensional morphology in the Stranski–Krastanov growth mode [34]. First, the film material can grow in the first few monolayers in a crystallographic structure that differs appreciably from its own bulk structure. The Stranski–Krastanov growth mode will then be accompanied by crystallographic changes to the bulk lattice structure of the film. The second possible reason is the formation of strain reliefs by increases to the surface area (e.g., surface buckling). When the strain energy is high, the increase in surface area is more than compensated.

Film growth by physical deposition without assisted ion bombardment is based on individual atomic processes (see e.g., the comprehensive review [38]):

- deposition of atoms
- formation of dimers
- attachment of adatoms to form islands or dimers
- detachment of atoms from islands
- re-evaporation (desorption) of adatoms
- thermally stimulated diffusion into substrate
- coalescence of islands
- direct deposition onto islands.

Molecules or atoms are evaporated with flux J_A onto the substrate, where they are immediately thermalized. The subsequent surface diffusion is given by

$$D_s = \frac{a^2 \nu_o}{4} \exp(-E_{sd}/k_B T), \tag{10.25}$$

where E_{sd} is the activation energy for surface diffusion, a is the surface nearest neighbor distance (also called jump length), and ν_o is the vibrational frequency of an atom in order of 10^{13} Hz.

The extent to which the various processes occur is determined by the atom deposition fluence or deposition rate and the temperature, which control the diffusion processes and can be described by set of deterministic reaction–diffusion equations [39, 40]. It is assumed that the adatoms or monomers with a concentration N_1 are mobile, meanwhile dimers are stable and immobile. The first rate equation describes the atomic processes during the early stages of film formation that influence the concentration of adatoms and can be expressed by

$$\frac{dN_1}{dt} = J_A - 2\sigma_1 N_1^2 - D_s \sum_{s \geq 2} \sigma_s N_1 N_s. \tag{10.26}$$

The first term on the right-hand side of (10.26) denotes the temporal increase of monomer density due to the deposition flux J_A (see 10.2). The second and third terms on the right-hand side of (10.26) represent the decrease of the monomer density due to the concurrence of two diffusing adatoms (which create a dimer) and the decrease due to capture of a monomer by a stable island, where N_s is the concentration of islands with s atoms ($s \geq 2$). D_s is the surface diffusion coefficient of adatoms. The factor 2 in the second term describes the fact that the formation of a dimer corresponds to the loss of two monomers. In (10.26), σ_1 and σ_s represent the dimensionless capture numbers of the monomers and islands, respectively. The capture numbers characterize the reaction cross-section of monomers and islands and can be assumed to be constant in a first approximation. It should be noted that the dissociation of dimers, re-evaporation of monomers, coalescence of islands and the direct impingement of atoms on islands were ignored. However, the right-hand side of (10.26) can be extended by two additional terms that describe the decrease in monomer density caused by direct impingement onto islands and monomers.

A second rate equation describes the temporal evaluation of the stable island concentration for $s \geq 2$ in a similar manner

$$\frac{dN_s}{dt} = D_s\sigma_{s-1}N_1N_{s-1} - D_s\sigma_s N_1 N_s, \tag{10.27}$$

where the first and the second term on the right-hand side of (10.27) refer to the increase of island density N_s due to the formation of dimers, either by the encounter between two diffused monomers (first term) or upon direct deposition onto an adatom (second term). A further term, $-2N_s(J_A\text{-}dN_1/dt)$, is added if the coalescence must be taken into account [41]. Introducing the average concentration of stable islands N_m and the average capture number σ, (10.26) and (10.27) can be simplified to

$$\frac{dN_1}{dt} = J_A - 2\sigma_1 N_1^2 - D_s\sigma N_1 N_m \tag{10.28}$$

and

$$\frac{dN_m}{dt} = D_s\sigma N_1^2. \tag{10.29}$$

The equation system (10.25) and (10.26) with an unlimited number of equations is thereby reduced to a solvable system of two coupled partial differential equations. Equations (10.26) and (10.28) both provide quantitative information about island density and island size distribution.

The time evolution of monomer and island densities can be calculated by integration of the rate equations (10.28) and (10.29). At beginning of the film deposition, the adatoms with a density of N_1 diffuse on the surface of the substrate with D_s until these adatoms meet other adatoms and create a dimer. The density of dimers N_2 increases linearly until $N_2 \sim N_1$. Then, the probability that a diffusing monomer will encounter another monomer or a dimer is approximately equal. Consequently, the density of stable nuclei N_s, where $s \geq 2$, increases until this density attains saturation at a coverage for $\overline{\theta} \approx 0.15 \ldots 0.2$ monolayers (ML). The mean free path of the diffusing adatoms, λ, is then comparable to the mean distance between islands L. Villain et al. [42] derived a simple expression for the relation between this distance, L, and the ratio D_s/J_A. They obtained

$$\frac{D_s}{J_A} \approx \frac{L^6}{\ln(L^2)}, \quad \text{and because the denominator is small,} \quad L \approx \left(\frac{D_s}{J_A}\right)^{1/6}, \tag{10.30}$$

meaning that the distance between the island depends upon or can only be controlled by the ratio D_s/J_A. The growth of islands is thereby preferred, because the adatoms reach these islands and are attached for $\theta > 0.2$ ML. The further growth of a monolayer is now characterized by island coalescence and percolation.

10.2.2 Deposition Under Assisted Low-Energy Ion Bombardment

The effect of additional low-energy ion bombardment during deposition on the growth of thin films was studied both experimentally and with computer simulation methods. In addition to the atomic process without ion bombardment, other impact-induced processes additionally influence the growth of thin films. Figure 10.11 presents the main elementary processes, based upon the ballistic interaction of the incidence particle with the atoms of the surface substrate and the growing film. These ion impact processes influence the growth of thin films during the film buildup. Figure 10.13 (c.f. Fig. 9.2) presents a summary of these non-thermal ion impact processes (without deposition of atoms and adsorption of residual gas atoms) that must be taken into consideration (partly in addition to the processes in Fig. 10.11).

The deposition of atoms with thermal energy from the vapor phase and their contribution to film growth (surface diffusion, formation of nuclei, coalescence of islands, etc. (see Fig. 10.12)), as well as non-thermal ion impact processes enforce the growth of thin films. The increased mobility and impact-induced attachment lead to film growth but also lead to the formation of adatoms and vacancies (processes X, Fig. 10.13). The adatoms can fill up vacancy sites (XIII) or lead to the growth of islands (XIV). However, the ion impact can also lead to fragmentation of the growing layer (process XI) or detachment of individual adatoms (process XII). Low-energy ion bombardment can also add to film depletion by desorption (XV). The ion beam-induced mixing provokes a gradual transition between the film and the substrates, resulting in an enhanced adhesion of the film (XVI). When reactive ions are used, a controlled phase formation in the growing film can be achieved (XVII).

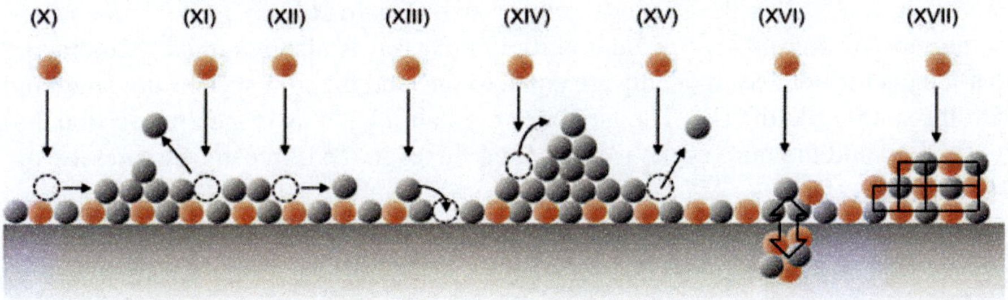

Fig. 10.13 Schematic representation of the non-thermal ion impact processes on film growth during low-energy ion bombardment (ions are colored orange and the deposited atoms are colored gray): (X) impact-induced attachment of adatoms (incl. diffusion) at steps or edges of terraces, (XI) fragmentation of islands (ion impact can cause the breakup of an island into several pieces), (XII) impact-induced detachment (steps or dimers are dissolved and isolated atoms (adatoms) on the surface are generated), (XIII) filling up a vacancy or vacancy island with single adatom (recombination), (XIV) pileup near descending steps (impact below steps can involve growth on the upper terrace), (XV) ion beam-induced desorption or sputtering, (XVI) ion beam-induced mixing, (XVII) ion beam-induced phase formation

Due to the complexity and large number of contributing processes, only very few experimental studies are known to exist regarding the initial stages in the formation of nuclei and islands under conditions of low-energy ion bombardment (see Sects. 9.2 and 10.3).

10.2.2.1 Average Film Composition

A thin film may be deposited by IBAD if the material deposition rate dominates over the sputter removal rate of the growing film. Conversely, film etching is observed. It is assumed that the film consists of two components with the composition A_xI_{1-x}. On the one hand, a component, I, made up of energetic particles with the energy E and the ion current density (flux) J_I, and on the other hand, an atomic component, A, which is deposited with the flux J_A. According to Berg et al. [43], a net deposition rate will be obtained only when the fractional coverage at the surface given by

$$x = \frac{J_A s_s}{J_I Y - J_A(s - s_s)}, \tag{10.31}$$

where Y is the sputtering yield, s is the sticking coefficients of the atoms to form the film onto the substrate and s_s is the sticking coefficient of atoms onto the substrate material. A net deposition can be realized only, if $x \geq 1$, i.e. if

$$J_I Y \leq J_A s. \tag{10.32}$$

The concentration distribution of the incorporated ion species in films deposited by concurrent material deposition can be approximately determined in a simple manner. According to (10.22), the impingement or arrival ratio $R = J_I/J_A$ is characterized by the ratio of both of the incident particle fluxes. It is also assumed that sticking coefficients for both components are equal to one and the film species are sputtered with the sputter yield Y(E). The number of remaining atoms is then proportional to $J_A - J_A Y(E)$ and the ratio of the incorporated fluxes in the film can be expressed by

$$\frac{J_I^f}{J_A^f} = \frac{J_I}{J_A - J_I Y(E)} = \frac{R}{1 - RY(E)}, \tag{10.33}$$

where this ratio is proportional to the concentration ratio of both components in the layer. Frequently, the sticking coefficient, s, of the incorporated ions is smaller than one. Then, (10.33) must be modified to

$$\frac{J_I^f}{J_A^f} = \frac{J_I}{J_A - J_I Y(E)} s = \frac{Rs}{1 - RY(E)}. \tag{10.34}$$

In more detailed consideration of the IBAD process, Van Vechten et al. [17] have taken into account that the ion reflection coefficient R_N (energetic particles can be reflected on the surface, see Sect. 5.10) and that an ion beam contains charged molecules in different charge states and neutrals (see Fig. 10.3) must be also considered. Then the incorporated flux ratio in the deposited film is given by

$$\frac{J_I^f}{J_A^f} = \frac{R_c(1 - \langle R_N \rangle)}{1 - R_c \langle Y(E) \rangle}, \tag{10.35}$$

where R_c is the corrected impingement ratio given by (10.23), $\langle R_N \rangle = \sum_i n_i \varepsilon_i R_{N,i} / \sum_i n_i \varepsilon_i$ is the weighted average reflection coefficient, and $\langle Y \rangle = \sum_i n_i \varepsilon_i Y_i / \sum_i n_i \varepsilon_i$ is the weighted average sputter yield (subscript i refers to I or A). The reflection coefficients and sputtering yields are expressed with r_i and Y_i, where n_i is the fractional component of each species, and ε_i is the number of atoms per ion species (see (10.1). The composition of a two-component film, A_xI_{1-x}, expressed as the atom fraction of the ionic species is then [14]

$$x = \frac{R}{R + [1 - R\langle Y(E) \rangle / (1 - \langle R_N \rangle)]} = \frac{R_c}{1 - R_c} \tag{10.36}$$

The reflection coefficient, $\langle R_N \rangle$, and the sputtering yield, $\langle Y \rangle$, are dependent upon ion energy and the incident angle.

10.2.2.2 Depth-Dependent Film Composition

In the last subchapter, the average concentration in films prepared by IBAD was considered in dependence of the sputtering yields, the reflection coefficients, and the ion-to-atom arrival ratio. The well-known fact that these films deposited on the substrate are not homogeneous remained disregarded. In particular, fluctuation of the film concentration at the interface between the substrate and the deposited layer and at the surface of the growing films was observed. Figure 10.14 shows an example of the concentration distribution of a BN film prepared by nitrogen ion beam-assisted deposition versus the depth on Si substrate. At the beginning of the deposition process under concurrent ion bombardment, the nitrogen ions are implanted into the substrate material. The modified depth in the substrate is dependent on the mean projected range of the applied ions. Intensive atomic redistribution by ballistic or collisional mixing of all participated species can be assumed. At high temperatures (homologous temperature of substrate > 0.2), the redistribution process is strengthened via radiation enhanced diffusion and thermal diffusion. As result, an intermixed buried interface is created. With increasing film thickness, the incoming ions are no longer capable of reaching the interface. A stationary growth then begins, which is characterized by a film stoichiometry proportional to the fluxes of both of the participated components (c.f. Fig. 9.3). In the near-surface region of the growing film, the stoichiometric

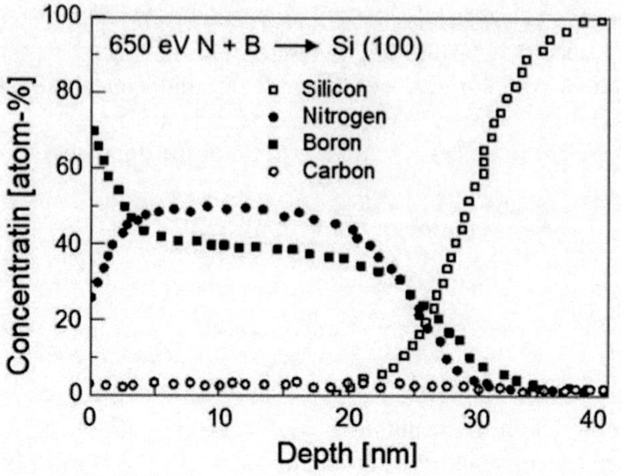

Fig. 10.14 Concentration distribution in a thin BN film on Si after boron deposition and concurrent 650 eV N ion bombardment at a substrate temperature of 350 °C and an arrival ratio of R = 1. The concentration distribution is measured by ERDA (210 MeV J^+-ions). For the depth calibration, a density of 2.1 g/cm^3 is assumed

composition can be changed by preferential sputtering of film components, diffusion processes, and the different manners of component deposition (thermal atoms are deposited on the surface and energetic ions are implanted into the film). TRIDYN simulations [44] confirm the considerable collisional broadening of the interface region, the stationary composition of film originated from the balance of boron deposition, the nitrogen implantation, and the boron sputtering.

The growth rate v_g of the IBAD films is given by the sum of the deposition rate, $J_A s_A/N + J_I s_I/N$, and the erosion rate $-v_s$. Assuming that the sticking of the atomic species, the trapping of the ionic species, and the sputtering all influence growth velocity, then the net growth velocity can be obtained by

$$v_g = \frac{J_A s_A + J_I s_I}{N} - v_s, \tag{10.37}$$

where s is again the sticking probability of the depositing atoms, s_I is trapping (sticking) probability of the incident ions in the film, Y(E) is the sputtering yield of the film material, and N is the atomic number density of the film.

Assuming that the sticking probabilities will be close to one ($s_A \cong s_I \cong 1$) and the sputtering yield only determines the erosion velocity and is given by

$$v_s = \frac{J_I Y(E)}{N} \tag{10.38}$$

(see Sect. 5.1) Thus, the growth velocity of an IBAD layer dependent on the sputtering yield is given by

$$v_g = \frac{J_A + J_I(1 - Y(E))}{N}. \tag{10.39}$$

Consequently, the thickness of a film deposited during the deposition time t_d is

$$h_f = \frac{J_A + J_I(1 - Y(E))}{N} t_d. \tag{10.40}$$

The concentration distribution of the incorporated ionic species in the film by the IBAD process is strongly dependent on the instantaneous surface position. As schematically shown in Fig. 10.15, the ion flux with an ion current density J_I and the flux of atoms with the flux density J_A concurrently impinge onto the surface of a semi-infinite substrate ($0 < x < \infty$). These two fluxes of material cause film growth in the x-direction, which means that the surface is simultaneously shifted during the deposition process. In the simplest case, in which only the sputtering slows down the growth process, the ion distribution function (see Appendix F, (F.1)) is given by $f(x)$ from the instantaneous surface x below the initial surface at $x = 0$ at the time $t = 0$. An ordinary differential equation then describes the incorporated ion distribution (the atom fraction of the incorporated ionic species) by

$$\frac{dN_I}{dt} = v_g \frac{dN_I}{dx} = J_I Y(E) f(x) \tag{10.41}$$

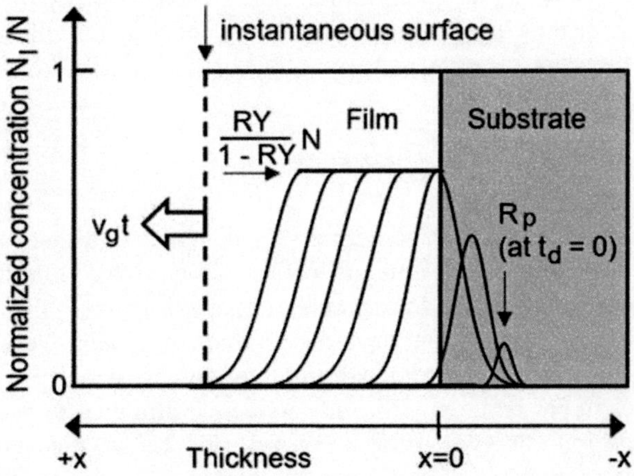

Fig. 10.15 Schematic representation of the concentration distribution of the incorporated ion species in relation to initial surface ($x = 0$) and the growing film ($x > 0$). It is assumed that the probability function f(x) is the same in the substrate and the deposited film (R_p is the projected range of the implanted ions, t_d is the deposition time)

with the solution

$$N_I = \frac{J_I Y(E)}{v_g} \int_0^\infty f(x)dx = \frac{[J_I Y(E)]N}{J_A - J_I Y(E)} \int_0^\infty f(x)dx = \frac{RY(E)}{1 - RY(E)} N \int_0^\infty f(x)dx, \tag{10.42}$$

where the factor $RY(E)/[1 - RY(E)]$ determines the maximum of the additional atom fraction of the ions incorporated in the film (Fig. 10.15). More complex models (see e.g., [45, 48]) have been proposed to describe the process of ion beam-assisted film growth, including radiation induced diffusion, mixing at the interface, trapping of the ionic species, and the sticking of the atomic species.

10.3 Thin Film Growth Under Assisted Ion Beam Bombardment

At or near the thermodynamic equilibrium, i.e. at high substrate temperatures and low deposition fluxes, three different growth modes can be distinguished (see Sect. 10.2.1). The interaction of the deposited thin films with the substrate at a certain temperature and supersaturation during thermal film deposition ultimately determines the growth morphology. For ion beam-assisted deposition, additional influencing factors have to be taken into account, such as energy of the particles, ion-to-atom ratio, or the chemical nature of the ion species controlling film growth.

10.3.1 Influence of the Ion Energy

A number of different studies have definitively demonstrated that low-energy ion bombardment strongly influences the nucleation and growth of thin films (e.g., see reviews by Greene et al. [35, 47], Petrov et al. [48], and Ensinger [49]), albeit neither a general concept nor a theory is known to exist that comprehensively describes the influence of assisting ion bombardment on film growth.

Two processes in particular significantly influence film growth and morphology in dependence on ion energy. On the one hand, it is known that during thin film growth up to ion energies of a few hundred electron-volts, ion-irradiated surfaces preferentially generate vacancies and adatoms. At higher ion energies, in addition to these point defects, one-dimensional defects (dislocations, dislocation loops, etc.) can be created. On the other hand, assisted ion bombardment during film growth causes an increase in the surface diffusion, referred to as ion beam-enhanced surface diffusion, which frequently favors layer-by-layer two-dimensional growth.

For the IBAD technologies, ion energies from several electron-volts up to some keV are used.

Ions with *thermal energies* (< 1 eV) can be physisorbed on the surface, dissociate and chemisorb to the surface (activated dissociative chemisorption), or scatter at the surface atoms, with some loss in their kinetic energy (direct inelastic scattering) [50]. For example, such experiments allow the examination of the reactivity at the surface of molecules [51]. For kinetic energies up to about 10 eV, the attractive chemical interaction becomes more important. This energy range is relevant for soft-landing deposition of polyatomic ions [52]. Above this energy, in the range of *hyperthermal energies* between 10 eV and 100 eV, the transferred energy at collisions is comparable to or larger than typical chemical binding energies. Consequently, surface modification is an emerging application of hyperthermal ion beams. Among other things, concurrent ion bombardment with hyperthermal energies and *low-energy ions* (up to about 1000 eV) during film deposition is able to strongly influence the growth of thin films [53]. In Fig. 10.11, the most important ion bombardment-induced processes are schematically represented.

When the incoming ions have a kinetic energy higher than the threshold displacement energy of the target atom ($E_d \approx 20\,\text{eV} \ldots 45\,\text{eV}$, see Sect. 4.1), this can knock out atoms of their lattice positions in the bulk. In dependence on the energy of the primary knock-on atoms, a more-or-less strong atomic motion along the path of this ion results in a rearrangement of the lattice. Radiation damage is created, including point defects, interstitials, and vacancies. The damage is partially annealed during the interaction process, depending on the nature of the target material and the temperature. Radiation defects can agglomerate to two- and three-dimensional defects. Additionally, bombardment with hyperthermal and low-energy ions induces an increased mobility of atoms in the bulk, referred to as radiation-enhanced diffusion (see Sect. 4.7) and phase formation or phase transformation (details are discussed in Sect. 6.2.2).

For low-energy incident ions, collision processes are limited to the interaction with atoms at the surface or in the near-surface region. If the energy transferred during collisions is high enough to overcome the surface binding energies, U_s (see Sect. 5.4.1), atoms can be removed from the surface and the near-surface region (sputtering). This detachment process leads to a reduction in growth velocity and may also leads to changes in film composition when different constituents of the film have different sputtering yields. When surface atoms are unable to leave the surface, they remain on the surface as adatoms. Additionally, adatoms leave a free lattice site at the surface, i.e. a surface vacancy is generated. Adatoms and vacancies are both capable of influence the morphology of film growth by providing additional nucleation sites (increased nucleation density) during thin film formation (in contrast to thermal deposition). In addition, an enhanced surface adatom diffusion can be expected at the surface, resulting in the breakup/fragmentation of islands or detachment of adatoms from islands (see Fig. 10.13, processes X and XI). The newly created adatoms then diffuse across the surface and can be re-evaporated or captured by islands, i.e. the island size and island density are influenced by this ion beam bombardment process. Few experimental studies are known that have investigated the influence of ion energy (frequently together with other parameters, such as temperature, etc.) on the early stages of film formation after ion beam assisted deposition.

- For example, experiments involving the densification of thin films, Yehoda et al. [54] found that these were governed by the average transferred ion energies while Rossi et al. [55] found the damage energy during ion bombardment to be the governing factor. These results can be interpreted as a combination of ballistic effects leading to densification and enhanced adatom mobility.
- Low-energy ion bombardment produces adatoms by sputtering in large numbers (Fig. 10.13). The surface vacancies generated by ion impact can enhance the island area density directly, by serving as nucleation points for the adatoms, or indirectly, by reducing the diffusion. As an example, Fig. 10.16 shows an STM image of silicon surface defects following 4.5 keV He ion bombardment of Si(001) with two different ion fluences [56]. The surface is characterized by individual adatoms, dimers, and adatom clusters (islands). These adatoms form adatom islands with an area density higher than the island density generated by thermal deposition. Because the number of adatoms and adatom islands is comparable with the number of vacancies and vacancy clusters, it can be expected that the influence of the vacancies on the nucleation in the ion beam-assisted deposition is significant.
- The kinetics of the damage formation and subsequent annealing process of Si(111) 7×7 surfaces bombarded with 1000 eV Ar ions were directly studied with STM [57]. The bombarded surface was characterized by adatoms removed by the ion bombardment and surface vacancies left behind by the removed adatoms. Upon annealing, the surface vacancies disappeared according to a bi-exponential first-order kinetics. The first annealing stage, with an activation energy of 0.83 eV, was attributed to the decomposition of large vacancy islands into smaller ones. The second stage, with an activation energy of 0.51 eV, was attributed to the reconstruction of the Si 7×7 surface arrangement. Both processes were assumed to be limited by surface diffusion of Si atoms. In contrast, reconstructed Au(111) and Pt(001) surfaces after bombardment with 600 eV Ar ions at 300 K showed

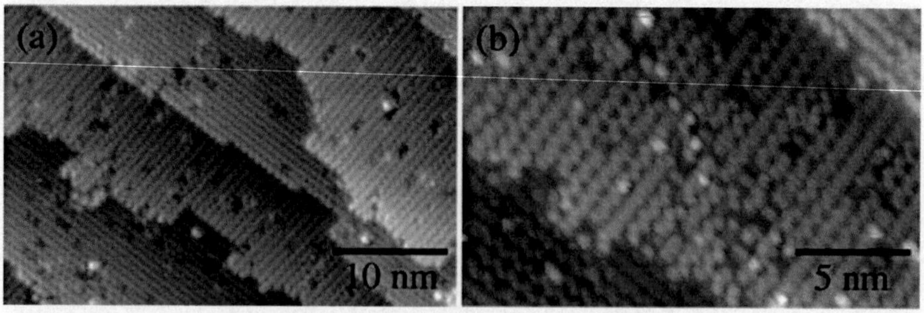

Fig. 10.16 STM image of Si(001) bombarded by 4.5. keV He ions at 180 K with **a** 1.7×10^{13} ions/cm^2 and **b** with 3.4×10^{13} ions/cm^2. Adatoms, dimers, and adatom clusters are visible as bright spots about 1 nm in diameter (figure adapted from [56])

no traces of vacancy or adatom islands [58]. The only visible defects were two-dimensional dislocation dipoles and individual dislocations.

- Using STM, Michely and Comsa [59] observed the formation of adatom islands for Pt(111) bombarded with 600 eV Ar ions at temperatures between 350 and 625 K. They assumed that each ion removes approximately one atom and generates a large local density of adatoms around the ion impact spot. From a theoretical point of view, Harrison et al. [60] predicted this process of ion bombardment-induced surface atom replacement and described it as radiation enhanced diffusion. These adatoms then condense into a large number of islands. As the result of this surface replacement process, subsurface vacancies are formed, which can be annealed for homologous temperatures > 0.3 eV [61].
- Consequently, this type of ion bombardment can successfully be applied to induce 2D growth above the critical homologous temperature. This was demonstrated by Rosenfeld et al. [62] for the deposition of Ag onto Ag(111) and Cu onto Cu(111) at room temperature and with short Ar ion pulses during the initial stages of the film growth. Under periodic ion bombardment, a layer-by-layer growth was observed. When the ion bombardment was carried out continuously, the growth mode was not affected, since adatoms were also created on top of growing islands, reducing the chance that deposited adatoms could descend from the island top.
- These results could be confirmed by Dvurechenskii et al. [63, 64] for the ion beam-assisted deposition of Ge/Si nanodots. Their studies demonstrated that the size distribution can be narrowed using pulsed ion beam-assisted deposition, as compared to a distribution of conventional MBE growth as well as the continuous IBAD (Fig. 10.17). Pulsed ion bombardment leads to smaller islands and a narrower size distribution. Nevertheless, this tendency is not so strongly developed in cases of continuous ion irradiation. This effect can be described by lowering

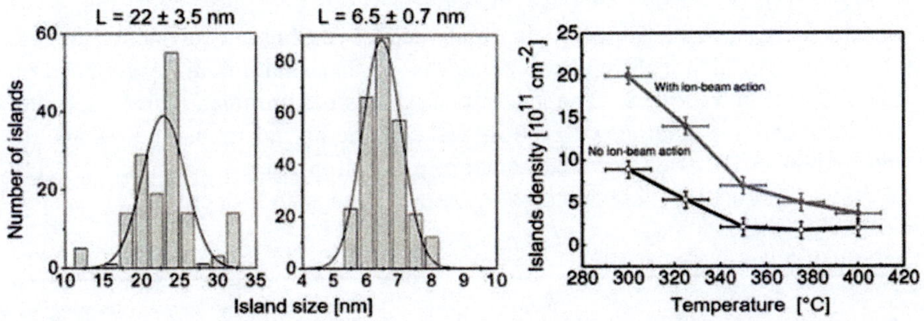

Fig. 10.17 Comparison of the experimentally measured size distribution of Ge islands on Si(1000) deposited at 350 °C. The nominal thickness is 5 monolayers. Left: conventional MBE without ion bombardment. Middle: pulsed Ge ion-assisted molecular beam epitaxy. The energy of the Ge ions was 100 eV and the deposition rate was 1 ML/s. L is the average size of the islands (figures adapted from [63]). Right: Comparison of the island area density prepared with and without pulsed ion bombardment at temperatures between 300 °C and 400 °C (figure adapted from [64])

the strain energy of these nanodots, corresponding to the strain relaxation caused by ion bombardment.

- Riekki and Koponen [65] have described this size selection growth in IBAD using a reactive kinetic model, including reversible adatom attachment and detachment processes. They found that an optimal stationary state with narrow size distribution is given only for low adatom fluxes in IBAD experiments. In agreement with the experimental results [63, 64], increasing the duration or the strength of the ion bombardment does not enhance the size selection, but instead enlarges the size dispersion.
- Likewise, Frantz et al. [66] studied the microscopic IBAD processes of the detachment of adatoms and the island dissolution by MD simulations with 25 eV Co ions onto Co islands of an Ag(100) surface. They found that the irradiation-induced detachment from the island is usual, while dissociation events of the island are very rare.
- Esch et al. [67] observed a drastic increase in the island area density in cases of simultaneous Ar ion bombardment (400 eV and 4000 eV) during the deposition of Pt films on Pt(111) at temperatures $\geq$ 200 K. The higher the ion energy, the stronger the increase. According to the authors, the increase is caused by nucleation at the ion impact-induced adatom clusters.
- In good agreement with the results of the simulation (see Sect. 10.3.4), low-energy ion bombardment during thin film deposition creates a larger area density of islands with smaller diameters than with thermal deposition or MBE. This can be considered the primary precondition for two-dimensional layer-by-layer growth. Nikzad and Atwater [68] studied microstructure evolution in the early stages of growth of polycrystalline Ge films on amorphous silicon dioxide by ion beam-assisted deposition and compared the results to conventional thermal film growth. The experimental results in Fig. 10.18 indicate that with increasing energy from 150 eV up to 350 eV, the Ge island size decreases, while the island density increases with respect to thermally deposited films at equivalent substrate coverages. An identical trend was observed by these authors in dependence on the ion-to-atom ratio [68]. The number of islands per unit area increases, while the sizes become smaller with the increase of the arrival ratio. The results are interpreted in terms of ion beam adatom desorption from islands (see process XVII in Fig. 10.13). These adatoms can re-attach to islands or create new islands

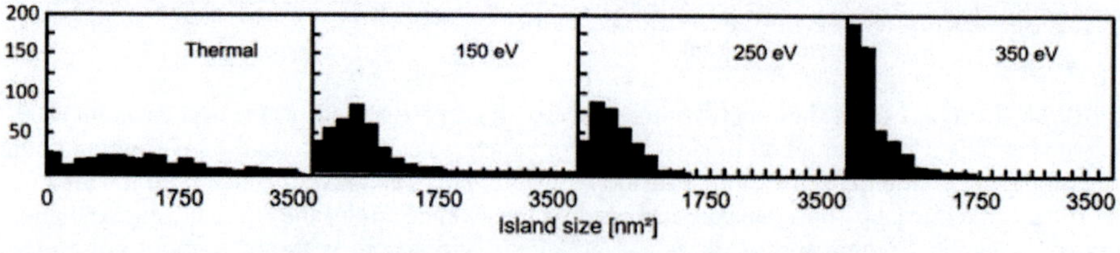

Fig. 10.18 Island size distribution histograms of thermally grown Ge films on amorphous SiO_2 surfaces and IBAD Ge films on SiO_2 prepared with different Ar ion energies (adapted from [68])

and consequently, this supersaturation of adatoms onto the substrate surfaces results in an increased island density and smaller island size.

In Sect. 4.1, it was established that a threshold displacement energy is necessary to generate permanent point defects. This energy is constant for a given crystallographic direction. However, beam-assisted deposition studies (e.g., [1]) and measurements of surface diffusion (e.g., [69]) have hinted that temperature may directly affect the dynamics of defect formation when ion energies are $\leq$ 100 eV. Wang and Seebauer [70] performed MD simulations of noble gas ion bombardment of adatoms on Si(111) and Ge(111) surfaces, where the Si and Ge self-interaction was determined by means of a Stillinger–Weber potential and the interaction with the bombarding noble gas ions and the adatoms by means of a universal potential. At the end of the simulations, the probability of adatom formation, sputtering, and bulk vacancy formation per ion impact was recorded. The simulations suggest that energy thresholds linearly decrease as temperature increases (about 0.1 eV/K). Wang and Seebauer [70] explained the reason that the number of defects should increase with temperature at a constant ion energy on the basis of energy threshold measurements for ion-enhanced surface diffusion of indium on silicon and germanium. As the ion energy decreases, the ions penetrate less close to the target atoms. Due to the lower penetration, the net repulsion potential appears more uniform when the ions come closest together. Thus, the target atoms must move farther away from their lattice sites to allow defect formation. The effect of the threshold energy decreasing as the temperature increases may result in specific defect formation at different temperatures during ion beam-assisted deposition.

10.3.2 Influence of Temperature on Ion Beam-Assisted Thin Film Growth

The strong influence of temperature on the quality of the epitaxial growth of thin films during thermal deposition is an accepted fact. It can now be assumed that an assisted low-energy ion bombardment during deposition can be used as a partial substitute for increased substrate temperature in the epitaxial growth process [69, 71], because the ion bombardment promotes the atomic arrangement processes and enhances the diffusion at the surface of the growing film.

Several experimental studies have shown that the optimal temperature for epitaxial growth of films prepared by IBAD can be significantly reduced. For example, Babaev et al. [72] reported on a decrease of the epitaxial temperature from 230 °C to 150 °C for the deposition of Sb films on rock salt under 400 eV noble gas ion bombardment. Shimizu et al. [73] produced high-quality epitaxial InP layers between 210 °C and 420 °C under concurrent 100 eV and 200 eV phosphorous ion bombardment. Of special application-oriented interest is the deposition of epitaxial Si and Ge layers by ion beam-assisted deposition as well as direct ion deposition (see Chap. 9). Itoh et al. [74] found a decrease of epitaxial temperature in single silicon films on Si substrates after partially ionized Si vapor deposition, and Wehner et al. [75] deposited epitaxial Si films on Si substrates under 30 eV ion bombardment for temperatures < 300 °C. Park et al. [76] reported a strong reduction of the growth temperature of $Si_{0.5}Ge_{0.5}$ alloy films on Si(001) formed by ion beam-assisted deposition. The temperature for an epitaxial thin film growth was about 200 °C lower than in conventional molecular beam epitaxy.

The influence of the substrate temperature becomes evident upon comparison of the growth morphology of IBAD films prepared at different temperatures. As an example Fig. 10.19 shows an atomic-resolution ADF-STEM image of GaN film on 6H-SiC deposited at 700 °C by ion beam-assisted molecular beam epitaxy. This temperature was found to be the optimal temperature for growing dense GaN thin films with high crystalline quality [77]. The nitrogen ion to gallium atom arrival ratio was 1.3 during the entire deposition process. The presence of two different GaN polytypes, cubic and hexagonal, is visible and was also confirmed by FFT analysis (insets in Fig. 10.19) [78]. The first nanometers of the GaN film consist of the w-GaN phase, followed by several nanometers of z-GaN. In Fig. 10.19, above the z-GaN layer, only w-GaN is present. This stacking sequence of the different regions, consisting of w-GaN, z-GaN and w-GaN layers. A variety of defects was found in the cubic and hexagonal GaN domains close to the GaN-SiC interface. Most of them were identified as basal-plane stacking faults (SF), grain, and staking mismatch boundaries. The w-GaN layer on top of the z-GaN layer has a significantly lower density of defects

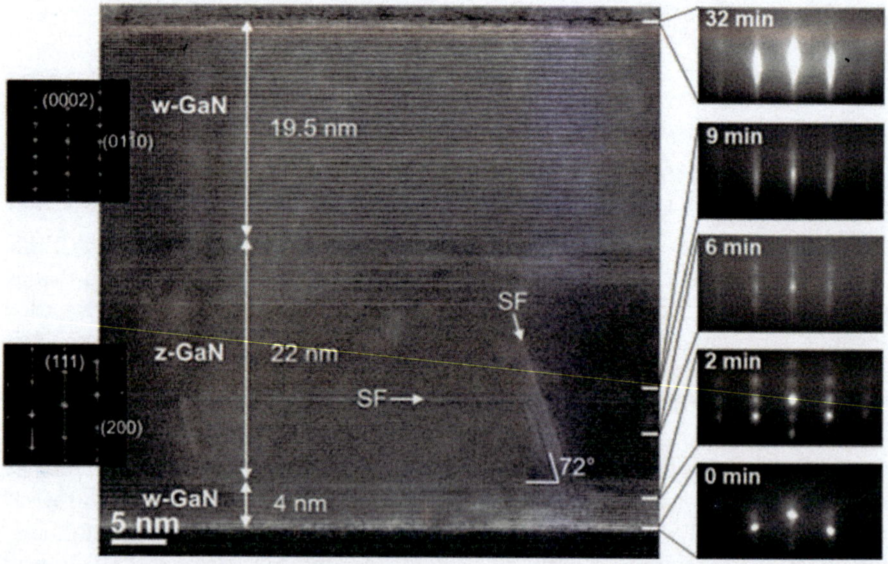

Fig. 10.19 Atomic-resolution ADF-STEM image of GaN film on 6H-SiC substrate prepared by molecular beam ion beam-assisted molecular beam epitaxy at 700 °C. Two representative FFT images are given (left). The STEM image shows z-GaN and w-GaN regions with various defect types and corresponding RHEED patterns obtained in-situ during different growth stages

than the w-GaN and z-GaN layers close to the interface. The RHEED patterns in Fig. 10.19, taken in-situ during the film deposition, represent the evolution of the crystalline surface structure. In the first three minutes of the ion beam-assisted deposition process, the spotty pattern of w-GaN islands was clearly visible. Within a few minutes, the spotty reflections continuously changed into streaky reflections and it was no longer possible to distinguish the different patterns of w-GaN and z-GaN. The transition from spots to streaks indicates the coalescence of the GaN film islands, resulting in a rather smooth, two-dimensional surface topography. After 9 min, no spotty reflections were observable, meaning that the film was completely coalesced at this stage of the deposition process. These RHEED streaks persisted throughout the entire remaining duration of the deposition (32 min). The growth morphology under these deposition conditions is therefore characterized by the existence of two GaN polytypes and crystal defects (see e.g., 72° stacking faults in Fig. 10.19).

In contrast to the film deposited at 700 °C, in Fig. 10.20, a cross-section TEM image of GaN films is shown, which was deposited on 6H-SiC under identical conditions, with the exception that the deposition temperature was reduced from 700 °C (Fig. 10.19) to 630 °C. Only the wurtzite GaN polytype was formed (Fig. 10.20, left) [79]. The perfection of the crystal lattice is obvious and at the interface between gallium nitride film and substrate, stresses associated with the lattice mismatch of GaN and SiC are relieved within approximately 2 nm. In contrast to higher deposition temperatures, only horizontally running stacking faults are generated (two of them can be seen in the GaN film at about 10 nm above the substrate).

The significant influence of the temperature during the IBAD process on the epitaxial growth of layers remains when other substrate materials are chosen. Figure 10.20, right, demonstrates the growth of GaN films on Si(001) substrates [80], where it is known that the lattice mismatch between both GaN lattices and the Si(001) surface is substantial. The RHEED patterns indicate that after deposition at 650 °C and 700 °C, the cubic polytype is the dominating GaN phase, and after 700 °C and after 750 °C is w-GaN the preferential phase. The topography is characterized by many ashlar-shaped crystallites with steps in their surfaces (see SEM image in Fig. 10.20) and some hexagonal-shaped crystallites.

The experimental results confirm that the temperature is a key factors governing the growth modes of thin films under assisted low-energy ion bombardment. In contrast to thermal deposition, it can be summarized that the temperature required for epitaxial growth is significantly lower for many materials under low-energy ion bombardment. Consequently, ion bombardment seems to strengthen the surface diffusion during the IBAD process. Unfortunately, the ion-enhanced diffusion is difficult to determine. Ditchfield and Seebauer [69] applied optical second harmonic microscopy to determine the surface diffusion coefficient of thermally deposited germanium diffusion on silicon surfaces under assisted noble gas ion bombardment between 10 eV and 65 eV and an ion incidence angle of 60°. According to these authors, two temperature regimes can be distinguished for surface diffusion under low-energy ion bombardment when conventional Arrhenius behavior for the ion-enhanced surface diffusion coefficient

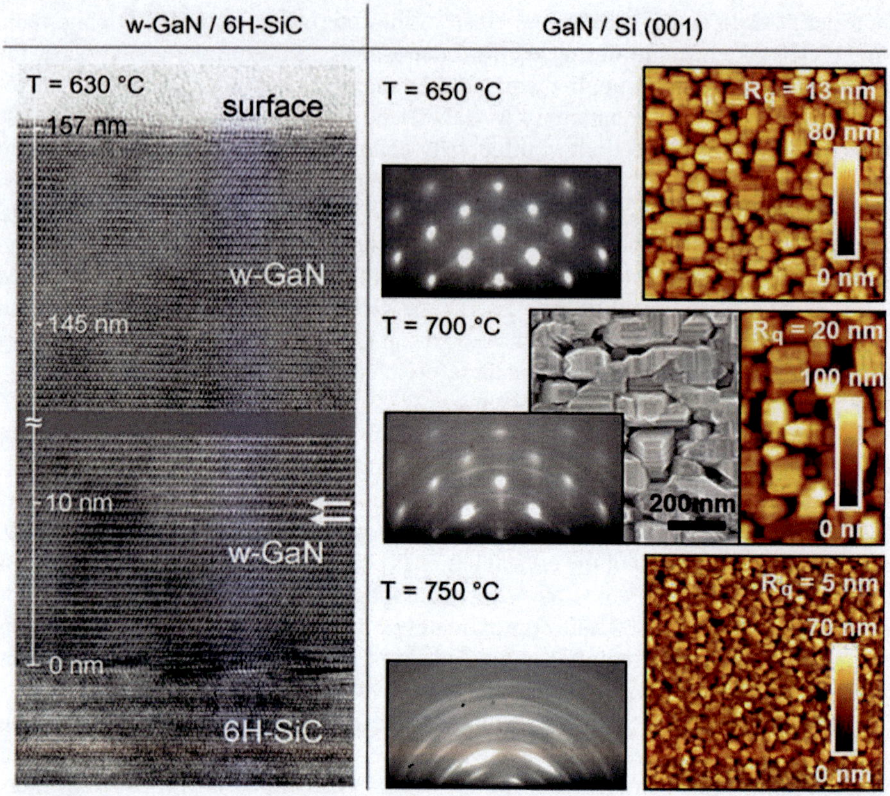

Fig. 10.20 Left: Cross-section, high-resolution TEM image of GaN film on 6H-SiC substrate prepared by ion beam assisted molecular beam epitaxy [79]. The deposition temperature is 630 °C. The zone axis is $[2\bar{1}\bar{1}0]$. Horizontal stacking faults are indicated by arrows. Right: RHEED diffraction patterns and the corresponding AFM images ($1.5 \times 1.5\ \mu m^2$) of GaN films onto Si(001) substrates deposited by ion beam assisted molecular beam epitaxy at different temperature (R_q is the root mean square roughness). The ion-to-atom ratio is 1.9 in all cases. The RHEED electron beam is parallel to z-GaN[110] direction. An SEM image is also shown for 700 °C

$$D_s = D_o \exp(-E_{sd}/k_B T) \tag{10.43}$$

is assumed (D_o is the pre-exponential factor of the diffusion, E_{sd} the activation energy of the surface diffusion, and k_B the Boltzmann constant). They found that for temperatures below 730 °C (low-temperature regime), E_{sd} is constant (identical to the non-bombardment case) and D_o remains constant up to an energy of 15 eV but then increases with energy by a factor of up to five at 65 eV (see Fig. 10.21, left). In contrast, above 730 °C (high-temperature regime) both E_{sd} and $\log(D_o)$ decrease linearly with ion energy for ion energies > 25 eV. With increased the ion mass, the described effects are more pronounced, i.e. D_o increases proportionally to the square root of ion mass and E_{sd} remains constant in the low-temperature regime. In the

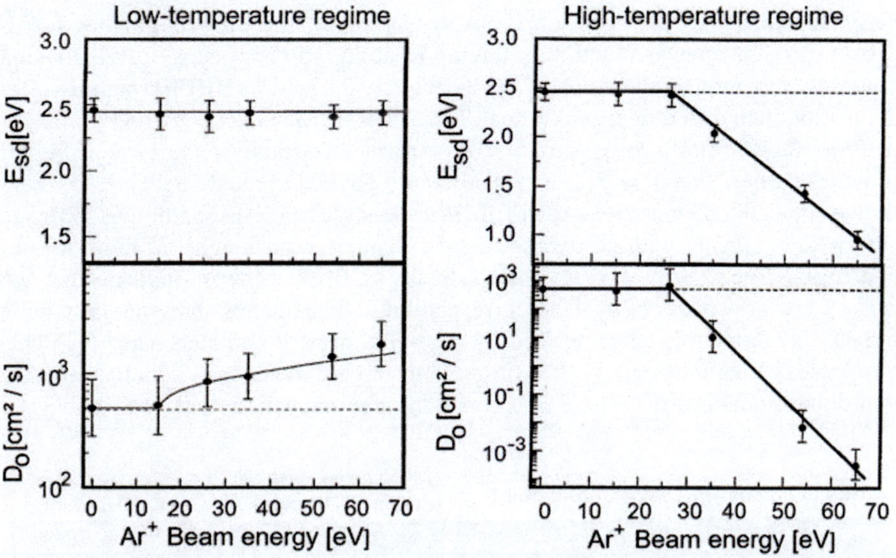

Fig. 10.21 Activation energy of diffusion E_a and pre-exponential factor D_o of the Ge surface diffusion on Si(111) under low-energy Ar ion bombardment as function of the ion energy. Left: for the low-temperature regime, right: for the high-temperature regime, i.e. temperatures >730 °C (adapted from [69])

high-temperature regime, E_{sd} decreases from about 2.5 eV (mass of He) to about 0.5 eV (mass of Xe), while D_o decreases by more than eight orders of magnitude.

Because in the low-temperature regime, the ion-enhanced diffusion increases (the pre-factor only decreases), Ditchfield and Seebauer [69] suggest that this effect is caused by a steady-state balance between a spatially-extended ion-induced surface modification and thermally activated annealing of this modification. For the high-temperature regime, it was suggested by the authors that the effect results from the inability of knocked-in surface atoms to incorporate into the surface and anneal the long-lived surface vacancies. These vacancies then serve as a preferential sink for diffusing adatoms and slow their motion.

10.3.3 Influence of the Ion-to-Atom Arrival Ratio

The influence of the arrival ratio on the evolution of the crystal structure at the surface during initial stages of ion beam-assisted deposition can be effectively observed in-situ by RHEED. RHEED patterns were obtained from the substrate prior to deposition and from the deposited films on the substrate. To this purpose, the substrate holder was rotated by the azimuthal angle to align the high symmetry alignment crystal axes with the fixed electron beam direction. For example, Neumann et al. [81] studied the early steps of 25 eV nitrogen ion beam-assisted GaN film growth on smooth 6H-SiC(0001) substrates as function of the ion-to-atom ratio by means of RHEED at a constant deposition temperature of 700 °C. Figure 10.22 shows RHEED patterns after deposition with different arrival ratios [82]. The RHEED patterns and STM images in Fig. 10.22 reveal a large variation in surface topography among the different growth regimes. For $R = 3.2$, a typical spotty RHEED pattern indicates a three-dimensional, purely wurtzitic growth of GaN islands. The corresponding STM image shows GaN islands with an average island diameter and height of about 40 nm and 6 nm, respectively. For $R \leq 1.6$, streaky RHEED patterns indicate that the surface has grown smoothly. The corresponding STM images show laterally wide terrace-like structures characterized by step-flow growth and step bunching. This morphology characterizes the transition from three-dimensional to predominantly two-dimensional growth. The RHEED pattern after deposition with $R = 1$ shows a

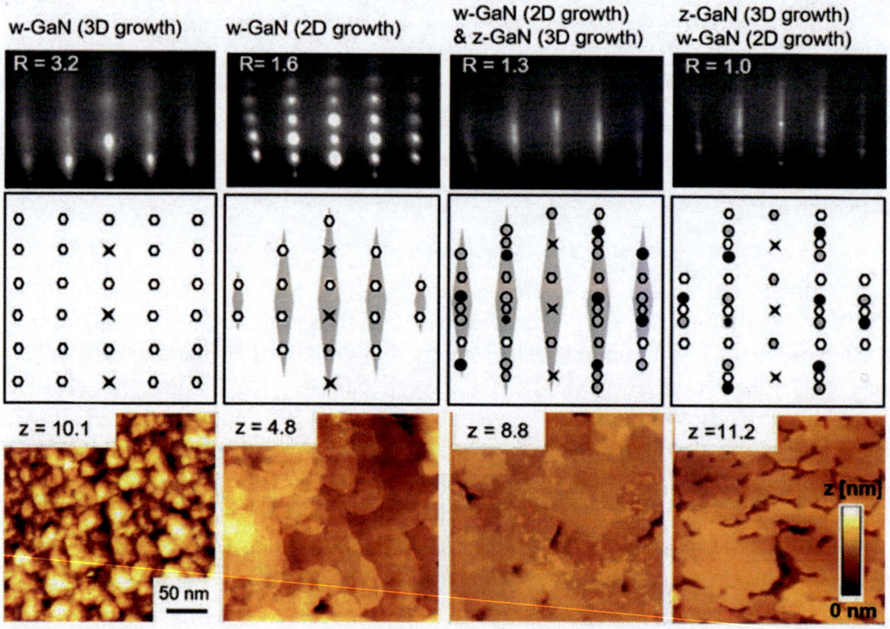

Fig. 10.22 First row: RHEED patterns from GaN deposited on 6H-SiC(0001) as function of the ion-to-atom ratio R. The GaN films were deposited by ion beam-assisted molecular beam epitaxy with nitrogen ion energy of 25 eV at a substrate temperature of 700 °C. The nominal film thickness is 11 nm (corresponds to a constant deposition time of 300 s). All figures are taken from [82]. Second row: Schematic reciprocal projections of the RHEED patterns. For an incidence direction of the electron beam parallel with $[2\bar{1}\bar{1}0]$ of (0001) oriented hexagonal polytype or parallel with $[1\bar{1}0]$ directions for (111) oriented cubic GaN, the theoretical diffraction patterns contain unequivocal reflexes to distinguish between the different GaN modifications (white rhombus—hexagonal, grey circle—cubic, black circle—cubic (twinned), grey rhombus—coincidental reflection from hexagonal and cubic polytypes and forbidden reflections) Third row: Corresponding STM images

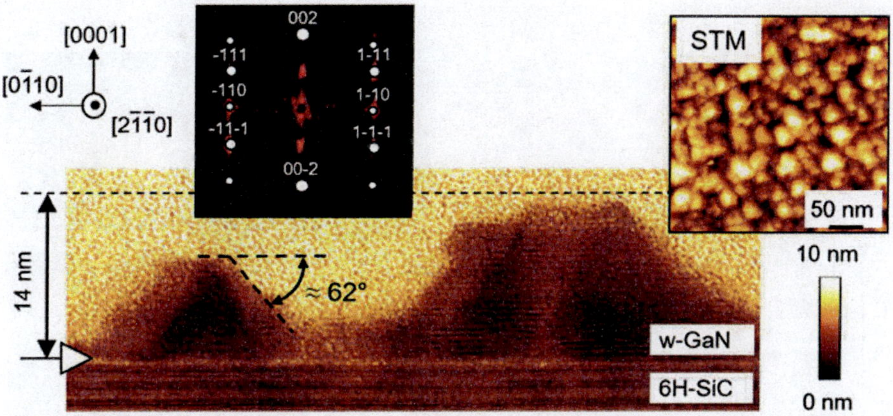

Fig. 10.23 Cross-section high-resolution TEM image, a Fast Fourier Transform (FFT) reflection pattern (diffractogram) and a STM image of w-GaN on 6H-SiC(0001) prepared by ion beam-assisted molecular beam epitaxy (IBA-TD). The ion energy was about 25 eV, the ion-to-atom ratio 3.1 and the deposition temperature 700 °C). The zone axis is parallel to the a-axis of the w-GaN unit cell. The lattice mismatch is about 3.6 %

superposition of streaks and double spots, the latter corresponding to (111) oriented z-GaN. This pattern is consistent with a locally smooth topography.

The electron microscopic studies provide local information about growth morphology in dependence on the ion-to-atom ratio R during the initial stage of growth. The high-resolution cross-sectional TEM image (Fig. 10.23) shows GaN islands prepared on a 6H-SiC substrate at a high ion-atom ratio of 3.1 [77]. The FTT pattern in Fig. 10.23 was obtained from such island and can be assigned to a single crystalline grain of w-GaN. The w-GaN islands are epitaxially grown to a 6H-SiC substrate surface. The interface between 6H-SiC and GaN is about 1–2 ML thick and contains no steps. Areas between the islands are covered with an approximately 2 nm (8 ML) thick layer. This is typical for a Stranski–Krastanov growth morphology. The angle between the side faces and the roof area of the mesa-like islands is about 62°. This corresponds to the angle between the (0001) and the (10$\bar{1}$1) planes of the hexagonal GaN crystal structure. The growth morphology changes significantly when the excessive supply of nitrogen is strongly reduced. The ion beam-assisted deposition of GaN on 6H-SiC(0001) surfaces under the condition of a less excessive supply of nitrogen (ion-to-atom ratio $R = 1.6$) leads to two-dimensional layer-by-layer growth during the initial stages of the deposition (Fig. 10.24). A perfect lattice matching on the atomic scale without formation of an interface between substrate and film could be established. The GaN crystal structure grows almost without the formation of dislocations. In Fig. 10.24, the cross-sectional TEM image shows the successive growth of two GaN polytypes: cubic GaN (z-GaN) and hexagonal GaN (w-GaN) are unambiguously distinguishable [77]. The FFT patterns can be assigned to single crystalline w-GaN and z-GaN layers, respectively. The hexagonal GaN film primarily grows up to a thickness of about 3.6 nm. This thickness corresponds to

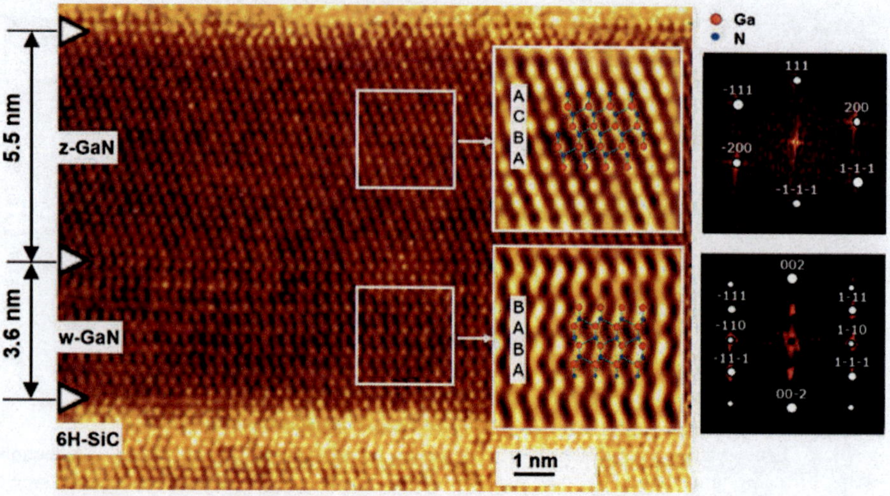

Fig. 10.24 Cross-section high-resolution TEM image, corresponding Fast Fourier Transform (FFT) reflection patterns (diffractograms), and STM images of w-GaN on a 6H-SiC(0001) surface prepared by ion beam-assisted molecular beam epitaxy (IBA-TD). The ion energy was about 25 eV, the ion-to-atom ratio 1.6, and the deposition temperature 700 °C). The zone axis is parallel to the a-axis of the w-GaN unit cell

the thickness of coalescence measured by RHEED [48]. A cubic GaN film is then formed by changing the stacking sequence (demonstrated in Fig. 10.24, details on the changing of the stacking sequence are provided in [82]).

These studies reveals that the ion-to-atom ratio has a major impact on the properties of the resulting film during the initial growth stage. The growth mode, surface roughness, and polytype mixture depend significantly on the arrival ratio. The desired layer-to-layer growth morphology for most IBAD films is reached when the ion-to-atom ratio is reduced below a critical magnitude.

10.3.4 Simulation of IBAD Thin Film Growth

Molecular Dynamic and Kinetic Molecular Dynamic Simulations

The influence of the deposition temperature on the crystallinity of layers prepared by IBAD was one of the first applications of the standard molecular dynamics (MD) methods in this field. Strickland and Roland [83] simulated the molecular beam deposition of Si films without and under concurrent low-energy Ar ion bombardment. Figure 10.25 shows vertical cross-section images of the deposited Si film without (left) and with (right) 10 eV Ar ion bombardment at a temperature of 300 K, obtained by MD simulations.

These figures demonstrate that the amount of crystalline material obtained at low temperatures and under ion bombardment is significantly higher than without

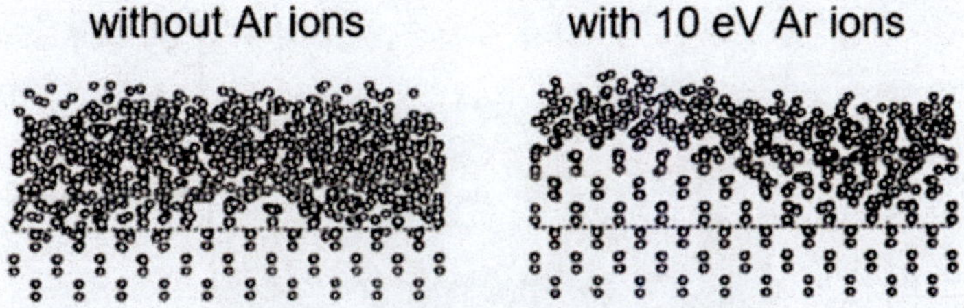

Fig. 10.25 Vertical cross-section images of Si films deposited without (left) and with (right) 10 eV Ar ion bombardment at 300 K obtained by MD simulations. The dashed line marked the initial substrate surface and the arrival ratio R = 1 (adapted from [83])

assisted ion bombardment. Only for very low ion energies (5 eV) and low ion fluxes could no significant increase of the amount of crystalline material could be proven, while for high ion energies and fluxes the crystalline fraction was enhanced.

Conventional computer simulations allow, depending on the size of the system, to follow the physical processes for a few fractions of a millisecond at most, while the time required for the growth of a monolayer prepared with IBAD is a few seconds. Backwell et al. [84] used an on-the-fly kinetic Monte Carlo method to study the growth of different materials via ion beam-assisted deposition in real time. As an example, Fig. 10.26 shows the film growth of Ag {111} under concurrent 100 eV Ar ion bombardment, where an arrival ratio of one was assumed. Differences in the evolution of film morphology after deposition with and without ion bombardment could be observed. First, with the same number of atoms, more monolayers are formed after deposition without assisted ions than in deposition with additional ion irradiation. This confirms that ion irradiation results in a densification of the film through the transfer of kinetic energy to the atoms, enabling increased surface diffusion. Second, stacking faults (ABAB stacking) were observed only in the films grown by deposition without assisted ion beam, whereas the films deposited by IBAD are characterized by preferred ABC stacking, because the assisted ion beam provides enough kinetic energy to the film to eradicate stacking faults. Third, interface mixing is observed between atoms from the original substrate and the newly deposited atoms only when the ion beam assist is included.

Detailed analysis of particular aspects of the microscopic IBAD and IBD (see Chap. 9) processes were carried out over longer periods of time (from picoseconds to seconds). Chason and Kellerman [85] studied the growth in the ion beam-assisted processes via Monte Carlo simulations. In particular, the effect of the ion-induced defects on the evolution of two-dimensional islands was investigated. According to these authors, a simple mechanism of independently diffusing surface vacancies and adatoms leads to more rapid development of two-dimensional islands than does thermal annealing alone. However, the coarsening of the islands is limited by the creation of new defects under ion bombardment, which leads to a continuous nucleation of new islands. For direct ion deposition (see Chap. 9), Jacobson et al. [86] used the kinetic Monte Carlo simulation to study the influence of ion energy on growth and to obtain information about the optimal energy for layer-to-layer growth.

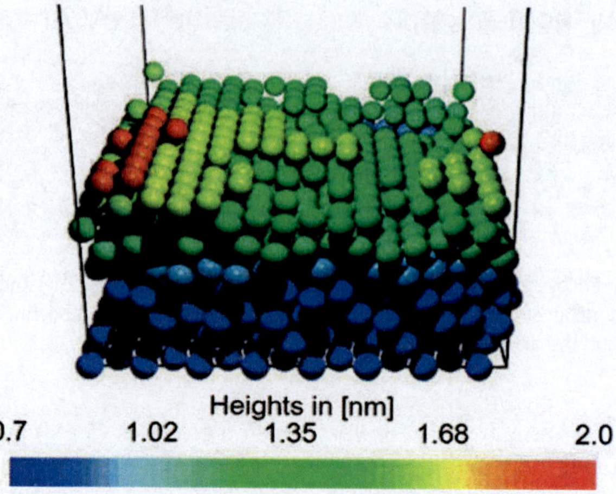

Fig. 10.26 Ag {111} oriented films prepared by 100 eV Ar ion-assisted deposition after 0.38 s of real time. Atoms are colored by heights in nanometer according to the color bar (adapted from [84])

They observed a strong influence of the ion energy on the density of islands and steps. In particular, the growth can be distinguished between layers fabricated only by thermal deposition and layers produced by ion beam-assisted deposition. The application of energetic ions increases the step density needed for layer-by-layer growth and leads to smoother surfaces at low temperatures. In Fig. 10.27, the Ag coverage is compared in each layer above a Ag(111) surface at 60 K after thermal and after energetic deposition (25 eV). It is shown that the growth generated by energetic particles leads to a smoother surface than that generated by thermal deposition. After thermal deposition of 3 ML, the growth of the sixth layer is still not complete. With the

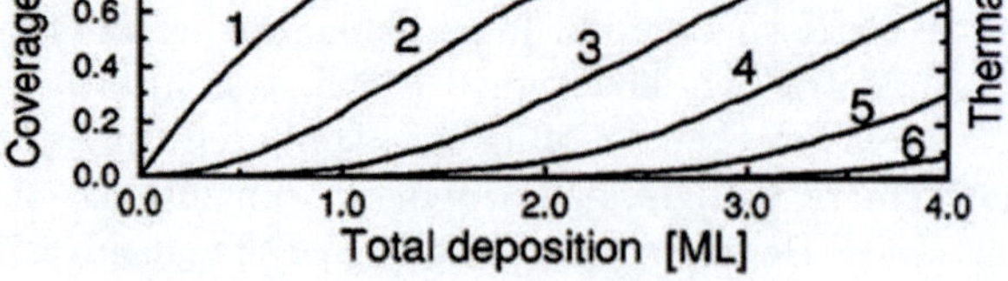

Fig. 10.27 Coverage in each growing Ag layer (layer marked at each graph) above the initial Ag(111) surface after 25 eV Ag ion deposition (above) and after thermal Ag deposition at 60 K. The plot was obtained by kinetic Monte Carlo simulation (adapted from [86])

25 eV deposition, the third layer is almost complete. These results are consistent with the notion that energetic ion-induced bombardment results in an increased mobility of atoms on the surface (see Fig. 10.13). Experimental studies [67] at temperatures > 200 K have shown that ion bombardment considerably increases the average island density, while average island size is simultaneously reduced. At low temperatures (< 70 K), the island density decreases due to ion impact induced adatom mobility.

Bunnik et al. [87] performed MD simulations to study the growth of Cu films on Mo(110) by thermal deposition and Ar ion beam-assisted deposition. They demonstrated that the influence of the ion impact is significant and leads to different crystal film structures. The growth mode of the Cu films goes through an unusual transition from a 2-dimensional to a 3-dimensional and to back to 2-dimensional growth mode. Subsequently, a stationary 2-dimensional growth mode is reached at about 4 nm thickness.

MD simulations by Zhou et al. [88] have demonstrated that the temperature during deposition and the ion-to-atom arrival ratio also strongly influence the growth behavior of IBAD films. For example, GaN films with higher crystallinity and lower defect concentrations could be achieved at substrate temperatures above 450 K. The best stoichiometric films were obtained at an arrival ratio between 2.6 and 2.8.

Study of IBAD Thin Film Growth by Rate Equations

Another possibility for describing the evolution of thin films deposited by IBAD and, in particular, the layer-by-layer growth, is modeling by rate equations. The advantage of this approach is its simplicity, which also allows a rapid numerical solution.

Initially, the kinetics of layer growth during the ion beam-assisted deposition of thin metallic films were studied using rate equations by Petzold and Heard [89], and Funsten et al. [90], with the aim of optimizing the FIB technology used to repair thin film lithography masks. These rate equations take into account various deposition parameters, including the incident ion flux, coverage of the substrate, film sputtering, and energy deposition by incident ions. The ion-to-atom arrival ratio was identified as the critical parameter of the ion beam-assisted deposition process [90].

Carter [91] used the classic rate equations for island nucleation and included the effects of monomer adatom sputtering, island sputtering, and island dissociation to study the influence of ion bombardment on the surface monomer adatom diffusion, the removal of adatoms from islands by sputtering, and the island dissociation by ion bombardment. Not included is the contribution of vacancy (aggregation, recombination with interstitials and adatoms), because the complexity of the defining equations increases enormously. Under these conditions, the rate equations for the monomer density $N_1(t)$ and the island density $N_s(T)$ at the time t, given by (10.26) and (10.27), must be extended. Carter was able to demonstrate that adatom sputtering alone can decrease island density and, in the later stage of island aggregation, increase island size, while island sputtering and dissociation processes can lead to increased island density.

Koponen et al. applied this method to island growth in the sub-monolayer region [92] and also for layer-by-layer growth [93]. Optimal conditions for sub-monolayer island growth are given when, on the one hand, adatom islands are small enough to prevent the onset of three-dimensional growth and, on the other hand, ion-induced surface erosion is avoided. In modeling growth during ion beam-assisted deposition, the processes of the formation of vacancies (see process (VII) in Fig. 10.11), the recombination of vacancies at adatoms islands, and the recombination of adatoms at vacancy islands (see Fig. 10.13) are included. The rate equations have been solved to obtain the size distribution of adatom and vacancy islands as a function of the coverage.

Also the layer-by-layer growth during IBAD was studied by Koponen [93] under inclusion of adatom diffusion, island diffusion, detachment and breakup, coalescence of large islands, and interlayer transitions of adatoms. It can be deduced that two requirements are necessary to achieve optimal layer-by-layer growth. First, an enhanced detachment of adatoms on the existing islands due to ion bombardment. Further, the detachment process must be assisted by a rapid downward transition of the adatom flux to underlying layers. If this mass transport process is not fast enough, then the onset of layer-by-layer growth is prevented. In order to obtain optimum conditions for layer-by-layer growth, both requirements must be satisfied. In practice, the energy and mass of the bombarding ions can be used to tune the breakup and detachment rate.

For example, Powell et al. [94] applied rate equations to develop a precursor-mediated chemisorption model for the description of GaN film growth kinetics. They determined the growth rate as

$$v_g = 2fRJ_A\overline{\theta}, \tag{10.44}$$

where f is the fraction of the incident ions or molecules that are accommodated and bond to chemisorbed atoms, R is the ion-to-atom arrival ratio (= J_I/J_A), and $\overline{\theta}$ is the coverage of chemisorbed atoms. In spite of the large number of arbitrary parameters, the model acceptably reproduces the growth trends of thin GaN film as a function of the temperature and the arrival ratio. In Fig. 10.28, the graphs calculated on the basis of this growth model are compared with the measured growth rates as a function of the arrival ratio and the temperature. The growth rate increases with the arrival ratio and decreases with the substrate temperature (GaN film deposition could not be observed for temperatures of 800 °C).

Figure 10.29 shows a comparison of the evolution of GaN films prepared by conventional MBE and hyperthermal ion beam-assisted molecular beam epitaxy (kinetic energy of the nitrogen ions < 25 eV). Both evolution processes were studied with the help of rate equations [95]. The height distribution functions for different deposition times are shown extracted from the size distribution function. Three-dimensional growth could be identified after conventional MBE (blue curves), i.e. the radius and height of GaN clusters increase at approximately the same rate. When the cluster radii become sufficiently large, so that the substrate surface is completely covered by a GaN film, the height of this film amounts to several tens of nanometers. In ion beam-assisted MBE (black curves in Fig. 10.29), the nitrogen ion bombardment plays a key role in the growth processes. As described above, nitrogen ion irradiation causes detachment of atoms from the top of GaN clusters, leading to a considerable

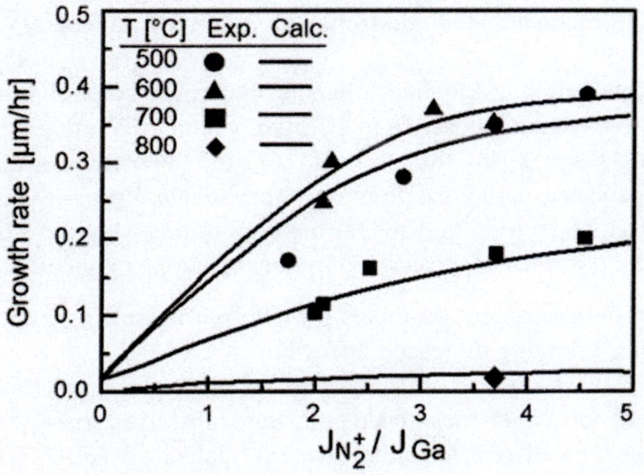

Fig. 10.28 Growth rate of GaN as a function of the ion-to-atom arrival ratio $J_{N_2^+}/J_{Ga}$ at four different temperatures. The ion energy of the N_2^+ molecules was 35 eV and the Ga atom fluence rate was 6×10^{14} cm^{-2} s^{-1} (figure adapted from [93])

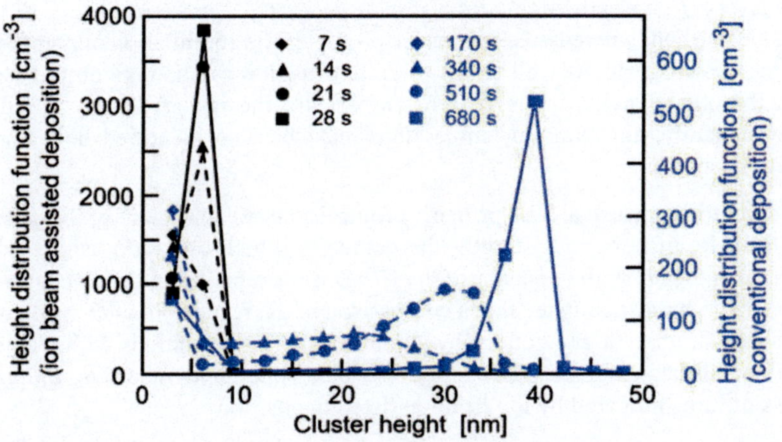

Fig. 10.29 Height distribution functions in dependence on the deposition time of GaN films prepared by ion beam assisted MBE (black) and conventional MBE (blue). The solid lines represent the height distribution for the total closed films and the dashed lines for depositions times before the total coverage

decrease of the height growth rate. Simultaneously, Ga and N atoms are displaced to the edges of a cluster, which results in an increased cluster radius growth. This means that the rate of radius growth is much greater than that of height growth. Consequently, a two-dimensional growth mode can be expected, in contrast to conventional MBE without ion irradiation. All clusters have different radii, but an approximately constant height of about 0.6 nm after a deposition time of only 28 s, when the entire surface is covered by clusters, so that an ultrathin GaN film with a thickness of about 2 ML is formed.

Molecular dynamic and kinetic molecular dynamic simulation, along with rate equations methods, make it possible to effectively study the interesting microscopic processes taking place during the ion beam-assisted deposition of thin films and to use these approaches as a tool for planning experiments. From the results presented in this subchapter, some prerequisites for the promotion of layer-by-layer growth by deposition in the presence of an assisted ion beam can be deduced:

- Assisted ion bombardment promotes growth by transferring more energy to the surface atoms, allowing increased diffusion.
- Ion energy plays a key role in the growth of high-quality crystalline films. For the inter- and intralayer mass transport, the transferred energy must exceed a minimum energy (between ten and a few tens of electron-volts). When incident or transferred energies are too high, permanent residual point defects are generated, which significantly influence the film properties. An optimum energy window for the bombarded ions must be identified (in general, smaller than 100 eV).
- Impact-induced physical effects during film growth, such as enhanced detachment of adatoms, fast downward diffusion, and the breakup of adatom islands, support the layer-by-layer growth mode.
- On the one hand, increased temperature promotes the annihilation of point defects in the growing film as well as the thermal diffusion of adatoms on the surface. On the other hand, with increased temperature the island density is reduced. Consequently, an optimum temperature must be also identified here for each material system.

The ion-to-atom arrival ratio primarily controls the stoichiometry of the deposited films. In spite of numerous studies, the degree to which this ratio can be used in determining the crystalline quality of the IBAD films remains an unsolved question.

Generally, it must be stated that a comparison of the results obtained by computer simulations or analytic predictions by application of rate equations with the limited experimental data available allow only restricted statements regarding the growth process of films prepared by ion beam-assisted deposition.

10.3.5 Epitaxial Growth by Ion Beam-Assisted Deposition

Within the scope of film growth prepared by ion beam-assisted deposition, epitaxial film growth is of particular importance, because epitaxial layer growth methods are capable of producing films with extreme purity, structural perfection, and homogeneity.

Epitaxy refers to the growth of (mono)-crystalline films on the surface of a monocrystalline substrate, where the crystalline orientation of the substrate enforces a crystalline order onto the deposited film. A distinction is made between homoepitaxy, where the film is on a substrate of the same composition, and heteroepitaxy, where the composition of the deposited film and the substrate is different. Notwithstanding the identical composition of film and substrate, large mismatches in the lattice parameters can occur. What is then expected is a straining of the film lattice without generation of lattice defects or an interface with a high density of dislocations to grade the strains from the substrate to the film, which can frequently result in a polycrystalline structure. A sufficient temperature is the prerequisite for epitaxial growth, in order to achieve significant mobility of adatoms.

It is generally accepted that epitaxial growth begins with an enhanced seeding of the substrate surfaces with oriented nuclei. These nuclei are formed by deposition of a small amount of material at an optimal temperature, referred to as epitaxial temperature. Heating of the surface leads to the formation of isolated oriented small crystallites that serve as seeds for the formation of oriented films during deposition. As has already been discussed, concurrent ion bombardment enhances this growth process.

Detailed experimental studies were carried out for epitaxial films prepared exclusively by nitrogen ion beam-assisted deposition of metal films. These studies focused on the determination of optimal conditions for the ion beam-assisted deposition of epitaxial films and the resulting epitaxial orientation relationships. In particular, these relationships describe the orientation of crystallographic planes and directions of the substrate and the film to each other, which can be determined by diffraction methods.

Petrov et al. [48] and Hultmann et al. [96] performed detailed studies on the influence of assisted-ion bombardment by reactive sputtering in an Ar-N_2 mixture on the epitaxial growth of TiN and $Ti_{0.5}Al_{0.5}N$ films on SiO_2 and steel, respectively. Preferred orientations and changes of the grain orientations were observed in dependence on film thickness for ion energies smaller than 20 eV and between 80 eV and 400 eV as well as for variations of the arrival ratio. The heteroepitaxy of GaN films on Al_2O_3 and MgO substrates prepared by reactive ion beam-assisted molecular beam epitaxy were investigated by Powell et al. [94]. They found that high growth temperatures, low ion energies, and low ion-to-atom arrival ratios lead to a significant improvement of the epitaxial film quality. GaN films with the lowest extended defect number densities (threading dislocation density $\approx 10^{10}\text{cm}^{-2}$) and the smallest X-ray diffraction rocking curve widths (5 arcmin) were obtained using $Al_2O_3(0001)$ substrates, temperatures $\geq$ 650 °C, $R \geq 3.5$, and N_2 ion energies of 35 eV. Higher nitrogen energies during deposition resulted in increased residual defect densities.

The influence of the ion incidence energy on the crystalline quality and the epitaxial matching is demonstrated in Fig. 10.30. The XRD (0002) rocking peak after 150 eV nitrogen assisted deposition has a FWHM of 3.8°, meaning that the medium tilt of crystallites around the sample normal is extremely large in comparison to single-crystalline material. Decreasing the ion energy results in a reduction of the FWHM. Consequently, for an ion energy of 25 eV, the GaN films exhibit a high degree of crystalline perfection. A threshold obviously exists between ion energies of 25 eV and 50 eV, particularly for the improvement of the crystalline quality. The fact that not only the polar spread of crystallites as measured by the rocking curves, but the azimuthal spread as well is affected by the ion energy is proven by the GaN $(10\bar{1}1)$ pole figures (Fig. 10.30 right). The pole figure of a film grown with an ion

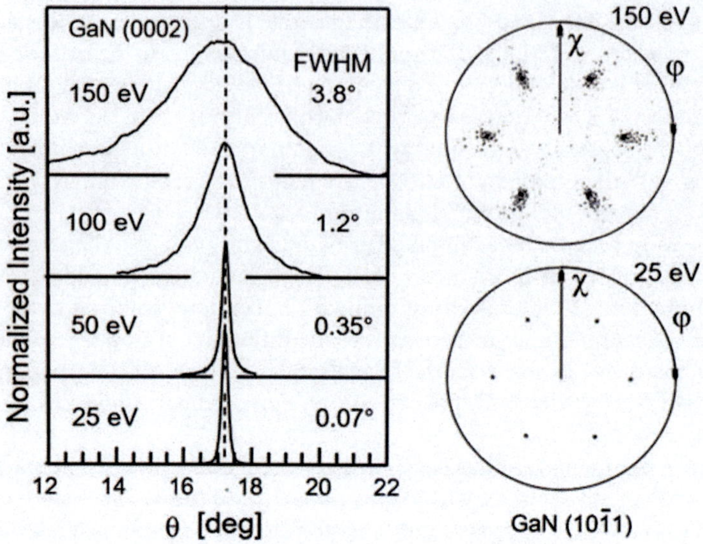

Fig. 10.30 XRD (0002) rocking peak (left) and (10$\bar{1}$1) XRD pole figures (right) of GaN films prepared with different nitrogen ion energies on c-plane sapphire substrates with an ion-to-atom ratio of 0.5. Dotted line indicates the theoretically expected positions [97]

energy of 150 eV (above) exhibits six broad and distorted pole density maxima with a high underground signal. The polar and azimuthal spread is about 3°. In contrast, the pole figure of a film deposited with ion energy of 25 eV (Fig. 10.30, right) is characterized by much narrower spreads of less than 0.07°.

Table 10.2 shows a summary of the orientation relationships of selected film-substrate combinations prepared by ion beam-assisted deposition. The nitrogen ion energy in these experimental studies ranges between 15 eV and 150 eV.

It is obvious from this table that although large lattice mismatch and/or differences of the linear thermal expansion coefficients between deposited film material and substrate exist, an epitaxial growth of films can still be realized (see also Fig. 10.20).

In order to understand the effect of ion energy in the growth of epitaxial thin films at low temperatures, the presence of optimum ion energy for the film growth is signi-ficant [71, 98, 99]. Rabalais et al. [98] reported on the optimal energy window of Si ions at the homoepitaxy of single crystalline silicon films prepared by mass-selected ion-assisted deposition (see also Sect. 9.3.2 and Fig. 9.20). The results are interpreted as a competition between lattice displacement energy, surface diffusion, and local atomic rearrangements. Layer-by-layer epitaxial growth was observed down to 160 °C. An approximate boundary line can be drawn between the regions where layer-by-layer epitaxial growth and island growth of defective, strained, and/or amorphous films have been observed. A distinct minimum is observed near 20 eV and 160 °C.

Mohajerzadeh et al. [99] also found an optimal energy between 40 eV and 50 eV for the deposition of Si-Ge compound films on Si(001) by IBA-TD. At lower

Table 10.2 Epitaxial orientation relationships between film and substrates prepared by ion beam-assisted deposition (z-GaN is the metastable zinc-blende structure of GaN, w-GaN is the stable wurtzite structure of GaN, SiC is the 6H-SiC polytype of SiC, and $LiAlO_2$ is γ-$LiAlO_2$)

Film/substrate	Lattice mismatch [%]	Ratio of lin. thermal expan. coefficients	Epitaxial orientation relationship	Refs.
GdN/MgO	+18.7		GdN(100) $\parallel$ MgO(100) with GdN[100] $\parallel$ MgO[100]	1
ScN/MgO	+7.0	0.3	ScN(001) $\parallel$ MgO(001) with ScN[100] $\parallel$ [100]MgO	2
w-GaN/MgO	+24.2 (along a-axis) −23.2 (along c-axis)	0.41 (along a-axis) 0.23 (along c-axis)	GaN(001) $\parallel$ MgO(001) with GaN[001] $\parallel$ MgO[001]	3
z-GaN/Si	+16.9	0.88	GaN(100) $\parallel$ Si(100) with GaN[010] $\parallel$ Si[010]	4
z-GaN/Al_2O_3	+5.2 (along a-axis) −4.2 (along c-axis)	0.43 (along a-axis) 0.37 (along c-axis)	GaN(111) $\parallel$ $Al_2O_3(0001)$ with GaN$[\bar{1}\bar{1}0]$ $\parallel$ $Al_2O_3[1\bar{1}00]$ and GaN$[1\bar{1}0]$ $\parallel$ $Al_2O_3[1\bar{1}00]$	5
w-GaN/Al_2O_3	+16.0 (along a-axes) +20.2 (along c-axes)	0.75 (along a-axes) 0.37 (along c-axes)	GaN(0001) $\parallel$ $Al_2O_3(0001)$ with GaN$[2\bar{1}\bar{1}0]$ $\parallel$ $Al_2O_3[1\bar{1}00]$ and GaN$[1\bar{1}00]$ $\parallel$ $Al_2O_3[1\bar{2}10]$ GaN$(2\bar{1}\bar{1}0)$ $\parallel$ $Al_2O_3(01\bar{1}2)$ with GaN[0001] $\parallel$ $Al_2O_3[1\bar{1}10]$ and GaN$[0\bar{1}00]$ $\parallel$ $Al_2O_3[2\bar{1}\bar{1}0]$	3
			GaN(0001) $\parallel$ $Al_2O_3(0001)$ with GaN$[2\bar{1}\bar{1}0]$ $\parallel$ $Al_2O_3[1\ \bar{1}00]$	6
w-GaN/Si	−17.4 (along a-axis) +4.5 (along c-axis)	1.56 (along a-axis) 0.88 (along c-axis)	GaN(0001) $\parallel$ Si(111) with GaN$[2\bar{1}\bar{1}0]$ $\parallel$ Si$[01\bar{1}]$	4

(continued)

Table 10.2 (continued)

Film/substrate	Lattice mismatch [%]	Ratio of lin. thermal expan. coefficients	Epitaxial orientation relationship	Refs.
w-GaN/SiC	−3.5 (along a-axes) −2.8 (along c-axes)	1.33 (along a-axes) 0.68 (along c-axes)	GaN(0001) || 6H-SiC(0001) with GaN$[10\bar{1}0]$ || 6H-SiC$[10\bar{1}0]$	7
w-GaN/$LiAlO_2$	+1.7 (along a-axes) +0.3 (along c-axes)	0.37 (along a-axes) 0.45 (along c-axes)	GaN$(1\bar{1}00)$ || $LiAlO_2(100)$ with GaN$[2\bar{1}\bar{1}0]$ || $LiAlO_2[001]$ and GaN[0001] || $LiAlO_2[010]$	8
TiN/Al_2O_3	+10.9 (along a-axis) +2.1 (along c-axis)	1.25 (along a-axis) 1.10 (along c-axis)	TiN(111) || $Al_2O_3(0001)$ with TiN$[\bar{1}10]$ || $Al_2O_3[1\bar{1}00]$	9
CrN/MgO(001)			(001)CrN || (001)MgO with [10]CrN || [100] MgO	10

The lattice mismatch is given by the lattice parameter of the substrate-lattice parameter of film divided by the lattice parameter of substrate, multiplied by 100%. The ratio of the linear thermal expansion coefficient is given by the thermal coefficient of the film material divided by coefficient of the substrate (ratio <1 indicates an in-plane thermal compression stress and ratio >1 indicates an in-plane thermal tensile stress)

[1]J.W. Gerlach, et al., Appl. Phys. Lett. 90 (2007) 061919
[2]D. Gall, et al., J. Appl. Phys. 84 (1998) 6034
[3]R.C. Powell, et al., J. Appl. Phys. 73 (19993) 189
[4]A. Finzel, Thesis University Leipzig, 2016
[5]J.W. Gerlach et al., Physica B 308–310 (2001) 81
[6]J.W. Gerlach, et al., Surf. Coating Technol. 128–129 (2000) 286
[7]L. Neumann, Thesis, University Leipzig, 2013
[8]Poppitz, Thesis, University Leipzig, 2015
[9]J.W. Gerlach et al., Thin Solid Films 459 (2004) 13
[10]D. Gall et al., J. Appl. Phys. 91 (2002) 3589

(< 30 eV) and at higher Si ion energies (> 70 eV), defects in the grown film and substrate were observed.

Atomic-resolution images of the GaN-SiC interface have shown that comparably low nitrogen ion bombardment (ion energy between 5 eV and 25 eV) during deposition leads to nearly perfect epitaxial growth of the GaN film (Fig. 10.31). Only the interface region (interface is marked with an arrow) between the substrate and the film was characterized by a changing of stacking sequence within the first tree monolayer of the deposited film (see insets in Fig. 10.31) [78].

These experimental results demonstrate that the energy of the incident ions during the ion beam-assisted deposition process is of exceeding importance for the epitaxial growth process [96]. On the one hand, it can be expected that processes at the surface as well as in the near-surface region of the growing thin film, are strengthened when the energy of the ions is higher than the thermal energy. On the other hand, the probability for a significant build-up of the defect concentration with increasing ion energy is well known (see Sect. 4.8). Consequently, Brice et al. [100] and later, Ma et al. [101] anticipated that the displacement damage energy of the surface is of prime importance for the explanation of epitaxial growth under hyperthermal and low-energy ion bombardment conditions.

Brice et al. [100] developed an analytical model (MD simulation), with the aim of separating the ion beam-induced surface and bulk defect processes. In this model, different threshold energies are assumed for the onset of the atomic displacement at the surface or in the bulk, where it is taken into account that the description of the damage processes at the surface and the bulk are identical. A second assumption is

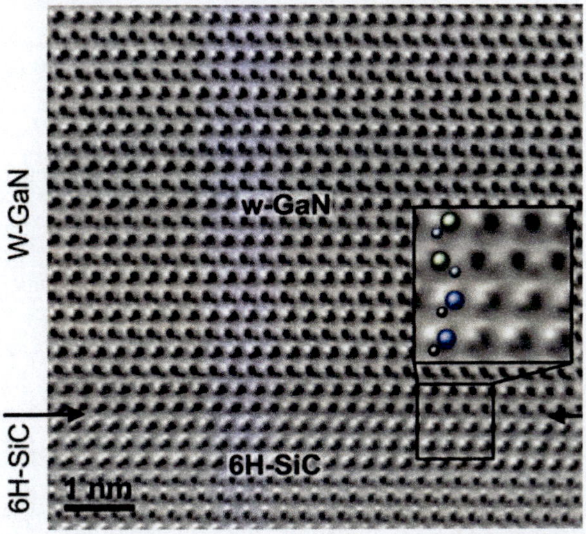

Fig. 10.31 Atomic-resolution ABF-STEM image of GaN-SiC interface. The inset shows a magnified image of the GaN-SiC interface. The arrow indicates the interface between substrate and film (adapted from [78])

that of a binary collision approximation rather than the application of a many-body collision specification. The most restricted condition in this model is the requirement that the surface displacement energy $E_d^{(s)}$ is one-half of the bulk displacement energy $E_d^{(b)}$, i.e. $E_d^{(s)} = \frac{1}{2}E_d^{(b)}$, based on the reasonable supposition that the coordination number of a surface atom is approximately half that of the bulk. Brice et al. [100] have shown that the damage energy (see Sect. 4.2) deposited into the surface and the underlying bulk can be given by

$$\nu^{(s)}\left(E, E_d^{(s)}\right) = \int\limits_{E(x)}^{E} \frac{S_d\left(\xi, E_d^{(s)}\right)}{[S_n(\xi) + S_e(\xi)]} d\xi \tag{10.45}$$

and

$$\nu^{(b)}\left(E, E_d^{(b)}\right) = \int\limits_{E=0}^{E(x)} \frac{S_d\left(\xi, E_d^{(b)}\right)}{[S_n(\xi) + S_e(\xi)]} d\xi, \tag{10.46}$$

where E is the ion energy, E(x) is energy of the ion in depth x from the surface, E_d is the threshold displacement energy (Sect. 4.1), ξ is the energy-integral variable, and denominators in (10.45) and (10.46) are nuclear and electronic stopping power (Chap. 3). $S_d(E, E_d)$ is the energy-loss rate that gives the correlation between the atomic displacement energy and the spatial rate of the damage energy deposition, and can be expressed by

$$S_d(E, E_d) = N \int\limits_{T_{min}}^{T_{max}} T\left(\frac{d\sigma}{dT}\right) dT, \tag{10.47}$$

where N is the atomic number density, T is the energy transfer from the ion to the target atom (2.35–2.37), and dσ/dT is the energy-transfer differential cross-section (2.76).

Ma et al. [101] have extended this model by the derivation of a reasonable approximation of the Kalbitzer nuclear stopping power (3.16), $s_n(\varepsilon) = 1.7\,\varepsilon^{1/2}$, in the region from a few electron-volts to some hundred electron-volts, where $s_n(\varepsilon)$ and ε are the reduced stopping power and reduced energy, respectively (see Sect. 3.1.1). They have obtained a general expression for the deposited energy in the surface layer with a thickness of x_s by

$$\nu^{(s)}\left(E, E_d^{(s)}\right) = S_{ne}(E)\left[E - E(x) - 4\left(E_d^{(s)}/\gamma\right)^{3/4}\left(E^{1/4} - E(x)^{1/4}\right)\right] \tag{10.48}$$

and in the underlying bulk by

$$\nu^{(s)}\left(E, E_d^{(b)}\right) = S_{ne}(E)\left[E(x) - \frac{E_d^{(b)}}{\gamma} - 4\left(E_d^{(b)}/\gamma\right)^{3/4}\left(E(x)^{1/4} - \left(E_d^{(b)}/\gamma\right)^{1/4}\right)\right] \tag{10.49}$$

for $\left(E_d^{(b)}/\gamma\right)^{3/4} < \gamma E(x_s)$, otherwise $\nu^{(s)}\left(E, E_d^{(b)}\right) = 0$. The used magnitude $S_{ne}(E)$ can be determined by

$$S_{ne}(E) = S_n(E)/(S_n(E) + S_e(E)) \tag{10.50}$$

and γ is the transfer energy efficiency factor (2.36). The model enables the prediction of a window of kinetic energies in which the deposition must be performed when the crystalline quality of the growing film is to be improved by low-energy ion bombardment.

Figure 10.32 shows the energy deposited in the bulk and first monolayer (surface layer) of Ge (left) and GaN (right) in dependence on the energy of the incidence helium ions (left) and nitrogen ions (right), respectively. The threshold energy for the displacements on the GaN surface was calculated by the empirical formula given by Eckstein and Preuss (see 5.34), where the surface binding energy of binary compound targets is given by the average value of the heat of sublimation, ΔH_s, of the constituents. Three different energy regions can be distinguished in Fig. 10.32:

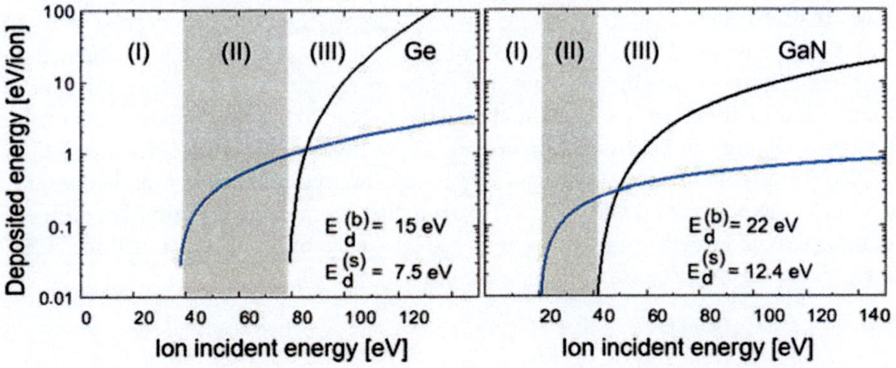

Fig. 10.32 Energy deposited in displacements of bulk and surface atoms in dependence on the ion incidence energy. Left: Ge bombarded by He ions under an incidence angle of 60° (adapted from [101]. Right: GaN prepared by perpendicularly bombarded by N ions (surface displacement energy calculated according to an empirical formula (see 5.34)

(I) $E < E_d^{(s)}/\gamma$: region of no displacements

In this region, the kinetic energy of the incident ions is not sufficient to cause displacements of atoms on the surface (i.e. in the first monolayer). Consequently, the ion energy does not have any significant effect on the growth process.

(II) $E_d^{(s)}/\gamma < E < E_d^{(b)}/\gamma$: region of surface displacements only (epitaxial window)

Ions have sufficient kinetic energy to displace atoms on the surface. For inducing displacements in the bulk, however, the kinetic ion energy is too low. With regard to high-quality crystal growth, this is the region in which the deposition process is ideally performed, because the ion irradiation enhances the ballistically driven diffusion of atoms on the surface. In this way, the atoms can more easily find and occupy energetically favorable positions and the number of lattice defects is decreased in comparison to the previously described region. In this context, it is important that displacements in the bulk material below the surface are not able to take place, so that the creation of additional buried point defects is avoided.

(III) $E > E_d^{(b)}/\gamma$: region of surface and bulk displacements

If the kinetic energy of the incident ions is larger than the threshold energy for atomic displacements in the surface and the bulk material, the ions penetrate into the material more deeply. Here, the crystal lattice is also damaged below the first monolayer.

For the examples chosen, the Brice model predicts this region of epitaxial growth (epitaxial window) to be between 40 eV and 80 eV for He ion bombarding of Ge (Fig. 10.32, left) and between 20 eV and 40 eV for N ion bombarding of GaN (Fig. 10.32, right).

It should be noted that this model does not take into account the influence of temperature on the diffusional behavior of atoms on the surface. Therefore, this model can be used to obtain only approximate values for the lower and upper limits of the ion kinetic energy in deposition processes where high-quality crystalline growth is yielded. If reliable information regarding the optimum range of ion kinetic energies is required, experimental verification is inevitable. Nonetheless, optimal conditions for the epitaxial growth of thin films prepared by ion beam-assisted deposition should be expected for hyperthermal ion bombardment ($E < 100$ eV).

10.4 Morphology of Thin Films Prepared by IBAD

Movchan and Demchishin [102] proposed a 'structure zone model' to describe the microstructure of evaporated coatings after vertical deposition without assisted ion bombardment. Three different types of morphologies can be distinguished, depending on the substrate temperature relative to the melting temperature of the film material (homologues temperature, T/T_M). At the lowest temperatures, the film possesses a columnar structure with a high density of voids in the grain boundaries (zone 1, $T/T_M < 0.3$). At higher temperatures, surface diffusion becomes relevant, resulting in densification of grain boundaries (zone 2, $0.3 < T/T_M < 0.5$). When bulk diffusion becomes important, the film consists of grains (zone 3, $T/T_M > 0.5$). Thornton [103] extended this model for thin metallic films prepared by magnetron, based on an additional parameter: the pressure during sputtering. In contrast to the IBAD process (no sputter gas), the additional sputter gas atoms under magnetron sputtering cause collisions with the sputtered atoms and change the kinetic energy of the ions arriving in the deposited thin film surface from random directions (increased beam divergence). The oblique component of the deposition flux provokes a self-shadowing process, which increases the density at grain boundaries. Moreover, the reduced kinetic energy of the sputtered particles leads to a limited adatom mobility. Consequently, these processes modify the morphology of deposited films. Thornton [103] described for the case of (magnetron) sputtering the influence of these processes on the microstructure by means of an additional transition zone (T) between zones 1 and 2. A different zone model, the revised structure zone model by Messier et al. [104], has been suggested to describe the morphology of films produced where the deposition process includes energetic particles. The introduction of ion energy rather than the pressure axis leads to an increase of the stability domain of zone T with increasing particle energy (Fig. 10.33, left). Denser thin films with a high degree of crystallinity can thereby be expected, due to the enhanced adatom mobility resulting from increased ion bombardment. As an example, Fig. 10.33 (right) shows a Mo_2N film prepared by IBA-TD at a homologous temperature of 0.35 (transition range between zones 1 and T). According to the structure zone model by Messier and coworker, the expected columnar morphology could be observed in the deposited film.

Different studies have definitively demonstrated that low-energy ion bombardment has a strong influence on the morphology of thin films (e.g., see reviews by

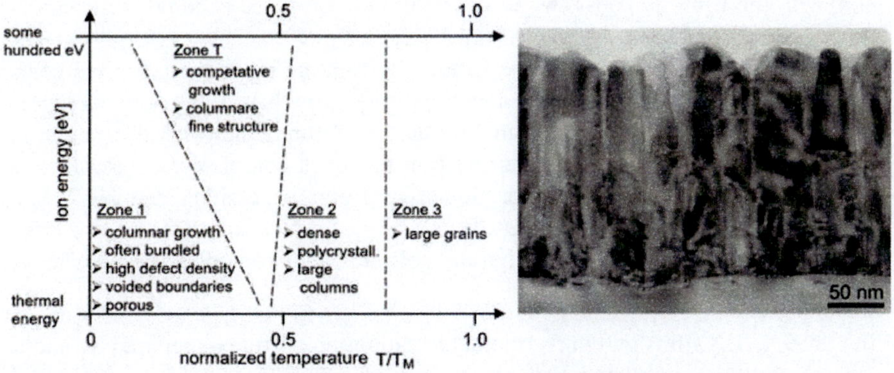

Fig. 10.33 Left: Schematic of the revised structure zone model by Messier et al. for ion beam-assisted deposition (modified version) Right: Cross-sectional X-TEM of polycrystalline Mo_2N film prepared by ion beam-assisted deposition at a homologous temperature of $T/T_M = 0.35$ and a nitrogen ion energy of about 100 eV

Greene [35], Petrov et. al. [48]). The morphology of the technologically important thin films of TiN, $Ti_{0.5}Al_{0.5}$ N and GaN prepared by reactive ion beam-assisted sputter deposition (reactive IBA-SD) has primarily been studied in detail. As is generally known, the energy and angular distribution of particle prepared by IBA-TD and IBA-PLD are unaffected due the operation under HV and UHV conditions. In contrast, methods based on IBA-SD (e.g., magnetron sputtering), which works at pressures some order of magnitude higher, the mean free path is significantly smaller than the target-substrate distance. It is thus expected that the thermalization of the sputtered particles (see Sect. 10.1.2.2) influences the morphology of the growing film:

- For instance, Adibi et al. [105] have deposited approximately 1 μm thick $Ti_{0.5}Al_{0.5}$ N films under N_2 ion bombardment at a pressure of 2.7 Pa on thermally oxidized Si(001) substrates and have studied the influence of the ion-to-atom arrival ratio R and the ion energy on the microstructure. By increasing the ion-to-atom arrival ratio R from 1 to 5.2 at a constant nitrogen molecule energy of about 20 eV (i.e. about 10 eV per incident nitrogen atom) a change from a porous (111) texture to a dense (002) oriented microstructure and an increase in the average column size to 35 nm was observed. Increasing the N_2 ion energy from 20 eV to 85 eV at $R = 1$ results in a decrease in the average column size to 25 nm and a reduction in intra-column porosity. Increasing R from 1 to 5.2 at a constant N_2 ion energy of 20 eV results in a change from a porous (111) texture to a completely dense (002) oriented microstructure, where an increase in the average column size to 35 nm could be determined. Consequently, pathways leading to microstructure and texture changes through variations in ion energy at constant arrival ratio and in arrival ratio at constant ion energy were found to be quite different.

- Petrov et al. [48] have shown that TiN films prepared on silicon oxide substrate under a pressure of 0.67 Pa are characterized by a continuous increase of the grain size with the film thickness, while the column boundaries become increasingly more open (Fig. 10.34).

  The columnar microstructure is formed by random nucleation, limited coarsening during coalescence, and a competitive growth. As the ion energy is increased to 120 eV (transition from the zone 1 to T-zone), the voids along column boundaries (arrows) disappear and the film becomes completely dense. This is accompanied, however, by incorporation of intra-granular residual damage. When the energy is increased above 160–200 eV, the defect density becomes so large that local epitaxial growth on individual columns is disrupted and re-nucleation occurs.

- In contrast, TiN films on c-axis orientated sapphire substrates prepared by about 100 eV ion beam-assisted thermal deposition (IBA-TD) under a significant lower working pressure of 1.5×10^{-3} Pa are characterized by a smooth surface (Fig. 10.35). The morphology of TiN films shows twisted and tilted grains.

  The $\pm 30°$ grain boundaries (see magnified section (A) in Fig. 10.35) consist of two grains, twisted or tilted to each other. The selected diffraction (SAD) pattern

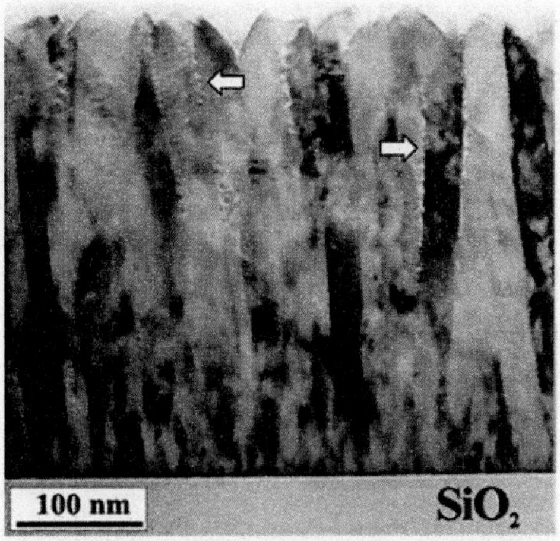

Fig. 10.34 Cross-sectional TEM micrograph of TiN deposited on amorphous SiO_2 at 300 °C with a total pressure of 0.67 Pa. The ion-to-atom arrival ratio was R = 1 and the ion energy $E_I = 20$ eV. Arrows indicate the voids at the grain boundaries (adapted from [48])

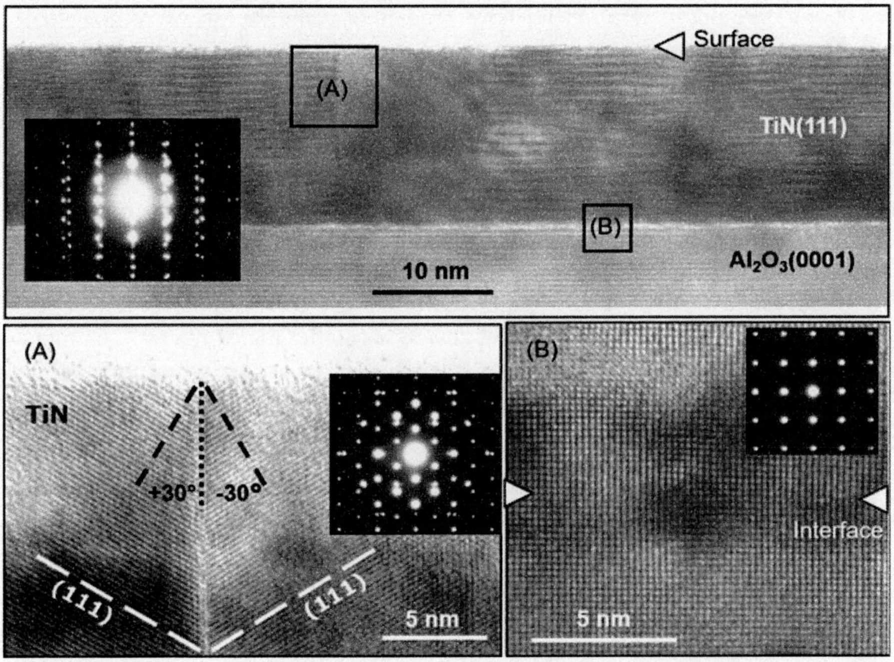

Fig. 10.35 Transmission electron micrographs and selected area electron diffraction patterns of the mosaic morphology of TiN film on sapphire substrate prepared by ion beam-assisted thermal deposition at 700 °C. The ion-to-atom arrival ratio was 0.66 and the ion energy of the nitrogen ions (N^+ and N_2^+ ions) was about 100 eV. The magnified sections (A) and (B) show a twisted grain boundary and the atomically smooth interface, respectively

grains in Fig. 10.35a represent an overlap of the diffraction patterns of the twisted grains. The magnified section (B) in Fig. 10.35 shows the atomically smooth interface between the substrate and the film without an amorphous interlayer. The SAD pattern in section (B) indicates that the grains are single crystals.

It is obvious that the morphology of films prepared by ion beam-assisted deposition methods is primarily affected by ion energy [106]. At ion energies that leave few or no permanent defects, the influence of ion-atom arrival ratio, temperature, and working pressure becomes increasingly important.

10.4.1 Roughness and Topography

To date, the evolution of surface topography and roughness in dependence on the IBAD parameters (ion energy, ion-to-atom ratio, substrate temperature, film thickness, etc.) have not been systematically examined. Some information exists about the influence of the ion energy on roughness. On the basis of detailed kinetic Monte Carlo molecular dynamic simulations for hyperthermal energies of the assisting ions (< 100 eV), Pomeroy et al. [107] have predicted a minimum rms roughness R_q (see Appendix L) as a function of ion energy near 25 eV for layer-by-layer growth of Cu films under hyperthermal energy Cu ion bombardment. This result could be confirmed in a study in which the influence of the hyperthermal deposition energy between 0.1 eV and 155 eV was studied on the roughness of a Co layer on Si(111) substrates [102]. When the ion energy is increased, the surface roughness decreases until it reaches a minimum at 25 eV, after which it again increases. Very thin GaN films prepared by ion beam-assisted molecular beam epitaxy with a nitrogen ion energy of 25 eV are characterized by a minimum of rms roughness. As an example, Fig. 10.36 shows a cross-sectional high-resolution image and a top-view image for a 13 nm thick GaN film [77]. The rms roughness is smaller than 1 nm (< 4…5 monolayers (ML)) for nitrogen ion energies < 25 eV). The GaN islands are formed on the substrate surface and grow individually until coalescence takes place. For low ion-to-atom arrival ratios (R < 1) the final growth was two-dimensional (see Fig. 10.36) and the average coalescence thickness was only about 10 ML, i.e. about 2.5 nm [81]. The roughness minimum can be attributed to effects induced by ion energy. First, the ion energy reinforces the mobility of the deposited adatoms and second, the incoming ions partially destroy existing islands. The result of this last process is a higher island density and, thereby, a higher probability of the attachment of the diffusing adatoms on these smaller islands. Consequently, these effects enhance the layer-by-layer growth and lead to smoother surfaces.

For higher energies (e.g., Co films on Si substrates, energies > 25 eV [108]), the defect formation and the sputtering by incoming ions result in an increased roughening of the surface.

For ion energies > 100 eV, an increase in the roughness can be observed, as a result of the grain growth and the increasing sputtering. For example, increasing

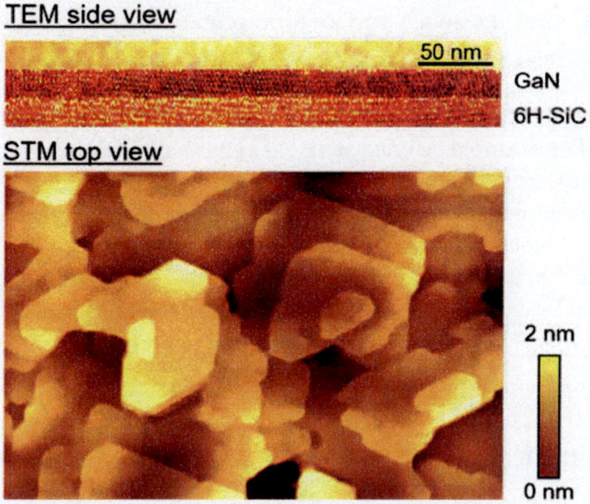

Fig. 10.36 High-resolution cross-sectional transmission electron microscopic image (above) and scanning tunneling microscopic image (bottom) of a 13 nm GaN film on 6H-SiC substrate prepared by nitrogen ion beam assisted molecular beam epitaxy

roughness as a function of an increase of the ion energy was observed for different metal films [109], boron carbon nitride films [110], and MgO films [111]. The influence of the energy of the assisting nitrogen ions on the surface roughness is illustrated in Fig. 10.37. The power spectral density (see Appendix L) is given for TiN films prepared with three different ion energies (all other deposition parameters were constant). It is obvious from this that the increase in roughness occurs in the mid-frequency range, i.e. in a range of spatial wavelengths between 10 nm and 100 nm.

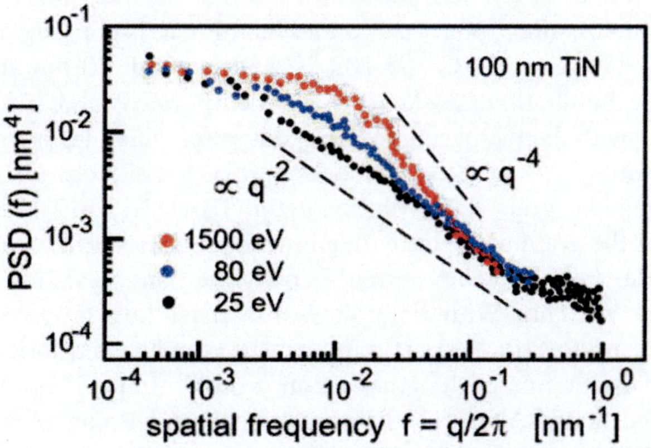

Fig. 10.37 Power spectral densities for TiN films prepared by IBAD with different nitrogen ion energies determined from 2 μm × 2 μm AFM measurements. The thickness of the films is about 100 nm

The PSD function of the rough surfaces are proportional to q^{-2} at low ion energies and q^{-4} at higher energie, where the slopes q^{-2} and q^{-4} correspond to the ballistic smoothing and the smoothing caused by ion-induced viscous flow, respectively (details see Sect. 7.2.2). As expected, roughness also increases with increased film thickness. For example, Miyata et al. [112] have studied the evolution of roughness under 100 eV oxygen ion beam-assisted sputter deposition of MgO films. The roughness increases by a factor of about 10 for films with a thickness of < 100 nm. It was observed that independent of the deposition rate, the surface roughness showed a steep increase in the roughness at a thickness of about 4 nm. According to the authors, this behavior can be explained by an increase in the extent of the crystallized regions of the film.

10.4.2 Grain Size

The grain size in polycrystalline films prepared by IBAD is dependent on the deposition temperature and post-annealing conditions, the ion flux (ion-to-atom arrival ratio), and the ion energy. Marinov [103] was the first to demonstrate the influence of additional ion bombardment on the average grain diameter of Ag films. It was assumed that the ion bombardment would increase the adatom mobility. Moreover, the higher the ion current, the greater the diameter of the Ag grains. In contrast, Huang et al. [114] observed that the average grain size decreases with an increase in the average energy per deposited atom for Ag films prepared by IBAD.

The grain growth in Cu films was studied as function of the *ion-to-atom arrival ratio*, the ion energy, and the annealing temperature. Roy et al. [115] observed a significant reduction of the grain size with an increase of the Ar ion to Cu atom arrival ratio up to 0.02 for two ion energies (62 eV and 600 eV) and different temperatures (between 60 °C and 230 °C), but independent of the substrate material. Above this R value, the grain size after 600 eV Ar ion bombardment is constant at about 30 nm for temperatures between 62 °C and 103 °C and is about 40 nm at 230 °C. It is assumed that the bombardment gas ions are incorporated increasingly at the Cu grain boundaries with the increasing ion-to-atom ratio, which leads to grain growth. The ion beam-induced lattice defect concentration, on the other hand, limits the grain growth. In contrast, the grain size evolution in thin IBAD Ni and TiN films showed a steady growth of the grain diameter with an increasing arrival ratio [116–118]. TiN films grown under low ion beam current density are nanocrystalline, with a grain size typically below 10 nm. With increasing ion-assisted flux, the size of crystallites increases rapidly up to 80 nm. The grain growth is connected to a change in the crystallographic orientation of the grains, from a preferred [111] orientation typical of evaporated fcc material to a [100] orientation [116]. Rajan et al. [118] found that the average grain diameter of about 20 nm without ion bombardment ($R = 0$) increases to about 45 nm after 1000 eV Ar ion bombardment during Ni deposition and with an arrival ratio $R = 0.45$. This growth evolution is associated with the changing of the main texture component to [220]. On the basis of these results, a

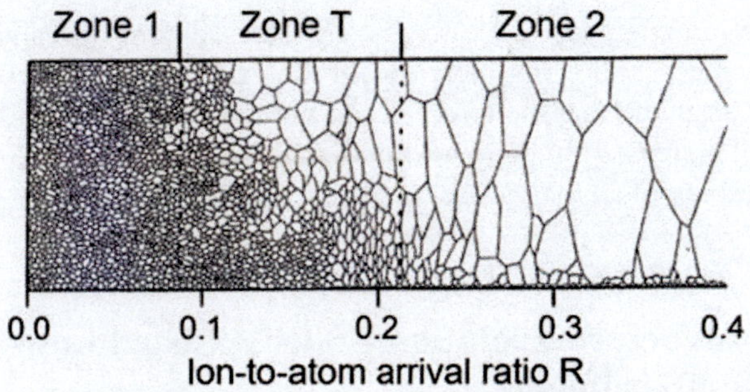

Fig. 10.38 Modified structure zone model for thin film grain morphology in terms of the ion-to-atom arrival ratio R instead of the homologous temperature (adapted from [118] in Chap. 9)

modified structure zone model has been proposed (Fig. 10.38), based on the ion-to-atom arrival ratio instead of the homologous temperature, to describe the variation of the microstructure (grain size and texture) as a function of the ion-to-atom arrival ratio at room temperature (compare with Fig. 10.33). It can be observed that, in general, with increasing arrival ratio for a given energy of ions, the grain size also increases.

Ma et al. [119] have studied the influence of an *ion energy* between 50 eV and 1200 eV on Cu grain growth. For the as-deposited films, the grain size increases from about 60 nm after deposition without ion bombardment to about 90 nm after 200 eV. Above this energy, the grain size gradually decreases until a grain size of about 70 nm is reached at 1200 eV. Upon annealing at 400°C, 500 eV IBAD Cu films show significant grain growth with final grain size close to the film thickness of 500 nm. Interestingly, no obvious grain growth is observed in Cu films bombarded with an ion energy of 1000 eV.

The influence of *film thickness* and *growth mode* on the grain size has also been studied. In thick TiN films prepared IBAD, both the roughness and the grain size increase with increasing film thickness [120]. The evolution of the grain sizes in GaN thin films in dependence on the film thickness is strongly correlated to the growth mode [37]. In the early stages of GaN film prepared by IBAD, grains with a diameter smaller than 50 nm can be observed for 3D growth (island growth) of (0001)-oriented w-GaN (hexagonal GaN). For GaN films deposited under conditions that induce a 2D layer-by-layer growth, (111)-oriented z-GaN (cubic GaN) grains with an average diameter smaller than 100 nm are formed. These grains appear in twinned form, i.e. they are rotated against each other around the [111] direction.

According to which process is dominant, it can be established that grain size and surface roughness can be increased or decreased in dependence on the ion energy. At low ion energies (between 20 eV and 50 eV), the formation of a high density of small, randomly oriented grains can be observed, because the incident ions disrupt existing islands at the early stages of film growth. Thus, the island density is increased and the attachment of adatoms on the periphery of small islands is significantly strengthened. The films are characterized by small grains and a very smooth surface. With an increase of the ion energy (> 20 eV... 50 eV), the adatom diffusivity during film growth also increases, allowing increases in the grain size and surface roughness. Harper [1] has assumed that the effect of ion bombardment at such low energies is similar to that of increased temperature. For higher ion energies (> 100 eV), sputtering processes contribute to an increasing roughness via surface erosion.

10.5 Microstructure Evolution Under Assisted Low-Energy Ion Bombardment

10.5.1 Texture Development

The crystal orientation in thin films may markedly affect physical properties. Two basic procedures permit the specific control of the orientation in thin crystalline films. On the one hand, the growth of a crystalline material during thin film deposition is referred to as epitaxial growth when the deposited layer grows with a defined orientation to the underlying substrate. A distinction is made between homoepitaxial growth, i.e. the growth of a crystalline film on a substrate of the same material, and heteroepitaxial growth, i.e. the growth of a crystalline film on a substrate of a dissimilar composition. Consequently, the growth process is often associated with the development of a preferred orientation or crystalline texture. The texture is the most important parameter in describing the anisotropy of polycrystalline materials. On the other hand, the ion beam-assisted deposition can be used to fabricate highly textured thin films on amorphous or crystalline substrate surfaces. Biaxial crystal alignment or preferred out-of-plane and in-plane crystallographic orientation can be achieved by low-energy off-normal ion bombardment, where the orientation can be controlled by the deposition methods and their associated processing parameters.

10.5.1.1 Pole Figures and Pole Figure Measurement

Thin polycrystalline films deposited by thermal and electron beam evaporation, pulsed laser ablation, or sputtering techniques are frequently characterized by a fiber- or columnar-like morphology with one preferred orientation normal to the surface, while in the azimuthal direction, the orientations of grains are randomly distributed. To characterize the thin film texture quantitatively, the orientation density function (ODF), f(**g**), of a crystalline phase is defined by

$$f(\mathbf{g})dg = \frac{dV}{V}, \tag{10.51}$$

where V is the sample volume and dV the volume of all crystallites with orientations **g** within the angular element dg. The ODF can be approximated by texture components. These are simple, bell-shaped functions in the orientation space, which can be interpreted as preferred orientations (peak components) or directions (fiber components). The fractions of these components can be determined by means of a quantitative texture description [121]. Due to this high compression of texture information, processes and properties correlated with texture can be more readily understood. The ODF cannot be directly measured. It is constructed by single orientation measurements or approximated by inversion of the pole figure. The two-dimensional pole density $P_{hkl}(\chi,\varphi)$ can be expressed by

$$P_{hkl}(\chi,\varphi)d(\chi,\varphi) = 4\pi\frac{dV(\chi,\varphi)}{V}, \quad (10.52)$$

where $d(\chi,\varphi) = \sin(\chi)d\chi d\varphi$ and the azimuth angle φ and the polar angle χ, respectively, indicate sample orientation. The pole density of each crystallographic direction can be plotted on a pole sphere or projected stereographically into the plane, i.e. each direction, given by φ and χ, is associated with a pole density value, $P_{hkl}(\chi,\varphi)$, according to (10.52), which quantifies the volume fraction dV/V of the crystallites. The measured intensity $I_{hkl}(\chi,\varphi)$ of the pole figure measurement is given by

$$I_{hkl}(\chi,\varphi) = N_{hkl}P_{hkl}(\chi,\varphi), \quad (10.53)$$

where N_{hkl} is a normalization factor. The main characterization methods are based on the diffraction of X-rays or electrons. When the Bragg condition is fulfilled, the geometrical position of sample and/or X-ray sources as well as the detector could be manipulated to perform different types of measurements to obtain various kinds of information on crystalline thin film structure, where a distinction is made between two well-defined orientations of the grains in a polycrystalline film. The *out-of-plane orientation* describes the orientation of grains perpendicular to surface (lattice planes parallel to surface), while the *in-plane orientation* gives the orientation of grains parallel to the surface (lattice planes perpendicular to surface). Some routinely performed measurement procedures include:

(i) The *symmetrical* $2\theta - \omega$ *scan* to qualitatively study the crystal structure. The measurement can be carried out by symmetrically varying the incident angle ω and the 2θ angle simultaneously. The value of 2θ is kept twice as large as ω in the overall measurement. It is equivalent to changing the length of the scattering vector while maintaining the fixed direction. This type of measurement provides information on the crystal orientation and out-of-plane interplanar spacing for crystal planes parallel to the film surface.

(ii) To assess the *out-of-plane tilt distribution* of crystallites, a *rocking curve* measurement (or ω-scan) can be performed by keeping the positions of incident beam (ω) and detector (2θ) fixed (keeping the length of scattering vector constant) at the Bragg's angle of planes of interest (in this case, planes parallel to surface of the film) and then rocking the film around the Bragg's angle. The full-width-half-maximum (FWHM) of the acquired peak is the measure of mosaic crystallite distribution.

(iii) *Pole figure* measurements are performed for a texture analysis of oriented polycrystalline and epitaxial film [122]. From such measurements, e.g., the epitaxial relationships between layer and substrate, i.e. the arrangement of the crystallite lattice with respect to the lattice of the single crystalline substrate, can be determined. The term in-plane pole figure is often used, since the in-plane detector arm is employed in the diffractometer for the pole figure measurement. An in-plane pole figure offers some advantages as compared to classical pole figure measurements with the parallel beam reflection method. The disadvantage is that the tilt angle and Bragg angle of the sample no longer correspond to individual goniometer axes. The pole figures are measured by a fixed diffraction angle 2θ and the varying of two geometrical parameters, such as the χ-angle (tilt angle from the surface normal) and the φ-angle (given by the rotation around the sample surface normal). The diffracted intensity obtained is then plotted as a function of these two angles χ and φ on a Wulff net. For a quantitative analysis, normalization of the intensity is required (obtained from a randomly oriented sample as reference). The relative intensity after normalization is referred to as pole density.

10.5.1.2 Preferred Crystal Orientation

Preferred orientation is a property of thin films after ion beam-assisted deposition that corresponds to the orientation tendency of the grains in grown films to orient themselves in certain directions according to a preferred crystallographic plane. This results in a higher fraction of occurrence of these planes and leads to a change in the relative intensities of the reflections from other crystallographic planes. The changing of the intensity in the peaks is dependent on the conditions of the film deposition.

The orientation of grains within a polycrystalline thin film is usually quite random. Randomly orientated grains or crystallites in a thin film are schematically shown in Fig. 10.39, above. The small cuboids (crystallites) represent the lattice plane arrangement within grains and show the crystallographic orientation of the lattices. The film consists of randomly oriented crystallites without a preferred orientation. Under certain conditions, however, the lattice planes of the crystallites may be oriented (Fig. 10.39, middle). A polycrystalline film is preferentially orientated or shows an out-of-plane orientation when a substantial fraction of the grains has an orientation [hkl] perpendicular to surface (represented by arrows in Fig. 10.39, white arrows indicate the preferred grain orientation). This arrangement is also called fiber texture. The fiber texture is characterized by a circular symmetry around a reference (substrate) axis. The crystals in the thin film are aligned with a high degree of orientation in this preferred direction. After subtraction of the background, the degree of preferred orientation of thin films can be determined by calculation of the orientation factor

$$T_c = \frac{I_{hkl}/I_{0,hkl}}{\frac{1}{n}\left[\sum\left(I_{hkl}/I_{0,hkl}\right)\right]}, \tag{10.54}$$

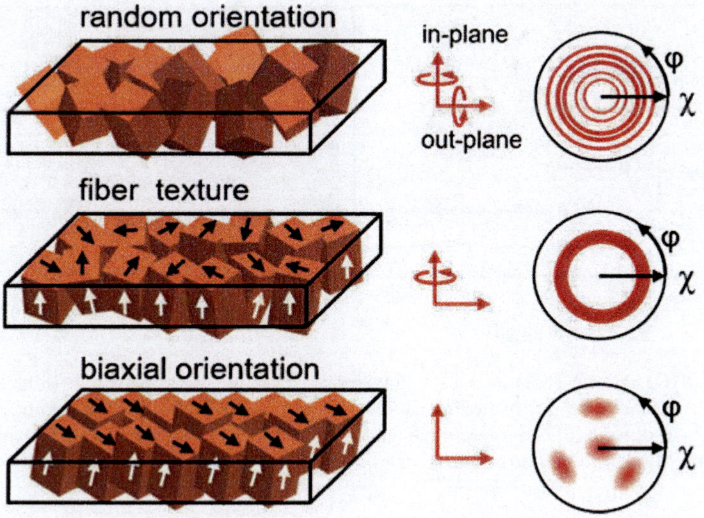

Fig. 10.39 Schematic representation of the crystal orientation in a randomly oriented film (at the top), in a film with fiber texture (middle), and one with a biaxial orientation (below). White arrows and black arrows indicate the out-of-plane and in-of-plane orientations of the grains, respectively. Examples for corresponding diffraction patterns are also schematically shown

where n is the number of all considered growth directions, I_{hkl} is the measured intensity of the hkl plane and $I_{0,hkl}$ the corresponding recorded intensity from the JCPDA data file.

The development of the preferred orientation during the deposition by (magnetron) sputtering and evaporation techniques in a gas environment has been studied in detail (see e.g., [123–125]). Here, the presentation is solely focused on films prepared by ion beam-assisted deposition.

10.5.1.3 Preferred Crystal Orientation by Ion Beam-Assisted Deposition

In early experiments, Trillat et al. [126] and Ogilvie and Thomson [127, 128] observed that after bombardment with Ar ions, single-crystalline Ag films were transformed into polycrystalline film, where the disorientation of the created crystallites was strongly dependent on the orientation of the host crystal surface, the ion energy, and the ion current density. Later, van Wyk and Smith [129] studied the preferred orientation in Cu films by self-ion bombardment and explained this phenomenon by means of a channeling model (see Sect. 10.5.1.5). In the subsequent decades, many studies have been published regarding the controlled approach to the preferred orientation in polycrystalline thin films by ion bombardment *after* film deposition (see e.g., [124, 125]).

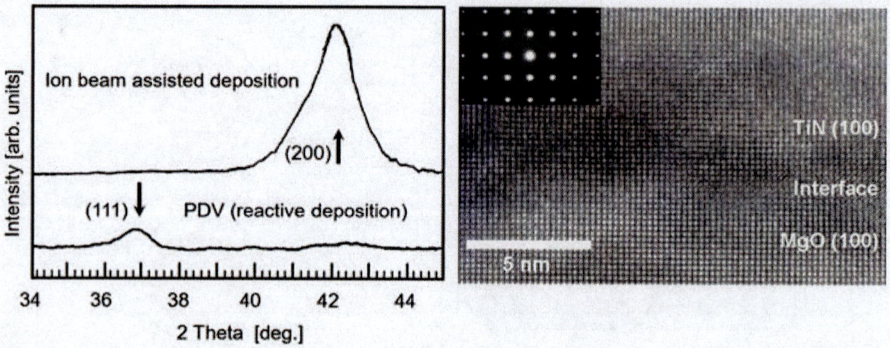

Fig. 10.40 XRD spectra (left) of a TiN film prepared by reactive deposition (left) and by Ar ion-assisted deposition (right) of titanium within a nitrogen environment at room temperature on silicon. High-resolution XTEM micrograph (right) of TiN after nitrogen ion beam assisted titanium deposition on MgO at room temperature

In contrast, ion beam-assisted deposition is characterized by deposition of a film material generated by evaporation, sputtering, or laser ablation and the bombardment of the film *during* the growth process. The first experiments to study the preferred orientation in thin films prepared by ion beam-assisted deposition were carried out by Dobrev [123]. While Ag films deposited at 300 °C without ion irradiation did not exhibit preferential orientation, these Ag films deposited under the same conditions and simultaneous 10 keV Ar ion irradiation were characterized by a [110] fiber texture. Later, Yu et al. [125] observed a preferred [110] orientation of sputtered Nb films. The polycrystalline Nb were irradiated with 200 eV Ar ions at 70° with respect to the surface normal during the sputter deposition. The degree of the fiber orientation was scaled by the ion-to-atom arrival ratio, independent of the temperature. As an example, Fig. 10.40 compares the diffractograms of TiN prepared by conventional PVD (left, below) and by Ar ion-assisted titanium deposition within a nitrogen atmosphere (left, above). The films after PVD are characterized by a columnar grain morphology and a [111] fiber texture, i.e. the [111] direction of the TiN crystallites is parallel to the surface normal of the substrate material and normal to the surface of the substrate, respectively. Figure 10.40, right, shows the crystalline microstructure of the MgO substrate, TiN film and interface. After IBA-TD, a preferred orientation of the formed titanium nitride crystallites can be observed. The TiN crystallites exhibit the expected fiber texture, i.e. the TiN grains show the tendency to have their surface normal perpendicular to the normal axis of the film [130]. Thus, one direction of the TiN crystallites is fixed (here the [111] direction), but the orientation of the crystallites to each other is free (the azimuth angle φ is not fixed, see Fig. 10.39). A number of studies have reported on the observation of the preferred orientation during the deposition of polycrystalline films (some examples are given in Table 10.3).

Preferred orientations after ion beam-assisted thermal deposition could be proved with different metals as well as oxides and nitrides. Obviously, the preferred orientation has been established for all crystal structures. Frequently, [111] and [110]

Table 10.3 Thin films prepared by ion beam-assisted deposition with a preferred orientation

Film material	Crystal structure	Preferred orientation	Alignment dependence	Refs.
Ag	fcc	[110]	ion fluence	1
		[111]		2
		(111)	arrival ratio, incidence angle	3
Nb	bcc	[110]	arrival ratio	4
		[110]	film thickness	5
Cu	fcc	(111)	arrival ratio	6
			ion energy	7
Pt		(111)	arrival ratio	8
AlN	hex	(0002) and (1000)	ion energy	9
			ion energy and arrival ratio	10
		(0002)	substrate temperature	11
		(0001)	substrate temperature	12
BN	Cubic	[110]		13
TiO_{2-x}	Tetragonal (rutile)	(200)		14
TiO_{2-x}	Tetragonal (anatase)	(101)	annealing temperature	15
TiN	fcc	(111)	arrival ratio	16
				17
		[001]		18
ZnO	hex	(0002)		19
			ion incidence angle	20
ITO	Cubic	(222) and (400)	type of the substrate	21
			film thickness	22
		(222)	arrival ratio	23
		(200)		24

(continued)

Table 10.3 (continued)

Film material	Crystal structure	Preferred orientation	Alignment dependence	Refs.
MoS_2	hex	[001]		25

The preferred orientation is given by [hkl] parallel to the surface normal or by (hkl) parallel to the surface

[1] D. Dobrev, Thin Solid Films 92 (1982) 41
[2] T.C. Huang, et al., J. Vac. Sci Technol. A3 (1985) 2161
[3] T. Feng et al., Appl. Surf. Sci. 254 (2008) 1565
[4] L.S. Yu et al., Appl. Phys. Lett.47 (1985) 932
[5] H. Ji, G.S. Was, Nucl. Instrum. Methods in Phys. Res. B 148 (1999) 880
[6] R.A. Roy et al., J, Vac. Sci. Technol. A6 (1988) 1621
[7] H. Ma, et al., Acta Mater. 98 (2015) 17
[8] B.J. Jiang, et al., Thin Solid Films 483 (2005) 411
[9] J.H. Edgar, et al., J. Mater Sci: Mater. Electr. 7 (1996) 247
[10] X. Wang, et al., Surf. Coat. Technol. 103–104 (1998) 334
[11] I.-H. Kim et al., Thin Solid Films 253 (1994) 47
[12] I. Gablech, et al., Thin Solid Films, 670 (2019) 105
[13] A.K. Ballal et al., Thin Solid Films 224 (1993) 46
[14] C. Li et al., Nucl. Instr. Meth. In Phys. Res. B 169 (2000) 21
[15] C. Yang et al., Appl. Surf. Sci. 254 (2008) 2685
[16] N. Nakagawa, et al., Nucl. Instrum. Methods in Phys. Res. B, 80–81 (1993) 1380
[17] T. Sato et al., Nucl. Instrum. Methods in Phys. Res. B, 19–20 (1987) 644
[18] I. Gablech, et al., Thin Solid Films, 638 (2017) 57
[19] D.H. Zhang, D.E. Brodie, Thin Solid Films, 213 (1992) 109
[20] D. Köhl et al., J. Phys. D: Appl. Phys. 43 (2010) 205301
[21] Y. Zhinong, et al., Thin Solid Films 517 (2009) 5395
[22] C. Liu, et al., J. Appl. Phys. 93 (2003) 2262
[23] L. Meng et al., Phys. Stat. Sol. (a) 205 (2008) 1961
[24] K. Thiele et al., J. Mater. Res. 18 (2003) 443
[25] D.N. Dunn et al., J. Mater Res. 13 (1998) 3001

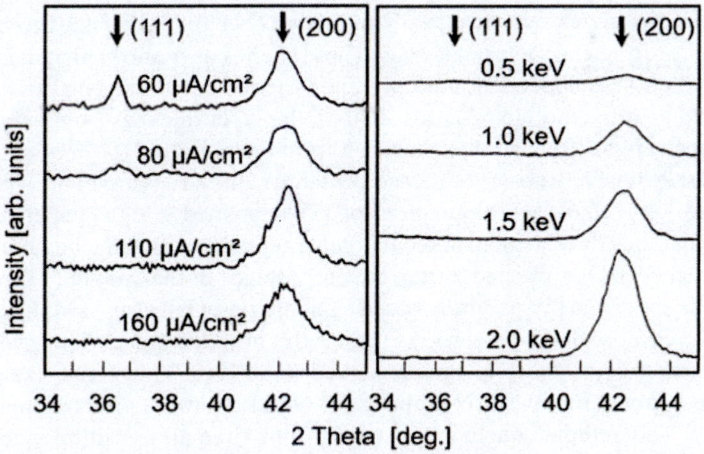

Fig. 10.41 XRD spectra of TiN films on Si(111) prepared by argon ion-assisted titanium deposition within a nitrogen atmosphere at room temperature in dependence on the ion current *density* (left, the ion energy is 1 keV) [130] and in dependence on the ion energy (right, the ion current density is 100 $\mu A/cm^2$) [131]

preferred orientations were observed for both face- and body-centered cubic structures, while for hexagonal structures, a [0001] orientation could be determined. According to the results obtained, summarized in Table 10.3, the preferred orientation can be controlled by the ion energy and the arrival ratio as well as by the film thickness.

Observations on the dependence of these accentuated orientations on the parameters of the ion beam-assisted thin film deposition have revealed a direct link to the existence of preferential orientations. In Fig. 10.41, left, the transition from the (111) orientation to the (200) preferred orientation is illustrated in dependence on the ion current density and the ion energy. With increased ion current density, the preferred orientation changes from the (111) to the (100) orientation (indicated by the existence of the TiN (200) peak). This effect can be also interpreted as the influence of an increased ion-to-atom arrival ratio. A similar behavior can be observed when studying the crystalline orientation in TiN films as a function of ion energy. Figure 10.41, right, shows that (100) preferred orientation is increasingly dominant even at ion energies below 500 eV. The above results for argon ion beam-assisted deposition also hold true for TiN films prepared by nitrogen ion-assisted deposition of titanium.

In addition to the ion-to-atom ratio and the ion energy, the substrate temperature and the film thickness significantly influence the texture during the IBAD process. For example, Hultman et al. [132] studied the effects of an arrival ratio with 20 eV N_2 ions on the texture of polycrystalline TiN films prepared by UHV reactive magnetron sputtering. While (111) and (002) orientations coexist during the early stages of the film growth, (111) orientated grains gradually overgrow the (002) oriented grains at a film thickness < 150 nm. Ultimately, the film has a complete (111) texture.

Table 10.4 summarizes some results of orientation changes during ion-beam-assisted deposition for frequently used materials. These studies have shown that the changing of the orientation is dependent not only upon the ion-to-atom arrival ratio and ion energy, but, in some individual cases, also on the layer thickness, temperature, and ion incidence angle. To date, this alteration from the preferred (001) orientation has been almost exclusively observed in face-centered cubic structures of the space group $Fm\overline{3}m$ (see Table 10.4). MD simulations of TiN deposited at high temperatures with hyperthermal ions have demonstrated that misoriented grains at the beginning of the deposition are (001) reoriented during the early stages of the growth [133].

For films deposited by sputtering based technologies, Gall et al. [134] have studied in detail the competition between (111) and (002) oriented grains. The studied film-substrate system consisted ScN films deposited on MgO (001) surfaces. Experimental studies have shown that the ScN grains grow with both (002) orientation and (111) orientation. The authors concluded that ScN films (like all transition metal nitride films) evolve toward a (111) orientation because the diffusion length of cations on the (001) surface is longer and the potential energy of adatoms on the (111) surface is lower than a possible preference of the (001) surface for adatom flow from the (002)- to the (001)-oriented grains.

10.5.1.4 Biaxial Orientation

In addition to the out-of-plane orientation of grains, the orientation of the grains parallel to the substrate can also be restricted by low-energy ion bombardment during thin film deposition. Consequently, nearly all the grains are aligned parallel to the substrate surface. These films are then biaxially oriented (see Fig. 10.13).

The first study to report on the biaxial orientation of grains in thin films prepared by ion beam-assisted deposition was published in 1992 [124]. A biaxial orientation is characterized by a totally fixed orientation of the crystallites produced (see Fig. 10.39). Iijima et al. [124] reported on the formation of biaxially oriented buffer layers of yttria-stabilized zirconia (YSZ) on polycrystalline Ni-based alloy (Hastelloy) by ion beam-assisted deposition. Subsequently, basal-plane-aligned $YBa_2Cu_3O_{7-x}$ films were successfully deposited on the biaxially oriented YSZ. In subsequent decades, biaxial alignment of grains in several materials after ion beam-assisted deposition could be proven (see e.g., [135]). As an example, the biaxial arrangement in polycrystalline TiN film is shown in Fig. 10.42. XRD pole figures obtained from titanium nitride films prepared by reactive deposition without ion irradiation (Ti deposited by PVD in nitrogen atmosphere) are shown in Fig. 10.42 (left). The typical (111) fiber texture at $\chi = 70.5°$ was obtained. The orientation distribution of the titanium nitride crystallites is given by a rotational symmetry with respect to the surface normal. This is a consequence of the non-perpendicular incidence of evaporating Ti atoms to the surface and is typical for films developing a columnar microstructure. Figure 10.42 (middle) shows a representative (111) pole figure of a TiN film prepared under perpendicular 2 keV nitrogen ion bombardment. The preferred orientation in the TiN films changes from (111) to (100), but a fiber

Table 10.4 Changing of the orientation in thin films during ion beam-assisted deposition

Type of structure/space group	Film material	Change of orientation	Change of orientation as function of the parameter	Refs.
fcc /Fm$\bar{3}$m	TiN	(111) → (200)		1–13
	TaN	(111) → (200)	Arrival ratio	14
	VN	(111) → (200)		15
	ZrN	(111) → (200)		15
	ScN	(111) → (200)	Ion energy	16
	CeO_2	(111) → (200)	Substrate temperature	17
		(111) → (200)		18
		(111) → (220) (111) → (113)		19,20
		(111) → (200)	Ion incident angle	21
	MgO	(111) → (200)		22
	Al	(111) → (200)		23
fcc /Fd$\bar{3}$m	TiC	(111) → (200)		24
Tetragonal/P42/mmm	TiO_2 (rutile)	(110) → (200)		25
fcc/Fm$\bar{3}$m	$Ti_{0.5}Al_{0.5}N$	(111)&(002) → (111)	Arrival ratio and film thickness	26

[1] Y. Andoh et al., Nucl. Instr. Meth. In Phys. Res. B 39 (1989) 158
[2] M. Satou et al., Nucl. Instr. Meth. In Phys. Res. B 39 (1989) 166
[3] I. Petrov et al., J. Vac Sci. Technol. A10 (1992) 265
[4] W. Ensinger, B. Rauschenbach, Nucl. Instr. Meth. In Phys. Res. B 80–81 (1993) 1409
[5] K. Baba, R. Hatada, Surf. Coatings Technol. 66 (1994) 368
[6] H. Kheyrandish et al., J. Vac. Sci Technol. A 12 (1994) 2723
[7] J.E. Green et al., Appl. Phys. Lett. 67 (1995) 2928
[8] H. Jiang et al., Thin Solid Films 258 (1995) 51
[9] W. Ensinger, Nucl. Instr. Meth. In Phys. Res. B 106 (1995) 142
[10] H. Wengenmair et al., Appl. Surf. Sci. 99 (1996) 313
[11] J.-H. Huang et al., Scripta Mater. 42 (2000) 573
[12] H.J. Shin et al. Nucl. Instr. Meth. In Phys. Res. B 190 (2002) 807
[13] G. Abadias et al., J. Appl. Phys. 99 (2006) 113,519
[14] C.-S. Shin, et al., J. Appl. Phys. 92 (2002) 5084
[15] C.-H-Ma, et al. Surf, Coat. Technol. 133–134 (2000) 289
[16] D. Gall, et al., J. Appl. Phys. 84 (1998) 6034
[17] M.Q. Huang, et al., Nucl. Instr. Meth. In Phys. Res. B 148 (1999) 793
[18] S. Zhu, et al. Appl. Phys. Lett. 65 (1994) 2012
[19] S. Kabakaraju et al., Thin Solid Films 305 (1997) 191
[20] C. Mansilla, Solid State Sci. 11 (2009) 1456
[21] Z.-N. Yu et al., Surf. Coat. Technol. 162 (2002) 11
[22] J. Wang et al., Surf. Coat. Technol. 158–159 (2002) 548
[23] S. Masaki et al., Nucl. Instr. Meth. In Phys. Res. B 59–60 (1991) 292
[24] D.E. Wolfe J. Singh, Surf. Coatings Technol. 124 (2000) 142
[25] F. Zhang et al., J. Appl. Phys. 83 (1998) 4101
[26] L. Hultmann et al., J. Appl. Phys. 93 (2003) 9086

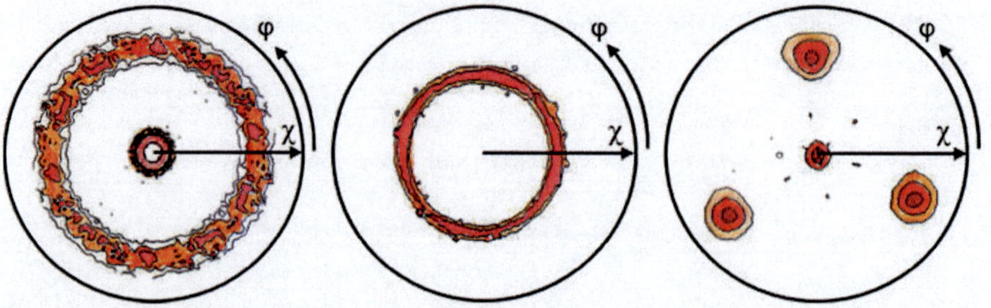

Fig. 10.42 XRD (111) pole figures of TiN film prepared by reactive deposition of titanium within a nitrogen environment (left), by perpendicular 2 keV nitrogen ion beam-assisted Ti deposition (middle) and by 2 keV nitrogen ion beam-assisted Ti deposition at an ion incidence angle of 55° (right)

texture at $\chi = 54.7°$ with a full width of half maxima (FWHM) of about 7° is also observed. Crystallites oriented in [100] direction parallel to the incident direction of ions grew preferentially with respect to arbitrarily aligned crystallites. Figure 10.42 (right) shows a (111) XRD pole figure of an ion beam-assisted deposited TiN film with a nitrogen ion incidence angle of 55°. This defined incident angle was selected because the angle between the (100) and (111) planes of the cubic unit cell is 54.7°. The experimental arrangement is schematically depicted in Fig. 10.43. The film is also (111) oriented, but instead of the fiber texture, a biaxial alignment of the crystallites was obtained, because the pole figure is characterized by the four-pole density maxima at $\chi = 0°$ and $\chi = 70.5°$, with $\varphi = 90°$, 210° or 330°. The surface normal of the cubic cell is now oriented parallel to the ion beam, so that a totally fixed orientation of the crystallites follows. It should be noted, however, that only an ion incidence angle of 54.7° allows both (111) orientation to sample surface normal and (100) orientation to the ion beam.

The possibility of controlling the biaxial orientation of thin films by means of ion beam-assisted deposition and to subsequently use these films as templates for the epitaxial growth of other materials has led to intensive studies of this phenomenon. Table 10.5 provides a summary of materials with a biaxial orientation that could be

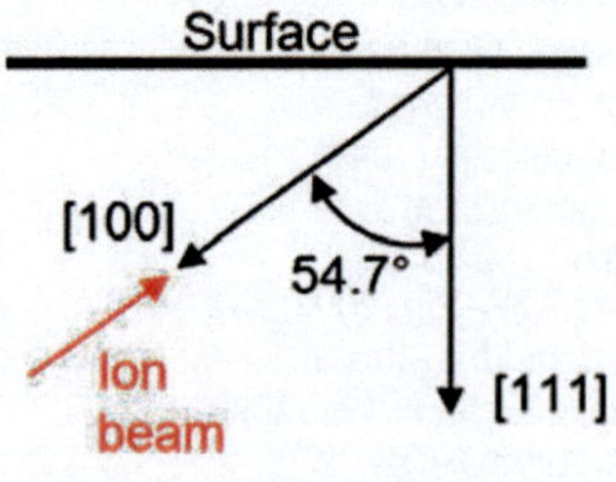

Fig. 10.43 Schematic depiction of the relation between the ion beam and the [100] and [111] directions of the TiN crystallites. Ions impinge the growing film at an angle of incidence of 54.7°; it is therefore aligned parallel to one (111) axis of the growing film

Table 10.5 Thin film materials with a biaxial orientation after ion beam-assisted thermal deposition (ion species without superscript characters—charge state not specified by the authors, YSZ—yttria-stabilized zirconia, GZO—$Gd_2Zr_2O_7$)

Space group	Film	Ion species	Energy [eV]	Epitaxial orientation or epitaxial orientation relationship	Refs.
$Fm\overline{3}m$	TiN	N	2000	[111] parallel to sample normal [100] parallel to the ion beam	1–3
		N	2000	(100) biaxial aligned (thickness ≤ 50 nm) (111) biaxial aligned (thickness > 50 nm)	4
		Ar + N	400–800	[111] parallel to sample normal	5
		N	100–1000	(111) parallel to surface ($E_I \leq$ 200 eV) (220) parallel to surface ($E_I \geq$ 350 eV)	6
		Ar	14,000	(100) biaxial orientation	7
	YSZ	O + Ar	300	[100] parallel with substrate normal [111] parallel with ion beam axis	8,9
		O, Ar	50–200	(100) biaxial orientation	10
			250	(100) parallel to substrate	11
		Ar	300	[111] parallel to sample normal [100] parallel to the ion beam	12
		O + Ar	75, 300	[111] or [110] biaxial orientation	13
		O + Ar	75, 300	(220) biaxial orientation	14
		Ar + O	300	(100) biaxial orientation	15
		Ar	300–400	(100) biaxial orientation	16
		Ar	300	(100) biaxial orientation (T ≤ 250 °C)	17
				(111) biaxial orientation (T > 250 °C)	
	MgO	Ar	700	(100) in-plane orientation	18
		Ar	750	[220] parallel to the ion beam	19
		Ar	750	[110] parallel to the ion beam	20
		Âr + O_2	750	[220] parallel to the ion beam	21
	CaF_2	Ar	800	[100] parallel to the ion beam	22
	MoS_2			(100) biaxial orientation	23
	CeO_2	Ar + O	250–500	(100) in-plane orientation	24
		Ar	100–300	(002) biaxial orientation	25
		Ar	200	[111] parallel to sample normal [100] parallel to the ion beam	26

(continued)

Table 10.5 (continued)

Space group	Film	Ion species	Energy [eV]	Epitaxial orientation or epitaxial orientation relationship	Refs.
		Ar + O	500–600	(100) in-plane orientation	[27]
		Ar^+, O	100–300	(200) out-of-plane orientation	[28]
Fd$\bar{3}$m	BN	N	96	(111) parallel to surface	[29]
	c-BN	N + Ar		[111] parallel to surface normal	[30]
	GZO	Ar, Ne		[111] parallel to sample normal [100] parallel to the ion beam	[31]

[1] J.W. Gerlach, et al., Appl. Phys. Lett. 68 (1993) 2360
[2] B. Stritzker, et al., Mat. Res. Soc. Symp. Proceed. 647 (2001) O9.1.1
[3] B. Rauschenbach, J.W. Gerlach, Cryst. Res. Technol. 35 (2000) 675
[4] M. Zeitler et al., Appl. Phys. Lett. 70 (1997) 1254
[5] R. Hühne, et al., Appl. Phys. Let. 85 (2004) 2744
[6] C.-H. Ma, et al., Thin Solid Films 446 (2004) 184
[7] L. Alberts, et al., Surf. Coat. Technol. 84 1996) 443
[8] Y. Iijima, et al., Appl. Phys. Lett. 60 (1992) 769
[9] Y. Iijima, et al., J. Mater. Res. 12 (1997) 2913
[10] R.P. Raede, et al., Appl. Phys. Lett. 61 (1992) 2231
[11] X.D. Wu, et al., Appl. Phys. Lett. 65 (1994) 1961
[12] H.C. Freyhardt, et al., IEEE Trans. Appl. Supercond. 7 (1997) 1426
[13] K.G. Ressler, et al., J. Amer. Ceram. Soc. 80 (1997) 2637
[14] N. Sonnenberg, et al., J. Appl. Phys. 74 (1993) 1027
[15] B. Holzapfel, et al., IEEE Trans. Appl. Supercond. 9 (1999) 1479
[16] N. Savvides, S. Gnanarajan, Physica, C 387 (2003) 328
[17] J. Diick, et al., Mat. Res. Soc. Symp. Proceed. Vol. 585 (2000) 55
[18] C.P. Wang, et al., Appl. Phys. Lett. 71 (1997) 2955
[19] J.R. Groves, et al., IEEE Trans. Appl. Supercond. 9 (1999) 1964
[20] R.T. Brewer, H.A. Atwater, Appl. Phys. Lett. 80 (2002) 3388
[21] T.P. Weber et al., Thin Solid films 476 (2005) 79
[22] J. R. Groves, et al., Energy Environ. Sci. 5 (2012) 6905
[23] D.N. Dunn, et al., J. Mater. Res. 13 (1998) 3001
[24] Y.J. Mao, et al., Surf. Coat. Technol. 103–104 (1998) 78
[25] S. Gnanarajan, N. Savvides, Thin Solid Films 350 (1999) 124
[26] S. Zhu, et al., Appl.Phys. Lett. 65 (1994) 2012
[27] X. Xiong, D. Winkler, Physica C 336 (2000) 70
[28] J. Wang, et al., Surf. Coat. Technol. 158–159 (2002) 548
[29] D. Litvinov, et al., Appl. Phys. Lett. 74 (1999) 955
[30] D.L. Medlin, et al., J. Appl. Phys. 79 (1996) 3567
[31] Y. Iijima, et al., Physica C 426–431 (2005) 899

prepared by ion beam-assisted deposition. It is noteworthy that the biaxial orientation was observed almost exclusively in crystalline films whose crystal structure can be assigned to the face-centered space groups Fm$\bar{3}$m (the space group Fd$\bar{3}$m differs from the space group Fm$\bar{3}$m only in the fact that the mirror plane is substituted with a sliding mirror plane). It stands to reason that in cubic materials, the two different directions imply the complete determination of the crystallographic orientation of the grain.

The biaxial orientation generated by ion beam-assisted deposition is also dependent on other various factors, namely, the ion-to-atom arrival ratio, the film thickness, and ion energy. Only a few known investigations exist with regard to the influence of the temperature and the substrate on the biaxial orientation of the deposited films. One example is the study of the change of the biaxial orientation yttria-stabilized zirconia layers deposited under an assisted Ar ion beam on metallic substrates as a function of the temperature [136]. The deposited YSZ films are amorphous after deposition below −130 °C. From −120 °C to 250 °C, (100) orientated films could be observed. Between 250 °C and 300 °C, an abrupt change of orientation to a (111) orientation was proven. A specific feature of the ion beam-induced biaxial orientation of YSZ could be observed by Ressler et al. [137]. Independent of the incident ion angle, the temperature during IBAD, and the arrival ratio, they found either a [111] or a [110] biaxial orientation in YSZ- and LCMO-films ($La_{1-x}Ca_xMnO_3$). It is assumed that the resistance to irradiation damage is different for both planes. The higher sputtering yield of oxygen from the (110) plane in contrast to the (111) plane leads to a preferred erosion of the (110) plane. Consequently, the (111) biaxial orientation is observed at higher ion incident energies and higher arrival ratios, while (110) could be found at lower ion energies and arrival ratios.

10.5.1.5 Texture Models

The commonly accepted process for the formation of preferred orientation (fiber texture) in growing films is the evolutionary selection model by van der Drift [138]. In this model, the difference in growth rates between several crystal planes at the surface of the film causes the fiber texture. Crystal grains oriented with their slower growing direction normal to the surface are terminated, while faster growing directions are preserved as they intersect the grain boundary. Figure 10.44 illustrates the van der Drift growth mechanism on the base of growth of two-dimensional crystallites with a square facetted morphology ($t = \Delta t$). The (01) crystallites grow only in the direction [01] direction so that the square morphologies are retained ($t = 25\Delta t$). In this case, the maximum growth rate is given for the [11] direction (arrows in Fig. 10.44). It is obvious that crystals whose faces are parallel to the substrate surface ((01)-oriented crystals) are overgrown by crystallites which are initially (11)-oriented (crystallites with a corner protruding in the growth direction in Fig. 10.44). Note that this mechanism leads to a preferential deposition of atoms at low-energy surfaces in order to minimize the surface energy [139]. Unfortunately, the influence of energetic ions

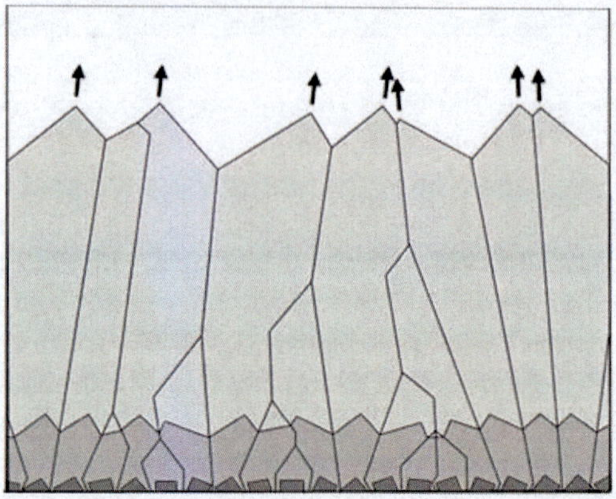

Fig. 10.44 Schematic representation of the van der Drift two-dimensional crystal growth, assuming randomly (10) oriented crystallites and infinite surface diffusion. The growth front is shown for different times of growth (colored differently). The evolutionary selection leads to crystals whose [11]-direction (arrows) are aligned almost perpendicular to the substrate

is not taken into consideration in this model. Consequently, several models have been proposed to explain the preferred orientation of films grown under additional low-energy ion bombardment.

Regrowth or Channeling Models

The first model to include the influence of the ion beam on the preferred orientation during thin film growth was proposed by van Wyk and Smith [129] and refined by Dobrev [123]. They suggest that the texture evolution from a strong [111] to a [220] correlates with the channeling direction of the crystalline grains of the studied metals. Ions travelling in channeling directions are characterized by a reduced nuclear energy deposition, i.e. the most open lattice direction aligned with the ion beam suffers the lowest nuclear energy loss. Consequently, the damage of such aligned grains is very small. These grains serve as seeds for epitaxial regrowth of ion-damaged misaligned grains. The development of the preferred orientation is attributed to the phenomenon of the recrystallization of oriented grains having the lowest nuclear energy deposition. Figure 10.45 shows a schematic view along the surface normal in the ion beam direction for the (100), the (110), and the (111) planes for crystal structures of the space group Fm$\bar{3}$m (e.g., TiN, MgO, etc.). It is obvious that under normal ion bombardment, the (111) plane seems more close-packed than the (100) plane or the (110) plane. Consequently, the (111) planes suffer greater ion damage. All crystallites growing in a direction deviant from the [110] direction will be dissolved under this ion bombardment. The surviving crystals grown in [110] directions serve as seeds for epitaxial regrowth of the surrounding grains.

Kiuchi et al. [140] simulated the damaging process of TiN (100) and (111) and predicted that in the [100] direction of TiN, the N ions penetrate the open channel, losing their kinetic energy primarily by electronic stopping, while ions lose their energy primarily by nuclear stopping in the (111) planes, leading to the amorphization

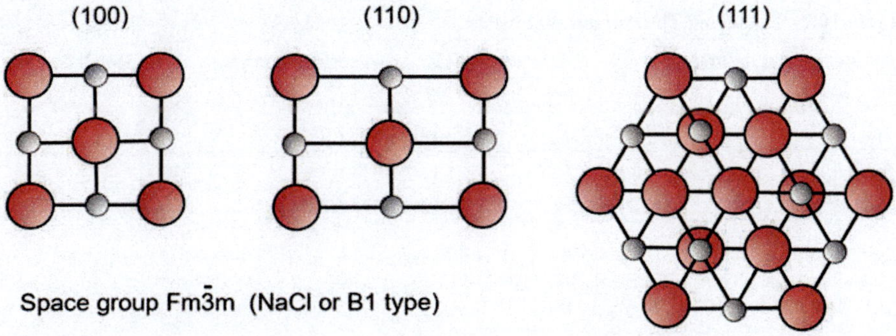

Fig. 10.45 View of the atomic arrangement parallel to surface and normal to the ion beam direction for a crystal from the space group Fm$\overline{3}$m

of the crystal. Brewer and Atwater [141] have also shown that anisotropic ion damage is responsible for the nucleation of biaxially textured MgO film.

Preferential Sputtering Model

The preferential sputtering models are directly associated with the above-discussed channeling models and models of surface energy minimization (see below). On the basis of low-angle ion sputtering experiments during film growth [42], Bradley et al. [142] proposed a selection model for texture evolution based on the changing of the sputtering yield with the grain orientation. They suggested that crystal grains with the lowest sputtering yield have the highest chance of surviving under ion bombardment. Consequently, due to their low sputtering yield, grains with crystal orientations normal to the ion flux direction remain preferred during film growth and serve as seeds for epitaxial regrowth. They also demonstrated that the degree of grain alignment increases with increasing ion flux (ion-to-atom arrival ratio).

It is well known (see Chap. 5) that the sputtering yield Y is inversely proportional to the surface energy U_s and directly proportional to the nuclear stopping power $S_n(E)$. Consequently, the lowest sputtering yield can be obtained for crystal planes that have the lowest density. Table 10.6 summarizes the relative plane density for a few different structure types. According to this table, the following preferred orientations can be expected when the ion beam is parallel to the surface normal of the bombarded film surface:

Table 10.6 Relative plane density of the fcc, bcc and hcp structure types (a and c represent the basal and height parameters of the hcp lattice cell)

Relative plane density		
fcc structure	bcc structure	hcp structure (for c/a = 1.6)
(110): (100): (111)	(111): (100): (110)	$(11\bar{2}0)$: $(10\bar{1}0)$: (0001)
0.612: 0.866: 1	0.409: 0.707: 1	0.625: 0.937: 1

Table 10.7 Calculated TiN surface energy

Surface energy [J/m^2]			Relative surface energy	Refs.
$U_{s,100}$	$U_{s,110}$	$U_{s,111}$	$U_{s,100}$: $U_{s,110}$: $U_{s,111}$	
2.3	2.6	4.0	1.0: 1.13: 1.74	1
4.94	6.99	8.536	1.0: 1.41: 1.73	2
1.06	2.59	4.59	1.0: 2.44: 4:33	3

[1] J. Pelleg, et al., Thin Solid Films 197 (1991) 117
[2] J.P. Zhao, et al., J. Phys. D: Appl. Phys. 30 (1997) 5
[3] M. Marlo, V. Milman, Phys. Rev. B 62 (2000) 2899

[111] for body-centered cubic crystals,
[110] for face-centered cubic crystals,
[11$\bar{2}$0] for hexagonal close-packed crystals.

It should be noted that the relative plane density directly correlates with the planar channeling directions [143].

Dong and Scrolovitz [144] studied the damage in fcc bicrystal consisting of [111] and [110] oriented grains irradiated with 100 eV ions by 3D-MD simulations. The [111] oriented crystal sustained significantly more damage than the [110] oriented crystal. It could be demonstrated that (i) the film texture changed from [111] and [110] to purely [110] during growth, and (ii) both ion beam-induced atomic sputtering and ion beam-induced damage are significantly reduced when the ion beam is oriented along the channeling directions of the crystals.

Surface Energy Minimization Model

The tendency for the lowest possible surface energy was proposed as the driving force for development of the preferred orientation in thin films, especially in TiN films [1, 145, 146]. An evolution of the texture from a random distribution of growing crystallites at the beginning toward a definite crystalline orientation of the crystallites can be observed. This process is characterized by a self-arrangement of the deposited atoms into the crystalline planes with the lowest surface energy parallel to the surface of the thin film. For example, the surface energy for different orientations in the case of TiN is summarized in the following table.

The ratio of the surface energies in different planes of the TiN can then be estimated to be $U_{s,100} < U_{s,110} < U_{s,111}$, i.e. the (200) plane has the lowest surface energy and can be predicted to be the preferred orientation. These results correspond well to the experimental results for TiN given in Tables 10.5 and 10.6. For example, Knuyt et al. [145] quantitatively modelled the evolution from the random orientation at the film-substrate interface toward a texture. The thickness at which the film is transformed from a random to a textured distribution of the growing crystallites and the time dependence of the development of the crystallites in the growing film could be determined.

Elastic Deformation Energy Minimization Model

Rickerby et al. [146] reported a high intrinsic stress in TiN films prepared by sputtering techniques. They assume that this stress, induced by ion bombardment, is connected with the formation of a preferred orientation and that this stress induces a considerable amount of strain energy in the thin film. Pelleg et al. [147], and later Oh and Je [148], predicted that the preferred orientation is determined by a competition between two thermodynamic parameters: the surface free energy and the strain energy. They have argued that at small thicknesses, the film shows an orientation corresponding to that of the lowest surface energy. At larger film thicknesses, the strain energy will become greater than the surface energy because the strain energy in the film increases with the thickness whereas the surface energy does not vary with film thickness (c.f. Figure 10.47). Accordingly, the preferred orientation is determined by the competition between strain energy and surface energy.

Consider a growing thin film under ion bombardment during deposition. The film will expand or contract according to the intrinsic stress or strain induced by the ion bombardment. For example, Windischmann [149] and Davis [150] have both proposed models to quantitatively describe the intrinsic stress during ion bombardment (see Sect. 10.6.7). The ion beam-induced expansion results in a thin film under biaxial strain ε_{ij} (Fig. 10.46). The uniform biaxial strain is then given by

$$\varepsilon_{ij} = \begin{pmatrix} \varepsilon_1 & 0 & 0 \\ 0 & \varepsilon_1 & 0 \\ 0 & 0 & \varepsilon_3 \end{pmatrix}. \tag{10.55}$$

When the symmetry of the film is high, the strain in both directions is isotropic, i.e. $\varepsilon_{11} = \varepsilon_{22} = \varepsilon_1$. Then, the stress tensor can be described by

$$\sigma_{ij} = \begin{pmatrix} \sigma_1 & 0 & 0 \\ 0 & \sigma_2 & 0 \\ 0 & 0 & 0 \end{pmatrix}, \tag{10.56}$$

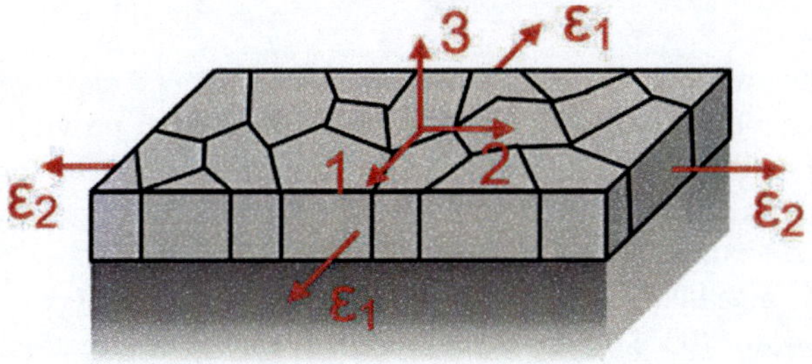

Fig. 10.46 Ion beam-induced expansion of film on a substrate results in a film under biaxial strain

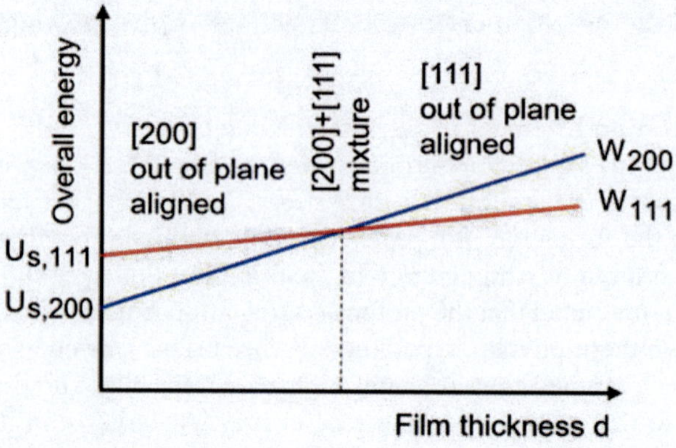

Fig. 10.47 Variation of the overall energy of a TiN film with increasing film thickness for a film grown without ion energies > 400 eV (figure is adapted from [157] and modified)

under the reasonable approximation that the film surface is unconstrained, i.e. $\sigma_{33} = 0$. The description of the elastic properties is based on Hooke's law, $\varepsilon_{kl} = s_{ijkl}\sigma_{ij}$, where ε_{kl} is the second-order strain tensor, s_{ijkl} is the fourth-compliance tensor, and σ_{ij} is the second-order stress tensor. According to the derivation of the stress in thin films under uniform biaxial strain and plane stress by Doerner and Nix [151], the strain energy density can be expressed by

$$W = \frac{1}{2}\varepsilon_{ij}\sigma_{ij} = Y_{hkl}\varepsilon_1^2, \tag{10.57}$$

where Y_{hkl} is the orientation-dependent effective elastic modulus. It is of interest to find that the effective biaxial elastic modulus is correlated with a specific direction within a thin film. According to Nye [152], these modules are given by

$$\begin{aligned} Y_{100} &= 1/s_{11} \\ Y_{110} &= 1/\left\{s_{11} - \frac{1}{2}\left[(s_{11} - s_{12}) - \frac{1}{2}s_{44}\right]\right\} \\ Y_{111} &= 1/\left\{s_{11} - \frac{2}{3}\left[(s_{11} - s_{12}) - \frac{1}{2}s_{44}\right]\right\} \end{aligned} \tag{10.58}$$

for cubic crystals, where s_{11}, s_{12}, and s_{44} are the compliance elastic constants. The ratios between the different effective elastic biaxial moduli provide a measure of the degree of departure from isotropy in cubic thin films. For TiN, for example, the experimental and calculated elastic constants have been summarized by Abadias and Tse [153]. Based on these values, the Zener anisotropic ratio $AR = 2(s_{11} - s_{12})/s_{44}$ is smaller than 1 (note that AR = 1 for an isotropic crystal). The biaxial modulus Y_{hkl} is then maximum for (001) direction and minimal for (111) direction. This implies that the elastic strain energy in a biaxially textured layer or thin film is maximum when there is a (100) orientation or minimum when there is a (111) orientation. Consequently, for TiN films, the ratio of the biaxial moduli and the strain energy density, W, are

$$Y_{111} < Y_{110} < Y_{200} \quad \text{and} \quad W_{111} < W_{110} < W_{200},$$

respectively. These relations were experimentally confirmed by Valvoda and Musil [154] as well as Zhao et al. [155]. It should be noted that this result is contrary to the results obtained from the above-discussed model of surface energy minimization (preferred (200) orientation is expected).

McKenzie et al. [156] have used a similar procedure, the Gibbs free energy minimization, to predict the preferred orientation in TiN thin films. The preferred orientation can be determined with the parameter $\delta = s_{11} - s_{21} - 1/2s_{44}$. Where $\delta < 0$, the Gibbs free energy is minimized when the [111] direction is normal to the biaxial stress plane, while when $\delta > 0$, the [100] is normal. This result is in good agreement with the observed preferred orientation of cubic TiN films.

Overall Free Energy Mode

The previously discussed models by Pelleg et al. [147] and Oh and Je [148] are based on the assumption that the sum of surface energy, $U_{s,hkl}$, and total strain (deformation) energy (given by intrinsic strain energy multiplied by the film thickness d), denoted as the overall free energy, E_{hkl}, determine the orientation of the crystallites.

Zhao et al. [155] extended the expression of the overall free energy by the addition of the stopping power S_{hkl} (energy contribution by energy loss of the impinged ion in the direction [hkl]). Then, the overall energy is then given by

$$U_{s,hkl} + W_{hkl}d + S_{hkl} = E_{hkl}. \tag{10.59}$$

The contributions are strongly dependent on the crystal direction. The preferred orientation of a thin film can be predicted by observing the overall free energy resulting from the minimization of the sum in (10.59) for the different crystal planes. Because the surface energy is independent of the film thickness, the stopping energy is depth-limited, and the strain energy increases linearly with the film thickness, a strong dependence of the preferred orientation on the film thickness can be expected (Fig. 10.47). Consequently, for small film thicknesses, the first term in (10.59), with a minimum surface energy, controls the preferred orientation, i.e. the (100) orientation can be observed (compare with the models of the surface energy minimization and the elastic deformation energy minimization, above). For larger film thicknesses, the strain energy term becomes more and more significant. With it, the (111) preferred orientation can ultimately be obtained (Fig. 10.47). Thus, a critical layer thickness above the strain energy density is expected to control the orientation of the layer crystallites. Determination of the critical film thickness is exceedingly difficult because the elastic moduli of the deposited material and the real surface energy must be known. The majority of studies on the ion beam-assisted growth process of thin films have verified a change of the preferred orientation from the (111) to the (100) orientation (see Table 10.3).

Overgrowth Model

Mahieu et al. [157] have proposed a mechanism for biaxial alignment in sputter-deposited thin films, based on a mechanism of aligned grains overgrowing misaligned grains (c.f. model by van der Drift [137]). It is known that grains oriented to the fastest geometrically growing direction perpendicular to the substrate will slowly envelop and overgrow the other out-of-plane oriented grains. For the growth conditions in zone T (see Fig. 10.33), a sufficiently high degree of mobility is assumed. The off-normal deposited adatoms are then capable of diffusing along the surface in the direction from which they were sputtered. Such biased or directional diffusion along the projection of the incoming material flux is based on the conservation of adatom momentum (see Sect. 9.2.2.2). This directional diffusion will result in an anisotropic growth rate of out-of-plane oriented crystallites, depending upon its in-plane alignment (Fig. 10.48). The growing crystallites have a different adatom capture probability depending upon its in-plane orientation. This probability can be estimated from the capture length, which is the projection of the crystallite cross-section of a plane perpendicular to the direction of the incoming flux. For example, several [001] out-of-plane oriented crystallites are formed at the initial stage of the deposition (black squares in Fig. 10.48). These grains will grow laterally toward the incoming material flux (grey area in Fig. 10.48). The [001] out-of-plane oriented crystallites have a tendency to keep their equilibrium shape during the lateral growth. It is obvious that the [001] out-of-plane oriented crystallites with $\alpha = 45°$ (angle between the projection of the incoming material flux and a [110] direction) will have a much larger area than the other in-plane aligned crystallites. A certain biaxial alignment will thus already exist at the moment of coalescence.

For example, with this model, Mahieu et al. [157] were able to explain the biaxial orientation in TiN and YSZ films deposited by sputtering at temperatures typical for the film growth condition in the zone T. Whether this model can be applied to explain the biaxial alignment in thin films prepared by IBAD must still be tested.

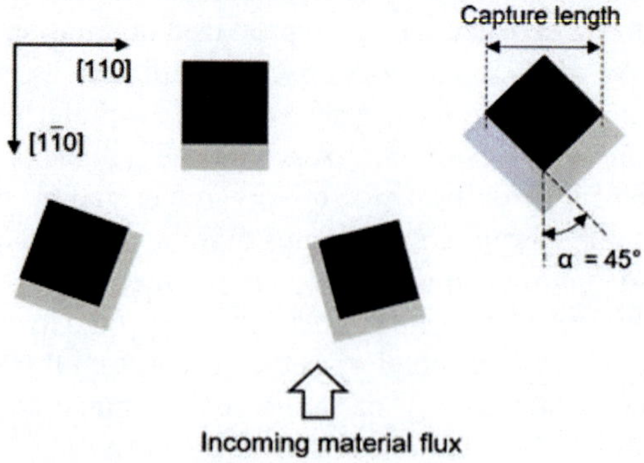

Fig. 10.48 Plane view of the [001] out-of-plane oriented grains with their lateral growth toward the incoming material flux. The grey region represents the lateral growth of the crystallites (figure is adapted from [157] and modified)

10.6 Densification, Stress, and Adhesion

The growth process of thin films by deposition at homologous temperatures $T_H <$ 0.2–0.3 without assisted low-energy ion irradiation is associated with the formation of small voids and columnar microstructures (e.g., see Figs. 10.33, left and 10.34), i.e. the density of such films is lower with respect to the bulk material. In general, these films are in a state of tensile stress parallel to the substrate–film interface. Consequently, a biaxial stress state is characterized by a stress in the film plane and no stress in the direction normal to film plane. It has been frequently observed that this ion beam-induced thin film compacting can increase/decrease the stress state, depending upon the ion-to-atom arrival ratio, temperature, and ion energy. A distinction should be made between extrinsic stress (caused by external forces, thermal induced changes of the film dimension, etc.) and intrinsic stress (caused by structural changes).

The densification and stress evolution in thin films were attributed to several factors, including the formation of thermal spikes, recoil implantation, impurity incorporation, surface diffusion, and the transferred momentum at the collision of incident ions with atoms of the growing film.

10.6.1 Transferred Momentum

Within the framework of ion beam-assisted deposition, the elastic collisions between the incident ions and the atoms of the growing film are approximately head-on collisions. The efficiency of the energy transfer from the incident ion to the atoms of the condensed film can be described by the transfer energy efficiency factor γ (see Sect. 2.2.1). Applying the constraints of energy and momentum conservation, the maximum momentum transfer per incident ion with the mass M_I to a the target atom with the mass M_A under the condition of elastic head-on collisions is given by

$$P_{\max} = \sqrt{2M_A \gamma E_I}, \tag{10.60}$$

where E_I is the kinetic energy of the incident ion, $\gamma = 4M_I M_A/(M_A + M_I)^2$ and M_A is the mass of the condensed atom (note that in contrast to the orginal publication, the mass of the film atoms must be used in this relationship). Targove and Macleod [158] introduced this consideration to quantify the properties of thin film deposited by ion beam-assisted deposition (refractive index, in particular) as a function of the total momentum transfer given by

$$P_o = J_I \cdot P_{\max} = J_I \cdot \sqrt{2M_A \gamma E_I}, \tag{10.61}$$

where J_I is the ion current density and the total momentum is given in $[\mu A \cdot cm^{-2} \cdot (amu)^{-1/2} \cdot (eV)^{1/2}]$. The dependence of the ion beam-assisted deposition process on the fluence of the deposited atoms can be considered by introduction of the normalized momentum as

$$P_n = R\sqrt{2M_A \gamma E_I}, \tag{10.62}$$

where R is the arrival ratio.

Consequently, direct thermal deposition (e.g., by evaporation with $\gamma = 1$) leads to a normalized momentum of

$$P_n = \sqrt{2M_A E_A}, \tag{10.63}$$

where E_A is the thermal energy of the deposited atoms. Combining both equations, the normalized momentum in ion beam-assisted deposition by evaporation of the atomic species can be expressed as

$$P_n = \sqrt{2M_A E_A} + R\sqrt{2M_A \gamma E_I} \tag{10.64}$$

where sputtering of the growing film and reflection on the film surface are insignificant. If sputtering and reflections must be considered, then (10.64) must be rewritten to

$$P_n = \sqrt{2M_A E_A} + R_c\sqrt{2M_A \gamma E_I}. \tag{10.65}$$

Because the kinetic energy of the incident ions is significantly higher than the thermal energy of the condensed atoms, the first term in (10.64) or (10.65) can be approximated to be generally negligible. If the film material is generated by sputtering, the relation of the normalized momentum must be modified to

$$P_n = R_c\sqrt{2M_A \gamma E_I} + R_{c,sp}\sqrt{2M_{sp}\gamma_{sp}\langle E_{sp}\rangle}, \tag{10.66}$$

where $\langle E_{sp}\rangle$ is the average kinetic energy of the sputtered atoms (see Appendix J).

10.6.2 Densification in Ion Beam-Assisted Thin Films

Only a few experimental studies have dealt with the densification of films prepared by ion beam-assisted vapor deposition. As far back as 1972, Mattox et al. [159] demonstrated the densification of tantalum films by ion bombardment during sputter deposition. The film density could be increased from 14 g/cm^3 without ion bombardment to 16.3 g/cm^3 after 500 Ar ion bombardment. The densification of deposited amorphous Ge films under low energy Ar ion bombardment was investigated by Yehoda et al. [54]. It was observed that the change of the void fraction depends more on the ion-to-atom arrival ratio than on the energy of the incident ions. It was

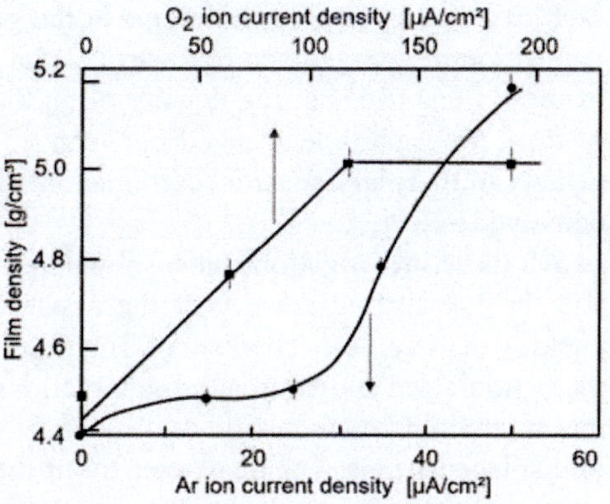

Fig. 10.49 Film density of oxygen and argon ion beam-assisted ZrO_2 films as function of the argon ion and molecular oxygen ion current density (presentation based on the measurement data of Martin et. al. [160])

also found that the densification of Ge films is highest when the Ar ion energy per deposited atom is about 5 eV.

The density of ZrO_2 films could be significantly raised from 4.4 g/cm^3 by Zr deposition without assisted ion bombardment to a maximum film density of 5.17 g/cm^3 after concurrent Ar ion irradiation and 5.03 g/cm^3 after O_2 ion-assisted bombardment, respectively [160]. Figure 10.49 illustrates the change in density as a function of the ionic current density.

In Fig. 10.50, it is quantitatively shown, that the density of a titanium film depends also on the ion-to-atom arrival ratio R and the Ar ion energy calculated by two-dimensional MD simulations [161]. With increasing arrival ratio and kinetic energy of the incident ions, the films become more and more compacted. At R $\geq$ 0.16, the Ti film attains maximum density.

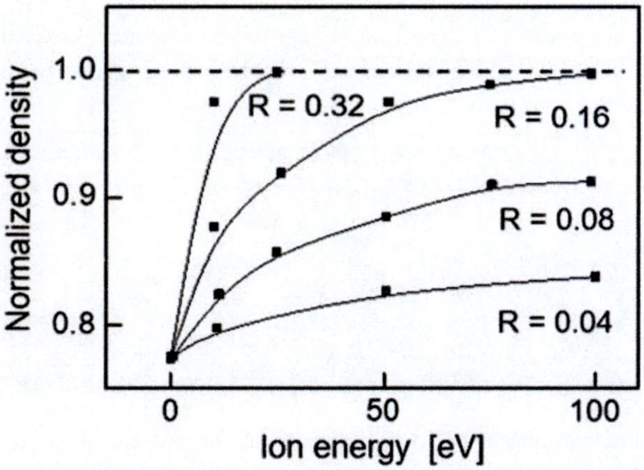

Fig. 10.50 Density of titanium films calculated by two-dimensional MD simulation as a function of the energy of assisted Ar ions for different ion-to-atom arrival ratios R (adapted from [161])

Molecular dynamic simulations of the growth of diamond-like films prepared by deposition of energetic carbon particels have shown, that the the maximum density of these films (95% of the density of bulk diamond) can be achieved if the energie of the carbon particles ranged between 40 eV and 70 eV, while, for example, the density of thermal deposited carbon atoms is smaller than 60% of the bulk diamond density [162].

All these investigations have shown that the ion-to-atom arrival ratio, ion energy and the ion current density of the assisted atoms significantly influence the film packing density. A mechanism of film densification by ion-assisted deposition based on the ion-atom and atom–atom interactions was proposed, where atoms are pushed into interstitial positions of the subsurface layer during the evolution of the cascades of displaced atoms. The displacement of the target atoms results in the formation of vacancies at or close to the surface, i.e. the depleted region at near-surface (about 2–3 monolayers), and a densification of the film at greater depths. The depleted region is then subsequently refilled during film growth by arriving atoms.

Müller used three-dimensional Monte Carlo simulations [163] to determine the densification in terms of the recoil implantation of surface atoms, ion incorporation, and sputtering. The film densification has been found to be approximately proportional to the ion-to-atom arrival ratio, where, the deposited atoms refill vacancies near the surface. In a two-dimensional MD simulation, Müller illustrated that ion bombardment during film growth, where there is low adatom mobility, causes the refilling of voids and induces homoepitaxial growth [164]. Figure 10.51 illustrates the influence of the ion-assisted ion bombardment on the microstructure of the growing film. The deposition with thermal atoms with a kinetic energy of 0.1 eV causes the formation of voids and microcolumns (Fig. 10.51, left). When additional Ar ions of 50 eV strike the growing film, the number of voids is strongly reduced, whereas the film packing density increases with an increasing the ion-to-atom ratio (Fig. 10.51, middle and right).

Based on the dependence of the density on the ion-to-atom arrival ratio by low-temperature ion beam-assisted processes, Grigorov et al. [165] proposed a model

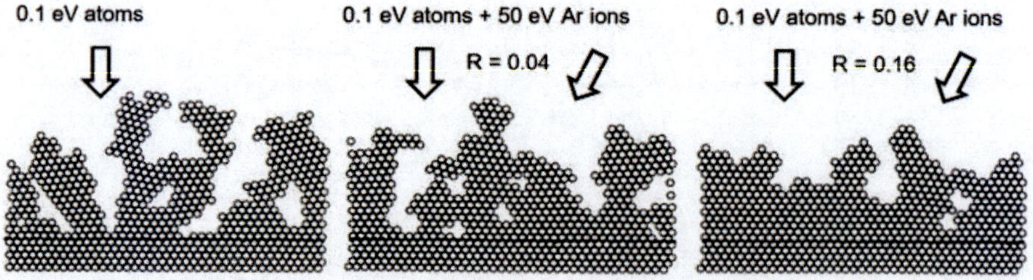

Fig. 10.51 Microstructure after deposition of thermal atoms with an energy of 0.1 eV arriving under normal incidence. Left: without ion bombardment, Middle: with 50 eV Ar ion bombardment and an arrival ratio of 0.04. Right: with 50 eV Ar ion bombardment and an arrival ratio of 0.16. The ion impingement angle relative to the substrate surface is 30°. The sputtering yield was 0.35 (figure adapted from [164])

allowing the quantitative determination of the optimum ion current density required to produce a film with maximum density. It becomes apparent that the most densely packed film can be produced when the ion bombardment triggers a structural rearrangement of about 80% of the deposited particles. Consequently, an average energy for maximum densification could be determined, which ranged between 34 eV and 87 eV for various metals. This study also implies that the average energy for maximum densification is about 2.2–2.7 times higher than the displacement energy E_d. The model by Grigorov et al. [166] allows an estimation of the optimum current density attainable by ion-induced film densification in a very thin surface film. If the mean energy (per film atom) deposited by the ions is of the order of the cohesive forces in this layer, a film of maximum density will be grown.

Carter [167] proposed a model to explain densification and stress evolution in thin films deposited with ion bombardment assistance, in which a constant ion current density J_I and an atom flux density, J_A, simultaneously impinge the surface of the growing film. The densification of voided films is then explained by forward sputtered recoils with the sputtering yield Y_{fr}, where the fractional densification is approximately proportional to $(J_I/J_A)Y_{fr} = RY_{fr}$. By increasing the product of the ion-to-atom arrival ratio and ion energy, $(J_I/J_A)E$, a saturation of the densification is expected. This is in good agreement with the experimental results from Martin et al. [160], where the densification first increases linearly and then, with increasing R, saturates for constant ion energy (see Fig. 10.49).

10.6.3 Residual Stress in Ion Beam-Assisted Thin Films

The goal of depositing thin films using IBAD is often to ensure that these films are tightly adherent to the substrate, free of defects and delamination, have a density comparable to that of bulk material and are almost free from stress. Without regard to extrinsic stress (e.g., interaction of the film with the environment), the residual stress in thin films prepared by simultaneous deposition and low-energy ion bombardment consists of two parts:

(i) The thermal stress, σ_{th}, is provoked by the difference of the thermal expansion coefficients of film and substrate materials during the deposition of the film at temperatures higher/lower than the temperature during the stress measurement.
(ii) The intrinsic stress, σ_{in}, is caused by ion beam-induced structural modifications (e.g., densification, grain growth, coalescence of grain boundaries, substrate–film misfit matching, insertion of excess atoms or ions, etc.) during film growth. The magnitude of the intrinsic stress strongly correlates to the ion irradiation and atom deposition parameters.

Consequently, the magnitude of the residual stress can be obtained by

$$\sigma = \sigma_{th} + \sigma_{in}. \tag{10.67}$$

According to Hooke's law, the residual stress in a thin film for a biaxial stress state is given by

$$\sigma = \left(\frac{E_f}{1 - \upsilon_f}\right)\varepsilon, \tag{10.68}$$

where ε is the strain, and ν_f and E_f are the Poisson's ratio and the Young's modulus of the film, respectively.

10.6.4 Measurement of Stress in Thin Films

The most common methods for the determination of stress, especially the biaxial stress in thin films prepared by ion beam-assisted deposition, are the measurement of the radius of curvature of the thin film-covered substrate and the X-ray diffraction technique. The choice of the method applied is dependent upon the measuring situation (in-situ or ex-situ), substrate and film material, and the sample dimension.

Stress Measurement Using the X-Ray Diffraction Technique
In principle, the measurement of the residual stress is based on the determination of the strain in crystal lattice and the assumption that the associated stress can be determined from the elastic constants, assuming a linear elastic distortion of the crystal lattice plane. For this purpose, the $\sin^2\psi$ X-ray diffraction technique (see e.g., [168]) is usually applied. By scanning hkl reflection at a selected sample orientation, described by the tilt angle ψ (Fig. 10.52), the lattice spacing can be determined for different groups of crystallites in the thin layer possessing various orientations with respect to the surface normal. The lattice strain along the direction ψ for in-plane isotropic material is then given by

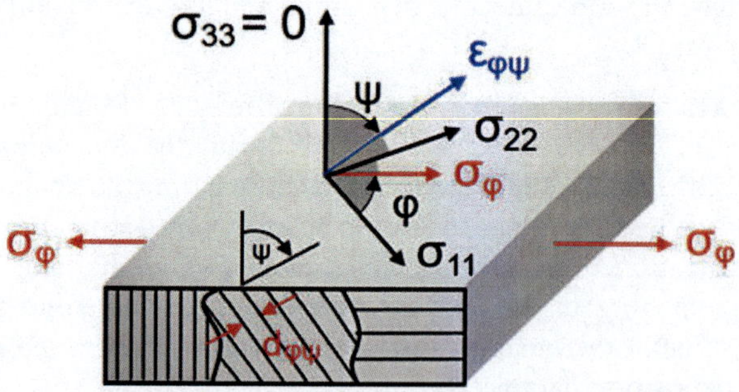

Fig. 10.52 Schematic depiction of the diffraction planes parallel to the surface. Note that stress components σ_{11} and σ_{22} both lie in the plane of the sample surface

$$\varepsilon_{\varphi\psi} = \frac{d_{\varphi\psi} - d_o}{d_o}, \tag{10.69}$$

where d_o is the stress-free lattice spacing and $d_{\varphi\psi}$ is the lattice spacing of the diffracting plane for a given {hkl} reflection measured in the direction (φ, ψ). According to the linear elasticity theory, $\varepsilon_{\varphi\psi}$ is connected to the stress tensor through the elastic constants of the material. A sample coordinate system (Fig. 10.52) is assumed, with the σ_{33} direction normal to the film surface and the σ_{11}, σ_{22} directions orthogonal in the plane of the film. Consequently, the shear components σ_{13} and σ_{23} are zero. Stress measurement is performed for both positive and (pseudo-) negative ψ angle tilts, where the pseudo-negative tilts are carried out by rotating the sample by 180° in the φ plane and then tilting in the same ψ direction. A plot of d_ψ as a function of $\sin^2\psi$ for both positive and (pseudo-) negative ψ angles should leads to a linear dependence (see e.g., Fig. 10.53 in the next subchapter). For elastic anisotropic material, it is necessary to introduce an additional dependence on the azimuthal angle φ. In the isotropic case, the volume averaged in-plane stress σ_{11} can be determined according to

$$d_\psi = d_o - \frac{\upsilon_f}{E_f}(\sigma_{11} + \sigma_{22})d_o + \frac{1 + \upsilon_f}{E_f}\sigma_\varphi d_o \sin^2\psi, \tag{10.70}$$

where ν_f and E_f are again the Young's modulus and Poisson's ratio of the thin film, respectively. Because

$$\frac{\partial d_{\varphi\psi}}{\partial sin^2\psi} = \frac{1 + \upsilon_f}{2E_f} d_o \sigma_\varphi, \tag{10.71}$$

the macroscopic stress σ_φ can be calculated from the dependence of $\sin^2\psi$ on d_o,

$$\sigma_\varphi = \frac{1 + \upsilon_f}{E_f} \frac{1}{d_o} \frac{\partial d_{\varphi\psi}}{\partial \sin^2\psi}, \tag{10.72}$$

where d_o and the elastic constants of the film material should be known.

Stress Measurement Using the Wafer Curvature Technique
This measuring technique is based on detection of the bending of the substrate that is deposited with a thin film. With tensile residual stress, the force of the substrate bonded to the film causes a concave curvature, whereas with compressive residual stress, the force of the substrate bonded to the film causes a convex curvature. Several in-situ configurations are based on the cantilever plate technique, allowing monitoring of the stress evolution during the ion beam-assisted deposition process. The substrate is held fixed at one side and its displacement from the zero-stress level due to the bending can be measured. The radius of the substrate curvature is commonly determined optically by measurement of the reflection angle of a laser beam [169] or by measuring the substrate displacement with the capacitance method [170]. The

smallest detectable deflection of the capacitance method is 1 nm. This corresponds to a measurement of the force per unit length of about 0.001 N/m. A similar resolution can be obtained using STM tips for measuring the substrate bending.

Stoney derived a now widely used formula for the evaluation of the stresses in thin films deposited on thick substrates [171]. This relationship gives the macroscopic stress in the thin film from the measured curvature κ of the substrate and can be expressed by

$$\kappa = 6\left(\frac{1-\upsilon_s}{E_s}\right)\frac{F}{h_s^2}, \tag{10.73}$$

where h_s, ν_s and E_s are the thickness of the substrate, the Poisson's ratio and the Young's modulus of the substrate, respectively. The formula is without regard to the film properties. Because the force F per unit length along the interface is proportional to the film stress σ_f given by $F = \sigma_f h_f$, the film stress can be obtained by the Stoney equation to be

$$\sigma_f = \frac{1}{6r_c}\left(\frac{E_s}{1-\upsilon_s}\right)\frac{h_s^2}{h_f}, \tag{10.74}$$

where h_f is the film thickness and r_c is the radius of curvature. According to Freund and Suresh [172], this relationship is applicable under the following conditions: (i) film and substrate thicknesses are small compared with the lateral dimensions, (ii) $h_f \ll h_s$, (iii) the substrate material is homogeneous, (iv) the influence of edge effects of the bended samples can be neglected, and (v) the lateral tilts and rotations of the deposited sample (substrate and film) are infinitesimally small.

For in-situ measurement of the stress in the vacuum, i.e. determination of the stress state in dependence on the deposited film thickness, the curvature is followed continuously by measurement. For this purpose, the (10.74) must be modified by

$$\sigma_f = \frac{1}{6}\left(\frac{E_s}{1-\upsilon_s}\right)\frac{h_s^2}{h_f}\left(\frac{1}{r_t}-\frac{1}{r_o}\right), \tag{10.75}$$

where r_o is the radius of curvature at the initial state (at the beginning of the experiment, t = 0) and r_t is the radius of curvature at the time t during the bending (deposition) experiment.

10.6.5 Thermal Stress

As mentioned above, the thermal stress σ_{th} in a film rigidly connected with a substrate is caused by the difference of the thermal expansion coefficients of the film, α_f, and the substrate, α_s, if the film is deposited on the substrate at a temperature T_s that is lower/higher than the temperature of the substrate during the measurement T_{sm}. The induced biaxial strain in the film can then be expressed by

$$\epsilon = (\alpha_s - \alpha_f)(T_{sm} - T_s). \tag{10.76}$$

The thermal stress is directly obtained by combining this equation with Hooke's law for the biaxial stress state, (10.68). This equation, also referred to as the general thermal stress equation and is given by

$$\sigma_{th} = \left(\frac{E_f}{1 - \upsilon_f}\right)(\alpha_s - \alpha_f)(T_{sm} - T_s) = \left(\frac{E_f}{1 - \upsilon_f}\right)\Delta\alpha\Delta T, \tag{10.77}$$

under the condition that no plastic deformation takes place during the temperature change. Provided that the measurement is performed at room temperature and the ion beam-assisted deposition is carried out at a higher temperature, $T_{sm} - T_s < 0$, a compressive stress can be determined for $\sigma_{th} < 0$ and a tensile stress for $\sigma_{th} > 0$.

As example, Fig. 10.53 (left) shows the results of X-ray diffraction technique measurements of the lattice spacing at a given temperature obtained during sample heating up and cooling down of 220 nm thick GaN films deposited on (0001) sapphire. These thin films were prepared by 25 eV nitrogen ion-assisted deposition of Ga onto the sapphire substrate. The strain measurements were performed by scanning the GaN reflections in a temperature cycle starting at room temperature (25 °C), heating to 520 °C, and then cooling back down to room temperature. Negative and positive slopes of the lattice parameter c versus $\sin^2\psi$ indicate compressive and tensile stress states in the GaN layer, respectively. Maximal thermal stresses of 446 MPa were found [173]. By fitting the temperature dependence of the slopes of linear dependencies from Fig. 10.53, left, the temperature at which a change from the compressive to the tensile stress state (or vice versa) occurs can be extrapolated (Fig. 10.53, right).

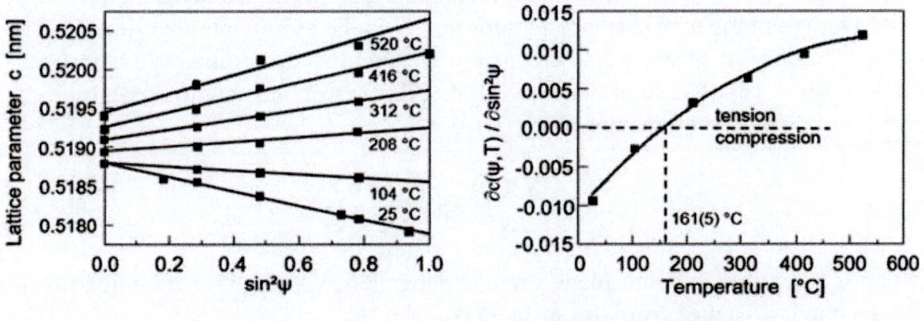

Fig. 10.53 Left: Lattice parameters c of deposited hexagonal GaN by nitrogen ion-assisted deposition as a function of sample tilt angle ψ measured in the temperature range between 25 and 520 °C. Each X-ray diffraction technique measurement at a specific angle ψ corresponds to a certain hk.l reflection. The negative and positive linear dependencies represent compressive and tensile stress states in the thin film, respectively. Right: Temperature dependence of the slopes of linear dependencies from the figure on the left. The temperature of the stress-free state (161 °C) in the thin film is extrapolated. A horizontal dashed line represents the stress-free state in the GaN thin film

This temperature of the stress-free state was found to be 161 °C. The temperature of the stress-free state in GaN thin film depends mainly upon the deposition conditions. The specific parameters of the thin film growth method (e.g., flux, growth rate, buffer material, temperature, etc.) and the deposition temperature pre-determine the stress states. Therefore, the resulting value is strongly dependent on the deposition and annealing conditions applied and cannot be generalized.

10.6.6 Intrinsic Stress by Low-Energy Ion Irradiation

It can be expected that the incorporation (implantation) of the implanted atoms in the lattice and the formation of implantation-generated Frenkel defects influences the stress state of the implanted film or near-surface region (expansion for an implanted particle or generated interstitial and dilatation for a formed vacancy). The volumetric extension of an implanted atom is given by $\Delta V/V = C_I \Delta V/\Omega_I$, where C_I is the concentration of implanted atoms and $\Delta V/\Omega$ is the specific volume extension per implanted atom (relaxation volume), expressed in units of the atom volume Ω_I. The strain, ε, due to the volume extension $\Delta V/V$ of such particles can be given by $\varepsilon = 1/3\ \Delta V/V$. According to Hooke's law (see (10.68)) the change of the stress per implanted atom can be expressed by

$$\sigma = \frac{E}{1-\upsilon}\varepsilon = \frac{1}{3}\frac{E}{1-\upsilon}C_I\frac{\Delta V}{\Omega_I}, \tag{10.78}$$

where C_I is identified with the concentration of the implanted ions (see Chap. 3). An extended view is that each implanted ion contributes to the displacement of other lattice atoms, i.e. the concentration is interpreted as the concentration of the displaced atoms, where $C_I = N_d(E)/N$ is the fraction (concentration) of displaced atoms, $N_d(E)$ is the average number of displaced atoms and N is the atomic number density. The average number of atoms per incident ion is given by a modified Kinchin-Pease relation (see 4.13). The total dpa over all recoils corresponds approximately to the fraction of displaced atoms dpa $\cong N_d(E)/N$ i.e.

$$C_I = \frac{N_d(E)}{N} \cong 0.4\frac{\nu(E)}{E_d N}. \tag{10.79}$$

Consequently, the total in-plane stress in dependence on the time of bombardment and the depth (distance from the surface) is given by

$$\sigma(z,t) = \frac{1}{3}\frac{E}{1-\upsilon}\left[C_I(x,t)\frac{\Delta V}{\Omega_I} + C_i(x,t)\frac{\Delta V}{\Omega_i} + C_v(x,t)\frac{\Delta V}{\Omega_v}\right], \tag{10.80}$$

where C_i and C_v are the concentrations of interstitials and vacancies, respectively, and $\Delta V/\Omega_i$ and $\Delta V/\Omega_v$ are the specific volume extensions per interstitial und vacancy, respectively. Hence, the stress evolution caused by incorporation of the implanted particles can be approximately determined by calculation of the depth distribution of the point defects (e.g., with the help of computer simulations, see Sect. 3.6) and the assumption that the specific volume of Frenkel defects is in order of magnitude of about 1 and 2 atomic volumes.

It is known that ion bombardment in the MeV and high keV energy ranges frequently leads to an increasing compressive stress, followed by a decrease and then saturation in the compressive stress with increasing fluence. This stress evolution is explained by ion beam-induced viscous flow (see e.g., [169]) and plastic deformation (see e.g., [174]).

However, little is known about irradiation-induced stresses, particularly for surfaces irradiated by low-energy ions. The mechanism of stress generation by direct low-energy ion bombardment of solids or deposited films has been investigated in dependence on the ion fluence (time of ion bombardment) [175–180], ion species [175], temperature [177], and ion energy [175, 176, 178].

Several interpretations have been published with the objective of explaining the stress evolution and relaxation by low-energy ion bombardment *without* additional material deposition:

- Dahmen et al. [175] explained the steady state in terms of a model that included compressive stress due to expansion of the lattice around the implanted ion and the subsequent removal of these implanted ions by sputtering. Because it is assumed that the defects are immobile, the steady-state stress is only dependent upon the energy-dependent implantation depth and the sputtering yield.
- Kalyanasundaram et al. [176] calculated the stress induced in Si by sub-keV Ar ions using MD simulations and found a compressive stress that reaches a steady-state of 1.6 GPa for 700 eV Ar ions, while much lower values of the stress were found experimentally. They found that the evolution of the compressive stress is proportional to the number of implanted particles and that the thickness of the stressed region corresponds to the range of these implanted particles.
- Ishii et al. [180] observed a rapid increase of compressive stress at the beginning of ion bombardment of silicon and then a steady-state that does not change with continued ion bombardment. The model by these authors explains the dependence of the steady-state stress on the ion flux by ion radiation induced fluidity (see also Sect. 6.5.6) and includes the relaxation generates by ion-induced defects. The analysis has shown that the defect recombination is the dominant mechanism controlling defect concentration both during and following the cessation of ion irradiation.
- Chan et al. [178] demonstrated that for heavy layer material (Pt) bombardment with heavier ions, a tendency to generate tensile stress can be observed and have explained this result by the fact that tensile stress is caused by the melting along the ion track and compressive stress by the incorporation of interstitials into the film. When the ion beam is turned off, a stress relaxation (lower compressive stress) can be detected, whereby this relaxation process can be small [180] or significant (stress relaxes to a tensile state) [177, 181].

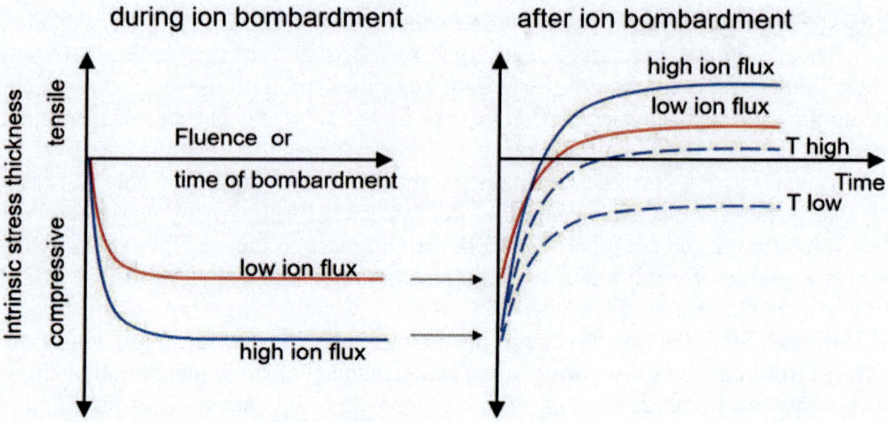

Fig. 10.54 Qualitative dependence of the intrinsic stress-thickness on the fluence (or time of ion bombardment) and the ion current density during low-energy ion bombardment (left) and the relaxation process immediately after the completion of the ion beam (right). The influence of temperature on relaxation process for the case of low ion flux is also shown schematically (dashed blue lines).

Based on experimental in-situ studies on the development of intrinsic stresses after 700 eV Ar ion bombardment of Cu layers, a continuum model was proposed by Chan et al. [177] to explain the stress evolutionunder ion bombardment. In general, the stress evolution during low-energy ion bombardment is characterized by a strong increase of compressive stress at the beginning and a steady-state level of saturation (see schematic Fig. 10.54, left), where this level is dependent on the ion current density (ion flux). In Fig. 10.54, the stress averaged over the depth of the implanted region (referred to as intrinsic stress thickness) is shown. After the ion bombardment is stopped, the compressive stress immediately starts to relax (Fig. 10.54, right), establishing a steady state level depending on the temperature and again the ion current density. Frequently, a transition from a compressive stress to a tensile stress can be observed.

The stress evolution after low-energy ion bombardment was explained by Chan et al. [177] as being the result of the incorporation of the implanted atoms as well as the generation of mobile point defects. The proposed model includes the temperature and flux dependence on the stress caused during ion bombardment and subsequent stress relaxation. Molecular dynamic simulations were used [178] to reveal that the dependence of irradiation-induced stress on the initial stress in the film can be derived from anisotropic interstitial diffusion in a uniform stress field.

10.6.7 *Intrinsic Stress in Thin Films Prepared by Ion Beam-Assisted Deposition*

In general, deposited thin films contain many defects of different types, which induce discrepancies from the ideal crystal structure and morphology and are capable of acting as sources of intrinsic stress. Common stress contributors in the film are grain boundaries, domain walls, impurities, etc. in the film-substrate interface, e.g., the mismatching or solid-state reactions between film and substrate and, in the film-vacuum interface, the surface defects and residual gas contaminations. According to Doerner and Nix [151], all thermally deposited films (e.g., by thermal evaporation) are generally under tensile stress. The application of energetic particles during the deposition dramatically changes the stress state. D'Heurle was the first to recognize that compressive stress can be attributed to ion beam-induced lattice distortion [181]. Later, Thornton and Hoffman (see e.g., [182, 183]) extensively studied the influence of energetic particles on the compressive stress of sputtered films. The stress evolution in thin deposited films bombarded with low-energy ions has also been studied (see previous chapter).

Over the past several decades, numerous studies have been published that deal with stress evaluation in deposited films under the assistance of low-energy ions. As an example, Kuratani et al. [184] studied the stress evolution of Si and Ni films on Si(100) by simultaneous Ar ion irradiation in dependence on the Ar ion energy and the arrival ratio. The stress state of the films was determined by the curvature technique after deposition. Without ion irradiation, the films exhibit a tensile stress of approximately 0.8 GPa. At ion bombardment with low energies, the development of strong compressive stress could be observed. With increasing ion energy, compressive stress was reduced and a transition to tensile stress could be detected. Figure 10.55, for example, shows the influence of the ion energy (left), the ion-to-atom arrival ratio (middle) and the deposition temperature (right) on the stress evolution of cubic BN prepared by nitrogen ion beam-assisted deposition of boron [6, 170, 185]. The

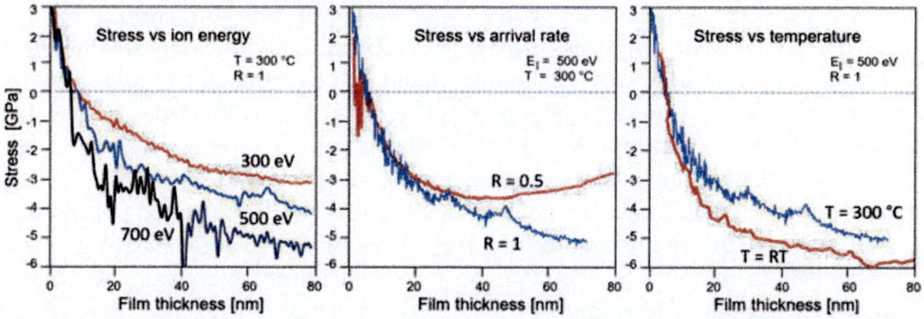

Fig. 10.55 In-situ stress evolution in boron nitride films on silicon in dependence on ion energy E_I (left), temperature of the substrate T (middle), and the ion-to-atom arrival ratio *R* (right) during deposition. The substrate deflection in-situ was determined with a highly sensitive capacitance measurement bridge with three capacitor plates (figures adapted from [185])

stress evolution is characterized by an increase of the compressive stress, increasing with ion energy and arrival ratio and decreasing with increased temperature. Table 10.8 provides a summary of selected studies on the field of stress evolution in films prepared by low-energy IBAD, together with the most important conditions for the ion bombardment and any achieved results.

The following is a summary of some generalized results that contribute significantly to the understanding of the origin of intrinsic stress and to the description of the stress state of films prepared by low energy ion beam-assisted bombardment:

- There are contradictory discussions about the influence of the incorporated bombarding gas on the development of the stress. On the one hand, for example, to explain the compressive stress in aluminum coatings [186], it is assumed that the gas inclusion supports the buildup of stresses, on the other hand, it was found that the variation of the gas content does not change the elastic strain of sputtered chromium coatings [187].
- The compressive intrinsic stress was found to be proportional to the momentum of the incident ion energy ($P \propto E^{1/2}$, see (10.62)).
- The tensile stress reached a maximum in films prepared by assisted ion bombardment at low ion energies and then the tensile stress decreased up to complete compensation of the tensile stress. This process is particularly dependent upon ion current density and ion-to-atom ratio.
- The transition of the tensile stress to a compressive stress can be expected for very small film thicknesses.
- In some cases, the stress relaxation under ion bombardment was attributed to an annealing effect by thermal spikes.

Based on the studies by Windischmann [149] and Pauleau [188] and selected results presented in Table 10.8, the stress evolution in thin films prepared by deposition under assistance of energetic ion bombardment can be schematically summarized in the following figure (Fig. 10.56, left). The stress state of as-deposited film without assisted ion bombardment (red points in the figures) is frequently characterized by tensile stress and a columnar growth morphology. The magnitude of the tensile stress is dependent on the kinetic processes active during film formation. It can be expected that the tensile film stress in as-deposited films is significantly smaller than the intrinsic stress generated by ion bombardment. For example, the tensile stress in as-deposited fcc metal films was determined to be $\leq$ 40 MPa [189]. Depending on the ion energy, the stress development can be divided into three typical ranges in terms of deposited energy. For low ion energies (range A, < 1 eV/atom), the impressed momentum is only small and the induced tensile stress reaches a maximum. As the ion energy increases, the intrinsic tensile stress decreases and a compressive stress is increasingly built up (range B, 1–20 eV/atom). For high ion energies, residual stresses decrease because the yield strength of the deposited material is reached (range C, > 25 eV/atom).

The typical behavior of the development of intrinsic stresses as a function of the energy of the incident ions could be confirmed both experimentally and by means of simulations. Based on MD simulations, Hoffman [190] demonstrated the increase

Table 10.8 Summary of stress measurements of films deposited by ion beam-assisted thermal, electron, and laser-beam evaporation (E_I—ion energy, RT—room temperature, s.s.—stainless steel, DLC—diamond-like carbon, P_n—normalized momentum, see Sec. 10.6.1)

Film	Ion species/Ion energy/Arrival ratio/Temperature	Results	Refs.
Cr	Xe- and Ar-ions/4.5 and 11.5 keV/10^{-4} to 10^{-2}/60–300 °C	• Tensile-compressive stress transition is a function of ion-to-atom arrival ratio • Stress is a function of temperature • Stress is correlated with momentum transfer	1
	Ar-ions/100–800 eV/360 °C	• Compressive stress increases with ion energy • Tensile-compressive stress transition is a function of ion energy and ion-to-atom arrival ratio	2
Al	Kr or Xe ions/0–200 eV	• Stress is a function of gas content up to 1% Xe and 3% Kr	3
	Ar ions/	• Compressive stress increases with increasing ion momentum transfer	4
Nb	Ar ions/100–800 eV/RT-400 °C	• Stress is a function of temperature and ion fluence • Stress maximum occurs at lower ion fluences and higher temperatures • Compressive stress varied oxygen content in the film	5
Cu	Ar ions/62 eV, 125 eV & 600 eV/60–230 °C/	• Tensile stress is increased with substrate temperature and decreases with ion flux • Tensile-compressive stress transition	6
W	Ar ions/200–600 eV/150–750 °C	• Maximum tensile stress = 1 GNm^{-2} @ 400 eV Ar-ions • Tensile-compressive stress transition is a function of temperature and ion current density • Stress is proportional to $(\text{ion energy})^{1/2}$	7,8
Ag	Ar ions/	• Tensile-compressive stress transition at about an ion transfer momentum of 0.2 $eV^{1/2}$	4
Mo	Ar-ions/ < 100 eV/RT	• Stress is function of the total energy per metal atom • Tensile-compressive stress transition for 40–50 eV Ar ions bombardment	9
Ge	Ar ions/200 – 400 eV/0 – 0.05/RT – 400 °C	• Strain is increased with arrival ratio and Ar ion energy • Linear relationship between perpendicular strain and arrival ratio	10
Ni	Ar-ions/60–800 eV/0.5–6.4/RT	• Compressive stress buildup proportional to $(\text{ion energy})^{1/2}$ • Tensile-compressive stress transition at small ion energies • Experimental data are consistent with Davis model	11

(continued)

Table 10.8 (continued)

Film	Ion species/Ion energy/Arrival ratio/Temperature	Results	Refs.
Ni and Si	Ar-ions/100–10,000/ RT	• Tensile stress of 0.8 GPa without ion bombardment • Stress state is strongly dependent on ion energy and the arrival ratio • Tensile-compressive stress transition occurs between 100 and 200 eV	12
Nb	Ar-ions/100–800 eV/ RT-400 °C	• Stresses that were strongly influenced by substrate temperature and ion dose • At 400 °C, high tensile stress changed to compression with increased bombardment • For low ion energies, an atom-to-ion arrival ratio of 0.03–0.1 ions/atom causes a stress transition • Compressive stress was measured without ion irradiation between RT and 200 °C • Ar-ion bombardment induced a compressive-tensile stress transition	13
c-BN	Ar/N-ions/1000 eV/ 400–500 °C	• Compressive stress between −0.15 and −5.5 GPa • Nearly the same elastic strain for c-BN and sp^2-bonded BN	14
	N/Ar ions/ 300–600 eV/ 0–2/RT-600 °C	• Compressive stress depends on ion energy, arrival ratio and temperature • Stress evolution depends on the film thickness • Tensile-compressive stress transition between 4 and 6 nm film thickness • Compressive stresses up to 3.5 GPa	15
DLC	CH_4 or C_4H_{10} ions	• Compressive stress determined by Ar ion energy • Contribution to the intrinsic stress by complex carbon-hydrogen species in the discharge	16
a-Al_2O_3	Ne, Ar or Kr ions/325 – 700 eV/ 0.003–0.19/RT	• Stress is a function of the ion energy and the noble gas ion species • Tensile-compressive stress transition between 1 and 3 eV/atom • Maximum compressive stress = −0.4 GPa at 20 eV/atom • Consistent with atom peening model	17
a-TiO_2	neutralized Ar-ions/ 300 V/RT & 150 °C	• Amorphous films show compressive stress • Minimum of compressive stress after annealing at 200 °C	18
SiO_2	O_2 and Ar ions/ 80–150 eV/RT-285 °C	• Compressive stress varied with the normalized momentum, linearly	19

(continued)

Table 10.8 (continued)

Film	Ion species/Ion energy/Arrival ratio/Temperature	Results	Refs.
$WSi_{2.2}$	Ar-ions/100–400 eV/ RT, 350 °C & 500 °C	• Tensile stress without ion bombardment • Tensile-compressive stress transition at room temperature with ion bombardment • Enhanced diffusion and local atomic arrangement are responsible for stress evolution • Stress influenced by compound formation during ion-assisted deposition	20
Si_xGe_{1-x}	Ar or Xe ions/ 50–1200 eV/ 200–450 °C/0.1–0.01	• Large strain changes up to 1.5% in and up to 0.5% in Si independence of ion energy, temperature and arrival ratio	21
B_xN_y	N_2 ions 0.25–2 keV/0–2	• Compressive stress decreases with increased ion energy for nitrogen-poor films • Compressive stress increases with ion energy for nitrogen-rich films	22
MgF_2	Ar ions/ $\leq$ 250 eV/RT	• Formation of stoichiometric MgF_2 films • For $P_n \leq 55$ $g^{1/2}$ $mol^{-1}eV^{1/2}$ films exhibit a reduced mass density and high intrinsic and extrinsic stresses • For $P_n > 60$ $g^{1/2}$ $mol^{-1}eV^{1/2}$ mass density is near the mass density of bulk material	23
Ni–Fe alloy	Ar-ions/150–1500 eV/ RT/0.022–2.2	• Tensile stress without ion bombardment • Tensile-compressive stress transition depends on ion current density (strongly) and ion energy (weakly)	24

[1] R.W. Hoffman, M.R. Gaerttner, J. Vac. Sci. Technol. 17 (1980) 425
[2] J.J. Cuomo, et al., J. Vac. Sci. Technol. (1986)
[3] V. Dietz et al., Nucl. Instr. Meth. in Phys. Res. B 59/60 (1991) 284
[4] C. K. Hwangho, et al., Appl. Optics, 28 (1989) 2769
[5] J.J. Cuomo, et al., J. Vac. Sci. Technol. 20 (1982) 349
[6] R.A. Roy et al., J. Vac. Sci. Technol. A6 (1988) 1621
[7] R.A. Roy, et al., Mater Res. Soc. Symp. Proceed. Vol. 128 (1989) 23
[8] R.A. Roy, et al., J. Mater. Res. 6 (1991) 80
[9] K. Morgan et al., IEEE Trans. Appl. Supercond. 25 (2015) 2,101,505
[10] C.J. Tsai, et al., Appl. Phys. Lett. 57 (1990) 2305
[11] K.O. Schweitz, et al., Nucl. Instr. Meth. In Phys. Res. B 127/128 (1997) 809
[12] N. Kuratani, et al., Surf. Coat. Technol. 66 (1994) 310–312
[13] J. J. Cuomo, et al., J. Vac. Sci. Technol. 20 (1982) 349 and J. Vac. Sci. Technol. A 6 (1988) 1621
[14] G.F. Cardinale, et al., Diamond Rel. Mater. 5 (1996) 1295
[15] M. Zeitler et al., J. Va. Sci. Technol. A 17 (1999) 597
[16] D. Nir, Thin Solid Films 146 (1987) 27
[17] L. Parfitt, et al., J. Appl. Phys. 77 (1995) 3029
[18] C.-C. Lee, et al., Appl. Optics, 44 (2005) 2995
[19] J.Y. Robic, et al., Thin Solid Films 290–291 (1996) 34
[20] D.S. Yee et al., J. Vac. Sci. Technol. A3 (1985) 2121
[21] C.J. Tsai, et al., J. Cryst. Growth 111 (1991) 931
[22] V. Stambouli, et al., Surf. Coat. Technol. 43/44 (1990) 137
[23] L. Dumas et al., Thin Solid Films 382 (2001) 61
[24] P.H. Wocjciechowski, J. Vac. Sci. Technol. A6 1988) 1924

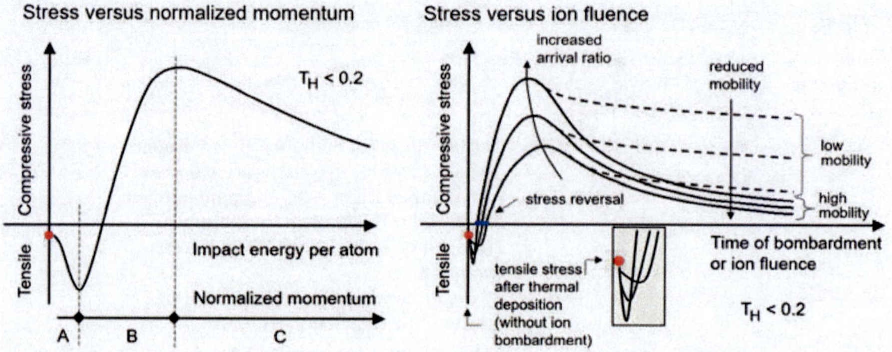

Fig. 10.56 Left: Schematic representation of the stress evolution in thin films prepared by ion beam-assisted deposition as a function of the ion energy at low temperatures (i.e. normalized temperature < 0.2). Right: Schematic representation of the stress evolution in thin films prepared by ion beam-assisted deposition as a function of the ion fluence (time of ion bombardment) at low temperatures (i.e. normalized temperature < 0.2). The inset shows more detail of the development of the tensile stress in the early stage of the IBAD as a function of the arrival ratio (black point indicates the tensile stress after only thermal deposition)

of tensile stress in Ni films deposited under Ar ion bombardment (arrival ratio was 0.16) from a non-zero tensile stress after deposition without ion bombardment to a maximum of the tensile stress after an ion energy of about 25 eV. Further increases in the ion energy lead to a decrease of the tensile stress. This behavior has been interpreted as densification (voids in the Ni after deposition without ion bombardment are closed by the assisted ion irradiation). 2D molecular dynamic simulations by Müller [164] confirmed the densification process under low-energy ion bombardment (see Fig. 10.51).

A similar behavior was very often observed for the intrinsic stress as a function of the duration of ion beam-assisted deposition (film thickness). Figure 10.56 (right) demonstrates the stress evolution as a function of the ion fluence (or duration of ion bombardment) in case that the energy per atom is low. With application of low ion energies, the tensile stress is increased up to a maximum, whereby the greater the ion-to-atom arrival ratio, the more particularly pronounced is the maximum (see inset in Fig. 10.56). The columnar or fibrous microstructure is probably strengthened at the beginning of the ion beam-assisted deposition process before densification takes place. Initial ion bombardment generates surface vacancies and bulk interstitials near the surface, with no other significant damage to the remaining material. According to Nix and Clemens [191], the maximum tensile stress corresponds to the onset of the island coalescence to form a continuous film. With advancing ion fluence, the transition from tensile stress to compressive stress and an increasing densification up to the maximum of the compressive stress can be observed.

The conditions required for the zero-stress state (stress reversal, see Fig. 10.56, right, blue points) were determined by Wolf [192] in dependence on the composition of the generated film and the ion energy. For example, Cr films require 100 eV per Cr atom for stress reversal, whereas carbon or boron ions require an order of magnitude less energy, because of their weaker type of bonding. According to Windischmann [149], stress reversal on films prepared by IBAD occurs at the normalized momentum P_n (see (10.62)) of $(0.3\ldots0.9)\sqrt{eV}/atom$, i.e. (5...15) $\times\ 10^{23}$ kgm/s/atom.

Some experimental studies have shown that the maximum value of compressive stress correlates with an increase of the ion-to-atom arrival ratio (see Fig. 10.56)). After the stress maximum is achieved, the stress evolution is characterized by a monotonous decrease of the stress, caused by the dominance of relaxation processes (see next chapter). The relaxation of the compressive stress (the intrinsic stress decreases with increasing ion fluence or irradiation time) is significantly pronounced in materials with a high atomic mobility for temperatures $T \leq T_H$, such as the fcc metals Ag, Cu, Al, etc. (see solid lines in Fig. 10.56). In contrast, a slight relaxation of the compressive stress can be observed in refractory materials such as Mo, Cr, W, with a low mobility for $T \leq T_H$ (see dashed line in Fig. 10.56, right). The compressive stress after high-energy assisted deposition is attributed to the dominance of interstitial defects as compared to vacancies.

10.6.8 Models of Stress Evolution During Ion Beam-Assisted Thin Film Growth

Various models have been proposed to explain stress evolution during thin film growth under assisted low-energy ion irradiation, whereby these models have focused on two aspects. First, the mechanism by which the stress is generated and second, stress evolution as a function of the ion energy.

Thermal Spike Annealing Model

Hirsch and Varga [193] proposed that the improved adhesion of deposited films can be explained by the annealing described within the thermal spike concept of incoming ions during the film deposition. Based on this observation, it has been assumed that the enhancement of film adhesion can be explained by an annealing effect caused by atomic rearrangement in ion beam-induced thermal spikes for ion energy between 65 eV and 3000 eV. This mechanism begins to be effective at a critical ion current density sufficient to ensure that all the deposited material atoms are subjected to rearrangement. It could be demonstrated that this critical ion current density was inversely proportional to the ion energy $\left(J_{I,c} \propto E_I^{-3/2}\right)$. Based on the binary collision approximation (Monte Carlo simulations), Brighton and Hubler [194] predicted a critical ion-to-atom arrival ratio R^{crit} above which stress annealing in the cascade volume can be expected causing the stress in the film. For example, for Ge films deposited under an assisted Ar ion beam, the critical ion-to-atom arrival ratio

$$R^{crit} = \begin{cases} 150E_I^{-1.59} & for \quad 0.2\,\text{keV} < E_I < 2\,\text{keV} \\ 4760E_I^{-2.04} & for \quad 2\,\text{keV} < E_I < 5\,\text{keV} \end{cases} \tag{10.81}$$

is strongly dependent on the ion energy E_I.

Atomic Peening or Forward Sputtering Model

Based on the peening mechanism by d'Heurle [181], Windischmann developed a quantitative model [149] to describe the ion beam-induced stress in thin films. In this model, the forward momentum of the bombarding ion is transferred to the lattice, which results in the generation of compressive stresses in the film. The linear cascade theory by Sigmund (see Sect. 5.2) is used to calculate the fractional number of displaced atoms, which should be proportional to distortional strain according to the Hooke's law. The model is based on assumptions:

(i) Energetic particles are capable to displace atoms from their lattice positions. This process results in a relative volumetric distortion (strain), ε, proportional to the number of atoms displaced from the equilibrium position, n_i/N, i.e.

$$\varepsilon = \frac{\Delta V}{V} = \frac{n_i}{N}K, \tag{10.82}$$

where N is the atomic number density, n_i is the number of implanted atoms into metasatabile lattice positons per unit volume and K is a proportionality factor.

(ii) The films are deposited at low temperatures (homologous temperature < 0.25). Then, it can be expected that the mass transport and defect mobility are sufficiently low, that deformation of the material is instantaneously frozen.

The volumetric distortion was determined by Windischmann [149] using (i) Sigmund's linear knock-on cascade theory of forward sputtering, (ii) a Born–Mayer potential for the interaction between the incident particle and the film atoms, and (iii) stopping cross-sections for the interatomic collision according to Wilson et al. [195] (see Sect. 3.1.1). Then, the strain is given by

$$\varepsilon = 4.79\delta_{MZ}\left(\frac{K\mathrm{J}_I}{N}\right)\sqrt{E_I}, \tag{10.83}$$

where J_I ion incident flux, U_s is the surface binding energy. δ_{MZ} contains the atomic number and mass of the interaction particles and is given by

$$\delta_{MZ} = \frac{M_2^{1/2}(Z_1Z_2)^{1/2}}{U_s(M_1+M_2)^{1/2}\left(Z_1^{2/3}+Z_2^{2/3}\right)}. \tag{10.84}$$

The intrinsic stress in the film can be obtained by substituting ε into Hooke's law for a biaxial stress state (see 10.68),

$$\sigma_{in} = \left(\frac{E_f}{1-\upsilon_f}\right)\varepsilon = 4.79\delta_{MZ}\left(\frac{E_f}{1-\upsilon_f}\right)\left(\frac{K}{N}\right)\mathrm{J}_I\sqrt{E_I}. \tag{10.85}$$

Replacing the atomic density with the Avogadro number N_A and the mass density ρ (in [g/cm^3]), then the compressive intrinsic stress can then be expressed by

$$\sigma_{in} = 4.79\delta_{MZ}\left(\frac{E_f}{1-\upsilon_f}\right)\frac{M_2}{\rho}\left(\frac{K}{N_A}\right)J_I\sqrt{E_I} = 4.79\delta_{MZ}E_{el}\left(\frac{K}{N_A}\right)J_I\sqrt{E_I}, \tag{10.86}$$

where N_A is the Avogadro number ($= NM_2/\rho$), ρ is the mass density, E_f is Young modulus of the film material, ν_f is the Poisson's ratio. The term E_{el} represents the stored elastic energy per mole.

Equation (10.86) predicts that the compressive stress is proportional to the square root of ion energy and the momentum $P_n = \sqrt{2M_2\gamma E_I}$ of the incident particle. The formation of an isotopic collision cascade was a precondition for the applicability of the Sigmund theory. This is no longer valid for very low incident ion energies and very light projectile ions. Therefore, a low energy limit can be expected for the application of this stress model.

The comparison of (10.82) and (10.86) also show that temporal change of the number of displaced atoms per unit volume into the film during deposition is approximately related to ion flux and energy, i.e.

$$\frac{dn_i}{dt} \propto J_I\sqrt{E_I}. \tag{10.87}$$

Based on an existing stress model for thermal deposition, Chason et al. [196] proposed a kinetic model to explain stress generation in polycrystalline thin films grown under additional ion bombardment (e.g., at sputter-induced deposition) by implementation of point defects generated by atomic peening. Two additional processes are included. The first is related to collision-induced densification of the structure near the grain boundary and the second is given by the introduction of mobile defects in the bulk of the film. The ballistically induced defects in the near-surface region can be incorporated as excess atoms at the grain boundary or annihilate at the free surface. Consequently, a complex dependency of the steady-state stress on the grain size, the growth rate, and the energies of the incoming particle flux could be determined.

Stress Model Based on Subplantation

Davis [150] and Robertson [197] independently devised relationships for the energy dependence of the stress (Davis formula) or the density increment (Robertson formula), respectively, based on the subplantation concept of Lifshitz et al. [198] to explain ion energy dependence of intrinsic stress.

The basic idea of the subplantation model is that the kinetic energy of the incident ions is transferred in linear collision processes with the film atoms within a time scale of about 10^{-13} s (first stage). It can be suggested that the lower the displacement energy of an atom, the higher the probability of its displacement. Consequently, the outcome is, for example, that sp^2-bonded atoms with a lower displacement energy are preferentially displaced, in contrast to sp^3-bonded atoms with a higher displacement energy. In addition, a fraction of the displaced atoms become stabilized in metastable sites with high displacement energies. According to Lifshitz et al. [198], in the second stage, a thermal spike is generated, which leads to thermalization of the excited atoms within a time scale of about 10^{-11} s. In the final stage (third stage, according to Lifshitz et al.), relaxation processes, such as diffusion, phase transformations, and chemical reactions can be expected over a long duration of between 10^{-11} s and 1 s.

On the one hand, it is assumed in the subplantation model that the compressive stress is caused by incident ions, which induce linear collision processes in compliance with the forward sputtering model by Windischmann [149, 192]), and on the other hand, the thermal spike after ion bombardment is assumed to reduce the stress as result of an annealing process. The incident ions impinging on the surface of the growing film cause atoms to be incorporated into the compact film. Consequently, the growing film expands. However, the expansion is limited by the contact plane of the film with the substrate, i.e. in the film plane parallel to the substrate surface (interface). These film atoms are thereby arranged in the aforementioned metastable sites.

Based on the assumption by Seitz and Koehler [199] that the thermal spike is characterized by intensive local heating in very small volume regions and the transfer of this energy to the film atoms, it can be expected that the thermal spike energy Q (see Sect. 4.6.2) is sufficiently high to release the film atoms from their positions. According to these authors, the number, n_T, of atoms participating in this rearrangement process during the life time of the thermal spike is given by

$$n_T = 0.016\delta(Q/E_s)^{5/3}, \tag{10.88}$$

where E_s is the activation energy (excitation energy) of the atomic rearrangement in the thermal spike (see Sect. 4.6.2), Q is the fraction of incident energy in the spike that is converted in lattice vibration (thermal energy), and δ is a material-dependent parameter and should lie in order of one. For low ion energy bombardment, the thermal energy in the spike Q is comparable to E_I, i.e. $Q \approx E_I$ [193]. The projected range of ions with such low energy will be small and most of the displaced atoms will be located close to the surface (or diffuse to the surface), i.e. a majority of atoms released from the metastable positions will end up on the free surface of the film. The rate of this relaxation process dn_R/dt can be expressed as

$$\frac{dn_R}{dt} = n_T \frac{N_I}{N} J_I = 0.016\,\delta \frac{N_I}{N} J_I (E_I/E_s)^{5/3}, \tag{10.89}$$

where N_I number of implanted atoms per unit volume.

Further, Davis [150] has also assumed that the steady-state is characterized by a balance between the defect production rate, dN_d/dt, (see (10.87) and the rate of relaxation (see (10.89)) and can be described by

$$\frac{dn_I}{dt} - \frac{dn_R}{dt} = \frac{N_I}{N} J_A, \tag{10.90}$$

where $\frac{N_I}{N} J_A$ is the net rate per unit area of atoms that are deposited on the surface or incorporated in the film. For ion beam-assisted deposition, the ratio between ion fluence, J_I, and the atoms deposited per united area and time, J_A, is expressed as the ion-to-atom arrival ratio $R = J_I/J_A$. Combining (10.87), (10.89) and (10.90) the fraction of incorporated atoms, n/N, the film is given by

$$\frac{N_I}{N} \propto \frac{\sqrt{E_I}}{(J_A/J_I) + 0.016\delta(E_I/E_s)^{5/3}} = \frac{\sqrt{E_I}}{1/\mathrm{R} + 0.016\delta(E_I/E_s)^{5/3}}. \tag{10.91}$$

Using (10.85) for the stress–strain relation in thin films (Hooke's law), the stress is found to be

$$\sigma_{in} \propto \left(\frac{E_f}{1-\upsilon_f}\right) \frac{\sqrt{E_I}}{1/\mathrm{R} + 0.016\delta(E_I/E_s)^{5/3}}, \tag{10.92}$$

where the stress $\sigma_{in} \propto E_I^{-7/6}$, i.e. the intrinsic stress decreases with increasing ion energy. With regard to stress development, a distinction can be made between two cases. At low ion flux, where $1/\mathrm{R} > 0.016\delta(E_I/E_s)^{5/3}$ and at high ion flux, where $1/\mathrm{R} < 0.016\delta(E_I/E_s)^{5/3}$ the intrinsic stresses are given by

$$\sigma_{in} \propto \left(\frac{E_f}{1-\upsilon_f}\right) R\sqrt{E_I} \quad \text{and} \quad \sigma_{in} \propto \frac{E_s^{\frac{5}{3}}}{0.016\delta E_I^{\frac{7}{6}}} \left(\frac{E_f}{1-\upsilon_f}\right), \tag{10.93}$$

respectively. With it, the intrinsic stress is proportional to J_I/J_A and $\sqrt{E_I}$ as well as to the normalized momentum P_n, whereas the atomic peening model provides a linear dependence of the intrinsic stress on J_I and $\sqrt{E_I}$ (compare (10.86) and (10.93)).

Figure 10.57 shows the stress in AlN films [150] and MgF_2 films [200] prepared by ion beam assisted deposition (measuring points) and after fitting (solid line). For example, the data for AlN were fitted by (10.92), assuming that $E_s = 11$ eV and $\delta = 1$ and $R = 1$. Based on experimental data and fitting of these data, the figure shows the expected behavior of the intrinsic film stress proportional to the square root of the incident ion energy (left) and the normalized momentum (right). Figure 10.58, right, also shows that a certain normalized momentum, here $P_n \approx 55$ $(\text{g·mol}^{-1}\ \text{eV})^{1/2}$, is necessary for the transition from a compressive to a tensile stress state. The maximum intrinsic stress is predicted for an ion incident energy of

$$E_I^{\max} \approx \left(\frac{3J_A}{7J_I \cdot 0.016\delta E_s^{-\frac{5}{3}}}\right)^{3/5} \approx 15\, R^{-3/5} E_s, \tag{10.94}$$

with the parameter $\delta = 1$. Consequently, energy E_I^{max} ranges between some ten electron-volts and less than 200 eV for typical activation energy E_s.

This model does not account for (i) relaxation processes not associated with the collisional cascade, (ii) that tensile stresses can be generated without ion bombardment during deposition, and (iii) that diffusion of defects must be considered.

A generalization of the qualitative subplantation model by Lifshitz et al. [198] was introduced by Boyd et al. [201]. This semiquantitative model describes the contributions of ion penetration, defect production, and radiation-enhanced diffusion in terms of analytical equations and provides quantities of the concentration retained in

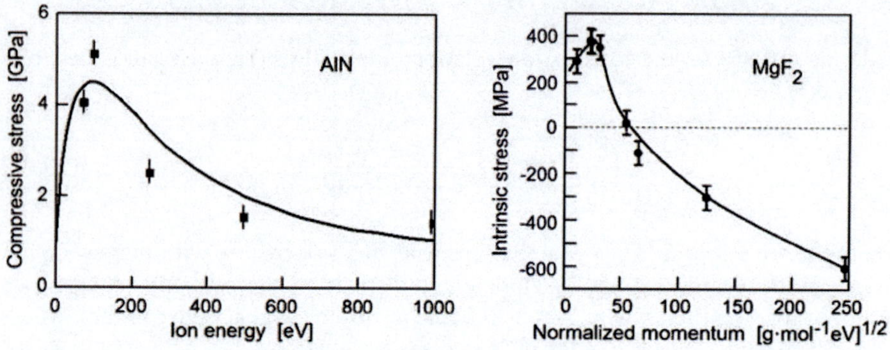

Fig. 10.57 Left. Compressive stress in AlN films produced by nitrogen ion-assisted deposition versus ion energy (figure adapted from [150]). Right: Stress in MgF_2 films produced by Ar ion-assisted deposition versus normalized momentum. The normalized momentum is calculated by (10.62); figure adapted from [200])

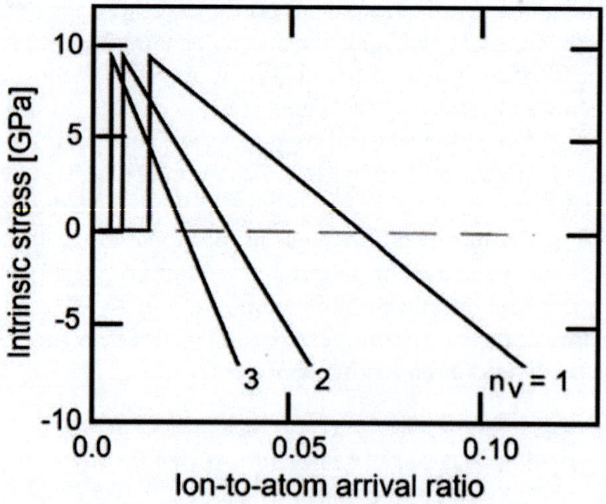

Fig. 10.58 Intrinsic stress as function of the ion-to-atom arrival ratio for three values of n_v where $d = 100$ nm, $f_{an} = 1$, and $\Delta_{shad} = 2.5$ nm (figure adapted from [202] and modified)

the subsurface layer, displacement energy thresholds and diffusion rates. This model was initially developed to describe the formation of carbon phases by direct ion beam deposition (see Sect. 9.3.1). In this model, a function of the experimental parameters is defined to allow estimation of the film properties (e.g., the c-BN content or the sp^3-bond fraction). In a simple phenomenological version of this model (the so-called static model), analytical approximations are derived of the ion energy dependence of the penetration and the damage production. It is assumed that the densification is the result of the collision process and that the relaxation is a result of a long-term radiation-enhanced diffusion process. Consequently, thermal spike relaxation was not taken into account.

Model of Transition from Tensile to Compressive Stress

It was demonstrated that stress evolution in the growing film is frequently characterized by a buildup of tensile stress in the initial stage up to a maximum (see Fig. 10.56). Subsequently, a sharp decrease in tensile stress and, at higher layer thickness, a transition from tensile to compressive stress can be observed. Knuyt et al. [202] have proposed a model that describes this behavior and tensile-compressive stress transition in films prepared by ion beam-assisted deposition. Assuming a typical columnar morphology of the deposited films, the lateral stress in a column is given by

$$\sigma_{in} = \left(\frac{E_f}{1-\upsilon_f}\right)\left(\frac{d_{gr}-d_{gr,eq}}{d}\right), \tag{10.95}$$

where d_{gr} is the lateral dimension of a single grain (defined as the mean distance between atoms on opposite planes of a column), $d_{gr,eq}$ is the equilibrium lateral dimension of the column, (i.e. the dimension in the unstressed state), and d is the total dimension of an assumed square column surface. A realistic expression for the elastic energy density in the film was found, under the condition that the mean width of the volume remaining between two neighboring columns d_{depl} can be minimized (i.e. $d_{depl} \approx (d_{depl})_{min}$). The stress is then given by

$$\sigma_{in} = \left(\frac{E_f}{1-\upsilon_f}\right)\frac{1}{d}\left[\Delta-\left(d_{depl}\right)_{\min}\right], \tag{10.96}$$

where the parameter Δ can be considered as the dimensional deficiency of a column within the film. This parameter consists two terms, i.e. $\Delta = \Delta_{shad} - \Delta d_{gr,def}$, whereby the first parameter represents the effect of the mutual shadowing of neighboring columns (see Chap. 11) including surface diffusion and thermal relaxation, and the second term describes the lateral column swelling caused by ion bombardment. Knuyt et al. [202] calculated these terms and were able to obtain an approximate expression for the lateral stress,

$$\sigma_{in} = \left(\frac{E_f}{1-\upsilon_f}\right)\left[\frac{1}{d}\left(\Delta-d_{depl,0}\right)-\frac{1}{3}f_{an}n_{\upsilon}R\right], \tag{10.97}$$

where f_{an} is a dimensionless factor that takes into account possible swelling anisotropy effects (i.e. for isotropic volume swelling expansion $f_{an} = 1$), n_v is a dimensionless parameter dependent on ion energy and point defect density that describes the mean number of surviving interstitials generated per impinging ion, and R is the ion-to-atom arrival ratio. It is further assumed that the minimum of elastic energy will be attained for $(d_{depl})_{min} \approx d_{depl,o}$. In general, the first contribution to the intrinsic stress caused by shadowing in (10.97) is positive (contribution of the tensile stress) because $\Delta_{shad} \geq d_{depl,o}$. The second contribution is negative (contribution of the compressive stress) and is dependent upon the arrival ratio and the ion energy. As an example, lateral stress as a function of the arrival ratio R is shown in Fig. 10.58 for three values of n_v (the mean volume swelling of the material due to the impingement of an ion). The stress-versus-arrival ratio curves reproduce the expected, frequently observed behavior (see Sect. 10.6.7 and Fig. 10.56, right). Initially, the columns grow in isolation from each other, hence $\sigma_{in} \approx 0$ (only in cases of strong shadowing). When the shadowing is weaker (Δ becomes smaller), the columns have an increasing effect on each other and tensile stress builds up. The maximum tensile stress is not dependent upon the arrival ratio. When shadowing effects are absent, the stress will always be compressive and will start at zero value for $R = 0$. Further, the stress curves will vary horizontally in accordance with n_v. The arrival ratio leading to zero intrinsic stress is the smallest for the largest value of n_v. The sharp maximum in the tensile stress may shift to negative (and hence unobservable) values of n_vR. The stress then evolves from tensile to compressive stress for increasing n_vR, but shows no maximum in the tensile region.

It can be asserted that the residual stress in films grown under additional ion bombardment has been studied in detail. However, a general understanding of the stress evolution results from the different experimental conditions is not comprehensively available. An explanation of the individual aspects of stress evolution is nonetheless already possible.

10.6.9 Adhesion

Adhesion is the tendency of dissimilar surfaces to cling to one another. With identical materials, this tendency is referred to as cohesion. Intermolecular forces (Van der Waals, electrostatic or chemical forces) are responsible for this effect. Here, adhesion is to be understood as the adherence of a thin film to a substrate, where the degree of adhesion is determined by the work which is necessary to separate atoms or molecules at the interface. Different types of interface (mechanical, chemical-bonding, diffusion and pseudo-diffusion interfacial layers) are known to cause the adhesion between the surfaces of two different materials [203]. For example, materials can be merged at the point of contact by diffusion. The condition for this is that both species are mobile (temperature is high enough) and capable of being soluble in each other.

The adhesion energy, W_{ad}, (in units of force or energy per surface area) is given by

$$W_{ad} = \gamma_s^F + \gamma_s^S - \gamma_s^{FS}, \tag{10.98}$$

where γ_s^F, γ_s^S and γ_s^{FS} are the specific surface energy of the film, the specific surface energy of the substrate and the specific interfacial energy, respectively [203]. The adhesion energy is the result of integration of the adhesion force over the distance x between the film and the substrate (thickness of the interface) and given by

$$W_{ad} = A \int F_{ad}(x)dx, \tag{10.99}$$

where the F_{ad} is the specific adhesion force and A is the contact area. The force F_{ad} is low (high), if two materials with low (high) surface energies are in contact.

Adhesion can be measured by (i) a scratch test (a hard stylus, e.g., a diamond, is moved under a load over the deposited sample and the critical load for spalling off the deposited layer is measured), (ii) a pull test (the force to pull off a stud fixed on the surface of the deposited film is measured), or (iii) a bending test of the substrate–layer combination.

At the beginning of the film growth under IBAD, an interface mixing between the film and the substrate atoms can be expected, driven by ion bombardment (Fig. 10.13, process XVI). This process leads to a significant broadening of the interface. Based on this result, an enhanced adhesion (anchoring) of the deposited films can be expected. Unfortunately, only a small number of systematic investigations in the area of the adhesion of layers prepared by IBAD are known. In comparison to films deposited without an assisted ion beam, a significant enhancement in adhesion could be proven after ion beam-assisted deposition for AlN and Si films on steel substrates [204], thin crystalline hydroxyapatite films on Ti-6Al-4 V [205, 206], Ag films on Mo substrates [207], and Cr-Ni coatings on steel substrates [208]. For example, Kellock et al. [209] were able to measure a strong adherence of Au films on GaAs after IBAD at low temperatures (120 °C–150 °C). The authors assume that the increased adhesion is caused by ion radiation-induced chemical displacement of As by Au at the GaAs surface. Hirsch and Varga [193] experimentally investigated the adhesion of amorphous Ge films under Ar ion bombardment. They interpreted the improved adhesion as an annealing within the thermal spike, caused by the incident ions. Above a critical ion current density, it is expected that all film atoms will be rearranged in the thermal spike. An empirical correlation was found between the critical ion current density and the ion energy, which is similar to that predicted by the Seitz-Koehler theory (see Sect. 4.7.2).

It is noteworthy that with the increasing energy of the assisting ion beam the adhesive strength increases throughout (partially up to a factor of ten, see e.g., [207]). It could also be shown that the increase in adhesion corresponds to a broadening of the interface region (interface mixing) between the film and the substrate (see e.g., [194]). Consequently, it can be assumed that the change in adhesion after ion beam-assisted deposition correlates with the dimension of the interface broadening. This so-called ion beam-induced mixing effect, which causes a highly effective process

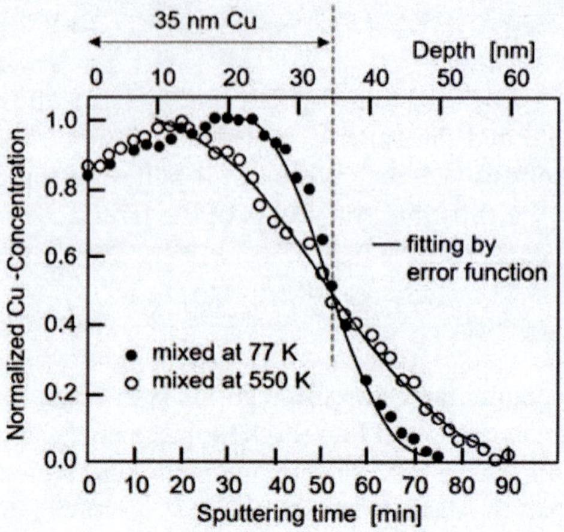

Fig. 10.59 Concentration distribution of a Cu top layer on titanium after 7.5 × 10^{15} ions/cm^2 at 77 and 550 K, measured by AES and RBS

of compositional changes at the interface region, has been studied intensively over the past several decades (see e.g., [210, 211]).

For example, Fig. 10.59 illustrates the mixing of Cu/Ti bilayers at two different temperatures after Xe ion beam irradiation [212]. The main feature of these concentration profiles is the progressive broadening of the interface region with increasing temperature. The measured concentration profiles suggest the migration of Cu into Ti and vice versa.

Initially, low-temperature ion mixing was believed to be controlled solely by ballistic two-body collisions. The mixing in a collision cascade is characterized by multiple relocations of atoms at low energies. The total relocation Δx within the cascade mixing can be described with the first and second moment of relocation probability distribution (see 4.32 or Appendix F), i.e. the mean spread is given by $\Omega^2_{mix} = \langle(\Delta x - \langle\Delta x\rangle)^2\rangle$. According to Sigmund and Gras-Marti [213], the mean spread is proportional to the ion fluence, $\Omega_{\mathrm{mix}} \propto \Phi^{1/2}$, i.e. a linear dependence of the mean spread on the fluence can be expected. It is obvious that this expression is similar to the thermal random walk process, where this process is $\propto t^{1/2}$. Consequently, an effective diffusion coefficient is given by $\Omega^2{}_{\mathrm{mix}} = 4Dt$, or the mixing rate can be defined as d(4Dt)/dΦ. However, it was found that the interface broadening exceeds the predictions of the ballistic consideration of the mixing process. Experimental studies have shown an increase of the mixing rate with increasing (negative) mixing heat and decreasing cohesive energy [210, 211].

According to the discussion above the adhesive force should be proportional to the mixing rate, $F_{ad} \propto d(4Dt)/d\Phi$, i.e. proportional to the broadening of the interface. This relationship is valid exclusively for small film thicknesses.

10.7 Thin Film Synthesis by Concurrent Low Energy Ion Bombardment

Compound synthesis in the reactive IBAD process is accomplished by adding one or more of the compound components in the vapor flux and adding one or more of the components in the ion beam flux. The generally available ion sources based on the ionization of simple gaseous components, therefore, the synthesis of nitrides and oxides, have been intensively studied.

The use of reactive IBAD to form nitrides and oxides offers a high degree of control over the film composition while also maintaining the ability to induce IBAD modification of the microstructure. Reactive ion beam-assisted deposition provides an opportunity for a precisely controllable synthesis. On the other hand, the applied IBAD techniques often lead to compound layers far from thermodynamic equilibrium, which are characterized by specific microstructures with distinctive properties. Aisenberg and Chabot [214] and Weissmantel [215] carried out the first experiments focused on the compound formation under low ion energy bombardment.

10.7.1 Synthesis of Nitrides by IBAD

Nitrides are a very interesting class of compounds because they have a unique combination of physical properties, such as a high degree of hardness, high melting temperature, and high thermal and electrical conductivities. These properties have attracted considerable interest for technical applications.

All IBAD configurations, IBA-TD, IBA-SD, and IBA-PLD, are used to synthetize binary and ternary nitride compounds. The fabrication of TiN, GaN, and cubic BN thin films by IBAD has been studied in detail. But also the nitride formation of metals (Al, Cr, Fe, Gd, V, Mo, Sc, etc.), Si_3N_4, and amorphous C:N have also been investigated depending on the conditions of the IBAD.

Example: Boron nitride

Boron nitrides are formed in four distinct crystalline phases: the hexagonal boron nitride phase (h-BN), the cubic boron nitride (c-BN) phase, the rhombohedral boron nitride phase, and the wurtzitic boron nitride phase (w-BN). The first two polytypes of boron nitride are of special interest due to their properties. The h-BN phase has a layered graphite-like structure with a two-dimensional sp^2 bond B-N of six atoms, and the cubic boron nitride phase is sp^3-bonded, similar to diamonds. c-BN is an especially interesting film material for protective layers, due its very high degree of hardness. The high band gap and the ability to dope c-BN with both n- and p-type dopants open additional electronic and optical applications. Comprehensive overviews of the synthesis and properties of boron nitride films have been published (e.g., see [216, 217]).

In addition to other manufacturing methods (pulsed laser ablation, ion plating, plasma enhanced CVD, etc.) ion beam-assisted deposition offers a possibility for controllably synthesizing c-BN films. On the one hand, boron is deposited by electron beam evaporation or sputtered by noble gas ion bombardment, and on the other hand, a low-energy ion source delivers a mixture of nitrogen and noble gas (particularly Ar) ions. The formation of the different BN polytypes can be detected by infrared spectroscopy and X-ray diffraction techniques or by a combination of both analysis methods. Figure 10.60 (left) demonstrates that low-energy ion bombardment is a necessary precondition for the formation (nucleation) of the c-BN phase.

This figure shows X-ray spectra of films prepared by the electron beam deposition of boron with and without simultaneous bombardment of a mixture of nitrogen and argon ions, using an RF ion source. In addition to diffraction peaks in the h-BN phase, after assisted ion bombardment, peaks in the c-BN can be identified, indicating the formation of the c-BN phase. High-resolution TEM studies by Kester et al. [218] have demonstrated that the microstructure of a boron nitride film is characterized by a sequential arrangement of different boron nitride layers. In Fig. 10.60 (right), the high-resolution TEM image demonstrates this typical arrangement. The boron nitride film has an approximately 4 nm thick amorphous mixed layer (a-BN) on a silicon substrate with a thickness in order of the ion range, followed by a sp^2-bonded h-BN film. Detailed electron diffraction studies (SAD studies) of these boron nitride

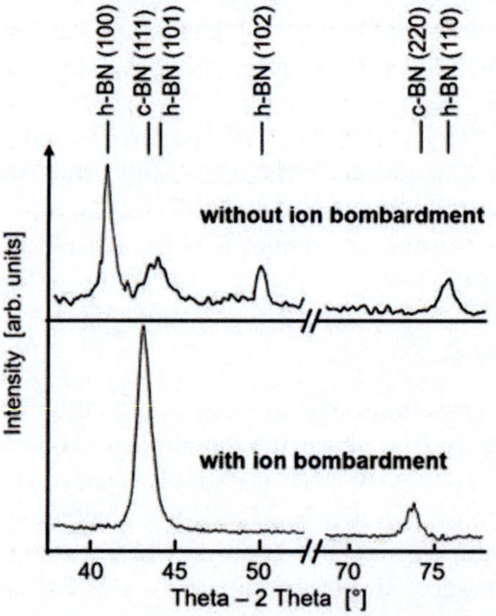

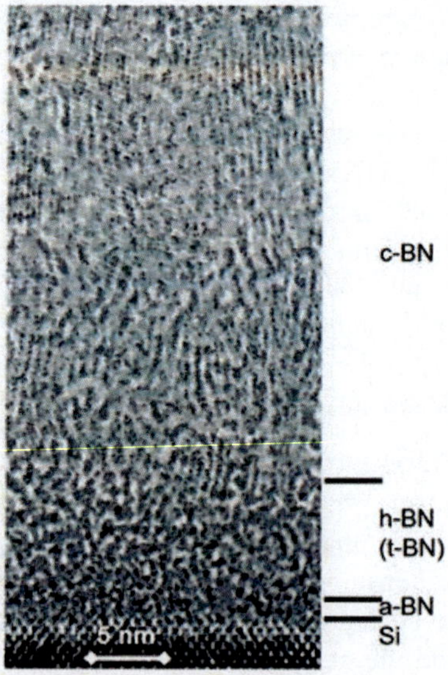

Fig. 10.60 Left: X-ray diffraction spectra of BN films prepared by electron beam evaporation of boron in a nitrogen atmosphere (working pressure of 1 × 10^{-2} Pa, substrate temperature of 600 °C) but without ion bombardment, and of a BN film prepared under 500 eV N/Ar ion bombardment. Right: High resolution XTEM image of a boron nitride film prepared by 500 eV N/Ar ion bombardment at 300 °C and an arrival ratio of 1.4 (figure taken from [185])

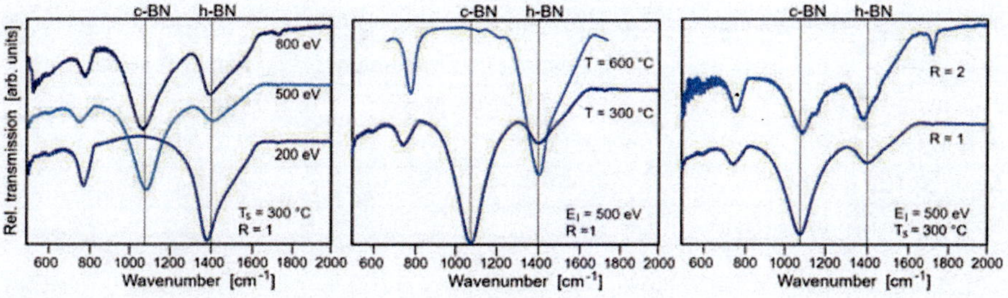

Fig. 10.61 FTIR spectra of boron nitride films on silicon in dependence on ion energy E_I (left), temperature of the substrate during deposition T_s (middle), and the ion to atom arrival ratio R (right)

interlayers indicate a turbostratic structure (labeled t-BN), which is characterized by the absence of ordering in the third dimension [219]. A top layer of polycrystalline c-BN then follows. This c-BN layer consists of small columnar grains (diameter up to several 10 nm), where the individual crystallites exhibit a preferential crystallographic orientation [216, 220]. It has been frequently observed (see e.g., [221, 222], not shown in Fig. 10.60), that the surface is covered by an sp^2-bonded BN film whose thickness approximately corresponds to the ion range.

The volume fraction of the c-BN phase in the deposited films can be quantified by comparison of the normalized FTIR intensities after transmission (reflection), i.e. ratio of the transmitted (reflected) intensity to the incident intensity. As an example, Fig. 10.61 shows FTIR spectra of boron nitride films on silicon prepared by ion beam-assisted deposition in dependence on the ion energy, substrate temperature, and arrival ratio [185] The FTIR spectra exhibit an absorption band at 1065 cm^{-1}, corresponding to the sp^3-bonded c-BN phase, and absorption bands at about 1380 and 780 cm^{-1}, corresponding to the sp^2-bonded h-BN. The shift to higher wavenumbers is commonly discussed as being due to the compressive stress in the BN films. It is obvious that the relative c-BN content varies in dependence on ion energy, the substrate temperature, and the ion-to-atom arrival ratio. For example, the c-BN content is about 60% for ion energy of 800 eV, $\geq$ 80% for ion energy of 500 eV, and not detectable for ion energy of 200 eV when the temperature during deposition is 300 °C and the R = 1. On the other hand, the relative c-BN content is reduced from about 80% after deposition at 300 °C up to about 5% after deposition at 600 °C under the condition that the ion energy is 500 eV and R = 1. A maximum c-BN amount (> 80%) was obtained at an ion-to-atom arrival ratio R = 1, where temperature during the IBAD process is 300 °C and E_I = 500 eV [169].

Several parameters can be identified that significantly influence the synthesis of cubic boron nitride films: (i) ion energy, (ii) ion-to-atom arrival ratio, (iii) ion mass, and (iv) substrate temperature [214]. According to Kulisch and Ulrich [217], the ranges for these parameters on the basis of numerous experimental results published in the literature are given in Table 10.9. The influence of the substrate remained disregarded.

Table 10.9 Decisive parameters of the c-BN films synthesis [217]

Parameter	Minimum	Maximum	Standard process
Ion energy [eV]	≈ 55	3500	500
Arrival ratio	≈ 0.6	130	
Ion mass [amu]	20 (Ne)	131 (Xe)	40 (Ar)
Substrate temperature [°C]	≈ 120	>1000	400

Low-energy ion bombardment during the growth of c-BN is obviously connected to an enormous level of compressive stress, and vice versa [169, 223]. This stress scales with the fraction of c-BN in the film.

It has been attempted to explain the formation of sp^3-bonded boron nitride films under low-energy ion bombardment on the basis of models that had already been used to discuss the evolution of compressive stresses, the generation of thermal spikes, and the processes of subplantation and resputtering. Nearly all models used to interpret the c-BN formation were initially developed to describe the diamond-like and amorphous carbon film formation (Sect. 9.3.1). Mirkarimi et al. [216] have divided these models into four categories:

Models Based on the Thermal Spike

It is assumed that the extreme temperatures and pressures within the thermal spike for extremely short periods of time (see Sect. 4.5) cause the transformation to c-BN. A large number of atomic rearrangements within a thermal spike are assumed to be triggered by bombardment with single low-energy ions [224]. Consequently, (i) a cylindrically symmetric Gaussian distribution of the deposited energy can be assumed and (ii) only the energy (thermal energy Q, where $Q < E_I$, see Sect. 4.7.2) deposited into phonon excitations is taken into consideration. Based on the algorithms by Vineyard [225] (see Sect. 4.5.2), the number of rearrangement processes can be estimated by [224]

$$n_T \approx \frac{0.042}{L}\left(\frac{Q}{E_s}\right)^2, \tag{10.100}$$

where L is length of the spike (the mean ion range in [nm]) and E_s ($\approx 3\ _{\mathrm{eV}}$) is activation energy of diffusion in the thermal spike. The effect of the rearrangement process in the spike becomes obvious if n_T (the total number of rearrangement processes during a thermal spike) is compared with the total number of atoms within the spike, n_s. When $n_T \geq n_S$ or $n_T/n_s \geq 1$, each atom in the spike will be rearranged one or multiple times and can form new chemical bonds. When $n_T < n_S$ or $n_T/n_s < 1$, the atoms in the spike become only partially rearranged. The ratio n_T/n_s is therefore a key parameter in the characterization of the resulting film structure. In c-BN formation, the ratio n_T/n_s is larger than one for ion energies between about 50 eV and 2 keV. The optimal conditions for the formation of sp^3-bonded BN can therefore be expected within this energy range. In good accordance with numerous experimental studies, the highest sp^3 bond fraction should be realized between 50 and 500 eV.

Models Based on Preferential Sputtering

Reinke et al. [223] developed a model for the formation of c-BN in dependence on the ion energy, substrate temperature, and ion mass, in terms of sputtering and desorption processes. It is assumed that both hexagonal sp^2-bonded and cubic sp^3-bonded BN phases are formed simultaneously during the ion beam-assisted deposition. Since the sputtering yield of c-BN is lower than that of h-BN, a preferential sputtering of the sp^2-bonded phase is expected under low-energy ion bombardment. Unfortunately, this model is not capable of describing the formation of c-BN at low ion energies (< 100 eV) and high energies (> 1 keV).

Stress Models

Ion bombardment induced stress rises quickly during the initial period of ion bombardment (see schematic stress evolution under ion bombardment in Fig. 10.57). The stress model (see Sect. 10.6.7) is based on the assumption that ion bombardment generates a high defect density, which induces (i) the buildup of high hydrostatic pressures and (ii) a pressure-driven phased transformation from the h-BN phase into the c-BN phase [227]. For example, a compressive stress >4…6 GPa should be necessary to create the sp^3-bonded cubic BN phase [228]. It is suggested that the ion bombardment generates point defects during nuclear collisions. These point defects produce volumetric strain and, thus, stress in the constrained film. If the stress is sufficiently high enough, sp^2-bonded BN is transformed into sp^3-bonded BN. Consequently, the defects produced during ion bombardment determine this transformation process, which is given by the balance between formation and recombination rates.

First Windischmann [149], and later Davis [150], derived analytic relations of ion-bombardment-induced stress (see Sect. 10.6.7). Windischmann [149] found that the intrinsic stress is proportional to $J_I E_I^{1/2}$ (see 10.93), where J_I and E_I are the ion flux and the ion energy, respectively. In this model, the temperature, and thereby the annealing of defects, are neglected. Kester and Messier [229] found the same ion energy dependence. They used the relationship for the momentum transferred from an ion to the film atoms developed by Tagove and Macleod [158] to define a c-BN formation threshold, which is proportional to $(2M_I \gamma E_I)^{1/2}$, where M_I is the ion mass and γ is the transfer energy efficiency factor (note the remarks of Mirkarimi et al. [228] and the comment in Sect. 10.6.7 on this relationship). A threshold value of about 200 $(\text{amu eV})^{1/2}$ for the c-BN nucleation could be proven. Davis [150] examined the thermally activated defect relaxation induced by spherical thermal spikes and found that the compressive intrinsic stress is proportional to the ion energy, $\sigma_{in} \propto E_I^{-7/6}$ for an arrival ratio of $R = 1$, i.e. in contrast to numerous experimental results, the stress decreases as energy increases.

Subplantation Models

Lifshitz et al. [198] have proposed a subplantation model in which the preferential displacement of the weakly bonded sp^2-bounded atoms by incident ions in deposited carbon films results in the accumulation of sp^3-bonded atoms (see Sect. 10.6.7). The application of this model is limited when the displacement threshold energy of both phases, e.g., the c-BN and the h-BN phases, are very similar. In contrast to the three-stage Lifshitz model, Robertson [197, 230] has suggested that two processes, local densification by penetration of individual ions and density relaxation during the thermal spike evolution, determine the portion of the sp^3-bonded boron nitride phase. The first process, densification, results in an increase of the ion bombardment induced sp^3 hybridization, while the second process, relaxation, leads to a diffusion of interstitials to the surface (annealing during the thermal spike). If the particle energy is too high, the resulting thermal spike induces a relaxation process, connected with a reduction of the ion bombardment induced high density. When the energy of the ions is too low, the penetration of depth of the ions is small and therefore an increase in density cannot be expected. The Robertson variant of the subplantation model quantifies the c-BN growth and predicts that the energy range will be only between 50 eV and 500 eV.

An extension of the subplantation model, the semiquantitative subplantation model [201], was used preferentially to describe the growth of amorphous carbon layers [231] and homoepitaxial silicon films [232]. However, with a suitable choice of the parameters in this phenomenological model, the growth of BN layers should be also writable.

It must be noted that none of the models presented is capable of satisfactorily and comprehensively explaining c-BN thin film growth under energetic particle bombardment.

10.7.2 Synthesis of Oxides by IBAD

Metal oxides have many excellent qualities that give them promising potential for a variety of significant applications. These same qualities are also expected from the thin films in this material class. Consequently, numerous deposition methods have been used, with the objective of preparing thin oxide films that show properties identical to those of the bulk material. Frequently, the deposition, in particular the thermal evaporation of oxide films, results in films with a significant porosity, depending upon the crystal structure, chemical composition, packing density, substrate temperature, deposition rate, and oxygen pressure [233]. The application of these methods is hence considerably limited. For example, the reduced packing density could be overcome only if the deposition is realized at elevated substrate temperatures to enhance the surface mobility of the atoms being deposited [234]. On the other hand, a reduced packing density of deposited titanium oxide film leads to a shift of the spectral characteristic of oxide films when they are exposed to air or changing conditions of humidity. For precision optics, this so-called vacuum-to-air shift is unacceptable.

It appears that ion beam-assisted deposition is a method that can be used to prepare thin films with properties comparable to the bulk material [1–6]. Consequently, binary and ternary oxide films have been synthesized either by oxygen ion-assisted deposition or by noble ion beam-assisted deposition into an oxygen atmosphere. Taking this route, oxide films of metals (Al, Fe, Zr, Ti, Hf, Ta, etc.), transparent oxide films (e.g., ITO and ZnO), and isolated films (e.g., SiO_2) could be produced and characterized.

Example: Titanium dioxide (titania)
Titanium dioxide, TiO_2, has an amorphous phase and three different crystalline phases brookite, anatase, and rutile, whereby brookite films were not observed for films deposited under high-vacuum conditions. Rutile is the most common and stable form. Its structure is based on a slightly distorted hexagonal close-packing of oxygen atoms with the titanium atoms occupying half of the octahedral interstices. Anatase is based on cubic packing of the oxygen atoms, but the coordination of the titanium is again octahedral.

Titanium dioxide is a very attractive material for several applications because this material has a high refractive index, is transparent over the wavelength range from the visible up to the near infrared, and is very hard. Rutile, in particular, is an attractive thin film material for numerous optical applications. For example, titanium dioxide is one component of alternating layers of high and low refractive index materials that, according to the principle of optical interference, tailor the spectral reflectance profile of the thin-film optics. This technique can be used to make optical filters, reflectors, and antireflection coatings.

Anatase is also an especially excellent photocatalyst for the sunlight-energized oxidation of most organic compounds. At the surface of titanium dioxide films, titanium dioxide transforms the energy of ultraviolet light into chemical energy. The free radicals, in turn, can react with almost all organic molecule and destroy them.

Numerous methods exist to prepare titanium dioxide films using ion beam-assisted deposition. The aim of these studies was to identify the existence range of the different titanium dioxides in dependence on the experimental parameters of ion beam-assisted deposition and to verify the suitability of the titanium dioxide films for specific applications.

In general, titanium oxide films deposited by IBAD at substrate temperatures lower than 200 °C are amorphous. Increasing the substrate temperature promotes the formation of crystalline titanium dioxide phases [235]. The application of substrate temperatures between 200 °C and over 600 °C leads to the formation of anatase (see e.g., [236]).

The duration and temperature of the annealing process determine the recrystallization process. There are only a few known studies on this topic. Jaing et al. [237] proved that an annealing temperature of 250 °C is necessary for the transformation of the amorphous phase into the crystalline anatase phase. Yang et al. [238] observed XRD peaks of anatase after annealing at 450 °C for one hour for titania films deposited by IBAD at temperatures between 50 °C and 300 °C. The rutile phase was formed by high-energy (40 keV) noble gas ion bombardment during film growth in an oxygen atmosphere at a substrate temperature of 250 °C [239]. The thermal stability of TiO_2 films deposited by IBAD at temperatures between 150 °C and 250 °C have been studied at recrystallization temperatures up to 300 °C. The results reveal that the

TiO_2 films are more stable after thermal annealing than in the initial state and that this is especially true for the films deposited at a substrate temperature of 150 °C [240]. Deposition at higher temperatures leads to the formation of polycrystalline titanium oxide films. For example, Yokota et al. [240] deposited polycrystalline TiO_2 films on Si(100) at 600 °C by titanium metal evaporation and assisted bombardment with oxygen ions and neutral oxygen molecules with an energy of 500 eV, where the anatase phase was formed under high Ti fluxes and the rutile phase under high oxygen fluxes.

Martev [241] studied the influence of the oxygen and mixed oxygen/argon ion-to-atom arrival ratio on the recrystallization of titanium dioxide films prepared by IBAD at deposition temperatures < 100 °C and ion beam energies of 520 eV and 630 eV, respectively. It was found that the transformation of the amorphous TiO_x compound into the crystal phase begins at low temperatures under ion bombardment and can be reinforced by increasing the ion-to-atom arrival ratio.

The stoichiometry of TiO_2 films prepared by oxygen ion beam-assisted deposition of TiO_2 for optical application has been investigated as a function of ion energy between 30 eV and 500 eV and current density up to 300 $\mu A/cm^2$ [242]. Low-energy ion bombardment improves stoichiometry, while high-energy bombardment is detrimental. Titanium dioxide films deposited by IBA-SD under simultaneous oxygen ion irradiation during growth have also illustrated that the ion beam-induced interface mixing effect strongly influences the adhesion and stoichiometry of the films [243].

The decisive experimental parameters in the field of the synthesis of crystalline titanium dioxides films by ion beam-assisted deposition are summarized in Table 10.10, based on published findings. It must be taken into consideration that major uncertainties exist with regard to the experimental parameters due to the somewhat limited number of detailed investigations.

A number of reviews of the optical properties of thin films deposited under assisted low-energy ion bombardment have been published [243, 244]. Of particular interest are the density and optical indices of TiO_2 thin films prepared by IBAD. Martin et al. [234, 245] studied the optical properties of crystalline TiO_2 films prepared by ion beam-assisted deposition. They demonstrated that the packing density of TiO_2 coatings is significantly higher compared to other conventionally produced TiO_2 coatings. This results in a reduction in the adsorption of moisture when the films are exposed to a humid atmosphere. Several authors have demonstrated that the density

Table 10.10 Decisive parameters of the crystalline titanium dioxide thin film synthesis by IBAD

Parameter	Minimum	Maximum	Standard process
Ion energy [eV]	≈ 30	40,000	≥ 500
Arrival ratio	0.04	1	≈ 0.25
Transferred ion energy per substrate atom [eV/atom]			> 50
Substrate temperature during deposition [°C]	RT	610 °C	400
Temperature during annealing [°]	≈ 150	650	500

of TiO_2 films can be significantly increased using oxygen ion-assisted deposition, in contrast to films prepared by thermal deposition (e.g., from 3.87 g/cm^3 to 4.02 g/cm^3 [246], where the density of anatase is 3.79 g/cm^3 and that of rutile is 4.23 g/cm^3).

Bennett et al. [247] and Tang et al. [248] have prepared TiO_2 films with refractive indices from 2.344 to 2.611 and from 2.322 to 2.495, respectively, at a wavelength of 550 nm. Yamada et al. [246] have produced titania films with a refractive index of 2.51 at deposition temperatures less than 100 °C. Williams et al. [249] have investigated the refractive indices of titania films deposited using IBAD at low temperatures (up to 100 °C) as function of the ion current density (ion-to-atom arrival ratio). Initially, with increases in ion current density, an increase of the refractive indices was observed. Films deposited at bombardment levels greater than the critical ion current density exhibit refractive indices that are less than those obtained at current densities below the critical value. The critical value is reduced by an increase of the ion energy.

Blood compatibility of the titanium oxide films prepared by IBAD were evaluated by clotting time measurement, platelet adhesion investigation, and hemolysis analysis. The results revealed that the blood compatibility of films was improved by a coating of non-stoichiometric titanium oxide film [250]. Zhang et al. [251] have also studied the blood compatibility of titanium oxide film prepared by ion beam-assisted deposition. They found that the clotting time of titanium oxide films is longer than that of low-temperature isotropic pyrolytic carbon, which is used in artificial heart valves in clinics.

10.8 List of Symbols

Symbol	Notation
C_i	Concentration of the component i
D_s	Surface diffusion coefficient
D_o	Pre-exponential factor of diffusion
E, E_I	Ion incident energy
E_A	Kinetic energy of atoms (molecules)
E_B	Atomic binding energy
E_{ev}	Kinetic energy of evaporated atoms
E_d	Displacement energy
$E_d^{(b)}$	Bulk displacement energy
$E_d^{(s)}$	Surface displacement energy
E_s	Activation energy of the atomic rearrangement in the thermal spike
E_{sp}	Kinetic energy of sputtered atoms
E_{sd}	Activation energy of the surface diffusion
I_{hkl}	Measured X-ray intensity

(continued)

(continued)

Symbol	Notation
J_A	Flux of atoms (impingement rate)
J_I	Ion current density (ion flux)
J_N	Flux of neutralized particles
J_R	Residual gas flux
L	Mean distance between islands
M_i	Mass of particle i
N_i	Atomic density of the component i
N_1, N_s	Concentration of the monomers and islands
P	momentum
Q	Fraction of incident energy converted into thermal energy (thermal spike energy)
R	Atom-to-atom ratio (impingement ratio, arrival ratio)
R_c	Corrected arrival ratio
R_N	Ion reflection coefficient
T_{max}	Maximum transferred energy
$S_e(E)$	Electronic stopping power
$S_n(E)$	Nuclear stopping power
U_s	Surface binding energy
W	Strain energy density
Y	Sputtering yield
Z_i	Atomic number of the element i
d	Lattice spacing
e	Electron charge
h_f, hs	Thickness of film, substrate
n	Number of collisions
p	Pressure
r_c	Radius of curvature
s	Sticking coefficient (trapping probability)
$s_n(\varepsilon)$	Reduced nuclear stopping power
t_d	Deposition time
v_A	Deposition rate (deposition velocity) of deposited atoms
v_e	Velocity of the outer shell electrons of the target atom
v_g	Net thin film growth velocity
v_I	Velocity of the projectile atom
v_s	Erosion rate
Θ	Contact angle
Λ	Distance between sputter source or crucible and the sample
Ω_i	Atom volume of the component i

(continued)

(continued)

Symbol	Notation
$\Omega^2{}_{mix}$	Mean spread (ion beam mixing)
α_f, α_s	Thermal expansion coefficient of film, substrate
γ	Mass transfer coefficient
γ_i	Surface tension (energy) of the component i
$\langle\delta\rangle$	Average charge per atom
ε	Reduced energy / strain
ζ_A	Tooling factors for the atomic flux measurement
ζ_I	Tooling factor for the ion beam flux measurement
s	Sticking coefficient
θ	Angle of ion incidence/Bragg angle
$\overline{\theta}$	Coverage
θ_{lab}	Scattering angle in the lab-system
κ	Curvature
λ	Mean free path
ν	Damage energy
ν_o	Jump frequency
σ	Collision cross-section/stress
σ_1, σ_s	Capture numbers of the monomers and islands
σ_{in}	Intrinsic stress
σ_{ij}	Biaxial stress
σ_{ce}	Cross-section of charge exchange
$\sigma_{sp,g}$	Collision cross-section between sputtered particles and particles of the background gas
φ, χ	Azimuthal and polar angles (pole figure measurement)
φ_{div}	standard deviation angle of the ion beam
φ^*	Angle between the center of the ion source and a point on the substrate surface
ψ	Tilt angle of the sample with respect to the beam (stress measurement)

References

1. G.K. Hubler, Fundamentals of ion beam-assisted deposition: Technique and film properties, Mater. Sci. Engng. A **115**, 181–192 (1989). Particle bombardment effects in thin film deposition, in *Plasma-surface interactions and processing of materials*, ed. by. O. Auciello, A. Gras-Marti, J.A. Valles-Abarca, D.L. Flamm (Kluwer Academic Publishers, Dordrecht, 1990), pp. 251–280
2. T. Itoh (ed.), *Ion Beam Assisted Film Growth* (Elsevier Science Publications, Amsterdam, 1989)
3. F.A. Smidt, Use of ion beam-assisted deposition to modify the microstructure and properties of thin films. Intern. Mater. Rev. **35**, 61–128 (1990)

4. J.K. Hirvonen, Ion beam assisted thin film deposition. Mater. Sci. Rep. **6**, 215–274 (1991)
5. G.K. Wolf, W. Ensinger, Ion bombardment during thin film deposition and its influence on mechanical and chemical surface properties. Nucl. Instr. Meth. Phys. Res. B **59–60**, 173–181 (1991)
6. B. Rauschenbach, Ion-assisted deposition of thin layers, in *Advances in Solid State Physics*, ed. by F. Kramer, Vol. 40, (Vieweg & Sohn, Braunschweig 2000), pp. 439–452
7. A.R. Gonzalez-Elipe, F. Yubero, J.M. Sanz, *Low Energy Ion-assisted Film Growth* (Imperial College Press, London, 2003)
8. D.M. Mattox, Fundamentals of ion plaTiNg. J. Vac. Sci. Technol. **10**, 47–52 (1973)
9. L. Pranevicius, Structure and properties of deposits grown by ion beam-activated vacuum deposition techniques. Thin Solid Films **63**, 77–85 (1979)
10. C. Weissmantel, G. Reisse, H.-J. Erler, F. Henny, K. Bewilogua, U. Ebersbach, C. Schürer, Preparation of hard coaTiNgs by ion beam methods. Thin Solid Films **63**, 315–325 (1979)
11. R. Gago, I. Jiménez, J.M. Albella, Thin film growth by ion beam-assisted deposition techniques, in *Material Surface Processing by Directed Energy Techniques*, ed. by Y. Pauleau, (Elsevier, Amsterdam, 2006), p. 345–382
12. M. Knudson, Das Cosinusgesetz in der kinetischen Gastheorie. Ann. Phys. **48**, 1113–1121 (1915)
13. D.M. Mattox, *Handbook of Physical Vapor Deposition (PVD) Processing* (Noyes Publications, Westwood, 1998)
14. H.R. Kaufman, R.S. Robinson, *Operation of broad-beam sources* (Commenwealth Sientific Corporation, Alexandria, 1987)
15. D. Van Vechten, G.K. Hubler, E.P. Donovan, F.D. Correll, Fundamentals of ion-atom-assisted deposition. I. Model of process and reproducibility of film composition, J. Vac. Sci. Technol. A **8**, 821–830 (1990)
16. R. Becker, W.B. Herrmannsfeldt, IGUN—A program for the simulation of positive ion extraction including magnetic fields. Rev. Sci. Instrum. **63**, 2756–2758 (2008)
17. D. Van Vechten, G.K. Hubler, E.P. Donovan, Characterization of a 3 cm Kaufman ion source with nitrogen feed gas. Vacuum **36**, 841–845 (1986)
18. J.W. Gerlach, P. Schumacher, M. Mensing, S. Rauschenbach, I. Cermak and B. Rauschenbach, Ion mass and energy selective hyperthermal ion beam-assisted deposition setup. Rev. Sci. Instrum. **88**, 063306 (2017)
19. M.W. Thompson, Physical mechanisms of sputtering. Phys. Rep. **69**, 335–371 (1981)
20. V.V. Serikov, K. Nanbu, Monte Carlo numerical analysis of target erosion and film growth in three-dimensional sputter chamber. J. Vac. Sci. Technol. A **14**, 3108–3123 (1996)
21. R.S. Robinson, Energetic binary collisions in rare gas plasmas, J. Vac. Sci. Technol. A **16**, 185–188 (1979). J.M.E. Harper, J.J. Cuomo, R.J. Gambino, H.R. Kaufman, R.S. Robinson, Mean free path of negative ions in diode sputtering. J. Vac. Sci. Technol. A **15**, 1597–1600 (1978)
22. A. Gras-Marti, J.A. Valles-Abarca, Slowing down and thermalization of sputtered particles fluxes: energy distributions. J. Appl. Phys. **54**, 1071–1075 (1983)
23. E. Fermi, *Nuclear physics, course notes*, ed. by J. Orear, A.H. Rosenfeld, R.A. Schluter (Midway Reprint, 1949 &1950)
24. F. Keywell, A mechanism for sputtering in the high vacuum based upon the theory of neutron cooling. Phys. Rev. **87**, 160–161 (1952). Measurements and collision—radiation damage theory of high-vacuum sputtering. Phys. Rev. **97**, 1611–1619 (1955)
25. K. Meyer, I.K. Schuller, C.M. Falco, Thermalization of sputtered atoms. J. Appl. Phys. **52**, 5803–5805 (1981)
26. W.D. Westwood, Calculation of the deposition rates in diode sputtering systems. J. Vac. Sci. Technol. **15**, 1–9 (1978)
27. A. Kirsch, W. Morokoff, Chr. Werner, Selfconsisted simulation of sputter deposition with Monte Carlo method. J. Appl. Phys. **75**, 2278–2285 (1994)
28. N. Bohr, The penetration of atomic particles through matter, Mat Fys. Medd. Dan. Vid. Selsk. XVIII, no. 8 (1948)

29. S. Sakabe, Y. Izawa, Cross sections for resonant charge transfer between atoms and their positive ions: collision velocity ≤ 1 .au. Atomic Data and Nucl. Data Table **49**, 257–314 (1991)
30. S. Schiller, U. Heisig, S. Panzer, *Electron Beam Technology* (Wiley, New York, 1982)
31. B. Lewis, J.C. Anderson, *Nucleation and Growth of Thin Films* (Academic, New York, 1978)
32. H. Neddermeyer, STM studies of nucleation and the initial stages of film growth. Crit. Rev. Solid State Mater. Sci. **16**, 309–335 (1990)
33. I.V. Markov, *Crystal Growth for Beginners: Fundamentals of Nucleation and Early Growth Stages and Epitaxy* (World Scientific, Singapore, 1995)
34. H. Brune, Microscopic view of epitaxial metal growth: nucleation and aggregation. Surf. Sci. Rep. **31**, 32–229 (1998)
35. J.E. Greene, The role of low-energy ion/surface interaction effects during crystal growth from vapor phase: nucleation, and growth, microstructure evolution, and defect annihilation, in *Handbook of Crystal growth*, ed. by D.T.J. Hurle, Vol. 1 (Elsevier, Amsterdam 1993), pp. 639–681
36. W.D. Wilson, I.G. Haggmark, J.P. Biersack, Calculations of nuclear stopping, ranges, and straggling in the low-energy region. Phys. Rev. B **15**, 2458–2468 (1977)
37. E. Bauer, Phänomenologische Theorie der Kristallabscheidung an Oberflächen. Zeitsch. f. Kristallogr. **110**, 372–394 (1952)
38. J.A. Venables, *Introduction to surface and thin film processes* (Cambridge University Press, Cambridge 2000) and H. Lüth, *Surfaces and Interfaces of Solid Surfaces* (Springer, Berlin, 1995)
39. J.A. Venables, Rate equation approaches to thin film nucleation kinetics. Phil. Mag. **27**, 697–738 (1973)
40. J.G. Amar, F. Family, P.-M. Lam, Dynamic scaling of the island-size distribution and percolation in a model of submonolayer molecular-beam, epitaxy. Phys Rev. B **50**, 8781–8799 (1994)
41. J.A. Venables, G.D.T. Spiller, M. Hanbücken, Nucleation and growth of thin films. Rep. Prog. Phys. **47**, 399–459 (1984)
42. J. Villain, ConTiNuum models of crystal growth from atomic beams with and without desorption. J. Phys. I France **1**, 19–42 (1991)
43. S. Berg, C. Nender, B. Gelin, Ion-assisted selective deposition of thin films. Vacuum **38**, 621–625 (1988)
44. W. Möller, D. Bouchier, O. Burat, V. Stambouli, Computer simulation of boron nitride deposition by ion beam-assisted evaporation. Suf. Coat. Technol. **45**, 73–81 (1991)
45. G. Carter, Peening ion ion-assisted thin-film deposition: a generalized model, J. Phys. D: Appl. Phys. **27**, 1046–1055 (1994). M.G. McLaren, G. Carter, The species dependence of inert gas ion incorporation in the in-induced deposition of tungsten. J. Appl. Phys. **86**, 2889–2895 (1999)
46. J.-K. Kim, J.S. Colligon, S.-H. Jeong, Mathematical modeling for the composition prediction of compound films grown by ion-assisted deposition technique and its application to TiN_x film. Nucl. Instr. Meth. In Phys. Res. B **225**, 509–515 (2004)
47. J.E. Greene, S.A. Barnett, J.-E. Sundgren, A. Rockett, Low-energy ion/surface interactions during film growth from the vapor phase: effect on nucleation and growth kinetics, defect structure, and elemental incorporation probabilities, in *Plasma-Surface Interactions and Processing of Materials*, ed. by O. Auciello, A. Gras-Marti, J.A. Valles-Abarca, L. Flamm (Kluwer Academic Publishers, Dordrecht 1990), pp. 281–311
48. I. Petrov, P.B. Barna, L. Hultman, J.E. Greene, Microstructural evolution during film growth. J. Vac. Sci. Technol. A **21**, S118–S128 (2003)
49. W. Ensinger, Low energy ion-assisted during deposition – an effective tool for controlling thin film microstructure. Nucl. Instr. Meth. Phys. Res. B **127**(128), 796–808 (1997)
50. L. Hanley, S.B. Sinnott, The growth and modification of materials via ion-surface processing. Surf. Sci. **500**, 500–522 (2002)

51. V. Grill, J. Shen, C. Evens, R.G. Cooks, Collisions of ions with surfaces at chemically relevant energies: Instrumentation and phenomena. Rev. Sci. Instr. **72**, 3149–3179 (2001)
52. S. Rauschenbach, R. Vogelsang, N. Malinowski, J.W. Gerlach, M. Benyoucef, G. Costan-TiNi, Z. Deng, N. Thontasen, K. Kern, Electrospay ion beam deposition: soft landing and fracmentation of functional molecules at solid surfaces. ASC Nano **3**, 2901–2910 (2009)
53. W. Ensinger, Ion bombardment effects during deposition of nitride and metal films. Surf. Coat. Technnol. **99**, 1–13 (1998)
54. J.E. Yehoda, B. Yang, K. Vedam, R. Messier, Investigation of the void structure in amorphous germanium film thin films as function of low-energy ion bombardment. J. Vac. Sci. Technol. A **6**, 1631–1635 (1988)
55. F. Rossi, B. André, A. van Veen, P.E. Mijnarends, H. Schut, F. Labohm, H. Dunlop, M.P. Delplancke, K. Hubbard, Physical properties of a-C: N films produced by ion beam-assisted deposition. J. Mater. Res. **9**, 2440–2449 (1994)
56. K. Kyuno, D.G. Cahill, R.S. Averbeck, J. Tarus, K. Nordlund, Surface defects and bulk defect migration produced by ion bombardment of Si(001). Phys. Rev. Lett. **83**, 4788–4791 (1999)
57. K. Yoneyama, K. Ogawa, Scanning tunneling microscope studies on recovery processes of sputter-induced surface defects on Si(111)-7x7. Jpn. J. Appl. Phys. **35**, 3719–3723 (1996)
58. O. Rodriguez de la Fuente, M.A. González, J.M. Rojo, Ion bombardment of reconstructed metal surfaces: from tow-dimensional dislocations dipoles to vacancy pits. Phys. Rev. B **63**, 085420 (2001)
59. T. Michely, G. Comsa, Generation and nucleation of adatoms during ion bombardment of Pt(111). Phys. Rev. B **44**, 8411–8444 (1991)
60. D.E. Harrison, Jr., P.W. Kelly, B.J.- Garrison, N. Winograd, Low energy ion impact phenomena on sigle crystal surfaces. Surf. Sci. **76**, 311–322 (1978)
61. T. Michely, J. Krug, *Islands, Mounds, and Atoms*, Vol. 42 (Springer Series in Surface Sience. Springer, Berlin 2004)
62. G. Rosenfeld, N.N. Lipkin, W. Wulfhekel, J. Kliewer, K. Morgenstern, B. Poelsema, G. Comsa, New concept for controlled homoepitaxy. Appl. Phys. A **61**, 455–466 (1995)
63. A.V. Dvurechenskii, J.V. Smagina, R. Groetschel, V.A. Zinovyev, V.A. Armbrister, P.L. Novikov, S.A. Teys, A.K. Gutakovskii, Ge/Si quantum dot nanostructures grown with low-energy ion beam assisted epitaxy. Surf. Coat. Technol. **196**, 25–29 (2005)
64. J.V. Smagina, V.A. Zinovyev, P.L. Novikov, V.A. Armbrister, E.S. Koptev, A.V. Dvurechenskii, Ge/Si nanostructures with quantum dots grown by ion beam assisted heteroepitaxy. J. Phys.: Conf. Series **245**, 012071 (2010)
65. K.A. Riekki, I.T. Koponen, Size selected growth of nanodots: Ion beam-assisted deposition and transition from 2D to 3D growth. Nucl. Instr. Meth. Phys. Res. B **290**, 48–53 (2012)
66. J. Frantz, M.O. Jahma, K. Nordlund, I.T. Koponen, Low-energy deposition of Co onto Co islands on Ag(100): effect on submonolayer growth. Phys. Rev. B **71**, 075411 (2005)
67. S. Esch, M. Breeman, M. Morgenstern, T. Michely, G. Comsa, Nucleation and morphology of homoepitaxial Pt(111)-films grown with ion beam-assisted deposition. Surf. Sci. **365**, 187–204 (1996)
68. A. Nikzad, H.A. Atwater, Microstructure modifications during early stages of ion-beam-asssited film growth of germanium om SiO_2. Suf. Coat. Technol. **51**, 243–246 (1992)
69. R. Ditchfeld, E.G. Seebauer, Semiconductor surface diffusion: Effects of low-energy ion bombardment, Phys. Rev. B **63**, 125317 (2001). Direct measurement of ion-influenced surface diffusion. Phys. Rev. Lett. **82**, 1185–1188 (1999)
70. Z. Wang, E.G. Seebauer, Extraordinary temperature amplification in ion-stimulated surface processes at low energies. Phys. Rev. B **66**, 205409 (2002). Temperature-dependent energy threshold for ion-stimulated defect formation in solids. Phys. Rev. Lett. **95**, 015501 (2005)
71. H.A. Atwater, Low energy ion-solid interactions during epitaxial growth. Solid State Phenom. **27**, 67–106 (1992)
72. V.O. Babaev, Ju. V. Bykov, M.B. Guseva, Effect of ion irradiation on the formation, structure and properties of thin metal films. Thin Solid Films **38** (1976), 1–8

73. S. Shimizu, O. Tsukakoshi, S. Komiya, Y. Makita, A molecular and ion beam epitaxy system for the growth of III-V compound semiconductors using a mass-separated low-energy group-V ion beam. Jap. J. Appl. Phys. **24**, 1130–1140 (1985)
74. T. Itoh, T. Nakamura, M. Muromachi, T. Sugiyama, Low temperature silicon epitaxy by partially ionized vapor deposition. Jap. J. Appl. Phys. **16**, 553–557 (1977)
75. G.K. Wehner, R.M. Warner Jr., P.D. Wang, Y.H. Kim, SubstituTiNg low-energy (< 30 eV) ion bombardment for elevated temperatures in silicon epitaxy. J. Appl. Phys. **64**, 6754–6760 (1988)
76. S.W. Park, J.Y. Shim, H.K. Baik, Low temperature epitaxial growth of $Si_{05}Ge_{05}$ alloy layer on Si(001) by ion beam-assisted deposition. J. Electro. Mater. **25**, 1399–1406 (1995)
77. L. Neumann, in *Anfangsstadium des ionenstrahlgestützten epitaktischen Wachstums von Galliumnitrid-Schichten auf Siliziumkarbid*. Dissertation, University of Leipzig
78. D. Poppitz, A. Lotnyk, J.W. Gerlach, J. Lenzner, M. Grundmann, B. Rauschenbach, An aberration corrected STEM study of structural defects in epitaxial GaN thin films grown by ion beam assisted MBE. Micron **73**, 1–8 (2015)
79. S. Sienz, J.W. Gerlach, T. Höche, A. Sidorenko, B. Rauschenbach, Ion beam-assisted molecular-beam epitaxy: a method to deposit gallium nitride films with high crystalline quality. Thin Solid Films **458**, 63–66 (2004)
80. A. Finzel, Ionenstrahlgestützte Molekularstrahlepitaxie von Galliumnitrid-Schichten auf Silizium. Dissertation, University Leipzig, 2016
81. L. Neumann, J.W. Gerlach, B. Rauschenbach, Initial stages of the ion beam assisted epitaxial GaN film growth on 6H-SiC(0001). Thin Solid Films **520**, 3936–3945 (2012)
82. B. Rauschenbach, A. Lotnyk, L. Neumann, D. Poppitz, J.W. Gerlach, Ion beam-assisted deposition of thin epitaxial GaN films. Materials **10**, 690–703 (2017)
83. B. Strickland, C. Roland, Low-temperature growth and ion-assisted deposition. Phys. Rev. B **51**, 5061–5064 (1995)
84. S. Blackwell, R. Smith, S.D. Kenny, J.M. Walls, Modeling evaporation, ion beam assist, and magnetron sputtering of thin metal films over realistic time scales. Phys. Rev. B **86**, 035416 (2012)
85. E. Chason, B.K. Kellerman, Monte Carlo simulations of ion-enhanced island coarsening. Nucl. Instr. Meth in Phys. Res. B **127**(128), 225–229 (1997)
86. J. Jacobson, B.H. Cooper, J.P. Sethna, Simulations of energetic beam deposition: from picoseconds to seconds. Phys. Rev. B **58**, 15847–15865 (1998)
87. B.S. Bunnik, C. de Hoog, E.F.C. Haddeman, B.J. Thijsse, Molecular dynamics study of Cu deposition on Mo and the effects of low-energy ion irradiation. Nucl. Instr. Meth B **187**, 57–65 (2002)
88. X.W. Zhou, D.A. Murdick, B. Gillespie, H.N.G. Wadley, Atomic assembly during GaN film growth: molecular dynamics simulation. Phys. Rev. B **73**, 045337 (2006)
89. H.-C. Petzold, P.J. Heard, Ion-induced deposition for X-ray mask repair: Rate optimization using a time-dependent model. J. Vac. Sc. Technol. B **9**, 2664–2669 (1991)
90. H.O. Funsten, Model for beam-induced deposition of thin metallic films, Nucl. Instr. Meth Phys. Res. B **72**, 183–196 (1992), and H.O. Funsten, J.W. Boring, R.E. Johnson, W.L. Brown, Low-temperature beam-induced deposition of thin films. J. Appl. Phys. **71**,1475–1484 (1992)
91. G. Carter, Island-density and size dynamics in ion-assisted atomic deposition. Vacuum **55**, 235–247 (1999)
92. I.T. Koponen, M. Rusanen, J. Heinonen, Island size distributions in submonolayer growth with mobile islands and breakup. Phys. Rev. E **58**, 4037–4040 (1998). M.O. Jahma, I.T. Koponen, T. Ala.Nissila, Optimal conditions for submonolayer nonoisland growth in ion beam-assisted deposition. Surf. Sci. **601**, 5628–5634 (2007)
93. I.T. Koponen, Modeling layer.by-layer growth in ion beam-assisted deposition of thin films, Nucl. Instr. Meth in Phys. Res. B **171**, 314–324 (2000)
94. R.C. Powell, N.E. Lee, Y.W. Kim, J.E. Greene, Heteroepitaxial wurtzite and zincblende structure GaN grown by reactive ion molecular beam epitaxy: Growth kinetics, microstructure, and properties J. Appl. Phys. **73**, 189–204 (1993)

95. Yu. V. Truskin, D.V. Kulikov, K.L. Safonov, J.W. Gerlach, Th. Höche, B. Rauschenbach, Atomic assembly during ion beam assisted growth: Kinetic modeling. J. Appl. Phys. **103**, 114904 (2008)
96. L. Hultmann, W.-D. Münz, J. Musil, K. Kadlec, I. Petrov, J.E. Greene, Low-energy (~100 eV) ion irradiation during growth of TiN deposited by reactive magnetron sputtering: effects of ion flux on film microstructure. J. Vac. Sci. Technol. A **9**, 434–438 (1991)
97. J.W. Gerlach, R. Schwertberger, D. Schrupp, S. Sienz, W. Attenberger, B. Rauschenbach, Influence of the ion irradiation during low-energy nitrogen ion-assisted deposition of wurtzitic gallium nitride films on sapphire, Proceed. Intern. Workshop on Nitride semiconductors, IPAP Conference Series **1**, 202–205 (2000)
98. J.W. Rabalais, A.H. Al-Bayati, K.J.Boyd, D. Marton, J. Kulik, Z. Zhang, W.K. Chu, Ion-energy effects in silicon ion beam epitaxy. Phys. Rev. B **53**, 10781–10792 (1996)
99. S. Mohajerzadeh, C.R. Selvakumar, D.E. Brodie, M.D. Robertson, J.M. Corbett, An ion beam vapor deposition technique for epitaxial growth of Si-Ge films on Si substrate. J. Vac. Sci. Technol. A **14**, 1963–1972 (1996)
100. D.K. Brice, J.Y. Tsao, S.T. Picraux, Partitioning of ion-induced surface and bulk displacements. Nucl. Instr. Meth. In Phys. Res. B **44**, 68–78 (1989)
101. Z.Q. Ma, Y.F. Zheng, B.X. Liu, Ion-induced surface and bulk displacement threshold for epitaxial growth. Phys. Stat. Sol. (A) **169**, 239–248 (1998). Sketching of preferred energy regime for ion beam assisted epitaxy, App. Surf. Sci. **137**, 184–190 (1999)
102. B.A. Movchan, A.V. Demchishin, Study of the structure and properties of thick vacuum condensates of nickel, titanium, tungsten, aluminum oxide and zirconium dioxide. Phys. Met. Metallogr. **28**, 83–90 (1969)
103. J.A. Thornton, Influence of apparatus geometry and deposition conditions on the structure and topography of thick sputtered coaTiNgs. J. Vac. Sci. Technol. **11**, 666–670 (1974)
104. R. Messier, A.P. Giri, R.A. Roy, Revised structure zone model for thin film physical structure. J. Vac. Sci. Technol. A **2**, 500–503 (1984)
105. F. Adibi, I. Petrov, J.E. Greene, L. Hultman, J.-E. Sundgren, Effects of high-flux low-energy (20–100 eV) ion irradiation during deposition on microstructure and preferred orientation of $Ti_{0.5}Al_{0.5}N$ alloys grown by ultrahigh-vacuum reactive magnetron sputtering. J. Appl. Phys. **73**, 8580–8589 (1993)
106. J.W. Gerlach, S. Sienz, W. Attenberger, B. Rauschenbach, Competition of epitaxy and ion beam irradiation-determined texture during ion beam-assisted deposition of gallium nitride films on r-plane sapphire. Phys. Stat. Sol. (c) **1**, 161–165 (2002)
107. J.M. Pomeroy, J. Jacobson, C.C. Hill, B.H. Cooper, J.P. Sethna, Kinetic Monte Carlo-molecular dynamics investigations of hyperthermal copper deposition on Cu(111). Phys. Rev. B **66**, 235412 (2002)
108. K. Vanormelingen, B. Degroote, H. Pattyn, A. Vantomme, Thin film growth using hyperthermal ions: a surface morphology study. Surf. Sci. **561**, 147–153 (2004)
109. T. Sikola, J. Spousta, L. Dittrichova, L. Benes, V. Perina, D. Rafaja, Ion beam-assisted deposition of metallic and ceramic thin films. Nucl. Instr. Meth. Res. Phys. B **127**(128), 673–676 (1997)
110. X.M. Yang, X.M. Wu, L.J. Zhuge, F. Zhou, Influence of assisted ion energy on surface roughness and mechanical properties of boron carbon nitride films synthesized by DIBSD. Appl. Surf. Sci. **255**, 4279–4282 (2009)
111. Z.H. Li, E.S. Cho, S.J. Kwon, Analysis of a MgO protective layer deposited with ion beam-assisted deposition in an AC PDP. J. Korean Phys. Soc. **49**, 2332–2337 (2006)
112. S. Miyata, A. Ibi, T. Izumi, Y. Shioara, Surface roughness of MgO thin film and its critical thickness for optimal biaxial texturing by ion beam-assisted deposition. J. Appl. Phys. **109**, 113922 (2011)
113. M. Marinov, Effect of ion bombardment on the initial stages of thin film growth. Thin Solid Films **46**, 267–274 (1977)
114. T.C. Huang, G. Lim, F. Parmigiani, E. Kay, Effect of ion bombardment during deposition on the x-ray microstructure of thin silver films. J. Vac, Sci. Technol. A **3**, 2161–2166 (1988)

115. R.A. Roy, J.J. Cuomo, D.S. Yee, Control of microstructure and properties of copper films using ion-assisted deposition. J. Vac. Sci. Technol. A **6**, 1621–1626 (1988)
116. W. Ensinger, B. Rauschenbach, Microstructural investigations on titanium nitride formed by medium energy ion beam-assisted deposition. Nucl. Instr. Meth. in Phys. Res. B **80**(81), 1409–1414 (1993)
117. W. Ensinger, The influence of the ion flux density on the properties of molybdenum films deposited from the vapour phase under simultaneous argon ion irradiation. Thin Solid Films **272**, 54–57 (1996)
118. K. Rajan, R. Roy, J. Trogolo, J.J. Cuomo, Low energy ion beam assisted grain size evolution in thin film deposition. J. Electr. Mater. **26**, 1270–1273 (1997)
119. H. Ma, Y. Zou, A.S. Sologubenko, R. Spolenak, Copper thin films by ion beam assisted deposition: strong texture, superior thermal stability and enhanced hardness. Acta Mater. **98**, 17–28 (2015)
120. J.M. López, F.J. Gordillo-Vázquez, O. Böhme, J.M. Albella, Low grain size TiN thin films obtained by low energy ion beam-assisted deposition. Appl. Surf. Sci. **173**, 290–295 (2001)
121. K. Helming, B. Rauschenbach, R.A. Schwarzer, Analysis of crystallographic texture in small sample areas. Textures Microstructures **26–27**, 111–124 (1996)
122. B. Rauschenbach, K. Helming, Implantation-induced texture. Instr. Meth. in Phys. Res. B **42**, 216–223 (1989)
123. D. Dobrev, Ion beam-induced texture formation in vacuum-condensed thin metal films. Thin Solid Films **92**, 41–53 (1982)
124. Y. Iijima, N. Tanabe, O. Kohno, Y. Ikeno, In-plane aligned $YBa_2Cu_3O_{7-x}$ thin films deposited on polycrystalline metallic substrates. Appl. Phys. Lett. **60**, 769–771 (1992)
125. L.S. Yu, J.M.E. Haper, J.J. Cuomo, D.A. Smith, Alignment of thin films by glancing angle ion bombardment during deposition. Appl. Phys. Lett. **47**, 932–933 (1985)
126. J.J. Trillat, L. Tertain, Etude des structures superficielles par diffraction électronique et bombardement ionique combinés. Applications. J. Chim. Phys. Phys.-Chim. Biol. **53**, 570–578 (1956)
127. G.J. Ogilvie, The structure of silver crystals after argon-ion bombardment. J. Phys. Chem. Solids **10**, 222–228 (1959)
128. G.J. Ogilvie, A.A. Thomson, Influence of temperature and bombardment rate on disorientation of silver single crystals by ion bombardment. J. Phys. Chem. Solids **17**, 203–209 (1961)
129. G.N. van Wyk, N.J. Smith, Ion bombardment induced preferential orientation in polycrystalline Cu Targets. Rad. Eff. **38**, 245–247 (1987)
130. B. Rauschenbach, J.W. Gerlach, Texture development in titanium nitride thin films grown by low-energy ion deposition. Cryst. Res. Technol. **35**, 675–688 (2000)
131. H. Wengenmair, J.W. Gerlach, U. Preckwinkel, B. Stritzker, B. Rauschenbach, Photon and ion beam-assisted deposition of titanium nitride. Appl. Surf. Sci. **99**, 313–318 (1996)
132. L. Hultman, J.-E. Sundgren, J.E. Greene, D.B. Bergstrom, I. Petrov, High-flux low-energy (20 eV) N^+_2 ion irradiation during TiN deposition by reactive magnetron sputtering: effects on microstructure and preferred orientation. J. Appl. Phys. **78**, 5395–5403 (1995)
133. D. Edström, D.G. Sangiovanni, L. Hultman, I. Petrov, J.E. Greene, V. Chirita, TiN film growth on misoriented TiN grains with simultaneous low-energy bombardment: restructuring leading to epitaxy. Thin Solid Films **688**, 137380 (2019)
134. D. Gall, I. Petrov, L.D. Madson, J.-E. Sundgren, J.E. Greene, Microstructure and electronic properties of the refractory semiconductor ScN grown on MgO(001) by ultra-high-vacuum reactive magnetron sputter deposition. J. Vac. Sci. Technol. A **16**, 2411–2417 (1998)
135. J.W. Gerlach, U. Preckwinkel, H. Wengenmair, T. Kraus, B. Rauschenbach, Biaxial alignment of TiN films prepared by ion beam-assisted deposition. Appl. Phys. Lett. **68**, 2360–2362 (1996)
136. J. Dzick, S. Sievers, J. Hoffmann, K. Thiele, F. Garcia-Moreno, A. Usoskin, C. Jooss, H.C. Freyhardt, Biaxially textured buffer layers on large-area polycrystalline substrates. Mat. Soc. Symp. Proc. **595**, 55–66 (2000)
137. K.G. Ressler, N. Sonneenberg, M.J. Cima, The development of biaxial alignment in yttria-stabilized zirconia films fabricated by ion beam-assisted deposition. J. Electr. Mater. **25**,

35–43 (1996). Mechanism of biaxial alignment of oxide thin films during ion beam-assisted deposition J. Am. Ceram. Soc. **80**, 2637–264 (1997)
138. A. van der Drift, Evolutionary selection, a principle governing growth orientation in vapour-deposited layers. Philips Res. Rep. **22**, 267–288 (1967)
139. C.W. Thompson, Structure evolution during Processing of polycrystalline films. Annu. Rev. Mater. Sci. **30**, 159–190 (2000)
140. M. Kiuchi, A. Chayahara, Y. Horino, K. Fujii, M. Satou, H.J. Kang, Y.W. Beag, Y. Kimura, R. Shimizu, Preferentially orientated crystal growth in dynamic mixing process—an approach by Monte Carlo simulation. Jap. J. Appl. Phys. **29**, 2059–2065 (1990)
141. R.T. Brewer, H.A. Atwater, Rapid biaxial texture development during nucleation of MgO thin films during ion beam-assisted deposition. Appl. Phys. Lett. **80**, 3388–3390 (2002)
142. R.M. Bradley, J.M.E. Harper, D.A. Smith, Theory of Thin Film orientation by ion bombardment during deposition. J. Appl. Phys. **60**, 4160–4164 (1986)
143. B.R. Appelton, G. Foti, Channeling, *in Ion Beam Handbook for Material Analysis.* ed. by J.W. Mayer, E. Rimini, (Academic Press, New York, 1977)
144. L. Dong, D.J. Srolovitz, Texture development mechanism in ion beam-assisted deposition. J. Appl. Phys. **84**, 5261–5269 (1998)
145. G. Knuyt, C. Quaeyhaegens, J. D'Heans, L.M. Stals, A quantitative model for the evolution from random orientation to unique texture in PVD thin film formation. Thin Solid Films **258**, 159–169 (1995)
146. D.S. Rickerby, A.M. Jones, B.A. Bellamy, Internal stress in titanium nitride coatings: modelling of complex stress systems. Surf. Coat Technol. **36**, 661–674 (1988)
147. J. Pelleg, L.Z. Zervin, S. Lungo, N. Croitoru, Reactive-sputter-deposited TiN films on glass substrates. Thin Solid Films **197**, 117–128 (1991)
148. U.C. Oh, J.H. Je, Effects of strain energy on the preferred orientation of TiN thin films. J. Appl. Phys. **74**, 1692–1696 (1993)
149. H. Windischmann, An intrinsic stress scaling law for polycrystalline thin films prepared by ion beam sputtering, J. Appl. Phys. **62**, 1800–1807 (1987). Intrinsic stress in sputter-deposited thin films. Crit. Rev. Solid State Mat. Sci. **17**, 547–596 (1992)
150. C.A. Davis, A simple model for the formation of compressive stress in thin films by ion bombardment. Thin Solid Films **226**, 30–34 (1993)
151. M.F. Doerner, W.D. Nix, Stress and deformation processes in thin films on substrates, CRC Crit. Rev. Solid St. Mater. Sci. **14**, 225–268 (1988)
152. J.F. Nye, *Physical Properties of Crystals* (University Oxford Press, Oxford, 1957)
153. G. Abadias, Y.Y. Tse, Diffraction stress analysis in fiber-textured TiN thin films grown by ion beam sputtering: application to (001) and mixed (001)+(111) texture. J. Appl. Phys. **95**, 2414–2428 (2004)
154. V. Valvoda, J. Musil, X-ray analysis of strain in titanium nitride layers. Thin Solid Films **149**, 49–60 (1989)
155. J.P. Zhao, X. Wang, Z.Y. Chen, S.Q. Yang, T.S. Shi, X.H. Liu, Overall energy model for preferred growth of TiN films during filtered arc deposition. J. Phys. D: Appl. Phys. **30**, 5–12 (1997)
156. D.R. McKenzie, Y. Yin, W.D. McFall, N.H. Hoang, The orientation dependence of the elastic strain energy in cubic crystals and its application to the preferred orientation in titanium nitride thin films. J. Phys.: Condens. Matter **8**, 5883–5890 (1996)
157. S. Mahieu, P. Ghekiere, D. Depla, R.D. Gryse, Biaxial alignment in sputter deposited thin films, Thin Solid Films **515**, 1229–1249 (2006). S. Mahieu, D. Depla, Reactive sputter deposition of TiN layers: modelling the growth by characterization of particle fluxes towards the substrate. J. Phys. D: Appl. Phys. **42**, 053002 (2009)
158. J.D. Targove, H.A. Macleod, Verification of momentum transfer as the dominant densifying mechanism in ion-assisted deposition. Appl. Optics **27**, 3779–3781 (1988)
159. D.M. Mattox, G.J. Kominiak, Structure modification by ion bombardment during deposition. J. Vac. Sci. Technol. **9**, 528–532 (1972)

160. P.J. Martin, R.P. Netterfield, W.G. Sainty, Modification of the optical and structural properties of dielectric ZrO_2 films by ion-assisted deposition. J. Appl. Phys. **55**, 235–241 (1984)
161. K.H. Guenther, B. Loo, D. Bruns, J. Edgell, D. Widham, K.-H. Müller, Microstructure analysis of thin films deposited by reactive evaporation and reactive ion-plating, Proc. SPIE **1019**, 73–83 (1988), see also J. Vac. Sci. Technol. A **7**, 1436–1445 (1989)
162. H.-P. Kaukonen, R.M. Nieminen, Moleclar-dynamics simulation of the growth of diamond.like films by energetic carbon-atom beams. Phys. Rev. Lett. **68**, 620–622 (1992)
163. K.-H. Müller, Modelling ion-assisted deposition of CeO_2 films. Appl. Phys. A **40**, 209–213 (1986)
164. K.-H. Müller, Ion beam-induced epitaxial vapour-phase growth: A molecular-dynamics study. Phys. Rev. B **35**, 7906–7913 (1987)
165. G.I. Grigorov, I.N. Martev, J.-P. Langeron, J.-L. Vignes, A choice of the optimum density of ion bombardment by ion-assisted physical vapor deposition of films. Thin Solid Films **161**, 249–256 (1988)
166. G.I. Grigorov, K.G. Grigorov, R. Sporken, R. Caudano, Ion-induced densification of pvd films–a choice of the optimum density of ion bombardment. Appl. Phys. A **63**, 399–401 (1996)
167. G. Carter, Peening ion-assisted thin-film deposition: a generalized model. J. Phys. D: Appl. Phys. **27**, 1046–1055 (1994)
168. I.C. Noyan, J.B. Cohen, *Residual stress: Measurement by Diffraction and Interpretation* (Springer, Berlin, 1987)
169. C.A. Volkert, Stress and plastic flow in silicon during amorphization by ion bombardment. J. Appl. Phys. **70**, 3521–3527 (1991)
170. M. Zeitler, S. Sienz, B. Rauschenbach, Study of stress evolution of boron nitride films prepared by ion-assisted deposition. J. Vac. Sci. Technol. A **17**, 597–602 (1999)
171. G.G. Stoney, The tension of metallic films deposited by electrolysis. Proc. Royal Soc. London, Ser. A **82**, 172–175 (1909)
172. L.B. Freund, S. Suresh, *Thin Film Materials: Stress, Defect Formation and Surface Evolution* (Cambridge University Press, Cambridge, 2003)
173. J. Keckes, J.W. Gerlach, R. Averbeck, H. Riechert, S. Bader, B. Hahn, H.-J. Lugauer, A. Lell, V. Härle, A. Wenzel, B. Rauschenbach, Temperature dependence of stresses in GaN thin films grown on (0001) sapphire: Modeling of thermal stresses. App. Phys. Lett. **79**, 4307–4309 (2001)
174. H. Trinkaus, A.I. Ryazanov, Viscoelastic model for the plastic flow of amorphous solids under energetic ion bombardment. Phys. Rev. Lett. **74**, 5072–5074 (1995)
175. K. Dahmen, M. Giesen, J. Ikonomov, K. Starbova, H. Ibach, Steady-state surface stress induced in noble gas sputtering. Thin Solid Films **428**, 6–10 (2003)
176. N. Kalyanasundaram, M.C. Moore, J.B. Freund, H.T. Johnson, Stress evolution due to medium-energy ion bombardment of silicon, Acta Mater. **54**, 483–491 (2006). N. Kalyanasundaram, M. Wood, J.B. Freund, H.T. Johnson, Stress evolution to steady state in ion bombardment of silicon. Mechan. Res. Comm. **15**, 50–56 (2008)
177. W.L. Chan, E. Chason, C. Iamsumang, Surface stress induced in Cu foils during and after low energy ion bombardment, Nucl.Instr. Meth.in Phys. Res. B **257**, 428–432 (2007). W.L. Chan, E. Chason, Stress evolution and defect diffusion in Cu during low energy ion irradiation: experiments and modeling. J. Vac. Sci. Technol. A **26**, 44–51 (2008)
178. W.L. Chan, K. Zhao, N. Vo, Y. Ashkenazy, D.G. Cahill, R.S. Averback, Stress evolution in platinum thin films during low-energy ion irradiation. Phys. Rev. B **77**, 205405 (2008)
179. T. Edler, S.G. Mayr, Mechanisms of stress generation during bombardment of Ge with keV ions: experiments and molecular dynamics simulations. New J. Phys. **9**, 325 (2007)
180. Y. Ishii, C.S. Madi, M.J. Aziz, E. Chason, Stress evolution in Si during low energy ion bombardment. J. Mater. Res. **29**, 2942–2948 (2014)
181. F.M. d'Heurle, Aluminium films deposited by rf sputtering. Metall. Mater. Trans. **1**, 725–732 (1970)

182. J.A. Thornton, D.W. Hoffman, Stress-related effects in thin films. Thin Solid Films **171**, 5–31 (1989)
183. D.W. Hoffman, J.A. Thornton, Compressive stress and inert gas in Mo films sputtered from a cylindrical-post magnetron with Ne, Ar, Kr, and Xe. J. Vac. Sci. Technol. **17**, 380–383 (1980)
184. N. Kuratani, O. Imai, A. Ebe, S. Nishiyama, K. Ogata, Internal stress in thin films prepared by ion beam and vapor deposition. Surf. Sci. Technol. **66**, 310–312 (1994)
185. M. Zeitler, Synthese and Spannungsanalyse von Bornitridschichten, hergestellt mittels ionengestützter Deposition. Disseration, University of Augsburg (1999)
186. V. Dietz, P. Ehrhart, D. Guggi, H.-G. Haubold, W. Jäger, M. Prieler, W. Schilling, Influence of ion bombardment during deposition on mustructure of evaporated aluminum films. Nucl. Inst. Meth. in Phys. Res. B **59**(60), 284–287 (1991)
187. W. Hoffman, M.R. Gaerttner, Modification of evaporated chromium by concurrent ion bombardment. J. Vac. Sci. Technol. **17**, 425–428 (1980)
188. Y. Pauleau, Generation and evolution of residual stress in physical vapour-deposited thin films. Vacuum **61**, 175–181 (2001)
189. R. Abermann, R. Kramer, J. Mäser, Structure and internal stress in ultra-thin silver films deposited on MgF_2 and SiO. Thin Solid Films **52**, 215–229 (1978)
190. R.W. Hoffman, Stress in thin films: The relevance of grain boundaries and impurities. Thin Solid Films **34**, 185–190 (1976)
191. W.D. Nix, B.M. Clemens, Crystallite coalescence: A mechanism for intrinsic tensile stresses in thin films. J. Mater Res. **14**, 3467–3473 (1999)
192. G.K. Wolf, Modification of chemical properties by ion beam-assisted deposition. Nucl. Instr. Meth. In Phys. Res. B **46**, 369–378 (1990)
193. E.H. Hirsch, I.K. Varga, Thin film annealing by ion bombardment. Thin Solid Films **69**, 99–105 (1980)
194. D.R. Brighton, G.K. Hubler, Binary collision cascade prediction of critical ion-to-atom arrival ratio in the production of thin films with reduced intrinsic stress. Nucl. Instr. Meth. in Phys. Res. B **28**, 527–533 (1987)
195. W.D. Wilson, L.G. Haggmark, J.P. Biersack, Calculations of nuclear stopping, range, and straggling in the low-energy region. Phys. Rev. B **15**, 2458–2468 (1977)
196. E. Chason, M. Karlson, J.J. Colin, D. Magnfält, K. Sarakinos, G. Abdadias, A kinetic model for stress generation in thin films grown from energetic vapor fluxes. J. Appl. Phys. **119**, 145307 (2016)
197. J. Robertson, Deposition mechanisms for promoting sp^3 bonding in diamond-like carbon. Diamond Rel. Mater. **2**, 984–989 (1993)
198. Y. Lifshitz, S.R. Kasi, J.W. Rabalais, Subplantation model for film growth from hyperthermal species. Phys. Rev. B **41**, 10468–10480 (1990) and Subplantation model for film growth from hyperthermal Species: application to diamond. Phys. Rev. Lett. **62**, 1290–1293 (1989)
199. F. Seitz, S. Koehler, Displacement of Atoms during Irradiation, *in Solid State Physics*, ed. by F. Seitz, D. Turnbull, Vol. 2, (Academic Press, New York, 1956), pp. 307–442
200. L. Dumas, E. Quesnel, J.Y. Robic, Y. Pauleau, Characterization of magnesium fluoride thin films produced by argon ion beam-assisted deposition. Thin Solid Films **382**, 61–68 (2001)
201. K.J. Boyd, D. Marton, J.W. Rabalais, S. Uhlmann, T. Frauenheim, Semiquantitative subplantation model for low energy ion interactions with surfaces. I. Noble gas ion-surface interactions. J. Vac. Sci. Technol. A **16**, 444–462 (1998)
202. G. Knuyt, W. Lauwerens, L.M. Stals, A unified theoretical model for tensile and compressive residual film stress. Thin Solid Films **370**, 232–237 (2000)
203. V.l. Popov, R. Pohrt, Q. Li, Strength of adhesive contacts: Influence of contact geometry and material gradients. Friction **5**, 308–325 (2017)
204. G.K. Wolf, M. Barth, W. Ensinger, Ion beam-assisted deposition for metal finishing. Nucl. Instr. Meth. in Phys. Res. B **37**(38), 682–687 (1989)
205. T.N. Kim, Q.L. Feng, Z.S. Luo, F.Z. Cui, J.O. Kim, Highly adhesive hydroxyapatite coatings on alumina substrates prepared by ion beam-assisted deposition. Surf. Coat. Technol. **99**, 20–23 (1998)

206. J.-M. Choi, H.-E. Kim, I.-S. Lee, Ion beam-assisted deposition (IBAD) of hydroxyapatite coating layer on Ti-based metal substrate. Biomaterials **21**, 469–473 (2000)
207. J. Zhu, Y. Hu, M. Xu, W. Yang, L. Fu, D Li, L. Zhou, Enhancement of the adhesive strength between Ag films and Mo substrate by Ag implanted via ion beam-assisted deposition. Materials **11**, 762 (2018)
208. L.-H. Tian, B. Tang, D.-X. Liuc, X.-D. Zhua, J.-W. He, Interfacial reactions duringIBAD and their effects on the adhesion of Cr–N coatings on steel. Surf. Coat. Technol. **191**, 149–154 (2005)
209. A.J. Kellock, J.E.E. Baglin, T.T. Bardin, J.G. Pronko, Adhesion improvement of Au on GaAs using ion beam-assisted deposition. Nucl. Instr. Meth in Phys. Res. B **59**(60), 249–253 (1991)
210. Y.-T. Cheng, Thermodynamic and fractal geometric aspects of ion-solid interactions. Mater. Sci. Rep. **5**, 54–97 (1990)
211. W. Bolse, Ion beam-induced atomic transport through bi-layer interfaces of low- and medium-Z metals and their nitrides. Mater. Sci. Engng. R **12**, 53–122 (1994)
212. B. Rauschenbach, M. Posselt, R. Grötschel, E. Brecht, G. Linker, O. Meyer, Ion mixing of Cu/Ti and Cu/Fe bilayers. Nucl. Instr. Meth in Phys. Res. B **69**, 277–280 (1992). B. Rauschenbach, W. Erfurth, G. Linker, O. Meyer, Ion beam mixing of Cu/Ti bilayers. Nucl. Instr. Meth in Phys. Res. B **68**, 438–442 (1992)
213. P. Sigmund, A. Gras-Marti, Theoretical aspects of atomic mixing by ion beams. Nucl. Instr. Meth. **182**(183), 25–41 (1981)
214. S. Aisenberg, R. Chabot, Ion beam deposition of thin films of diamondlike carbon. J. Appl. Phys. **42**, 2953–2958 (1972)
215. C. Weissmantel, Reactive film preparation. Thin Solid Films **31**, 11–18 (1976)
216. P.B. Mirkarimi, K.F. McCarty, D.L. Medlin, Review of advances in cubic boron nitride film synthesis. Mater. Sci. Engng. R **21**, 47–100 (1997)
217. W. Kulisch, S. Ulrich, Parameter spaces for nucleation and the subsequent growth of cubic boron nitride films. Thin Solid Films **423**, 183–195 (2003)
218. D.J. Kester, K.S. Ailey, R.F. Davis, Deposition and characterization of boron nitride thin films. Diamond Rel. Mater. **3**, 332–336 (1994)
219. J. Thomas, N.E. Weston, T.E. O'Connor, Turbostratic boron nitride, thermal transformation to order-layer-lattice boron nitride. J. Am. Chem. Soc. **84**, 4619–4622 (1963)
220. D.V. Shtansky, O. Tsuda, Y. Ikuhara, T. Yoshida, Crystallography and structural evolution of cubic boron nitride films during bias sputter deposition. Acta mater. **48**, 3745–3759 (2000)
221. D. Bouchier, G. Sené, M.A. Djouadi, P. Möller, Effect of noble gas ions on the synthesis of c-BN by ion beam-assisted deposition. Nucl. Instrum. Meth. in Phys Res. B **89**, 369–372 (1994)
222. T.A. Friedmann, P.B. Mirkarimi, D.L. Medlin, K.F- McCarty, E.J. Klaus, D.R. Boehme, H.A. Johnsen, M.J. Mills, D.K. Ottesen, J.C. Barbour, Ion-assisted pulsed laser deposition of cubic boron nitride films. J. Appl. Phys. **76**, 3088–3101 (1994)
223. S. Reinke, M. Kuhr, W. Kulisch, R. Kassing, Recent results in boron nitride deposition in light of sputter model. Diamond Rel. Mater. **4**, 272–283 (1995)
224. H. Hofsäss, H. Feldermann, R. Merk, M. Sebastian, C. Ronning, Cylindrical spike model for the formation of diamond like thin films by ion deposition. Appl. Phys. A **66**, 153–181 (1998)
225. G. H. Vineyard, Thermal spikes and activated processes. Rad. Effects Defects in Solids **29**, 245–248 (1976)
226. S. Reinke, M. Kuhr, W. Kulisch, Mechanism in ion induced c-BN growth. Diamond Rel. Mater. **3**, 342–345 (1994)
227. D.R. McKenzie, W.D. McFall, W.G. Sainty, C.A. Davies, R.E. Collins, Compressive stress induced formation of cubic boron nitride. Diamond Rel. Mater. **2**, 970–976 (1993)
228. P.B. Mirkarimi, K.F. McCarty, D.L. Medlin, W.G. Wolfer, T.A. Friedmann, R.J. Klaus, G.F. Cardinale, D.G. Howitt, On the role of ions in the formation of cubic boron nitride films by ion-assisted deposition. J. Mater. Res. **9**, 2925–2938 (1994)
229. D.J. Kester, R. Messier, Phase control of cubic boron nitride thin films. J. Appl. Phys. **72**, 504–513 (1992)

230. J. Robertson, The deposition mechanism of diamondlike a-C and a-C:GH. Diamond Rel. Mater. **3**, 361–368 (1994)
231. D. Marton, K.J. Boyd, I.W. Rabalais, Y. Lifshitz, Semiquantitative subplantation model for low energy ion interactions with surfaces. II. Ion beam deposition of carbon and carbon nitride. J. Vac. Sci. Technol. A. **16**, 455–462 (1998)
232. K.J. Boyd, D. Marton, J.W. Rabalais, S. Uhlmann, T. Frauenheim, Semiquantitative subplantation model for low energy ion interactions with surfaces. III. Ion beam homoepitaxy of Si. J. Vac. Sci. Technol. A **16**, 463–471 (1998)
233. H.K. Pulker, G. Paesold, E. Ritter, Refractive indices of TiO_2 films produced by reactive evaporation of various titanium-oxygen phases. Appl. Optics **15**, 2986–2991 (1976)
234. P.J. Martin, Ion based methods for optical thin film deposition. J. Mater Sci. **21**, 1–25 (1986)
235. H.-C. Chen, K.-S. Lee, C.-C. Jaing, C.-C. Lee, Enhancement of the stability of ion-assisted TiO_2 thin films by using re-annealing with different substrate temperatures. Proc. SPIE **7067**, 70670O (2008)
236. G. Wang, J.R. Brewer, F. Namavar, R.F. Sabirianov, H. Haider. K. L. Garvin, C.L. Cheung, Structural study of titanium oxide films synthesized by ion beam-assisted deposition. **30**, 59–64 (2008)
237. C.-C. Jaing, H.-C. Chen, C.-C. Lee, Effects of thermal annealing on titanium oxide films prepared by ion-assisted deposition. Opt. Rev. **16**, 396–399 (2009)
238. C. Yang, H. Fan, Y. Xi, J. Chen, Z. Li, Effects of depositing temperatures on structure and optical properties of TiO_2 film deposited by ion beam assisted electron beam evaporation. Appl. Surf. Sci. **254**, 2685–2789 (2008)
239. F. Zhang, Z. Zheng, Y. Chen, D. Liu, X. Lu, Study on the effect of ion beam bombardment during deposition on preferred orientation in rutile-type titanium dioxide. J. Appl. Phys. **83**, 4101–4105 (1998)
240. K. Yokota, Y. Yano, K. Nakamura, M. Ohnishi, F. Miyashita, Effects of oxygen ion beam application on crystalline structures of TiO_2 films deposited on Si wafers by an ion beam-assisted deposition. Nucl. Instr. Meth. B **242**, 393–395 (2002)
241. I.N. Martev, Oxygen-ion-assisted deposition of TiO films. Vacuum **58**, 327–334 (2000)
242. J.R. McNeil, A.C. Barron, S.R. Wilson, W.C. Herrmann, Ion-assisted deposition of optical thin films: low energy vs high energy bombardment. Appl. Optics **23**, 552–559 (1984)
243. A.M. Ektessabi, Ion beam-assisted sputter deposition of thin oxide films. Surf. Coat. Technol. **68–69**, 208–216 (1994)
244. S. Mohan, M.G. Krishan, A review of ion beam-assisted deposition of optical thin films. Vacuum **46**, 645–659 (1995)
245. P.J. Martin, H.A. MacLeod, R.P. Netterfield, C.G. Pacey, W.G. Sainty, Ion beam-assisted deposition of thin films. Appl. Optics **22**, 178–184 (1983)
246. Y. Yamada, H. Uyama, T. Murata, H. Nozoye, Low temperature deposition of titanium-oxide films with refractive indices by oxygen-radical beam assisted evaporation combined with ion beams. J. Vac. Sci. Technol. A **19**, 2479–2482 (2001)
247. J.M. Bennett, E. Pelleier, G. Albarnd, J.P. Borgogno, C. Carniglia, R.A. Schmell, T.H. Allen, T. Tuttle-Hart, K.H. Guenther, A. Saxer, Comparison of properties of tirtanium dioxide films prepared by various techniques. Appl. Optics **28**, 3303–3317 (1989)
248. Q. Tang, K. Kikuchi, S. Ogura, A. Macleod, Mechanism of columnar microstructure growth in titanium oxide thin films deposited by ion beam-assisted deposition. J. Vac. Sci. Technol. A **17**, 3379–3384 (1999)
249. F.L. Williams, D.E. Reicher, C.B. Jung, J.R. McNeil, Metal dioxides using ion-assisted deposition at low temperature. J. Vac. Sci. Technol. **7**, 2286–2288 (1989)
250. H. Nan, Y. Ping, C. Xuan, L. Yongxang, Z. Xiaolan, C. Guangjun, Z. Zihong, Z. Feng, C. Yuanru, L. Xianghuai, X. Tingfei, Blood compatibility of amorphous titanium oxide films synthesized by ion enhanced deposition. Biomaterials **19**, 771–776 (1998)
251. F. Zhang, X. Liu, Y. Mao, N. Huang, Y. Chen, Z. Zheng, Z. Zhou, A. Chen, Z. Jiang, Artificial heart valves: improved hemocompatibility by titanium oxide coatings prepared by ion beam-assisted deposition. Surf. Coating Technol. **103–104**, 146–150 (1998)

Chapter 11
Ion Beam Sputtering Induced Glancing Angle Deposition

Abstract The method of ion beam sputtering under glancing angle conditions in combination with an additional rotation of the sample holder allows the growth of almost arbitrarily designed nano- and microstructures of all material classes on surfaces. The self-shadowing and the surface diffusion essentially govern the structure evolution. It is demonstrated that by varying the particle incidence angle, the temperature, azimuthal rotation frequency, and the beam divergence of the sputtered particles, a wide variety of nanostructure morphologies (e.g., slanted and vertical columns, screws, spirals, or zigzag columns) can be generated. Ballistic simulations are preferably used to simulate the growth of these structures. It can be shown that two basic alternatives of ballistic simulations, off-lattice simulations and on-lattice simulations, are available to successfully model growth. A remarkable result of all experimental investigations and computer simulations is that the column tilt angle is always smaller than the incidence angle. Various explanations are known to explain this fact. These models will be presented and it will be shown that especially the competition model is able to describe a relation between the tilt angle and the angle of incidence for the complete range of material incidence angles. For various applications, patterning of the substrate prior to growth is required to fabricate arrays for highly regular nanostructures. This fabrication is demonstrated and the application of these structures for the realization of biosensors and magnetic nanotubes is shown.

A fascinating ion sputtering method for fabricating nano- and microstructures is based on GLancing Angle Deposition (GLAD). This technique provides high-precision control over nanostructures and microstructures ranging from simple vertical posts to polygonal microstructures. This method, a constructive way to fabricate nanostructures, features the deposition of vapor-deposited or sputtered particles on a substrate at a large deposition angle with respect to the normal of the substrate surface, and also provides the ability to rotate the substrate about an axis perpendicular to the substrate. By controlling the polar and azimuthal rotation of the substrate, a wide variety of dimensions and shapes of nanostructures can be realized. In Fig. 11.1, the method is schematically presented. The most important mechanism in the GLAD process is ballistic shadowing between the initial nucleation points. This shading prevents the deposition of material in regions, called shadow regions, behind the

B. Rauschenbach, *Low-Energy Ion Irradiation of Materials*, Springer Series in Materials Science 324, https://doi.org/10.1007/978-3-030-97277-6_11

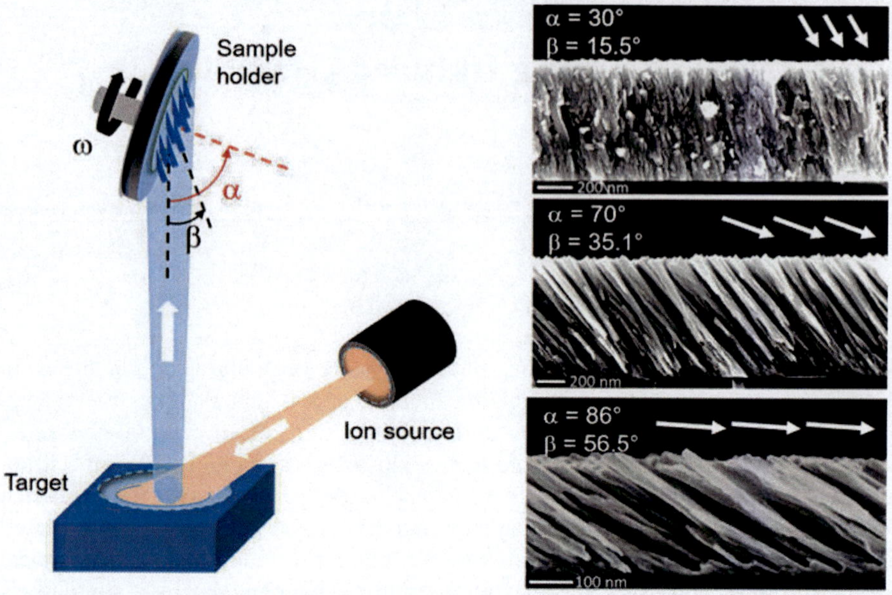

Fig. 11.1 Left: Schematic configuration of sputter-induced glancing angle deposition. The angles α and β characterize the tilt angle (≡ incidence angle of sputtered particles) and the growth angle of generated columnar structures with respect to the surface normal, respectively. The rotation frequency ω describes the number of rotation around the surface normal per unit time. Right: SEM cross-section image of Mo film on Si substrate deposited by ion sputtering induced glancing angle deposition at different incidence angles α of sputtered particles

initial nucleation sites. In the absence of adatom diffusion, the incident material is deposited at the point of impact, creating oblique columnar nanostructures and highly porous films. Additional rotation of the sample holder allows vertical growth of nanocolumns or other nanostructures.

The GLAD process is a physical vapor deposition (PVD) process, i.e. the desired material is first evaporated or sputtered, then transported through vacuum and finally deposited on a substrate surface at oblique angles. The most commonly used vapor deposition methods are resistive heating or electron beam bombardment of the material in a high vacuum. Sputtering techniques to deposit the material are also commonly used. In particular, DC (diode or direct current) and RF (radio frequency) magnetron configurations are used to produce GLAD coatings. Sputter-induced GLAD is a simple, cost-effective, and flexible technique for fabricating nano- and microstructures and highly porous films on large surfaces, regardless of the material type. The major drawbacks of these sputtering techniques are the high process pressure (> 0.1 Pa) and the large flux divergence of the sputtered material. Comprehensive reviews of the GLAD method can be found in [1–8].

This chapter is exclusively focused on the *ion beam sputtering method under glancing angle conditions* (Fig. 11.1) to produce nanostructures and highly porous films. In this case, the GLAD process can be carried out under high vacuum conditions and the divergence of the sputtered particle flux is almost comparable to the evaporation process. In the following, ion sputtering induced glancing angle deposition is introduced and some selected results and potential applications are presented.

11.1 Basic Mechanisms of Oblique Deposition

The conventional deposition technique is usually characterized by perpendicular impingement of vapor particles on the surface of a substrate. The morphology of these layers is characterized by a columnar structure and can be described by the three-zone model proposed by Movchan-Demchishin [9]. Within this model, a columnar morphology (see Fig. 11.1, right) can be observed for homologous temperatures $T/T_M < 0.3$, which are relevant for the deposition experiments at very oblique angles (T is the temperature of the substrate and T_M is the melting temperature of the deposited material, both given in unit of [K]).

If the vapor particles or sputtered particles are incident at a larger angle (angle of incidence $\alpha > 70°$) to the surface normal of the substrate, the formation of columns can be expected. This type of physical vapor deposition, characterized by large angles of incidence α, is called Oblique Angle Deposition (OAD). The processes involved in the growth of oblique columns are shown schematically in Fig. 11.1. The incident particle flux arrives the substrate at an angle α and forms randomly distributed film nuclei on the surface (Fig. 11.2, left). These nuclei grow and form columns tilted in the direction of the incident particles. As a result, the tilted columns shadow neighboring columns from the incident particle stream (dashed lines, Fig. 11.2, middle and right). This self-shadowing at the atomic level has the consequence that no further particles penetrate into the shadowed areas and thus they remain unfilled. Consequently, a porous morphology of the deposited layers can be expected. With increasing layer thickness, smaller columns are completely shadowed, i.e. the growth of these columns is stopped (Fig. 11.2, right). Figure 11.2 also demonstrates that the columns grow with a certain tilt angle β, where this tilt angle is defined as the angle between the substrate normal and the column center axis. A remarkable result of all experimental investigations and computer simulations is that the column tilt angle, β, is always smaller than the incidence angle α, i.e. $\beta < \alpha$ (for discussion see Sect. 11.4).

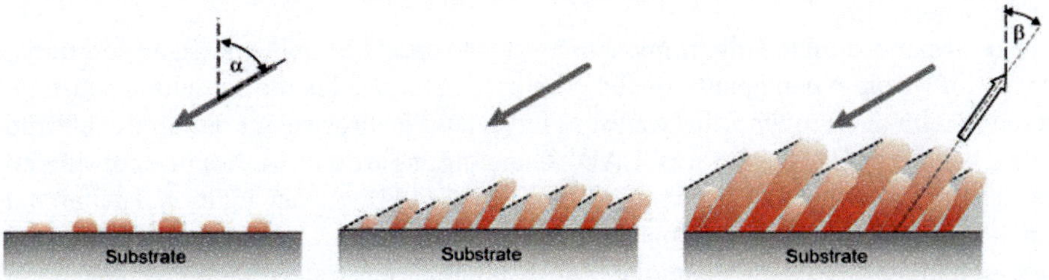

Fig. 11.2 Schematic processes of film growth by oblique angle deposition (OAD)

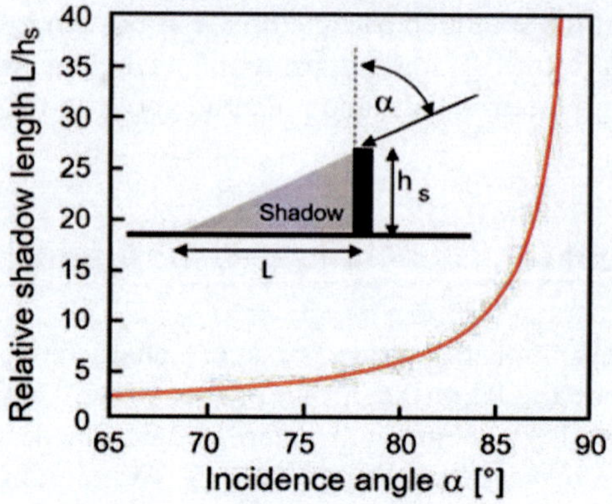

Fig. 11.3 Relative shadowing length, L/h_S, versus the angle of incidence α

Assuming a flat surface, the geometric shadowing length L of an isolated structure is then given by

$$L = h_s \cdot \tan \alpha, \tag{11.1}$$

where h_s is height of the structure casting the shadow. Figure 11.3 depicts that a nonlinear increase in the intrinsic shadow length can be observed with increasing angle of incidence. For incidence angles $\alpha > 75°\ldots80°$, extremely long shadow lengths must be expected, which can be larger than the dimension of the microstructures, h_s, and also the surface diffusion length of deposited particles. It should be noted that deposition also leads to the formation of adatoms, which can diffuse on the surface of the deposited film. It is well known that the mobility of adatoms is strongly dependent on temperature. With increasing temperature, this diffusion process increasingly determines the shape of the columns (see Sect. 11.3.3).

Kundt [10] first established that the optical properties of obliquely deposited metal films depend on the angle of incidence. Later, König and Helwig [11] recognized the self-shadowing effect in OAD, and Nieuwenhuizen and Haanstra [12] made the first hypothesis for a correlation between the tilt angle and the incidence angle based on SEM studies of OAD thin films. As shown above, the angle of material incidence influences the tilt angle of the columns (Figs. 11.1 and 11.2), as well as the porosity of the whole film.

To further design the thin films with respect to specific applications, an additional degree of freedom can be introduced. The parameter used is the azimuthal substrate rotation with the frequency ω (see Fig. 11.1, left). The process is commonly referred to as glancing angle deposition (GLAD). Glancing angle deposition can be considered as an extension of the OAD technique, combining OAD with an additional azimuthal rotation around the surface normal of the substrate (Fig. 11.1). Young and Kowal [13] first combined the OAD technique with azimuthal substrate rotation at deposition angles between 30° and 60° to realize fluorite polarization filters for visible light on glass surfaces. However, this technique remained almost unnoticed until the mid-1990s. Then, Brett's group [14–16] demonstrated the possibility of transforming the columnar OAD structures into new nano- and microstructures with previously unknown shapes by appropriate substrate rotation.

11.2 Experimental Realization of Sputter-Induced OAD and GLAD

In general, OAD and GLAD sputter experiments were performed in ultra-high or high vacuum load-locked deposition chambers with a setup as sketched in Fig. 11.4. The deposition chamber is equipped with a low-energy broad beam ion source for sputtering the target and a heated substrate holder that can be moved along all three spatial coordinates. The deposition angle α between target normal and substrate normal can be continuously adjusted. In GLAD experiments, the substrate holder rotates azimuthally around the substrate normal at variably controllable speeds. To reduce the angular scattering of the sputtered particles (beam divergence), an aperture can be also installed between the target and the substrate. By varying the tilt angle of the substrate and the azimuthal rotation frequency, a wide variety of nanostructure morphologies can be generated. Specifically, tilting with respect to the surface normal of the substrate affects the tilt-dependent morphology, while rotation about the surface normal of the substrate causes changes in the helical morphology. On this basis, simple and precise control of the shape and dimension of nano- and microstructures is possible [17]. For example, Fig. 11.5 shows SEM cross-sectional images of different nano- and microstructures deposited by sputter-induced deposition under grazing incidence at a fixed material incidence angles between $\alpha = 78°$ and $\alpha = 85°$ without and with various substrate rotation frequencies, where ε is the rotation

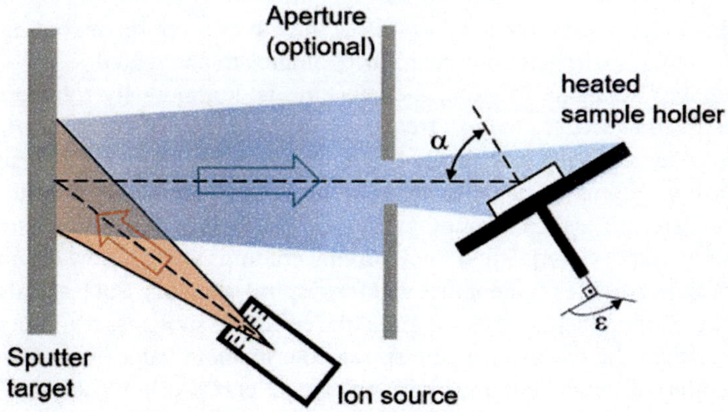

Fig. 11.4 Sketch of the experimental setup used for ion beam sputter OAD and GLAD. ε is the rotation angle around the surface normal

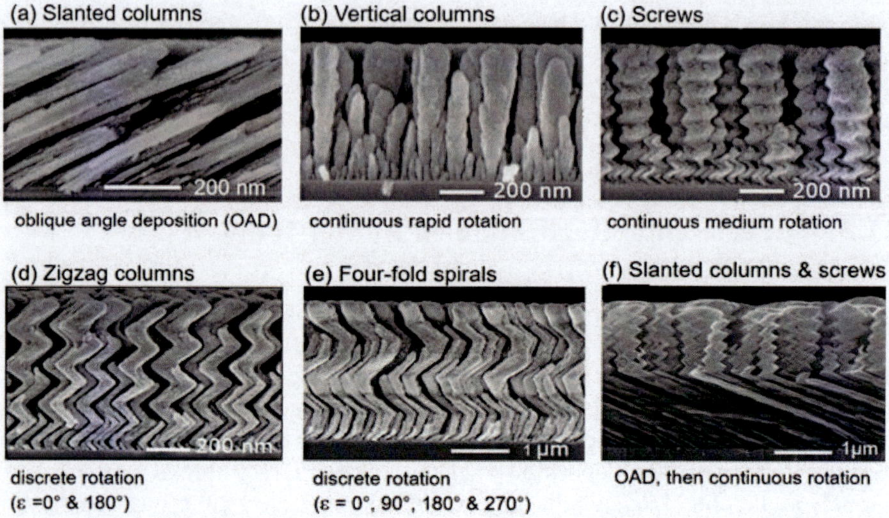

Fig. 11.5 Thin sculptured Si films on Si substrates deposited by direct sputter technique under grazing incidence of the material flux

angle (further structures see e.g., [18, 19]). The rotation angle, ε (see Fig. 11.4), varied between 0° and 360° and can be adjusted controllably. For example, for the production of zigzag structures, two angles, 0° and 180° are selected and material is deposited under these two settings. The fabrication of slated Si columns on Si substrates without sample rotation (OAD) is also illustrated in Fig. 11.5 after sputter-induced deposition under an incidence angle of 83°, but without sample rotation.

Variation of the Rotation Frequency

Rotation of the substrate about its normal changes the apparent direction of the incident particle stream, which in turn affects the shaded regions on the substrate. The evolving columns are expected to follow the direction of the incoming particle stream, allowing a variety of columnar morphologies to be created.

In *continuous* substrate rotation, the substrate is continuously rotated about its normal at a fixed substrate rotation frequency ω (in units of [revolutions/time]). In this process, the ratio v/ω, in units of [thickness/rev], between the deposition rate of the deposited atoms v, in units of [thickness/time], and the substrate rotation frequency ω determines the morphology of the grown structures. If this ratio is large (slow substrate rotation frequency), the evolving columns try to follow the constantly changing direction of the incident flux and form spiral columns. Such spirals have an open core, i.e. the height gained by the spiral per complete substrate revolution (called pitch) is less than the diameter of the spiral. Intermediate values of ratio v/ω lead to the formation of helical columns with a compact core. At low values of the ratio (fast substrate rotation frequency), the particle flow appears omni-azimuthal from the perspective of growing columns, triggering the formation of vertical columns. As an example, Fig. 11.5 shows SEM cross-sectional images of vertical columns

(Fig. 11.5 b) and screws (Fig. 11.5 c) deposited by sputter-induced GLAD at a fixed material incidence angle of $\alpha = 83°$ with different substrate rotation frequencies (further structure see e.g., [18, 19]).

A second type of substrate rotation is *discrete* substrate rotation, in which the sample is rotated stepwise through a specific azimuth angle. Once the rotation angle is reached, the rotation stops and deposition of the material occurs under these angular conditions. This process aims to generate diverse, tailored n-fold columnar morphologies, where n denotes the number of arms deposited per complete substrate revolution. This gradual rotation of the substrate allows successive deposition of a series of structure arms pointing in different directions, leading, for example, to zigzag chevrons or quadruple helical structures [4, 19, 20]. Figure 11.5. demonstrates the growth of twofold columnar structures (zig-zag, Fig. 11.5, d) after deposition of material from two discrete angular positions (0° and 180°) and the deposition of four-fold spirals (Fig. 11.5, e) from four discrete angular positions (0°, 90°, 180°, and 270°) [21].

A third variant is *combination* of deposition on a stationary substrate followed by deposition with a fast continuous substrate rotation or vice versa. For example, tilted columns are grown without substrate rotation, then the substrate is rapidly rotated through the desired angle, and spirals or screws are deposited directly on the first arms, etc. (Fig. 11.5f).

Mircostructure of Deposited Columns

According to Hara et al. [22] and van Kranenburg and Lodder [1], the GLAD process leads to bundling of columns because the column spacing is larger in the plane given by the particle incidence direction and the substrate normal than in the direction perpendicular to it. Therefore, the bundling of the columns is observed in the direction perpendicular to the plane of incidence (i.e. in the direction perpendicular to the particle flux). The bunching process has been studied, especially for thin metal films after electron beam evaporation, using high-resolution electron microscopy and tomographic techniques (see, e.g., [7, 23–30]). For example, the SEM image in Fig. 11.6 (left) illustrates the layer morphology of randomly arranged Si spirals sputtered at a Si particle incidence angle of 88°. In the first ten nanometers of microscopic bundling, an increase in the structure diameter can be observed. In addition to the SEM image, the TEM cross-sectional image, Fig. 11.6 (right), also shows a nucleation layer of vertical posts with an average diameter of about 20 nm at the interface to the silicon substrate at the beginning of the structure growth [26]. As the sputter deposition progresses, some of the nucleation structures (fibers) agglomerate to form the actual spiral, which has an average diameter of 150 nm at the tip, meaning that about 60 up to 70 originally formed nucleation centers (fibers) form a structural unit (spiral). The spirals consist of a large number of individual fibers growing close together, with constant diameter and spacing along the spiral. It is likely that there is branching within the spiral arms, resulting in an increased number of fibers within the arms [27]. Electron diffraction studies have shown that the spirals consist of amorphous silicon after deposition at room temperature.

Fig. 11.6 Cross-section SEM micrograph (left) and cross-section TEM bright field micrograph (right) Si spirals deposited by ion sputter-assisted GLAD

11.3 Oblique Thin Film Growth

It can be concluded that the morphology of obliquely deposited thin films is the result of shadowing phenomena during the manufacturing process. In particular, changes in temperature during deposition and deposition rate affect the morphology [21].

11.3.1 Modeling of the Oblique Film Growth

Computer simulations have become an important tool for understanding the physics of thin film deposition. In general, numerical methods are required to describe film growth as a result of OAD or GLAD, since analytical models describing the growth processes with all their facets are not available. Small-scale effects on time scales down to microseconds can be studied for oblique film growth using molecular dynamics and Monte Carlo simulations. These methods are powerful tools that can directly investigate the influence of parameters such as substrate temperature, deposition rate or the used material. However, these methods are not able to process a large number of particles due to the high computational effort (see Sect. 3.6). Some effects, in particular the self-shadowing effect, are long range effects that require the inclusion of larger parts of the sample and thus larger amounts of particles.

Ballistic models, first used by Vold [31] to simulate sediment formation, have been successfully applied to growing films, assuming (i) linear trajectories for the incoming particles and (ii) the particles stuck to the position of their first contact with the surface (for details on ballistic approaches, see, e.g., Meakin et al. [32–34]). Ballistic simulations (ballistic Monte Carlo simulations) are also commonly used approaches to describe the morphological evolution of deposited layers in the context of OAD and GLAD, where short-range effects such as surface diffusion (small number of particles acting over nanometers) and long-range processes such as self-shadowing (billions of particles acting over micrometers) must be combined within the concept of collision detection [35]. Ballistic models are based on sequential deposition. Starting from the idea of König and Helwig [11] that shadowing of incident atoms by the growing film is responsible for the anisotropic structure after oblique incidence, Henderson et al. [36] and Dirks and Leamy [37, 38] first applied a two-dimensional ballistic deposition simulation to describe film growth under GLAD conditions.

In ballistic simulations, two basic alternatives can be chosen for the trajectories of the incoming particles: (i) the particle positions and trajectories are treated as floating-point numbers (off-lattice simulations), and (ii) the particles are constrained to specific grid points, which are usually treated as integers (on-lattice simulations). Mostly, despite the limiting condition, the latter type of simulations is used because the computational effort of handling integers is much lower than for floating-point numbers. To ensure absolute isotropy of the simulation results, ballistic simulations outside the lattice can be used [39, 40]. Mathematically, spherical particles are assumed, which can reside at freely chosen floating point coordinates. Based on the simplification of an absolutely parallel beam, mathematical simplifications can be applied that minimize the computational effort, so that a lattice-independent simulation of column growth can be realized (for details, see the Appendix in [41]). Figure 11.7 shows a cut through a three-dimensional off-lattice simulation cell and, for comparison, a SEM image of a Si film deposited at an angle of 82° by OAD. The general morphology of the simulated OAD films agrees well with experimental observations. However, collision detection for spherical particles that can be are freely positioned is very computationally expensive, and the spatial partitioning of the simulation cell is often not helpful due to the fractal space-filling behavior of growing films [41]. Therefore, simulations on the lattice are frequently used (e.g., in [42–44]). In such ballistic Monte Carlo simulations on the grid, the simulation volume is formed by a lattice (Fig. 11.8), for which a cubic geometry is typically used. In this, a particle is represented by a single point of the lattice, so that the

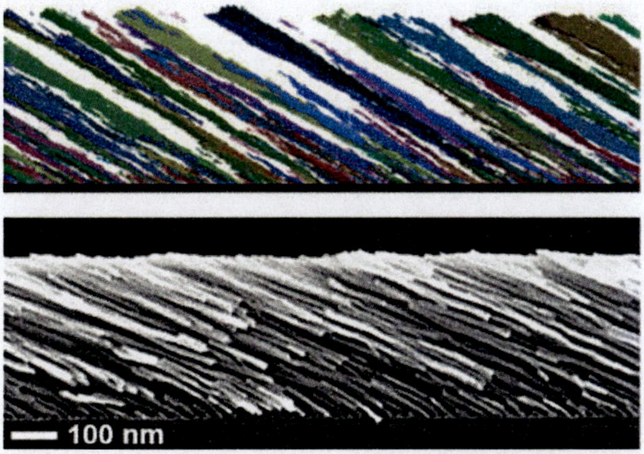

Fig. 11.7 Image of a simulated film (above) and a cross-section SEM image (below) of a Si film deposited at an angle of 82° for comparison. The different colors indicate the individual nanostructures

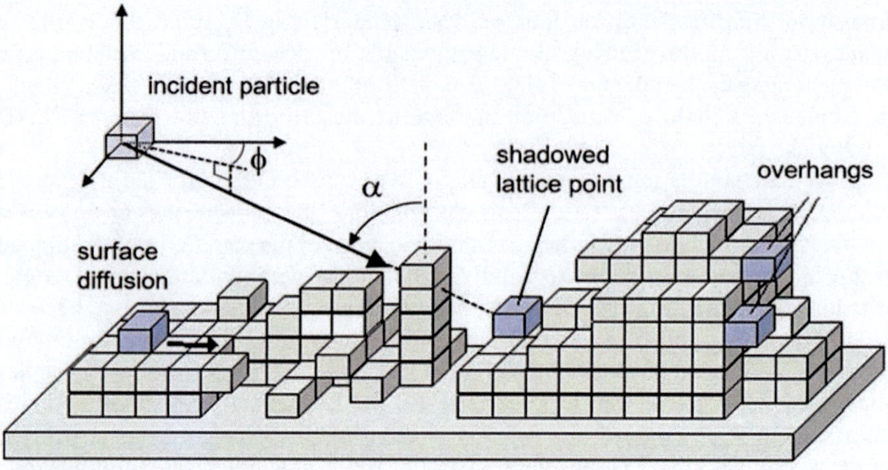

Fig. 11.8 Basic processes in the on-lattice 3-dimensional Monte Carlo ballistic simulations. Atoms move towards the surface with the angle α and ϕ then sticking to the surface, then it is re-emitted or can diffuse. Some surface points are shadowed from the incident fluxes of particles due to the nearby higher surface features (figure is taken from [45] and colored)

particle positions are constrained to the lattice points (this is called lattice effect). Due to the restriction to an absolutely parallel particle flow, the simulation outside the grid cannot be used to investigate the rotation of the substrate (GLAD process) or the influence of beam divergence. With such approach, the simulation cell is divided into regular cubic lattice coordinates. This is achieved by restricting the particle positions to integer coordinates (see e.g., [42, 44, 46, 47]). However, the lattice effect, which is inherent in all such lattice simulations, leads to the artificial anisotropies. Therefore, all simulations on the lattice involving oblique deposition, substrate motion (such as rotation), or beam divergence are questionable [48].

Ballistic approaches on the lattice aiming at describing the oblique film growth are based on the assumption that (i) the particles are hard spheres or cubes in a three-dimensional simulation (or hard circles or squares in a two-dimensional simulation), (ii) the particles are launched at a randomly chosen position above the substrate, (iii) the emitted particles move with straight-line trajectories towards the film surface, and (iv) the particles become part of the growing film at the point of first contact. The description of the growth behavior by $(2+1)$ dimensional ballistic Monte Carlo simulations is based on the following sequence of steps (Fig. 11.8, for details see e.g., [49, 50]):

- each incident particle is represented with the dimension of a lattice point,
- depending on the deposition method, a certain angular distribution is chosen for the angle of incidence of the particles, whereas a fixed grazing angle of incidence α is chosen for the simulations of deposition under oblique angle,
- at each simulation step, a particle is moved to a randomly chosen grid point on the substrate surface, moving in a straight line until it hits the surface,

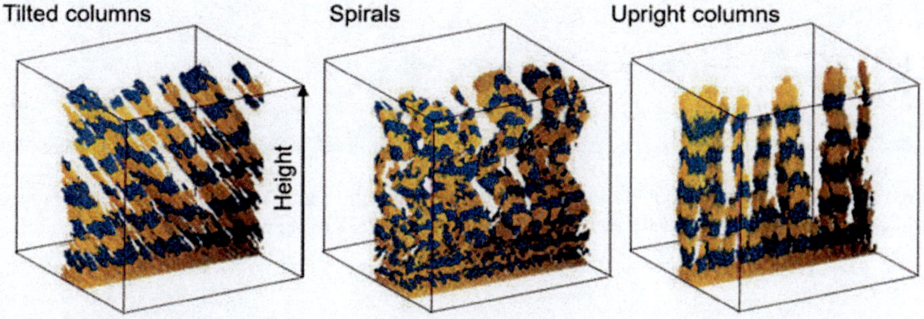

Fig. 11.9 3-dimensional on-lattice simulations of the growth of tilted columns, spirals and upright columns [48, 51]

- in general, a constant sticking coefficient is used, and the direction of the reemitted particles is determined by the cosine distribution (see Appendix K),
- the shadowing effect is taken into account, where the particle trajectory may be cut off on its way by large surface features,
- after each deposition effect the next particle is started.

For example, Fig. 11.9 exhibits simulated structures grown with increasing substrate rotation speed, leading to the growth of tilted columns (no rotation), spirals (low rotation speed), and upright columns (high rotation speed) [41]. In comparison with experimental results (c.f. Fig. 11.5) the morphology of the different microstructures are precisely reproduced, the structures are separated and the growth competition can be observed as well.

Depending on the temperature during oblique vapor deposition, surface diffusion of adatoms can play an important role in film growth. For example, Sánchez et al. [52] have shown that diffusion processes lead to a transformation of the porous into a compact packing of the deposited layers already at homologous temperatures $T/T_M > 0.19 \ldots 0.25$. Consequently, several approaches to describe the oblique growth of microstructures have included surface diffusion.

Frequently, the influence of the surface diffusion on the shape of micro- and nanostructures grown by OAD and GLAD is modeled as a ballistic Monte Carlo process, where adatom mobility (diffusion on the surface) must be integrated in addition to film growth phenomena such as nucleation, coalescence, self-shadowing, reemission, etc. [49, 50, 53–55]. For this purpose, the strict condition for a ballistic process is intermitted, i.e. after the first contact of the particle with the surface, a motion of this particle is allowed. Surface diffusion can be considered as thermally driven hopping process, where the jump probability exhibits an Arrhenius-type temperature dependence [56]. The activation energy of this process is given by the sum of the energies of the bonds that must be broken for a certain jump [57]. Unfortunately, definitive values for the surface self-diffusion length, the surface self-diffusion coefficient and the activation energy are not available or are very uncertain. However,

Fig. 11.10 Two-dimensional ballistic simulations of a Ni film for different substrate temperatures and comparable depositions times (between 3.6 and 4.3 s). The angle of incidence is 45° and the deposition rate is 1 nm/s (figures are taken from [61])

approximate values for the surface diffusion length [21] and the surface activation energy [58–60] are known.

Figure 11.10 depicts a sequence of 2-dimenional ballistic simulations including thermally induced diffusion of an obliquely deposited Ni film in dependence of substrate temperature. With increasing the surface mobility (temperature during deposition), a transition takes place from a porous film with columnar microstructure to a densely packed film. It is also obvious that the film thickness decreases due the agglomeration [57, 61]. The study by Müller [61] also shows that the transition temperatures to achieve densely packed films increases with increasing the deposition rate. In detail, Meakin and Jullien [33, 34] studied the change of film density in an off-lattice model for random deposition of particles in two dimensions in dependence on the incidence angle. Two different cases were compared. In the first case, the arriving particles are assumed to adhere irreversibly to a growing deposit, and in a second case, the particles are allowed to diffuse as many times as necessary until they finally reach a local minimum position. The obtained results show that the angle of incidence can have a significant effect on the film density. For an incidence angle of $\alpha = 0°$, the normalized mean density is about 0.357 and 0.723 for the first and second models, respectively. In both cases, the density decreases with increasing angle of incidence, as expected (the mean density is zero at $\alpha = 90°$).

11.3.2 Growth on Isolated Seed Points

Of particular importance for the interpretation of the growth mode are studies on the growth of fan-like structures at isolated points on the surface under the condition of ballistic deposition. Assume a surface with precast seed points (see Sect. 11.5.1) or microscopic elevations that rotates about the surface normal during deposition. Ballistic deposition conditions are applied, i.e. the particles move parallel to each other through the vacuum and stick at the first contact point on the surface. Such an aggregation leads to the formation of a fan-like structure (see e.g., [62, 63]). This is remarkable because no columnar microstructure is observed in the direction of the incident beam. The first experimental evidence was provided by Ye and Lu [46]. They have produced silicon fan microstructures on cylindrical tungsten pillars. The results could be experimentally confirmed by Tanto et al. [47, 64] and Grüner et al.

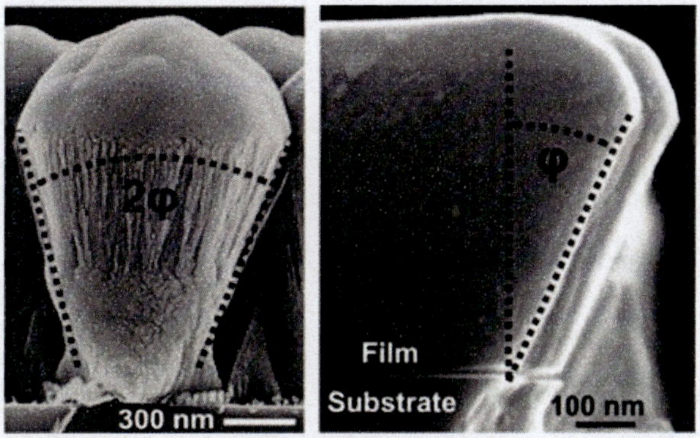

Fig. 11.11 SEM image of a Ge fan structure deposited at normal incidence onto a pre-patterned substrate (left, fan angle $\varphi \approx 19.5°$) and an image of an Si overhang of an deposited Si film on a flat Si surface (right, fan angle $\varphi \approx 24°$) deposited at normal incidence onto an edge

[41]. Figure 11.11, left, shows two cross-section SEM images of Ge fan-like structures. The material is deposited at normal incidence onto a pre-patterned deposited substrate. The fan shape of the individual structures, which develop from protruding nucleation sites, is the result of lateral structure broadening caused by the sticking of overhang particles. Rotation during deposition results in a conical structure capped by a curved dome (see e.g., Fig. 11.11, left). A similar behavior can be observed at the edge of a flat substrate (overhang in Fig. 11.11, right). The fan is characterized by the fan angle, φ, given between the vertical growth direction and the boundary of the grown fan structure (see also Fig. 11.22). Prior to experimental evidence of fan-like structures, numerous theoretical studies (simulations) have addressed the formation of fan-like structures [39, 62, 63, 65–68] and their scaling properties [65–67]. Preferably, simulations were performed in $1 + 1$ dimensions [39, 41, 61, 65, 68] (in exceptional cases also in $2 + 1$ dimensions [66]). For example, Fig. 11.12 shows the results of a two-dimensional off-lattice simulation where well-defined substrates were introduced into the simulation cell. The simplest substrate is a single, isolated particle (nucleation point) at a fixed position (Fig. 11.12, left). After an assumed normal incidence of particles, the expected fan-like structure with a curved growth front is formed. Regardless of the dimension of the nucleation point, the shape of the growth front is approximately a spherical calotte. Ramandal and Sander [39] have given a first order approximation for the curvature of the calotte. The radius of the curved growth front can be described by means of the relation, $r(\varphi) = vt\sqrt{\cos\varphi}$, where v is the constant growth rate (or deposition rate) and t is the time of deposition. Since perfect point-like nucleation points (Fig. 11.12, left) are not achievable in real deposition experiments, Fig. 11.12 also shows the growth of fan structures on a small spherical area (middle) and a plateau (right). The morphology of the grown layer on the plateau reflects the experimentally observed growth of layers (overhang, c.f. Fig. 11.11, right).

Fig. 11.12 Off-lattice simulated fan-like structure on an isolated point (left), am large spherical seed (middle) and a plateau [41]. The concentric circles are added to highlighted the nearly spherical growth front on the top of the fan

11.3.3 Growth of Sculptured Thin Films

The morphology of thin films generated by oblique deposition is driven by the shadowing effect. According to the selection model by van der Drift [69], only the structures (columns, chevrons, spirals, etc.) with the fastest vertical growth component will be successful in the competitive growth, because those columns will overgrow the slower growing columns (details see Sect. 10.5.1.5). The result of this process is that the obliquely deposited thin films will consist of structures with different sizes. This also means that a transition from nano- to microscopic structures can occurs during film growth. An example is shown in Fig. 11.13, left. The detail of a SEM image exhibits a two-fold chevron Si structure deposited by ion beam sputter deposition. The diameter of the chevron arms, w, increases with the height h_s. As result of the competitive growth, the growth of smaller nano- and microstructures are stopped (red arrows in Fig. 11.13). In particular, for vertical columnar structures grown with GLAD and substrate rotation, it is known that as a consequence of column extinction and column coalescence in the early growth stages, average column diameters and column-to-column separations increase [2, 17, 70]. Under highly oblique deposition conditions and low temperatures, shadowing exceeds the diffusion-mediated smoothing process [71]. This self-shadowing favors the growth of larger structures over smaller, adjacent structures in the beginning of growth (Fig. 11.13, left). Thus, the number of columns is continuously reduced during the advancing deposition. When the surviving structures increase their lateral size while the incoming particle flux is held constant, the column-to-column separation λ also increases. These circumstances are reflected in the PSD presentation (see Appendix L) of films that consisting vertical Si columns with increasing of the time of sputter deposition (Fig. 11.13, right) [72]. Within the time interval from 10 min to 200 min the maximum peaks positions q_{max} in the PSD curves decrease, resulting in a shift of the column-to-column separation from about 22 nm to about 84 nm. This indicates

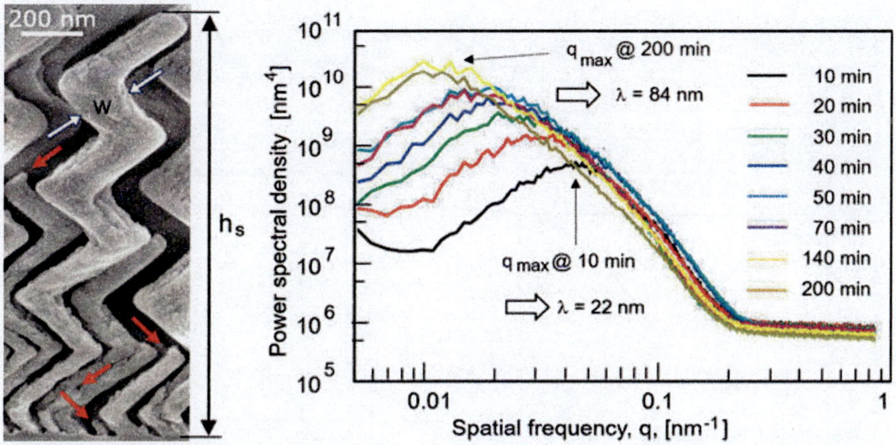

Fig. 11.13 Left: Cross-sectional SEM micrograph of Si chevrons (zig-zag-structures) on a planar Si substrate prepared by ion beam assisted sputter deposition. Red arrows indicate separate structures that stopped growing due to enhanced shadowing by adjacent structures. Right: Power spectral density representation calculated from top view SEM micrographs of glancing angle deposited vertical Si columns with different deposition times. The maxima of the PSD curves after 10 and 200 min deposition times are indicated by black arrows

that the growth of the glancing angle deposited Si nanostructures starts on seeds with lateral dimensions in the range of approximately 20 nm, and that column competition, extinction and merging during deposition lead to a reduction of the number of surviving structures which in turn increase their diameters and column-column distance.

Influence of Temperature

It is known that the competition between the shadow length, which is determined by the deposition angle, and the length of adatom mobility, which is determined by a thermally activated process, governs the shape of the structures fabricated with OAD or GLAD [60]. In general, nanostructures fabricated by OAD tend to broaden and compete with each other as height increases, leading to the extinction of some structures [2, 4, 73]. According to the zone model by Movchan-Demchishin [9], a columnar morphology should be observed for homologous temperatures $T/T_M < 0.3$, which are relevant for the most OAD and GLAD experiments. Figure 11.14a, b illustrates the influence of the temperature at a constant rotation frequency [72]. The SEM micrographs in Fig. 11.14a indicate that the Si film thickness decreases from about 540 nm to 330 nm, while width of the Si screws (diameter) increases from about 117 nm to about 140 nm when the temperature increases from room temperature (RT) to 360 °C during the sputter-induced deposition. The morphological shape of the individual screws also changes. At low temperatures between 200 °C and 300 °C, the individual spiral fibers merge to screw-like structures. Above this temperatures, broad screw-like structures are formed. With the aim of identifying the underlying mechanism of column growth, the scaling behavior of the Si structure diameter, w,

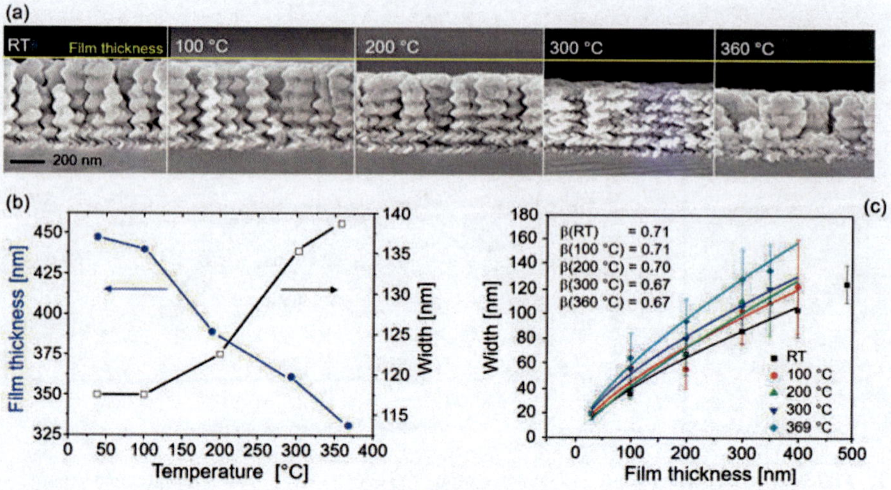

Fig. 11.14 **a** Influence of the deposition temperature on the growth of Si screws grown at a low rotational velocity of 0.035 rev/min. The deposition time was 155 min for each film and the angle of particle incidence was 85°. **b** Dependence of the film thickness and width of the microstructures o the temperature. The films were deposited under condition as in the case (**a**). **c** Average column width *w* data are plotted as a function of column length *h* for Si columns prepared at different temperatures and at a deposition rate of 17 nm/rev. Each data point is the average of measurements on up to 12 columns. Each error bar on the measured column width represents the standard deviation of the measured value

with increasing structure height (or deposition time) for vertical columnar structures can be investigated by determining the scaling exponent β (see Chap. 6 and note that the tilt angle is also referred to by β). Column width changes with the column length according to a power law, $w \propto h^{\beta}$ with the growth exponent $\beta = 0.67$–0.71 (Fig. 11.14c) for Si and $\beta = 0.59$ for Ge [75]. These relatively high growth exponents are to be expected if the contribution of surface diffusion to the growth of the microstructures is insignificant. Lower growth exponents between 0.28 and 0.34 (simulation) and between 0.3 and 0.5 (experiment) were determined for columnar metal structures [49].

At homologous substrate temperatures < 0.45, an expected improvement in the crystallinity of the films is observed, and the morphology of the films significantly changes. Increased temperature leading to longer surface diffusion of adatoms causes atypical growth at glancing angle conditions [74].

Influence of Rotation Frequency

Robbi et al. [10, 11] were the first who have applied the SEM to directly study the influence of the substrate rotation on the evolving morphology of obliquely deposited thin films. It is well known (Sect. 11.2) that the ratio of the deposition rate, v, to substrate rotation frequency, *ω*, determines the morphology of the growing structure, when the temperature during the deposition $T < 0.3T_M$ [2, 5, 18]. For homologous temperatures < 0.3 (e.g., $T/T_M \approx 0.18$ for Si at room temperature), only

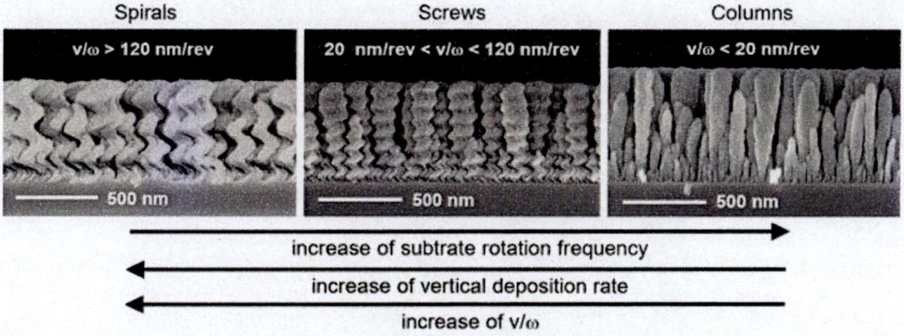

Fig. 11.15 Sculptured thin films comprising Si nanostructures deposited with GLAD by ion beam sputter deposition in dependence on the ratio of the deposition rate to the substrate rotational speed at constant temperature (room temperature)

a low adatom mobility is expected [9]. Therefore, it can be assumed that the surface diffusion length is not sufficient to affect the shape of the columnar microstructure. Figure 11.15 illustrates the different Si structural morphologies formed at different v/ω values when the substrate is continuously rotated [17, 18]. Similar observation for sculptured films of Ge [74–76] and various metals [23] have confirmed this evolutionary sequence. The v/ω-dependent morphological change is related to the growth front of the structures 'following' the particle flux under conditions of glancing angle deposition and low adatom mobility. Since most of the incoming particles stick to where they land, namely at the tops of the growing structures, a continuous change of the particle flux direction leads to the growth of helical structures like spirals and screws. As the columns grow tilted in the direction of the incoming particle flux, slow substrate rotation leads to the formation of spiral-like structures. In this process, the growing nanostructures follow the substrate rotation. At higher rotation speeds, the pitch of the spirals decreases so that the structures are more like screws. At high rotation frequencies, the influence of the individual revolutions is lost and the structures degrade into upright columns (c.f. Fig. 11.5). However, it has been found that this rather simple classification can only be seen regarded as rule of thumb and is valid only for limited structure heights, since the competitive character of the growth of adjacent structures favors both structure broadening and structure merging effects, which gradually lead to a change of the morphology of the structure with increasing structure height. Figure 11.16 demonstrates the morphological change of the Si nanostructures prepared by ion beam-assisted sputter deposition at glancing angles [17]. At a constant ratio of deposition rate and the substrate rotation frequency v/ω, the morphology changes from spirals to screws to columns with increase of the film thickness only as result of merging. Consequently, controlling the ratio v/ω and the thickness of the deposited film at different incidence angles α is a simple method to tune the shape of microstructure and the density of sculptured thin films. Further tuning of the films to achieve specific morphologies of sculptured

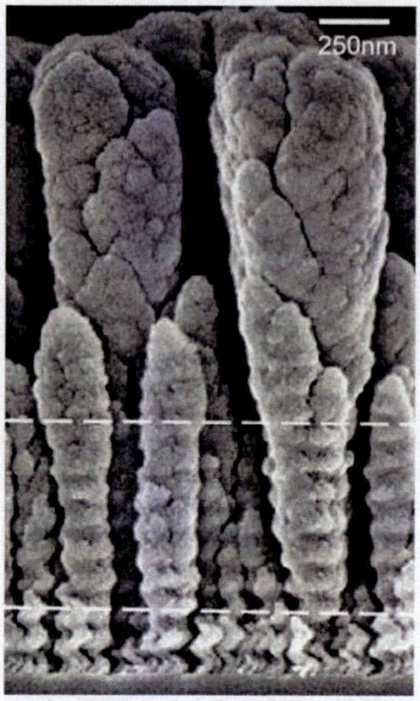

Fig. 11.16 Cross-section SEM micrograph of Si nanostructures prepared by ion beam assisted sputter deposition. The ratio of ratio of deposition rate to the substrate rotation frequency was 100 nm/rev, the angle of Si particle incidence was 83°. The film was deposited at room temperature

films may be possible if applying more complex sample movements as phi-sweep [77] or swing tilting-rotation [78].

Influence of Beam Divergence

Oblique angle experiments require a highly directional flux to ensure that the self-shadowing effect is not counteracted by atoms reaching the substrate at an angle significantly different from angle of incidence. As described in Appendix K, the flux of emitted particles is characterized by a $\cos^n$-distribution and the source of the particles is no ideal point source. Consequently, a significant fraction of the emitted particles strike the substrate at an angle that differs from the angle of incidence α. This broadening of the particle beam is described by the divergence angle φ_{div}, i.e. the particles reach the surface of the grown structures at an angle of incidence $\alpha \pm \varphi_{div}$, where φ_{div} is given by (6.57). It can be expected that the tilting angle β of the grown nanostructures vary in same order of magnitude. Consequently, the angular divergence of the beam is controlled using an aperture between target and substrate (cf. Fig. 11.4). This procedure leads to a reduced divergence, but also in lower flux density of atoms to be deposited. Figure 11.17 shows an aperture between the target and substrate with aim to reduce the beam divergence for the case of an incident angle $\alpha = 0°$. The atom flux density, J_a, at the substrate surface after an aperture with the radius r_a is given by

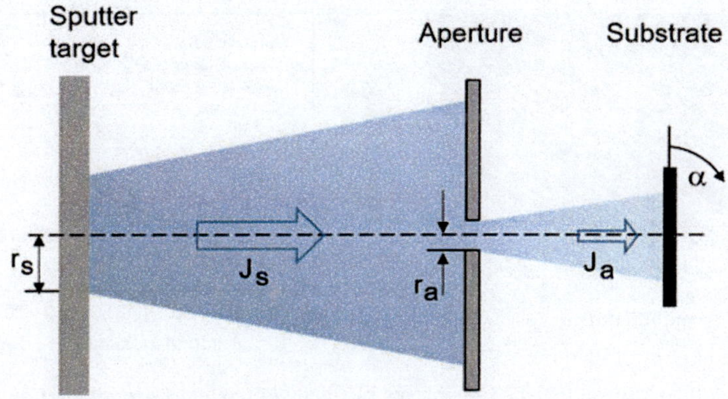

Fig. 11.17 Schematic representation of an aperture between the particle source (sputter target) and the substrate

$$\frac{J_a}{J_s} = \frac{r_s}{r_a}\exp\left[1 - \left(\frac{r_a}{\sigma\sqrt{2}}\right)^2\right], \tag{11.2}$$

where atom flux density without an aperture, J_s, is given by constant atom flux density of the sputtered particles, r_s is radius of the sputtering spot on the surface of sputter target and σ is the standard derivation of the Gaussian distribution of the particle beam. If the substrate is tilted at an angle α with respect to the surface normal of the substrate, the relationship given in Appendix K for current density of sputtered particles must be considered.

In most sputter experiments, a slit aperture is included between the sputter target and substrate to prevent the sputtered atoms emitted with larger deposition angles from reaching the substrate. In GLAD experiments to fabricate Si and Ge nanostructures, apertures are used to limit the divergence angle to $< 10°$ [17, 44, 79, 80]. It should be noted, however, that the deposition rate is significantly reduced by the aperture (or apertures).

11.4 Relation Between the Column Tilt Angle and the Angle of Particle Incidence

The incident flux of sputtered or evaporated particles arrives the substrate at an angle α. The morphology of the grown thin film deposited under off-normal conditions is determined by a geometric self-shadowing effect and surface diffusion. This generally leads to the formation of columns inclined by an angle β. For example, Fig. 11.18 (left) exhibits the growth of tilted Mo columns after deposition at an oblique incidence angle $\alpha = 86°$ on oxidized Si substrates. Deposition at this highly oblique angle

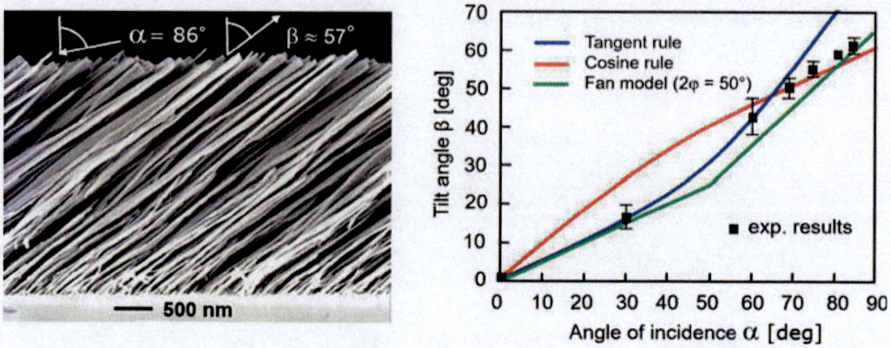

Fig. 11.18 Left: Cross-sectional SEM images of tilted Mo columns deposited at an incidence angle of 84° on thermally oxidized Si substrates at room temperature [81]. Right: Experimentally measured tilt angle of grown Si columns at oblique deposition (black squares) in dependence on the angle of incidence. For comparison the predictions of some common models for the tilt angles are shown [41]. The angle $\Phi = 2\varphi$, where φ is the fan angle

leads to the self-assembly of nanostructured thin films consisting of individual tilted columns inclined in the direction of the incident particle flux (growth or tilt angle β ≈ 57° with respect to the surface normal).

It has been known for decades that independent of the deposited material, the tilt angle of the growth direction is always smaller than the angle of the incident material, i.e. $\beta < \alpha$. In the past, various relationships have been proposed to describe the more than 50 years old experimentally confirmed finding that the growth angle is not identical to the deposition angle [12, 41, 64, 82–87].

Nieuwenhuizen and Haanstra [12] proposed the first empirical relation to describe the difference between the incident angle, α, and the growth (tilt) angle, β, called tangent rule, which is as follows

$$\tan\beta = \frac{1}{2}\tan\alpha. \tag{11.3}$$

This relationship is based on the assumption that surface diffusion of adatoms can be neglected [21]. Dirks and Leamy [37] have confirmed this equation by computer simulations. Moreover, the conditions of deposition and the properties of the deposited material are not taken into account. In particular, this relationship tends to overestimate the angle at incidence angles $\alpha > 60°$ (Fig. 11.18, right).

Based on a ballistic study of the shadowing of columns, Tait et al. [85] has derived the relationship,

$$\beta = \alpha - \arcsin\left(\frac{1-\cos\alpha}{2}\right) \tag{11.4}$$

called cosine rule. Comparison with experimental studies (an example is shown in Fig. 11.1) indicate that especially at incidence angles $\alpha < 60°$, the growth angles of the columns are overestimated and for $\alpha > 70°$ the growth angles are underestimated.

The discrepancies for both rules are not unexpected, because neither material properties nor deposition parameters are taken into account. Considering of material and deposition parameters, Lichter and Chen [82] derived an angular relationship in a continuum approach. They found the relation

$$\tan\beta = \frac{2}{3}\frac{\tan\alpha}{1+\epsilon\tan\alpha\sin} \quad \text{with } \varepsilon = \frac{4}{27}\frac{h_s}{D_{sd}}v \tag{11.5}$$

where v is the deposition (growth) rate, h_s is the height of the initial shadowing feature and D_{sd} is the adatom diffusion coefficient. The quantity ε varies in a limited range $0 < \varepsilon \leq 3.7$. But even this rule cannot satisfactorily explain the experimentally verified relationship between the tilt angle and the angle of incidence over the entire angular range.

Tanto et al. [64] proposed a semi-empirical model (Tanto model) based on geometrical shadowing effect due to the aggregation of columns under the obliquely incident material flux. As described in the last subsection, the grown overhang structure, called fan, is characterized by the fan angle φ, which depends on the material properties, temperature, deposition rate, and impurity content. This fan angle is used as the only input parameter in Tanto's model and can be determined experimentally on a fan grown at normal material incidence. In this model, Tanto et al. [64] assumed that isolated nucleation points are initially formed on the substrate, (Fig. 11.19, left). On this protruding nucleation points the individual fan structures begin to grow against the incoming material flow. After a specific time of growth, the upper edges of the fan structures start to cast shadows on the lower edges of the adjacent growing fans (Fig. 11.19, left). Thus, no further material can be deposited, i.e. the growth stops.

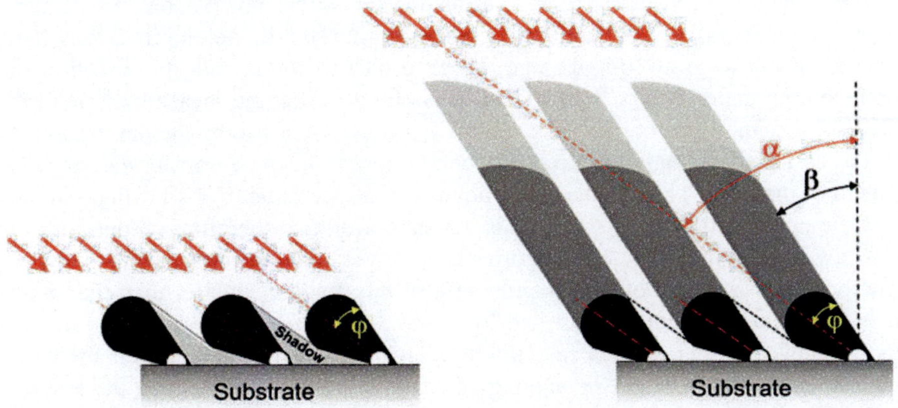

Fig. 11.19 Schematic of Tanto's fan model. Left: before the shadowing is active. Right: grown nanostructures (figure is taken from [51] and modified)

The shadow casted by the upper, non-shadowed edge of the fan structures reinforces the lower edges to grow parallel with the upper edges. Since the angle by which this upper edge grows away from the direction of the incident particle flux is the fan angle φ, the tilt angle β of the finally arising nanostructure is defined by [64]

$$\beta = \alpha - \varphi \quad \text{for } \alpha > 2\varphi. \tag{11.6}$$

For smaller incidence angles, Tanto et al. [64] derived the relation

$$\beta = \alpha - \arctan\left[\frac{\sin(2\varphi) - \sin(2\varphi - 2\alpha)}{\cos(2\varphi) + \cos(2\varphi - 2\alpha) + 2}\right] \quad \text{for } \alpha < 2\varphi. \tag{11.7}$$

In Fig. 11.18 (right) experimental results are compared with expected results according to the cosine, tangent and fan relationships. The predicted tilt angles β are in correct order of magnitude, but an accurate agreement with the experimental results cannot be established. Furthermore, the kink at $\alpha = 2\varphi$ (fan relationship by Tanto et al.) indicates that deeper lying physical processes are not taken into account.

Recently, Grüner et al. [41] proposed a competition model that use the classic Tanto relation and additionally considers some structural aspects. First, the curved shape of the fan structure is regarded, assuming an approximately spherical growth front of the fan structure. The shape of the growth front defines an isochrone surface for each time point of the fan structure growth.

Second, on the base of a detailed consideration of the time evolution of interacting fan structures [41] shows that the shadowing of the lower part of a fan does not cause its lower edge to grow parallel to the upper edge that casts the shadow. Therefore, the shadowed lower edge of each fan encounters the shadow-casting upper corner of the front-standing fan. The growth competition between these both edges defines the long-time behavior of the growing film. Therefore, Grüner et al. [41] proposed to divide the growth process on a tilted row of isolated seed points into three stages.

The substrate is considered to be an array of seed points, which in turn are located in an oblique, parallel particle flux. As in the original fan model, an individual fan structure starts to grow (initial stage (I)) on each seed point into the direction of the incoming material flux (Fig. 11.20). It is obvious, that the local tilt angle β_I is identical with the angle of incidence α. This growth stage lasts until the fan structures start to cast shadows onto the structures behind them. During a transition stage (II), growth of the lower parts of the fan structures is stopped due to shadowing. At this time the upper part of each structure is not shadowed and therefore it continues to grow away from the particle trajectories at an angle equal to the half of fan angle. The local growth angle of the structures β_{II} at this stage is smaller than the angle of incidence α and can be described by Tanto´s relation, (11.6), using the upper, non-shadowed corner of each fan structure. However, from the observed shape of the nanostructures, it can be concluded, that this is not the final stage of the growth process. In fact, the shadowed corner of the individual fan structures still grows faster than the upper corner. Due to this, the shadowed corner will catch up with this upper corner until both finally touch each other (Fig. 11.21). Finally, this former shadowed

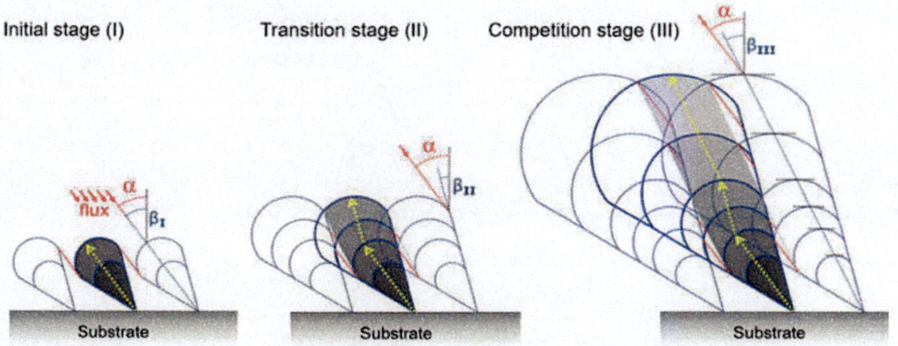

Fig. 11.20 Scheme of the three growth stages. The yellow arrows represent the local growth direction for each stage [41]

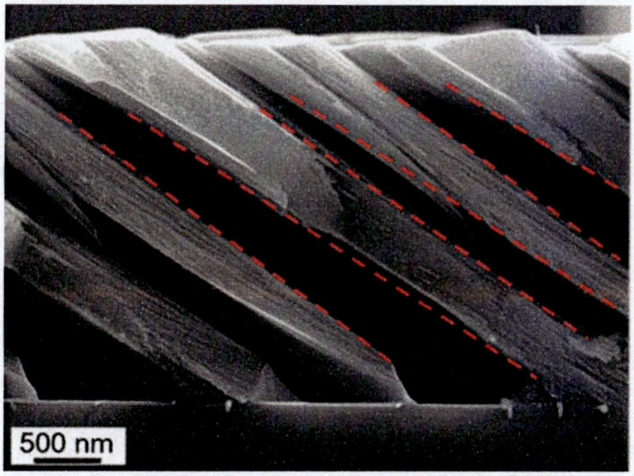

Fig. 11.21 SEM image of silicon nanostructures, obliquely deposited ($\theta = 80°$) onto a patterned substrate. The dotted red indicators highlight the converging corners of the nanostructures observed in the transition stage (II)

corner will start to cast a shadow onto the first corner, which is now left behind and therefore stops to grow. From this time on, both corners shadow each other in a competitive process, leading to the final growth stage (III). As a result, the columns at the back can partially shade those at the front. This can also be considered as the origin of the competitive growth which is inherent to all OAD and GLAD processes.

In order to calculate the morphological properties of the oblique fan structures, the angles and lengths as shown in Fig. 11.22 are used. Using the angles and lengths as shown in this figure the angle φ is given by $\tan\varphi = r/a$ and the tilt angle is given by $\sin\beta/a = \sin(\alpha - \beta)/r$. Combining both equations yields to a relation between the tilt angle and the angle of incidence

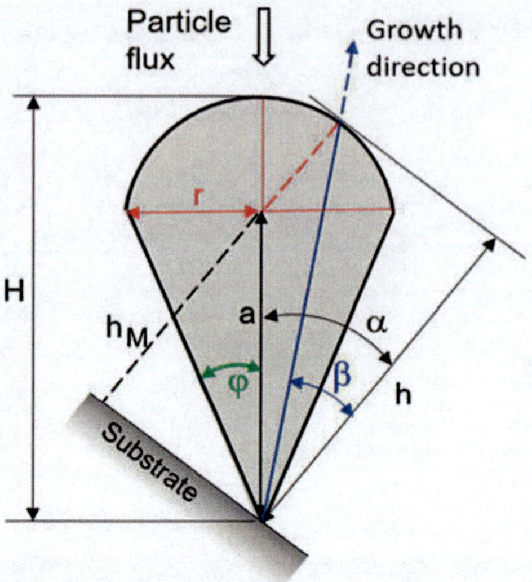

Fig. 11.22 Sketch of the used angles and length. The substrate is tilted by an angle α. Note that all red lines have the same length, due to the approximation of the spherical growth front. The blue line represents the connection of the fan's origin with its highest point of the fan and therefore the part of the fan structure that is never shadowed

$$\tan\beta = \frac{\sin\alpha}{\cos\alpha + \tan\varphi} \quad \text{or approximated to } \beta \approx \left(1 + \frac{\varphi}{90^o}\right) \tag{11.8}$$

for the complete range of material incidence angles, i.e. $0° \leq \alpha \leq 90°$, where φ is given in [°]. In addition to the analytical relationship between the angle of inclination of the growing structures and the angle of incidence of the incident material flow, (11.8), the relative growth rate and porosity, and thus the density of inclined thin films, can also be determined on the basis of this model [41].

(i) As shown in Fig. 11.22, the thickness of the obliquely deposited film, h, can be determined for an arbitrarily chosen deposition time with respect to the film thickness, H, of a film normally deposited film under the same conditions. It is obvious, that the normally deposited film is a superposition of vertically standing fan structures so that h = H. From Fig. 11.22 follows the geometric relations $h = h_M + r = a\cos\alpha + r$ and H = a + r. Using the above presented equation $\tan\varphi = r/a$, the relative thickness of the film (or growth rate), h/H, is given by

$$\frac{h}{H} = \frac{a\cos + r}{a + r} = \frac{\cos + \tan\varphi}{1 + \tan\varphi}. \tag{11.9}$$

(ii) Poxson et al. [88] derived that the porosity and relative thickness of such an obliquely deposited film are related, assuming that the adhesion coefficient of the incoming material is equal to one. At least for evaporation onto an unheated substrate, neither sputtering nor reflection or desorption are expected, so this assumption is valid. Following these authors, the porosity of the film can be expressed as $P = 1 - V_{mat}/V_{film}$, where V_{film} denotes the total volume of the porous film (film thickness times substrate area) and V_{mat} represents the fraction of the film actually filled with material. The amount of material reaching a tilted substrate is geometrically reduced by a factor of $\cos\alpha$ compared to normal deposition on a substrate with the same surface area. Thus, with $V_{mat} = V_{norm} \cdot \cos\alpha$, follows

$$P = 1 - \frac{V_{norm}}{V_{film}} \cos \alpha. \tag{11.10}$$

Since the same substrate area is assumed for both volumes, the consideration can be reduced to the film thicknesses. Therefore, the porosity of the film is given by the relative growth rate h/H and

$$P = 1 - \frac{H}{h} \cos \alpha. \tag{11.11}$$

Combing (11.9) and (11.11), then the film density can be expressed by

$$\rho_{film} = \rho_{bulk} \frac{\cos \alpha + \tan \varphi \cos}{\cos \alpha + \tan \varphi}. \tag{11.12}$$

According to (11.12) is the density of thin films prepared by OAD also depends on the fan angle.

Application

In summary, relations for the column tilt angle β, the relative growth rate h/H and the film density ρ are derived. In each of these expressions only one parameter, the fan angle φ is used. An advantage of this model is also that the specific properties of the deposition process (e.g., the divergence of the depositing particle beam), is taken into account.

Based on (11.8), the relationship between the angle of incidence and the growth angle of obliquely deposited films can be adequately described. For example, Fig. 11.23 shows results of a measurement of the tilt angle and the film thickness as a function of the incidence angle after oblique deposition of Si and Ge, respectively [41, 51]. Fitting the model to the experimentally determined tilt angles results in fan angles of 25.15° for Si (measured angle = 24°) and 21.5° for Ge (measured angle = 19.5°). The application of the model to the tilted deposition of various compounds is shown in Fig. 11.24. Measurements of tilt angles of MgF_2 nanostructures for the entire range of incidence angles have been reported by Tait et al. [85] and Messier et al. [89] (Fig. 11.24, left). The experimental values of these two groups fit well and confirm the predicted quasi-linear behavior. Poxson et al. [88] have experimentally determined the porosities of SiO_2 and ITO (indium tin oxide) films by optical

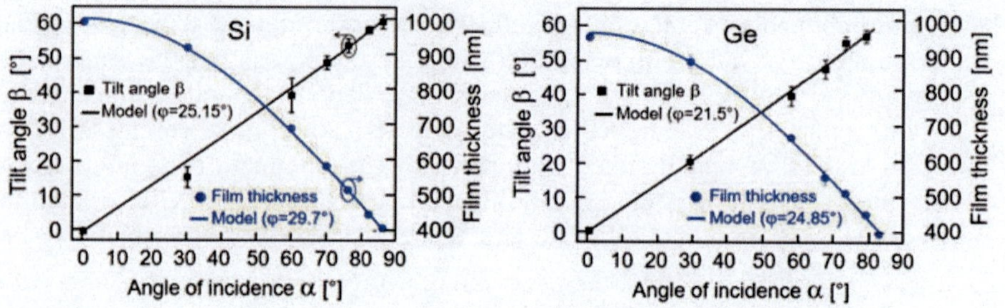

Fig. 11.23 Measured (points) and fitted tilt angles (solid lines) for varying angles of incidence for Si and Ge

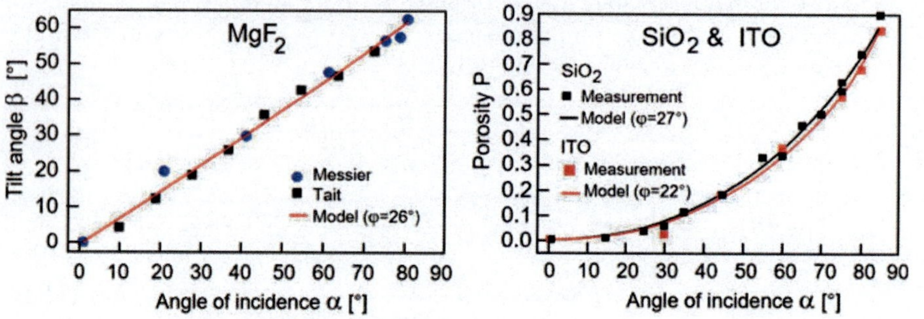

Fig. 11.24 Column tilt angles for MgF_2 taken from Tait et al. [85] and Messier et al. [89] (left) and the porosity of SiO_2 and ITO films taken from Poxson et al. [88] (right) compared with the results of the competition model (lines)

methods as a function of the material incidence angle (squares in Fig. 11.24, right). Comparison of the fitted fan angles for SiO_2 and ITO obtained from the tilt angle and porosity measurements shows good agreement, i.e. the use of the fan angle as input for porosity prediction illustrates the applicability of the model [51].

The proposed model, which is capable of predicting the angle of inclination for the entire range of incidence angles $0° \leq \alpha \leq 90°$ was also verified for metals. For example, Fig. 11.25 shows results of measuring the tilt angle and film thickness as a function of the incidence angle after oblique deposition of Mo [41]. Fitting the model to the experimentally determined tilt angles yields a fan angle of 38.6°. It is obvious that the experimental results can be described very well with the proposed model over the whole angular range.

11.5 Growth on Patterned Substrates

OAD and GLAD on plane surfaces are initially characterized by randomly distributed seed points (nuclei). These nuclei grow to islands with non-uniform diameters and are capable of shadowing the incoming particle flux from adjacent regions (shadowing

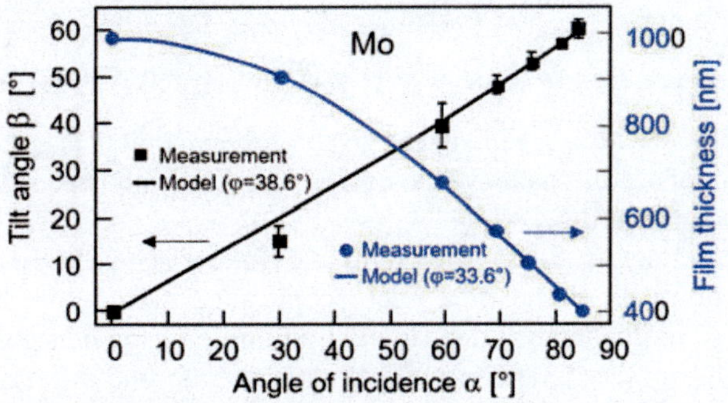

Fig. 11.25 Measured (points) and fitted tilt angles (solid line) for varying angles of incidence for molybdenum

effect, see Fig. 11.2). This leads to variations in the diameter, height, and arrangement of the grown nano- and microstructures. However, some applications require a lateral periodic arrangement of structures and precise control of structure diameters and the height (see Sect. 11.6). This cannot be readily achieved on bare substrates. It requires patterning of the substrate prior to OAD or GLAD process. Malac et al. [90] were the first to cover the substrate with hillock-like features that serve as seeds for the structures that grow in subsequent oblique deposition process.

The principle of the patterning of the substrate is shown in Fig. 11.26. The substrate is patterned using various techniques (next subchapter) to create an array of seed points characterized by the parameters of spacing, height, width, lattice type, and shape of the seeds [6, 19, 91–94]. If the shadowing length L cast by an (artificial) seed point of height h_s is larger than the spacing d, no condensation between the seeds should occurs. The section of a column or seed that can be still directly impacted by the incoming particle flux is the exposing height Λ. As the deposition progresses, this should result in the artificial nuclei capturing all incoming particles, creating structures that develop only on the provided nuclei and mimic the periodicity of the nuclei. The shadowing length is given by (11.1) and the exposing height by

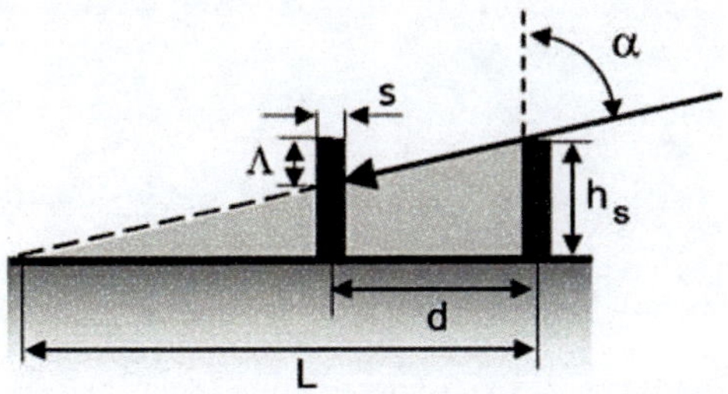

Fig. 11.26 Schematic diagram illustrating shadowing length L, exposing height Λ, the width s (diameter) and spacing between two pattering points d

$$\Lambda = \begin{cases} d/\tan\alpha & for\ d \le L \\ h_s & for\ d > L \end{cases}, \tag{11.13}$$

where the diameter (width, s, see Fig. 11.26) of the seed points is ignored. Then, the minimum height of seeds to achieve the nearest-neighbor full shadowing is given for $\Lambda = h_s$, i.e. $\alpha = \arctan(d/h_s)$, cf. (11.1).

The shadow effect is demonstrated in the following SEM image sequence. Figure 11.27 shows the effect of a microscopic dust particle on the distribution of material (here molybdenum) deposited at an angle of incidence $\alpha = 84°$ on an oxidized Si(001) surface [23]. The angle of rotation ε around the surface normal of the substrate was varied in defined angular steps, Δε (see also Fig. 11.4). At each of the selected angles ε, material is deposited onto the substrate. After that the substrate is quickly rotated by a desired small rotation angle Δε. The shadowed areas are caused from the dust particle in the center of each SEM image. Depending on the selected angular step size, various shadow areas are created. The shadowed areas in Fig. 11.27, right, are not visible, because of the small revolution angle.

Of particular importance is the issue of shadowing by a nucleation point on the neighboring points when the points are arranged periodically on the surface of the substrate [92].

Assuming a square arrangement of the nucleation points (Fig. 11.28) with a center given at (00). The seed points at this position (00) should have a shadowing length L (radius of the gray circle in Fig. 11.28). The seed points with indices (hk) inside the circle can have a shadowing effect on seed point (00), where the shadowing length between seed points (00) and (hk) is $d = \sqrt{h^2 + k^2}$ (a is the lattice constant of the square lattice). When the patterned substrate is rotated, the exposure height Λ changes due to the change of the shadow direction. All seed points (hk) within the shadow circle will shade the point (00). Ye et al. [92, 93] were able to find that all the seeds shading a seed at (00) are its nearest neighbors, second nearest neighbors, fourth nearest neighbors, seventh nearest neighbors, etc. Consequently, (11.13) can

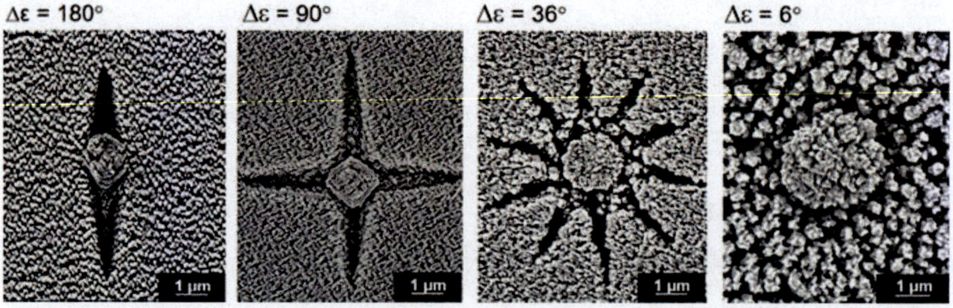

Fig. 11.27 Top-view SEM images of twofold (zigzag, Δε = 180°), (b) fourfold (Δε = 90°), tenfold (Δε = 36°) and 60-fold (Δε = 6°) shading structures after deposition of Mo at α = 84° and at room temperature on natively oxidized Si(001) substrates. The shadowed areas are caused from the dust particle in the center of each SEM image. ε is the revolution angle and Δε is the step width of the rotation angle

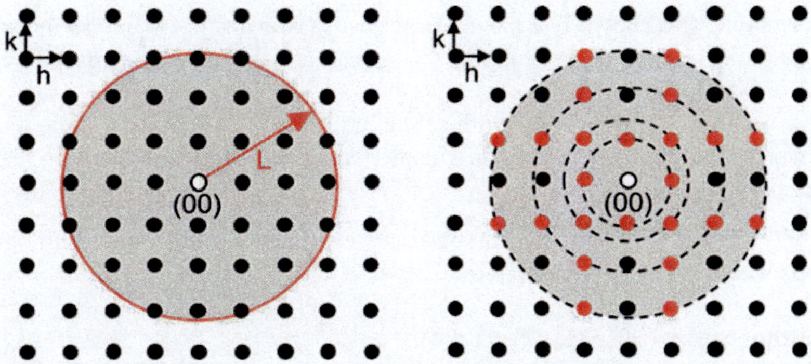

Fig. 11.28 Shadow conditions (left) and highlighted seed points (red) that have a shadowing effect on the seed at (00) (right) outlined for square lattice. Figure is taken from [93] and modified

be rewritten for a square lattice as follows

$$\Lambda = \frac{a\sqrt{h^2 + k^2}}{\tan\alpha} \tag{11.14}$$

or if the width (diameter) of the seed point, s, is included, as follows

$$\Lambda = \frac{a\sqrt{h^2 + k^2} - s}{\tan\alpha}. \tag{11.15}$$

Similar expressions were obtained for the hexagonal arrangement of seed points on flat surfaces [92]. The individual shape (cross-section) of the seed points affects the size of the shaded area. Pyramidal seed points, for example, provide different shades than cylindrical ones [5].

11.5.1 Pattering Techniques

Various patterning techniques are used to achieve regular arrays of seed points. General requirements on micro- and nanofabrication to produce seed patterns are:

- the patterned areas should be on order of square millimeters to a few square centimeters,
- the width of the individual seed should be between a few tens of nanometers and up to one micrometer,
- the distance between the seed point should vary between a hundred nanometers and few micrometers,

- in principle, the known five basic planar grids (Bravais grids such as the square, rectangular, centered rectangular, tetragonal and hexagonal grids) should be realized, and
- the average deviation of the width of the individual seed points and the distance between seed points should be less than 10% of the width and distance between seed points, respectively.

In order to create pre-patterned substrates on a large scale, various lithography methods were preferentially applied:

Photolithography (see e.g., [78, 91, 94])
This technique for transferring a pattern onto a substrate is based on the following approach. After cleaning the surface, a photoresist is applied, such as a spin coating. Then the photoresist is exposed to intense light, which causes chemical changes. Using a special solution, called developer, the exposed areas or the unexposed areas of the deposited substrate can be dissolved by the positive or negative photoresist, respectively. The patterned photoresist can be used as a starting point, or the resist pattern can be transferred to the underlying material by chemical (wet etching) or physical etching techniques (dry etching).

Advantages (for fabrication of OAD or GLAD seed layers):

- versatile and fast procedure,
- large area size (wafer),
- mass production.

Disadvantages:

- resolution is limited by diffraction at about the half of the used wavelength (Abbe criterion),
- expensive equipment, and
- each seed arrangement requires a special mask.

Electron Beam Lithography (EBL, see e.g., [20, 90, 95, 96])
Electron beam lithography is a method of producing submicron and nanoscale seed dots by exposing electron radiation sensitive surfaces (resists) to an electron beam. It exploits the fact that certain resists change their properties when irradiated with electrons. By computer control of the position of the electron beam, it is possible to write arbitrary structures on a surface, so that the original digital image can be transferred directly to the desired substrate. Figure 11.29 shows the application of EBL in combination with the generation of nanostructures by direct ion beam sputtering including glancing angle deposition.

Advantages:

- diameters $s < 100$ nm of individual seed points can be realized,
- process is mask-less,
- high flexibility in the arrangement of the seed points on substrate surface.

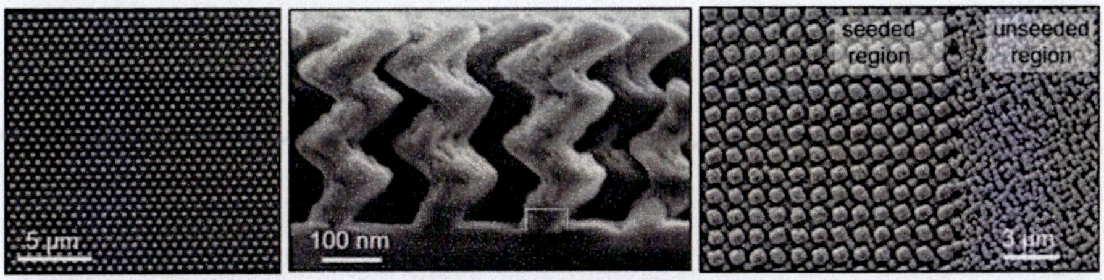

Fig. 11.29 SEM micrographs of structures prepared by direct ion beam sputtering -induced glancing angle deposition on substrates pre-patterned by electron beam lithography. Left: Pre-pattered Si substrate after EBL and before deposition. Middle: Silicon zig-zag structures on circular seed points (diameter of the seed point is about 80 nm). Right: Si columns grown on pre-pattered and non-patterned Si substrates

Disadvantages:

- very high production times (low throughput),
- expensive equipment and production costs.

Nano-sphere Lithography (NSL, see e.g., [44, 91, 97–100])
This lithographic technique is based on the coating the surface with a suspension containing monodisperse spherical colloids (e.g., polystyrene spheres). By self-assembly, monolayers or bilayers of hexagonal close-packed (hcp) nanospheres with controllable lattice constant are formed and used as lithographic mask (Fig. 11.30, left). After the polystyrene sphere mono- or bilayers are formed on substrate, metals (Au, Ag, etc.) or silicon are deposited through this mask, which contains triangular holes between the spheres. In contrast to masks of a monolayer of polystyrene spheres, where the deposited metal particles are arranged in a hexagonal two-dimensional lattice with a two-point basis (honeycomb structure), a primitive hexagonal arrangement of the metal clusters is obtained for polystyrene bilayers. After deposition, the nanospheres are removed from the surface, leaving metal seed points in the regions corresponding to the void between the nanospheres (Fig. 11.30, middle). Subsequently, the OAD or GLAD procedure can be performed on pre-pattered substrates (Fig. 11.30, left).

Advantages:

- cost-effective, simple and fast lithography tool,
- precisely controlled spacing,
- high resolution.

Disadvantages:

- the area of the ordered seed points is smaller than one square centimeter, and
- only two seed arrangements (hexagonal and honeycomb) are possible.

In addition to these commonly used techniques, the nanoimprint [101], the laser direct-write lithography [5, 77], the pattering with block copolymers [102] or the

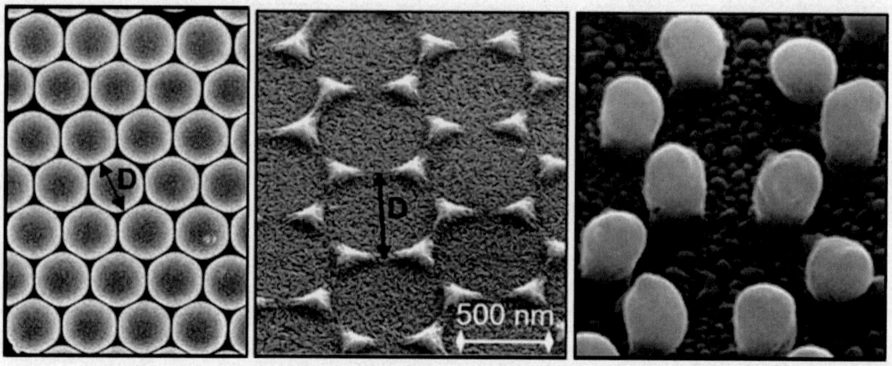

Fig. 11.30 SEM micrographs of the nanosphere preparation route (D is the diameter of the nanospheres). Left: top-view of a monolayer of polystyrene spheres on a Si surface. Middle: honeycomb arrangement of Si seed array after Si deposition and removing of the spheres. Right: amorphous Si nanostructures grown on the hexagonal arranged seed points after ion- beam-induced sputter deposition at an incidence angle of 86° and at room temperature

diffraction mask projection laser ablation [103] are applied to produce pre-patterned substrate surfaces.

11.5.2 Arrays of High-Regular Nanostructures and Its Design

It is obvious that the complete shadowing of nearest-neighbor is given by (11.13) and Fig. 11.26 for $\Lambda = h_s$, i.e. $\alpha = \arctan(d/h_s)$. If the diameter of the seed, s, is included in this consideration (see Fig. 11.26), no deposition of incident material between the seed points is expected if $d \leq h_s \tan\alpha + s$ [104]. With the aim of producing periodically ordered structures, this condition should be considered in seed deposition.

Periodically ordered nanostructures can be fabricated by ion beam sputtering on differently pre-patterned substrates. For this purpose, samples with square, honeycomb-like, and hexagonally closed packed (hcp) arrangement of the artificial seeds were prepared for subsequent deposition at oblique particle incidence.

For example, the following SEM images (Fig. 11.31) demonstrate the variety of the structures achievable with this method [19, 72]. Exact structural uniformity over large areas (on order of square centimeters) is possible when the design parameters, such as seed height, inter-seed spacing or periodicity, and seed arrangement are well controllable. In the following, the influence of various design parameters on the

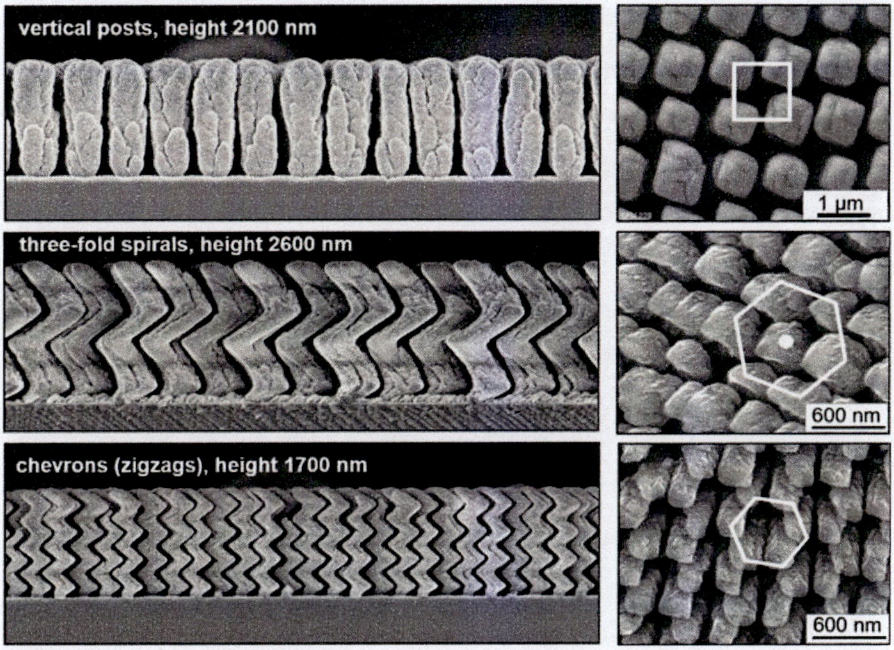

Fig. 11.31 Cross-section (left) and top-view (right) SEM micrographs of sculptured thin Si films prepared by ion-sputtering-induced glancing angle deposition [19]. Left: Si nanostructures on Si substrates with EBL templates, where the inter-seed distance was 150 nm and the seed height about 100 nm. Right: Si nanostructures on NSL pre-patterned Si substrates

morphological evolution of nanostructures prepared by sputtering-induced GLAD on patterned substrates is briefly discussed.

Arrangement of the Seed Points

Experimental studies [72, 91] and MC simulations [44] have shown that the column morphology is strongly influenced by the arrangement of the seed points on the underlying substrate. For example, Si columns were prepared by ion beam sputtering-induced GLAD onto square, honey-comb-like and hexagonally closed packed (hcp) patterns at a glancing incidence angle $\alpha = 85°$ (divergence: $9° \pm 3°$), and with a deposition rate of about (3.5 ± 0.2) nm/min. In Fig. 11.32 (top row), SEM micrographs of the three different patterns and the indicated in-plane distances [72, 91]. The bottom row of Fig. 11.32 shows SEM micrographs of columnar Si nanostructures deposited on the different seed patterns. Figure 11.32a shows that in the case of the square pattern and a sufficient height of the seed points, the resulting structures have a columnar shape with an almost circular cross-section. The situation is different, as shown in Fig. 11.32b, in the case of a honeycomb-like pattern on the substrate. GLAD with fast substrate rotation leads to the evolution of columnar structures with a triangular cross-section. According to Main et al. [105], this difference can be explained

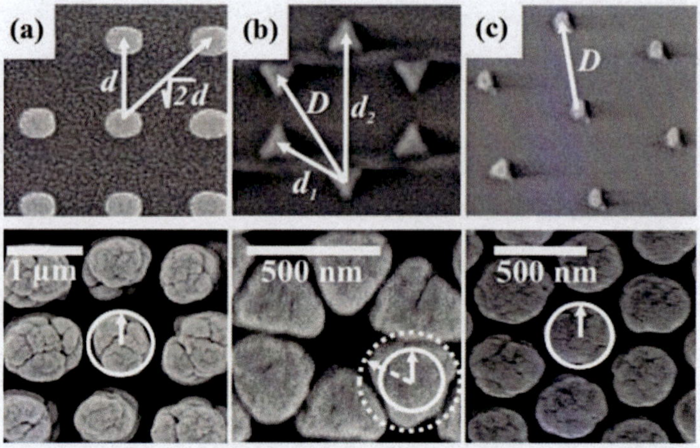

Fig. 11.32 Upper row: Top-view SEM micrographs of the patterned surfaces fabricated by EBL (**a**) and NSL (**b** and **c**), where D is the diameter of the nanosphere, d is the period, $d_1 = D\sqrt{3}$ is the distance of the three nearest neighbors and $d_2 = 2d_1 = 2D\sqrt{3}$ is the distance of the second-nearest neighbors. Bottom row: Top-view SEM micrographs of the grown Si columns on substrates with different pre-pattern. **a** square seed pattern: the white circle and arrow indicate the saturation radius R. **b** Honeycomb seed pattern: the full circle and arrow indicate the saturation radius R_1 in direction d_1 of the nearest neighbor, and the dotted circle and arrow indicate the different value of the *saturation radius* R_2 in direction d_2. **c** hcp seed pattern: the white circle and arrow indicate the saturation radius. Figures are taken from [91]

by the symmetry of the seed layer. In the case of the square seed design, the saturation radius R of the nearly cylindrical shaped structures grows exponentially with the distance of the nearest neighbors (period d), (see Sect. 6.1.2). In the case of a tetragonal pattern, each seed point is surrounded by eight neighboring seeds (four nearest neighbors and four second nearest neighbors) in a circle with a radius of $\sqrt{2}d$ (see also Fig. 11.28). Therefore, the growth front of the evolving structure meets almost the same 'free volume' in each growth direction. Consequently, the final structure (after reaching the saturation radius R) has an approximately circular cross-section. On honeycomb patterned substrates, each seed point has three nearest neighbors at a distance $d_1 = D/\sqrt{3}$d (as opposed to eight nearest neighbors in the case of the square pattern). Therefore, the shadowing effect of the surrounding nuclei (and the columns that develop from them) is not nearly multidirectional as in the tetragonal case. On the contrary, the growing structures will experience a strong shadowing effect in the direction toward the nearest neighbors, but will have more space to fill in the other directions. The second nearest neighbors are of little importance for the growth of the columnar structure on the examined seed point. Because of these widely varying distances from one seed to its neighbors, two growth directions occur in the case of a honeycomb template [97]. In the end, the final columnar structure has a cross-section resembling an equilateral triangle, with the sides pointing to the nearest neighbors and the tips pointing to the third nearest neighbors [91]. Hawkeye et al. [6] have therefore concluded that the larger the distance between the nucleation points, the more rotationally symmetric the resulting columnar structure.

In summary, the arrangement of the seed points themselves, i.e. the number, spacing, and relative position of the seed points to each other, has a strong influence on the development of structural morphology during GLAD.

Dimension of the Seeds

Studies have shown that the growth of nanocolumns is strongly influence by pattern seed width, s, and the period of the seed arrangement [75, 80, 91, 106], where distance d (c.f. Figure 11.26) now represents the periodicity. Basically, a distinction can be made between $s \ll d$, i.e. $s/d \ll 1$ (seed size is small compared to the period) and $s < d$, i.e. $s/d < 1$ (seed size is only slightly smaller than period). Figure 11.33 illustrates the growth for both cases. For $s \ll d$, closely bunched multiple nanocolumns evolve on each seed point [106]. This effect can be attributed to a pronounced column competition [94]. Thus, a broad feature width and a small feature spacing lead to growth of multiple nanocolumns on a single Si seed point. An example of a nanocolumn grown on such structure can be seen in the inset of Fig. 11.33, left, which grew as an individual entity. Si pattern formed after the patterning process is also distinctly visible under the grown nanocolumn. Moreover, the growth can also be observed in the spaces between the seeds and on the sidewalls of the Si pattern. This growth could be attributed to exposure height Λ [92]. Obviously, a larger inter-seed distance strongly influenced the growth on these structural features and allowing the formation of isolated columns. In Fig. 11.34, left, schematically summarized these growth processes [75, 106]. For $s < d$ (i.e. larger seed sizes relative to the period), the growth behavior is strongly altered (Fig. 11.33, middle and right). In this case, increased growth is observed at the nucleation boundaries. These structures tend to broaden with increasing height [2, 107], leading to extinction of some structures. Structures with an open center emerge. This behavior is expected when $s \geq d/2$ at incidence angles $\alpha > 70°$.

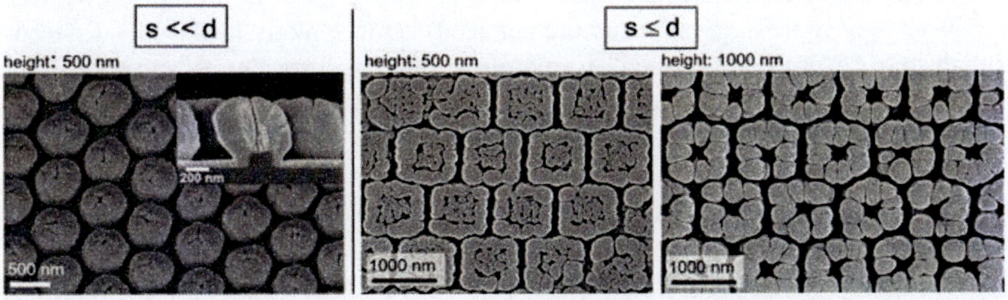

Fig. 11.33 Top-view SEM micrographs of Ge columns grown on patterned Si substrates by ion beam sputter -induced glancing angle deposition. Left: 500 nm columns grew on hcp Si array under the condition that $s/d \approx 0.28$ ($s = 200$ nm and $d = 722$ nm, i.e. $s \ll d$). The inset represent corresponding cross-section SEM micrograph. Right: Ge columns with a height of 500 nm and 1000 nm grown on hcp Si arrays under the condition that $s < d$

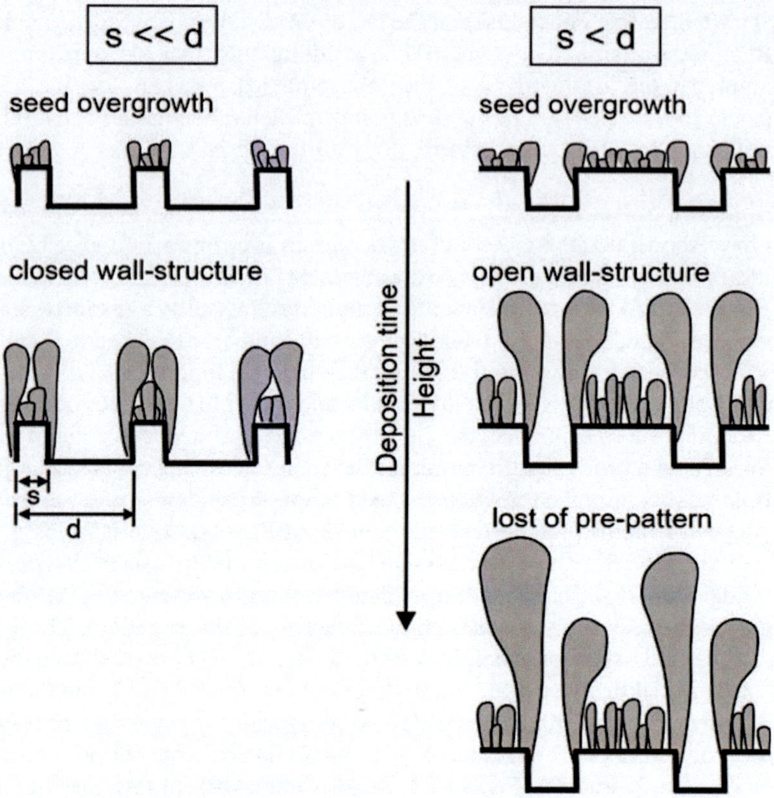

Fig. 11.34 Schematic diagram illustrating competitive growth for s ≪ d (left) and s < d (right), where s is the seed size and d is the distance between two seed points

It is evident that in competitive growth, when the seed size is too large relative to the equilibrium structure size on planar substrates, the growth of multiple structures (i.e. survival of more than one structure per seed) is more likely [107]. The different growth steps (overgrowth of seeds, formation of open wall structures, and loss of the pre-patterned structure) are shown schematically on the right in Fig. 11.33 [75, 106].

Influence of the Periodicity

In the following, the distance between next-neighboring pattern points is referred to as pattern periodicity. When using NSL technique and honeycomb patterns, the periodicity d is given by $d = D/\sqrt{3}$, where D is the diameter of the spheres, i.e. periodicity is definable within the pre-pattern process. Figure 11.35 presents the influence of the periodicity on the shape of Ge nanocolumns for different angles of incidence of sputtered Ge particles [106, 108]. As the angle of incidence increases, the structures become broader and are closer together. As a result, the cross-section shape increases, corresponding to threefold honeycomb symmetry. In addition, the cross-section size increases with periodicity. However, the tendency of the cross-section shape to reflect pattern symmetry decreases. The structure shape changes from triangular to hexagonal/round with increasing pattern periodicity and decreasing incidence angle. The shape evolution can be explained by nearest neighbors shadowing. The formation of the cross-section shape is determined by the ratio of the average interspace distance z (Fig. 11.35) and the periodicity d, i.e. z/d. As the periodicity decreases, this ratio increases. Thus, the nanocolumns are exposed to a higher flux of material from the main incidence (see arrows in Fig. 11.35). As a result, the edge planes directed toward the third nearest neighbors become more pronounced, resulting in a triangular cross-sectional shape.

Consequently, cross-section shape can be precisely adjusted from triangular at low honeycomb periodicity and high glancing angle incidence to hexagonal at high honeycomb lattice periodicity and small glancing angle.

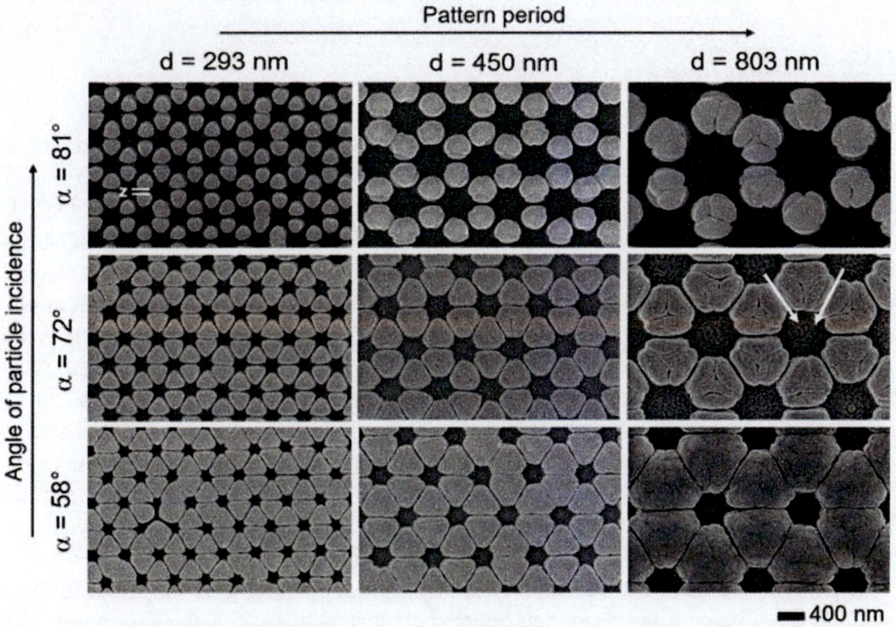

Fig. 11.35 SEM top views of Ge nanocolumns prepared by ion beam sputter-induced glancing angle deposition in honeycomb arrangement. The seed points are deposited using NSL. The arrows indicate the main directions of incidence for the formation between the planes of each structure (figures are taken from [106])

11.6 Applications of Films Prepared by Ion Sputter Induced Glancing Angle Deposition

Numerous comprehensive reports have extensively reported on the applications of thin films prepared by OAD or GLAD (see, e.g., reviews [2, 4–7]). It has been shown that nanostructured GLAD and OAD films using conventional PVD methods such as electron beam evaporation, laser ablation or magnetron sputtering are promising candidates for the fabrication of mechanical and optical devices, magnetic data storage devices, sensors, etc. In this chapter, two examples (fabrication of biosensor and magnetic nanotubes) are used to demonstrate applications of micro- and nanostructured thin films deposited by sputter-induced glancing angle deposition.

11.6.1 Biosensors

Plasmon-based thin-film biosensors have been developed for the detection of a variety of target samples such as pesticides, contaminants, viruses, and diseased tissue [109, 110]. The use of effects such as surface-enhanced fluorescence (SEF), surface-enhanced Raman scattering (SERS), and surface-enhanced infrared absorption (SEIRA) enables the detection of minimal amounts of these substances. Sculptured thin films are also suitable for biochip applications using a 2D array of small pixels (< 0.1 mm^2 per pixel) on a substrate to detect a large number of analytes in parallel (lab-on-a-chip systems). The signal from Raman active molecules is enhanced by many orders of magnitude when these molecules are brought close to metal surfaces (< a few nm) [109, 110]. This effect is typically attributed to the strong enhancement of the electric field near light-induced surface plasmons and can be quantified by SERS [111, 112] and used to detect and measure a variety of different samples such as DNA, bisphenol-A-polycarbonate, hemoglobin, viruses, etc. (see e.g., [109]).

The extent of amplification is influenced by several parameters, such as the surface area and the distance between neighboring nanostructures. On the one hand, the larger the surface area, the more molecules can be adsorbed and the more molecules contribute to the signal. On the other hand, the smaller the distance between nanostructures, the higher the field intensity. The shape and curvature of the cap of the nanostructures also play an important role, because higher curvature leads to higher field strength. The porosity of the nanostructured films plays another crucial role in sensor efficiency, as the total surface area of the film and the distance between individual nanostructures are strongly related to it. The simplest way to change the properties of the sculptured films is to vary the angle of incidence.

Frequently, silver is used by ion beam sputter-induced GLAD to fabricate biosensors based on surface-enhanced fluorescence or Raman scattering [115–117] because it can be passivated with a thin surface layer to become stable against oxidation and

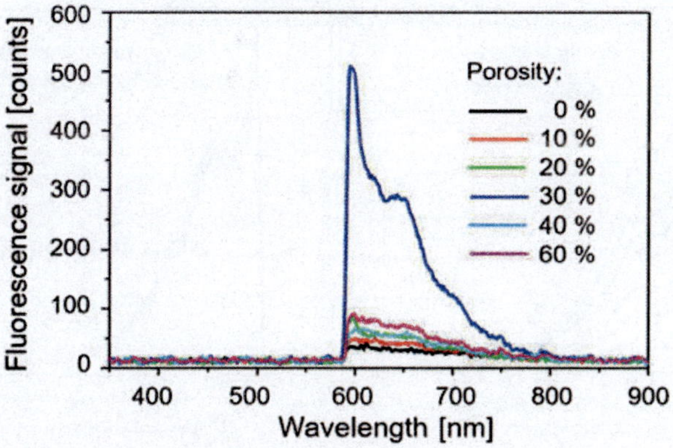

Fig. 11.36 Dependence of the surface enhanced fluorescence signal strength on the porosity for nanostructured 350 nm thick Ag films

offers high enhancement factors. As is also well known, a more oblique incidence enhances the shadowing effect and leads to more separated nanostructures and higher film porosity (see Sect. 11.1). Figure 11.36 illustrates the influence of the porosity of sculptured thin Ag films on the fluorescence signal. A high sensitivity of fluorescence signal intensities from the film porosity was found, with a maximum at a porosity of around 30% [8, 113]. Finally, the surface of the nanostructures (roughness, composition, crystal structure, etc.) also has an influence on the chemical stability and behavior of the material during film growth [114].

Generally, various functionalization procedures are carried out to allow, first, a Raman-active agent to react with the surface of the nanostructure and, second, the sensor to selectively respond to specific molecules [113, 118].

Quantification of Glycated Hemoglobin [118]

Diabetes is one of the major health concerns worldwide and is mainly monitored by measuring of immediate fasting glycemic index of glucose either in the saliva or blood plasma. Glycated hemoglobin (HbA1c) results from the glycosylation of hemoglobin (Hb) in the blood. HbA1c is a more reliable measure of diabetes than Hb because glycemic index remains constant over a period of three months and has therefore become one of the important tests for confirming and controlling diabetes. Acceptable HbA1c levels in human blood generally range from 4 % to 8 % of total human Hb [119]. As a result, a GLAD based sensor for HbA1c detection was developed that has increased intrinsic fluorescence and does not require any fluorescent tags.

This sensor was designed over 4-ATP layered carbonized Ag sculptured thin films by immobilizing a self-assembled monolayer of anti-HbA1c (anti-glycated hemoglobin) antibody over it. The non-specific binding sites were blocked by bovine serum albumin (BSA) to prevent the occurrence of any false signals (for details see [8, 118]). Figure 11.37, left, shows the intrinsic fluorescence spectra obtained from the sensor chip for different concentrations of HbA1c [118]. With an increase in the HbA1c concentration, a trend of decrease in the fluorescence intensity was

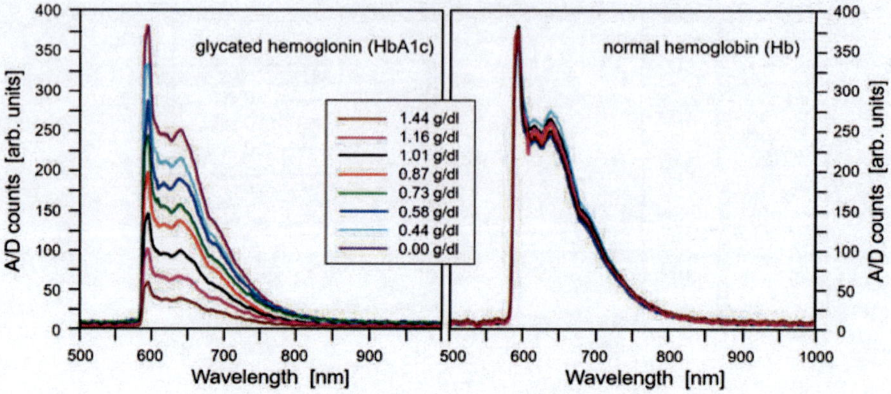

Fig. 11.37 Enhanced intrinsic fluorescence spectra for varying concentrations of HbA1c (left) and, normal Hb-concentrations (right) on the carbonized nano-sculptured Ag film sensor chips

observed. The reason for decrease in fluorescence intensity is the blocking of the fluorescence sites by the addition of HbA1c molecules on the sensor chip. Negative control experiments on Hb were performed to ensure the specificity of the present sensor. Figure 11.37, right, presents the intrinsic fluorescence spectra for different concentrations of Hb. It can be seen that the fluorescence intensity does not change with increasing Hb concentration. The total SEF enhancement per molecule was estimated to be 1.06×10^3, while that on the tip of nanocolumns was 1.45×10^4. In addition, negative control experiments were performed with Hb to ensure the specificity of the present sensor. To date, fluorescence enhancements of up to three orders of magnitude have been achieved for quite complicated geometries fabricated using traditional methods [120].

Applications for Detection of Biomarkers [113]

The detection of biomarkers generated by ion beam sputtering induced GLAD is demonstrated using endocrine-disrupting compounds as example. A biosensor chip is developed for the detection of a protein biomarker for endocrine disrupting compounds, vitellogenin, in aquatic environment. The aim is to quantify the tiniest amounts of this compound. Endocrine system is an ensemble of glands that secrete various hormones directly into the blood stream to maintain homeostasis. The endocrine system regulates many vital processes in a living organism, such as growth, reproduction and metabolism. Any disturbance in the function of the endocrine system results in impaired metabolism, birth rate, growth, and many other complex processes. Since these compounds can accumulate in organisms over time, even the existence of traces of these substances in the environment is hazardous. Their direct detection is difficult, not only because of their low concentration, but also because their spatial and temporal distribution is usually unknown.

A better choice is to utilize the accumulation of the endocrine disruptors in organisms and to check directly if they are affected. The protein vitellogenin (Vg), which is an egg yolk precursor and synthesized only by the female individuals of vertebrates,

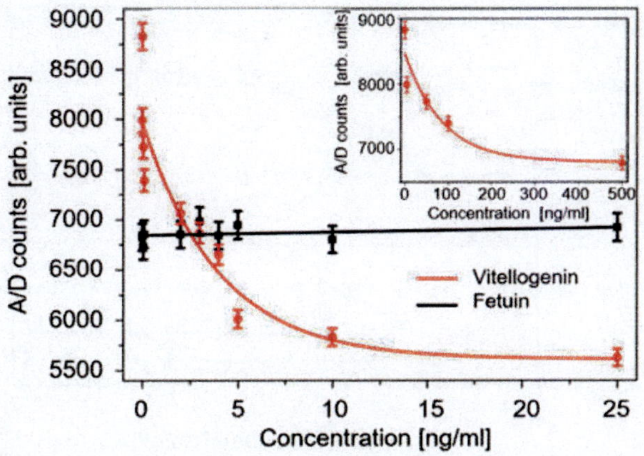

Fig. 11.38 Comparison of change of SERS signal intensity of the 1077 cm^{-1} peak for different concentrations of vitellogenin and fetuin (figure is taken from [113])

but not by their male counterparts. If Vg is found in male individuals, it is direct evidence that they are affected by an estrogenic endocrine-disrupting compound.

The sensor is prepared in a similar manner as for the detection of glycosylated hemoglobin, but now with an anti-Vg antibody (see [113] for details). The reduction of the intensity of the Raman signal of 4-ATP at a shift of 1077 cm^{-1} is taken as a measure of the Vg concentration in the test solution (Fig. 11.38). Even a concentration as low as 5 pg/ml leads to a intensity of around 10 %, which can be easily detected. To show the selectivity of the sensor, different concentrations of another protein (fetuin) were applied to it. No change in the Raman signal intensity could be found in this case (Fig. 11.38). Sensors prepared as described can be stored for at least 3 months, without the loss of response efficiency.

Detection of Bacteria Concentrations [121]

With the objective to detect bacteria concentrations down to one bacterium per 10 μl, a bacteriophage functionalized sensor can be used, where the use of bacteriophages enables to distinguish between living and dead bacteria [122]. The preparation of the sensor by ion beam-induced sputter GLAD is described in detail in [113, 121]. After the formation of the 4-ATP layer a glutaraldehyde cross-linking layer is added to serve as substrate for the bacteriophages. When living bacteria come close to the bacteriophages, DNA is injected into them, forcing the bacteria to produce more phages. During this process the intensity of the observed Raman signals is increased.

This process allows selective detection of, for example, Escherichia coli (E. coli) bacteria down to a concentration of 150 cfu ml^{-1} (cfu-colony forming units). Based on recorded SERS spectra the Raman enhancement versus concentration at 1077 cm^{-1} is shown in Fig. 11.39 for E. coli B and E. coli μX. The measurement requires 10 μl of the test medium, which corresponds to a single peak. The peak was selected because of the maximum gain at this band, which corresponds to the greatest sensitivity and dynamic range of the sensor. Specificity can also be investigated. For this purpose,

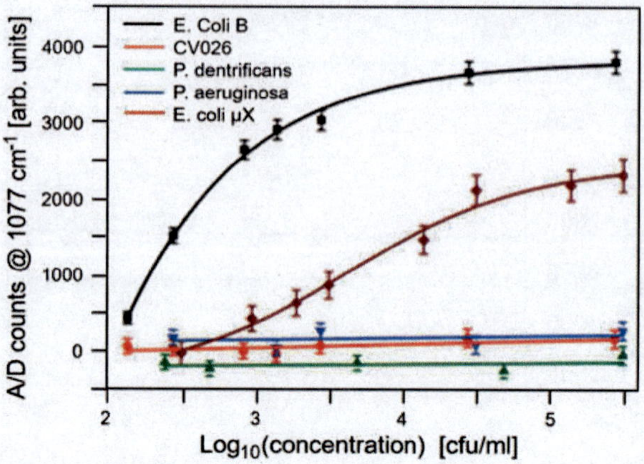

Fig. 11.39 SERS intensity of the 1077 cm^{-1} peak in dependence on the concentration (in cfu/ml)] for five different bacteria strains (figure taken from [121])

different bacterial strains (Chromobacterium violaceum, Paracoccus denitrificans, Pseudomonas aeruginosa) were applied to the sensor and no change in Raman signal intensity was detected, as shown in Fig. 11.38.

Therefore, this type of sensor allows detection of a single bacterium in a mixture of different strains, as normally occurs in clinical or environmental cultures.

11.6.2 Magnetic Nanotubes

The combination of glancing angle sputter deposition and atomic layer deposition (ALD) offers the possibility to prepare magnetic nanotubes whose magnetic behavior can be tailored [123, 124]. For this purpose, sculptural thin Si films can be deposited, consisting of either vertical columnar structures deposited with glancing angle deposition and fast substrate rotation (structure height about 520 nm) or inclined columnar structures deposited by oblique angle deposition (structure height about 600 nm). These columns are coated with Fe_2O_3 using of ALD, resulting in tubular structures with defined wall thicknesses surrounding the Si columns. Subsequent reduction in Ar/H_2 atmosphere converts the Fe_2O_3 tubes into ferrimagnetic Fe_3O_4 tubes (details of the preparation see [123]). At a deposition rate of nominally 0.02 nm/rev., 500 ALD cycles resulted in Fe_2O_3 coating of the Si pillars with a thickness of about 10 nm. Figure 11.40 illustrates the tilted Si nanostructure grown at different incident angles and a TEM micrograph (Fig. 11.40d) showing a distinct Fe_3O_4 layer of constant thickness covering the Si nanocolumn, with the 5 nm thick SiO_2 protective insulating layer preventing aerobic re-oxidation of the magnetic Fe_3O_4 material.

To evaluate the magnetic response of the Fe_3O_4-coated thin Si films, the changes in the coercive field H_C (i.e. the field strength required to demagnetize the material after magnetization in a field of strength H) were measured and angular-dependent

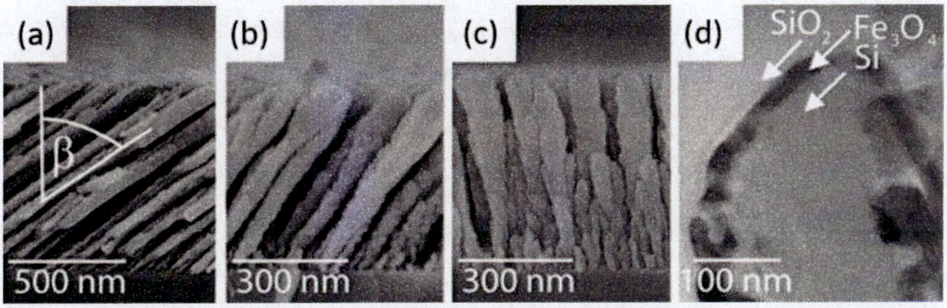

Fig. 11.40 SEM micrographs of inclined Si columns prepared by sputter -induced glancing angle deposition with different tilt angles: (**a**) $\beta = (57 \pm 2)°$, (**b**) $\beta = (47 \pm 2)°$, and (**c**) $\beta = (0 \pm 2)°$. The TEM bright-field micrograph (**d**) of the top of a GLAD-grown Si structure after ALD indicates the conformal coverage of the Si template with 20 nm Fe_3O_4 and 5 $_{nm}$ SiO_2 [123]

recorded (with respect to the substrate normal) using SQUID at T = 300 K. For example, Fig. 11.41, left, shows a recorded hysteresis loop for extracting the coercivity H_C [123, 124]. The measurement of H_C is repeated at every angle δ (angle between the surface normal of the sample and the tilt direction the sample). Drawing $\Delta H_C = H_C(\delta) - H_C(0°)$ as a function of δ, remarkable differences are observed, as shown in Fig. 11.41 (right). While H_C remains approximately constant in the case of the vertical columnar structures, ΔH_C increases monotonically from $\delta = 0°$ (perpendicular alignment of the projection of the long axis of the Fe_3O_4 tubes on the substrate surface with respect to the direction of H) to $\delta \approx 90°$ (parallel alignment of the directions of H and the projection of the long tube axis on the substrate surface) in the case of inclined columns.

It can be concluded that for the tilted tubular ferrimagnetic Fe_3O_4 structures, the easy magnetic axis is in the direction of the long axis of the tube. In contrast, for vertical columns, the coercivity H_C is independent of the angle. Consequently,

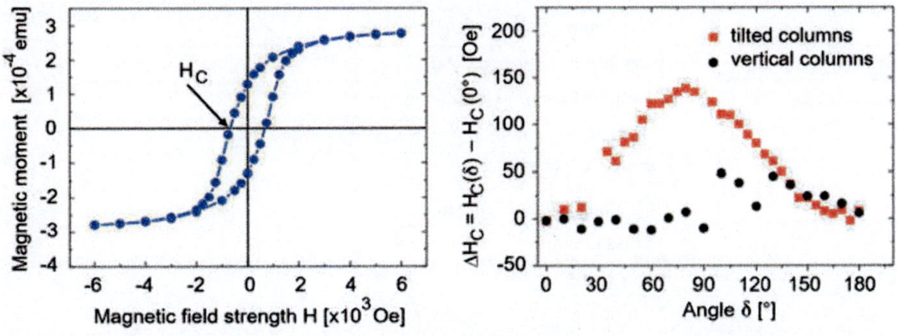

Fig. 11.41 Left: Example of a magnetization hysteresis loop to extract the coercivity H_C of a sculptured thin film consisting of inclined tube-like Fe_3O_4 structures ($\delta = 57°$). Right: Dependency of H_C on the angle δ between magnetic field H and direction of column growth in terms of $\Delta H_C = H_C(\delta) - H_C\ (0°)$

tubular magnetic nanostructures can be deposited whose magnetic behavior can be tailored.

11.7 List of Symbols

Symbol	Notation
D	Diameter of a sphere
D_{sd}	Coefficient of the self-diffusion
H	Film thickness after perpendicular deposition
H_c	Coercivity
J_a	Atom flux density behind an aperture
J_s	Sputtered atom flux density
L	Shadow length
P	Porosity
T	Temperature
T_M	Melting temperature
d	Distance between two pattering points (periodicity)
h	Film thickness after oblique deposition
h_s	Structure height
r_s	Radius of the sputter spot on the target (beam radius)
$r(\varphi)$	Radius of the calotte curvature
s	Seed diameter
v	Growth velocity, deposition rate
w	Diameter of the nanostructure
z	Average interspace distance
Λ	Exposing height at OAD
Φ	Double fan angle ($= 2\varphi$)
α	Angle of particle incidence
β	Tilt angle, growth angle or scaling exponent
ε	Rotation angle
λ	Column-column distance
ρ	Mass density
φ	Fan angle
φ_{div}	Angle of beam divergence
ω	Azimuthal rotation frequency

References

1. H. van Kranenburg, C. Lodder, Tailoring growth and local composition by oblique-incidence deposition: a review and new experimental data. Mater. Sci. Eng. R **11**, 295–354 (1994)
2. A. Lakhtakia, R. Messier, *Sculptured thin films: nanoengineered morphology and optics* (SPIE Press, Bellingham, 2005)
3. T. Karabacak, T.-M. Lu, Shadowing growth and physical self-assembly of 3D columnar structures, in *Handbook of Theoretical and Computational Nanotechnology*, ed. by M. Rieth, W. Schommers (American Scientific Publishers, Stevenson Ranch, 2005), Chap. 69, p. 729
4. M.M. Hawkeye, M.J. Brett, Glancing angle deposition: fabrication, properties, and applications of micro-and nanostructured thin films. J. Vac. Sci. Technol. A **25**, 1317–1335 (2007)
5. M.T. Taschuk, M.M. Hawkeye, M.J. Brett, Glancing angle deposition, in *Handbook of deposition technologies for films and coatings*, ed. by P.M. Martin, (Elsevier, 2010), pp. 621–678
6. M.M. Hawkeye, M.T. Taschuk, M.J. Brett, *Glancing angle deposition of thin films: engineering the nanoscale* (Wiley, New York, 2014)
7. A. Barranco, A. Borras, A.R. Gonzalez-Elipe, A. Palmero, Perspectives on oblique angle deposition of thin films: From fundamentals to devices. Progr. Mater. Sci. **76**, 59–153 (2016)
8. C. Grüner, I. Abdulhalim, B. Rauschenbach, Glancing angle deposition for biosensing applications, in *Encyclopedia of Interfacial Chemistry*, ed. by K. Wandelt (Elsevier, Oxford, 2018), pp. 129–137
9. B.A. Movchan, A.V. Demchishin, Study of structure and properties of thick condensates of nickel, titanium, tungsten, aluminum oxide and zirconium oxide in vacuum. Fiz. Metal. Metalloved. **28**, 653–660 (1969)
10. A. Kundt, Ueber Doppelbrechung des Lichtes in Metallschichten, welche durch Zerstäuben einer Kathode hergestellt sind. Ann. Phys. Chem. **263**, 59–71 (1886)
11. H. König, G. Helwig, Über die Struktur schräg aufgedampfter Schichten und ihr Einfluss auf die Entwicklung submikroskopischer Oberflächenrauhigkeiten. Optik **6**, 111–124 (1950)
12. J.M. Nieuwenhuizen, H.B. Haanstra, Microfractography of thin films. Philips Tech. Rev. **27**, 87–91 (1966)
13. N.O. Young, J. Kowal, Optically active fluorite films. Nature **183**, 104–105 (1959)
14. K. Robbie, M.J. Brett, A. Lakhtakia, First thin-film realization of a helicoidal bianisotropic medium. J. Vac. Sci. Technol. **13**, 2991–2993 (1995)
15. K. Robbie, L.J. Friedrich, S.K. Dew, T. Smy, M.J. Brett, Fabrication of thin films with highly porous microstructures. J. Vac. Sci. Technol. **13**, 1032–1035 (1995)
16. K. Robbie, M.J. Brett, A. Lakhtakia, Chiral sculptured thin films. Nature **384**, 616–616 (1996)
17. C. Patzig, A. Miessler, T. Karabacak, B. Rauschenbach, Arbitrarily shaped Si nanostructures by glancing angle ion beam sputter deposition. Phys. Stat. Sol. B **247**, 1310–1321 (2010)
18. C. Patzig, B. Rauschenbach, Temperature effect on the glancing angle deposition of Si sculptured thin films. J. Vac. Sci. Technol. A **26**, 881–886 (2008)
19. B. Rauschenbach, C. Patzig, Periodic nanoscale Si structures by ion beam induced glancing angle deposition, in *Proceedings of IEEE 2nd International Nanoelectronic Conference*, Shanghai (2008), pp. 1084–1088
20. C. Patzig, B. Rauschenbach, W. Erfurth, A. Milenin, Ordered silicon nanostructures by ion beam induced glancing angle deposition. J. Vac. Sci. Technol. B **25**, 833–838 (2007)
21. L. Abelmann, C. Lodder, Oblique evaporation and surface diffusion. Thin Solid Films **305**, 1–21 (1997)
22. K. Hara, M. Kamiya, T. Hashimoto, K. Okamoto, H. Fujiwara, Oblique-incidence anisotropy of the iron films evaporated at low substrate temperatures. J. Mag. Mag. Mat. **73**, 161–166 (1988)
23. S. Liedtke-Grüner, in *Growth of obliquely deposited metallic thin films*. Dissertation, University Leipzig (2019)

24. T. Brown, K. Robbie, Observations of self-assembled microscale triangular-shaped spikes in copper and silver thin films. Thin Solid Films **531**, 103–112 (2013)
25. K. Itoh, F. Ichikawa, Y. Takahashi, K. Tsutsumi, Y. Noguchi, K. Okamoto, Columnar grain structure in cobalt films evaporated obliquely at low substrate temperatures. Jpn. J. Appl. Phys. **45**, 2534–2538 (2006)
26. E. Schubert, J. Fahlteich, Th. Höche, G. Wagner, B. Rauschenbach, Chiral silicon nanostructures. Nucl. Instr. Meth in Phys. Res. B **244**, 40–44 (2006)
27. M. Malec, R.F. Egerton, Observations of the microscopic growth mechanism of pillars and helices formed by glancing-angle thin-film deposition. J. Vac. Sci. Technol. A **19**, 158–166 (2001)
28. Z. Dohnálek, G.A. Kimmel, D.E. McCready, J.S. Young, A. Dohnálková, R.S. Smith, B. D. Kay, Structural and chemical characterization of aligned crystalline nanoporous MgO films grown via reactive ballistic deposition. J. Phys. Chem. B **106**, 3526–3529
29. K.M. Krause, D.W. Dick, M. Malac, M.J. Brett, Taking, a little off the top: nanorod arry morphology and growth studied by focused ion beam topography. Langmuir **26**, 17558–21756 (2010)
30. S. Liedtke, C. Grüner, A. Lotnyk, B. Rauschenbach, Glancing angle deposition of sculptured thin metal films at room temperature.Nanotechnology **28**, 385604 (2017)
31. M. J. Vold, A numerical approach to the problem of sediment volume. J. Colloid Sci. **14**, 168–174 (1959) and Sediment volume and structure in dispersions of anisometric particles. J. Phys. Chem. **63**, 1608–1612 (1959)
32. P. Meakin, Models for colloidal aggregation. Ann. Rev. Phys. Chem. **39**, 237–267 (1988)
33. P. Meakin, R. Jullien, Restructuring effects in the rain model for random deposition. J. Physique **48**, 1651–1662 (1987)
34. P. Meakin, R. Jullien, Simple ballistic deposition models for the formation of thin films. Proc. SPIE **0821** (1988)
35. M. Teschner, S. Kimmerle, B. Heidelberger G. Zachmann L. Raghupathi A. Fuhrmann, M.-P. Cani, F. Faure, N. Magnenat-Thalmann, W. Strasser, P. Volino Collision detection for deformable objects. Computer Graphics Forum **24**, 61–81 (2005)
36. D. Henderson, M.H. Brodsky, P. Chaudhari, Simulation of structural anisotropy and void formation in amorphous thin films. Appl. Phys. Lett. **25**, 641–643 (1974)
37. A.G. Dirks, H.A. Leamy, Columnar microstructure in vapor-deposited thin films. Thin Solid Films **47**, 219–233 (1977)
38. H.A. Leamy, G.H. Gilmer, A.G. Dirks, The microstructure of vapor deposited thin films, in *Current Topics in Materials Science*, ed. by E. Kaldis, Vol. 6 (North-Holland, 1980), pp. 301–344
39. P. Ramanlal, L.M. Sander, Theory of ballistic aggregation. Phys. Rev. Lett. **54**, 1828–1831 (1985)
40. R. Jullien, P. Meakin, Simple three-dimensional models for ballistic deposition with restructuring. Europhys. Lett. **4**, 1385–1390 (1987)
41. C. Grüner, S. Liedtke, J. Bauer, S.G. Mayr, B. Rauschenbach, Morphology of thin films formed by oblique physical vapor deposition. ACS Appl. Nano Mater. **1**, 1370–1376 (2018)
42. P. Meakin, P. Ramanlal, L.M. Sander, R.C. Ball, Ballistic deposition on surfaces. Phys Rev. A **34**, 5091–5103 (1986)
43. D.X. Ye, T.M. Lu, Ballistic aggregation on two-dimensional arrays of seeds with oblique incident flux: growth model for amorphous Si on Si. Phys. Rev. B **75**, 23540 (2007)
44. C. Patzig, T. Karabacak, B. Fuhrmann, B. Rauschenbach, Glancing angle sputter deposited nanostructures on rotating substrates: experiments and simulations. J. Appl. Phys. **104**, 094318 (2008)
45. T. Karabacak, Y.-P. Zhao, G.-C. Wang, T.-M. Lu, Growth-front roughening in amorphous silicon films by sputtering. Phys. Rev. B **64**, 085322 (2001)
46. D.X. Ye, T.M. Lu, Fanlike aggregations on seeds by parallel ballistic flux: experimental results and Monte Carlo simulations of the growth of three-dimensional Si structures. Phys. Rev. B **75**, 115420 (2007)

47. B. Tanto, C.F. Doiron, T.M. Lu, Large artificial anisotropic growth rate in on-lattice simulation of obliquely deposited nanostructures. Phys. Rev. E **83**, 016703 (2011)
48. C. Grüner, S. Grüner, S.G. Mayr, B. Rauschenbach, Avoiding anisotropies in on-lattice simulations of ballistic deposition. Phys. Stat. Sol. (b) **258**, 2000036 (2021)
49. T. Karabacak, J.P. Singh, Y.-P. Zhao, G.C. Wang, T.-M. Lu, Scaling during shadowing growth of isolated nanocolumns. Phys. Rev. B **68**, 125408 (2003)
50. T. Karabacak, G.-C. Wang, T.-M. Lu, Physical self-assembly and the nucleation of three-dimensional nanostructures by oblique angle deposition. J. Vac. Sci. Technol. A **22**, 1778–1784 (2004)
51. C. Grüner, in *Oblique angle deposition of thin films—theory, modelling and application.* Dissertation, University Leipzig (2019)
52. P.A. Sánchez, T. Sintes, J.H. E. Cartwright, O. Piro, Influence of microstructure on the transitions between mesoscopic thin-film morphologies in ballistic-diffusive models. Phys. Rev. E **81**, 011140 (2010)
53. T. Smy, D. Vick, M.J. Brett, S.K. Dew, A.T. Wu, J.C. Sit, K.D. Harris, Three-dimensional simulation of film microstructure produced by glancing angle deposition. J. Vac. Sci. Technol. A **18**, 2507–2512 (2000)
54. M. Suzuki, Y. Taga, Numerical study of the effective surface area of obliquely deposited thin films. J. Appl. Phys. **90**, 5599–5606 (2001)
55. Y. Saito, S. Omura, Domain competition during ballistic deposition: Effect of surface diffusion and surface patterning. Phys. Rev. E **84**, 021601 (2011)
56. W. Jost, *Diffusion in Solids, Liquids Gases* (Academic Press, New York, 1970)
57. S. Müller-Pfeiffer, H. van Kraneneburg, J.C. Lodder, A two-dimensional Monte Carlo model for thin film growth by oblique evaporation: simulation of two-component systems for the example of Co-Cr. Thin Solid Films **213**, 143–153 (1992)
58. G. Ehrlich, Molecular processes at the gas-solid interface, in *Structure and Properties of Thin Films,* ed. by C.A. Neugebauer, J.B. Newkirk, D.A. Vermilyea (Wiley, New York, 1959), pp. 423–475
59. G. Neumann, W. Hirschwald, Mechanism of surface self diffusion. Z. Phys. Chem. B **81**, 63–176 (1972)
60. S. Liedtke-Grüner, C. Grüner, A. Lotnyk, J.W. Gerlach, B. Rauschenbach, Biaxially textured titanium thin films by oblique angle deposition: conditions and growth mechanisms. Phys. Stat. Sol. (a) **217**, 1900636 (2019)
61. K.-H. Müller, Dependence of thin-film microstructure on deposition rate by means of a computer simulation. J. Appl. Phys. **58**, 2573–2576 (1985)
62. D. Bensimon, B. Shraiman, S. Liang, On the ballistic model of aggregation. Phys Lett. **102** A, 238–240 (1984)
63. M. Pelliccione, T.-M. Lu, Self-shadowing in ballistic fan formation from point seeds. Phys. Rev. B **75**, 245431 (2007)
64. B. Tanto, G. Ten Eyck, T.-M. Lu, A model for column angle evolution during oblique angle deposition. J. Appl. Phys. **108**, 026107 (2010)
65. S. Liang, L.P. Kadanoff, Scaling in a ballistic aggregation model. Phys. Rev. A **3**, 2628–2630 (1985)
66. J. Krug, P. Meakin, Columnar growth in oblique incidence ballistic deposition: faceting, noise reduction, and mean-field theory. Phys. Rev. A **43**, 900–919 (1991)
67. F. Porcú, F. Prodi, Ballistic accretion on seeds of different sizes. Phys. Rev. A **44**, 8313–8315 (1991)
68. A.V. Limaye, R.E. Amritkar, Theory of growth of ballistic aggregates. Phys. Rev. A **34**, 5085–5090 (1986)
69. A. van der Drift, Evolutionary selection, a principle governing growth orientation in vapour-deposited layers. Philips Res. Rep. **22**, 267–288 (1967)
70. B. Rauschenbach, C. Patzig, Dünne Schichten durch Deposition unter streifenden Einfall. Vakuum für Forschung & Techn. **22**, 14–19 (2010)

71. A. Amassian, K. Kaminska, M. Suzuki, L. Martinu, K. Robbie, Onset of shadowing-dominated growth in glancing angle deposition. Appl. Phys. Lett. **91**, 173114 (2007)
72. C. Patzig, in *Glancing angle deposition of silicon nanostructures by ion beam sputtering.* Dissertation, University Leipzig (2009)
73. C. Khare, J.W. Gerlach, M. Weise, J. Bauer, T. Höche, B. Rauschenbach, Growth temperature altered morphology of Ge nanocolumns. Phys. Stat. Sol. A **208**, 851–856 (2011)
74. C. Khare, J.W. Gerlach, T. Höche, B. Fuhrmann, H.S. Leipner, B. Rauschenbach, Effects of annealing on arrays of Ge nanocolumns formed by glancing angle deposition. Appl. Surf. Sci. **258**, 9762–9769 (2012)
75. C. Khare, J.W. Gerlach, B. Fuhrmann, B. Rauschenbach, Influence of substrate temperature on glancing angle deposited Ag nanorods. J. Vac. Sci. Technol. A **28**, 1002–1009 (2010)
76. C. Khare, Gowth of Ge, Ag and multilayered Si/Ge nanostructures by ion beam sputter glancing angle deposition. Disserataion, University Leipzig (2012)
77. M.O. Jensen, M.J. Brett, Porosity engineering in glancing angle deposition thin films. Appl. Phys. A **80**, 763–768 (2005)
78. D.X. Ye, T. Karabacak, B.K. Lim, G.C. Wang, T.M. Lu, Growth of uniformly aligned nanorod arrays by oblique angle deposition with two-phase substrate rotation. Nanotechnology **15**, 817–821 (2004)
79. C. Khare, R. Fechner, J. Bauer, M. Weise, B. Rauschenbach, Glancing angle deposition of Ge nanorod arrays on Si patterned substrates. J. Vac. Sci. Technol. A **29**, 041503 (2011)
80. C. Khare, B. Fuhrmann, H.S. Leipner, J. Bauer, B. Rauschenbach, Optimized growth of Ge nanorod arrays on Si patterns. J. Vac. Sci. Technol. A **29**, 051501 (2011)
81. S. Liedtke-Grüner, C. Grüner, A. Lotnyk, J.W. Gerlach, M. Mensing, P. Schumacher, B. Rauschenbach, Crystallinity and texture of molybdenum thin films obliquely deposited at room temperature. Thin Solid Films **685**, 6–16 (2019)
82. S. Lichter, J. Chen, Model for columnar microstructure of thin solid films. Phys. Rev. Lett. **56**, 1396–1399 (1986)
83. R. Fiedler, G. Schirmer, Säulenwachstum bei aufgedampften Schichten. Thin Solid Films **167**, 281–289 (1988)
84. Y. D. Fan, X. P. Li, J. Yang, J. P. Li, Microscopic model for columnar growth of thin films. Phys. Stat. Sol. (a) **134**, 157–166 (1992)
85. R.N. Tait, T. Smy, M.J. Brett, Modelling and characterization of columnar growth in evaporated films. Thin Solid Films **226**, 196–201 (1993)
86. Hodgkinson, I., Q.h. Wu, J. Hazel, Empirical equations for the principal refractive indices and column angle of obliquely deposited films of tantalum oxide, titanium oxide, and zirconium oxide. Appl. Optics **37**, 2653–2659 (1998)
87. H. Zhu, W. Cao, G. K. Larsen, R. Toole, Y. Zhao, Tilting angle of nanocolumnar films fabricated by oblique angle deposition. J. Vac. Sci. Technol. B **30**, 030606 (2012)
88. D. J. Poxson, F. W. Mont, M. F. Schubert, J. K. Kim, E. F. Schubert, Quantification of porosity and deposition rate of nanoporous films grown by oblique-angle deposition. Appl. Phys. Lett. **93**, 101914 (2008)
89. R. Messier, T. Gehrke, C. Frankel, V.C. Venugopal, W. Otaño, A. Lakhtakia, Engineered sculptured nematic thin films. J. Vac. Sci. Techn. **15**, 2148–2152 (1994)
90. M. Malac, R.F. Egerton, M.J. Brett, B. Dick, Fabrication of submicrometer regular arrays of pillars and helices. J. Vac. Sci. Technol. B **17**, 2671–2674 (1999)
91. C. Patzig, C. Khare, B. Fuhrmann, B. Rauschenbach, Periodically arranged Si nanostructures by glancing angle deposition on patterned substrates. Phys. Stat. Sol. (b) **247**, 1322–1344 (2010)
92. D.-X. Ye, C. L. Ellison, B.-K. Lim, T.-M. Lu, Shadowing growth of three-dimensional nanostructures on finite size seeds. J. Appl. Phys. **103**, 103531 (2008)
93. D.-X. Ye, in *Shadowing growth by physical vapor deposition.* Dissertation, Rensselaer Polytechnic Institute Troy, New York (2006)
94. M.W. Horn, M.D. Pickett, R. Messier, A. Lakhtakia, Blending of nanoscale and microscale in uniform large–area sculptured thin–film architectures. Nanotechnology **15**, 303–310 (2004)

95. S. Kesapragada, P. Sotherland, D. Gall, Ta nanotubes grown by glancing angle deposition. J. Vac. Sci. Technol. B **26**, 678–681 (2008)
96. M. Summers, B. Djurfors, M. Brett, Fabrication of silicon submicrometer ribbons by glancing angle deposition. J. Micro/Nanolithography, MEMS and MOEMS **4**, 033012 (2005)
97. C. Patzig, B. Rauschenbach, B. Fuhrmann, H.S. Leipner, Growth of Si nanorods in honeycomb and hexagonal-closed-packed arrays using glancing angle deposition, J. Appl. Phys. **103**, 024313 (2008)
98. C.M. Zhou, D. Gall, Growth competition during glancing angle deposition of nanorod honeycomb arrays. Appl. Phys. Lett. **90**, 093103 (2007)
99. C. Zhou, D. Gall, Surface patterning by nanosphere lithography for layer growth with ordered pores. Thin Solid Films **516**, 433–437 (2007)
100. A. Pawar, I. Kretzschmar, Patchey particles by glancing angle deposition. Langmuir **24**, 355–358 (2008)
101. B. Dick, J.C. Sit, M.J. Brett, I.M.N. Votte, C.W.M. Bastiaansen, Embossed polymetric relief structures as a template for the growth of periodic inorganic microstructures. Nano Lett. **1**, 71–73 (2001)
102. S. Krishnamoorthy, C. Hinderling, H. Heinzelmann, Nanoscale patterning with block copolymers. Mater. Today **9**, 40–47 (2006)
103. M. Mäder, T. Höche, J.W. Gerlach, R. Böhme, K. Zimmer, B. Rauschenbach, Large area metal dot matrices made by diffraction mask projection laser ablation. Phys. Stat. Sol. (RRL) **2**, 34–36 (2008)
104. M.O. Jensen, M.J. Brett, Periodically structured glancing angle depsoition thin films. IEEE Trans. Nanotechnol. **4**, 269–277 (2005)
105. E. Main, T. Karabacak, T.M. Lu, Continuum model for nanocolumn growth during oblique angle deposition. J. Appl. Phys. **95**, 4346–4351 (2004)
106. M. Weise, in *Dreidimensionale Germanium- und Siliziumstrukturen.* Unpublished Diploma thesis, Universität Leipzig (2010)
107. B. Dick, M.J. Brett, T. Smy, M. Belov, M.R. Freeman, Periodic submicrometer structures by sputtering. J. Vac. Sci. Technol. B **19**, 1813–1819 (2001)
108. J. Bauer, M. Weise, B. Rauschenbach, N. Geyer, B. Fuhrmann, Shape evolution in glancing angle deposition of arranged germanium nanocolumns. J. Appl. Phys. **111**, 104309 (2012)
109. I. Abdulhalim, M. Zourob, A. Lakhtakia, Surface plasmon resonance for biosensing: a mini-review. Electromagnetics **28**, 214–242 (2008)
110. A. Shalabney, I. Abdulhalim, Sensitivity-enhancement methods for surface plasmon sensors. Laser & Photonics Rev. **5**, 571–606 (2011)
111. M. Fleischmann, P.J. Hendra, A.J. McQuillan, Raman spectra of pyridine adsorbed at a silver electrode. Chem. Phys. Lett. **26**, 163–166 (1974)
112. M.G. Albrecht, J.A. Creighton, Anomalously intense Raman spectra of pyridine at a silver electrode. J. Am. Chem. Soc. **99**, 5215–5217 (1977)
113. S.K. Srivastava, A. Shalabney, I. Khalaila, C. Grüner, B. Rauschenbach, I. Abdulhalim, SERS biosensor using metallic nano-sculptured thin films for the detection of endocrine disrupting compound biomarker vitellogenin. Small **10**, 579–3587 (2014)
114. I. Abdulhalim, A. Karabchevsky, C. Patzig, B. Rauschenbach, B. Fuhrmann, E. Eltzov, R. Marks, J. Xu, F. Zhang, A. Lakhtakia, Surface-enhanced fluorescence from metal sculptured thin films with application to biosensing in water. Appl. Phys. Lett. **94**, 063106 (2009)
115. Q. Zhou, Z. Li, Y. Yang, and Z. Zhang, Arrays of aligned, single crystalline silver nanorods for trace amount detection. J. Phys. D: Appl. Phys. **41**, 152007 (2008)
116. J.R. Sánchez-Valencia, J. Toudert, A. Borras, C. López-Santos, A. Barranco, I.O. Feliu, A.R. González-Elipe, Tunable in-plane optical anisotropy of Ag nanoparticles deposited by DC sputtering onto SiO_2 nanocolumnar films. Plasmonics **5**, 241–250 (2010)
117. A. Karabchevsky, C. Khare, B. Rauschenbach, I. Abdulhalim, Microspot sensing based on surface-enhanced fluorescence from nanosculptured thin films, J. Nanophotonics **6**, 061508 (2012)

118. S.K. Srivastava, C. Grüner, D. Hirsch, B. Rauschenbach, I. Abdulhalim, Enhanced intrinsic fluorescence from carboxidized nano-sculptured thin films of silver and their application for label free dual detection of glycated hemoglobin. Opt. Express **25**, 4761–4772 (2017)
119. S.Y. Song, Y.D. Han, Y.M. Park, C.Y. Jeong, Y.J. Yang, M.S. Kim, Y. Ku, H.C. Yoon, Bioelectrocatalytic detection of glycated hemoglobin (HbA1c) based on the competitive binding of target and signaling glycoproteins to a boronate-modified surface. Biosens. Bioelectron. **35**, 355–362 (2012)
120. Y. Wang, L. Wu, T.I. Wong, M. Bauch, Q. Zhang, J. Zhang, X. Liu, X. Zhou, P. Bai, J. Dostalek, B. Liedberg, Directional fluorescence emission co-enhanced by localized and propagating surface plasmons for biosensing. Nanoscale **8**, 8008–8016 (2016)
121. S.K. Srivastava, H.B. Hamo, A. Kushmaro, R.S. Marks, C. Grüner, B. Rauschenbach, I. Abdulhalim, Highly sensitive and specific detection of *E. coli* by a SERS nanobiosensor chip utilizing metallic nanosculptured thin films. The Analyst **140**, 3201–3209 (2015)
122. B. Van Dorst, J. Mehta, K. Bekaert, E. Rouah-Martin, W. De Coen, P. Dubruel, R. Blust, J. Robbens, Recent advances in recognition elements of food and environmental biosensors: a review. Biosens. Bioelectron. **26**, 1178–1194 (2010)
123. O. Albrecht, R. Zierold, C. Patzig, J. Bachmann, C. Sturm, B. Rheinländer, M. Grundmann, D. Görlitz, B. Rauschenbach, K. Nielsch, Tubular magnetic nanostructures based on glancing angle deposited templates and atomic layer deposition. Phys. Stat. Sol. (b) **247**, 1365–1371 (2010)
124. O. Albrecht, R. Zierold, S. Allende, J. Escrig, C. Patzig, B. Rauschenbach, K. Nielsch, D. Görlitz, Experimental evidence for an angular dependent transition of magnetization reversal modes in magnetic nanotubes, J. Appl. Phys. **109**, 093910 (2011)

Appendix A

A.1 Thomas–Fermi Approximation for an Isolated Atom

The idea underlying the Thomas–Fermi theory [1, 2] is based on the replacement of complicated terms in the electron–electron repulsion energy distribution by a simple functional of the electron density. In this semi-classical theory, a simplification of the many-body problem is assumed by the approximation that an atom can be described as a uniformly distributed (negatively charged) electron cloud around a nucleus. In the Thomas–Fermi approximation, the total energy is given as a functional of electron density, where the density variation in the electron cloud is determined by the Coulomb potential of the nucleus and the electron gas. According to Eliezer et al. [3], the Thomas–Fermi approximation can be used under the assumptions that the system is in its lowest quantum state, the number of electrons is large and the change in the potential energy is small compared to the total energy of an electron.

The electron energy of a free Fermi electron gas, in a 3-dimensional real space with side length L and volume L^3 is given by

$$E = \frac{1}{2}\left(\frac{\pi\hbar}{L}\right)^2 \left(n_x^2 + n_y^2 + n_z^2\right) = \frac{1}{2}\left(\frac{\pi\hbar}{L}\right)^2 R^2, \quad n_x, n_y, n_z = 1, 2, 3, \ldots \tag{A.1}$$

where $\hbar = h/2\pi$ and h is the Planck's constant. The radius R of a sphere in the space n_x, n_y, n_z includes all occupied states to which the Fermi energy E_F (maximum energy of electrons) is assigned. The number of energy levels within the maximum energy value at zero temperature can be expressed by

$$N_F = \frac{1}{2^3}\frac{4\pi}{3}R^3 = \frac{1}{3\pi^2}\left(\frac{L}{\hbar}\right)^3 (2m_e E_F)^{3/2}, \tag{A.2}$$

B. Rauschenbach, *Low-Energy Ion Irradiation of Materials*, Springer Series in Materials Science 324, https://doi.org/10.1007/978-3-030-97277-6

where m_e is the electron mass. It must be noted that because of the Pauli principle, two identical Fermions can never occupy the same quantum mechanical state. However, due to this condition, only a few electrons have the minimum possible energy. The other electrons must occupy higher energetic levels. Thus, some electrons in the Fermi gas can have a very high energy when the gas is extremely cold. The highest energy is called the Fermi energy E_F.

Then, the number of states in three dimensions (density of states) is defined as

$$D(E) = \frac{dN_F}{dE}\left[\frac{1}{3\pi^2}\left(\frac{L}{\hbar}\right)^3 (2m_e E_F)^{3/2}\right] = \frac{L^3}{3\pi^2}\left(\frac{2m_e}{\hbar^2}\right)^{3/2}\sqrt{E}. \tag{A.3}$$

The state of density is proportional the root of energy, is zero at the absolute zero point, and proportional to the volume L^3.

The electrostatic potential of the electrons and nucleus inside the atom is V(r), where r is the distance from the nucleus. The maximum energy of an electron $E = E_F$ is assumed to be $-$ V(r). Then, the density of the charges (electrons per volume) is given by

$$\rho(r) = \frac{N_F}{L^3} = \frac{1}{3\pi^2}\frac{(2m_e)^{3/2}}{\hbar^3}[-V(r)]^{3/2}. \tag{A.4}$$

By considering the density of charge as continuous, it satisfies the Poisson's equation

$$\frac{d^2V(r)}{dr^2} = -4\pi e^2\rho(r). \tag{A.5}$$

Combining the (A.4) and (A.5), the Poisson equation for the potential V(r) is given by

$$\frac{d^2V(r)}{dr^2} = \frac{4e^2}{3\pi}\frac{(2m_e)^{3/2}}{\hbar^3}[V(r)]^{3/2}. \tag{A.6}$$

The boundary conditions by the solution of (A.6) are

(i) for $r \to 0$, the potential V(r) becomes the Coulomb field Ze^2/r of the nucleus and
(ii) for $r \to \infty$, the potential is zero, which indicate that the atom as whole is uncharged.

The interactions between charged particles are almost always affected by the presence of other charged particles in the vicinity, and in the process are usually reduced in strength or screened out (see Chap. 2). The screened Coulomb potential is given by

$$V(r) = \frac{-Ze^2}{r}\chi(r), \tag{A.7}$$

where the term $\chi(r)$ represents the screening of the nuclear charge by atomic electrons and Z an average atomic number of the two interacting particles.

The first and second derivations of the (A.7) are determined to

$$\frac{2}{r}\frac{dV(r)}{dr} = Ze^2\left(\frac{2\chi(r)}{r^3} - \frac{2}{r^2}\frac{d\chi(r)}{dr}\right) \tag{A.8}$$

and

$$\frac{d^2V(r)}{dr^2} = Ze^2\left(\frac{2}{r^2}\frac{d\chi(r)}{dr} - \frac{2\chi(r)}{r^3} - \frac{1}{r}\frac{d^2\chi(r)}{dr^2}\right), \tag{A.9}$$

where the first derivation is multiplied by 2/r. After addition of both derivations, the potential can be expressed by

$$\frac{d^2V(r)}{dr^2} = -\frac{Ze^2}{r}\frac{d^2\chi(r)}{dr^2}. \tag{A.10}$$

Using the Poisson equation (A.6), (A.10) can be rewritten to

$$\frac{d^2\chi(r)}{dr^2} = -\frac{4e^2r}{3\pi\hbar^3}\frac{[2m_eV(r)]^{3/2}}{Z}. \tag{A.11}$$

This equation can be made dimensionless by writing $r = a_{TF}\chi$ and $dr/d\chi = a_{TF}$. Then, (A.11) can be transformed to

$$\frac{d^2\chi}{dx^2} = \frac{4e^2}{3\pi\hbar^3}Z^{1/2}(2m_ea_{TF})^{3/2}\frac{\chi^{3/2}}{x^{1/2}}, \tag{A.12}$$

where the screened potential of (A.7) is used in form of $V(r) = \frac{-Ze^2}{a_{TF}x}\chi(r)$ and a_{TF} is the Thomas–Fermi screening length. Using the definition of Thomas–Fermi screening length

$$a_{TF} = \frac{1}{2}\left(\frac{3\pi}{4}\right)^{3/2}\frac{\hbar^2}{Z^{1/2}e^2m_e} = 0.8853\frac{a_0}{Z^{1/3}}, \tag{A.13}$$

where a_0 is the Bohr radius as outlined in (2.7). Then (A.12) can transformed into a dimensionless, non-linear second-order differential Thomas–Fermi equation

$$\frac{d^2\chi}{dx^2} = \frac{\chi^{3/2}}{x^{1/2}}, \tag{A.14}$$

where the boundary conditions for the screening functions of the neutral Thomas–Fermi atom are $\chi(0) = 1$ and $\chi(\infty) = 0$. This equation is a universal equation depending on the number of electrons (Z). On the one hand, the Thomas–Fermi differential equation cannot be solved in closed form, but on the other hand, the Thomas–Fermi approximation make this method to a very useful approach in atomic physics. Many approximated solutions of the non-linear Thomas–Fermi differential equation have been published [4]. As an example, the solution of this equation is given by

$$\chi(x) = \left(1 + 0.02747x^{1/2} + 1.243x - 0.1486x^{3/2} \right. \\ \left. +0.2302x^2 + 0.007298x^{5/2} + 0.006944x^3\right)^{-1} \tag{A.15}$$

with an accuracy of 0.3% [5].

References

1. L. H. Thomas, The calculations of atomic fields, Proc. Cambridge Phil. Soc. **23**, 524 (1927)
2. E. Fermi, Eine statistische Methode zur Bestimmung einiger Eigenschaften des Atoms und ihre Anwendung auf die Theorie des periodischen Systems der Elemente Z. Physik **48**, 73 (1928)
3. S. Eliezer, A. Ghatak, H. Hora, *Fundamentals of Equations of State*, (World Scientific Publ., Singapore 1986)
4. E. Hille, Some aspects of the Thomas–Fermi equation, J. Anal. Math. **23**, 147 (1970)
5. R. Latter, Atomic energy levels for the Thomas–Fermi and Thomas–Fermi-Dirac potential, Phys. Rev. **89**, 510 (1955)

Appendix B

B.1 Particle Movement in a Central Force Field

Two particles are assumed to interact during the collision process by the central collision force, i.e. the force acts on the line connecting the centers of these particles (Fig. B.1). In a central force field, the particle is only accelerated in the radial direction. The motion of a particle in the x–y-plane of motion can be described by polar coordinates $x = r$ cos and $y = r$ sin (see Appendix C). The trajectory of the particle with the reduced mass M_r is then given by specifying each position of the trajectory with the time and depends on the impact parameter b and the angle $\varphi = \varphi(t)$. Consequently, the motion of the particle can be determined by two time-dependent functions r(t) and (t).

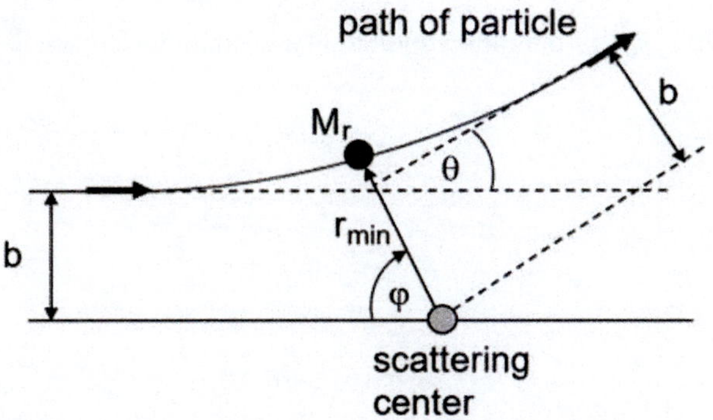

Fig. B.1 Trajectory of a particle with the reduced mass M_r in a central force field. At any time the position of the particle is given by distance r(t) between and the angle φ(t) The angle φ(t) is determined from the initial direction of the particle. The impact factor b and the asymptotic scattering angle are shown

B. Rauschenbach, *Low-Energy Ion Irradiation of Materials*, Springer Series in Materials Science 324, https://doi.org/10.1007/978-3-030-97277-6

On the base of the first and second time derivatives of the position coordinates

$$\begin{aligned}
\frac{dx}{dt} &= \frac{dr}{dt}\cos\varphi - r\frac{d\varphi}{dt}\sin\varphi,\\
\frac{d^2x}{dt^2} &= \frac{d^2r}{dt^2}\cos\varphi - 2\frac{dr}{dt}\frac{d\varphi}{dt}\sin\varphi - r\left(\frac{d\varphi}{dt}\right)^2\cos\varphi - r\frac{d^2\varphi}{dt^2}\sin\varphi,\\
\frac{dy}{dt} &= \frac{dr}{dt}\sin\varphi + r\frac{d\varphi}{dt}\cos\varphi,\\
\frac{d^2y}{dt^2} &= \frac{d^2r}{dt^2}\sin\varphi + 2\frac{dr}{dt}\frac{d\varphi}{dt}\sin\varphi - r\left(\frac{d\varphi}{dt}\right)^2\sin\varphi + r\frac{d^2\varphi}{dt^2}\cos\varphi,
\end{aligned} \tag{B.1}$$

the polar velocity components are given by

$$\begin{aligned}
v_r &= \frac{dx}{dt}\cos\varphi + \frac{dy}{dt}\sin\varphi,\\
v_\theta &= \frac{dy}{dt}\cos\varphi - \frac{dx}{dt}\sin\varphi.
\end{aligned} \tag{B.2}$$

Using these equations, the radial and angular components of the particle acceleration can be determined to

$$\begin{aligned}
a_r &= \frac{d^2x}{dt^2}\cos\varphi + \frac{d^2y}{dt^2}\sin\varphi,\\
a_\theta &= \frac{d^2y}{dt^2}\cos\varphi - \frac{d^2x}{dt^2}\sin\varphi.
\end{aligned} \tag{B.3}$$

With the help of (B.1) the radial and angular velocities and accelerations can be obtained to

$$\begin{aligned}
v_r &= \frac{dr}{dt}\\
v_\theta &= r\frac{d\varphi}{dt}\\
a_r &= \frac{d^2r}{dt^2} - r\left(\frac{d\varphi}{dt}\right)^2,\\
a_\theta &= 2\frac{dr}{dt}\frac{d\varphi}{dt} + r\frac{d\varphi^2}{dt^2},
\end{aligned} \tag{B.4}$$

where the expressions $r\left(\frac{d\varphi}{dt}\right)^2$ and $2\frac{dr}{dt}\frac{d\varphi}{dt}$ correspond to the centrifugal and Coriolis accelerations, respectively.

The generalized equation of motion $M_r a = -dV/r$ for a particle with a mass M_r is used to describe the motion of particle in the central force field. Together with (B.4) this yields to

$$M_r a_r = M_r\left[\frac{d^2r}{dt^2} - r\left(\frac{d\varphi}{dt}\right)^2\right] = -\frac{dV}{dr}. \tag{B.5}$$

An important result is that the acceleration of the particles in the central field force in the radial direction (the angular acceleration) is zero.

With respect to the origin, the vector of the total angular momentum can be expressed by

$$\boldsymbol{L} = \boldsymbol{r} \times M_r \frac{d\boldsymbol{r}}{dt}. \tag{B.6}$$

It is obvious that both the vector of the distance **r** and the velocity vector d**r**/dt lie in the plane of motion. Consequently, the direction of the angular momentum is time-independent and therefore this angular momentum can be reduced to

$$L = M_r r^2 \frac{d\varphi}{dt}. \tag{B.7}$$

Consequently, the radial component in (B.5) can be transformed into

$$M_r \frac{d^2r}{dt^2} - M_r r\left(\frac{d\varphi}{dt}\right)^2 + \frac{dV(r)}{dr} = 0 = M_r \frac{d^2r}{dt^2} - \frac{L^2}{M_r r^2} + \frac{dV(r)}{dr}. \tag{B.8}$$

On this base the equation of the motion of the particles in the central force field is given by

$$M_r \frac{d^2r}{dt^2} = -\frac{d}{dr}\left(V(r) + \frac{L^2}{2M_r r^2}\right). \tag{B.9}$$

After multiplying this equation by dr/dt and integrating, the following is obtained,

$$\int dt\left[M_r \frac{d^2r}{dt^2}\frac{dr}{dt} + \frac{d}{dr}\left(V(r) + \frac{L^2}{2M_r r^2}\frac{dr}{dt}\right)\frac{dr}{dt}\right]$$
$$= \int d\left(\frac{dr}{dt}\right) M_r \frac{dr}{dt} + \int dr \frac{d}{dr}\left(V(r) + \frac{L^2}{2M_r r^2}\frac{dr}{dt}\right) = 0, \tag{B.10}$$

where the relationships $dt\frac{dr}{dt} = dr$ and $dt\frac{dr}{dt}\frac{d^2r}{dt^2} = d\left(\frac{dr}{dt}\frac{dr}{dt}\right)$ are used.

Using again (B.7) to solve the integrals in (B.10). Under the condition of the law of conservation of energy, this equation can be formulated as follows

$$\frac{1}{2}M_r\left(\frac{d^2r}{dt^2}\right)^2+\frac{1}{2}M_r\left(\frac{d\varphi}{dt}\right)^2+V(r)=0\equiv E_r \text{ or}$$

$$E_r=\frac{1}{2}M\left(\frac{d^2r}{dt^2}\right)_r^2+\frac{L^2}{2M_r r^2}+V(r), \tag{B.11}$$

where E_r represents the total energy (sum of kinetic and potential energy).

The positive potential implies a repulsive interaction. Equation (B.11) can be rewritten to

$$\frac{dr}{dt}=\pm\sqrt{\frac{2}{M_r}[E_r-V(r)]-\left(\frac{L}{M_r r}\right)^2}=\pm v\sqrt{1-\frac{V(r)}{E_r}-\left(\frac{b}{r}\right)^2}. \tag{B.12}$$

The sign is positive if r increases and is negative if r decreases. To describe the trajectory of the particle with the reduced mass as a function of the angle, the derivative of the angle φ with respect to r is sought using (2.38) and (B.12).

$$\frac{d\varphi}{dt}=\frac{d\varphi}{dt}\frac{dt}{dr}=\frac{vb}{r^2}\frac{1}{v\sqrt{1-\frac{V(r)}{E_r}-\left(\frac{b}{r}\right)^2}}=\frac{b}{r^2\sqrt{1-\frac{V(r)}{E_r}-\left(\frac{b}{r}\right)^2}}. \tag{B.13}$$

By integration, a relation between the angle φ and the impact parameter b can be determined as follows

$$\varphi=\int_{r_{\min}}^{\infty}\frac{b}{r^2\sqrt{1-\frac{V(r)}{E_r}-\left(\frac{b}{r}\right)^2}}dr, \tag{B.14}$$

where $r_{\min}$ is the the closest approach between the particle with the reduced mass and the scattering center.

After integration of the radial velocity equation of motion, (B.12), the so-called time integral can be determined to

$$t=\int\frac{dr}{\sqrt{\frac{2}{M_r}[E_r-V(r)]-\left(\frac{L}{M_r r}\right)^2}}+const. \tag{B.15}$$

That gives an implicit time dependence of the distance between the particle with the mass M_r and the scattering center.

Appendix C

C.1 Polar, Cylindrical and Spherical Coordinates

From the practical point of view, the application of polar, cylindrical and spherical coordinates can result in significant mathematical simplifications. For example, the polar coordination system is a two-dimensional coordination system in which each point P in a plane is determined by a distance r from a reference point and the azimuthal angle φ from a reference direction (Fig. C.1, left). In contrast, the spherical coordinates (r, φ, α) of a point P denote (i) the distance of this point, r, from the origin of the coordinate system, (ii) the polar angle α determined the angle between the + z axis and the radius vector and (iii) the azimuth angle φ the angle measured between the + x axis and the projection of the radius vector onto the (x, y) plane, where r, φ, and α vary from zero to ∞, from zero to 2π and from zero to π, respectively (Fig. C.1).

The relations between the Cartesian and the polar, cylindrical and spherical coordinates are given by the following relationships:

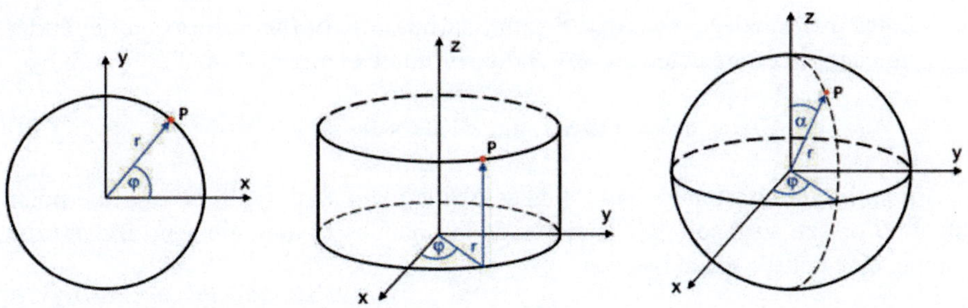

Fig. C.1 Polar, cylindrical and spherical coordinates

B. Rauschenbach, *Low-Energy Ion Irradiation of Materials*, Springer Series in Materials Science 324, https://doi.org/10.1007/978-3-030-97277-6

(i) Cartesian coordinates ↔ polar coordinates:

$$x = r\cos\varphi, r = \sqrt{x^2 + y^2} \tag{C.1}$$

$$y = r\sin\varphi, \quad \varphi = arctan\left(\frac{y}{x}\right) \text{ for } x > 0, y \geq 0 \tag{C.2}$$

(ii) Cartesian coordinates ↔ cylindrical coordinates:

$$x = r\cos\varphi, \quad r = \sqrt{x^2 + y^2} \tag{C.3}$$

$$y = r\sin\varphi, \quad \varphi = arctan\left(\frac{y}{x}\right) \tag{C.4}$$

$$z = z, \quad z = z \tag{C.5}$$

(iii) Cartesian coordinates ↔ spherical coordinates:

$$x = r\sin\alpha\cos\varphi, \quad r = \sqrt{x^2 + y^2 + z^2} \tag{C.6}$$

$$y = r\sin\alpha\cos\varphi, \quad \varphi = \begin{cases} arctan\frac{y}{x} & for\ x > 0, y \geq 0 \\ arctan\frac{y}{x} + 2\pi & for\ x < 0, y \geq 0 \\ arctan\frac{y}{x} + \pi & for\ x < 0, y < 0 \end{cases} \tag{C.7}$$

$$z = r\cos\alpha, \quad \alpha = \arccos\frac{z}{\sqrt{x^2 + y^2 + z^2}} = \arccos\frac{z}{r} \tag{C.8}$$

Infinitesimal small changes of the azimuthal angle dφ on the surface of the cylinder leads to changes of area element dA and the volume element dV and are given by

$$dA = rdrd\varphi \text{ and } dV = rdrdzd\varphi. \tag{C.9}$$

Similarly, infinitesimal small changes of the azimuthal angle dφ and the polar angle dθ on the surface of a sphere leads to changes of area element dA and the volume element dV given by

$$dA = r^2\sin\alpha d\alpha d\varphi = r^2 d\Omega \text{ and } dV = r^2\sin\alpha drd\alpha d\varphi, \tag{C.10}$$

where $d\Omega \equiv \sin\alpha\ d\alpha\ d\varphi$ is the infinitesimal element of solid angle, which vary from zero to 4π.

Appendix D

D.1 Differential Rutherford Scattering Cross-Section

The derivation of the differential Rutherford scattering cross-section is based on the following assumptions:

(1) Collison partners are considered as point masses and charges.
(2) The electrical repulsion between the charges is the only force that is present in the scattering event.
(3) The target atom is infinitely heavy and is at rest.
(4) The energy of the incident particle is not changed during the scattering process.
(5) The lab-system coincides with the cm-system, consequently.

The effect of a repulsive Coulomb force between two equal charges is illustrated schematically in Fig. D.1. The path of the incident particle is hyperbolic and can be described by the relative distance r between the colliding particles and the angle φ.

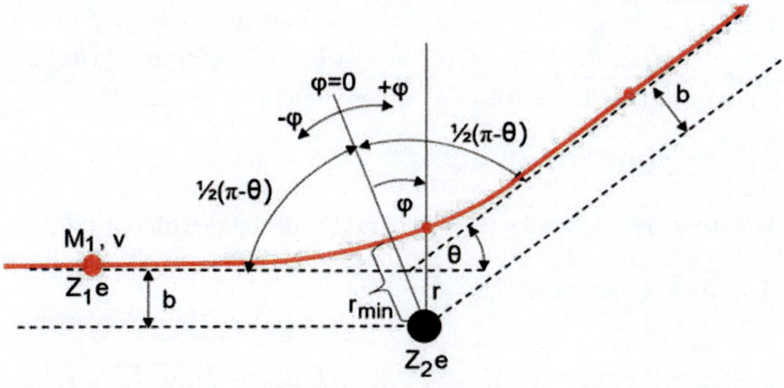

Fig. D.1 Classical Rutherford scattering of the projectile Z_1e by the target atom Z_2e. r and φ are the polar coordinates, b is the impact parameter, r_{min} is the distance of closest approach

B. Rauschenbach, *Low-Energy Ion Irradiation of Materials*, Springer Series in Materials Science 324, https://doi.org/10.1007/978-3-030-97277-6

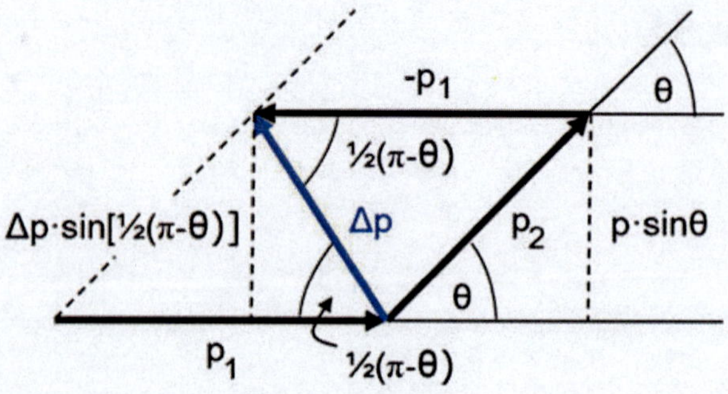

Fig. D.2 Momentum transfer Δp in elastic scattering at a fixed target atom. p_1 and p_2 are the incident and scattered particle momenta

The total change in the linear momentum of the incident particle can be determined from the momentum generates by the electrostatic force between the incident particle and the target particle to

$$\Delta \boldsymbol{p} = \boldsymbol{p}_2 - \boldsymbol{p}_1 = \int \boldsymbol{F} dt. \tag{D.1}$$

In Fig. D.2 is schematically shown the momenta relationships during the scattering event.

According to Fig. D.2 and assuming that the target atom does not move during the scattering (energy of the incident particle is the same before and after the scattering process), the momentum transfer can be expressed as follows

$$\Delta p = \sin\left(\frac{\pi - \theta}{2}\right) = p \sin\theta = M_1 v \sin\theta \tag{D.2}$$

or after applying trigonometric relations as follows

$$\begin{aligned}\Delta p = \frac{M_1 \sin\theta}{\sin\left(\frac{\pi-\theta}{2}\right)} &= \frac{2M_1 v \sin\frac{\theta}{2}\cos\frac{\theta}{2}}{\sin\frac{\pi}{2}\cos\frac{\theta}{2} - \cos\frac{\pi}{2}\sin\frac{\theta}{2}} = \frac{2M_1 v \sin\frac{\theta}{2}\cos\frac{\theta}{2}}{\cos\frac{\theta}{2}} \\ &= 2M_1 v \sin\frac{\theta}{2},\end{aligned} \tag{D.3}$$

where M_1 is the mass and v is the velocity of the incident projectile. On the other side, the momentum transfer can also be described by the force components. According Fig. D.3 the force components can be written by

$$|\Delta \boldsymbol{p}| = \int F \cos\varphi dt + \int F \sin\varphi dt. \tag{D.4}$$

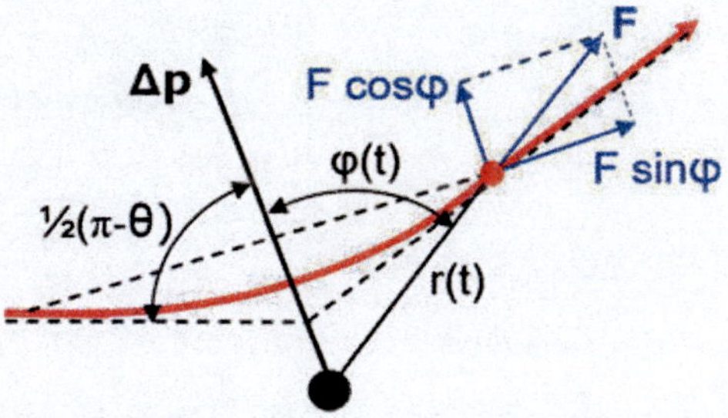

Fig. D.3 Transfer momentum vector and vector components of the Coulomb force during the scattering process. The vectors demonstrate the instantaneous directions during the motion of the incident particle

It is obvious that the angle φ between the Coulomb force with the transfer momentum at any orbital position leads to an overall change in linear momentum. In general, the very small contribution of the second term in (D.4) is negligible. The lower and upper integration limits are given for the case that $\varphi = -\infty$ and $\varphi = +\infty$, respectively. According to Fig. D.1, theses integration limits are $\pm \frac{1}{2}(\pi - \theta)$. Thus, (D.4) can be rearranged together with (D.2) to give

$$|\Delta \boldsymbol{p}| = 2M_1 v \sin\frac{\theta}{2} = \int\limits_{-\frac{1}{2}(\pi-\theta)}^{+\frac{1}{2}(\pi-\theta)} F\cos\varphi dt = \int\limits_{-\frac{1}{2}(\pi-\theta)}^{+\frac{1}{2}(\pi-\theta)} F\cos\varphi\frac{dt}{d\varphi}d\varphi. \quad \text{(D.5)}$$

The conservation of angular momentum can be stated as

$$M_1 r^2\frac{d\varphi}{dt} = M_1 vb \quad \text{or} \quad \frac{d\varphi}{dt} = \frac{vb}{r^2}. \quad \text{(D.6)}$$

Substituting (D.6) in the integral of (D.5) leads to

$$2M_1 v^2 b\sin\frac{\theta}{2} = \int\limits_{-\frac{1}{2}(\pi-\theta)}^{+\frac{1}{2}(\pi-\theta)} r^2 F\cos\varphi d\varphi. \quad \text{(D.7)}$$

As well-known, the electrostatic Coulomb force is given by

$$F = \frac{1}{4\pi\varepsilon_o}\frac{Z_1 Z_2 e^2}{r^2}, \quad \text{(D.8)}$$

where the term $1/4\pi\varepsilon_o$ in (D.8) is usually set equal 1 (Gaussian system of units). The upper (D.7) can be rearranged with (D.8) to

$$2M_1 v^2 \mathrm{b} \sin\frac{\theta}{2} = Z_1 Z_2 e^2 \int\limits_{-\frac{1}{2}(\pi-\theta)}^{+\frac{1}{2}(\pi-\theta)} \cos\varphi d\varphi. \tag{D.9}$$

The integration of this equation yields to

$$\frac{2M_1 v^2}{Z_1 Z_2 e^2} b \sin\frac{\theta}{2} = \int\limits_{-\frac{1}{2}(\pi-\theta)}^{+\frac{1}{2}(\pi-\theta)} \cos\varphi d\varphi = 2\sin\frac{1}{2}(\pi-\theta) = 2\cos\frac{\theta}{2}. \tag{D.10}$$

Thus, the scattering angle θ can be determined as follows

$$\frac{\cos\frac{\theta}{2}}{\sin\frac{\theta}{2}} = \cot\frac{\theta}{2} = \frac{M_1 v^2}{Z_1 Z_2 e^2} b \tag{D.11}$$

or the impact parameter b to

$$b = \frac{Z_1 Z_2 e^2}{M v^2} \cot\frac{\theta}{2} = \frac{Z_1 Z_2 e^2}{2E} \cot\frac{\theta}{2}. \tag{D.12}$$

Equation (2.58) is used for the calculation of the differential scattering-cross section in the case of the Rutherford scattering. It should be noted that the relationships for the scattering cross-section in laboratory and in the cm-system are identical under the formulated assumptions

$$\frac{d\sigma}{d} = \frac{b}{\sin\theta}\left|\frac{d\mathrm{b}}{d\theta}\right|, \tag{D.13}$$

where dΩ is the differential solid angle. With this equation the differential Rutherford scattering cross-section per solid angle can be expressed as

$$\begin{aligned} \frac{d\sigma}{d\Omega} &= \frac{1}{\sin\theta}\frac{Z_1 Z_2 e^2}{2E}\cot\frac{\theta}{2} \cdot \frac{Z_1 Z_2 e^2}{2E}\frac{1}{2\sin^2\frac{\theta}{2}} \\ &= \frac{1}{2}\left(\frac{Z_1 Z_2 e^2}{2E}\right)^2 \frac{\cot\frac{\theta}{2}}{\sin\theta}\frac{1}{\sin^2\frac{\theta}{2}} \\ &= \left(\frac{Z_1 Z_2 e^2}{4E}\right)^2 \frac{1}{\sin^4\frac{\theta}{2}} \end{aligned} \tag{D.14}$$

and using (2.56) the differential Rutherford scattering cross-section can be given by

$$\frac{d\sigma}{d\theta} = 4\pi \left(\frac{Z_1 Z_2 e^2}{4E} \right)^2 \frac{\cos\frac{\theta}{2}}{\sin^3\frac{\theta}{2}}. \tag{D.15}$$

The Rutherford scattering formulae were derived under the condition that the target atom (scatter) is infinitely heavy and at rest. In this particular case, the formula for the differential Rutherford scattering cross-section is identical in both systems, i.e.

$$\frac{d\sigma}{d\theta} = \frac{d\sigma_{lab}}{d\theta} = \frac{d\sigma_{cm}}{d\theta}. \tag{D.16}$$

Appendix E

E.1 Reduced Stopping Power Cross-Section

In Chap. 2, (2.75), is shown that the differential cross-section for a screened Coulomb potential can be approximately expressed by [1]

$$d\sigma = -\frac{\pi a_{TF}^2}{2t^{3/2}} f\left(t^{1/2}\right) dt. \tag{E.1}$$

Using (3.3) the nuclear stopping power is given by

$$\begin{aligned} \left(\frac{dE}{dx}\right)_n &= N \int\limits_{T_{\min=0}}^{T_{\max}} T \frac{d\sigma}{dT} dT \\ &= -\pi a_{TF}^2 N \int\limits_0^{T_{\max}} T \frac{f\left(t^{1/2}\right)}{2t^{\frac{3}{2}}} dt \\ &= -\frac{\pi a_{TF}^2 N}{\varepsilon^2} T_M \int\limits_0^{T_{\max}} T f\left(t^{1/2}\right) dt^{1/2}, \end{aligned} \tag{E.2}$$

where $t = \varepsilon^2 (T/T_{max})$, see (2.70). By introduction of the reduced length [2]

$$\rho = N M_2 4\pi a_{TF}^2 \frac{M_1}{M_1 + M_2} x, \tag{E.3}$$

the normalized stopping cross-section in the lab-system [1] can be obtained by

$$s_n(\varepsilon) \equiv \frac{d\varepsilon}{d\rho}. \tag{E.4}$$

B. Rauschenbach, *Low-Energy Ion Irradiation of Materials*, Springer Series in Materials Science 324, https://doi.org/10.1007/978-3-030-97277-6

Substituting the stopping power $S_n(E)$ from (E.2) into (3.7), the normalized stopping power can be expressed by

$$S_n(\varepsilon) = \frac{1}{\varepsilon}\int_0^{\varepsilon} f(t^{1/2})dt^{1/2}, \tag{E.5}$$

where $T_{max} = \gamma E$, see (2.36), and $f\left(t^{1/2}\right) = \lambda_m t^{1/2-m}$, see (2.74). For the integration of (E.5) the substitution $y = t^{1/2}$ is used. The power law approximation for the normalized nuclear stopping cross-section is then given by

$$s_n(\varepsilon) = \frac{\lambda_m}{\varepsilon}\int_0^{\varepsilon} y^{1-2m}dy = \frac{\lambda_m}{\varepsilon}\frac{y^{1-2m}}{2(1-m)}\bigg|_0^{\varepsilon} = \frac{\lambda_m}{2(1-m)}\varepsilon^{1-2m}. \tag{E.6}$$

The relation between the normalized nuclear stopping cross-section $s_n(\varepsilon)$ and $S_n(E)$ can be obtain by using (3.1) in the form [3]

$$NS(E) = \frac{dE}{dx} = \frac{d\varepsilon}{d\rho}\left(\frac{dE}{d\varepsilon}\frac{d\rho}{dL}\right) \tag{E.7}$$

with

$$\frac{d\varepsilon}{d\rho} = \frac{dE}{dx}\left(\frac{d\varepsilon}{dE}\frac{dL}{d\rho}\right). \tag{E.8}$$

This algebraic transformation allows to formulate the relation between $s_n(\varepsilon)$ and $S_n(E)$ as follows

$$s_n(\varepsilon) = \frac{\varepsilon}{\pi a_{TF}^2\gamma E}S_n(E). \tag{E.9}$$

References

1. J. Lindhard, V. Nielsen, M. Scharff, Approximation method in classical scattering by screened Coulomb fields – Notes on atomic collision I, Mat. Fys. Medd. Dan. Vid. Selsk., **36** no. 10 (1968)
2. J. Lindhard, M. Scharff, H. E. Schiott, Range concepts and heavy ion Ranges, – Notes on atomic collision II, Mat. Fys. Medd. Dan. Vid. Selsk., **33** no. 14 (1963)
3. M. Nastasi, J. W. Mayer, J. K. Hirvonen, *Ion–Solid Interaction: Fundamentals and Application*, (Cambridge Univ. Press, Cambridge, 1996)

Appendix F

F.1 Concentration Distribution After Ion Implantation

Gaussian Distribution

The vertical concentration distribution after ion implantation can be approximately described by a Gaussian distribution function (see Fig. F.1). This function represents the probability density function of a normally distributed random variable x with variance σ^2 and the expected value μ and can be expressed by

$$f(x) = \frac{1}{\sigma\sqrt{2\pi}} \exp\left[-\frac{1}{2}\left(\frac{x-\mu}{\sigma}\right)^2\right], \qquad \text{(F.1)}$$

where this function satisfies the condition $\int_{-\infty}^{\infty} f(x)dx = 1$. Then, the concentration distribution can be determined by $N(x) = f(x)\cdot\Phi$, where N(x) is the number of atoms in the depth x and Φ is the ion fluence.

It is possible to define the characteristic parameters of the concentration distribution after ion implantation directly by their probability density function f(x). The projected ion range R_p and the standard deviation ΔR_p are usually taken by

$$R_p = \int_{-\infty}^{\infty} x \cdot f(x)dx \quad \text{and} \quad \Delta R_p = \sqrt{\int_{-\infty}^{\infty} (x - R_p)^2 \cdot f(x)dx}. \qquad \text{(F.2)}$$

Then, the concentration distribution is given by

$$N(x) = \frac{\Phi}{\sqrt{2\pi}\,\Delta R_p} \exp\left[-\frac{1}{2}\frac{(x-R_p)^2}{(\Delta R_p)^2}\right]. \qquad \text{(F.3)}$$

B. Rauschenbach, *Low-Energy Ion Irradiation of Materials*, Springer Series in Materials Science 324, https://doi.org/10.1007/978-3-030-97277-6

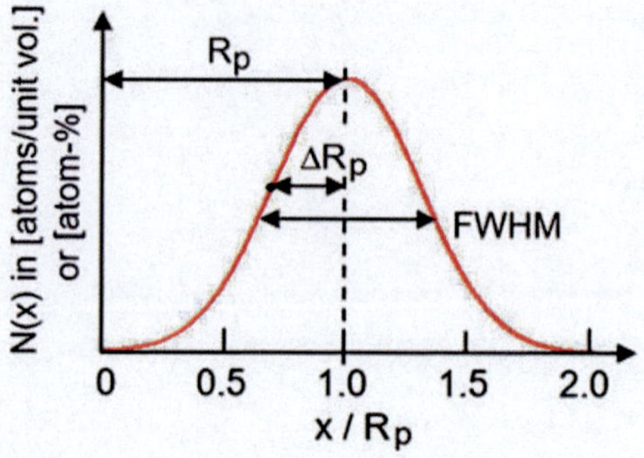

Fig. F.1 Gaussian concentration distribution with the characteristic parameters (R_p is the projected ion range, ΔR_p is the standard deviation and FWHM is the full width of the half maximum)

The integration is performed under the assumption of the integration limits range between $-\infty$ and $+\infty$. This leads to a small error in the determination of the concentration distribution. The corrected relation of the concentration distribution is given by

$$N(x) = \frac{2\Phi}{\sqrt{2\pi}\,\Delta R_p\left[1 + erf\left(\frac{R_p}{\sqrt{2\pi}\,\Delta R_p}\right)\right]} \exp\left[-\frac{1}{2}\frac{\left(x - R_p\right)^2}{\left(\Delta R_p\right)^2}\right]. \tag{F.4}$$

In general, the difference between the distribution calculated with (F.3) and with (F.4) is insignificant and can be neglected. Here, the symmetrical Gaussian shape for a concentration profile is a first order approximation to the actual profile.

High-Order Moments of Gaussian Distribution

In practice, the two parameters R_p and ΔR_p are frequently not sufficient to reproduce the measured concentration distributions of implanted species. To reproduce the profiles with higher accuracy, higher-order Gaussian moments are required.

An arbitrary distribution N(x) can be characterized in terms of its moments. The moment μ_i for an arbitrary index i can be determined by

$$\mu_i = \int_{-\infty}^{\infty} \left(x - R_p\right)^i N(x)dx. \tag{F.5}$$

The first-order moment μ_1 corresponds to the mean projected ion range R_p and the second-order moment $\Delta R_p = \sqrt{\mu_2}$ is associated with the standard deviation of the projected range [see (F.2)].

The third-order moment, called skewness, characterizes the degree of asymmetry of a distribution around R_p and can be described by

$$\gamma = \frac{\mu_3}{\Delta R_p^3} = \frac{\int_{-\infty}^{\infty} (x - R_p)^3 \cdot f(x)dx}{\Delta R_p^3}. \tag{F.6}$$

A positive value of skewness means a distribution with an asymmetric tail extending toward greater depths ahead of R_p and a negative value means a distribution whose tail extends towards depths behind R_p.

The fourth-order moment, called kurtosis, is given by

$$\beta = \frac{\mu_4}{\Delta R_p^4} = \frac{\int_{-\infty}^{\infty} (x - R_p)^4 \cdot f(x)dx}{\Delta R_p^4} \tag{F.7}$$

and describes the sharp of the profile. The Gaussian distribution is characterized by $\gamma = 0$ and $\beta = 3$. For $\beta > 3$, the peak of the concentration distribution is sharper and for $\beta < 3$ the distribution has a gentler peak.

Joined Half-Gaussian Distribution

The concentration profile described with a Gaussian distribution is symmetrical with respect to R_p. Frequently, asymmetrical profiles are measured after ion implantation. Gibbons and Mylroil [1] have proposed a joined half-Gaussian distribution, using two Gaussian profiles with different standard deviations and a joined peak position. Then, the concentration distribution can be expressed by

$$N(x,t) = \begin{cases} N_{max} exp\left[-\frac{1}{2}\frac{(x-R_p)^2}{\Delta R_{pf}^2}\right] & for\ x < R_p \\ N_{max} exp\left[-\frac{1}{2}\frac{(x-R_p)^2}{\Delta R_{pb}^2}\right] & for\ x \geq R_p, \end{cases} \tag{F.8}$$

where R_p is the peak position, ΔR_{pf} is the standard deviation of the near-surface region and ΔR_{pb} is the standard deviation of the deeper region.

The maximum concentration can be determined in the same manner as before [see (3.40)] and is given by

$$N_{\max} = \sqrt{\frac{\pi}{2}} \frac{\Phi}{\Delta R_{pf} + \Delta R_{pb}} \approx 1,25 \frac{\Phi}{\Delta R_{pf} + \Delta R_{pb}} \tag{F.9}$$

and the standard deviations can be evaluated by integrations

$$\Delta R_{pf}^2 = \frac{\int_{-\infty}^{R_p} (x - R_p)^2 N(x)dx}{\int_{-\infty}^{R_p} N(x)dx}, \ \Delta R_{pb}^2 = \frac{\int_{R_p}^{\infty} (x - R_p)^2 N(x)dx}{\int_{R_p}^{\infty} N(x)dx}. \tag{F.10}$$

Pearson IV Distribution

Other distribution functions, for example the Pearson IV function [2], are also used to reproduce the measured concentration function with higher accuracy [3].

If higher order moments are expected to significantly affect the concentration distribution, a Pearson distribution function f(ξ) can also be used by solving the following differential equation

$$df(\xi) = \frac{(x-a)f(\xi)}{b_o + b_1x + b_2x^2}dx, \tag{F.11}$$

where $\xi = (x - R_p)/\Delta R_p$. Also in this case, the Pearson distribution is unit-normalized, i.e. $\int_{-\infty}^{\infty} f'(x)dx = 1$, and the higher-order moments are formed in the same kind as in the case of a Gaussian distribution. The four parameters in (F.11) are related to the expressions

$$b_0 = -\frac{\Delta R_p^4(3\gamma^2 - 4\beta)}{10\beta - 12\gamma^2 - 18}, b_1 = a = -\frac{\gamma\Delta R_p^2(\beta+3)}{10\beta - 12\gamma^2 - 18} \text{ and } b_2 = -\frac{6 - 3\gamma^2 - 2\beta}{10\beta - 12\gamma^2 - 18}. \tag{F.12}$$

Under the condition that $0 < b_1^2/4b_0b_2 < 1$, the concentration distribution can be represented by Pearson IV distribution to

$$N(x) = \frac{\Phi}{\sqrt{2\pi}\Delta R_p} exp\left[-\frac{1}{2b_2}\ln\left(b_o + b_1x + b_2x^2\right) - \frac{2b_2a + b_1}{b_2\sqrt{4b_2b_0 - b_1^2}} arctan\left(\frac{2b_2a + b_1}{\sqrt{4b_2b_0 - b_1^2}}\right)\right] \tag{F.13}$$

with high exactness. For example, the Pearson IV distribution is used to calculate the concentration distribution in the SUPREM computer program (see Sect. 3.6).

References

1. J. F. Gibbons, S. Mylroie, Estimation of impurity profiles in ion-implanted amorphous targets using joined half-Gaussian distributions, Appl. Phys. Lett. **22**, 568 (1973)
2. W. K. Hofer, Concentration profiles of boron implanted in amorphous and polycrystalline silicon, Philips Res. Rep. Vol. Suppl. **8**, 41 (1975)
3. K. Suzuki, *Ion implantation and Activation,* Vol. 1, (Bentham Sci. Publ. 2013)

Appendix G

G.1 Influence of a Subsequent Annealing on the Implanted Concentration Profiles

Assuming a Gaussian concentration distribution after ion implantation is given by (3.35) to

$$N(x) = \frac{\Phi}{\sqrt{2\pi}\,\Delta \mathrm{Rp}} \exp\left[-\frac{1}{2}\frac{\left(x - R_p\right)^2}{\Delta R_p^2}\right]. \tag{G.1}$$

Considering the concentration distribution as an instantaneous diffusion source, this concentration profile can be described mathematically by

$$N(x,t) = \frac{\Phi}{2\sqrt{\pi Dt}} \int\limits_{-\infty}^{\infty} d\langle x\rangle N(\langle x\rangle) \exp\left[-\frac{1}{2}\frac{(x - \langle x\rangle)^2}{2\mathrm{Dt}}\right], \tag{G.2}$$

where D is the diffusion coefficient, t is the diffusion time and x is the average distance of the diffuse particle. The concentration distribution N(x,t) according to this equation can only be determined by numerical approaches. However, it is possible to identify the source concentration $N(x)$ with a Gaussian distribution, [see (G.1)], then the concentration profile can be expressed by

$$N(x,t) = \frac{\Phi}{2\pi\,\Delta \mathrm{R}_p\sqrt{2Dt}} \int\limits_{-\infty}^{\infty} d\langle x\rangle exp\left[-\frac{1}{2}\frac{\left(\langle x\rangle - R_p\right)^2}{\Delta R_p^2}\right] exp\left[-\frac{1}{2}\frac{(x - \langle x\rangle)^2}{2\mathrm{Dt}}\right]. \tag{G.3}$$

B. Rauschenbach, *Low-Energy Ion Irradiation of Materials*, Springer Series in Materials Science 324, https://doi.org/10.1007/978-3-030-97277-6

Combining the both exponential terms in (G.3) to

$$\left[\frac{(\langle x\rangle - R_p)^2}{2\Delta R_p^2}\right] + \left[\frac{(x - \langle x\rangle)^2}{4Dt}\right] = \left[\frac{\langle x\rangle^2 - 2\langle x\rangle + R_p^2}{2\Delta R_p^2}\right] + \left[\frac{x^2 - 2x\langle x\rangle + \langle x\rangle^2}{4Dt}\right]$$
$$= \frac{2Dt\left(\langle x\rangle^2 - 2\langle x\rangle + R_p^2\right) + \Delta R_p^2(x^2 - 2x\langle x\rangle + \langle x\rangle^2)}{4Dt\Delta R_p^2}$$
$$= \frac{\langle x\rangle^2\left(\Delta R_p^2 2Dt\right) - 2\langle x\rangle\left(2R_p Dt + \Delta R_p^2 x\right) + \left(2R_p^2 Dt + \Delta R_p^2 x^2\right)}{4Dt\Delta R_p^2} \tag{G.4}$$

and create a new variable

$$\alpha = \frac{\langle x\rangle}{2}\left(\frac{\Delta R_p^2 + 2Dt}{Dt}\right)^{1/2}. \tag{G.5}$$

Then (G.4) can be rewritten by

$$\left[\frac{(\langle x\rangle - R_p)^2}{2\Delta R_p^2}\right] + \left[\frac{(x - \langle x\rangle)^2}{4Dt}\right] = \alpha^2 + 2\alpha\left[\frac{2R_pDt + \Delta R_p^2 x}{2\Delta R_p\left(\Delta R_p^2 + 2Dt\right)^{1/2}}\right] + \frac{2DtR_p^2 + \Delta R_p^2 x^2}{4Dt\Delta R_p^2}$$
$$= \left[\alpha - \frac{2R_pDt + \Delta R_p^2 x}{2\Delta R_p\left(\Delta R_p^2 + 2Dt\right)^{1/2}}\right]^2 - \frac{4D^2t^2R_p^2 + 4R_pDt\Delta R_p^2 x}{4Dt\left(\Delta R_p^2 + 2Dt\right)^{1/2}}$$
$$+ \frac{2DtR_p^2 + \Delta R_p^2 x^2}{4Dt\Delta R_p^2}\left(\frac{2\Delta R_p^2}{2\Delta R_p^2} + \frac{2Dt}{2Dt}\right)$$
$$= \left[\alpha - \frac{2R_pDt + \Delta R_p^2 x}{2\Delta R_p\left(\Delta R_p^2 + 2Dt\right)^{1/2}}\right]^2 - \frac{x^2 - 2xR_p + R_p^2}{2\left(\Delta R_p^2 + 2Dt\right)}. \tag{G.6}$$

Substituting of (G.6) into (G.3), then follows

$$N(x, t) = \frac{\Phi}{\pi\sqrt{2(\Delta R_p^2 + 2Dt)}} \exp\left[-\frac{1}{2}\frac{(x - R_p)^2}{\Delta R_p^2 + 2Dt}\right]$$
$$\int_{-\infty}^{\infty} d\alpha \exp\left[-\left(\alpha - \frac{2R_pDt + \Delta R_p^2 x}{2\Delta R_p\sqrt{Dt(\Delta R_p^2 + 2Dt)}}\right)^2\right]. \tag{G.7}$$

The integral is known and has the value of $\sqrt{\pi}$ (see e.g., [1]).

Consequently, follows the concentration distribution after implantation and subsequent annealing to

$$N(x,t) = \frac{\Phi}{\sqrt{2\pi\left(\Delta R_p^2 + 2Dt\right)}} \exp\left[-\frac{1}{2}\frac{(x - R_p)^2}{\Delta R_p^2 + 2Dt}\right]. \qquad (G.8)$$

Reference

1. J. Crank, *The Mathematics of Diffusion,* (Oxford University Press, Oxford 1975)

Appendix H

H.1 Threshold Displacement Energy of Different Materials

Threshold displacement energy of graphite, carbon, diamond, fullerene and highly ordered pyrolytic graphite (HPOG)

Material	Experiment/simulation	Notes	E_d [eV]	References
Graphite	Electron irradiation, single crystals	[0001] @ 15 K	60 ± 10	1
	MD simulation including ZBL potential	296 K	25	2
		900 K	60	
	Electron irradiation, polycryst. graphite	323 K	25.8 ± 0.9	3
	Electron microscopy	330–600 K	24	4
	MD simulation, Brenner potential	300 K	44.5	5
		1800 K	42	
	Density functional theory calculations		25	6
	MD simulation	AA stacking, $E_{d(avg}$	34	7
		AB stacking, $E_{d(avg}$	34.5	
	Transmission electron microscopy	[0001]	34	8
	Electron irradiation	[0001] @ RT	31	9
	Electron irradiation	[0001]	27	10
	Auger electron spectroscopy	[0001]	34	11

(continued)

B. Rauschenbach, *Low-Energy Ion Irradiation of Materials*, Springer Series in Materials Science 324, https://doi.org/10.1007/978-3-030-97277-6

(continued)

Material	Experiment/simulation	Notes	E_d [eV]	References
	Transmission electron microscopy		20–30	12
	Transmission electron microscopy		10–20	13
	Measurement of resistivity versus temperature	[0001]	23–31	14
		$[11\bar{2}0]$	30–42	
	Transmission electron microscopy		30	15
	Ne ion irradiation		35.3 ± 1	16
	Electron irradiation		15	17
	Electron irradiation		25	18
			30	19
Carbon	MD simulation	$E_{d(min)}$	25 ± 5	20
		$E_{d(av)}$	69 ± 14	
	Electron irradiation		10	21
Diamond	Electron irradiation & Transmission electron microscopy	[100]	37.5 ± 1.2	22
		[110]	47.6 ± 1.3	
		[111]	45.0 ± 1.3	
	Transmission electron microscopy		80	15
	Measurement of resistivity	15 K	35	23
	Determination of swelling	~ 370 K	< 55	24
	Measurement of resistivity	300 K	< 80	25
	Electron irradiation		35	17
	MD simulation	[001], [011]	50	26
		[111]	60	
	MD simulation	$E_{d(min)}$	30	27
		$E_{d(av)}$	70	
	MD simulation	[100]	52	28
		[110]	60	
		[111]	45	
	Electron irradiation	[110]	32	29
		[100]	27	
		[111]	34	
			40	19

(continued)

(continued)

Material	Experiment/simulation	Notes	E_d [eV]	References
HOPG	Transmission electron microscopy		12	[30]
C_{60}	Transmission electron microscopy		7.6–15.7	[21]
			< 15	[31]
Multi-wall C nanotube	Transmission electron microscopy	@ 300 K	≈ 15 to 20	[20]

[1] M. W. Lucas, E. W. J. Mitchell, Carbon 1 (1964) 401
[2] A. J. McKenna, et al., Carbon 99 (2016) 71
[3] D. T. Eggen, Techn. Rep. NAA-SR-69, North American Aviation, Los Angeles 1950
[4] S. M. Ohr, et al., in Electron Microscopy and Structure of Materials (ed. G. Thomas), Univ. of California Press, Berkeley, 1981, p. 964
[5] B. D. Hehr, et al., Nucl. Technol. 160 (2007) 251
[6] O. V. Yazyev, et al., Phys. Rev. B 75 (2007) 115,418
[7] R. Smith, K. Beardmore, Thin Solid Films, 272 (1996) 255
[8] R. F. Egerton, Phil. Mag. 35 (1977) 1425
[9] G. L. Montet, G. E. Myers, Carbon 9 (1971) 179
[10] H. Abe, et al., MRS Symp. Proceed. Vol. 373 (1995) p. 383
[11] D. Marton, et al., Phys. Rev. B48 (1993) 6757
[12] G. Lulli, et al., Ultramicroscopy 60 (1995) 187
[13] M. Zaiser, F. Banhart, Phys. Rev. Lett, 79 (1997) 3680
[14] T. Iwata, T. Nihira, J. Phys. Soc. Japn. 31 (1971) 1761
[15] L. W. Hobbs, in Hren, J. J., Goldstein, J. I., Joy, D. C. (Eds.), Introduction to Analytical Electron Microscopy, Plenum Press, New York, 1987, p. 399
[16] H. J. Steffens, et al., Phys. Rev. Lett. 68 (1992) 1726
[17] F. Banhardt, J. Appl. Phys. 81 (1997) 3440
[18] J. W. Corbett, Electron Radition Damage, in Semiconductors and Metals, Academic Press, New York 1966
[19] S. J. Zinkle, C. Kinoshita, *J. Nucl. Mater.,* 251 (1997) 200
[20] A. Yu. Konobeyev, et al., Nucl. Energy Technol. 3 (2017) 169
[21] F. Banhardt, Rep. Prog. Phys.62 (1999) 1181
[22] J. Koike et al., Appl. Phys. Lett. 60 (1992) 1450
[23] J. C. Bourgoin, B. Massarani, Phys. Rev. b 14 (1976) 3690
[24] J. F. Prim, et al., Phys. Rev. B 34 (1986) 8870
[25] C. D. Clark, et al., Discuss. Faraday Soc. 31/1961) 96
[26] W. Wu, S. Fahy, Phys. Rev. B 49 (1994) 3030
[27] D. Delgado, R. Vila, J. Nucl. Mater. 419 (2011) 32
[28] R. Kalish, et al., Phys. Stat. Sol. (a) 174 (1999) 83
[29] J. W. Steeds, Nucl. Instr. Meth. in Phys Res. B 269 (2011) 1702
[30] K. Nakai, et al., Ultramicroscopy, 39 (1991) 361
[31] T. Füller, F. Banhardt, Chem. Phys. Lett. 254 (1996) 372

Threshold displacement energy of silicon

Experiment/simulation	Notes	E_d [eV]	References
DFT-MD simlation	$E_{d(av)}$	36 ± 2	[1]
	[100]	20 ± 2	
	[111]	12.5 ± 1.5	
Recommendation		20.5	[2]
Electron irradiation	[111]	12 ± 0.6	[3]
Electron irradiation		21	[4]
MD-simulation	[100]	18.6	[5]
	[111]	16.2	
Low-energy electron irradiation and comparison with computational results	[100]	30	[6]
	[110]	35	
	[111]	22	
MD-simulation	[111]	15.8	[7]
	[100]	12.2	
Computer simulation		10–26	[8]
Quantum–mechanical calculation	[100] and [111]	9.75 ± 0.25	[9]
MD simulation	[100]	23	[10]
	surface E_d	18	
MD simulation using different forms of potential	[100]	9.75–23.5	[11]
	[111]	13.25–20.25	
	[− 1− 1− 1]	11.75–17.25	
MD simulation	[111] (at 1100 K and E_I < 100 eV)	9 ± 2 (for adatoms)	[12]
		18 ± 1 (for knock-in´s)	
		19 ± 1 (for sputtering)	
		21 ± 1 (bulk vacancies)	
Si, n-type	300 K	11 ± 1.6	[13]
Si, n-type	80 K, 245 K	22 ± 1	[14]
Si, n-type	[111], 290 K	22	[15]
Si, n-type	300 K	≈ 13	[16]
	80 K	≈ 20	
Si, p-type	[111], ≈ 290 K	12	[17]
Si, p-type	77 K–300 K	≈ 20.7	[18]
Si, n-type	77 K–300 K	≈ 14.4	
Si, p-type	300 K	≈ 13	

[1] E. Holmström, et al., Phys. Rev B78 (2008) 045,202
[2] P. J. Griffin, Report SAND 2016-1941C
[3] J. Loferski, P. Rappaport, Phys. Rev. 111 (1958) 432

[4] J. Corbett, G. D. Watkins, Phys. Rev. 138 (1965) A555
[5] L. A. Miller, et al., Phys. Rev. B 49 (1994) 16,953
[6] P. Hemment, P. Stevens, J. Appl. Phys. 40 (1969) 4893
[7] L. Miller, et al., Radiat. Eff. Defects Solids 129, (1994) 127
[8] M. Sayed, et al., Nucl. Instrum. Methods Phys. Res. B 102 (1995) 232
[9] W. Windl, et al, Nucl. Instrum. Methods Phys. Res. B 141 (1998) 61
[10] S. Uhlmann, et al., Radiat. Eff. Defects Solids 141 (1997) 185
[11] M. Mazzarolo, et al. Phys. Rev. B 63 (2001) 195,207
[12] Z. Wang, E. G. Seebauer, Phys. Rev. B 66 (2002) 205,409
[13] N. N. Gerasimenko, et al., Soviet Phys. Semicond. 5 (1971) 1439
[14] P. L. F. Hemment, P. R. C. Stevens, in Atomic Collision Phenomena in Solids, Eds. D. W. Palmer, et al., North- Holland, Amsterdam, (1970) p. 217
[15] P. C. Bunbury, in Radiation Effects in Semiconductors, Ed. F. L. Vook, Plenum Press, New York (1968), p. 280
[16] T. I. Kolomenskaya, et al., Soviet Phys. Semicond. 1 (19,679 652
[17] J. A. Grimshaw, Phys. Lett. 22 81,966) 372
[18] R. L. Novak, Bull. Amer. Phys. Soc. 8 (1963) 235
[19] V. S. Vavilov, et al., Soviet Phys. Semicond. 2 (1960) 1301

Threshold displacement energy of iron

Experiment/simulation	Notes	E_d [eV]	References
ASTM standard practice for Fe		40	1
MD simulation with 11 different potentials	$E_{d(av)}$	22–53.5	2
	[100]	9–21	
	[110]	25–47	
	[111]	17–35	
Simulation	[100]	17.95	3
	[111]	26.17	
Electron irradiation, single crystal	[100]	17 ± 1	4
	[110]	> 30	
	[111]	20 ± 1.5	
High-energy electron irradiation	[100]	17	5
	[110] and [111]	30	
MD simulation, low-energies, Finnis–Sinclair potential	[100] @ 0 K	≈ 18	6
	[100] @ 100 K	20 ± 5	
	[110]	≈ 31	
	[111]	> 70	
MD simulation for three different strains and two temperatures (300 K and 500 K)	$E_{d(av)}$ @ 300 K	50.0[a)]/35.8[b)]/46.2[c)]	7
	$E_{d(av)}$ @ 500 K	57.4[a)]/44.4[b)]/51.7[c)]	
	$E_{d(min)}$ @ 300 K	20.6[a)]/15.7[b)]/19.6[c)]	
	$E_{d(min)}$ @ 500 K	19.6[a)]/14.6[b)]/17.4[c)]	

(continued)

(continued)

Experiment/simulation	Notes	E_d [eV]	References
Dynamic calculations with different potentials	[100]	≈ 17	[8]
	[110]	≈ 34	
	[111]	≈ 38	
Electron irradiation	@ 100 °C	16.7	[9]
High-energy electron irradiation	@ 20.4 K	24	[10]
MD-simulation	[100] @ 36 K	16	[11]
	[110] @ 36 K	30	
	[111] @ 36 K	30	
	[100] @ 300 K	15	
	[110] @ 300 K	37	
	[111] @ 300 K	27	
Electron irradiation	[110]	30	[12]
	[100]	20	
	[111]	25	
Electrical resistance	$E_{d(av)}$	24	[13]
MD simulation	$E_{d(av)}$	44	[14]
		17	[15]
Combination of MD and kinetic MC simulation	[100]	19	[16]
	[110]	27	

[(a)] unstrained, [(b)]2% hydrostatic tensile strain, [(c)]5% Bain strain (tetragonal shear)

[1] ASTM Standard E693-94 (1994)
[2] K. Nordlund et al., Nucl. Instr. Meth B 246 (2006) 322
[3] V. M. Agranovich, V. V. Kirsanov, Sov. Phys. – Solid State 12 (1971) 2147
[4] F. Maury, et al., Phys. Rev. B 14 (1976) 5303
[5] J. N. Lomer, M. Pepper, Phil. Mag. 16 (1967) 11
[6] D. J. Bacon, et al., J. Nucl. Mater. 205 (1993) 52
[7] B. Beeler, et al., J. Nucl. Mater 474 (2016) 113
[8] C. Erginsoy et al., Phys. Rev. 133 (1964) A595
[9] R. W. Series, et al., Rad. Eff. (1977) 81
[10] P. G. Lucasson, R. M. Walker, Diss. Faraday Soc. 31 (1961) 57
[11] K. Zolnikov, et al., Nucl. Instr. Meth. B 352 (2015) 43
[12] P. Vajda, Rev. Mod. Phys. 49 (1977) 481
[13] P. G. Lucasson, R. M. Walker, Phys. Rev. 127 (1962) 485
[14] P. Lucasson, in M. T. Robinson, F. N. Young (Eds.), Fundamental aspects of radiation damage in metals, Oak Ridge Nat. Lab. (1975) p. 42
[15] P. Jung, Phys. Rev. B 23 (1981) 664
[16] N. Soneda, T. Diaz de la Rubia, Phi. Mag. A 78 (1998) 995

Appendix I

I.1 Impact Parameter, Mean Free Path and Collision Number

It is assumed that n gas particles per unit volume are randomly distributed (for details see [1–3]). The particles are considered as hard spheres moving through the gas with mean velocity v. A collision occurs when two particles approach to a distance $b_{max} = r_1 + r_2$ (= d), where r_1 and r_2 are radii of the particles, b is the distance of the line perpendicular to the initial directions of motion of the two colliding particles (see Fig. I.1). Thus, b_{max} is the maximum value of the impact parameter at which a collision can be expected. The target area or collision cross-section σ for such a collision is given by

$$\sigma = \pi (r_1 + r_2)^2 = \pi d^2 = \pi b_{max}^2. \tag{I.1}$$

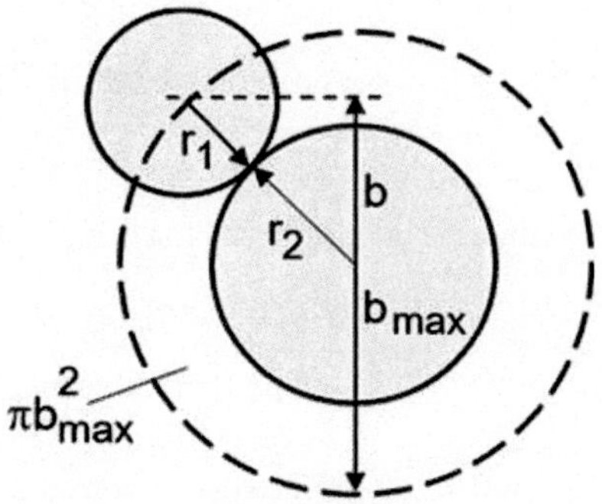

Fig. I.1 Schematic representation of the collision of two particles with the radii r_1 and r_2

B. Rauschenbach, *Low-Energy Ion Irradiation of Materials*, Springer Series in Materials Science 324, https://doi.org/10.1007/978-3-030-97277-6

The average distance between two collisions is the mean free path λ. The number of collisions for particles moving in a straight line within the radial distance d is given by a collision cylinder of the dimension b_{max} × travel distance × N, where N is number of particles per unit volume, called atomic number density. Consequently, for a collision distance corresponding to a travel distance λ, the mean free path length follows to be

$$1 = \pi d^2 \cdot N \cdot \lambda = \sigma \cdot N \cdot \lambda \quad \text{or} \quad \lambda = \frac{1}{N\pi d^2} = \frac{1}{N\sigma} = \frac{1}{\sigma}\frac{k_B T}{p}, \tag{I.2}$$

$$\text{with} \;\; N = \frac{V}{n} = \frac{p}{k_B T},$$

where p is the pressure and n is the total number of particles. In the same way, the average time τ between two collisions can be defined to

$$\tau = \frac{\lambda}{\langle v \rangle} = \frac{1}{N\sigma}\sqrt{\frac{\pi M}{8 k_B T}}, \tag{I.3}$$

where $\langle v \rangle$ is the mean velocity and M is mass of the particle.

A condition for the deduction of (I.2) is that all particles except one particle are stationary particles. In reality, the motion of the other particles cannot be neglected. Consequently, a more realistic description, under the condition that an average relative velocity, $\langle v \rangle_r$, is introduced for an encounter between two particles, is necessary. The relative velocity can be expressed by vector sum of the two colliding particles $\boldsymbol{v}_r = \sqrt{\boldsymbol{v}_1 - \boldsymbol{v}_2}$ and the scalar product of the relative velocity as $v_r = \sqrt{v_1 v_2}$. Then, the relative velocity is given by

$$v_r = \sqrt{(\boldsymbol{v}_1 - \boldsymbol{v}_2) \cdot (\boldsymbol{v}_1 - \boldsymbol{v}_2)} = \sqrt{(\boldsymbol{v}_1 \cdot \boldsymbol{v}_1) - 2\boldsymbol{v}_1 \cdot \boldsymbol{v}_2 + (\boldsymbol{v}_2 \cdot \boldsymbol{v}_2)} \tag{I.4}$$

and the average relative velocity is given by

$$\langle v_r \rangle = \sqrt{\langle v_1 \cdot v_1 \rangle - 2\langle v_1 \cdot v_2 \rangle + \langle v_2 \cdot v_2 \rangle}. \tag{I.5}$$

Since the middle term of the last equation is zero (both vector velocities are random and uncorrelated), the relationship between the average relative and the mean velocity is

$$\langle v \rangle_r = \sqrt{\boldsymbol{v}_1^2 + \boldsymbol{v}_2^2} = \sqrt{2}\langle v \rangle. \tag{I.6}$$

The average relative velocity is slightly larger than the mean velocity. Thus, the mean free path length is given by (compare with I.2)

$$\lambda = \frac{1}{N\sqrt{2}\sigma}. \tag{I.7}$$

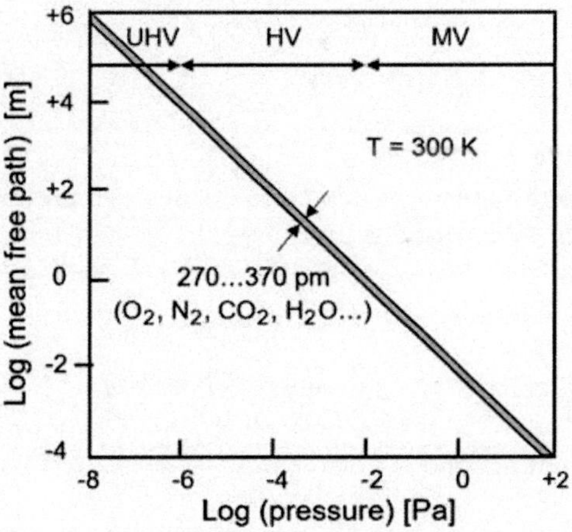

Fig. I.2 Mean free paths in dependence on the pressure at room temperature. The diameters of the gas molecules are assumed to be between 270 and 370 pm, which is typical for the gases as nitrogen, oxygen, rare gases, etc. (UHV—ultra-high vacuum, HV—high vacuum, MV—medium vacuum)

Fig. I.2 shows the mean free path in dependence of the pressure for several gases at room temperature

Taking into account the movement of gas particles (e.g., generated by sputtering or evaporation) through a background gas, the mean free path length between collisions of sputtered or evaporated gas particles with each other and collisions between gas particles and particles of the background gas must be determined. Under the condition of such a gas mixture, where the number density N_{sp} of the first gas component (produced by sputtering or evaporation) is much lower than the density N_g of the second component (background gas), the mean free path according to Jeans [1] is given by

$$\lambda_{sp,g} = \frac{1}{\pi\sqrt{2}N_{sp}d_{sp,sp}^2 + \pi\sqrt{1+\frac{M_{sp}}{M_g}}N_g d_{sp,g}^2} \quad \text{and}$$

$$\lambda_{g,sp} = \frac{1}{\pi\sqrt{2}N_g d_{g,g}^2 + \pi\sqrt{1+\frac{M_g}{M_{sp}}}N_{sp} d_{sp,g}^2}, \tag{I.8}$$

where $d_{sp,sp}$ is the collision diameter of binary collisions between *sp* type particles and $d_{sp,g}$ the collision diameter between *sp* and *g* type particles ($d_{sp,g} = (d_{sp} + d_g)/2$), M_{sp} and M_g are the masses of the sputtered or evaporated particles and background gas particles, respectively. For this binary gas mixture and for low concentration (dilute mixture) $N_{sp} \ll N_g$, the last equation can be reduced to

$$\lambda_{sp,g} = \frac{1}{\pi\sqrt{1+\frac{M_{sp}}{M_g}}N_g d_{sp,g}^2} \quad \text{and} \quad \lambda_{g,sp} = \frac{1}{\pi\sqrt{2}N_g d_{g,g}^2} \tag{I.9}$$

and is thus comparable to (I.7).

The mean collision frequency z describes the average number of collisions over time in a well-defined system and is given by

$$z = N \cdot \pi d^2 \cdot \sqrt{2} \cdot \langle v \rangle = N\sqrt{2}\sigma \langle v \rangle. \tag{I.10}$$

where the average velocity is given by $v = \sqrt{8RT/\pi M}$.

The total number of collisions, called collision rate, is N times z, but divided by 2 to avoid double counting, and is given by

$$Z = \frac{zN}{2} = N^2\frac{\sqrt{2}}{2} \cdot \pi d^2 \cdot \langle v \rangle = N^2\frac{\sqrt{2}}{2}\sigma\sqrt{\frac{8RT}{\pi M}}. \tag{I.11}$$

References

1. J. Jeans, *The dynamical theory of gases, Dover*, (New York 1954) and E. J. Davis, Aerosol Sci. Technol. **2**, 121 (1983)
2. S. Chapman, T. G. Cowlings, *The Mathematical Theory of Non-Uniform Gases*, (Cambridge Univ. Press, Cambridge 1990)
3. E. W. McDaniel, *Collision Phenomena in Ionized Gases*, (Wiley New York 1964)

Appendix J

J.1 Mean Energy of Sputtered Atoms

Falcone [1] proposed an energy distribution of the sputtered particles which is characterized by a cutoff in the energy distribution of sputtered atoms. The sputtering yield according to Falcone is then given by (c.f. 5.48)

$$Y_F(E) = Y\left(1 - \frac{E_{th}}{E}\right)^2, \tag{J.1}$$

where E is the energy of the incident ion, Y is the sputtering yield and E_{th} is the sputtering threshold energy (see Sect. 5.4.1). The mean energy of sputtered atoms is defined by

$$\langle E \rangle_{sp} = \frac{1}{Y(E)} \int_0^{E_M} dEEF(E_{sp}), \tag{J.2}$$

where $F(E_{sp})$ is the kinetic energy distribution of the sputtered atoms according to Thompson [see (10.6)]. The upper integration limit is given by a maximum energy (cut-off energy) $E_M = \gamma E - U_s$, where γ is the transfer energy efficiency factor [see (2.36)] and U_s is the surface binding energy. Under the condition that $E \gg E_{th}$, the integration of (J.2) yields the average energy of the sputtered particles to be

$$\langle E \rangle_{sp} = 2U_s\left[\ln\left(\frac{E}{E_{th}}\right) - \frac{3}{2}\right]. \tag{J.3}$$

The choice of a cut-off yield energy distribution is partly discussed critically [2].

Another way to determine the average energy of the sputtered atoms is based on the definition

B. Rauschenbach, *Low-Energy Ion Irradiation of Materials*, Springer Series in Materials Science 324, https://doi.org/10.1007/978-3-030-97277-6

$$\langle E\rangle = \frac{\int\limits_0^\infty EF(E)dE}{\int\limits_0^\infty F(E)dE}, \tag{J.4}$$

where F(E) represents the distribution of the particles. For example, the Thompson distribution of the sputtered atoms [3, 4], given by (10.6) or (10.8), is

$$F(E_{sp}) = \frac{2U_s E_{sp}}{(U_s + E_{sp})^3}, \tag{J.5}$$

where E_{sp} is the energy of the sputtered particle and U_s is the surface binding energy. Equation (J.3) can be integrated assuming a Thompson distribution, (J.5), where the upper integration limit is given by T_{max} [5]. The maximum transferred energy T_{max}, given by (2.27), describes the maximal energy that can be transferred from the incident ion with the energy E to the target atom.

Assuming that $T_{max} \gg U_s$, the integration of the denominator [5] of (J.4) yields

$$\int\limits_0^{T_{max}} (E_{sp})FdE = -2U_s \frac{U_s + 2E_{sp}}{2(U_s + E_{sp})^2}\Bigg|_0^{T_{max}} = -U_s\frac{U_s + 2E_{sp}}{2(U_s + E_{sp})^2} + 1 \approx 1 \tag{J.6}$$

and the integration of the numerator leads to

$$\begin{aligned}\int\limits_0^{T_{max}} E_{sp}F(E_{sp})dE &= U_s^2\frac{3U_s + 4E_{sp}}{(U_s + E_{sp})^2} + 2U_s\ln(U_s + E_{sp})\Bigg|_0^{T_{max}} \\ &= U_s^2\frac{3U_s + 4E_{sp}}{(U_s + E_{rp})^2} + 2U_s\ln(U_s + T_{max}) - 3U_s - 2U_s\ln(U_s) \\ &\approx 2U_s\ln(T_{max}) - 3U_s - 2U_s\ln(U_s) = 2U_s\ln\left(\frac{T_{max}}{U_s}\right) - 3U_s.\end{aligned} \tag{J.7}$$

Consequently, the mean energy of the sputtered particles can be expressed by

$$\langle E_{sp}\rangle = 2U_s\left[\ln\left(\frac{T_{max}}{U_s}\right) - \frac{3}{2}\right]. \tag{J.8}$$

Remarkable is the similarity of (J.3 and J.8).

For comparison, it should be remembered that the average velocity of molecules, $\langle v\rangle = \sqrt{8k_BT/\pi M}$, in a volume is consistent with Maxwell-Boltzmann theory, while the average velocity of molecules coming out of the aperture of a Knudsen cell, $\langle v_K\rangle = (3\pi/4)\sqrt{2k_BT/\pi M}$, is approximately a factor of 1.178 greater [6].

Consequently, the average kinetic energy of the molecules in an emerging beam from a Knudsen cell can be given by

$$\langle E_{Knudsen}\rangle = \frac{M\langle v_K\rangle^2}{2} = \frac{9\pi}{16}k_B T. \tag{J.9}$$

References

1. G. Falcone, Sputtering transport theory: The mean energy, Phys. Rev. B **38**, 6398 (1988)
2. Z. L. Zhang, Comment on "Sputtering transport theory: The mean energy", Phys. Rev. B **71**, 026,101 (2005)
3. M. W. Thompson, II. The energy spectrum of ejected atoms during the high energy sputtering of gold, Phil. Mag. **18**, 377 (1968)
4. P. Sigmund, in *Sputtering by particle bombardment I*, ed. by R. Behrisch, Topics in Appl. Physics, Vol. 47, (Springer-Verlag Berlin Heidelberg, 1981) pp.9
5. J. Held, A. Hecimovic, A. von Keudell, V. Schulz von der Gathen, Velocity distribution of titanium neutrals in the target region of high power impulse magnetron sputtering discharges, Plasma Source Sci. Technol. **27**, 105,012 (2018)
6. V.A. Elyukhin, R. Pena-Sierra, G. Garcia-Salgado, Thermodynamic models of molecular beams, Superficies y Vacio **17**, 25 (2004)

Appendix K

K.1 Particle Impingement Flux and Source Emission Characteristic

Hertz-Knudsen Equation

In the context of deposition experiments, it is useful to known the number of particles (atoms, molecules) hitting a unit area and in a unit time. Assume that particles with a velocity v strike the element dA of the surface at the angle θ to the normal of dA (see Fig. K.1). These particles cover a distance of vdt in the time interval dt. Consequently, all particles in the volume dA × v dt × cos θ hit the surface element dA. In the case of a one-dimensional consideration, the number of particles in the velocity range between v and v + dv is given by the Maxwell–Boltzmann distribution function of particle velocities in a gas

$$f(v_i) = \sqrt{\frac{M}{2\pi k_B T}} \exp\left(-\frac{M v_i^2}{2k_B T}\right), \tag{K.1}$$

where M is the mass of the particle. This distribution has the form of a Gaussian distribution and is symmetric about the origin. The probability of finding a particle with the velocity v_i is proportional to $\exp\left(-M v_i^2/2k_B T\right)$. Then the particle flux J can be expressed by

$$J(v)d^3v = f(v)v\cos\theta d^3v \tag{K.2}$$

and the total flux per area dA and unit time dt is given by integration over all particles n which strike the surface

B. Rauschenbach, *Low-Energy Ion Irradiation of Materials*, Springer Series in Materials Science 324, https://doi.org/10.1007/978-3-030-97277-6

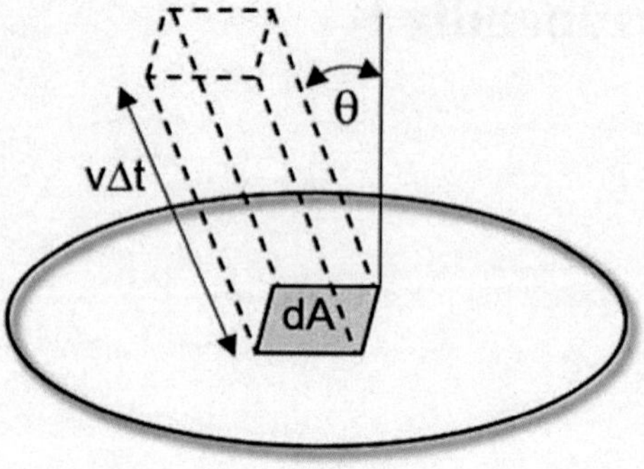

Fig. K.1 Schematic presentation of the particle deposition with a deposition angle θ

$$J = \frac{dn}{Adt} = n\int f(v)v\cos\theta d^3v, \tag{K.3}$$

where n is particles per volume (particle density). In spherical coordinates (see Appendix C) is $d^3v = v^2\, dv \sin\theta\, d\theta\, d\varphi$ and the flux

$$\begin{aligned} J &= \frac{dn}{Adt} \\ &= n\int f(v)v\cos\theta v^2\sin\theta d\theta\int_0^{2\pi} d\varphi \\ &= n\int_0^{\infty} f(v)v^3dv\int_0^{\pi/2}\cos\theta\sin\theta d\theta\int_0^{2\pi} d\varphi \\ &= \pi n\int_0^{\infty} f(v)v^3dv. \end{aligned} \tag{K.4}$$

After substitution of the mean velocity (for definition, see e.g., [1])

$$\langle v\rangle = \int_0^{\infty} f(v_i)v\cdot 4\pi v^2dv = 4\pi\int_0^{\infty} f(v_i)v^3dv, \tag{K.5}$$

the total particle flux is given by

$$J = \frac{dn}{Adt} = \frac{1}{4}n\langle v\rangle. \tag{K.6}$$

This total flux, the so-called impingement rate (particles per unit area and time), depends only on the number of particles and the average velocity [2]. Assuming that the particles exhibit the behavior of an ideal gas ($n = p/k_BT$), and by substituting the mean velocity, it follows that

$$\langle v\rangle = 4\pi\left(\frac{M}{2\pi k_B T}\right)^{3/2}\int_0^\infty \exp\left(-\frac{Mv^2}{2k_B T}\right)v^3 dv$$
$$= 4\pi\left(\frac{M}{2\pi k_B T}\right)^{3/2}\frac{1}{2}\left(\frac{M}{2k_B T}\right)^{-2} = \sqrt{\frac{8k_B T}{\pi M}}. \tag{K.7}$$

With that, the Hertz-Knudsen equation of the evaporation is given by

$$J = \frac{p}{\sqrt{2\pi M k_B T}}. \tag{K.8}$$

This equation is derived from the kinetic gas theory and specifies the number of particles strike a surface per unit area and unit time at equilibrium. This is called the impingement rate. The relation (K.8) represents the maximum possible evaporation particle flux density. If (K.8) is multiplied with the mass of particle M, the mass evaporation rate can be determined as follows

$$J \cdot M = \sqrt{\frac{M}{2\pi k_B T}}p. \tag{K.9}$$

In general, two corrections to modify (K.8) must be discussed. First, a pressure independent sticking coefficient, s (≤ 1), is introduced to account for the fact that particles can be reflected or desorbed from the surface. Second, in (K.8) is ignored the hydrostatic pressure of the evaporated material in the gas phase. This is correct, if the evaporation is realized under ideal vacuum conditions. If this prerequisite is not given, the difference between the equilibrium pressure p_e of the evaporant at the temperature T and the ambient hydrostatic pressure p must be considered. Consequently, the Hertz-Knudsen equation must be rewritten by

$$J = s\frac{p_e - p}{\sqrt{2\pi M k_B T}}. \tag{K.10}$$

This is consistent with the kinetic theory in which the impingement flux is proportional to pressure. Consequently, the maximum flux or impingement rate is determined by p_e and can only be achieved by $p = 0$.

Particle Emission Characteristic

The determination of the emission characteristics is based on the assumptions that (i) there are no collisions between the emitted particles and the particles of the residual gas and (ii) the sticking coefficient, $s = 1$, where this coefficient is defined as the ratio of number of particles sticking on the surface to the total number of particles impinging on the substrate surface in the same time period.

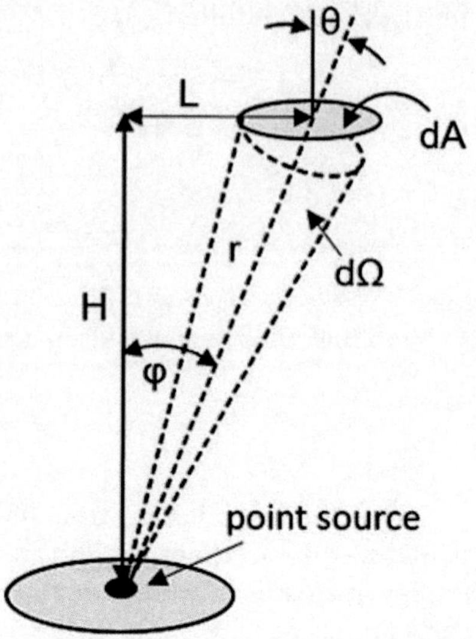

Fig. K.2 Schematic presentation of the particle flux within a solid angle $d\Omega$

The probability of a gas molecule leaving a solid surface in a given direction within a solid angle $d\Omega$ is proportional to cos θ $d\Omega$, where θ is the angle between the direction and the normal to the surface [3]. Figure K.2 schematically shows the deposition under an oblique angle, where dA is the area element of the substrate, φ is the angle between the surface normal and the line between emission source and area element dA of the substrate. Then, the solid angle $d\Omega$ is given by

$$d\Omega = \frac{\cos\theta}{r^2} dA. \tag{K.11}$$

and the amount $d\overline{M}$ of material passing through solid angle $d\Omega$ is given by

$$d\overline{M} = Cmd\Omega = Cm\frac{\cos\theta}{r^2} dS, \tag{K.12}$$

where m is the total mass of material emitted from the source in all directions and C is a constant.

According to Holland and Steckelmacher [4], a distinction should be made between two types of particle sources:

(i) the point source, which uniformly emits the particles in all spatial directions, i.e. the substrate surface can be assumed to be a sphere with the point source in the center (i.e. $\theta = 0$ and $dA = 2\pi \sin\varphi \, d\varphi$, see Appendix C) and
(ii) the small-surface source obeying cosine law of Knudsen [3], when the distance between the particle source and the substrate is relatively large compared to

the surface area of the particle source. In the second case, the small-surface source, the material is only emitted in the hemisphere.

Then, the integrals and the constants are given by

$$\int d\overline{M} = Cm \int_{\varphi=0}^{\varphi=\pi} 2\pi \sin\varphi d\varphi \text{ and } C = 1/4\pi \text{ for the point source} \tag{K.13}$$

and

$$\int d\overline{M} = Cm \int_{\varphi=0}^{\varphi=\pi/2} 2\pi \cos\varphi \sin\varphi d\varphi \text{ and } C = 1/\pi \text{ for the small-surface source.} \tag{K.14}$$

The solutions of the last two equations lead to

$$d\overline{M} = \frac{m}{4\pi} d\Omega = \frac{m}{4\pi r^2} \cos\theta dA \text{ for the point source} \tag{K.15}$$

and

$$d\overline{M} = \frac{m}{\pi} \cos\varphi d\Omega = \frac{m}{\pi r^2} \cos\varphi \cos\theta dA \text{ for the small-surface source.} \tag{K.16}$$

The deposited film thickness, d_f, can be determined with $d\overline{M} = \rho d_f \, dS$, where ρ is the mass density of the film material. For the point source follows the film thickness to

$$d_f = \frac{m}{4\pi\rho r^2} \cos\theta = \frac{m}{4\pi\rho} \frac{H}{r^3} = \frac{m}{4\pi\rho} \frac{H}{\left(H^2 + L^2\right)^{3/2}} \tag{K.17}$$

and for the small-surface source the film thickness is given by

$$d_f = \frac{m}{\pi\rho r^2} \cos\varphi \cos\theta = \frac{m}{\pi\rho} \frac{H^2}{\left(H^2 + L^2\right)^2}. \tag{K.18}$$

Experimentally, it was found out that the Knudsen laws do not fully reflect the measured particle emission data after evaporation or sputtering, since the angular profile is also dependent on the orifice of the particle source (difference between point source and real source).

Consequently, in a first approximation the cos θ - distribution (K.17) is modified by

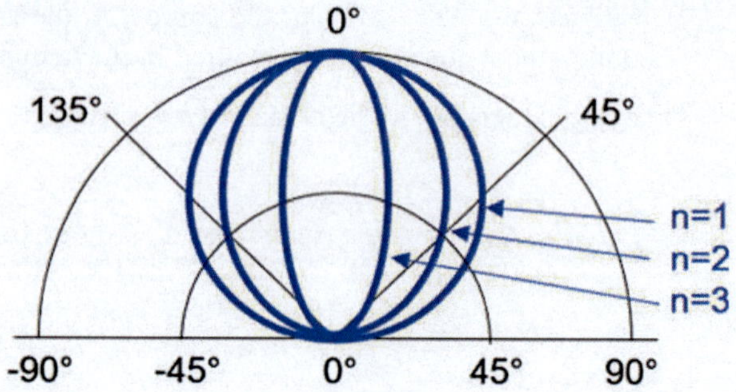

Fig. K.3 Angular distribution of the cosine distribution for n = 1, 2 and 3

$$d_f = \frac{m}{4\pi\rho r^2}\cos^n\theta \tag{K.19}$$

for the point source, where n is a real number [5]. Figure K.3 depicts the normalized angular distribution for n = 1, 2 and 3.

In practice, significantly larger particle sources are applied, i.e. that the distance between particle source and the substrate is comparable to the diameter (length) of the source area. Examples are boats or crucibles for thermal evaporation and electron beam evaporation, respectively. With the objective of using the formalism for small area sources, Villa and Pompa [6] replaced the equation for the film thickness [cf. (K.18)] in the application of larger sources by

$$d_f = \iint\limits_{P(x,y)} \frac{Q}{r^2}\cos^n\varphi\cos\theta dxdy. \tag{K.20}$$

where Q is an experimentally determinable coefficient and P(x,y) describes the form of the source area [6].

References

1. W. Kauzmann, *Kinetic Theory of Gases*, Thermal Properties of Matter, Vol. I, (Benjamin, Reading, MA, 1966)
2. R. Glang, in *Handbook of Thin Film Technology*, ed. by L. I. Maissel and R. Glang, (McGraw-Hill, New York 1970)
3. M. Knudsen, Das Cosinusgesetz in der kinetischen Gastheorie, Ann. der Physik, **48**, 1113 (1915)
4. L. Holland, W. Steckelmacher, The distribution of thin films condensed on surfaces by the vacuum evaporation method, Vacuum **2**, 346 (1952)
5. G. Deppisch, Schichtdickengleichmäßigkeit von aufgedampften Schichten in Theorie und Praxis, Vak.-Tech. **3**, 67 (1981)
6. F. Villa, O. Pompa, Emission pattern of real vapor sources in high vacuum: an overview, Appl. Optics **38**, 695 (1999)

Appendix L

L.1 Statistical Analysis of Roughness and Roughness Measurement Techniques

Roughness

The average roughness in the one-dimensional case is given by the mean value of the heights h(x) over the measured spatial size L by

$$\langle h\rangle = \frac{1}{L}\sum_{x=1}^{L}|h(x)|. \tag{L.1}$$

For two-dimensional surface, the average absolute deviation of the roughness from the mean line $\overline{h}$ is described by the *arithmetic average height*

$$\langle h\rangle = \frac{1}{L^2}\sum_{x=1}^{L}\sum_{y=1}^{L}\left|h(x,y)-\overline{h}(x,y)\right|. \tag{L.2}$$

where $\overline{h} = \frac{1}{L^2}\sum_{x=1}^{L}\sum_{y=1}^{L} h(x,y)$. The $\overline{h}$ - line divides the roughness profile so that the sum of the squares of the deviations of the profile height from it is equal to zero (Fig. L.1).

The arithmetic average height provides no information about the wavelength of the roughness and is not responsive to small changes in the profile. In general, it is assumed that the probability distribution function of heights at a position $\mathbf{r}$ = (x,y) can be described by a Gaussian distribution function (Fig. L.1)

$$f(h) = \frac{1}{\sqrt{2\pi}\,w}\exp\left(-\frac{h^2}{2w^2}\right), \tag{L.3}$$

B. Rauschenbach, *Low-Energy Ion Irradiation of Materials*, Springer Series in Materials Science 324, https://doi.org/10.1007/978-3-030-97277-6

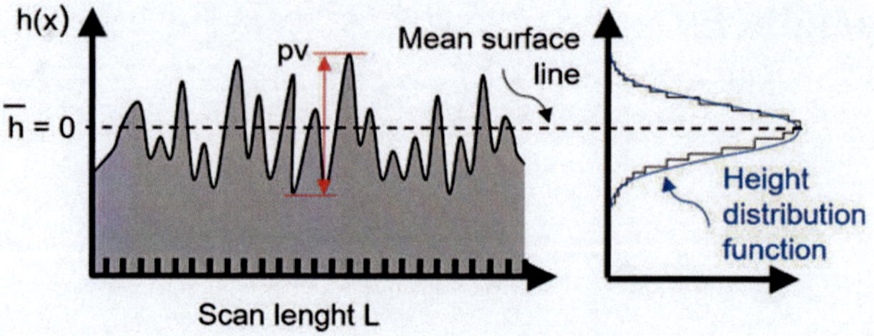

Fig. L.1 Schematic of a surface profile, the histogram of surface heights and the equivalent Gaussian function. The histogram represents the height distribution function

where w (or R_q) is the *root-mean-square (rms) roughness* or *interface width* and can be considered as the second-order moment of the Gaussian distribution of the heights h. The interface width is defined by

$$w = R_q = \frac{1}{L^2}\sqrt{\left\langle \sum_{x=1}^{L}\sum_{y=1}^{L}\left[h(x,y) - \bar{h}(x,y)\right]^2 \right\rangle} = \left(\left\langle\left[h(x,y) - \bar{h}(x,y)\right]^2\right\rangle\right)^{1/2}, \tag{L.4}$$

which gives the deviation of the surface height. The angular brackets $\langle\ldots\rangle$ represent the spatial average over the measured area. The time-dependent evolution of the roughness due to deposition or erosion processes can be expressed as follows

$$w(L,t) = \left(\left\langle[h(r,t) - \bar{h}(r,t)]^2\right\rangle\right)^{1/2}. \tag{L.5}$$

The measurement of the mrs-roughness is bandwidth-limited, i.e. surface spatial wavelengths smaller than the resolution of the measuring instrument are not recorded. The vertical distance between the maximum, h_{max}, and minimum height values, h_{min}, in the array of surface heights, labelled as *peak-to-valley value* (pv), can be expressed by (Fig. L.1).

$$\mathrm{pv} = h_{max} - h_{min}\,. \tag{L.6}$$

In addition to the first moment, h, and second moment, w, higher moments of the distribution characterize the asymmetry of height distribution. The third order moment (called *skewness*) is given by

$$R_s = \frac{1}{Lw^3}\sum_{x=1}^{L}\sum_{y=1}^{L-n}\left[h(x,y) - \bar{h}(x,y)\right]^3 \tag{L.7}$$

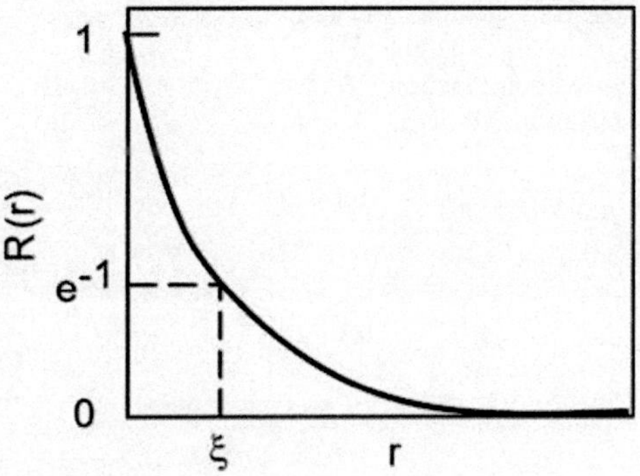

Fig. L.2 General behavior of the autocorrelation function R(r) (figure adapted from [1])

and the fourth order moment (called *kurtosis*) by

$$R_k = \frac{1}{Lw^4}\sum_{x=1}^{L}\sum_{y=1}^{L-n}\left[h(x, y) - \overline{h}(x, y)\right]^4, \tag{L.8}$$

where the skewness is a measure of the symmetry of the distribution with respect to the mean surface line (Fig. L.1) and the kurtosis characterized the sharpness of the distribution (see also Appendix F).

Autocorrelation Function

The vertical properties of a surface can be characterized by the interface width or the root-mean-square roughness. The *autocorrelation function* R(r) is used to measure the correlation between different heights on surfaces

$$R(r = x, y) = \frac{1}{L(L-n)}\frac{1}{w^2}\sum_{x=1}^{L}\sum_{y=1}^{L-n} h(x+n, y)h(x, y), \tag{L.9}$$

where n is the number of measuring points (pixels) used for calculation. The autocorrelation function decreases exponentially (Fig. L.2). On this base, a *lateral correlation length* ξ can be defined as the value of r at which $R(r) = e^{-1}$ [1], where $e \cong 2.718$ is the Euler number. The lateral correlation length can be viewed as a length that is an average measure of the distance beyond which the surface elevation variations are uncorrelated, implying that there is no characteristic long range length scale, i.e. this magnitude is the distance within the surface heights are significantly correlated.

Height-Height Correlation Function

The probability of finding a surface height difference between two points separated by distance r (see inset in Fig. L.3) is given by the *height-height correlation function*, which is an equivalent function to the autocorrelation function. If one applies a matrix of digitized height values (e.g., after characterizing the surface by STM, AFM or SEM), then the height-height-correlation function in one direction and averaged

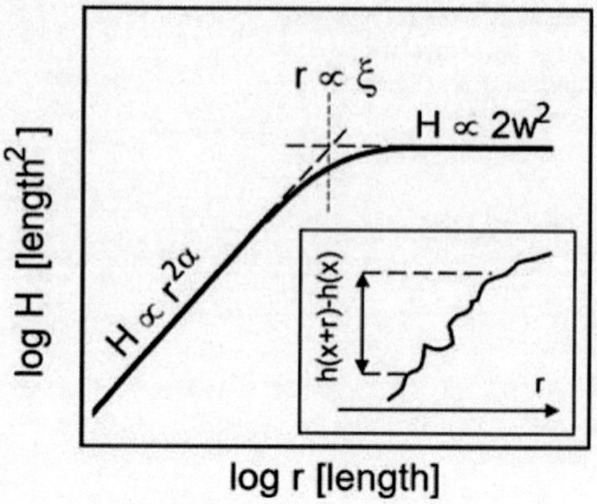

Fig. L.3 General behavior of the height-height correlation function H(r) for self-affine surfaces

over the other direction for self-affine surfaces is given by

$$H(r) = \frac{1}{L(L-n)} \sum_{x=1}^{L} \sum_{y=1}^{L-n} h(x+n, y)h(x, y) = \langle [h(x+r) - h(x)]^2 \rangle, \quad \text{(L.10)}$$

where x characterized a specific surface point. The assumption here is that the statistical behavior of the surface is independent of the orientation of the surface. On the one hand, the height-height correlation function for self-affine surfaces is simple an inversion and vertical translation of the autocorrelation function [1], expressed by

$$H(r) = \langle [h(x+r) - h(x)]^2 \rangle = 2w^2[1 - R(r)] \quad \text{(L.11)}$$

and on the other hand, (L.10) can be phenomenologically interpreted by [1]

$$H(r, t) = 2w^2 \left[1 - \exp\left(-\frac{r}{\xi} \right)^{2\alpha} \right], \quad \text{(L.12)}$$

where α is the roughness or Hurst exponent.

For r ≪ ξ is $H(r, t)2w^2 \left[1 - \left(1 - \left(\frac{r}{\xi} \right)^{2\alpha} \right) \right] \approx \frac{2w^2}{\xi^{2\alpha}} r^{2\alpha} \propto (mr)^{2\alpha}$ and for r ≫ ξ is $H(r, t) \propto 2w^2$, where m is the local slope and given by $m = \left(\sqrt{2w} \right)^{-\alpha} / \xi \propto w^{-\alpha}/\xi$. Therefore, height-height correlation function is characterized by [2].

$$H(r, t) \propto \begin{cases} (mr)^{2\alpha} & \text{for } r \ll \xi \\ 2w^2 & \text{for } r \gg \xi \end{cases}. \quad \text{(L.13)}$$

Figure L.3 shows the behavior of the height-height correlation function for a self-affine surface.

From Fig. L.3, it can be seen that (i) the roughness exponent α can be derived from the slope of the height-height correlation function at small values of r, (ii) the rms roughness w is obtained from the height-height correlation function after saturation (i.e. $r > \xi$) and the average value of ξ is obtained from the crossovers of the H(r)-function.

Power Spectral Density

The root-mean square roughness is sensitive only to vertical signals and ignores horizontal structures. In contrast, the *power spectral density* (PSD) function [3, 4] evaluates surface roughness as the spread of height deviations from a mean line and the lateral distribution over which the height variations occur.

The PSD can be obtained from Fourier transform of the surface profile, i.e. the PSD is a squared fast Fourier transform (FFT) of the measured topography and provides the spectrum of the spatial frequencies of that surface. The topography information h(x,y) is Fourier transform from the real space in the reciprocal space, where the wave-vectors $\mathbf{q} = (q_x, q_y)$ are the spatial frequencies in which the PSD is defined, and the power corresponds to the squared value of the roughness amplitude. From mathematical point of view, PSD is evaluated from Fourier transform (FT) of surface profile *h(x)*. For a one-dimensional profile h(x), measured e.g., by a mechanical profiler, the PSD is given by

$$PSD(q_x) = |FT(h(x)|^2 = \lim_{L\to\infty} \frac{2}{L^2} \left| \int_{-L/2}^{L/2} h(x) \cdot \exp[-2\pi i q_x] dx \right|^2. \quad \text{(L.14)}$$

The PSD is a measure describing the contributions of the surface frequencies (wavelengths) to roughness. The roughness R_q and average slope s are first and second statistical momentums of the PSD frequency spectra and are given by the following integrals

$$R^2_{q,1D} = \int_{q_{\min}}^{q_{\max}} PSD(q_x) dq_x \quad and \quad s^2 = \int_{q_{\min}}^{q_{\max}} (2\pi q_x)^2 PSD(q_x) dq_x. \quad \text{(L.15)}$$

For two-dimensional profiles h(x,y), measured e.g., by AFM or white-light interferometer, the PSD (dimension in units of [length^4]) can be expressed by

$$\mathrm{PSD}(q_x, q_y) = \lim_{L\to\infty} \frac{1}{L^2} \left| \int_{-L/2}^{L/2} \int_{-L/2}^{L/2} h(x, y) \cdot \exp\left[-2\pi i \left(q_x x + q_y y\right)\right] dx dy \right|^2$$

(L.16)

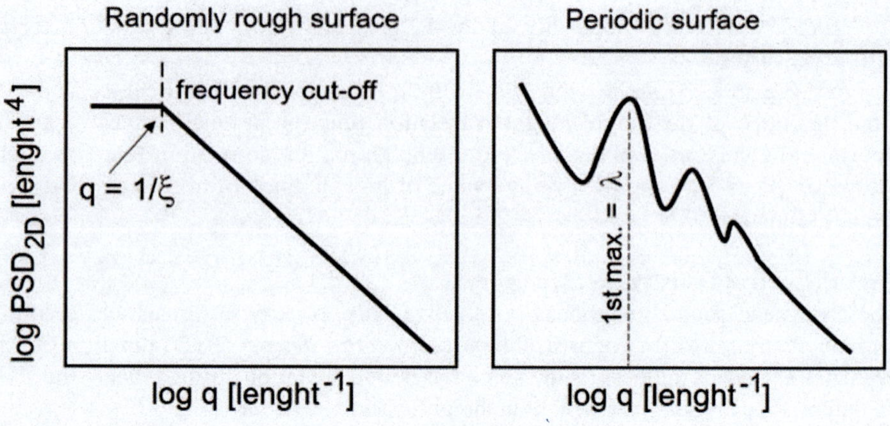

Fig. L.4 Chematic presentation of PSD of a rough surface (left) and a periodic surface (right) versus spatial frequency, where wave-vector $q = 2\pi/\lambda = 2\pi f$

and leads to information about the relative contributions of all surface spatial frequencies [L.4]. The rms-roughness for two-dimensional profiles can be obtained by integration of the relation

$$R_{q,2D}^2 = 2\pi \int_{q_{x,\min}}^{q_{x,\max}} \int_{q_{y,\min}}^{q_{y,\max}} PSD(q_x, q_y) q_x dq_x q_y dq_y. \tag{L.17}$$

In general, the PSD function is presented by plots of log PSD as function of log spatial frequency or the spatial wavelength λ (Fig. L.4).

If periodic behavior of surface structures on self-affine surfaces is present (e.g., ripples), the PSD function is characterized by a damped periodic oscillation (Fig. L.4, right), where the first maximum denotes the roughness amplitude and the first minimum denotes the wavelength of the periodic features (ripples).

The two-dimensional PSD function can be simplified, when an isotropic roughness distribution is assumed. Then, the PSD function is given by [5].

$$PSD(q) = \frac{1}{2\pi} \int_0^{2\pi} PSD(q, \varphi) d\varphi, \tag{L.18}$$

where $q = \sqrt{q_x^2 + q_y^2}$ and $\varphi = \tan^{-1}(q_x/q_y)$.

Measurement Methods of Roughness

The most common methods for metrological determination of roughness measurements are the mechanical or stylus profilometry, the atomic force microscopy, the optical profilometry, interferomety and angle-resolved light scattering due to their

Table L.1 Spatial resolution (black strip) and rms height resolution (blue strip) of methods used for typical measurements of surfaces

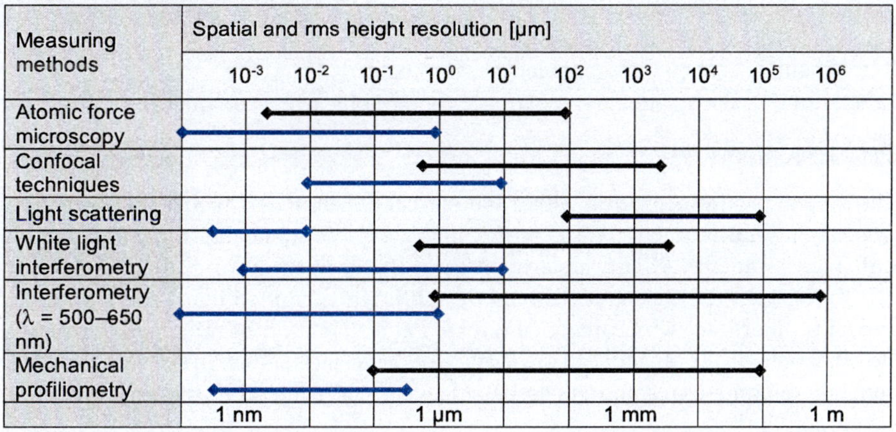

ability to quantitatively measure with micro- and nanoscale resolution. In Table L.1 are summarized the spatial resolution and rms height resolution for these measuring methods.

Atomic force microscopy (AFM)

This class of techniques is characterized by the controlled scanning a nanoscale tip (probe) over a sample to measure the surface topography with up to atomic resolution, where the tip can be without contact (non-contact AFM, scanning tunneling microscopy, STM), in contact (contact mode) or in intermittent contact (tapping mode). In AFM, the deflection of the cantilever is measured, which is proportional to the interaction force between the sample and the probe [7].

Scanning size: < 100 μm
Key benefit: high-resolution
Limitations: small measurement range, sensitive to vibration, great influence of the tip shape.

Light scattering

This measuring technique is based on the comparison of the intensity of incident and scattered light beam at different wavelengths (wavelength spectrum). Two typical methods can be distinguished [4]. The total integrated scattering (TIS) is defined as the ratio of the light intensity scattered into a hemisphere over the sample to the light intensity reflected from the specimen surface. The angle-resolved scattering (ARS) [8] is applied to analyze the in-plane angular intensity distribution of the scattered light to determine the surface roughness. The instrumentation of TIS is simpler than for ARS, but more statistical information can be with ARS than with the TIS method. However, all these scattering techniques do not measure the real-space topography.

Theories (models) are required to relate the measured spectrum of the scattered beam to the roughness spectrum of the surface.

Scanning size: ≤ 1 mm
Key benefit: fast and non-contact measurement
Limitations: does not obtain real-space topography, needs assumptions of models.

Optical profilometry

The optical profilometry uses either the optical coherence of white light (scanning white light interferometry, WLI) or the phase shifts of monochromatic light (phase shift interferometry, PSI) to determine the vertical position of each surface point [9]. Topography analysis is performed by splitting the emitted light beam inside the optical profiler into two beams, one directed at a standardized reference mirror and the other outside the objective at the surface of the tested sample. In the vertical direction, sub-nanometer resolution can be achieved, while lateral resolution is severely diffraction limited.

Scanning size: < 1–5 mm ($\leq$ 100 mm with stitching method)
Key benefit: non-contact sub-nanometer height resolution
Limitations: diffraction-limited spatial resolution, very rough surfaces and transparent thin film are difficulty measurable.

Mechanical or stylus profilometry

A sharp needle (radius of curvature between 1 and 10 μm) is dragged across the surface. The deflection of the needle correlates with the surface topography [10].

Scanning size: < 50 mm ($\leq$ 250 mm with stitching method)
Key benefit: sub-nanometer height resolution
Limitations: extremely robust measurement technique, can cause surface damages, tip shape influences the test reading.

Typically, the total PSD of a surface cannot be determined over the entire spatial frequency bandwidths. Therefore, the stitching method [5] is applied, which provides the PSD for individual sample regions over larger spatial frequency bandwidths. Subsequently, the results of the individual roughness measurements on the same specimen have to be combined.

References

1. M. Pelliccione, T.-M. Lu, *Evolution of Thin Film Morphology*, (Springer, Berlin 2008)
2. F. Family, T. Vicsek. Scaling of the active zone in the Eden process on percolation networks and the ballistic deposition model, J. Phys. A: Math. Gen. **18**, L75 (1985)
3. J. M. Elson, J. M. Bennett, Calculation of the power spectral density from surface profile data, Appl. Opt. **34**, 201 (1995)

4. J. M. Bennett, L. Mattson, *Introduction to Surface Roughness and Scattering*, (Optical Soc. of America, Washington 1993)
5. A. Duparré, J. Ferre-Borrull, S. Gliech, G. Notni, J. Steinert, J. M. Bennett, Surface characterization techniques for determining the root-mean-square roughness and power spectral densities of optical component, Appl. Opt. **41,** 154 (2002)
6. A.-L. Barabási, H. E. Stanley, *Fractal Concepts in Surface Growth*, (Cambridge University Press, Cambridge 1995)
7. W. C. Sanders, *Atomic Force Microscopy: Fundamental Concepts and Laboratory Investigations*, (CRC Press, Boca Raton, 2020)
8. J. M. Bennett, Recent development in surface roughness characterization, Meas. Sci. Technol. **3**, 1119 (1992)
9. J. P. de Groot, Interference microscopy for surface structure analysis, in *Handbook of Optical Metrology*, ed. by T. Yoshizawa, (CRC Press, Boca Raton 2015), pp. 791.
10. D. J. Whitehouse, *Handbook of Surface and Nanometrology*, (CRC Press, Boca Raton 2011)

Appendix M

M.1 Dynamic Scaling and Frequency Analysis

Dynamic scaling is a powerful method for predicting surface properties at various length scales. Originally, the scaling theory was based on self-affine surfaces. Later, this theory was applied to describe the evolution of more complex surfaces (e.g., mounded surfaces). For a self-affine surface, it is important that the horizontal and vertical directions of the surface can be scaled so that the new surface is statistically identical to the original surface (see, e.g., [1–4]).

Based on numerous experimental results, Family and Vicsek [1] have proposed a dynamic scaling hypothesis for the time-dependent evolution of rough surfaces. They assumed that the rms roughness, w(L,t), increases with time and eventually saturates at a certain time value, the saturation time τ_s, which depends on the selected section L of the fully measured surface. The function w(L,t), presented in Appendix L, is invariant when there are no characteristic times or length scales, i.e. w(L,t) is a homogenous function of only two variables [c.f. (K.5)]. Family and Vicsek [1] have reduced this function to a single variable given by

$$w(L,t) = \left(\left\langle [h(r,t) - \overline{h}(r,t)]^2 \right\rangle\right)^{1/2} \propto L^{\alpha} f\left(\frac{t}{L^z}\right), \tag{M.1}$$

where α is the roughness exponent, z is the dynamic exponent (z = α/β) and β is the growth exponent (see Sect. 6.1.2). The Family-Vicsek scaling function, f, is given by

$$f\left(\frac{t}{\tau_s}\right) \propto \begin{cases} \left(\frac{t}{\tau_s}\right)^{\beta} & \text{for} \left(\frac{t}{\tau_s}\right) \ll 1 \\ \text{constant for} \left(\frac{t}{\tau_s}\right) \gg 1 \end{cases}. \tag{M.2}$$

B. Rauschenbach, *Low-Energy Ion Irradiation of Materials*, Springer Series in Materials Science 324, https://doi.org/10.1007/978-3-030-97277-6

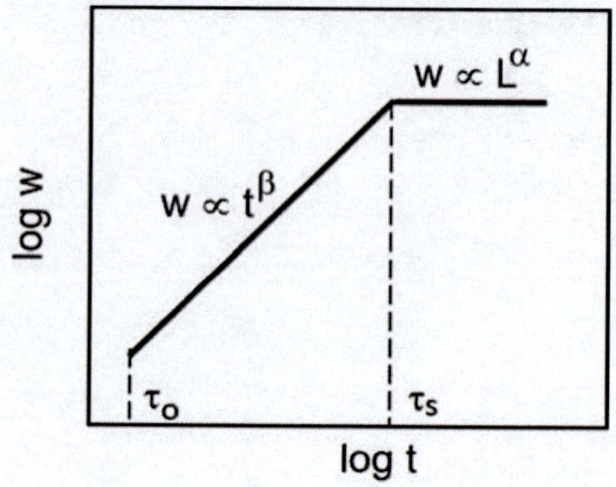

Fig. M.1 Schematic of the interface width w (or rms roughness R_q) as function of the time. α is the roughness exponent, β is the surface growth exponent, τ_s is the saturation time

This is based on the assumption that roughness increases as material is removed (or is deposited by sputtering or evaporation) from a flat substrate until it remains constant after the saturation time τ_s, i.e. $t < \tau_\varepsilon$, is $w \propto (t/\tau_s)^\beta \propto L^{\alpha-\beta z} t^\beta \infty t^\beta$ and when t >> τ_s is w independent of L. Consequently, the scaling relation, (M.1), can be expressed by

$$w(L,t) \propto \begin{cases} t^\beta & \text{for } \tau_o \\ L^\alpha & \text{for } t \gg \tau_S \end{cases}, \tag{M.3}$$

where the scaling exponents are defined in Chap. 6.

For a given dimension, the three scaling exponents defines a universality class, i.e. if two or more processes have the same values of these exponents, they have the same university class, which means that their underlying dynamics obey the same physical mechanisms. It should be noted that a different scaling behavior can be expected for non-self-affine surfaces. Based on this hypothesis, it is possible to scale a surface height profile in time using (M.3). Figure M.1 shows schematically the time dependence of roughness. The saturation time, τ_s, indicates the time when the lateral correlation length ξ is larger than r, or the time at which the surface roughness begins to saturate.

The time τ_0 terminates erosion process in the initial stage from the process of the roughness evolution. In the case of the growth of thin films, the exponents α and β take specific values depending on the growth mechanism. The most common growth mechanisms [1, 3] are the random deposition, random deposition with surface relaxation described by the Edwards-Wilkinson equation, nonlinear growth described by the Kardar-Parisi-Zhang equation and the diffusion-controlled surface evolution according to Mullins (c.f. Table 6.2). In principle, all three exponents can be determined experimentally [1, 3].

(i) the roughness exponent α from slope of the height-height correlation function for r < ξ (linear part of the log–log plot of $\sqrt{H(r,t)}$ in dependence of distance r),

(ii) the growth exponent β from the slope of roughness as function of the time for t < τ_o (linear fit of the log–log plot of roughness w in dependence of time t) and

(iii) the dynamic exponent from the time evolution of the lateral correlation length.

Frequency Analysis

Although experimental studies have demonstrated the existence of power laws, they do not prove that dynamical scaling is an adequate description of the ion beam-induced roughening processes. The underlying mechanism remains hidden.

Tong and Williams [5] have developed a method to model the contribution to the smoothing or roughening processes following low energy ion bombardment [see (7.23) and (7.24)]. They have proposed a stochastic rate equation to describe the evolution of surface morphology by a Fourier transform of spatial coordinates in reciprocal space given by

$$\frac{\partial \tilde{h}(\boldsymbol{q}, t)}{\partial t} = -\tilde{h}(|\boldsymbol{q}|, t) \sum_{j=1}^{n} a_j q^j + \eta(\boldsymbol{q}, t), \tag{M.4}$$

where $\tilde{h}$ is the Fourier transform of h and $a_j \geq 1$ are constants. Since the incident ion beam consists of discrete particles, there are inevitably statistical fluctuations around its mean value, which are modeled by addition of η to (M.4). The spatially and temporally uncorrelated noise η can cause profound changes in topography (c.f.

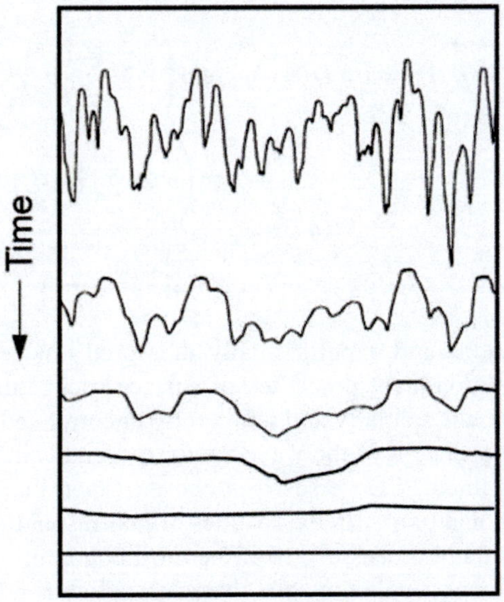

Fig. M.2 Simulation of a typical erosion dynamics for the case of a uniform ion bombardment of an initially rough surface in absence of the noise. Figure is taken from [7] and modified

Figure 6.1), while surfaces become flat after ion bombardment when no noise is assumed (Fig. M.2). The fluctuations are assumed to be Gaussian white noise and can be expressed by [6]

$$\langle \eta(x,t) \rangle = 0 \text{ and } \langle \eta(x,t)\eta(y,t') \rangle = 2\Omega\delta^d(x-y)\delta(t-t'), \tag{M.5}$$

where the angular brackets denote ensemble averaging, Ω is the fluctuation or white noise strength (intensity) and d is the spatial dimension (d = 1 or 2). This relationship yields a finite magnitude, if x = y and t = t′, i.e. sputter events that are separated in time and or space results in $\eta(x,t)\eta(y,t') = 0$. These events are called uncorrelated. The solution of (M.4) is given by

$$\tilde{h}(q,t) = \tilde{h}(q,0) \cdot \exp\left(-t\sum_{j=1}^{n} a_j q^j\right). \tag{M.6}$$

Based on frequency analysis, the spectral power density (PSD) was determined to be [5, 7, 8]

$$PSD(q,t) = \Omega(q)\frac{1-\exp\left(-2t\sum_{j=1}^{n} a_j q^j\right)}{\sum_{j=1}^{n} a_j q^j}, \tag{M.7}$$

if the surface is almost smooth. If a surface characterized by PSD (q,0) is rough before ion bombardment, the spectral power density can be expressed by [9]

$$PSD(q,t) = PSD(q,0) \cdot \exp\left(-2t\sum_{j=1}^{n} a_j q^j\right) + \Omega(q)\frac{1-\exp\left(-2t\sum_{j=1}^{n} a_j q^j\right)}{\sum_{j=1}^{n} a_j q^j}. \tag{M.8}$$

Using these equations and experimentally measured power spectral densities, one can conclude the dominant processes of surface roughening and smoothing. Assuming that the noise is spatially and temporally uncorrelated, the noise strength Ω is constant, i.e. independent of the wave-vector q. Usually, a log–log plot of the experimentally determined PSD versus q is chosen. Then the linear slope of the PSD curve is proportional to q^{-j}. Integer values of j correspond to different surface roughening and smoothing processes [10]. The most common surface mechanisms contributing to smoothing/roughening are characterized by j = 1 (viscos flow), j = 2 (desorption/ballistic smoothing/sputter erosion), j = 3 volume diffusion and j = 4 (surface diffusion). Figure M.3 schematically illustrates the PSD functions in Fourier

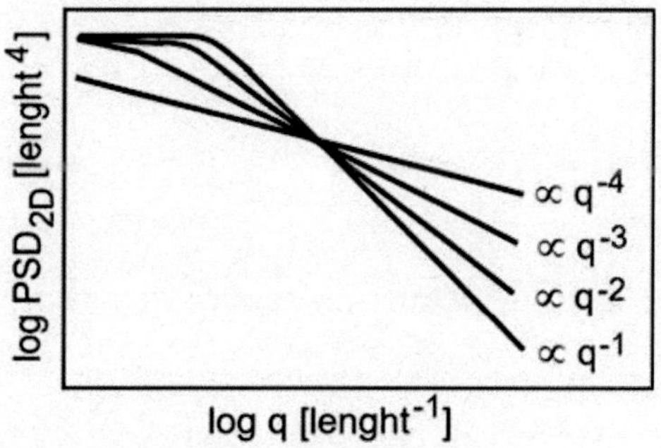

Fig. M.3 Characteristic behavior of the PSD function in Fourier space for the different surface mechanisms

space for the different surface mechanisms (details see Sect. 7.2.1). Note that several mechanisms can act simultaneously, characterized by the existence of different slopes of the PSD function versus the spatial frequency or spatial wavelengths.

References

1. M. Pelliccione, T.-M. Lu, *Evolution of Thin Film Morphology*, (Springer, Berlin 2008)
2. F. Family, T. Vicsek, Scaling of the active zone in the Eden process on percolation networks and the ballistic deposition model, J. Phys. A: Math. Gen. **18**, L75 (1985)
3. A.-L. Barabási, H. E. Stanley, *Fractal Concepts in Surface Growth*, (Cambridge University Press, Cambridge 1995)
4. P. Meakin, The growth of rough surfaces and interfaces, Phys. Rep. **235**, 189 (1983)
5. W. M. Tong, R. S. Williams, Kinetics of Surface growth: Phenomenology, scaling, and mechanisms of smoothening and roughening, Annual Rev. Phys. Chem. **45**, 401 (1994)
6. T. Viscek, *Fractal Growth Phenomena*, (World Sci. Publ., Singapore 1992)
7. G. S. Bales, R. Bruinsma, E. A, Eklund, R. P. U. Karunasiri, J. Rudnick, A. Zangwill, Growth and erosion of thin solid films, Science **249**, 264 (1990)
8. D. G. Stearns, Stochastic model for thin film growth and erosion, Appl. Phys. Lett. **62**, 1745 (1993)
9. S. G. Mayr, R. S. Averback, Surface smoothing of rough amorphous films by irradiation-induced viscous flow, Phys. Rev. Lett. **87**, 196,106 (2001)
10. W.M. Tong, R.S. Williams, Kinetics of surface growth: Phenomenology, scaling, and mechanisms of smoothing and roughening, Ann. Rev. Phys. Chem. **45**, 401 (1994)

Appendix N

N.1 Ehrlich-Schwoebel Barrier and Edge Step Diffusion

The adatoms generated (e.g., by ion bombardment) diffuse on the surface until they encounter adatom traps (e.g., another adatom or a step of a growing island). The adatoms attached to such traps cause the islands or clusters of adatoms to expand until they grow together into a complete monolayer. Then the layer-by-layer growth process begins anew. If an adatom approaches a step from above, it is prevented from descending by a step-edge barrier. This Ehrlich-Schwoebel barrier (Fig. N.1) causes the adatoms at the edge of an island to be partially reflected by the edge of the island, i.e. they remain on the surface of the island (see Fig. N.1). Consequently, additional activation energy, the energy E_{ES} for downward diffusion of adatoms, is required for adatoms to overcome the step. This specific growth instability develops in the presence of an ES barrier at a step edge and promotes the imbalance between upward and downward flows in stepped surfaces, leading to the formation of mounds. In their

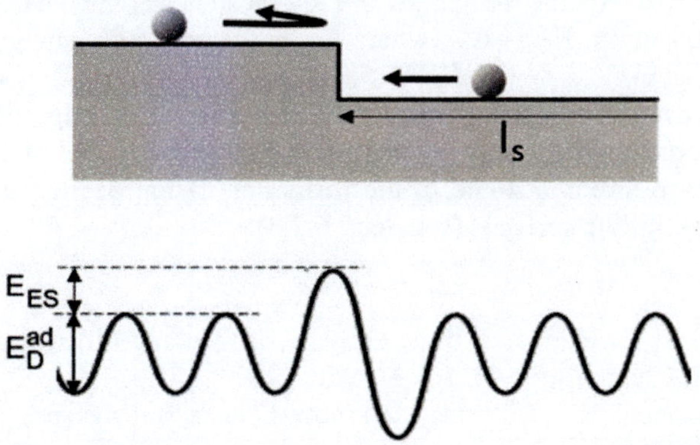

Fig. N.1 Scheme of the Ehrlich-Schwoebel (ES) barrier with the diffusing energy E_D^{ad} of the adatoms on the substrate and the energy barrier E_{ES}

B. Rauschenbach, *Low-Energy Ion Irradiation of Materials*, Springer Series in Materials Science 324, https://doi.org/10.1007/978-3-030-97277-6

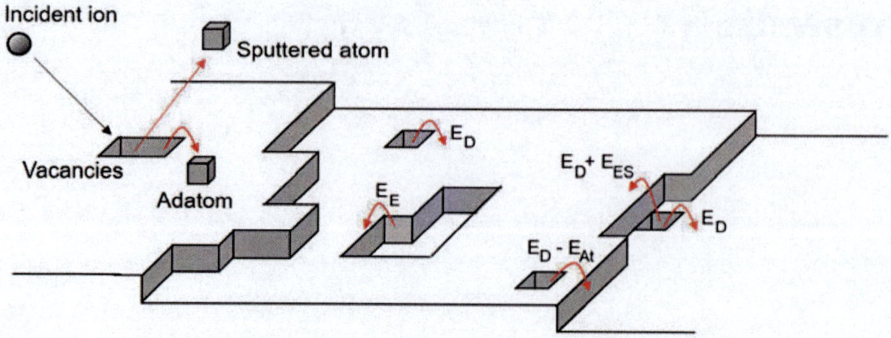

Fig. N.2 Schematic of the different mechanisms of vacancy diffusion on surfaces

seminal publications, Ehrlich et al. [1] and Schwoebel et al. [2] demonstrated the presence of this additional barrier to the diffusion of an adatom at the downward step edge and reported that adatoms located at the edge of terraces are more likely to diffuse backward in the same layer (or on the surface of the island) than to migrate to a lower step. Consequently, a higher adatom density and thus a higher probability of new island formation can be expected.

Politi and Villain [3] found that the lateral size of growth hills induced by the Ehrlich-Schwoebel barrier scales with l_c^2/l_s, where l_s is the Schwoebel length (Fig. N.1) and l_c is the nucleation length, defined as the typical width of a terrace just before a new terrace is formed on its step.

The formation of etch pits and walls is strongly dependent on the diffusion of vacancies (see Sect. 6.2.3). According to Villian [4], the different processes of vacancy diffusion are associated with different activation energies (Fig. N.2). As a result of ion bombardment, sputtered atoms (or molecules), adatoms and vacancies are generated. The activation energy of the individual vacancies diffusing on the terrace is E_D. For vacancies at the upper step edge, a higher activation energy, $E_D + E_{ES}$, is required to overcome the step, while for vacancies at the lower step edge, the barrier is slightly lower, $E_D - E_{At}$, where E_{At} is the activation energy of attraction of vacancies by steps.

It can be expected, that three processes (ES barrier, step-edge diffusion, step-vacancy attraction) cause jumps of adatoms (vacancies) in the uphill (downhill) direction and therefore contribute to the formation of ion beam induced pits and mounds onto crystalline surfaces (see Sect. 6.2.3).

References

1. G. Ehrlich, F. G. Hudda, Atomic view of surface self-diffusion: Tungsten on tungsten, J. Chem. Phys. **44**, 1039 (1966)
2. R. L. Schwoebel, E. J. Shipsey, Step motion on crystal surfaces, J. Appl. Phys. **37**, 3682 (1966)

3. P. Politi, J. Villain, Ehrlich-Schwoebel instability in molecular-beam epitaxy: A minimal model, Phys. Rev. B **54**, 5114 (1996)
4. J. Villain, Continuum models of crystal growth with and without desorption, J. Phys. I (France) **1**, 19 (1991)

Appendix O

O.1 Coefficients of the Surface Evolution Equations

Coefficients of the Bradley-Harper Equation in the Notation Proposed by Bradley and Harper [1]

$$\Gamma_x(\theta) = \frac{A}{B_1}\sin\theta - \frac{B_2}{2B_1}\left(1 + \frac{A^2}{2B_1}\right)\cos\theta - \frac{AC}{B_1^2}\left(3 + \frac{A^2}{B_1}\right)\cos\theta \tag{O.1}$$

$$\Gamma_y(\theta) = -\frac{\mu^2}{a^2}\left(\frac{1}{2}B_2 + \frac{AC}{B_1}\right)\cos\theta \tag{O.2}$$

$$A = \left(\frac{a}{\sigma}\right)^2 \sin\theta \tag{O.3}$$

$$B_1 = \left(\frac{a}{\sigma}\right)^2 \sin^2\theta + \left(\frac{a}{\mu}\right)^2 \cos^2\theta \tag{O.4}$$

$$B_2 = \left(\frac{a}{\sigma}\right)^2 \cos\theta \tag{O.5}$$

$$C = \frac{1}{2}\left[\left(\frac{a}{\mu}\right)^2 - \left(\frac{a}{\sigma}\right)^2\right]\sin\theta\cos\theta, \tag{O.6}$$

where θ is the angle of ion incidence, N is the atomic number density, *a* is the average ion range, σ is the longitudinal straggling, and μ is the lateral straggling (see Fig. 8.40).

B. Rauschenbach, *Low-Energy Ion Irradiation of Materials*, Springer Series in Materials Science 324, https://doi.org/10.1007/978-3-030-97277-6

Coefficients of the Anisotropic Kuramoto–Sivashinsky Equation in the Notation Proposed by Cuerno et al. [2]

$$v = \frac{F}{\sigma} c \tag{O.7}$$

$$v' = \frac{F}{\sigma} s\left(a_\sigma^2 c^2 - 1\right) \tag{O.8}$$

$$\lambda_x = \frac{F}{\sigma} c\{a_\sigma^1 \left(3s^2 - c^2\right) - a_\sigma^4 s^2 c^2 \tag{O.9}$$

$$\lambda_y = -\frac{F}{\sigma} c\left(a_\sigma^2 c^2\right) \tag{O.10}$$

$$v_x = \frac{F}{2} a_\sigma \left(2s^2 - c^2 - a_\sigma^2 s^2 c^2\right) \tag{O.11}$$

$$v_y = -\frac{F}{2} a_\sigma c^2 \tag{O.12}$$

$$F = \frac{JE}{\sqrt{2\pi}} \exp\left(-\frac{a_\sigma^2}{2}\right), \tag{O.13}$$

where $a_\sigma = a/\sigma$, $s = \sin\theta$, $c = \cos\theta$ and Λ is a constant [see E. (5.20)].

Coefficients of the Effective Surface Diffusion Model in the Notation Proposed Makeev and Barabási [3] (Under the Simplified Assumption that $\sigma = \mu$)

$$D_{xx} = \frac{Fa^2}{24a_\sigma}\left[a_\sigma^4 s^4 c^2 + a_\sigma^2 \left(6s^2 c^2 - 4s^4\right) + 3c^2 - 12s^2\right] \tag{O.14}$$

$$D_{yy} = \frac{Fa^2}{24a_\sigma} 3c^2 \tag{O.15}$$

$$F = \frac{JE\Lambda}{\sqrt{2\pi}} \exp\left(-\frac{a_\sigma^2 + a^2 s^2}{2}\right), \tag{O.16}$$

where $a_\sigma = a/\sigma$, $s = \sin\theta$, $c = \cos\theta$ and Λ is a constant [see E. (5.20)].

Coefficients of the Generalized Continuum Equation in the Notation Proposed by Makeev et al. [4]

(1) Coefficients of the Bradley-Harper equation

$$v = Fc \tag{O.17}$$

$$v' = F\frac{s}{f^2}\left[a_\sigma^2 a_\mu^2 c^2 (a_\sigma^2 - 1) - a_\sigma^4 s^2\right] \tag{O.18}$$

$$v_x = Fa\frac{a_\sigma^2}{2f^3}\left[2a_\sigma^4 s^4 - a_\sigma^4 a_\mu^2 s^2 c^2 + a_\sigma^2 a_\mu^2 s^2 c^2 + a_\mu^4 c^4\right] \tag{O.19}$$

$$v_y = -Fa\frac{c^2 a_\sigma^2}{2f} \tag{O.20}$$

(2) Coefficients of the dispersive non-linearities

$$\begin{aligned}\xi_x = Fa\frac{2a_\sigma^2 sc}{2f^5}&\left[-6s^6 a_\sigma^8 + a_\sigma^8 a_\mu^2 s^4 (4 + 3c^2) - a_\sigma^8 a_\mu^4 c^2 s^2 + a_\sigma^6 a_\mu^4 c^2 s^2 (4 - 6c^2)\right.\\ &+ a_\sigma^6 a_\mu^2 s^4 (-3 + 15s^2) + a_\sigma^4 a_\mu^4 c^2 s^2 (4 + 3s^2) - a_\sigma^4 a_\mu^6 3c^4 (1 + s^2)\\ &\left.+ a_\sigma^2 a_\mu^6 c^4 (9 - 3s^2) - 3a_\mu^8 c^6\right]\end{aligned} \tag{O.21}$$

$$\xi_y = Fa\frac{a_\sigma^2 sc}{2f^3}\left[-a_\sigma^4 a_\mu^2 c^2 + a_\sigma^4 s^2 (2 + c^2) - a_\mu^4 c^4 + a_\sigma^2 a_\mu^2 c^2 (3 - 2s^2)\right] \tag{O.22}$$

(3) Non-linear coefficients according Makeev, Cuerno and Barabási

$$\Omega_1 = -Fa^2\frac{3}{6}\frac{1}{f^2}\frac{s}{a_\mu^2}\left[f^2 - fa_\sigma^2 c^2 - (a_\mu^2 - a_\sigma^2)c^2(f - a_\sigma^4 s^2)\right] \tag{O.23}$$

$$\begin{aligned}\Omega_2 = Fa^2\frac{1}{6}\frac{1}{f^4}&\left[-3sf^2(f - a_\sigma^4 s^2) + a_\sigma^2 c^2 (3a_\sigma^2 sf + a_\sigma^6 s^3) f\right.\\ &\left.+2(a_\mu^2 - a_\sigma^2)c^2(3sf^2 + 6a_\sigma^4 s^3 + a_\sigma^8 s^5)\right]\end{aligned} \tag{O.24}$$

(4) Coefficients of the ion-induced effective surface diffusion (ESD):

$$\mathrm{D}_{xx} = F\frac{a^3}{24}\frac{1}{f^5}\left[-4f^2(3a_\sigma^2 s^2 f + a_\sigma^6 s^4) + a_\sigma^2 c^2 f (3f + 6a_\sigma^4 s^2 f + a_\sigma^8 s^4)\right.$$

$$+2(a_\mu^2 - a_\sigma^2)c^2(15a_\sigma^2 s^2 f^2 + 10a_\sigma^6 s^4 f + a_\sigma^{10} s^6)] \quad (O.25)$$

$$\mathrm{D}_{yy} = F\frac{a^3}{24}\frac{1}{f^5}\frac{3a_\sigma^2}{a_\mu^2} f^4 c^4 \quad (O.26)$$

$$\mathrm{D}_{xy} = F\frac{6a^3}{24}\frac{1}{f^5}\frac{f^2}{a_\mu^2}[-2f^2 a_\sigma^2 s^2 + a_\sigma^2 c^2(f^2 + a_\sigma^4 s^2 f)$$
$$+2(a_\mu^2 - a_\sigma^2)c^2(3a_\sigma^2 s^2 f + a_\sigma^6 s^4 f)] \quad (O.27)$$

(5) Coefficients of the Kardar-Parisi-Zhang non-linearities:

$$\lambda_x = F\frac{c}{2f^4}\{a_\sigma^8 a_\mu^2 s^4(3 + 2c^2) + 4a_\sigma^6 a_\mu^4 c^4 s^2 - a_\sigma^4 a_\mu^6 c^4(1 + 2s^2)$$
$$-f^2[2a_\sigma^4 s^2 - a_\sigma^2 a_\mu^2(1 + 2s^2)] - a_\sigma^8 a_\mu^4 s^2 c^2 - f^4\} \quad (O.28)$$

$$\lambda_y = F\frac{c}{2f^4}(a_\sigma^4 s^2 + a_\sigma^4 a_\mu^2 c^2 - f^2) \quad (O.29)$$

The parameters in these coefficients are given by

$$F = \frac{JE\Lambda a}{\sigma\mu N\sqrt{2\pi f}}\exp\left(-\frac{a_\sigma^2 a_\mu^2 c^2}{2f}\right) \quad (O.30)$$

$$f = a_\sigma^2 s^2 + a_\mu^2 c^2 \quad (O.31)$$

and $a_\sigma = a/\sigma$, $a_\mu = a/\mu$, s = sin θ, c = cos θ, and the material constant Λ given by (5.20).

Coefficients in the Notation (One-Dimensional Case) Proposed by Muñoz-García et al. [5]

$$v = \alpha_0\Phi \quad (O.32)$$

$$\gamma_x = -\alpha_o\Phi\alpha_{1x} \quad (O.33)$$

$$v_x = -\alpha_0\Phi a_{2x} + \alpha_0^2(1 - \Phi)\Phi\alpha_{1x}^2/\gamma_0$$

$$\Omega_x = \alpha_o[-\Phi\alpha_{3x} + (1 - \Phi)D/\gamma_o - \Phi R_{eq}\gamma_{2x})\alpha_{1x}] \quad (O.34)$$

$$\xi_x = -\alpha_0 \Phi \alpha_{5x} \tag{O.35}$$

$$\mathrm{B}_{xx} = D R_{eq} \gamma_{2x} + \alpha_o \left[\Phi \alpha_{4xx} - (1 - \Phi) D/\gamma_o - \Phi R_{eq} \gamma_{2x} \right) \alpha_{2x} \right] \tag{O.36}$$

$$\lambda_x^{(1)} = -\alpha_0 \Phi \alpha_{6x} \tag{O.37}$$

$$\lambda_{xx}^{(2)} = -\alpha_o (1 - \Phi) D/\gamma_o - \Phi R_{eq} \gamma_{2x}) \alpha_{6x}, \tag{O.38}$$

where α_0 is the excavation rate (see Sect. 8.2.1.2).

Coefficients in the Notation Proposed by Bradley [6, 7]

$$a_\sigma = a/\sigma \tag{O.39}$$

$$a_\mu = a/\mu \tag{O.40}$$

$$M_x^{(0)} = D \exp\left(-\frac{a_\sigma^2}{2}\right) \frac{2\pi}{a_\mu \sqrt{B_1}} \exp\left(\frac{A^2}{2B_1}\right) \tag{O.41}$$

$$M_x^{(1)} = \frac{aA}{B_1} M_x^{(0)} \tag{O.42}$$

$$\mathrm{C}_{11}(\theta) = -aM_x^{(0)} \left[\frac{A}{B_1} \sin\theta - \frac{\cos\theta}{2} \left(\frac{B_2}{2B_1} + \frac{6AC}{B_1^2} + \frac{A^2 B_2}{B_1^2} + \frac{2A^3 C}{B_1^3} \right) \right] \tag{O.43}$$

$$\mathrm{C}_{22}(\theta) = -M_x^{(0)} \left(\frac{B_2}{2} + \frac{AC}{B_1} \right) \cos\theta, \tag{O.44}$$

where a is the average ion range, σ is the longitudinal straggling, μ is the lateral straggling and A, B_1, B_2, C are the Bradly-Harper coefficients [see (O.3)–(O.6)].

References

1. R. M. Bradley, J. M. E. Harper, Theory of ripple topography induced by ion bombardment, J. Vac. Sci. Technol. A **6**, 2390 (1988)
2. R. Cuerno, A.-L. Barabási, Dynamic scaling of ion-sputtered surfaces, Phys. Rev. Lett. **74**, 4746 (1995)
3. M. A. Makeev, A.-L. Barabási, Ion-induced effective surface diffusion in ion sputtering, Appl. Phys. Lett. **71**, 2800 (1997)
4. M. A. Makeev, R. Cuerno, A.-L. Barabási, Morphology of ion-sputtered surfaces, Nucl. Instr. Meth. in Phys. Res. B **97**, 185 (2002)

5. J. Muñoz-García, R. Cuerno, M. Castro, Coupling of morphology to surface transport in ion-beam irradiated surfaces: Oblique incidence, Phys. Rev. B **78**, 205408 (2008)
6. R. M. Bradley, Exact linear dispersion relation for the Sigmund model of ion sputtering, Phys. Rev. B **84**, 075413 (2011)
7. M. P. Harrison, R. M. Bradley, Crater function approach to ion-induced nanoscale pattern formation: Craters for flat surfaces are insufficient, Phys. Rev. B **89**, 245401 (2014)

List of Materials, Substances, and Microorganism

It should be noted that the following list does not include ion species, contaminations and impurities.

(continued)

B. Rauschenbach, *Low-Energy Ion Irradiation of Materials*, Springer Series in Materials Science 324, https://doi.org/10.1007/978-3-030-97277-6

(continued)

(continued)

(continued)

(continued)

(continued)

(continued)

(continued)

(continued)

(continued)

(continued)

(continued)

Index

B. Rauschenbach, *Low-Energy Ion Irradiation of Materials*, Springer Series in Materials Science 324, https://doi.org/10.1007/978-3-030-97277-6

J

K

T